Marine Mammals: Evolutionary Biology

2nd Edition

Marine Mammals: Evolutionary Biology
2nd Edition

Annalisa Berta
Department of Biology
San Diego State University
San Diego, California

James L. Sumich
Biology Department
Grossmont College
El Cajon, California

Kit M. Kovacs
Biodiversity Programme
Norwegian Polar Institute
Tromsø, Norway

With illustrations by
Pieter Arend Folkens
A Higher Porpoise Design Group
Benecia, California

Peter J. Adam
Department of Biology
San Diego State University
San Diego, California

AMSTERDAM • BOSTON • HEIDELBERG • LONDON
NEW YORK • OXFORD • PARIS • SAN DIEGO
SAN FRANCISCO • SINGAPORE • SYDNEY • TOKYO
An imprint of Elsevier

ELSEVIER

Cover illustrations: Carl Buell

Academic Press is an imprint of Elsevier
30 Corporate Drive, Suite 400, Burlington, MA 01803, USA
525 B Street, Suite 1900, San Diego, California 92101-4495, USA
84 Theobald's Road, London WC1X 8RR, UK

This book is printed on acid-free paper. ⊗

Library of Congress Cataloging-in-Publication Data
Application submitted.

British Library Cataloguing in Publication Data
A catalogue record for this book is available from the British Library

ISBN 13: 978-0-12-088552-7
ISBN 10: 0-12-088552-2

Printed in China

10 11 12 13 14 10 9 8 7 6 5 4 3 2

CONTENTS

PREFACE

The second edition, like the previous one, *Marine Mammals: Evolutionary Biology,* is written for two audiences: as a text for an upper-level undergraduate or graduate-level course on marine mammal biology and as a source book for marine mammal scientists in research, education, management, and legal/policy development positions. One of our major goals is to introduce the reader to the tremendous breadth of topics that comprise the rapidly expanding interdisciplinary field of marine mammal science today. Our motivation for writing this book was the lack of a comprehensive text on marine mammal biology, particularly one that employs a comparative, phylogenetic approach. We have attempted, where possible, to demonstrate that hypotheses of the evolutionary relationships of marine mammals provide a powerful framework for tracing the evolution of their morphology, behavior, and ecology. This approach has much to offer but is limited, in many cases, by available comparative data. We hope that this book stimulates others to pursue marine mammal research in this exciting new direction.

ACKNOWLEDGMENTS

In preparing the second edition, we have been guided by the detailed, thoughtful, and constructive comments of colleagues and students. The many colleagues who contributed photographs and line drawings are identified in the captions. We appreciate the copyediting of Christian Lyderson and Fred Inge Prestenge for library assistance. The production and editorial staff at Academic Press have been very helpful in preparation of this book; we are especially grateful to our Developmental Editor, Kirsten Funk, and Senior Editor, Andrew Richford, as well as the Manager of Editorial Services at SPI, Christine Brandt. Finally, we thank friends and colleagues who provided inspiration by asking, "Why do phylogenies matter?" Although we have relied on existing published literature for information, the interpretations presented here are solely ours. In the spirit of improving this work, we would appreciate notification of any errors, either of omission or of fact.

Annalisa Berta
aberta@sunstroke.sdsu.edu

Jim Sumich
jim.sumich@gcccd.net

Kit Kovacs
kit.kovacs@npolar.no

1

Introduction

1.1. Marine Mammals—"What Are They?"

Some 100 living species of mammals (listed in the Appendix) depend on the ocean for most or all of their life needs. Living **marine mammals** include a diverse assemblage of species that have representatives in three mammalian orders. Within the order Carnivora are the **pinnipeds** (i.e., seals, sea lions, walruses), the sea otter, and the polar bear. The order Cetacea includes whales, dolphins, and porpoises, and the order **Sirenia** is composed of sea cows (manatees and dugongs). Marine mammals were no less diverse in the past and include extinct groups such as the hippopotamus-like **desmostylians,** the bizarre bear-like carnivore **Kolponomos,** and the aquatic sloth **Thalassocnus.**

1.2. Adaptations for Aquatic Life

Marine mammals are well adapted for life in the water though they differ in the degree to which they are adapted to this habitat. Pinnipeds, sea otters, and polar bears are amphibious, spending some time on land or ice to give birth and to molt, whereas cetaceans and sirenians are fully aquatic. A few major aquatic **adaptations** are briefly reviewed in this chapter and are covered in greater detail in subsequent chapters. Adaptations of the skin, specifically its increased insulation (through development of **blubber** or a dense fur layer) and **countercurrent heat exchange systems,** help them cope with the cold. Similarly, the eyes, nose, ears, and limbs of marine mammals have changed in association with their ability to live in a variety of aquatic environments, which include saltwater, brackish, and freshwater. Perhaps the most notable among sensory adaptations are the high frequency sounds produced by some whales for use in navigation and foraging. Other marine mammals (e.g., pinnipeds, polar bears, and sea otters) have an acute sense of smell; these same groups also possess well-developed whiskers with sensitive nerve fibers that serve as tactile sense organs. Pinnipeds have front and hind limbs modified as flippers that propel them both in the water and on land. In cetaceans and sirenians, the hind limbs are virtually absent and locomotion is accomplished by vertical movement of the tail. Most marine mammals cope with living in salt water by conserving water in their heavily lobulated kidneys, which are efficient at concentrating urine.

Many marine mammals are capable of prolonged and deep dives. Adaptations of the respiratory system, such as flexible ribs that allow the lungs to collapse and thickened tissue in the middle ear of pinnipeds and cetaceans, enable them to withstand the tremendous pressures encountered at great depths. The long dives of these animals are accomplished by a variety of circulatory changes including a slowed heart rate, reduced oxygen consumption, and shunting blood to only essential organs and tissues.

1.3. Scope and Use of This Book

Our goal for this second edition remains the same as for the first edition: to provide an overview of the biology of marine mammals with emphasis on their evolution, anatomy, behavior, and ecology. These topics are presented and discussed using, in so far as possible, an explicit phylogenetic context. In doing so we consider different ways of incorporating evolutionary history into comparative analyses of marine mammal biology. The phylogenetic approach advocated in this book is a young but vigorously developing research field that we believe has much to offer marine mammal science. Over the past six years, interest in this approach has grown and we are pleased to offer a number of new case studies that integrate a phylogenetic approach into studies of marine mammal biodiversity.

The book is divided into two major sections: *Part I: Evolutionary History* (Chapters 2–6) is where the origin and diversity of marine mammals are revealed, and *Part II: Evolutionary Biology, Ecology, and Behavior* (Chapters 7–15) is where we attempt to explain how this diversity arose by examining patterns of morphological, behavioral, and ecologic diversity. We have intended to explain these concepts, wherever possible, by example and with a minimum of professional jargon. Words and phrases included in the glossary appear in boldface type at their first appearance in the text. "Further reading" sections have been placed at the end of each chapter and are intended to guide the reader to more detailed information about a particular topic.

1.4. Time Scale

A historical discussion of marine mammals requires a standard time framework for relating evolutionary events. Figure 1.1 presents the geologic time scale that is used throughout this book (based on Harland *et al*., 1990). Our interest lies in the **Cenozoic Era,** the last 65 million years of earth history, during which time all marine mammals made their first appearance. Whales and sirenians were the first to appear, beginning approximately 50 million years ago (Ma) during the early **Eocene.** Pinnipeds trace their ancestry back between 29 and 23 Ma to the late **Oligocene.** The sea otter lineage goes back approximately 7 Ma to the late **Miocene,** although the modern sea otter is known in the fossil record only as far back as the early **Pleistocene** (1.6 Ma). Polar bears appear even later, during the late Pleistocene (0.5 Ma). The desmostylians, extinct relatives of sirenians, range from the early Oligocene through the late Miocene. The extinct carnivoran *Kolponomos is* known from a brief time interval during the early Miocene, and the extinct marine sloth *Thalassocnus* lived during the late Miocene–late **Pliocene** (7–3 Ma).

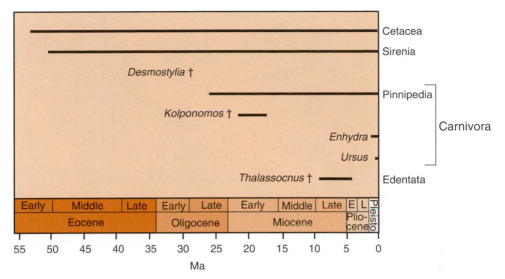

Figure 1.1. Chronologic ranges of marine mammal taxa. Solid bars show reported maximum ranges. Ma = million years ago. (Time scale and correlations are from Harland *et al.*, 1990, and Berggren *et al.*, 1995.)

1.5. Early Observations of Marine Mammals

The study of marine mammals probably began with casual observations of the appearance and behavior of whales in the 4th century B.C. Still, the knowledge and history of these animals themselves go much further back. Drawings of seals and dolphins on pieces of reindeer antler and in caves have been found from **Paleolithic** times. The Greek philosopher Aristotle (384–322 B.C.) in his *Historia Animalium* describes dolphins, killer whales, and baleen whales, noting that "the [latter] has no teeth but does have hair that resemble hog bristles." Unfortunately, Aristotle's observations were dismissed by many later workers because of his misclassification of dolphins as fish. Following Aristotle, the only other authority on whales in ancient times was Pliny the Elder (24–79 A.D.). In his 37-volume *Naturalis Historia,* he included a book on whales and dolphins in which he provided accounts based on Aristotle's findings and his own observations. Knowledge of marine mammals languished for a thousand years after Aristotle and Pliny during the Dark Ages. During the Renaissance, a rapid increase in exploration of the oceans was followed by the publication of scientific reports from various expeditions. The earliest of these was the *Speculum Regale,* an account of Iceland in the 13th century that considered whales the only truly interesting sight the island had to offer. Its author correctly distinguished between northern right whales and bowhead whales, which were still confused by many naturalists five centuries later. In the 16th century, explorers discovered the rich feeding grounds in the high Arctic and the large whale populations that these supported. In the mid-1500s, Konrad Gesner in his *Historia Animalium* presented illustrations of whales; among them was one so large that sailors mistook it for an island (Figure 1.2).

A walrus is also illustrated in Gesner's work (Figure 1.3a). Among the earliest drawings of seals, *Vitulus marinus* (Figure 1.3b) in Pierre Belon's *De Aquatilibus* (1553) is most remarkable for its accuracy, particularly in the detail of the hind limbs. In Guillaume Rondelet's *De Piscibus* (1554), two seals are illustrated, one probably

Figure 1.2.   Woodcut by Conrad Gesner, from *Historia Animalium,* first published between 1551 and 1558, shows a whale so large that sailors mistook it for an island.

Figure 1.3.   Early illustrations of pinnipeds. **(a)** Walrus from Conrad Gesner's *Historia Animalium,* probably taken from a drawing by Albert Dürer. **(b)** Seal from P. Belon, *De Aquatilibus* (1553). **(c)** Seal from Guillaume Rondelet, *De Piscibus* (1554). **(d)** Seal from Guillaume Rondelet, *De Piscibus* (1554). **(e)** "Sea lion" from R. Brookes, *The Natural History of Quadrupeds* (1763).

representing the common seal and the other the Mediterranean monk seal (Figure 1.3c, d; King, 1983). In another book, *The Natural History of Quadrupeds* (1763) by R. Brookes, it is obvious from the illustration and description of the male with a large snout or trunk that the elephant seal is depicted as a cheerful "sea lion" with a "seaweed tail" (Figure 1.3e; King, 1983).

In 1596, the Dutch navigator Wilhelm Barents discovered Spitzbergen (the largest island in the Svalbard Archipelago, north of Norway) and early in the 17th century commercial whalers were sent there by Dutch and English companies to establish a whaling town. Although these expeditions were concerned primarily with whale products, they also produced a number of publications that provided reasonably accurate descriptions of the external appearance of the most common kinds of whales. The best of these are found in *Spitzbergische oder Groenlandische Reisen Beschreibung* (1675) by Frederich Martens and *Bloyeyende Opkomst der Aloude en Hedendaagsche Groenlandsche Visschery* (1720) by C. G. Zorgdrager, both of which contained engravings that continued to be reproduced in books until the early 19th century. Georg Wilhem Steller, ship's naturalist and physician for Vitus Bering's second expedition to North America, was among the first Europeans to explore Alaska and the Aleutian and Commander Islands. His notes of marine mammals living in the Bering Sea, *The Beasts of the Sea* (1751), contained a natural history account of the sea otter, sea lion, fur seal, and the now extinct Steller sea cow, the only first-hand scientific observation of this species.

Another naturalist, Lacépéde, compiled a volume on whales (1804), in which most of the illustrations were copied from previous publications (Figure 1.4). Lacépéde acknowledged that not having ever seen a whale, he had made his descriptions from those of other naturalists. In the first half of the 19th century, additions to the

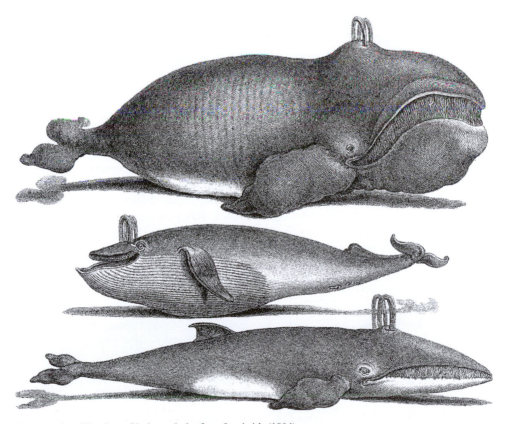

Figure 1.4.   Woodcut of baleen whales from Lacépéde (1804).

literature included Peter Camper's *Observations Anatomiques sur Plusiers Especes de Cétacés* (1820). The foremost European cetologist of the second half of the 19th century was P.-J. Van Beneden, a Belgian zoologist whose many monographs on whales and pinnipeds (including *Histoire Naturelle des Cétacés des Mers d'Europe,* 1889) were published in Brussels between 1867 and 1892. John Edward Gray, who became Keeper of the Zoology Department at the British Museum of Natural History, published his *Catalogue of Seals and Whales in the British Museum* in 1866. John Allen (1880), in his comprehensive monograph of North American pinnipeds, provided keys to the families and genera, described the North American species, and gave accounts of pinniped species in other parts of the world.

Meanwhile, the whaling industries of several countries were making other contributions to the study of whales. Whaling captains such as William Scoresby and Charles Scammon made their own observations in the field or collected those of their colleagues. Scoresby published *An Account of the Arctic Regions* (1820), which is still a valuable source of information on the northern right whale. Scammon's book, *The Marine Mammals of the North-Western Coast of North America,* was published in 1874 and has become a classic, particularly valued for its description of the natural history of the gray whale in California.

Land-based whaling stations used in more modern whaling provided the material for Frederick True's 1904 monograph *The Whalebone Whales of the Western North Atlantic* and Roy Chapman Andrews's 1916 monograph on the Sei whale in the Pacific.

Apart from whalers, the only people seriously interested in the study of whales (**cetology**) at this time were comparative anatomists (for a more detailed account of the beginnings of cetology see Matthews, 1978). Among their ranks were Rondelet, Bartholin, Camper, Cuvier, Hunter, and Owen. These pioneers in the study of cetacean anatomy made the most of specimens that came their way and the writings that many of them produced show that they made accurate observations. Cuvier in particular made several fundamental advances in cetology. His *Le Régne Animal* (1817) and *Recherches sur les Ossemens Fossiles* (1823) contain the original descriptions and illustrations of the three species of cetacean that he named (Cuvier's beaked whale, Risso's dolphin, and the spotted dolphin).

During this time, confusion over the affinities of another marine mammal group, the dugongs, led some to consider them an unusual tropical form of walrus. In a publication from 1800, the manatee is inaccurately shown as hog-nosed (Figure 1.5a). The earliest illustration of a sirenian to be published, the West Indian manatee from the 1535 edition of *La Historia General de la Indias* by Gonzalo Fernandez de Oviedo y Valdés, is little changed from this depiction more than two centuries later (Figure 1.5b).

Figure 1.5. Early illustrations of manatees. **(a)** An "American manatee" (species, unknown) from a lithograph (Reynolds and Odell, 1991). **(b)** West Indian manatee from the 1535 edition of *La Historia General de la Indias* by Gonzalo Fernandez de Oviedo y Valdes.

Descriptions of the anatomy of various pinnipeds followed including the walrus (Murie, 1870) and the Steller sea lion (Murie, 1872, 1874). Another accomplished anatomist, W. C. S. Miller (1888), dissected a variety of pinnipeds including the southern fur seal and southern elephant seal, recovered on the *H. M. S. Challenger* expedition to the Antarctic during the years 1873–1875. Thompson (1915) published the first account of the osteology of Antarctic seals including the Ross seal, the Weddell seal, and the leopard seal. Howell (1929) published his well-known comparative study of both phocids and otariids based on the California sea lion and the ringed seal. He followed this with a book on aquatic adaptations in mammals (Howell, 1930).

1.6. Emergence of Marine Mammal Science

Marine mammal science has emerged as a discipline in its own right only in the last 20 – 30 years. This increasing interest in marine mammals is clearly shown by the expansion of the literature dealing with these animals. J. A. Allen's bibliography of cetaceans and sirenians (1882), covering the 350 years from 1495 to 1840, contains 1014 titles, just under three publications per year. In the period from 1845 to 1960, between 3000 and 4000 articles were published, with a conservative estimate of about 28 titles a year (Matthews, 1966). By comparison, c. 24,000 papers on marine mammals were published between 1961 and 1998 according to the *Zoological Record,* a rate of 646 per year. From 1999 to 2004, marine mammal publications increased to a rate of more than 856 per year. Among the major influences that contributed to the birth of marine mammal science was the growing recognition that marine mammal populations were limited in numbers and that their exploitation had to be regulated (Boyd, 1993). The aim of many early studies was to obtain accurate information about the biology of these animals for use in establishing an effective management policy for sustainable exploitation. It is ironic that the decline in whale stocks heralded the beginning of the scientific study of marine mammals. As a result of concerns regarding stock viability, the *Discovery* investigations (1925–1951) were undertaken to examine the biology of whale stocks in the Southern Ocean. Not only was the biology of whales examined but also their food supplies and their distributions and abundances in relation to oceanographic conditions. For example, British scientists N. A. Mackintosh and J. F. G. Wheeler (1929) examined 1600 carcasses for gut contents in order to produce their report on blue and fin whales. Leonard Harrison-Matthews had comparable samples in his reports on the humpback whale, sperm whale, and southern right whale in 1938 (Watson, 1981).

In the 1950s, the theme of the *Discovery* investigations was continued by the Falkland Islands Dependencies Survey (later known as the British Antarctic Survey) when it established a research program on the southern elephant seal on South Georgia Island under the directorship of R. M. Laws. In parallel with these and other studies, with a focus on population ecology, there also was growing interest in the anatomy and physiology of marine mammals (Irving, 1939; Scholander, 1940; Slijper, 1962; Norris, 1966; Andersen, 1969; Ridgway, 1972; Harrison, 1972–1977). The establishment of various scientific committees (e.g., the International Whaling Commission's Scientific Committee in 1946 and the U.S. Marine Mammal Commission in 1972) to provide advice about the status of various marine mammal populations also required knowledge and data on the general biology of these animals and thus served to stimulate research. Since the early 1980s, the biology of various marine mammal species has been the subject of many notable books, beginning with Ridgway and

Harrison's series entitled *Handbook of Marine Mammals* (1981–1998). This has been followed by detailed separate accounts of the biology of the Pacific walrus (Fay, 1982), gray whale (Jones *et al.*, 1984), bowhead whale (Burns *et al.*, 1993), bottlenose dolphin (Leatherwood and Reeves, 1990; Reynolds *et al.*, 2000), Hawaiian spinner dolphin (Norris *et al.*, 1994), harbor porpoise (Read *et al.*, 1997) sperm whale (Whitehead, 2003), harp and hooded seals (Lavigne and Kovacs, 1988), elephant seals (Le Boeuf and Laws, 1994), and the northern fur seal (Gentry, 1998). Comprehensive treatments of marine mammal groups are available for pinnipeds (King, 1983; Bonner, 1990; Riedman, 1990; Renouf, 1991), for whales (Matthews, 1978; Gaskin, 1982; Evans, 1987; Mann *et al.*, 2000), for manatees and dugongs (Hartman, 1979; Reynolds and Odell, 1991), and for sea otters (Kenyon, 1969; Riedman and Estes, 1990). Valuable field identification guides for all marine mammals are found in Reeves *et al.* (2002), for pinnipeds and sirenians in Reeves *et al.* (1992), and for whales and dolphins in Leatherwood and Reeves (1983) and Carwardine (1995). Recent additions to the growing literature on marine mammal biology include edited books on health and medicine (Dierauf *et al.*, 2001), cell and molecular biology (Pfeiffer, 2002), conservation biology (Evans and Raga, 2001), evolutionary biology (Hoelzel, 2002), and even an encyclopedia on marine mammals (Perrin *et al.*, 2002).

Matthews (1966) wrote "the greatest revolution in the study of the Cetacea . . . has come with the possibility of keeping living cetaceans in oceanariums." However, one of the most significant advances in marine mammal science in recent years has undoubtedly been the move toward studying animals under wild, unrestrained conditions at sea. This is in large part the result of technological advances in microelectronics (e.g., satellite **telemetry** and **time-depth recorders**). For example, the application of microelectronics led to the discovery that elephant seals regularly dive to depths of 1000 m with consistently long dive durations, typically lasting 15 to 45 minutes. This feature of elephant seal biology, in addition to studies on a variety of other species, has forced physiologists to reexamine our understanding of the biochemical pathways used by these animals to maximize the efficiency of oxygen utilization. Studies with **crittercams** provide a visual record of everything that a marine mammal sees. For example, crittercams have revealed Wedell seals flushing prey from crevices in the ice.

Technological advances in molecular biology (e.g., analysis of DNA variation) have also provided unparalleled opportunities to examine interactions among populations and the roles of individuals within those populations. For example, using DNA fingerprinting and other techniques, it is possible to assess paternity and kinship among whales, animals for which this has previously been virtually impossible owing to the difficulty of observing them mating underwater. These techniques have also made it possible to measure effective population sizes and interpret historical events such as population bottlenecks. Molecular techniques also have contributed to our knowledge of the **systematics** and **taxonomy** of various marine mammal groups.

As pointed out by Watkins and Wartzok (1985), information and research about marine mammals range "from intensive to eclectic." Much of the available data is difficult to synthesize because techniques vary widely and sample sizes often are necessarily small. This is not a reflection of poor science but rather the environmental, practical, and legal complications implicit in marine mammal research. It is apparent that the database must be expanded. Even within a relatively homogeneous group like odontocete whales, one well-known species (the bottlenose dolphin, *Tursiops truncatus*) cannot be used reliably to characterize all toothed whales. With this in mind, we hope that as readers of this book you will be able to identify areas in which research must be done. We encourage you to pursue research on marine mammals—there are still many gaps in our knowledge of this diverse and unique assemblage of mammals.

1.7. Further Reading and Resources

There are a large number of Internet addresses with information about marine mammal programs and organizations; a few that we consider the most useful are listed here: http://www.marinemammalogy.org—Society for Marine Mammalogy (SMM), a professional international organization of marine mammal scientists, publishes a journal (quarterly) of original research on marine mammals: *Marine Mammal Science.* http://web.inter.NL.net/users/J.W.Broekema/ecs/index.htm—European Cetacean Society (ECS), professional biologists and others interested in whales and dolphins. http://www.earthwatch.org—Earthwatch Institute, offers opportunities for marine mammal enthusiasts to work as volunteers with research scientists.

Also, for career and hobbyist information about marine mammals see books by Glen (1997) *The Dolphin and Whale Career Guide,* Samansky (2002) *Starting Your Career as a Marine Mammal Trainer,* and *Strategies for Pursuing a Career in Marine Mammal Science* published by SMM and available online.

References

Allen, J. A. (1880). "History of the North American Pinnipeds, a Monograph of the Walruses, Sea-Lions, Sea-Bears, and Seals of North America." *U.S. Geol. Geogr. Surv. of the Territories,* Misc. Publ. No. 12, Government Printing Office, Washington, DC.

Allen, J. A. (1882). "Preliminary List of Works and Papers Relating to the Mammalian Orders Cete and Sirenia." *Bull. U.S. Geol. Geogr Surv. of the Territories* 6(3) (Art. 18): 399–562.

Andersen, H. T. (ed.) (1969). *The Biology of Marine Mammals.* Academic Press, New York.

Andrew, R. C. (1916). "Monographs of the Pacific Cetacea 2: The Sei Whale." *Mem. Amer. Mus. Nat. Hist. 1:* 291–388.

Belon, P. (1553). *Petri Bellonii Cenomani De aquatilibus: libro duo cum conibus ad viuam ipsorum effigiem, quoad eius fieri potuit, expressis.* Apud Carolum Stephanum, Typographum Regium, Paris.

Berggren, W. A., D. V. Kent, C. C. Swisher, Jr., and M. P. Aubry (1995). A Revised Cenozoic Geochronology and Chronostratigraphy. *In* "Geochronology, Time Scales and Global Stratigraphic Correlations" (W. A. Berggren *et al.,* eds.), pp. 129–212. SEPM Special Publication, No. 54.

Bonner, W. N. (1990). *The Natural History of Seals.* Christopher Helm, London.

Boyd, I. L. (1993). "Introduction: Trends in Marine Mammal Science." *Symp. Zool. Soc. London 66:* 1–12.

Brookes, R. (1763). *A New and Accurate System of Natural History (6 vols.)* Vol. 1 "The Natural History of Quadrupeds." Printed for J. Newbery, London.

Burns, J. J., J. J. Montague, and C. J. Cowles (1993). *The Bowhead Whale.* Special Publication, No. 2. Soc. Mar. Mammal. Allen Press, KS.

Camper, P. (1820). *Observations anatomiques sur la structure intèrieure et le squelette de plusieurs espèces de cètacès*; publie'es par son fils, Adrien-Gilles Camper; avec des notes par G. Cuvier. Gabriel Dufour, 1820 (A. Belin), Paris.

Carwardine, M. (1995). *Whales, Dolphins, and Porpoises.* D. K. Publishing, New York.

Cuvier, G. (1817). *Le regne animal distribue d'apres son organisation, pour servir de basea l'histoire naturelle des animaux et d'introduction a l'anatomie comparee.* Deterville, Paris.

Cuvier, G. (1823). *Recherches sur les ossemens fossiles: ou l'on rétablit les caractères deplusieurs animaux dont les révolutions du globe ont détruit les espèces.* Nouvelle Édition, entirement refondue, et considérablement augmentée. Dufour et d'Ocagne, 1821–1825. Paris.

Dierauf, L., and F. M. D. Gulland (eds.) (2001). *CRC Handbook of Marine Mammal Medicine.* CRC Press, Boca Raton, FL.

Evans, P. G. H. (1987). *The Natural History of Whales and Dolphins*. Christopher Helm, London/Facts on File, New York.

Evans, P. G. H., and J. A. Raga (eds.) (2001). *Marine Mammals: Biology and Conservation*. Kluwer Academic/Plenum Publishers, New York.

Fay, F. H. (1982). "Ecology and Biology of the Pacific Walrus, *Odobenus rosmarus divergens* Illiger." *U. S. Dept. Int. Fish Wild. Serv*. North American Fauna, No. 74.

Fernandez de Oviedo y Valdes, G. (1535). *Historia general y natural de la Indias. Edición y estudio preliminar de Juan Pérez de Tudela Bueso*. Ediciones Atlas, Madrid.

Gaskin, D. E. (1982). *The Ecology of Whales and Dolphins*. Heinemann, London.

Gentry, R. L. (1998). *Behavior and Ecology of the Northern Fur Seal*. Princeton University Press, Princeton, NJ.

Gesner, K. (1551–1587). *Conradi Gesneri Historiæ animalium*. C. Froschouerum, Tiguri.

Glen, T. B. (1997). *The Dolphin and Whale Career Guide*. Omega Publishing Company, Chicago.

Gray, J. E. (1866). *Catalogue of Seals and Whales in the British Museum*. 2nd ed. British Museum, London.

Hamilton, R. (1839). "The Naturalists Library (conducted by W. Jardine)." Mammalian. Vol. 8. *Amphibious Carnivora, Including the Walrus and Seals, also of the Herbivorous Cetacea*. W. H. Lizars, Edinburgh and W. Curry, Jun. and Co., Dublin.

Harland, W. B., R. L. Armstrong, A. V. Cox, L. E. Craig, A. G. Smith, and D. G. Smith (1990). *A Geologic Time Scale-1989*. Cambridge University Press, New York.

Harrison, R. J. (1972–1977). *Functional Anatomy of Marine Mammals*, Vols. 1–3. Academic Press, London.

Hartman, D. S. (1979). "Ecology and Behavior of the Manatee *(Trichechus manatus)* in Florida." *Am. Soc. Mammal. Special Publication,* No. 5, 1–153.

Hoelzel, A. R. (2002). *Marine Mammal Biology,* Blackwell Science, Oxford.

Howell, A. B. (1929). "Contributions to the Comparative Anatomy of the Eared and Earless Seals (Genera *Zalophus* and *Phoca)*." *Proc. U. S. Natl. Mus. 73:* 1–143.

Howell, A. B. (1930). *Aquatic Mammals*. Thomas, Springfield, IL.

Irving, L. (1939). "Respiration in Diving Mammals." *Physiol. Rev. 19* : 112–134.

Jones, M. L., S. L. Swartz, and S. Leatherwood (eds.) (1984). *The Gray Whale*. Academic Press, New York.

Kenyon, K. (1969). "The Sea Otter in the Eastern Pacific Ocean," North American Fauna No. 68, *Bur. Sport Fish. Wild*. U.S. Government Printing Office, Washington, DC.

King, J. E. (1983). *Seals of the World,* 2nd ed., British Museum of Natural History, London, and Cornell University Press, Ithaca, NY.

Lacépéde, B. (1804). *Histoire naturelle de Lacépède: comprenant les cétacés, les quadrupèdes ovipares, les serpents et les poissons*. Furne et cie, Paris.

Larson, L.M. Speculum Regale (Iceland 13th Century). *The King's Mirror: Translated from the Old Norwegian*. (Scandinavian monographs: 3). American-Norwegian Foundation, 1917, New York.

Lavigne, D. M., and K. M. Kovacs (1988). *Harps and Hoods*. University of Waterloo Press, Ontario, Canada.

Leatherwood, S., and R. R. Reeves (1983). *The Sierra Club Handbook of Whales and Dolphins*. Sierra Club Books, San Francisco, CA.

Leatherwood, S., and R. R. Reeves (eds.) (1990). *The Bottlenose Dolphin*. Academic Press, San Diego, CA.

Le Boeuf, B.J., and R. M. Laws (eds.) (1994). *Elephant Seals*. University of California Press, Berkeley.

Mackintosh, N. A., and J. F. G. Wheeler (1929). "Southern Blue and Fin Whales." *Discovery Report 1:* 257–540.

Mann, J., R. C. Connor, P. L. Tyack, and H. Whitehead (eds.) (2000). *Cetacean Societies: Field Studies of Dolphins and Whales*. University of Chicago Press, Chicago.

Martens, F. (1675). *Spitzbergische oder Groenlandische Reise Beschreibung gethan im Jahr 1671: aus eigner Erfahrunge beschrieben, die dazu erforderte Figuren nach dem Leben selbst abgerissen (so hierbey in Kupffer zu sehen) und jetzo durch den Druck mitgetheilet*. Auff Gottfried Schultzens Kosten gedruckt, Hamburg.

Matthews, L. H. (1966). Chairman's Introduction to First Session of the International Symposium on Cetacean Research. *In* "Whales, Dolphins and Porpoises." (K. S. Norris, ed.), pp. 3–6. University of California Press, Berkeley.

Matthews, L. H. (1978). *The Natural History of the Whales*. Columbia University Press, New York.

Miller, W. C. G. (1888). The myology of the Pinnipedia. *In* "Report on the Scientific Results of the Voyage of H.M.S. Challenger." 26(2): 139–240; appendix to Turner's report. Challenger Office, 1880–1895, Edinburgh.

Murie, J. (1870). "Researches Upon the Anatomy of the Pinnipedia. Part I. On the Walrus *(Trichechus rosmarus* Linn.)." *Trans. Zool. Soc. London 7:* 411–464.

Murie, J. (1872). "Researches Upon the Anatomy of the Pinnipedia. Part 2. Descriptive Anatomy of the Sea-Lion *(Otaria jubata)*." *Trans. Zool. Soc. London 7:* 527–596.

Murie, J. (1874). "Researches Upon the Anatomy of the Pinnipedia. Part 3. Descriptive Anatomy of the Sea-Lion *(Otaria jubata)*." *Trans. Zool. Soc. London 8:* 501–562.

Norris, K. (1966). *Whales, Dolphins, and Porpoises*. University of California Press, Berkeley, CA.

Norris, K. S., B. Wursig, R. S. Wells, and M. Wursig (1994). *The Hawaiian Spinner Dolphin*. University of California Press, Berkeley, CA.

Perrin, W. F., B. Wursig, and J. G. M. Thewissen (eds.) (2002). *Encyclopedia of Marine Mammals*. Academic Press, San Diego, CA.

Pfeiffer, C. J. (ed.) (2002). *Molecular and Cell Biology of Marine Mammals*. Krieger Publishing Company Malabar, FL.

Pliny the Elder. C. Plini Secundi Naturalis historiae libri XXXVII; post Ludovici Iani obitum recognovit et scripturae discrepantia adiecta edidit Carolus Mayhoff. Teubner, 1906–09, Lipsiae.

Read, A. J., P. R. Wiepkema, and P. E. Nachtigall (eds.) (1997). *The Biology of the Harbour Porpoise*. De Spil Publishers, Woerden, The Netherlands.

Reeves, R. R., B. S. Stewart, P. J. Clapham, and J. Powell (2002). *National Audubon Society Guide to Marine Mammals of the World*. Alfred A. Knopf, New York.

Reeves, R. R., Stewart, B. S., and Leatherwood, S. (1992). *The Sierra Club Handbook of Seals and Sirenians*. Sierra Club Books, San Francisco.

Renouf, D. (ed.) (1991). *Behaviour of Pinnipeds*. Chapman & Hall, New York.

Reynolds, J. E., III, and D. K. Odell (1991). *Manatees and Dugongs*. Facts on File, New York.

Reynolds, J. E., III, and S. A. Rommel (eds.) (1999). *Biology of Marine Mammals*. Smithsonian Institution Press, Washington, D.C.

Reynolds, J. E., III, R. S. Wells, and S. D. Eide (2000). *The Bottlenose Dolphin: Biology and Conservation*. University Press of Florida, Gainesviller, FL.

Ridgway, S. H. (ed.) (1972). *Mammals of the Sea*. Thomas, Springfield, IL.

Ridgway, S. H., and R. Harrison (ed.) (1981–1998). *Handbook of Marine Mammals,* Vols. 1–6. Academic Press, San Diego, CA.

Riedman, M. L. (1990). *The Pinnipeds*. University of California Press, Berkeley, CA.

Riedman, M.L., and J. Estes (1990). "The Sea Otter *(Enhydra lutris)*: Behavior, Ecology, and Natural History." *U.S. Dep. Int. Biol. Rep. 90*(14): 1–126.

Rondelet, G. (1554–1555). Libri de piscibus marinis, in quibus veræ piscium effigies expressæ sunt. apud Matthiam Bonhomme, Lugduni.

Samansky, T. S. (2002). "Starting Your Career as a Marine Mammal Trainer." DolphinTrainer.com

Scammon, C. M. (1874). *The Marine Mammals of the North-Western Coast of North America: Described and Illustrated: Together with an Account of the American Whalefishery*. John H. Carmany, San Francisco.

Scholander, P. F. (1940). "Experimental Investigations on the Respiratory Function in Diving Mammals and Birds." *Hvalrådets Skrifter. Det Norske Videnskaps-Akademi I Oslo 22:* 1–131.

Scoresby, W. (1820). *An Account of the Arctic Regions: With a History and Description of the Northern Whale-Fishery*. Edinburgh.

True, F. (1904). "Whale Bone Whales of the Western North Atlantic Compared with Those Occurring in European Waters" . . . . *Smithsonian Contrib. Knowledge*: 33. Washington, DC.

Slijper, E. (1962). *Whales*. Hutchinson, London.

Steller, G. W. (1751). "The Beasts of the Sea." *Novi Comm. Acad. Sci. Petropolitanae 2:* 289–398.

Thompson, R. B. (1915). "Osteology of Antarctic Seals." *Rep. Scient. Results Scott. Nam. Antarc. Exped. 4*(3): 17–31.

Van Beneden, P. J. (1889). *Histoire naturelle des cetaces des mers d'Europe*. Bruxelles.

Watkins, W. A., and D. Wartzok (1985). "Sensory Biophysics of Marine Mammals." *Mar Mamm. Sci. 1:* 219–260.

Watson, L. (1981). *Whales of the World*. Hutchinson, London.

Whitehead, H. (2003). *Sperm Whales: Social Evolution in the Ocean*. University of Chicago Press, Chicago.

Zorgdrager, C. G. (1720). *Bloeyende Opkomst der Aloude en Hedendaagsche Groenlandsche Visschery*. Johannes Oosterwyk, T'Amsterdam.

2

Systematics
and Classification

2.1. Introduction: Systematics—What Is It and Why Do It?

Systematics is the study of biological diversity that has as its emphasis on the reconstruction of **phylogeny,** the evolutionary history of a particular group of organisms (e.g., species). Systematic knowledge provides a framework for interpreting biological diversity. Because it does this in an evolutionary context it is possible to examine the ways in which attributes of organisms change over time, the direction in which attributes change, the relative frequency with which they change, and whether change in one attribute is correlated with change in another. It also is possible to compare the descendants of a single ancestor to look for patterns of origin and extinction or relative size and diversity of these groups. Systematics also can be used to test hypotheses of adaptation. For example, consider the evolution of the ability to hear high frequency sounds, or **echolocation,** in toothed whales. One hypothesis for how toothed whales developed echolocation suggests that the lower jaw evolved as a unique pathway for the transmission of high frequency sounds under water. However, based on a study of the hearing apparatus of archaic whales, Thewissen *et al.* (1996) proposed that the lower jaw of toothed whales may have arisen for a different function, that of transmitting low frequency sounds from the ground, as do several vertebrates including the mole rat. According to this hypothesis, the lower jaw became specialized later for hearing high frequency sound. In this way the lower jaw of toothed whales may be an **exaptation** for hearing high frequency sounds. An exaptation is defined as any adaptation that performs a function different from the function that it originally held. A more complete understanding of the evolution of echolocation requires examination of other characters involved such as the presence of a melon and the morphology of the middle ear and jaw as well as the bony connections between the ear and skull (see Chapter 11).

An understanding of the evolutionary relationships among species can also assist in identifying priorities for conservation (Brooks *et al.*, 1992). For example, the argument for the conservation priority of sperm whales is strengthened by knowing that this lineage occupies a key phylogenetic position as basal relative to the other species of toothed whales. These pivotal species are of particular importance in providing baseline comparative data for understanding the evolutionary history of the other species of toothed whales. Sperm whales provide information on the origin of various morphological characters that permit suction feeding and the adaptive role of these features in the early evolution of toothed whales.

Perhaps most importantly, systematics predicts properties of organisms. For example, as discussed by Promislow (1996), it has been noted that some toothed whales (e.g., pilot whales and killer whales) that have extended parental care also show signs of reproductive aging (i.e., pregnancy rates decline with increasing age of females), whereas baleen whales (e.g., fin whales) demonstrate neither extended parental care nor reproductive aging (Marsh and Kasuya, 1986). Systematics predicts that these patterns would hold more generally among other whales and that we should expect other toothed whales to show reproductive aging.

Finally, systematics also provides a useful foundation from which to study other biological patterns and processes. Examples of such studies include the coevolution of pinniped parasites and their hosts (Hoberg, 1992, 1995), evolution of locomotion and feeding in pinnipeds (Berta and Adam, 2001; Adam and Berta, 2002), evolution of body size in phocids (Wyss, 1994), evolution of phocid breeding patterns (Perry *et al.*, 1995) and pinniped recognition behavior (Insley *et al.*, 2003), and the evolution of hearing in whales (Nummela *et al.*, 2004). Male social behavior among cetaceans was studied using a phylogenetic approach (Lusseau, 2003), and Kaliszewska *et al.* (2005) explored the population structure of right whales, based on genetic studies of lice that live in association with these whales.

2.2. Some Basic Terminology and Concepts

The discovery and description of species and the recognition of patterns of relationships among them is founded on the concept of evolution. Patterns of relationships among species are based on changes in the features or **characters** of an organism. Characters are diverse, heritable attributes of organisms that include DNA base pairs, anatomical and physiological features, and behavioral traits. Two or more forms of a given character are termed the **character states.** For example, the character "locomotor pattern" might consist of the states "alternate paddling of the four limbs (quadrupedal paddling)," "paddling by the hind limbs only (pelvic paddling)," "lateral undulations of the vertebral column and hind limb (caudal undulation)," and "vertical movements of the tail (caudal oscillation)." Evolution of a character may be recognized as a change from a preexisting, or **ancestral** (also referred to as **plesiomorphic** or **primitive**), character state to a new **derived** (also referred to as **apomorphic**) character state. For example, in the evolution of locomotor patterns in cetaceans, the pattern hypothesized for the earliest whales is one in which they swam by paddling with the hind limbs. Later diverging whales modified this feature and show two derived conditions: (1) lateral undulations of the vertebral column and hind limbs and (2) vertical movements of the tail.

The basic tenet of **phylogenetic systematics,** or **cladistics** (from the Greek word meaning "branch"), is that shared derived character states constitute evidence that the species possessing these features share a common ancestry. In other words, the shared derived features or **synapomorphies** represent unique evolutionary events that may be used to link two or more species together in a common evolutionary history. Thus, by sequentially

linking species together based on their common possession of synapomorphies, the evolutionary history of those **taxa** (named groups of organisms) can be inferred.

Relationships among taxonomic groups (e.g., species) are commonly represented in
the form of a **cladogram,** or **phylogenetic tree,** a branching diagram that conceptually
represents the best estimate of phylogeny (Figure 2.1). The lines or branches of the
cladogram are known as **lineages** or **clades.** Lineages represent the sequence of ancestor-
descendant populations through time. Branching of the lineages at **nodes** on the cladogram represents **speciation events,** a splitting of a lineage resulting in the formation of
two species from one common ancestor. Trees can be drawn to display the branching
pattern only or in the case of molecular phylogenetic trees drawn with proportional
branch lengths that correspond to the amount of evolution (approximate percentage
sequence divergence) between the two nodes they connect.

The task in inferring a phylogeny for a group of organisms is to determine which characters are derived and which are ancestral. If the ancestral condition of a character or
character state is established, then the direction of evolution, from ancestral to derived,
can be inferred, and synapomorphies can be recognized. The methodology for inferring
direction of character evolution is critical to cladistic analysis. **Outgroup comparison** is
the most widely used procedure. It relies on the argument that a character state found in
close relatives of a group (the **outgroup**) is likely also to be the ancestral or primitive state
for the group of organisms in question (the **ingroup**). Usually more than one outgroup is
used in an analysis, the most important being the first or genealogically closest outgroup
to the ingroup, called the **sister group.** In many cases, the primitive state for a taxon can
be ambiguous. The primitive state can only be determined if the primitive states for the
nearest outgroup are easy to identify and those states are the same for at least the two
nearest outgroups (Maddison *et al.*, 1984).

Using the previous example, determination of the primitive cetacean locomotor pattern is based on its similarity to that of an extinct relative to the cetaceans, a group of four

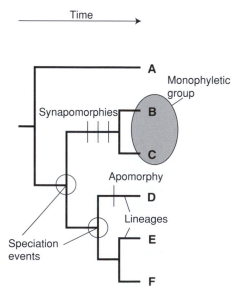

Figure 2.1.   A cladogram illustrating general terms discussed in the text.

legged mammals known as the **mesonychids** (i.e., an outgroup), which are thought to have swam by quadrupedal paddling. Locomotion in whales went through several stages. Ancestral whales (i.e., *Ambulocetus*) swam by pelvic paddling propelled by the hind limbs only. Later diverging whales (i.e., *Kutchicetus*) went through a caudal undulation stage propelled by the feet and tail. Finally, extinct dorudontid cetaceans and modern whales adopted caudal oscillation using vertical movements of the tail as their swimming mode (Figure 2.2; Fish, 1993).

Derived characters are used to link **monophyletic groups,** groups of taxa that consist of a common ancestor plus all descendants of that ancestor. In contrast to a monophyletic group, **paraphyletic and polyphyletic groups** (designated by quotation marks) include a common ancestor and some, but not all, of the descendants of that ancestor. A real example of a paraphyletic group is the recognition of an extinct group of cetaceans known as **"archaeocetes."** A rapidly improving fossil record and phylogenetic knowledge of whales now support the inclusion of "archaeocetes" as the ancestors of both baleen whales and toothed whales rather than as a separate taxonomic category (e.g., Thewissen *et al.*, 1996). In a **polyphyletic group,** taxa that are separated from each other by more than two ancestors are placed together without including all the descendants of their common ancestor. For example, recent molecular data supports river dolphins as a polyphyletic group because Indian river dolphins do not share the same common ancestor as other river dolphins (Figure 2.3).

Monophyletic groups can be characterized in two ways. First, a monophyletic group can be **defined** in terms of ancestry, and second, it can be **diagnosed** in terms of characters (see Appendix 3). For example, whales or cetaceans can be defined as including the common ancestor of *Pakicetus* (an extinct whale) and all of its descendants including

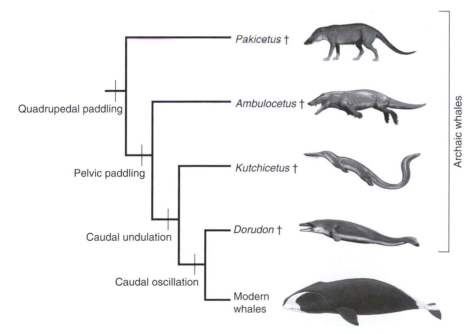

Figure 2.2. Distribution of character states for locomotor pattern among cetaceans. Reconstructions of the archaic whales *Pakicetus, Rodhocetus, Kutchicetus,* and *Dorudon* are illustrated by Carl Buell. The modern mysticete, the bowhead whale, *Balaena mysticetus,* is illustrated by P. Folkens.

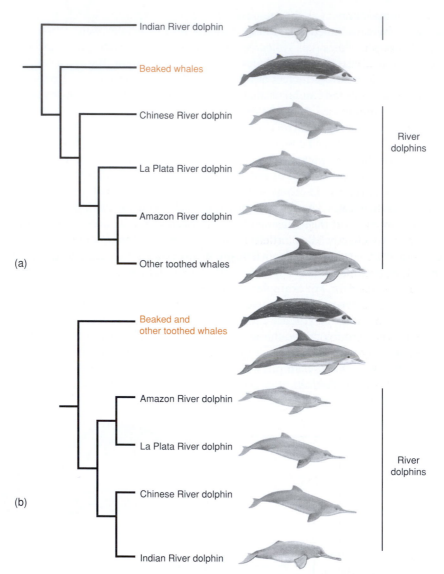

Figure 2.3. Alternative hypotheses for the phylogeny of river dolphins. **(a)** Molecular view supporting river dolphin polyphyly. **(b)** Morphologic view of river dolphin monophyly.

both modern toothed and baleen whales. Note that this definition is based on ancestry and does not change because there will always be a common ancestor for whales. On the other hand, cetaceans can be diagnosed by a number of characters (e.g., thick, dense auditory bulla and morphology of cusps on posterior teeth; see also Chapter 4). The usefulness of the distinction between definition and diagnosis is that, although the definition may not change, the diagnosis can be altered to reflect changes in our knowledge of the distribution of characters. New data, new characters, or reanalysis of existing characters can modify the diagnosis. For example, in the early 1990s discoveries of new fossil cetaceans (e.g., *Ambulocetus* and *Rodhocetus*) have provided new characters illuminating the transition between whales and their closest ungulate relatives. The definition of

Cetacea has not changed, but the diagnosis has been modified according to this new character information. A third term also used in this book, **characterization,** refers to a list of distinguishing features, both shared primitive and shared derived characters, that are particularly useful in field or laboratory identification of various species.

A concept critical to cladistics is that of **homology.** Homology can be defined as the similarity of features resulting from common ancestry. Two or more features are homologous if their common ancestor possessed the same feature. For example, the flipper of a seal and the flipper of a walrus are homologous as flippers because their common ancestor had flippers. In contrast to homology, similarity not due to homology is termed **homoplasy.** The flipper of a seal and the flipper of a whale are homoplasious as flippers because their common ancestor lacked flippers. Homoplasy may arise in one of two ways: **convergence (parallelism)** or **reversal.** Convergence is the independent evolution of a similar feature in two or more lineages. Thus, seal flippers and whale flippers evolved independently as swimming appendages; their similarity is homoplasious by convergent evolution. Reversal is the loss of a derived feature coupled with the reestablishment of an ancestral feature. For example, in phocine seals (e.g., *Erignathus, Cystophora,* and the Phocini) the development of strong claws, lengthening of the third digit of the foot, and deemphasis of the first digit of the hand are character reversals because none of them characterize phocids ancestrally but are present in terrestrial arctoid carnivores.

It is a common, but incorrect, practice to refer to taxa as being either primitive or derived. This is deceptive, because individual taxa that have diverged earlier than others may have undergone considerable evolutionary modification on their own relative to taxa that have diverged later in time. For example, otariid seals have many derived characters, although they have diverged earlier than phocid seals. In short, taxa are not primitive, although characters may be.

2.3. How Do You Do Cladistics?

Cladograms are constructed using the following steps:

1. Select a group whose evolutionary relationships interest you. Name and define all taxa for that group. Assume that the taxa are monophyletic.

2. Select and define characters and character states for each taxon.

3. Arrange the characters and their states in a data matrix (see example in Table 2.1).

4. For each character, determine which state is ancestral (primitive) and which is derived. This is done using outgroup comparison. For example, if the distribution of character #1, thick fat layers of the skin, is taken into consideration, two character states are recognized: "absent" and "present." In Table 2.1, the outgroup (bears) have the former condition, which is equivalent to the ancestral state. This same state is also seen in one of the ingroup taxa, the fur seals and sea lions. The other ingroup taxa have thick fat layers "present," which is a synapomorphy that unites walruses and seals to the exclusion of fur seals and sea lions.

5. Construct all possible cladograms by sequentially grouping taxa based on the common possession of one or more shared derived character states (circles around character states in Table 2.1) and choose the one that has the most shared derived character states distributed among monophyletic groups (Figure 2.4b). Note that the tree in Figure 2.4a shows no resolution of relationships among taxa, referred to as a **polytomy,** and that the

Table 2.1. Data Set for Analysis of Recent Pinnipeds Plus an Outgroup Showing Five Characters and Their Character States

	Character/Character states				
Taxon	1 Thick fat layers	2 Locomotor type	3 Pelage	4 Middle ear bones	5 Lacrimal bone
Outgroup	absent	forelimb + hind limb	abundant	small	present
Ingroup:					
Fur seals and Sea lions	absent	forelimb	abundant	small	absent
Walruses	present	hind limb	sparse	large	absent
Seals	present	hind limb	sparse	large	absent

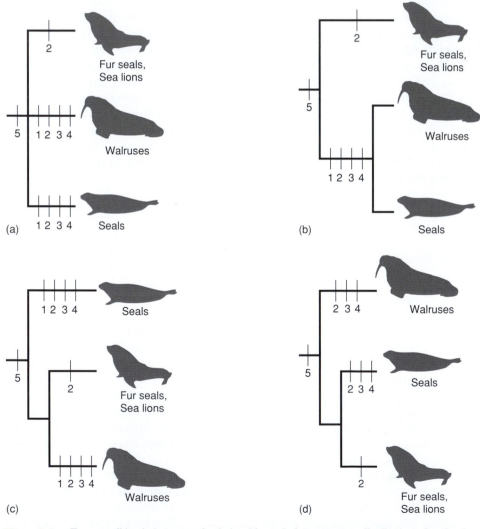

Figure 2.4. Four possible cladograms of relationship and character-state distributions for the three ingroups listed in Table 2.1. Part **b** has the most shared derived characters.

trees in Figure 2.4c and 24.d show mostly characters that are unique to one taxon and tell us nothing about relationships among different taxa.

The use of molecular characters (i.e., nucleotide sequence data) in cladistic analysis follows the same logic as other types of character data. Molecular data chosen should be nonrecombinant, maternally inherited alleles or fixed attributes. Next, generate sequences from these sources. The main repository for these sequences is the public nucleotide database (e.g., GenBank in the United States). Third, align the sequences. This is based on the assumption that sequence similarity equals sequence homology. This is a critical step and the identification of homologous nucleotide sequences can be as difficult in molecular phylogeny as it is in morphological studies. Finally, construct trees from the aligned sequence data.

2.4. Testing Phylogenetic Hypotheses

An important aspect of the reconstruction of phylogenetic relationships is known as the principle of **parsimony.** The basic tenet of the principle of parsimony is that the clado-gram that contains the fewest number of evolutionary steps, or changes between character states of a given character summed for all characters, is accepted as being the best estimate of phylogeny. For example, for all the possible cladograms for the data set of Table 2.1, the one (see Figure 2.4b) illustrated in detail in Figure 2.5a is the shortest because it contains the fewest number of evolutionary steps.

An alternative method to parsimony that is most often used with molecular data is **maximum likelihood.** This method is based on different assumptions about how characters evolve and a different method for joining taxa together. The approach begins with a mathematical formula that describes the probability that different types of nucleotide substitutions will occur. Given a particular phylogenetic tree with known branch lengths, a computer program can evaluate all possible tree topologies and compute the probability of producing the observed data, given the specified model of character change. This probability is reported as the tree's likelihood. The criterion for accepting or rejecting competing trees is to choose the one with the highest likelihood. One advantage of this approach is that by giving an exact probability for each tree this method facilitates quantitative comparison among trees. Closely related to likelihood methods are **Bayesian methods** for inferring phylogenies (Hulsenbeck *et al.*, 2001). Bayesian infer-ences of phylogeny employ a Markov chain Monte Carlo algorithm to solve the compu-tation aspects of sampling trees according to their posterior probabilities. The posterior probability of a tree can be interpreted as the probability that the tree is correct. To obtain posterior probabilities this approach requires a likelihood model and various parameters (e.g., phylogeny, branch lengths, and a nucleotide substitution model). One advantage of Bayesian inference is its ability to handle large data sets.

The methods used to search for the most parsimonious tree depend on the size and complexity of the data matrix. These methods are available in several computer programs [e.g., PAUP (Swofford, 2000); HENNIG86 (Farris, 1988); MacClade (Maddison and Maddison, 2000)]. The latter is particularly useful in visually assessing the evolution of characters. Recently, systematists have become concerned about the relative accuracy of phylogenetic trees (i.e., how much confidence can be placed in a specific phylogenetic reconstruction). Studies indicate that methods of phylogenetic

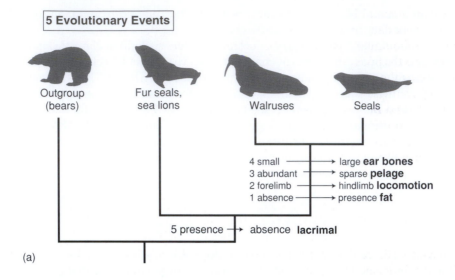

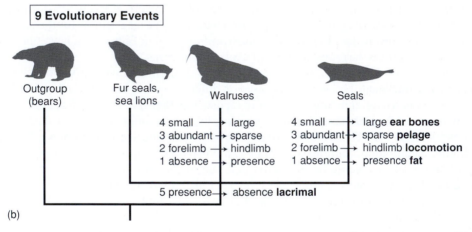

Figure 2.5.   Two of the four possible cladograms. **(a)** Most parsimonious cladogram. Note a total of five evolutionary events. **(b)** Alternative cladogram showing different relationships for taxa. Note that this cladogram requires nine evolutionary events, four more than the most parsimonious cladogram.

analysis are most accurate if sufficient consideration is given to such parameters as sampling, rigorous analysis, and computer capabilities (Hillis, 1995).

A related issue in systematics is how to evaluate different data sets (e.g., morphology, behavior, and DNA sequences), particularly whether they should be combined (also referred to as a "total evidence" approach) or analyzed separately (Bull *et al.*, 1993; Hillis, 1995). The results of a total evidence analysis can then be compared with the results of the separate analyses. Before data sets can be combined, it is necessary to determine if they are congruent, that is, the order of branching is not contradictory. Several statistical tests have been developed to test for significant incongruences among data sets (e.g., Hulsenbeck and Bull, 1996; Page, 1996). Having compared several or all possible trees often leads to the question: How good is the tree? If more than one tree is supported by the data, investigators typically examine the topologies of trees close to the optimal trees. Computer programs can evaluate multiple trees and create a **consensus tree** that represents the branching pattern supported by all of the nearly optimal trees.

Determining the accuracy and reliability of phylogenetic information in a given data set is an important aspect of phylogenetic analysis. There are several methods (i.e., **bootstrap analysis** and **Bremer support**) commonly employed that provide various ways to identify which portions of a tree are well supported and which are weak. If bootstrap support for a particular branch is high (i.e., 70% or higher), an investigator will usually conclude that it likely indicates a reliable grouping.

2.5. Going Beyond the Phylogenetic Framework: Elucidating Evolutionary and Ecological Patterns

Once a phylogenetic framework is produced, one of its most interesting uses is to elucidate questions that integrate evolution, behavior, and ecology. One technique used in this book to facilitate such evolutionary studies is **optimization,** or **mapping** (Funk and Brooks, 1990; Brooks and McLennan, 1991, 2002; Maddison and Maddison, 2000). Once a cladogram has been constructed, a feature or condition is selected to be examined in light of the phylogeny of the group. Examples included in this book include the evolution of body size, host-parasite associations, mating-reproductive behavior, hearing, feeding, and locomotor behavior. The condition of the terminal taxon (at the ends of branches) is identified and "mapped" onto the cladogram. There are various ways of mapping character changes onto the cladogram as discussed by Maddison and Maddison (2000). Hypothetical states are assigned to the nodes that reflect the most parsimonious arrangement of these conditions at each node. This allows one to determine the evolutionary trend of the condition in question. For example, consider the evolution of body size in phocid seals. One traditional assumption had been that small body size is the ancestral condition among phocids. This view is based on the assumption that seals of large body size represent an evolutionary advancement because they have a decreased surface area that in turn reduces body heat loss, an advantage in cold environments. This assumption, however, lacks historical evidence. When body size is mapped onto a phylogeny for seals and their relatives (walruses and sea lions; Figure 2.6), there is a more parsimonious explanation for the data (Wyss, 1994). Accordingly, large body size is the ancestral condition for seals. A decrease in body size evolved secondarily among phocine seals (e.g., harbor, ribbon, and spotted seal). This hypothesis led Wyss (1994) to question whether this decrease in size among phocids was correlated with any other pattern of character evolution. He discovered that phocines were characterized by massive character reversals and he hypothesized that these reversals might be related to shifts in timing during development (neoteny). In addition to a decrease in body size, a number of other characters among phocines provided evidence for developmental juvenilization (i.e., failure of certain regions of the skull to ossify, resulting in perforations in the basicranium and the lack of fusion of certain cranial bones). In this example, a phylogenetic approach provided a framework for questions regarding the relationship between the evolution of body size and the pattern of evolution of other characters. A developmental explanation for the observed body size pattern was then proposed and further evidenced by other characters.

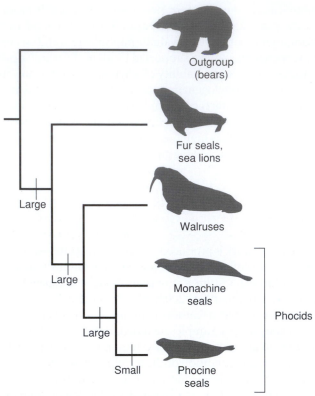

Figure 2.6.   Body size mapped onto pinniped phylogeny. (Based on Wyss, 1994, and Bininda-Emonds and Russell, 1996.)

Another growing area of interest in the comparative study of phylogenies is how to deal with different types of character change, such as discrete or categorical (e.g., presence or absence of limbs) versus continuously varying characters (e.g., amount of time spent foraging). Several different methods have been proposed to incorporate phylogenetic information into comparative analyses. Examples of these techniques include Felsenstein's (1985) method of independent contrasts and the spatial autocorrelation techniques of Chevrud *et al*. (1985). These methods are designed for use with primarily continuous characters and as such are beyond the scope of this text (see Felsenstein, 2004 for a recent review).

2.6. Taxonomy and Classification

In addition to phylogeny reconstruction an integral component of systematics is **taxonomy**, the description, identification, and classification of species. Although the taxonomy of mammals is relatively well known compared to other groups of organisms, we still are discovering previously unknown species of marine mammals. In the last decade, two new species of beaked whale were described (Reyes *et al*., 1991; Dalebout *et al*., 2002), another was resurrected (Van Helden *et al*., 2002), a new dolphin was reported (Beasley *et al.*, 2005), and evidence was presented for distinguishing three forms (probably subspecies) of killer whale (Pitman and Ensor, 2003). Among baleen whales a new species of balaenopterid was also reported (Wada *et al.*, 2003).

Recently, there has been recognition that DNA sequences can provide universal characters for taxonomic identification. This discovery has lead to the application of DNA or molecular taxonomy, the identification of specimens of known species (e.g., Baker *et al.*, 2003; Dalebout *et al.*, 2004). Such genetic characters are particularly useful for species in which morphological characters are subtle or difficult to compare because of rarity of specimens or widespread distributions. Given a database of "reference" sequences based on validated specimens (i.e., identified by experts for which diagnostic skeletal material or photographs are available), unknown "test" specimens can be identified to species based on their phylogenetic grouping with sequences from recognized species to the exclusion of sequences from other species. An example of the application of molecular taxonomy is the little known family Ziphiidae (beaked whales), which resulted in the correct identification of specimens involving animals previously misidentified from morphology (Dalebout *et al.*, 1998, 2002, 2004).

Nomenclature is the formal system of naming taxa according to a standardized scheme, which for animals is the International Code of Zoological Nomenclature. These formal names are known as scientific names. The most important thing to remember about nomenclature is that all species may bear only one scientific name. The scientific name is, by convention, expressed using Latin and Greek words.

Species names are always italicized (or underlined) and always consists of two parts, the genus name (always capitalized, e.g., *Trichechus*) plus the specific epithet (e.g., *manatus*). For this reason, species names are known as binomials and this type of nomenclature is called binomial nomenclature. Species also have common names. In the previous example, *Trichechus manatus* is also known in English by its common name, West Indian manatee.

Classification is the arrangement of taxa (e.g., species) into some type of hierarchy. Taxonomic ranks are hierarchical, meaning that each rank is inclusive of all other ranks beneath it. The major taxonomic ranks used in this book are as follows:

Major taxonomic ranks	*Example*
Order	Sirenia
Family	Trichechidae
Genus	*Trichechus*
Species	*manatus*

We need a system of classification so that we can communicate more easily about organisms. The two major ways to classify organisms are **phenetic** and **phylogenetic.** Phenetic classification is based on overall similarity of the taxa. Phylogenetic classification is that which is based on evolutionary history, or pattern of descent, which may or may not correspond to overall similarity. Phylogenetic systematists contend that classification should be based on phylogeny and should include only monophyletic groups. We have provided the most recent information on the classification and phylogeny of marine mammals. The classification of many marine mammal groups, however, is in a constant state of change due to new discoveries and information. Indeed, some systematists have offered compelling arguments for the elimination of taxonomic ranks altogether. In general, it is more important to know the names and characteristics of larger taxonomic groups like the Pinnipedia and the Sirenia than it is to memorize their rank.

2.7. Summary and Conclusions

A primary goal of systematics, the reconstruction of phylogenetic relationships, provides a framework in biology for interpreting patterns of evolution, behavior, and ecology. Relationships are reconstructed based on shared derived similarities between species, whether similarities in morphologic characters or in molecular sequences, that provide evidence that these species share a common ancestry. The direction of evolution of a character is inferred by outgroup comparison. The best estimate (most parsimonious) of phylogeny is the one requiring the fewest number of evolutionary changes. Phylogenetically based comparative analyses have proven to be a powerful tool for generating and testing ideas about the links between behavior and ecology. Taxonomy involves the description, identification, naming, and classification of species. Molecular taxonomy, the use of DNA sequences for identification of specimens of known species, is especially applicable for species in which morphological characters are difficult to observe or compare.

2.8. Further Reading

Readers are referred to texts by Wiley (1981), Wiley *et al.* (1991), Smith (1994), and Felsenstein (2004) for discussion of the principles and practice of phylogenetic systematics. Treatment of molecular data in phylogeny reconstruction is reviewed by Swofford *et al.* (1996), Graur and Li (2000), and Nei and Kumar (2000). Brooks and McLennan (1991, 2002), Harvey and Pagel (1991), Martins (1996), and Krebs and Davies (1997) provide examples of the use of phylogeny in studies of ecology and behavior.

Important websites with information on software programs related to phylogenetics are http://evolution.genetics.washington.edu created by Joe Felsenstein and the home pages of the Tree of Life Web project (http://tolweb.org/tree/phylogeny.html).

For a comprehensive reference data set to assist in the genetic identification of cetaceans see www.DNA-surveillance

References

Adam, P. J., and A. Berta (2002). "Evolution of Prey Capture Strategies and Diet in the Pinnipedimorpha (Mammalia: Carnivora)." *Oryctos 4:* 83–107.

Baker, C. S., M. L. Dalebout, S. Lavery, and H. A. Ross (2003). "www.DNA-Surveillance: Applied Molecular Taxonomy for Species Conservation and Discovery." *Trends Ecol. Evol. 18:* 271–272.

Beasley, I., K. M. Robertson, and P. Arnold (2005). "Description of a New Dolphin, the Australian Snubfin Dolphin *Orcaella heinsohni* sp. N. (Ceacea, Delphinididae). *Mar. Mamm. Sci. 21:* 365–400.

Berta, A., and P. J. Adam (2001). Evolutionary biology of pinnipeds. *In* "Secondary Adaptation to Life in Water" (J. M. Mazin and V. de Buffrenil, eds.), pp. 235–260. Verlag Dr. Friedrich Pfeil, München, Germany.

Bininda-Emonds, O. R. P., and A. P. Russell (1996). "A Morphological Perspective on the Phylogenetic Relationships of the Extant Phocid Seals (Mammalia: Carnivora: Phocidae)." *Bonner Zoologische Monographien, 41:* 1–256.

Brooks, D. R., R. L. Mayden, and D. A. McLennan (1992). "Phylogeny and Biodiversity; Conserving our Evolutionary Legacy." *Trends Ecol. Evol. 7:* 55–59.

Brooks, D. R., and D. A. McLennan (1991). *Phylogeny, Ecology, and Behavior*. University of Chicago Press, Chicago.

Brooks, D. R., and D. A. McLennan (2002). *The Nature of Diversity*. University of Chicago Press, Chicago.

Bull, J. J., J. P. Hulsenbeck, C. W. Cunningham, D. L. Swofford, and P. J. Waddell (1993). "Partitioning and Combining Data in Phylogenetic Analyses." *Syst. Biol. 42:* 384–397.

Chevrud, J. M., M. M. Dow, and W. Leutenegger (1985). "The Quantitative Assessment of Phylogenetic Constraints in Comparative Analyses: Sexual Dimorphism in Body Weights Among Primates." *Evolution 39:* 1335–1351.

Dalebout, M. L. C. S. Baker, J. G. Mead, V. G. Cockcroft, and T. K. Yamada (2004). "A Comprehensive and Validated Molecular Taxonomy of Beaked Whales, Ziphiidae." *J. Heredity 95:* 459–473.

Dalebout, M. L., J. G. Mead, C. S. Baker, A. N. Baker, and A. van Helden (2002). "A New Species of Beaked Whale *Mesoplodon perrini* sp. N. (Cetacea: Ziphiidae) Discovered Through Phylogenetic Analysis of Mitochondrial DNA Sequences." *Mar. Mamm. Sci. 18:* 577–608.

Dalebout, M. L., A. van Helden, K. Van Waerebeek, and C. S. Baker (1998). "Molecular Genetic Identification of Southern Hemisphere Beaked Whales (Cetacea: Ziphiidae)." *Mol. Ecol. 7:* 687–694.

Farris, J. S. (1988). *HENNIG86, Version 1.5*. Distributed by the author, Port Jefferson Station, NY.

Felsenstein, J. (1985). "Confidence Limits on Phylogenies: An Approach Using the Bootstrap. *Evolution 39:* 783–791.

Felsenstein, J. (2004). *Inferring Phylogenies*. Sinauer, Sunderland, MA.

Fish, F. (1993). "Influence of Hydrodynamic Design and Propulsive Mode on Mammalian Swimming Energetics." *Aust. J. Zool. 42:* 79–101.

Funk, V, and D. R. Brooks (1990). "Phylogenetic Systematics as the Basis of Comparative Biology." *Smithson. Contrib. Bot. 73:* 1–45.

Graur, D., and W-H. Li (2000). *Fundamentals of Molecular Evolution,* 2nd ed. Sinauer, Sunderland, MA.

Harvey, P., and M. D. Pagel (1991). *The Comparative Method in Evolutionary Biology*. Oxford University Press, Oxford.

Hillis, D. (1995). "Approaches for Assessing Phylogenetic Accuracy." *Syst. Biol. 44:* 3–16.

Hoberg, E. P. (1992). "Congruent and Synchronic Patterns in Biogeography and Speciation Among Seabirds, Pinnipeds, and Cestodes." *J. Parasitol. 78:* 601–615.

Hoberg, E. P. (1995). "Historical Biogeography and Modes of Speciation Across High Latitude Seas of the Holarctic: Concepts for Host-Parasite Coevolution Among the Phocini (Phocidae) and Tetrabothriidae (Eucestoda)." *Can. J. Zool. 73:* 45–57.

Hulsenbeck, J. P. F., and J. J. Bull (1996). "A Likelihood Ratio Test to Detect Conflicting Phylogenetic Signal. *Syst. Biol. 45:* 92–98.

Hulsenbeck, J. P. F. Ronquist, R. Nielsen, and J. P. Bollback (2001). "Bayesian Inference of Phylogeny and Its Impact on Evolutionary Biology." *Science 294:* 2310 2314.

Insley, S. J., A. V. Phillips, and I. Charrier (2003). "A Review of Social Recognition in Pinnipeds." *Aquat. Mamm. 29:* 181–201.

Kaliszewska, Z. A., J. Seger, V. J. Rowntree, S. G. Barco, et al. (2005). "Population Histories of Right Whales (Cetacea: *Eubalaena*) Inferred from Mitochondrial Sequence Divesities and Divergences of Their Whale Lice (Amphipoda: Cyamus)." *Mol. Ecol. 14*(10).

Krebs, J. R., and N. B. Davies (1997). *Behavioural Ecology: An Evolutionary Approach,* 4th ed. Blackwell Science, London.

Lusseau, D. (2003). "The Emergence of Cetaceans; Phylogenetic Analysis of Male Social Behavior Supports the Cetartiodactyla Clade." *J. Evol. Biol. 16:* 531–535.

Maddison, W., M. Donoghue, and D. Maddison (1984). "Outgroup Analysis and Parsimony." *Syst. Zool. 33:* 83–103.

Maddison, W. P., and D. R. Maddison (2000). *Mac Clade: Analysis of Phylogeny and Character Evolution, Version 4.0*. Sinauer, Sunderland, MA.

Martins, E. (ed.) (1996). *Phylogenies and the Comparative Method in Animal Behavior*. Oxford University Press, New York.

Marsh, H., and T. Kasuya (1986). "Evidence for Reproductive Senescence in Female Cetaceans." *Rep. Int. Whal. Comm., Spec. Issue 8:* 57–74.

Nei, M., and S. Kumar (2000). *Molecular Evolution and Phylogenetics*. Cambridge University Press, Cambridge.

Nummela, S. J., G. M. Thewissen, S. Bajapi, S. T. Hussain, and K. Kumar (2004). "Eocene Evolution of Whale Hearing." *Nature 430:* 776–778.

Page, R. D. M. (1996). "On Consensus, Confidence, and 'Total Evidence.'" *Cladistics 12:* 83–92.

Perry, E. A., S. M. Carr, S. E. Bartlett, and W. S. Davidson (1995). "A Phylogenetic Perspective on the Evolution of Reproductive Behavior in Pagophilic Seals of the Northwest Atlantic as Indicated by Mitochondrial DNA Sequences." *J. Mammal. 76*(1): 22–31.

Pitman, R. L., and P. Ensor (2003). "Three Forms of Killer Whales (*Orcinus orca*) in Antarctic Waters." *J. Cetacean Res. Management 5:* 131–139.

Promislow, D. E. L. (1996). Using comparative approaches to integrate behavior and population biology. *In* "Phylogenies and the Comparative Methods in Animal Behavior" (E. Martins, ed.), pp. 288–323. Oxford University Press, New York.

Reyes, J. C., J. G. Mead, and K. Van Waerebeek (1991). "A New Species of Beaked Whale *Mesoplodon peruvianus* sp. n. (Cetacea: Ziphiidae) from Peru." *Mar Mamm. Sci. 7*(1): 1–24.

Smith, A. B. (1994). *Systematics and the Fossil Record*. Blackwell Science, London.

Swofford, D. L. (2000). *PAUP*: Phylogenetic Analysis using Parsimony, Version 4*. Sinauer Associates, Sunderland, MA.

Swofford, D. L., G. J. Olsen, P. J. Waddell, and D. M. Hillis (1996). Phylogenetic inference. *In* "Molecular Systematics" (D. M. Hillis, C. Moritz, and B. Mable, eds.), 2nd ed., pp. 407–514. Sinauer Associates, Sunderland, MA.

Thewissen, J. G. M., S. I. Madar, and S. T. Hussain (1996). "*Ambulocetus natans,* an Eocene Cetacean (Mammalia) from Pakistan." *CFS. Cour. Forschungsinst. Senckenberg 191:* 1–86.

Van Helden, A. L., A. N. Baker, M. L. Dalebout, J. C. Reyes, K. Van Waerebeek, and C. S. Baker (2002). "Resurrection of *Mesoplodon traversii* (Gray, 1874), Senior Synonym of M, Bahamondi Reyes, Van Waerebeek, Cardenas and Yanez, 1995 (Cetacea: Ziphiidae)." *Mar. Mamm. Sci. 18:* 609–621.

Wada, S., M. Oishi, and T. Yamada. (2003). "A Newly Discovered Species of Living Baleen Whale. *Nature 426:* 278–281.

Wiley, E. O. (1981). *Phylogenetics: The Theory and Practice of Phylogenetic Systematics*. Wiley, New York.

Wiley, E. O, D. Siegel-Causey, D. R. Brooks, and V. A. Funk (1991*). The Complete Cladist: A Primer of Phylogenetic Procedures*. Univ. Kans. Mus. Nat. Hist., Spec. Publ., No. 19, Lawrence, KS.

Wyss, A. R. (1994). "The Evolution of Body Size in Phocids: Some Ontogenetic and Phylogenetic Observations." *Proc. San Diego Soc. Nat. Hist. 29:* 69–75.

3

Pinniped Evolution and Systematics

3.1. Introduction

Modern pinnipeds are aquatic members of the mammalian Order Carnivora and comprise three monophyletic families: the Otariidae (eared seals or fur seals and sea lions), the Odobenidae (walruses), and the Phocidae (true or earless seals). Pinnipeds comprise slightly more than one fourth (28%) of the diversity of marine mammals. Thirty-four to thirty-six living different species of pinnipeds are distributed throughout the world: 19 phocids, 14–16 otariids, and the walrus. Roughly 90% of an estimated 50 million individual pinnipeds are phocids; the remaining 10% are otariids and odobenids (Riedman, 1990; Rice, 1998). The fossil record indicates that extant pinnipeds represent only a small fraction of what was once a much more diverse group. For example, only a single species of walrus exists today, whereas no less than 10 genera and 13 species existed in the past (Deméré, 1994a). The earliest well-documented record of pinnipeds is from the late Oligocene (27 to 25 Ma; Figure 3.1), although a slightly earlier record (29 Ma) is less well substantiated.

New discoveries of fossil pinnipeds together with comparative studies of living taxa have enabled a more complete understanding of pinniped origin, diversification, and morphology. These topics are explored in this chapter. Characters defining major groups of pinnipeds are also listed for reference. Controversies regarding the relationship of pinnipeds to other carnivores, relationships among pinnipeds, and the alliance of an extinct pinniped group, desmatophocids, also are considered.

3.2. Origin and Evolution

3.2.1. Pinnipeds Defined

The name pinniped comes from the Latin *pinna* and *pedis* meaning "feather-footed," referring to the paddle-like fore- and hind limbs of seals, sea lions, and walruses, which they use in locomotion on land and in the water. Pinnipeds spend considerable amounts

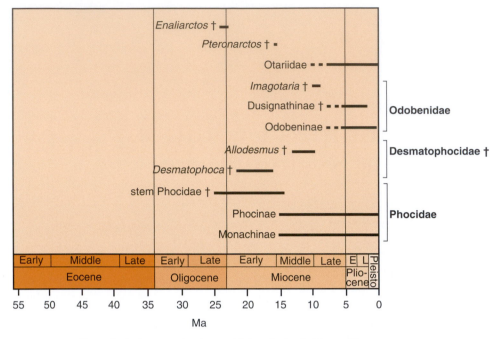

Figure 3.1.   Chronological ranges of extinct and living pinnipeds. Ma = million years ago.

of time both in the water and on land or ice, differing from cetaceans and sirenians, which are entirely aquatic. In addition to blubber, some pinnipeds have a thick covering of fur.

In seeking the origin of pinnipeds we must first define them. Is the group mono-phyletic or not? Although this question has been subject to considerable controversy during the last century (e.g., see Flynn *et al.*, 1988), the majority of scientists today agree that the Pinnipedia represent a natural, monophyletic group. Pinnipeds are diagnosed by a suite of derived morphological characters (for a complete list see Wyss, 1987, 1988; Berta and Wyss, 1994). All pinnipeds, including both fossil and recent taxa, possess the characters described later, although some of these characters have been modified or lost secondarily in later diverging taxa.

Some of the well known synapomorphies possessed by pinnipeds (enumerated in Figure 3.2 and illustrated in Figures 3.3–3.5) are defined as follows:

1. *Large infraorbital foramen.* The infraorbital foramen, as the name indicates, is located below the eye orbit and allows passage of blood vessels and nerves. It is large in pinnipeds in contrast to its small size in most terrestrial carnivores.

2. *Maxilla makes a significant contribution to the orbital wall.* Pinnipeds display a unique condition among carnivores in which the maxilla (upper jaw) forms part of the lateral and anterior walls of the orbit of the eye. In terrestrial carnivores, the maxilla is usually limited in its posterior extent by contact of several facial bones (jugal, palatine, and/or lacrimal).

3. *Lacrimal absent or fusing early in ontogeny and does not contact the jugal.* Associated with the pinniped configuration of the maxilla (character 2) is the great reduction or absence of one of the facial bones, the lacrimal. Terrestrial carnivores have

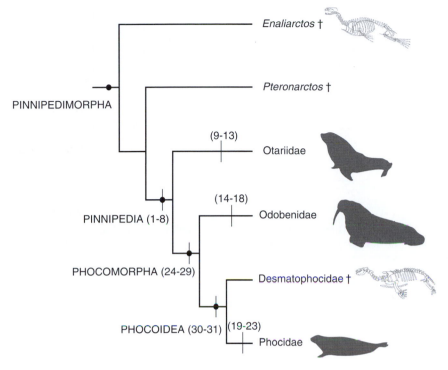

Figure 3.2.   A cladogram depicting the relationships of the major clades of pinnipeds. Numbers at nodes refer to synapomorphies listed in the text and † = extinct taxa; see also Figures 3.3, 3.4, and 3.5. For more detailed cladograms of individual families, see Figures 3.12, 3.19, and 3.20. (Modified from Wyss, 1988; Berta and Wyss, 1994; Deméré, 1994b.)

a lacrimal that contacts the jugal or is separated from it by a thin sliver of the maxilla and thus can be distinguished from pinnipeds.

4. *Greater and lesser humeral tubercles enlarged.* Pinnipeds are distinguished from terrestrial carnivores by having strongly developed tubercles (rounded prominences) on the proximal end of the humerus (upper arm bone).

5. *Deltopectoral crest of humerus strongly developed.* The crest on the shaft of the humerus for insertion of the deltopectoral muscles in pinnipeds is strongly developed in contrast to the weak development observed in terrestrial carnivores.

6. *Short and robust humerus.* The short and robust humerus of pinnipeds is in contrast to the long, slender humerus of terrestrial carnivores.

7. *Digit I on the hand emphasized.* In the hand of pinnipeds the first digit (thumb equivalent) is elongated, whereas in other carnivores the central digits are the most strongly developed.

8. *Digit I and V on the foot emphasized.* Pinnipeds have elongated side toes (digits I and V, equivalent to the big toe and little toe) of the foot, whereas in other carnivores the central digits are the most strongly developed.

3.2.2. Pinniped Affinities

Since the name Pinnipedia was first proposed by Illiger in 1811, there has been debate on the relationships of pinnipeds to one another and to other mammals. Two hypotheses

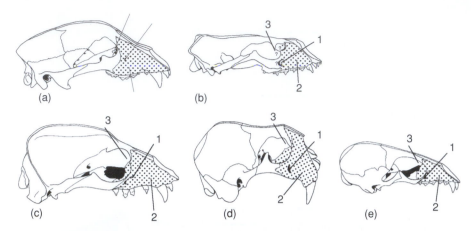

Figure 3.3. Lateral views of the skulls of representative pinnipeds and a generalized terrestrial arctoid. **(a)** Bear, *Ursus americanus.* **(b)** Fossil pinnipedimorph, *Enaliarctos mealsi.* **(c)** Otariid, *Zalophus californianus.* **(d)** Walrus, *Odobenus rosmarus.* **(e)** Phocid, *Monachus schauinslandi,* illustrating pinniped synapomorphies. Character numbers (see text for further description): 1 = large infra-orbital foramen; 2 = maxilla (stippled) makes a significant contribution to the orbital wall; 3 = lacrimal absent or fusing early and does not contact jugal. (From Berta and Wyss, 1994.)

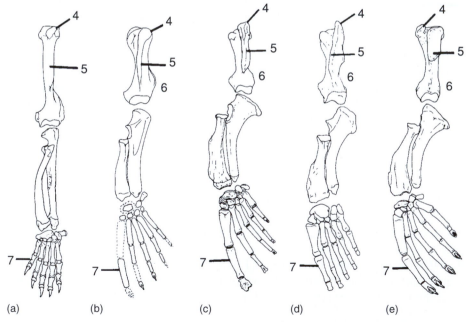

Figure 3.4. Left forelimbs of representative pinnipeds **(b–e)** and a generalized terrestrial arctoid **(a)** in dorsal view illustrating pinniped synapomorphies. Labels as in Figure. 3.3 plus character numbers (see text for further description): 4 = greater and lesser humeral tubercles enlarged; 5 = deltopectoral crest of humerus strongly developed; 6 = short, robust humerus; 7 = digit I on manus emphasized. (From Berta and Wyss, 1994.)

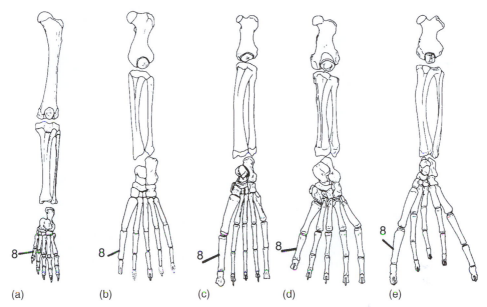

(a)   (b)   (c)   (d)   (e)

Figure 3.5. Left hind limbs of representative pinnipeds **(b–e)** and a generalized terrestrial arctoid **(a)** in dorsal view illustrating pinniped synapomorphies. Labels as in Figure 3.3. Character number (see text for further description): 8 = digit I and V on the foot emphasized. (From Berta and Wyss, 1994.)

have been proposed. The monophyletic hypothesis proposes that the three pinniped families share a single common evolutionary origin (Figure 3.6a). The **diphyletic** view (also referred to as pinniped diphyly; Figure 3.6b) calls for the origin of pinnipeds from two carnivore lineages, the alliance of odobenids and otariids being somewhere near ursids (bears) and a separate origin for phocids from the mustelids (weasels, skunks, otters, and kin).

Traditionally, morphological and paleontological evidence supported pinniped diphyly (McLaren, 1960; Tedford, 1976; Repenning *et al.*, 1979; Muizon, 1982). On the basis of his reevaluation of the morphological evidence, Wyss (1987) argued in favor of a return to the single origin interpretation. This hypothesis of pinniped monophyly has received considerable support from both morphological (Flynn *et al.*, 1988; Berta *et al.*, 1989; Wyss and Flynn, 1993; Berta and Wyss, 1994) and biomolecular studies (Sarich, 1969; Árnason and Widegren, 1986; Vrana *et al.*, 1994; Lento *et al.*, 1995; Árnason *et al.*, 1995; Flynn and Nedbal, 1998; Flynn *et al.*, 2000; Davis *et al.*, 2004).

All recent workers, on the basis of both molecular and morphologic data, agree that the closest relatives of pinnipeds are arctoid carnivores, which include procyonids (raccoons and their allies), mustelids, and ursids, although which specific arctoid group forms the closest alliance with pinnipeds is still disputed (see recent review Flynn and Wesley-Hunt, 2005). There is evidence to support a mustelid (Bininda-Emonds and Russell, 1996; Flynn and Nedbal, 1998; Bininda-Emonds *et al.*, 1999), ursid (Wyss and Flynn, 1993; Berta and Wyss, 1994), and ursid-mustelid (Davis *et al.*, 2004) ancestry.

Although both morphological and molecular data support pinniped monophyly there is still disagreement on relationships among pinnipeds. Most of the controversy lies in the debate as to whether the walrus is most closely related to phocids or to otariids. Some recent morphologic evidence for extant pinnipeds unites the walrus and phocids as sister

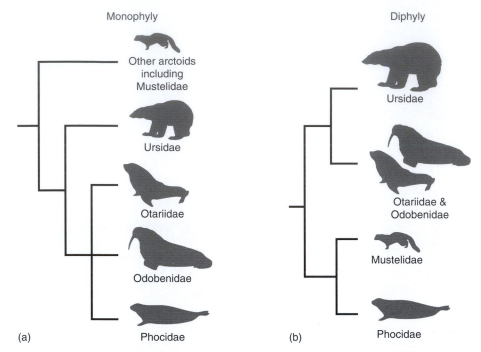

Figure 3.6.   Alternative hypotheses for relationships among pinnipeds. **(a)** Monophyly with ursids as the closest pinniped relatives. **(b)** Diphyly in which phocids and mustelids are united as sister taxa, as are otariids, odobenids, and ursids.

groups (Figure 3.7a; Wyss, 1987; Wyss and Flynn, 1993; Berta and Wyss, 1994) and is discussed later in this chapter. An alternative view based mostly on molecular data (Vrana *et al.*, 1994; Lento *et al.*, 1995; Árnason *et al.*, 1995; Davis *et al.*, 2004) but with support from total evidence analyses (e.g., Flynn and Nedbal, 1998) supports an alliance between the walrus and otariids (Figure 3.7b).

3.2.3. Early "Pinnipeds"

An understanding of the evolution of early "pinnipeds" necessitates a knowledge of certain fossil taxa. The earliest diverging lineage of "pinnipeds" actually are members of the Pinnipedimorpha clade and appear to have originated in the eastern North Pacific

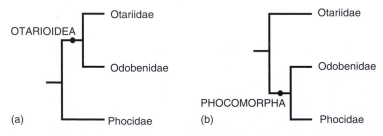

Figure 3.7.   Alternative hypotheses for position of the walrus. **(a)** "Otarioidea" clade. **(b)** Phocomorpha clade.

(Oregon) during the late Oligocene (27–25 Ma; see Figure 3.1). The earliest known pinnipedimorph, *Enaliarctos,* is represented by five species (Mitchell and Tedford, 1973; Berta, 1991). The ancestral pinnipedimorph dentition, exemplified by *E. barnesi* and *E. mealsi,* is **heterodont,** with large blade-like cusps on the upper cheekteeth well-adapted for shearing (Figure 3.8). These dental features together with those from the skull (when compared with terrestrial carnivores) indicate closest similarity in terms of derived characters with archaic bears (amphicynodonts; see Figure 3.8).

Other species of the genus *Enaliarctos* show a trend toward the decreasing shearing function of the cheekteeth (e.g., reduction in the number and size of cusps). These dental trends herald the development of simple peg-like, or **homodont,** cheekteeth characteristic of most living pinnipeds (Berta, 1991). The latest record of *Enaliarctos* is along the Oregon coast from rocks of 25–18 Ma in age. An "enaliarctine" pinniped also has been reported from the western North Pacific (Japan) in rocks of late early Miocene (17.5–17 Ma; Kohno, 1992), although the specimen needs further study before its taxonomic assignment can be confirmed.

The pinnipedimorph *E. mealsi* is represented by a nearly complete skeleton collected from the Pyramid Hill Sandstone Member of the Jewett Sand in central California (Figure 3.9; Berta *et al.,* 1989; Berta and Ray, 1990). The entire animal is estimated at 1.4–1.5 m in length and between 73 and 88 kg in weight, roughly the size and weight of a small male harbor seal.

Considerable lateral and vertical movement of the vertebral column was possible in *E. mealsi.* Also, both the fore- and hind limbs were modified as flippers and used in aquatic locomotion. Several features of the hind limb suggest that *E. mealsi* was highly capable of maneuvering on land and probably spent more time near the shore than extant pinnipeds (see also Chapter 8).

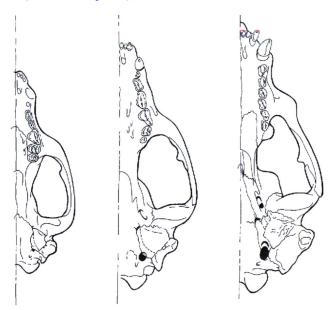

Figure 3.8.   Skulls and dentitions of representative pinnipeds and a generalized terrestrial arctoid in ventral view. **(a)** Archaic bear, *Pachcynodon* (Oligocene, France). **(b)** Fossil pinnipedimorph, *Enaliarctos mealsi* (early Miocene). **(c)** Modern otariid, *Arctocephalus* (Recent, South Atlantic). (From Tedford, 1976.)

(a)

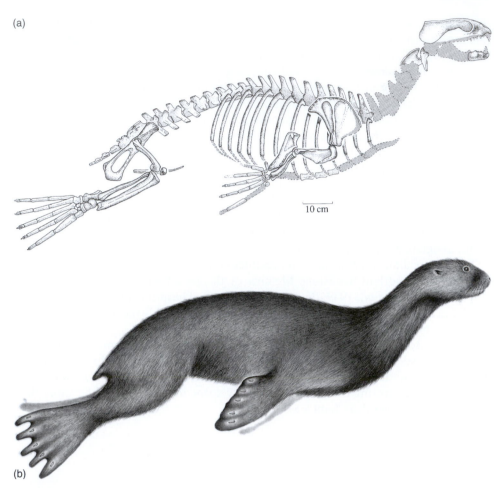

10 cm

(b)

Figure 3.9.   The pinnipedimorph, *Enaliarctos mealsi.* **(a)** Skeletal reconstruction. **(b)** Life restoration. Total estimated length, snout to tail, 1.4–1.5 m. Shaded areas are unpreserved bones. (From Berta and Ray, 1990.)

A later diverging lineage of fossil pinnipeds more closely allied with pinnipeds than with *Enaliarctos* is *Pteronarctos* and *Pacificotaria* from the early-middle Miocene (19–15 Ma) of coastal Oregon (Barnes, 1989, 1992; Berta, 1994; see Figure 3.1). A striking osteological feature in all pinnipeds is the geometry of bones that comprise the orbital region (Wyss, 1987). In *Pteronarctos,* the first evidence of the uniquely developed maxilla is seen. Also, in *Pteronarctos* the lacrimal is greatly reduced or absent, as it is in pinnipeds. A shallow pit on the palate between the last premolar and the first molar, seen in *Pteronarctos* and pinnipeds, is indicative of a reduced shearing capability of the teeth and begins a trend toward homodonty.

3.2.4. Modern Pinnipeds

3.2.4.1. Family Otariidae: Sea Lions and Fur Seals

Of the two groups of seals, the otariids are characterized by the presence of external ear flaps, or **pinnae,** and for this reason they are sometimes called eared seals (Figure 3.10).

Figure 3.10. Representative otariids. **(a)** Southern sea lion, *Otaria byronia* and **(b)** South African fur seal, *Arctocephalus pusillus,* illustrating pinna. Note also the thick, dense fur characteristic of fur seals. (Illustrations by P. Folkens from Reeves *et al.*, 1992.)

Another characteristic feature of otariids that can be used to distinguish them from phocids is their method of movement on land. Otariids can turn their hindflippers forward and use them to walk (described in more detail in Chapter 8). Otariids generally are smaller than most phocids and are shallow divers targeting fast swimming fish as their major food source. The eared seals and sea lions, Family Otariidae, can be diagnosed as a monophyletic group by several osteological and soft anatomical characters (Figures 3.2 and 3.11) as follows:

9. *Frontals extend anteriorly between nasals.* In otariids, the suture between the frontal and nasal bones is W-shaped (i.e., the frontals extend between the nasals). In other pinnipeds and terrestrial carnivores, the contact between these bones is either transverse (terrestrial carnivores and walruses) or V-shaped (phocids).

10. *Supraorbital process of the frontal bone is large and shelf-like, especially among adult males.* In otariids, the unique size and shape of the supraorbital process, located above the eye orbit, readily distinguishes them from other pinnipeds. The supraorbital process is absent in phocids and the modern walrus.

11. *Secondary spine subdivides the supraspinous fossa of the scapula.* A ridge subdividing the supraspinous fossa of the scapula (shoulder blade) is present in otariids but not in walruses or phocids.

12. *Uniformly spaced pelage units.* In otariids, pelage units (a primary hair and its surrounding secondaries) are spaced uniformly. In odobenids and phocids, the units are arranged in groups of two to four or in rows (see Chapter 7, Figure 7.10).

13. *Trachea has an anterior bifurcation of the bronchi.* In odobenids and phocids, the trachea divides into two primary bronchi immediately outside the lung (Fay, 1981; King, 1983a). In otariids, this division occurs more anteriorly, closer to the larynx and associated structures.

The Otariidae often are divided into two subfamilies, the Otariinae (sea lions) and the Arctocephalinae (fur seals). Five genera and species of sea lions are recognized: *Eumetopias, Neophoca, Otaria, Zalophus,* and *Phocarctos.* Sea lions are characterized and readily distinguished from fur seals by their sparse pelage (see Figure 3.10a). The fur seals, named for their thick dense fur, are divided into two genera. *Arctocephalus* (generic

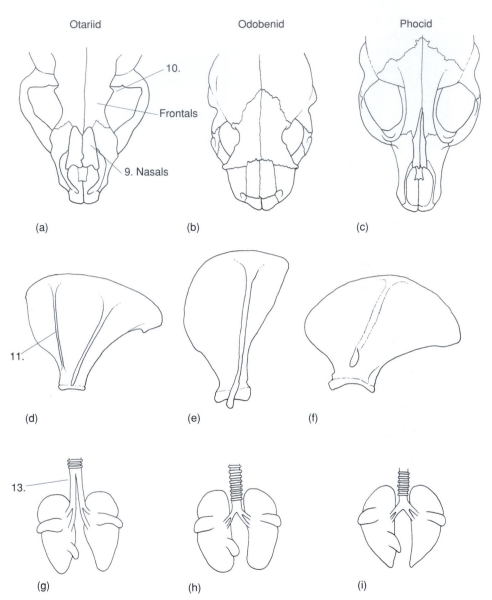

Figure 3.11. Otariid synapomorphies. (Illustrations by P. Adam.) Character numbers (see text for further description). **(a–c)** Skulls in dorsal view: 9 = frontals extend anteriorly between nasals, contact between these bones is transverse (walrus) or V-shaped (phocids); 10 = supraorbital process of the frontal bone is large and shelf-like, this process is absent in the modern walrus and phocids. **(d–f)** Left scapulae in medial view: 11 = secondary spine subdividing the supraspinous fossa of the scapula, this ridge is absent in the walrus and phocids. **(g–i)** Lungs in ventral view: 13 = trachea has an anterior bifurcation of the bronchi. (Modified from King, 1983b.) This division occurs immediately outside the lungs in the walrus and phocids.

name means bear head), or southern fur seals, live mostly in the southern hemisphere and a single species of northern fur seal, *Callorhinus ursinus* (generic name means beautiful nose), inhabits the northern hemisphere (see Figure 3.10b). Relationships among the otariids based on morphology (Berta and Deméré, 1986; Berta and Wyss, 1994) indicate that only the Otariinae are monophyletic with a sister group relationship suggested with *Arctocephalus*. *Callorhinus* and the extinct taxon *Thalassoleon* are positioned as sequential sister taxa to this clade (Figure 3.12). Another recent analysis (Bininda-Emonds *et al.*, 1999) suggested that both subgroups were monophyletic.

Molecular sequence data (Lento *et al.*, 1995, 1997; Wynen *et al.*, 2001) revealed paraphyly among both fur seals and sea lions. New Zealand fur seal *(Arctocephalus forsteri)* and the northern fur seal *(Callorhinus ursinus)*, both arctocephalines, are separated from each other by two sea lion lineages (Steller's sea lion, *Eumetopias jubatus,* and Hooker's sea lion, *Phocarctos hookeri*), and the two sea lions are no more closely related to each other than they are to other otariid taxa (i.e., the arctocephalines). A different arrangement among otariids is suggested by Árnason *et al.* (1995), but a limited number of species were sampled. Their study supports an alliance between *Arctocephalus forsteri* and the Antarctic fur seal, *Arctocephalus gazella,* and unites Steller's sea lion, *Eumetopias,* and the California sea lion, *Zalophus*. In addition to the extant fur seal genera *Callorhinus* and *Arctocephalus,* several extinct otariids are known. The earliest otariid is *Pithanotaria starri* from the late Miocene (11 Ma) of California. It is a small animal characterized by double rooted cheekteeth and a postcranial skeleton that allies it with other otariids. A second extinct late Miocene taxon (8–6 Ma), *Thalassoleon* (Figure 3.13), recently reviewed by Deméré and Berta (in press) is represented by three

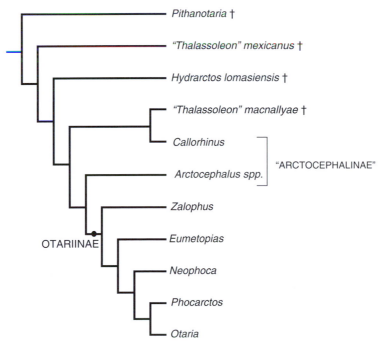

Figure 3.12. Phylogeny of the Otariidae based on morphologic data showing monophyletic Otariinae and paraphyletic "Arctocephalinae" with † = extinct taxa. (From Berta and Wyss, 1994, and Berta and Deméré, 1986.)

(a)

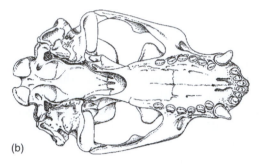

(b)

Figure 3.13. Skull of an early otariid, *Thalassoleon mexicanus,* from the late Miocene of western North America in **(a)** lateral and **(b)** ventral views. Original 25 cm long. (From Repenning and Tedford, 1977.)

species: *T. mexicanus* from Cedros Island, Baja California, Mexico, and southern California; *T. macnallyae* from California; and *T. inouei* from central Japan. *Thalassoleon* is distinguished from *Pithanotaria* by its larger size and in lacking a thickened ridge of tooth enamel at the base of the third upper incisor (Berta, 1994). A single extinct species of northern fur seal, *Callorhinus gilmorei,* from the late Pliocene in southern California and Mexico (Berta and Deméré, 1986) and Japan (Kohno and Yanagisawa, 1997) has been described on the basis of a partial mandible, some teeth, and postcranial bones. Several species of the southern fur seal genus *Arctocephalus* are known from the fossil record. The earliest known taxa are *A. pusillus* (South Africa) and *A. townsendi* (California) from the late Pleistocene (Repenning and Tedford, 1977).

The fossil record of sea lions is not well known. Only the late Pleistocene occurrences of *Otaria byronia* from Brazil (Drehmer and Ribeiro, 1998) and *Neophoca palatina* (King, 1983b) from New Zealand can be considered reliable (Deméré et al., 2003).

3.2.4.2. Family Odobenidae: Walruses

Arguably the most characteristic feature of the modern walrus, *Odobenus rosmarus,* is a pair of elongated ever-growing upper canine teeth **(tusks)** found in adults of both sexes (Figure 3.14b). A rapidly improving fossil record indicates that these unique structures evolved in a single lineage of walruses and that "tusks do not a walrus make." The modern walrus is a large-bodied shallow diver that feeds principally on benthic invertebrates, especially molluscs. Two subspecies of *Odobenus rosmarus* are usually recognized, *Odobenus r. rosmarus* from the North Atlantic and *Odobenus r. divergens* from the North Pacific. A population from the Laptev Sea has been described as a third subspecies, *Odobenus. r. laptevi* (Chapskii, 1940). Monophyly of the walrus family, the Odobenidae,

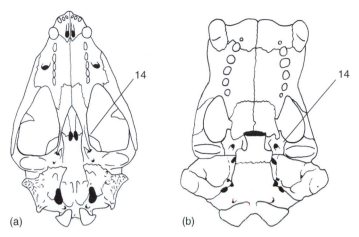

(a)   (b)

Figure 3.14.   A walrus synapomorphy. Skull of an **(a)** otariid and **(b)** walrus in ventral view illustrating differences in the pterygoid region. Character number: 14 = broad, thick pterygoid strut; in the otariid the pterygoid strut is narrow. (From Deméré and Berta, 2001.)

is based on four unequivocal synapomorphies (Deméré and Berta, 2001; Figures 3.2, 3.14, and 3.15):

14. *Pterygoid strut broad and thick*. The pterygoid strut is the horizontally positioned expanse of palatine, alisphenoid, and pterygoid lateral to the internal nares and hamular process. Basal pinnipedimorphs are characterized by having a narrow pterygoid strut, which in walruses is broad with a ventral exposure of the alisphenoid and pterygoid.

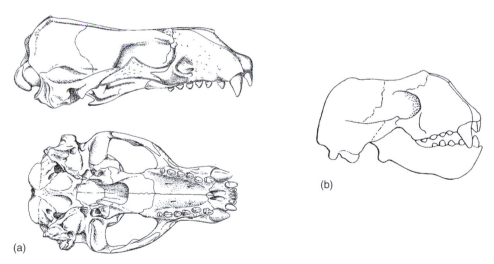

(b)

(a)

Figure 3.15.   Skulls of fossil odobenids. **(a)** Lateral and ventral views of *Imagotaria downsi* from the Miocene of western North America. Original 30 cm long. (From Repenning and Tedford, 1977.) **(b)** Lateral view of *Protodobenus japonicus* from the early Pliocene of Japan. Original 25 cm. (From Horikawa, 1995.)

15. *P4 protocone shelf strong and posterolingually placed with convex posterior margin*. In basal walruses (i.e., *Proneotherium, Imagotaria, and Prototaria*) the P4 protocone is a posteromedially placed shelf. That differs from *Enaliarctos,* which has a anterolingually placed protocone shelf. In later diverging walruses (i.e., *Dusignathus* and odobenines) the protocone shelf is greatly reduced or absent.

16. *M1 talonid heel absent*. The condition in walruses (absence of talonind heel) differs from other pinnipedimorphs in which a distinct cusp, the hypoconulid, is developed on the talonid heel.

17. *Calcaneum with prominent medial tuberosity*. In basal pinnipedimorphs, otariids and phocids, the calacaneal tuber is straight-sided. In walruses a prominent medial protuberance is developed on the proximal end of the calcaneal tuber.

Morphological study of evolutionary relationships among walruses has identified two monophyletic groups. The Dusignathinae includes the extinct genera *Dusignathus, Gomphotaria, Pontolis,* and *Pseudodobenus*. The Odobenidae includes in addition to the modern walrus, *Odobenus,* the extinct genera *Aivukus, Alachtherium, Gingimanducans, Prorosmarus, Protodobenus,* and *Valenictus* (Deméré, 1994b; Horikawa, 1995). Dusignathine walruses developed enlarged upper and lower canines, whereas odobenines evolved only the enlarged upper canines seen in the modern walrus.

At the base of walrus evolution are *Proneotherium* and *Prototaria,* from the middle Miocene (16–14 Ma) of the eastern North Pacific (Kohno *et al.*, 1995; Deméré and Berta, 2001). Other basal odobenids include *Neotherium* and *Imagotaria* from the middle-late Miocene of the eastern North Pacific (Figure 3.15). These archaic walruses are characterized by unenlarged canines and narrow, multiple rooted premolars with a trend toward molarization, adaptations suggesting retention of the fish diet hypothesized for archaic pinnipeds rather than the evolution of the specialized mollusc diet of the modern walrus. The dusignathine walrus, *Dusignathus santacruzensis,* and the odobenine walrus, *Aivukus cedroensis,* first appeared in the late Miocene of California and Baja California, Mexico. Early diverging odobenine walruses are now known from both sides of the Pacific in the early Pliocene. *Prorosmarus alleni* is known from the eastern United States (Virginia) and *Protodobenus japonicus* from Japan. A new species of walrus, possibly the most completely known fossil odobenine walrus, *Valenictus chulavistensis,* was described by Deméré (1994b) as being closely related to modern *Odobenus* but distinguished from it in having no teeth in the lower jaw and lacking all upper postcanine teeth. The toothlessness (except for tusks) of *Valenictus* is unique among pinnipeds but parallels the condition seen in modern suction feeding whales and the narwhal.

Remains of the modern walrus *Odobenus* date back to the early Pliocene of Belgium; this taxon appeared approximately 600,000 years ago in the Pacific.

3.2.4.3. Family Phocidae: Seals

The second major grouping of living seals, the phocids, often are referred to as the earless seals for their lack of visible ear pinnae, a characteristic that readily distinguishes them from otariids. Another characteristic phocid feature is their method of movement on land. The phocids are unable to turn their hindflippers forward and progression over land is accomplished by undulations of the body (described in more detail in Chapter 8). Other characteristics of phocids include their larger body size in comparison to otariids, averaging as much as 2 tons in elephant seal males. Several phocids, most notably the elephant seal and the Weddell seal, are spectacular divers that feed on pelagic, vertically migrating squid and fish at depths of 1000 m or more.

Wyss (1988) reviewed the following characters that support monophyly of the Family Phocidae (Figures 3.2, 3.16, and 3.17):

19. *Lack the ability to draw the hind limbs forward under the body due to a massively developed astragalar process and greatly reduced calcaneal tuber*. The phocid astragalus (ankle bone) is distinguished by a strong posteriorly directed process over which the tendon of the flexor hallucis longus passes. The calcaneum (one of the heel bones) of phocids

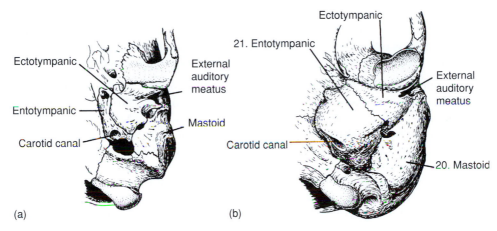

Figure 3.16. Phocid synapomorphies. Ventral view of ear region of **(a)** an otariid and **(b)** a phocid. Character numbers (see text for further description): 20 = pachyostotic mastoid bone—this is not the case in other pinnipeds; 21 = greatly inflated entotympanic bone—in other pinnipeds, this bone is flat or slightly inflated. (Modified from King; 1983b.)

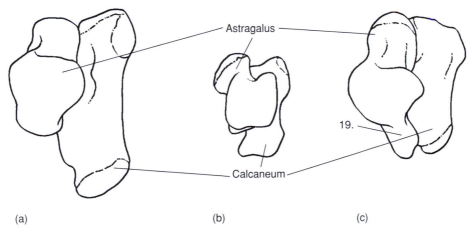

Figure 3.17. Phocid synapomorphies. Left astragalus (ankle) and calcaneum (heel) of **(a)** an otariid, **(b)** a walrus, and **(c)** a phocid. Character numbers: 19 = lack the ability to draw the hind limbs under the body due to a massively developed astragalar process and greatly reduced calcaneal tuber; these modifications do not occur in other pinnipeds. (Illustrations by P. Adam.)

is correspondingly modified. The calcaneal tuber is shortened and projects only as far as the process of the astragalus. This arrangement prevents anterior flexion of the foot, resulting in seals' inability to bring their hind limbs forward during locomotion on land.

20. *Pachyostic mastoid region.* In phocids, the mastoid (ear) region is composed of thick, dense bone **(pachyostosis),** which is not the case in otariids or the walrus.

21. *Greatly inflated entotympanic bone.* In phocids, the entotympanic bone (one of the bones forming the earbone or tympanic bulla) is inflated. In other pinnipeds, the entotympanic bone is either flat or slightly inflated.

22. *Supraorbital processes completely absent.* Phocids differ from other pinnipeds in the complete absence of the supraorbital process of the frontal (see Figure 3.11).

23. *Strongly everted ilia.* Living phocines, except *Erignathus,* are characterized by a lateral eversion (outward bending) of the ilium (one of the pelvic bones) accompanied by a deep lateral excavation (King, 1966). In other pinnipeds and terrestrial carnivores, the anterior termination of the ilium is simple and not everted or excavated.

Traditionally, phocids have been divided into two to four major subgroupings, monachines (monk seals), lobodontines (Antarctic seals), cystophorines (hooded and elephant seals), and phocines (remaining Northern Hemisphere seals). Based on morphologic data, Wyss (1988) argued for the monophyly of only one of these groups, the Phocinae, composed of *Erignathus* and *Cystophora* plus the tribe Phocini, consisting of *Halichoerus, Histriophoca, Pagophilus, Phoca,* and *Pusa.* According to Wyss, both the Monachinae" and the genus *"Monachus"* are paraphyletic with *"Monachus,"* in turn representing the outgroup to the elephant seals, *Mirounga,* and the lobodontines (including the Weddell seal, *Leptonychotes;* crabeater seal, *Lobodon;* leopard seal, *Hydrurga;* and the Ross seal, *Ommatophoca;* Figures 3.18, 3.19, and 3.20). Another morphology-based study found reasonable support for both the Monachinae and Phocinae, although with differing relationships among the taxa within each group (Bininda-Emonds and Russell, 1996). The basal position of *Monachus* and *Erignathus,* in the Monachinae and

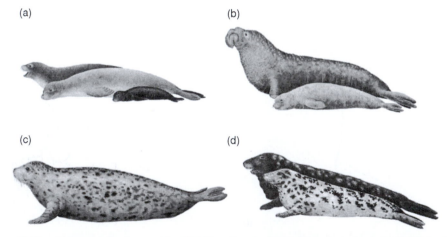

Figure 3.18. Representative "monachines" **(a)** Hawaiian monk seal, *Monachus schauinslandi;* **(b)** Northern elephant seal, *Mirounga angustirostris,* and phocines; **(c)** Harbor seal, *Phoca vitulina;* and **(d)** grey seal, *Halichoerus grypus.* (Illustrations by P. Folkens from Reeves *et al.,* 1992.)

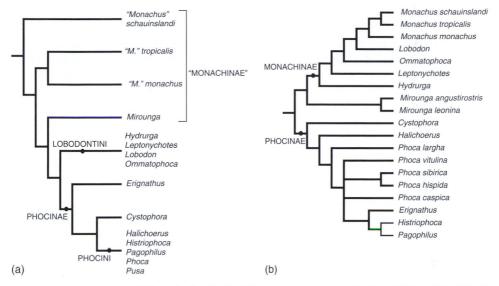

Figure 3.19. Alternative phylogenies for the Phocidae based on morphologic data. **(a)** From Wyss (1988) and Berta and Wyss (1994). **(b)** From Bininda-Emonds and Russell (1996).

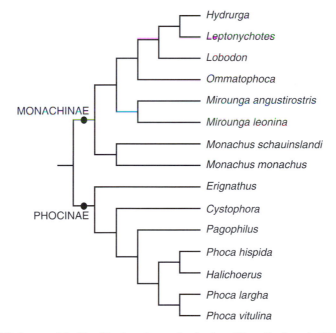

Figure 3.20. Phylogeny of the Phocidae based on molecular data. (From Davis *et al.*, 2004.)

Phocinae, respectively, was not supported. Instead, both taxa were recognized as later diverging members of their respective subfamilies rendering the Lobodontini and Phocini as paraphyletic clades (Bininda-Emonds and Russell, 1996; see Figure 3.19).

Several molecular studies provide data on phocid interrelationships (e.g., Árnason *et al.*, 1995; Carr and Perry, 1997; Mouchaty *et al.*, 1995; Davis *et al.*, 2004; Fyler *et al.*, 2005; see Figure 3.20). In the most inclusive study to date (Davis *et al.*, 2004), strong support was found for monophyly of both the Monachinae and Phocinae. There was also strong support for three monachine lineages: (1) the Hawaiian monk seal, *Monachus schauinslandi,* sister taxon to the Mediterranean monk seal, *Monachus tropicalis*, (2) monophyletic elephant seals (*Mirounga* spp.) sister taxon to (3) lobodontine seals (i.e., *Hydrurga, Lobodon, Ommatophoca,* and *Leptonychotes*). Among phocines, the bearded seal, *Erignathus,* was the most basal and the hooded seal, *Cystophora,* sister to the Phocini. Within the Phocini, *Pagophilus* was sister group of the *Phoca* species complex. The position of the grey seal, *Halichoerus grypus,* among species of the genus *Phoca* proposed earlier by Árnason *et al.* (1995), suggests that generic status of the grey seal is not warranted, a conclusion reached in other molecular (Mouchaty *et al.*, 1995; O'Corry-Crowe and Westlake, 1997; but see Carr and Perry, 1997, for a different view) and morphologic (Burns and Fay, 1970) studies. Included in this redefined *Phoca* complex, in addition to the grey seal are the ringed seal, *Phoca hispida;* the spotted seal, *Phoca largha;* and the harbor seal, *Phoca vitulina.*

Four subspecies of the harbor seal are currently recognized based on morphologic, molecular, and geographic differences (Stanley *et al.*, 1996): *P. v. vitulina* and *P. v. concolor* (the eastern and western Atlantic Ocean populations, respectively) and *P. v. richardsi* and *P. v. stejnegeri* (the eastern and western Pacific Ocean populations, respectively). A fifth subspecies, *P. v. mellonae,* a freshwater population from the area of Seal Lakes in northern Québec, Canada, is morphologically and behaviorally distinct from the others (Smith and Lavigne, 1994). Árnason *et al.*'s (1995) data support an alliance between the eastern Pacific harbor seal, *P. v. richardsi,* and the Eastern Atlantic harbor seal, *P v. vitulina,* with the spotted seal, *Phoca largha,* positioned as sister taxon to this clade as well as supporting a sister group relationship between the grey seal, *Halichoerus grypus,* and the ringed seal, *Phoca hispida* (see Figure 3.20).

The report of a fossil phocid from the late Oligocene (29–23 Ma) of South Carolina (Koretsky and Sanders, 2002), if its stratigraphic provenance is correct, makes it the oldest known phocid and pinniped (see Figure 3.1). Prior to this record, phocids were unknown until the middle Miocene (15 Ma) when both phocine and "monachine" seals became distinct lineages in the North Atlantic. The extinct phocine seal *Leptophoca lenis* and the "monachine" seal *Monotherium? wymani* are known from Maryland and Virginia during this time (Koretsky, 2001). *Leptophoca lenis,* or a closely related species, also is represented in the eastern Atlantic, from deposits in the Antwerp Basin, Belgium (Ray, 1976). Another addition to the fossil record of phocids is a new genus and species of phocine seal, based on an articulated skeleton (lacking the skull, neck, and part of the hind limbs) reported from the middle Miocene of Argentina (Cozzuol, 1996). Other fossil seals are represented by well-preserved skeletal material. For example, *Acrophoca longirostris* (Figure 3.21) and *Piscophoca pacifica* from the late Miocene and early Pliocene were found in the Pisco Formation of Peru (Muizon, 1981), and *Homiphoca capensis* was discovered in South Africa (Hendey and Repenning, 1972). Although originally considered "monachines" (i.e., lobodontine seals), ongoing study by Cozzuol (1996) indicates that *Acrophoca, Piscophoca,* and new material from the Pliocene of Peru (provisionally referred to two new genera and species) may be closer to phocines. Other phocines evolved in the vast inland Parathethys Sea from 14 to 10 Ma (Koretsky, 2001 see also Chapter 6) and differentiated mostly in the Pleistocene.

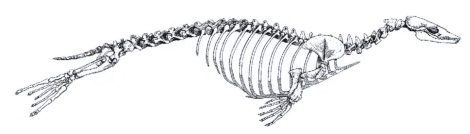

Figure 3.21. Skeleton of an archaic phocid, *Acrophoca longirostris,* from the Miocene of Peru. (From Muizon, 1981.)

3.2.5. Walruses and Phocids Linked?

As noted previously, an alliance between the walrus and phocid seals and their extinct relatives, the Phocomorpha clade, has been proposed based on morphologic data. Among the synapomorphies that unite these pinnipeds are the following (see Figure 3.2):

24. *Middle ear bones enlarged.* The middle ear bones of the walrus and phocids are large relative to body size, which is not the case in otariids and terrestrial carnivores.

25. *Abdominal testes.* The testes are abdominal (inguinal) in phocids and the walrus. In contrast, the testes of otariids and terrestrial carnivores lie outside the inguinal ring in a scrotal sac.

26. *Primary hair nonmedullated.* The outer guard hairs in the walrus and phocids lack a pith or medulla, which is present in otariids and other carnivores (see also Chapter 7).

27. *Thick subcutaneous fat.* The walrus and phocids are characterized by thick layers of fat; these layers are less well developed in otariids and terrestrial carnivores.

28. *External ear pinna lacking.* The walrus and phocids lack external ear pinnae, the presence of which characterizes otariids and terrestrial carnivores.

29. *Venous system with inflated hepatic sinus, well-developed caval sphincter, large intervertebral sphincter, duplicate posterior vena cava, and gluteal route for hind limbs.* The walrus and phocids share a specialized venous system that is related, in part, to their exceptional diving capabilities (see Chapter 10). In contrast, otariids and terrestrial carnivores have a less specialized venous system that more closely approximates the typical mammalian pattern.

The question of whether walruses are more closely related to phocids or to otariids involves further exploration of both morphological and molecular data sets. Study of basal walruses will likely provide additional characters at the base of walrus evolution that can then be compared for alliance with either otariids or phocids. The molecular results, which consistently support a link between otariids and the walrus, have been explained as a methodological problem or a long-branch attraction effect. When dealing with sequence data, branch lengths refer to the expected amounts of evolutionary change along that branch. The walrus lineage is a relatively long branch and it is unlikely that intraspecific variation among the walrus subspecies will aid in bisecting the branch, although this has yet to be examined. Because there is a tendency for longer branches to attract one another and thus give a misleading tree, it is possible that the walrus and otariid alliance may be incorrect. A more conservative interpretation of the molecular data is that the walrus is an early, but not first, independent divergence from the common pinniped ancestor (Lento *et al.*, 1995).

3.2.6. Desmatophocids: Phocid Relatives or Otarioids?

Study of pinniped evolutionary relationships has identified a group of fossil pinnipeds including *Desmatophoca* and *Allodesmus* (Figure 3.22) that are positioned as the common ancestors of phocid pinnipeds (Berta and Wyss, 1994). This interpretation differs from previous work that recognized desmatophocids as **otarioid** pinnipeds, a grouping that includes walruses (Barnes, 1989). The question of otarioid monophyly was examined in a comprehensive pinniped data set. The strict consensus tree that resulted by forcing otarioid monophyly was 34 steps longer than the preferred hypothesis (Berta and Wyss, 1994).

Desmatophocids are known from the early middle Miocene (23–15 Ma) of the western United States and Japan. Newly reported occurrences of *Desmatophoca* from Oregon confirm the presence of sexual dimorphism and large body size in these pinnipeds (Deméré and Berta, 2002). Allodesmids are known from the middle to late Miocene of California and more recently from Japan (Barnes and Hirota, 1995). They are a diverse group characterized, among other characters, by pronounced sexual dimorphism, large eye orbits, bulbous cheektooth crowns, and deep lower jaws.

A number of features are shared among phocids and their close fossil relatives *Allodesmus* and *Desmatophoca* (identified as the Phocoidea clade) and hence they support a close link between these taxa (see Figure 3.2). These synapomorphies include among others:

30. *Posterior termination of nasals posterior to contact between the frontal and maxilla bones*. In phocids and desmatophocids, the V-shaped contact between the frontal and maxilla bones is the result of the nasals extending posteriorly between these bones (see Figure 3.11).

31. *Squamosal jugal contact mortised*. A mortised or interlocking contact between the squamosal and jugal (cheekbones) distinguishes phocids and desmatophocids from other pinnipeds in which these bones overlap one another in a splint-like arrangement (see Figure 3.3).

3.3. Summary and Conclusions

Pinniped monophyly is a well-supported hypothesis based on both morphological and molecular data. The closest relatives of pinnipeds are arctoid carnivores, with most evidence supporting either a link between pinnipeds and ursids or pinnipeds and mustelids.

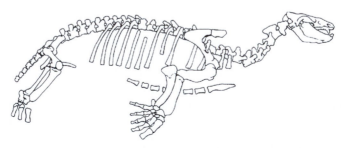

Figure 3.22.   Skeleton of the desmatophocid, *Allodesmus kernensis,* from the Miocene of western North America. Original 2.2 m long. (From Mitchell, 1975.)

The earliest pinnipedimorphs (i.e., *Enaliarctos*) appear in the fossil record approximately 27–25 Ma in the North Pacific. Modern pinniped lineages diverged shortly thereafter with the first appearance of the phocid seals in the North Atlantic. Phocids usually are divided into two to four subgroups including "monachines" and phocines. Although the monophyly of monachines has been questioned based on morphology, molecular data strongly support monachine and phocine monophyly. Walruses appeared about 10 million years later in the North Pacific. A rapidly improving fossil record indicates that enlarged tusks characteristic of both sexes of the modern walrus were not present ancestrally in walruses. Two monophyletic lineages of walruses are recognized (Dusignathinae and Odobeninae). The last pinniped lineage to appear in the fossil record, the otariids, are only known as far back as 11 Ma in the North Pacific. Morphologic data support monophyly of the sea lions (Otariinae) but not the fur seals (Arctocephalinae).

Molecular data indicates both fur seals and sea lions are paraphyletic. A remaining conflict is the position of the walrus. Most morphologic data support a phocid and walrus alliance, whereas the molecular data consistently supports uniting otariids and the walrus. Resolution of these conflicts will likely benefit from detailed exploration of both morphologic and molecular data sets.

3.4. Further Reading

Relationships among various arctoid carnivores and pinnipeds are reviewed in Flynn *et al.* (2000) and Davis *et al.* (2004). For descriptions of morphology see Berta (1991, 1994) and Barnes (1989, 1992) for basal pinnipedimorphs, Repenning and Tedford (1977) for fossil otariids and walruses, and Muizon (1981) and Koretky (2001) for fossil phocids. Reviews of the evolution and phylogeny of walruses include Deméré (1994a, 1994b), Kohno *et al.* (1995), and Deméré and Berta (2001). For alternative views on phocid phylogeny see the morphologically based studies of Wyss (1988), Bininda-Emonds and Russell (1996), and Bininda-Emonds *et al.* (1999) and the molecular studies of Árnason *et al.* (1995), Davis *et al.* (2004), and Fyler *et al.* (2005). Lento *et al.* (1995) and Wynen *et al.* (2001) provide molecular evidence for otariid relationships, but see Berta and Deméré (1986) for a different view based on morphologic data.

References

Árnason, U., K. Bodin, A. Gullberg, C. Ledje, and S. Mouchaty (1995). "A Molecular View of Pinniped Relationships with Particular Emphasis on the True Seals." *J. Mol. Evol. 40:* 78–85.

Árnason, U., and B. Widegren (1986). "Pinniped Phylogeny Enlightened by Molecular Hybridization Using Highly Repetitive DNA." *Mol. Biol. Evol. 3:* 356–365.

Barnes, L. G. (1989). "A New Enaliarctine Pinniped from the Astoria Formation, Oregon, and a Classification of the Otariidae (Mammalia: Carnivora)." *Nat. Hist. Mus. L.A. Cty., Contrib. Sci. 403:* 1–28.

Barnes, L. G. (1992). "A New Genus and Species of Middle Miocene Enaliarctine Pinniped (Mammalia: Carnivora) from the Astoria Formation in Coastal Oregon." *Nat. Hist. Mus. LA. Cty., Contrib. Sci. 431:* 1–27.

Barnes, L. G., and K. Hirota. (1995). "Miocene Pinnipeds of the Otariid Subfamily Allodesminae in the North Pacific Ocean: Systematics and Relationships." *Island Arc 3:* 329–360.

Berta, A. (1991). "New *Enaliarctos* (Pinnipedimorpha) from the Oligocene and Miocene of Oregon and the Role of "Enaliaretids" in Pinniped Phylogeny." *Smithson. Contrib. Paleobiol. 69:* 1–33.

Berta, A. (1994). "New Specimens of the Pinnipediform *Pteronarctos* from the Miocene of Oregon." *Smithson. Contrib. Paleobiol. 78:* 1–30.

Berta, A., and T. A. Deméré (1986). "*Callorhinus gilmorei* n. sp., (Carnivora: Otariidae) from the San Diego Formation (Blancan) and Its Implications for Otariid Phylogeny." *Trans. San Diego Soc. Nat. Hist. 21*(7): 111–126.

Berta, A., and C. E. Ray (1990). "Skeletal Morphology and Locomotor Capabilities of the Archaic Pinniped *Enaliarctos mealsi.*" *J. Vert. Paleontol. 10*(2): 141–157.

Berta, A., C. E. Ray, and A. R. Wyss (1989). "Skeleton of the Oldest Known Pinniped, *Enaliarctos mealsi.*" *Science 244:* 60–62.

Berta, A., and A. R. Wyss (1994). "Pinniped Phylogeny." *Proc. San Diego Soc. Nat. Hist. 29:* 33–56.

Bininda-Emonds, O. R. P., J. L. Gittleman, and A. Purvis (1999). "Building Large Trees by Combining Information: A Complete Phylogeny of the Extant Carnivora (Mammalia)." *Biol. Rev. 74:* 143–175.

Bininda-Emonds, O. R. P., and A. P. Russell (1996). "A Morphological Perspective on the Phylogenetic Relationships of the Extant Phocid Seals (Mammalia: Carnivora: Phocidae)." *Bonner Zool. Monogr. 41:* 1–256.

Burns, J. J., and F. H. Fay (1970). "Comparative Morphology of the Skull of the Ribbon Seal, *Histriophoca fasciata* with Remarks on Systematics of Phocidae." *J. Zool. Lond. 161:* 363–394.

Carr, S. M., and E. A. Perry (1997). Intra- and interfamilial systematic relationships of phocid seals as indicated by mitochondrial DNA sequences. *In* "Molecular Genetics of Marine Mammals" (A. E. Dizon, S. L. Chivers, and W. E. Perrin, eds.), Spec. Publ. No. 3, pp. 277–290. Soc. Mar. Mammal. Allen Press, Lawrence, KS.

Chapskii, K. K. (1940). "Raspostranenie morzha v moryakh Laptevykh I Vostochno Sibirkom." *Problemy Arktiki 1940*(6): 80–94.

Cozzuol, M. A. (1996). "The Record of the Aquatic Mammals in Southern South America." *Muench. Geowiss. Abh. 30:* 321–342.

Davis, C. S., I. Delisle, I. Stirling, D. B. Siniff, and C. Strobeck. (2004). "A Phylogeny of the Extant Phocidae Inferred from Complete Mitochondrial DNA Coding Regions." *Mol. Phylogenet. Evol. 33:* 363–377.

Deméré, T. A. (1994a). "The Family Odobenidae: A Phylogenetic Analysis of Fossil and Living Taxa." *Proc. San Diego Soc. Nat. Hist. 29:* 99–123.

Deméré, T. A. (1994b). "Two New Species of Fossil Walruses (Pinnipedia: Odobenidae) from the Upper Pliocene San Diego Formation, California." *Proc. San Diego Soc. Nat. Hist. 29:* 77–98.

Deméré, T. A., and A. Berta (2001). "A Reevaluation of *Proneotherium repenningi* from the Miocene Astoria Formation of Oregon and Its Position as a Basal Odobenid (Pinnipedia: Mammalia)." *J. Vert. Paleontol. 21:* 279–310.

Deméré, T. A., and A. Berta (2002). "The Pinniped Miocene *Desmatophoca oregonensis* Condon, 1906 (Mammalia: Carnivora) from the Astoria Formation, Oregon." *Smithson. Contrib. Paleobiol. 93:* 113–147.

Deméré, T.A., and A. Berta (in press). "New Skeletal Material of *Thalassoleon* (Otariidae: Pinnipedia) from the Late Miocene-Early Piocene (Hemphillian) of California." *Florida Mus. Nat. Hist. Bull.*

Deméré, T. A., A. Berta, and P. J. Adam (2003). "Pinnipedimorph Evolutionary Biogeography." *Bull. Amer. Mus. Nat. Hist. 279:* 32–76.

Drehmer, C. J., and A. M. Ribeiro (1998). "A Temporal Bone of an Otariidae (Mammalia: Pinnipedia) from the late Pleistocene of Rio Grande do Sul State, Brazil." *Geosci. 3:* 39–44.

Fay, F. H. (1981). Walrus: *Odobenus rosmarus. In* "Handbook of Marine Mammals" (S. H. Ridgway and R. J. Harrison, eds.), Vol. 1, pp. 1–23. Academic Press, New York.

Flynn, J. J., and M. A. Nedbal (1998). "Phylogeny of the Carnivora (Mammalia): Congruence vs. Incompatibility Among Multiple Data Sets." *Mol. Phylogenet. Evol. 9:* 414–426.

Flynn, J. J., M. A. Nedbal, J. W. Dragoo, and R. L. Honeycutt (2000). "Whence the Red Panda? *Mol. Phylogenet. Evol. 170:* 190–199.

Flynn, J. J., N. A. Neff, and R. H. Tedford (1988). Phylogeny of the Carnivora. *In* "The Phylogeny and Classification of the Tetrapods" (M. J. Benton, ed.), Vol. 2, pp. 73–116. Clarendon Press, Oxford.

Flynn, J. J., and G. D. Wesley-Hunt (2005). Carnivora. *In* "The Rise of Placental Mammals" (K. D. Rose and J. David Archibald, eds.), pp. 175–198. Johns Hopkins, Baltimore, MD.

Fyler, C., T. Reeder, A. Berta, G. Antonelis, and A. Aguilar. (2005). "Historical Biogeography and Phylogeny of Monachine Seals (Pinnipedia: Phocidae) Based on Mitochondrial and Nuclear DNA Data." *J. Biogeogr. 32:* 1267–1279.

Hendey, Q. B., and C. A. Repenning (1972). "A Pliocene Phocid from South Africa." *Ann. S. Afr. Mus. 59*(4):71–98.

Horikawa, H. (1995). "A Primitive Odobenine Walrus of Early Pliocene Age from Japan." *Island Arc 3:* 309–329.

Illiger, J. C. W. (1811). *Prodromus systematics Mammalium et Avium*. C. Salfeld, Berlin.

King, J. E. (1966). "Relationships of the Hooded and Elephant Seals (Genera *Cystophora* and *Mirounga*)." *J. Zool. Soc. Lond. 148:* 385–398.

King, J. E. (1983b). "The Ohope Skull-A New Species of Pleistocene Sea Lion from New Zealand." *N. Z. J. Mar Freshw. Res. 17:* 105–120.

King, J. E. (1983a). *Seals of the World*. Oxford University Press, London.

Kohno, N. (1992). "A New Pliocene Fur Seal (Carnivora: Otariidae) from the Senhata Formation, Boso Peninsula, Japan." *Nat. Hist. Res. 2:* 15–28.

Kohno, N., L. G. Barnes, and K. Hirota (1995). "Miocene Fossil Pinnipeds of the Genera *Prototaria* and *Neotherium* (Carnivora; Otariidae; Imagotariinae) in the North Pacific Ocean: Evolution, Relationships and Distribution." *Island Arc 3:* 285–308.

Kohno, N., and Y. Yanagisawa (1997). "The First Record of the Pliocene Gilmore Fur Seal in the Western North Pacific Ocean." *Bull. Natl. Sci. Mus. Ser C: Geol. (Tokyo) 23:* 119–130.

Koretsky, I. A. (2001). "Morphology and Systematics of Miocene Phocinae (Mammalia: Carnivora) from Paratethys and the North Atlantic Region." *Geol. Hung. Ser. Palaeontol. 54:* 1–109.

Koretsky, I., and A. E. Sanders (2002). "Paleontology from the Late Oligocene Ashley and Chandler Bridge Formations of South Carolina, 1: Paleogene Pinniped Remains; The Oldest Known Seal (Carnivora: Phocidae)." *Smithson. Contrib. Paleobiol. 93:* 179–183.

Lento, G. M., M. Haddon, G. K. Chambers, and C. S. Baker (1997). "Genetic Variation, Population Structure, and Species Identity of Southern Hemisphere Fur Seals, *Arctocephalus spp.*" *J. Heredity 88:* 28–34.

Lento, G. M., R. E. Hickson, G. K. Chambers, and D. Penny (1995). "Use of Spectral Analysis to Test Hypotheses on the Origin of Pinnipeds." *Mol. Biol. Evol. 12:* 28–52.

McLaren, I. A. (1960). "Are the Pinnipedia Biphyletic?" *Syst. Zool. 9:* 18–28.

Mitchell, E. D. (1975). "Parallelism and Convergence in the Evolution of the Otariidae and Phocidae." *Rapp. P.-v. Réun. Cons. Int. Explor. Mer. 169:* 12–26.

Mitchell, E. D., and R. H. Tedford (1973). "The Enaliarctinae: A New Group of Extinct Aquatic Carnivora and a Consideration of the Origin of the Otariidae." *Bull. Amer. Mus. Nat. Hist. 151*(3): 201–284.

Mouchaty, S., J. A. Cook, and G. F. Shields (1995). "Phylogenetic Analysis of Northern Hair Seals Based on Nucleotide Sequences of the Mitochondrial Cytochrome b Gene. *J. Mammal. 76:* 1178–1185.

Muizon, C. de (1981). *Les Vertébrés Fossiles de la Formation Pisco (Pérou). Part 1. Recherche sur les Grandes Civilisations,* Mem. No. 6. Instituts Francais d'Études Andines, Paris.

Muizon, C. de (1982). "Phocid Phylogeny and Dispersal." *Ann. S. Afr. Mus. 89*(2): 175–213.

O'Corry-Crowe, G. M., and R. L. Westlake (1997). Molecular investigations of spotted seals *(Phoca largha)* and harbor seals *(P. vitulina),* and their relationship in areas of sympatry. *In* "Molecular Genetics of Marine Mammals" (A. E. Dizon, S. J. Chivers, and W. F. Perrin, eds.), Spec. Pub. No. 3, pp. 291–304. Soc. Mar. Mammal, Allen Press, Lawrence, KS.

Ray, C. E. (1976). "Geography of Phocid Evolution." *Syst. Zool. 25:* 391–406.

Reeves, R. R., B. S. Stewart, and S. Leatherwood (1992). *The Sierra Club Handbook of Seals and Sirenians*. Sierra Club Books, San Francisco, CA.

Repenning, C. A., C. E. Ray, and D. Grigorescu (1979). Pinniped biogeography. *In* "Historical Biogeography, Plate Tectonics, and Changing Environment" (J. Gray and A. J. Boucot, eds.), pp. 357–369. Oregon State University, Corvallis, OR.

Repenning, C. A., and R. H. Tedford (1977). "Otarioid Seals of the Neogene." *Geol. Surv. Prof. Pap. (U.S.) 992:* 1–93.

Rice, D. W. (1998). *Marine Mammals of the World*. Soc. Mar. Mammal., Spec. Publ. No. 4, pp. 1–231.

Riedman, M. (1990). *The Pinnipeds. Seals, Sea Lions and Walruses*. University California Press, Berkeley, CA.

Sarich, V. M. (1969). "Pinniped Phylogeny." *Syst. Zool. 18:* 416–422.

Smith, R. J., and D. M. Lavigne (1994). "Subspecific Status of the Freshwater Harbor Seal *(Phoca vitulina mellonae):* A re-assessment." *Mar. Mamm. Sci. 10*(1): 105–110.

Stanley, H., S. Casey, J. M. Carnahan, S. Goodman, J. Harwood, and R. K. Wayne (1996). "Worldwide Patterns of Mitochondrial DNA Differentiation in the Harbor Seal *(Phoca vitulina)." Mol. Biol. Evol. 13:* 368–382.

Tedford, R. H. (1976). "Relationships of Pinnipeds to Other Carnivores (Mammalia)." *Syst. Zool. 25:* 363–374.

Vrana, P. B., M. C. Milinkovitch, J. R. Powell, and W. C. Wheeler (1994). "Higher Level Relationships of Arctoid Carnivora Based on Sequence Data and 'Total Evidence.'" *Mol. Phylogenet. Evol. 3:* 47–58.

Wynen, L. P., S. D. Goldsworthy, S. Insley, M. Adams, J. Bickham, J. P. Gallo, A. R. Hoelzel, P. Majluf, R. P. G. White, and R. Slade (2001). "Phylogenetic Relationships Within the Family Otariidae (Carnivora)." *Mol. Phylogenet. Evol. 21:* 270–284.

Wyss, A. R. (1987). "The Walrus Auditory Region and Monophyly of Pinnipeds." *Amer. Mus. Novit. 2871:* 1–31.

Wyss, A. R. (1988). "On "Retrogression" in the Evolution of the Phocinae and Phylogenetic Affinities of the Monk Seals. *Amer. Mus. Novit. 2924:* 1–38.

Wyss, A. R., and J. Flynn (1993). A phylogenetic analysis and definition of the Carnivora. *In* "Mammal Phylogeny: Placentals" (F. S. Szalay, M. J. Novacek, and M. C. McKenna, eds.), pp. 32–52. Springer-Verlag, New York.

4

Cetacean Evolution and Systematics

4.1. Introduction

The majority of marine mammals belong to the Order Cetacea, which includes whales, dolphins, and porpoises. Two major groups of extant whales are recognized—the Mysticeti, or baleen whales, and the Odontoceti, or toothed whales. Toothed whales are more diverse, with approximately 75 species known compared to 12–14 mysticete species.

Cetaceans together with sirenians are the earliest recorded marine mammals, appearing in the Eocene about 53–54 Ma (Figure 4.1). Cetaceans are also the most diverse mammalian group to adapt to a marine existence. New discoveries of fossil whales provide compelling evidence for both the phylogenetic connections of cetaceans as well as the evolutionary transformation from a terrestrial to a fully aquatic existence.

4.2. Origin and Evolution

4.2.1. Whales Defined

The mammalian order Cetacea comes from the Greek *ketos* meaning whale. Whales and sirenians (see Chapter 5) are the only marine mammals to live their entire lives in water. A thick layer of blubber, rather than hair or fur, insulates them. The hind limbs have been lost and they use the horizontal tail flukes for propulsion. Steering and maintenance of stability when moving is accomplished by a pair of paddle-shaped foreflippers.

Whales have traditionally been defined as a monophyletic group. Geisler (2001) provided 15 unequivocal derived characters to diagnose Cetacea (Figure 4.2) including the following basicranial and dental features:

1. *Mastoid process of petrosal not exposed posteriorly*. In cetaceans, the mastoid process is not exposed posteriorly, the lambdoidal crest of the squamosal is in continuous contact with exoccipital and basioccipital. In noncetacean mammals, the mastoid region is exposed on the outside of the skull (O'Leary and Geisler, 1999).

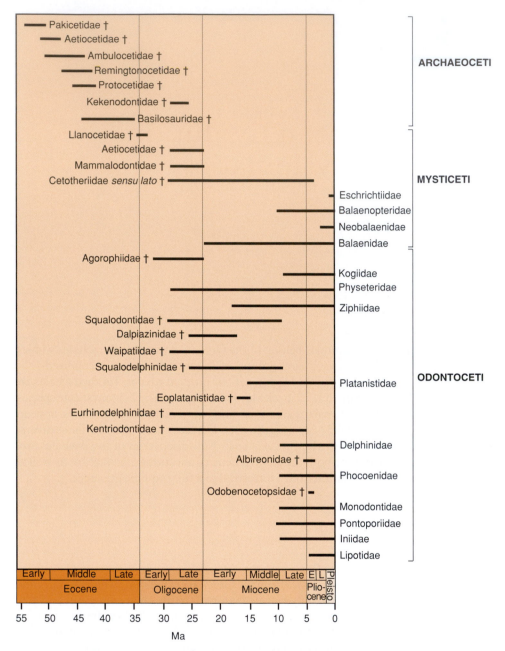

Figure 4.1.   Chronologic ranges of extinct and living cetaceans. Ma = million years ago.

2. *Pachyosteosclerotic bulla.* The auditory bulla of cetaceans consists of dense, thick (pachyostotic) and **osteosclerotic** (replacement of spongy bone with compact bone) bone, referred to as pachyosteosclerotic bone. Pachyosteosclerosis occurs in the ear region of all cetaceans and it is absent in noncetacean mammals (Thewissen, 1994; Luo and Gingerich, 1999).

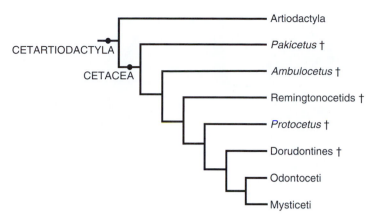

Figure 4.2. A cladogram depicting the relationships for cetaceans and their terrestrial relatives (Thewissen *et al.*, 2001).

3. *Bulla articulates with the squamosal via a circular entoglenoid process*. In cetaceans, a platform (entoglenoid process) is developed for articulation with the squamosal (Luo and Gingerich, 1999; O'Leary and Geisler, 1999). Although the bulla contacts the squamosal in archaic ungulates, a distinctive process is not developed.

4. *Fourth upper premolar protocone absent*. In fossil relatives of cetaceans, the protocone is present in contrast to the absence of this cusp in cetaceans (O'Leary, 1998; O'Leary and Geisler, 1999).

5. *Fourth upper premolar paracone height twice that of first upper molar*. In archaic cetaceans (e.g., *Pakicetus* and *Ambulocetus*), the upper fourth premolar has an anterior cusp (paracone) that is elevated twice as high as that of the first upper molar. In relatives of cetaceans, the paracone is not higher than in the first upper molar (Thewissen, 1994; O'Leary and Geisler, 1999).

4.2.2. Cetacean Affinities

4.2.2.1. Relationships of Cetaceans to Other Ungulates

Linnaeus, in an early edition of *Systema Naturae* (1735), included cetaceans among the fishes, but by the tenth edition he had followed Ray (1693) in recognizing them as a distinct group unrelated to fishes. Flower (1883) was the first to propose a close relationship between cetaceans and ungulates, the hoofed mammals. This idea has been endorsed on the basis of dental and cranial evidence by Van Valen (1966) and Szalay (1969) who argued for a more specific link between cetaceans and an extinct group of ungulates, mesonychian condylarths (Figures 4.3 and 4.4). Among fossil taxa, mesonychian condylarths are usually recognized as closely related to cetaceans, although recent work indicates that other ungulates are likely closer relatives (see Theodor *et al.*, 2005). Mesonychians had wolf-like proportions including long limbs, a **digitigrade** stance (walking on their fingers and toes), and probably hoofs. In addition, most genera had massive, crushing dentitions that differ from other ungulates in suggesting a carnivorous diet. A connection between cetaceans and mesonychians (referred to as Cete) comes from the skull, dentition, and postcranial skeletons of a rapidly increasing number of basal

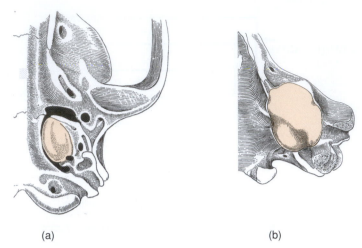

(a)   (b)

Figure 4.3.   Whale synapomorphies. **(a)** Basicranium of mesonychian condylarth, *Haplodectes hetangen-sis,* **(b)** Basicranium of archaic whale, *Gaviacetus razai* illustrating the difference in the ear region. Character number 2 (see text for more explanation) pachyostotic bulla; in the condylarth pachyostosis is absent. (From Luo and Gingerich, 1999.)

whales such as *Protocetus, Pakicetus, Rodhocetus,* and *Ambulocetus.* The hind limbs of these whales distally show a **paraxonic** arrangement, a condition in which the axis of symmetry in the foot extend about a plane located between digits III and IV (Figure 4.5). This paraxonic arrangement bears striking resemblance to that of mesonychian condylarths as well as that of the Artiodactyla (even-toed ungulates including deer, antelope, camels, pigs, giraffes, and hippos). Morphologic evidence in support of mesonychians as the sister group of the cetaceans is reviewed by O'Leary (1998), O'Leary and Geisler (1999), Luo and Gingerich, (1999) and O'Leary *et al.* (2003).

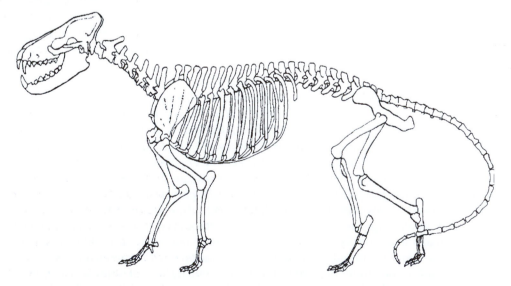

Figure 4.4.   Skeleton of *Mesonyx,* a mesonychian condylarth. (From Scott, 1888.)

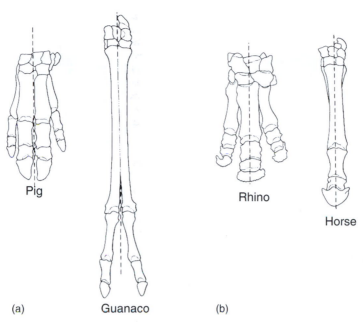

Pig

Rhino

Horse

Guanaco

(a)   (b)

Figure 4.5. Synapomorphy uniting Cete (cetaceans and mesonychian condylarths) + artiodactyls.
Paraxonic foot arrangement **(a)** in which the axis of symmetry runs between digits III and IV
(from MacFadden, 1992); in the primitive mesaxonic arrangement **(b)** the axis of symmetry
runs through digit III.

Among extant groups, artiodactyls are most commonly cited as the sister group of the
Cetacea based on morphologic data, and the majority of morphologically based studies
have found the Artiodactyla to be monophyletic (e.g., Thewissen, 1994; O'Leary, 1998;
O'Leary and Geisler, 1999; Geisler, 2001). Close ties between cetaceans, perissodactyls
(odd-toed ungulates), and phenacodontids proposed previously by Thewissen (1994),
Prothero (1993), and Prothero *et al.* (1988), respectively, are no longer tenable.

Like morphologic analyses, most molecular sequence data including that from both
combined and separate data sets (i.e., noncoding, protein coding, nuclear, mitochondrial DNA and transposons; Irwin and Árnason, 1994; Árnason and Gullberg, 1996;
Gatesy, 1998; Gatesy *et al.*, 1996, 1999a, 1999b, 2002; Shimamura *et al.*, 1997, 1999;
Nikaido *et al.*, 1999; Shedlock *et al.*, 2000; Murphy *et al.*, 2001; Árnason *et al.*, 2004) support the derivation of Cetacea from within a paraphyletic Artiodactyla and some of
these studies further suggest that cetaceans and hippopotamid artiodactyls are sister
taxa and united in a clade—Cetancodonta (Árnason *et al.*, 2000; Figure 4.6).

Until recently, morphologic data did not support molecular-based hypotheses that
supported close ties between artiodactyls and cetaceans. At issue was the morphology of
the ankle. Traditionally the ankle of artiodactyls, in which a trochlea is developed on the
distal part of the astragalas, had long been recognized as a unique feature that enabled
rapid locomotion. Recent discoveries of the ankle bones of archaic cetaceans show that
a trocheated or "double pulley" ankle is also present in basal cetaceans and supports a
close relationship between artiodactyls and cetaceans (Gingerich *et al.*, 2001; Thewissen
et al., 2001). If artiodactyls are paraphyletic, then either mesonychians are not closely
related to cetaceans (making many dental characters convergent), or the specialized heel

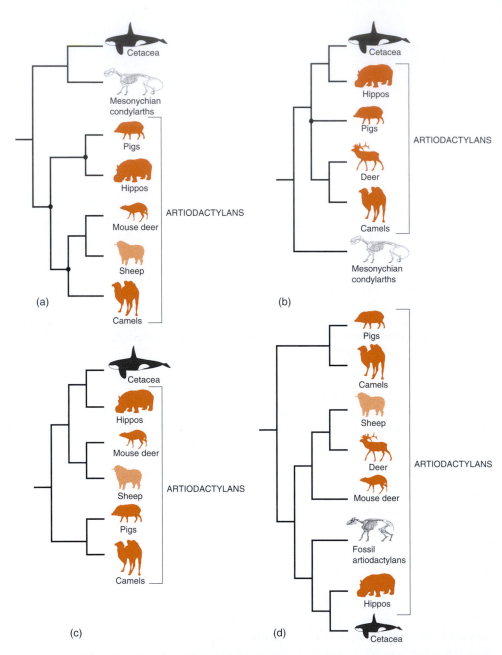

Figure 4.6.   Alternative hypotheses for relationships between cetaceans and various ungulate groups. **(a)** Morphologic data (O'Leary and Geisler, 1999; Geisler, 2001). **(b)** Morphologic data (Geisler and Uhen, 2003). **(c)** Molecular data (Gatesy *et al.*, 2002). **(d)** Combined molecular and morphologic data with mesonychian condylarths excluded. (O'Leary *et al.*, 2004).

morphology has evolved several times independently in artiodactyls or has been lost in the mesonychian/cetacean clade. Morphologic data presented by O'Leary and Geisler (1999) support a sister group relationship between Mesonychia and Cetacea with this clade as the sister group of a monophyletic Artiodactyla. Other morphologic studies support either a sister group relationship between artiodactyls and cetaceans or agree with the hippopotamid hypothesis (Gingerich *et al.*, 2001; Thewissen *et al.*, 2001; Geisler and Uhen, 2003). There is need for further exploration of evidence for a link between anthracotheres (pig-like extinct artiodactyls), hippos, and early cetaceans (see Gingerich, 2005; Boisserie *et al.*, 2005).

Controversy has ensued regarding the efficacy of morphologic vs molecular characters, analysis of extant vs extinct taxa, and analysis of data subsets (e.g., see Naylor and Adams, 2001; O'Leary *et al.*, 2003; Naylor and Adams, 2003; O'Leary *et al.*, 2004). More extensive phylogenetic analyses are necessary to clarify relationships among whales, artiodactylans, and their extinct relatives. Such analyses should include a better sampling of species and characters in combined analyses that include morphologic and molecular data as well as fossil and extant taxa. Toward this end, the most comprehensive study to date of whales, artiodactylans, and their extinct relatives (i.e., 50 extinct and 18 extant taxa) combined approximately 36,500 morphologic and molecular characters (O'Leary *et al.*, 2004). Because topologies were not well resolved given the instability of several taxa (i.e., Mesonychia) a subagreement tree summarized the maximum number of relationships supported by all minimum length topologies. This tree is consistent with a close relationship between cetaceans and hippopotamuses.

4.2.2.2. Relationships among Cetaceans

Prior phylogenetic analyses that used molecular data to support odontocete paraphyly, specifically a sister group relationship between sperm whales and baleen whales (Milinkovitch *et al.*, 1993, 1994, 1996), have been shown to be weakly supported (Messenger and McGuire, 1998). Recent molecular studies have consistently supported odontocetes as monophyletic (Gatesy, 1998; Gatesy *et al.* 1999a; Nikaido *et al.*, 2001). Several recent studies have made significant contributions to resolution of interrelationships among cetaceans by using comprehensive data sets (including both fossil and recent taxa) and rigorous phylogenetic methods (e.g., Messenger and McGuire, 1998; Geisler and Sanders, 2003).

4.2.3. Evolution of Early Whales—"Archaeocetes"

The earliest whales are archaeocetes, a paraphyletic stem group of cetaceans. Archaeocetes evolved from mesonychian condylarths. Archaeocete whales have been found from early to middle Eocene (52–42 Ma) deposits in Africa and North America but are best known from Pakistan and India. Archaeocetes have been divided into five or six families, the Pakicetidae, Protocetidae, Ambulocetidae, Remingtonocetidae, and Basilosauridae (Dorudontinae is sometimes recognized as a separate family) (Thewissen *et al.*, 1998; Thewissen and Williams, 2002; Uhen, 2004).

The Pakicetidae are the oldest and most basal cetaceans and include *Pakicetus, Nalacetus, Himalayacetus,* and *Icthyolestes* (see Thewissen and Hussain, 1998 and Williams, 1998 for taxonomic reviews). Pakicetids are known from the late early Eocene of Pakistan and India (e.g., Gingerich and Russell, 1981; Gingerich *et al.*, 1983; Thewissen and Hussain, 1998; Thewissen *et al.*, 2001). *Pakicetus* possessed a very dense and inflated auditory bulla that is partially separated from the squamosal (cheek bone), a feature suggesting their ears were adapted for underwater hearing (Gingerich and Russell, 1990; Thewissen and Hussain, 1993). However, pakicetids were predominantly land or freshwater animals and, except for features of the ear, had few adaptations consistent with aquatic life. Recent discoveries of pakicetid skeletons indicate that they had running adaptations (i.e., slender metapodials, heel bone with long tuber (Thewissen *et al.*, 2001).

The monophyletic Ambulocetidae include *Ambulocetus, Gandakasia,* and *Himalayacetus* (Thewissen and Williams, 2002). One of the most significant fossil discoveries is that of a whale with limbs and feet, *Ambulocetus natans,* also from the early Eocene of Pakistan (Thewissen *et al.*, 1994). The well-developed hind limbs and toes that ended as hooves of this so-called walking whale leave no doubt that they were used in locomotion. Thewissen *et al.* (1994) suggested that *Ambulocetus* swam by undulating the vertebral column and paddling with the hind limbs, combining aspects of modern seals and otters, rather than by vertical movements of the tail fluke, as is the case in modern whales (Figure 4.7; see also Chapter 8). The front limbs and hand of *Ambulocetus* also were well developed, with flexible elbows, wrists, and digits. Body size estimates suggest that *Ambulocetus* weighed between 141 and 235 kg and was similar in size to a female Steller's sea lion (Thewissen *et al.*, 1996). A second genus of ambulocetid whale, *Gandakasia,* is distinguished from *Ambulocetus* by its smaller size (Thewissen *et al.*, 1996).

A very diverse lineage of early whales, the Protocetidae, include *Rodhocetus, Artiocetus, Indocetus, Babicetus, Takracetus,* and *Gaviacetus* from India-Pakistan; *Protocetus* and *Eocetus* from Egypt; *Pappocetus* from Africa; *Georgiacetus;* and *Natchitochia* from the southeastern United States (Thewissen *et al.*, 1996; Uhen, 1998a; Hulbert *et al.*, 1998; Gingerich *et al.*, 2001; Thewissen *et al.*, 2001). Partial skeletons of *Rodhocetus* and *Artiocetus* suggest that protocetids swam using the robust tail as well as the fore limbs and hind limbs (Gingerich *et al.*, 2001) (Figure 4.8).

The Remingtonocetidae, a short-lived archaeocete clade (early middle Eocene of India-Pakistan) containing the genera *Remingtonocetus, Dalanistes, Andrewsiphius, Attockicetus,* and *Kutchicetus* are characterized by long, narrow skulls and jaws and robust limbs. Morphology of the jaws of remingtonocetids suggests a diet of fast-

(a)

(b)

Figure 4.7.   *Ambulocetus natans* (**a**) skeletal reconstruction (Thewissen, 2002) and (**b**) life restoration (Thewissen and Williams, 2002).

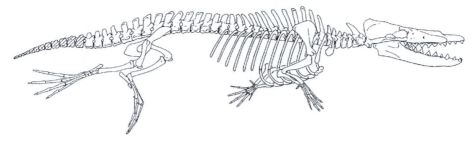

Figure 4.8.   Skeleton of *Rodhocetus kasrani*. Dashed lines and crosshatching show reconstructed parts. Original 2 m in length. (From Gingerich *et al.*, 2001.)

swimming aquatic prey (Thewissen, 1998). The middle ear is large and shows some specializations for underwater hearing (Bajpal and Thewissen, 1998; Gingerich, 1998). Although it has been suggested that remingtonocetids are ancestral to odontocetes, based on the presence of pterygoid sinuses in the orbits (air filled sacs in the pterygoid bone; Kumar and Sahni, 1986), this is now considered unlikely and they are recognized as a lineage of basal cetaceans (Thewissen and Hussain, 2000).

The paraphyletic Basilosauridae were late diverging archaeocetes and include one lineage of large species with elongated trunk vertebrae, the Basilosaurinae, and the Dorudontinae, a group of species without elongated vertebrae (see Uhen, 2004, for a recent taxonomic review). Some basilosaurines were gigantic, approaching 25 m in length, and are known from the middle to late Eocene and probably also from the early Oligocene in the northern hemisphere (Gingerich *et al.*, 1997). The several hundred skeletons of *Basilosaurus isis* are known from the middle Eocene of north central Egypt (Wadi Hitan, also known as the Valley of Whales or Zeuglodon Valley), which provide evidence of very reduced hind limbs in this species (Gingerich *et al.*, 1990; Uhen, 2004; Figure 4.9). Although it was suggested that *B. isis* used its tiny limbs to grasp partners during copulation (Gingerich *et al.*, 1990), the limbs could just as easily be interpreted as vestigial structures without function.

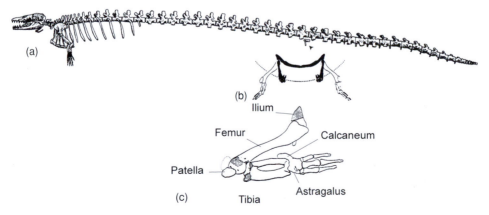

Figure 4.9.   Skeleton and hind limb of *Basilosaurus isis*. (From Gingerich *et al,* 1990.) **(a)** Skeleton in left lateral view and position of hind limb (*arrow*). **(b)** Hypothesized functional pelvic girdle and hind limb in resting posture (*solid drawing*) and functional extension (*open*). **(c)** Lateral view of left hind limb.

The dorudontines were smaller dolphin-like species that were taxonomically and ecologically more diverse than the basilosaurines. They are known from the late Eocene in
Egypt, southeastern North America, Europe, and New Zealand (Uhen, 2004). Among
the abundant fossil cetaceans from Egypt are the remains of *Dorudon atrox,* one of the
earliest and best known fully aquatic cetaceans (Uhen, 2004). *Dorudon* had short forelimb flippers, reduced hind limbs, and tail-based propulsion as in modern cetaceans
(Uhen, 2004). Also from this locality is a new genus and species of dorudontine,
Ancalecetes simonsi (Gingerich and Uhen, 1996), which differs from *D. atrox* in several
peculiarities of the forelimb including fused elbows that indicate very limited swimming
capability (Figure 4.10). Modern whales, including both odontocetes and mysticetes,
likely diverged from dorudontines (Uhen, 1998b).

4.2.4. Modern Whales

Estimates of the divergence time for the mysticete-odontocete split differ depending on
data (gene sequences, short interspersed element [SINE] insertions, or fossils) and
method (molecular clock, Bayesian). According to the fossil record, mysticetes and
odontocetes diverged from a common archaeocete ancestor about 35 Ma (Fordyce,
1980; Barnes *et al.*, 1985). On the basis of mitochondrial genomic analyses, Árnason
et al. (2004) postulated a 35-Ma age for the split between odontocetes and baleen whales
in agreement with the fossil record.

Modern whales differ from archaeocetes because they possess a number of derived
characters not seen in archaeocetes. Arguably one of the most obvious features is
the relationship of the bones in the skull to one another in response to the migration of
the nasal openings (blowholes) to the top of the skull. Termed **telescoping,** the modern
whale skull has premaxillary and maxillary bones that have migrated far posteriorly and
presently form most of the skull roof resulting in a long **rostrum** (beak) and dorsal nasal
openings. The occipital bone forms the back of the skull and the nasal, frontal, and parietal bones are sandwiched between the other bones (Figure 4.11).

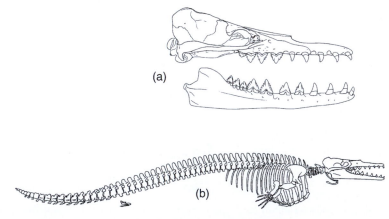

(a)

(b)

Figure 4.10.   Skeletal reconstructions of *Dorudon atrox.* **(a)** Skull and jaws (Uhen, 2002). **(b)** Skeleton in
right lateral view (Gingerich and Uhen, 1996).

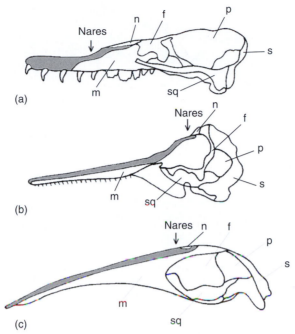

Figure 4.11. Telescoping of the skull in cetaceans. Note the posterior position of the nares and the different arrangement of the cranial bones in an archaic whale (archaeocete) **(a)**, a modern odontocete **(b)**, and mysticete **(c)**. Cranial bones: premaxilla (stippled), frontal (f), maxilla (m), nasals (n), parietal (p), squamosal (sq), supraoccipital (s). (Modified from Evans, 1987.)

Another derived feature of modern whales is a fixed elbow joint. The laterally flattened forelimbs are usually short and rigid with an immobile elbow. Archaeocetes have flexible elbow joints, capable of rotation.

4.2.4.1. Mysticetes

The baleen, or whalebone, whales are so named for their feeding apparatus: plates of **baleen** hang from the roof of the mouth and serve to strain planktonic food items. Although extant mysticetes lack teeth (except in embryonic stages) and possess baleen, this is not true for some fossil toothed mysticetes, as discussed later. Major evolutionary trends within the group include the loss of teeth, development of large body size and large heads, shortening of the intertemporal region, and shortening of the neck (Fordyce and Barnes, 1994).

Deméré *et al.* (in press) identified seven unequivocal synapomorphies to diagnose mysticetes (see Figure 4.11).

1. *Lateral margin of maxillae thin.* Mysticetes are distinguished from odontocetes in their development of thin lateral margins of the maxilla.

2. *Descending process of maxilla present as a broad infraorbital plate.* Mysticetes display a unique condition of the maxilla in which a descending process is developed as a broad plate below the eye orbit. Odontocetes lack development of a descending process.

3. *Posterior portion of vomer exposed on basicranium and covering basisphenoid/basioccipital suture.* Mysticetes are distinguished from odontocetes in having the posterior portion of the vomer exposed on the basicranium.

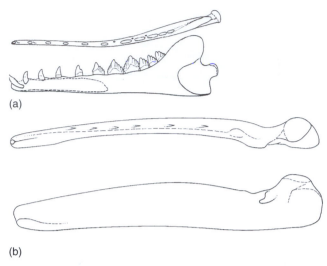

(a)

(b)

Figure 4.12.   Mysticete mandibular symphysis in dorsal and medial views illustrating ancestral
(a) Zygorhiza kochii and derived conditions **(b)** gray whale, *Eschrichtius robustus.* (From
Deméré, 1986 and unpublished manuscript.) Illustrated by M. Emerson.

4. *Basioccipital crest wide and bulbous.* The wide, bulbous basioccipital crest in mysticetes
is in contrast to the transversely narrow basioccipital crest in odontocetes.

5. *Mandibular symphysis unfused with only a ligamental or connective tissue attachment,
marked by anteroposterior groove.* Mysticetes display the unique condition of having an
unfused mandibular symphysis (Figure 4.12). Odontocetes possess a bony/fused mandibular symphysis.

6. *Mandibular symphysis short with large boss dorsal to groove.* Mysticetes are distinguished from odontocetes in having a short mandibular symphysis with a large boss dorsal to
the groove. In odontocetes, the mandibular symphysis is long with a smooth surface dorsal to
the groove.

7. *Dorsal aspect of mandible curved laterally.* Mysticetes possess a mandible that curves
laterally in dorsal view (see Figure 4.12). Most odontocetes possess a mandible that appears
straight when viewed dorsally; physeterids and pontoporiids are exceptions and possess
medially bowed mandibles due to a long fused symphysis.

4.2.4.1.1. Archaic Mysticetes

Archaic toothed mysticetes have been grouped into three families: the Aetiocetidae, the
Llanocetidae, and the Mammalodontidae. The Aetiocetidae includes four genera:
Aetiocetus (A. cotylalveus, A. polydentatus [Figure 4.13], *A. tomitai, A. weltoni*), *Chonecetus
(C. goedertorum, C. sookensis), Ashrocetus (A. eguchii)* and *Morawanocetus (M. yabukii)*
(Barnes *et al.*, 1995). *Aetiocetus* and *Chonecetus* possess multicusped teeth and nutrient
foramina (openings for blood vessels) for baleen. The oldest described mysticete is the
toothed *Llanocetus denticrenatus,* the only member of the family Llanocetidae. It is known
only from a fragment of large inflated mandible (Mitchell, 1989) of late Eocene or early
Oligocene age (Seymour Island, Antarctica). More complete material of the same species
(actually of the same specimen) was recovered and is under study (Fordyce, 1989). The holotype skull and skeleton represent a large individual with a skull length of about 2 m. The multicusped teeth of *Llanocetus denticrenatus* may have functioned in filter feeding, contrasting

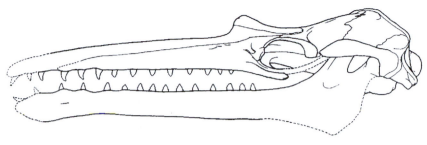

Figure 4.13. Skull and lower jaw restoration of an archaic mysticete whale, *Aetiocetus polydentatus,* from the late Oligocene of Japan. (From Barnes *et al.*, 1995.)

with the long pincer-like jaws and teeth typical of other fish-eating archaeocetes (Uhen, 2004). Another archaic toothed mysticete, *Mammalodon colliveri* (Figure 4.14), represents the Mammalodontidae from the late Oligocene or early Miocene in Australia, and has a relatively short rostrum, flat palate, and heterodont teeth. Only the holotype has been described (see Fordyce, 1984) but other late Oligocene specimens occur in the southwest Pacific (Fordyce, 1992).

Baleen-bearing mysticetes include several extinct lineages. The earliest known baleen-bearing mysticete *Eomysticetus whitmorei* (see Figure 4.14) was described from the late Oligocene of South Carolina (Sanders and Barnes, 2002). The "Cetotheriidae" is a large, diverse, nonmonophyletic assemblage of extinct toothless mysticetes that have been grouped together primarily because they lack characters of living mysticetes (see Figure 4.14). "Cetotheres" range in age from the late Oligocene to the late Pliocene of North America, South America, Europe, Japan, Australia, and New Zealand. At least 60 species of "cetotheres" have been named; however, many are based on noncomparable elements and the entire group is in clear need of systematic revision.

Most "cetotheres" were of moderate size, up to 10 m long, but some were probably as short as 3 m. Some fossil "cetotheres" have actually been found with impressions of baleen.

Kimura and Ozawa (2001) presented the first cladistic analysis that included eight "cetotheres," in addition to basal mysticetes (aetiocetids), and representatives of most extant families (*Caperea* was excluded). Their results supported "cetotheres" as more closely related to Balaenopteridae + Eschrichtiidae than to Balaenidae and identified two subgroups one of which is more closely related to these two modern lineages than it is to other "cetotheres." Geisler and Sanders (2003) included a more limited sample of "cetotheres" and their results supported inclusion of several Miocene "cetotheres" (*Diorocetus* and *Pelocetus*) together with extant mysticetes in a clade distinct from the eomysticetids.

4.2.4.1.2. *Later Diverging Mysticetes*
Relationships among the four families of modern baleen whales: Balaenopteridae (fin whales or **rorquals),** Balaenidae (bowhead and right whales), Eschrichtiidae (gray whale), and Neobalaenidae (pygmy right whale) have been contentious. Prior molecular studies did not sample all species nor did they yield well resolved relationships between the four major groups of mysticetes (Árnason and Gullberg, 1994; Árnason *et al.*, 1993). In more inclusive, better resolved molecular phylogenies of mysticetes, Rychel *et al.* (2004) and Sasaki *et al.* (2005) found evidence to support Balaenidae as the most basal mysticetes, and Neobalaenidae as the next diverging lineage and sister group to the balaenopterid-eschrichtiid clade (Figure 4.15).

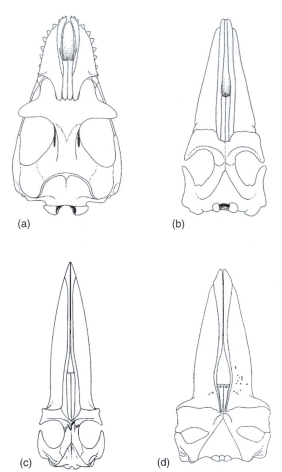

Figure 4.14. Archaic mysticete skulls in dorsal view. **(a)** *Mammalodon collivieri.* (From Fordyce and Muizon, 2001.) **(b)** "*Mauicetus" lophocephalus.* (From Fordyce and Muizon, 2001.) **(c)** *Eomysticetus whitmorei.* (From Sanders and Barnes, 2002.) **(d)** "Cetothere" *Agalocetus patulus.* (From Kellogg, 1968.) Not to scale.

Prior phylogenetic analyses of mysticetes based on morphology either failed to employ rigorous systematic methods or included limited taxon/character sampling (McLeod *et al.*, 1993; Geisler and Luo, 1996). Geisler and Sanders (2003) presented the first comprehensive morphological analysis that included significant numbers of extant and fossil mysticetes and odontocetes. Their most parsimonious tree divided extant mysticetes into two clades: Balaenopteroidea (Eschrichtiidae + Balaenopteridae) and Balaenoidea (Balaenidae + Neobalaenidae) (see Figure 4.15). Deméré *et al.* (in press) in a phylogenetic analysis of extinct and extant mysticetes confirmed strong support for both of these clades and provided limited resolution for a larger sample of basally positioned "cetotheres" (see Figure 4.15). This same result was also supported by total evidence analyses in this study. Future work should be directed toward clarifying the taxonomic status and evolutionary relationships among balaenopterid species (e.g., *B. brydei-edeni-borealis-omurai* complex), balaenids, and other mysticetes (gray, sei, and minke whales).

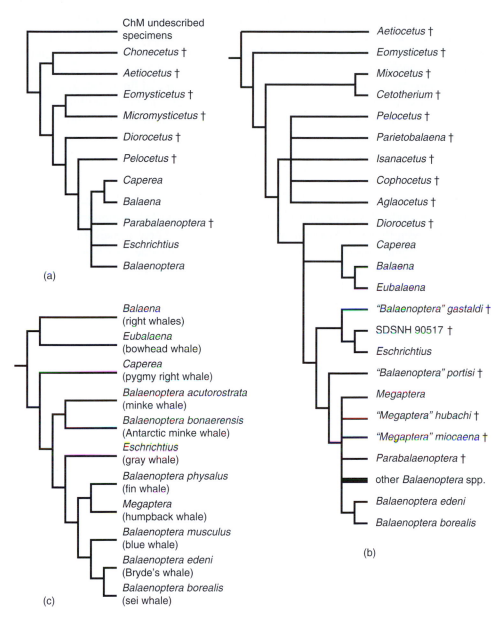

Figure 4.15. Relationships among mysticetes based on molecular and morphologic data. **(a)** Morphologic data (Geisler and Sanders, 2001); † = extinct taxa. **(b)** Morphologic data (Deméré *et al.*, in press). **(c)** Molecular phylogeny based on mitochondrial and nuclear sequence data (Rychel *et al.*, 2004).

Family Balaenopteridae The Balaenopteridae, commonly called the rorquals, which include fin whales and the humpback, are the most abundant and diverse living baleen whales. They include six to nine species ranging from the small 9-m minke whale, *Balaenoptera acutorostrata,* to the giant blue whale, *Balaenoptera musculus.* The blue whale has the distinction of being the largest mammal ever to have lived, reaching 33 m in length and weighing over 160 tons (Jefferson *et al.*, 1993). A new species, *Balaenoptera omurai,* was recently reported from Japan and distinguished from related species based on morphologic and molecular characters (Wada *et al.,* 2003).

Balaenopterids are characterized by the presence of a dorsal fin, unlike gray whales and balaenids, and by numerous throat grooves that extend past the throat region (Barnes and McLeod, 1984; Figure 4.16). The fossil record of the group extends from the middle Miocene and fossils are reported from North and South America, Europe, Asia, and Australia (Barnes, 1977; Deméré, 1986; McLeod *et al.*, 1993; Oishi and Hasegawa, 1995; Cozzuol, 1996; Dooley *et al.*, 2004; Deméré *et al.*, in press).

Family Balaenidae The family Balaenidae includes the right whales, *Eubalaena,* and the bowhead whale, *Balaena.* Some molecular data, however, do not support generic distinction between the two (Árnason and Gullberg, 1994). Three species (or subspecies according to some workers) of right whale are recognized, the North Atlantic right whale *(Eubalaena glacialis)* and the North Pacific right whale *(Eubalaena japonica)* and the South Atlantic right whale *(Eubalaena australis)*. Hunters called them the "right" whales to kill because they inhabited coastal waters, were slow swimming, and floated when dead. Balaenids are characterized by large heads that comprise up to one third of the body length. The mouth is very strongly arched and accommodates extremely long baleen plates (Figure 4.17).

The oldest fossil balaenid, *Morenocetus parvus,* is from the early Miocene (23 Ma) of South America (Cabrera, 1926). *M. parvus* has an elongated supraorbital process and a triangular occipital shield that extends anteriorly; both characters are developed to a lesser extent than in later balaenids (McLeod *et al.*, 1993). Relatively abundant fossils of later diverging balaenids are known, especially from Europe. Among Pliocene *Balaena* species is a nearly complete skeleton of a new bowhead from the Pliocene Yorktown Formation of the eastern United States (Westgate and Whitmore, 2002).

Family Neobalaenidae Traditionally, the small, 4-m long pygmy right whale, *Caperea marginata,* found only in the southern hemisphere, has been included in the Balaenidae (e.g., Leatherwood and Reeves, 1983). Its placement in a separate family, the Neobalaenidae, is supported by anatomical data (Mead and Brownell, 1993). In a molecular analysis that employed both mitochondrial and nuclear genes (Rychel *et al.*,

(a)   (b)

Figure 4.16. A representative of the Family Balaenopteridae (blue whale, *Balaenoptera musculus*). **(a)** Dorsal view of the skull. **(b)** Left side of body. (Illustrated by P. Folkens.) Note the dorsal fin and throat grooves. (From Barnes and McLeod, 1984.) Original skull length 6 m.

2004) the position of *Caperea* varied; it was either positioned as sister to balaenids or as diverging off the stem between balaenids and *Eschrichtius* (see Figure 4.15).

Caperea has a unique type of cranial architecture, distinguished from other mysticetes by a larger, more anteriorly thrusted occipital shield and a shorter, wider, and less arched mouth that accommodates relatively short baleen plates (see Figure 4.17). Other differences in the pygmy right whale in comparison with balaenids include the presence of a dorsal fin, longitudinal furrows on the throat (caused by mandibular ridges that might be homologous to throat grooves), coarser baleen, smaller head size relative to the body, a proportionally shorter humerus, and four instead of five digits on the hand (Barnes and McLeod, 1984).

No well-documented fossils of neobalaenids exist.

Family Eschrichtiidae The family Eschrichtiidae is represented by one extant species, the gray whale. It has a fossil record that goes back to the Pleistocene (100,000 years bp). The gray whale is now found only in the North Pacific although a North

(a)

(b)

(c)

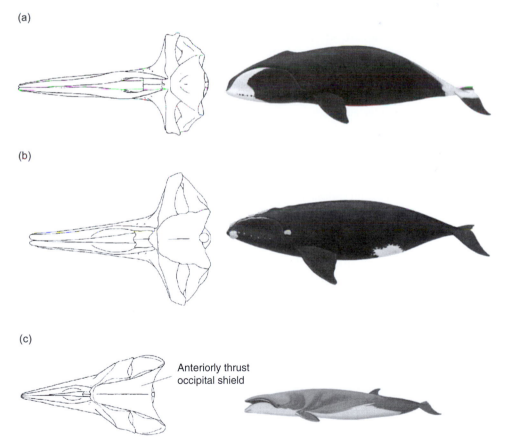

Anteriorly thrust
occipital shield

Figure 4.17. Representative balaenids and neobalaenid. Dorsal view of the skull and left side of the body. Note the large head and arched rostra. (Illustrated by P. Folkens.) **(a)** Bowhead, *Balaena mysticetus*. **(b)** Northern right whale, *Eubalena glacialis*. **(c)** Pygmy right whale, *Caperea marginata*. (From Barnes and McLeod, 1984.) Original skull lengths are 1.97 m, 3.27 m and 1.47 m, respectively.

Atlantic population became extinct in historic time (17th or early 18th century according to Bryant, 1995). There are two North Pacific subpopulations: the western North Pacific population migrates along the coast of Asia and is extremely rare. The much larger eastern North Pacific population was severely over exploited in the late 19th and early 20th centuries but has recovered sufficiently to be removed from the list of endangered species. Molecular analyses (Árnason and Gullberg, 1996; Hasegawa *et al.*, 1997; Rychel *et al.*, 2004; Sasaki *et al.*, 2005) position the gray whale as sister taxon to balaenopterids or nested within this lineage (see Figure 4.15).

The gray whale lacks a dorsal fin and is characterized by a small dorsal hump followed by a series of dorsal median bumps. Gray whales have two to four throat grooves in comparison to the numerous throat grooves of balaenopterids. The baleen plates differ from those of balaenids by being fewer in number, thicker, and white. A unique feature is the presence of paired occipital tuberosities on the posterior portion of the skull for insertion of muscles that originate in the neck region (Barnes and McLeod, 1984; Figure 4.18).

4.2.4.2. Odontocetes

The majority of whales are odontocetes, or toothed whales, named for the presence of teeth in adults, a feature distinguishing them from extant mysticetes. Odontocetes encompass a wide diversity of morphologies ranging from the large, deep-diving sperm whale, which has relatively few teeth and captures squid by suction feeding, to the smallest cetaceans, the porpoises, which have many spade-shaped teeth for seizing fish. Another useful distinction of odontocetes is a difference in telescoping of the skull in which the maxilla "telescopes," or extends posteriorly, over the orbit to form an expanded bony supraorbital process of the frontal (Miller, 1923; see Figure 4.11). In living odontocetes, this supraorbital process forms an origin for a facial (maxillonasolabialis) muscle (Mead, 1975a), which inserts around the single blowhole and associated nasal passages. The facial muscle complex and nasal apparatus generate the high frequency sounds used by living odontocetes for echolocation (see Chapter 11).

Among purported diagnostic features of odontocetes include two characters that have been specifically related to echolocation abilities: the presence of a **melon,** a region of adipose tissue on top of the skull with varying amounts of connective tissue within it, and **cranial** and **facial asymmetry**, a condition in which bones (= cranial asymmetry) and soft structures (= facial asymmetry) on the right side of the facial region are larger and more developed than equivalent structures on the left side. Cranial asymmetry is not universal among odontocetes, in either presence or extent. Both cranial and facial asymmetry are found in all modern representatives of the seven extant odontocete families, but fossil evidence indicates that cranial asymmetry is less pronounced in the more basal members of these groups and is totally absent in some extinct taxa. When present, the skew is always to the left side with the right side larger. Heyning (1989) argued that it is more likely that cranial asymmetry evolved only once.

Milinkovitch (1995) proposed another scenario in which facial asymmetry started to develop in the ancestor of all extant cetaceans and by chance was oriented to the left. It follows from his argument that left-oriented facial asymmetry might be an ancestral character for odontocetes that was lost or greatly reduced in baleen whales. Accordingly, cranial asymmetry would accompany facial asymmetry and would have been developed

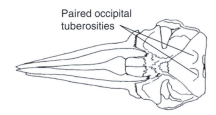

Paired occipital
tuberosities

Figure 4.18. The Family Eschrichtiidae (gray whale *Eschrichtius robustus*). Dorsal view of the skull illustrating the paired occipital tuberosities. (From Barnes and McLeod, 1984.) Original skull
length 2.33 m.

independently in two (possibly up to four) odontocete lineages. Geisler and Sanders
(2003) provide a test of this hypothesis in their evaluation of the distribution of asymmetry of the premaxilla in cetaceans. Their results suggest that asymmetry of the premaxilla evolved five times among odontocetes.

Regarding the presence of a melon, Milinkovitch (1995) noted that mysticetes possess
a fatty structure just anterior to the nasal passages that may be homologous to the melon
of odontocetes (Heyning and Mead, 1990). It has been suggested that the "vestigial"
melon of mysticetes might be a hint of more generalized **paedomorphism** of their facial
anatomy, seen for example in a fossil delphinoid that has dramatically reversed telescoping of the skull (Muizon, 1993b). Milinkovitch (1995) further suggested that presence of
a melon (along with facial and cranial asymmetry and echolocation abilities) might be
ancestral for all cetaceans and that baleen whales greatly reduced or lost this character.
Heyning (1997) disputed this interpretation, arguing that it assumes *a priori* that the
melon regressed from a larger melon in the common ancestor, a claim that lacks empirical evidence. In addition, study of the inner ear of an archaic mysticete, which more
nearly resembles the nonecholocating modern mysticetes than early fossil toothed
whales, offers little support for the suggestion that echolocation was present in ancestral
mysticetes and was lost secondarily in extant mysticetes (Geisler and Luo, 1996). In summary, Milinkovitch's alternative interpretations of odontocete morphological synapomorphies are less parsimonious interpretations of character transformations and they
lack supporting data.

The traditional monophyletic view of odontocetes is followed here based on a comprehensive reappraisal of both morphologic and molecular data (Messenger, 1995;
Messenger and McGuire, 1998). In a recent reevaluation of purported odontocete
synapomorphies, Geisler and Sanders (2003) identify 14 unequivocal synapomorphies,
a few of which are as follows (Figure 4.19):

1. *Nasals elevated above the rostrum*. The height of the nasals in odontocetes ranges
between 229–548% of rostral height. In the primitive condition seen in baleen whales
and artiodactyls, nasal height ranges between 92–139% of rostral height.

2. *Frontals higher than nasals*. In odontocetes the frontals are higher than the nasals.
Mysticetes and artiodactyls have frontals that are lower than the nasals.

3. *Premaxillary foramen present*. Odontocetes possess infraorbital or premaxillary
foramina of varying shapes and sizes. Neither mysticetes nor artiodactyls possess
foramina in the premaxillary bones.

4. *Maxillae overlay supraorbital process*. "Telescoping" of the skull in odontocetes
involves the presence of ascending processes of the maxillae that cover the supraor-

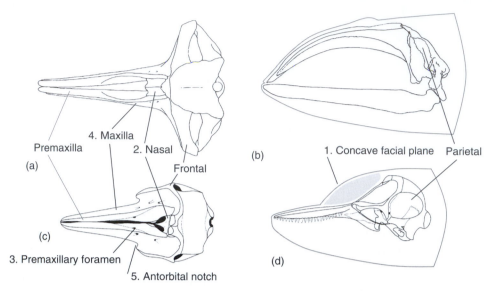

Figure 4.19.   Simplified outlines of cetacean skulls in dorsal and lateral views illustrating odontocete
synapomorphies. **(a)** and **(b)** a mysticete, *Balaena mysticetus,* **(c)** and **(d)** an odontocete,
Tursiops truncatus. (Modified from Fordyce, 1982.)

bital processes of the frontals. This condition is not seen in mysticetes or terrestrial
mammals.

4.2.4.2.1. *Basal Odontocetes*

The phylogenetic relationship of the generally accepted basal odontocetes (i.e.,
Agorphius, Xenorophus, and *Archaeodelphius*) from the Oligocene age (28–24 Ma) are
becoming better understood (e.g., Geisler and Sanders, 2003). According to these workers *Archaeodelphius* is the basal-most member from the a clade that includes *Xenorophus*
and related taxa. Geisler and Sanders (2003) mention an undescribed specimen that they
refer to *Agorophius pygmaeus* (late Oligocene, South Carolina), which was previously
represented by the holotype skull apparently now lost (Fordyce, 1981). The few known
skulls of these basal odontocetes demonstrate that these animals had only a moderate
degree of telescoping (the nares were anterior to the orbits) and that the cheekteeth had
multiple roots and accessory cusps on the crowns (Barnes, 1984a).

4.2.4.2.2. *Later Diverging Odontocetes*

Only one of the two major later diverging odontocete clades proposed by Geisler and
Sanders (2003), the Physeteroidea (Physeteridae + Ziphiidae), is generally accepted by
most workers. The Platanistoidea (river dolphins and their kin) plus the Delphinidae +
Monodontidae + Phocoenidae are more contentious.
   Molecular and morphologic phylogenies for odontocetes are presented in Figures
4.20 and 4.21. Geisler and Sander's (2003) morphologically based proposal of two major
odontocete clades: the Physeteroidea (Physeteridae + Ziphiidae) and the Platanistoidea
(river dolphins and their kin) plus the Delphinidae + Monodontidae + Phocoenidae differs from previous hypotheses. According to Heyning (1989, 1997) the Physeteroidea
(Physeteridae + Kogiidae) are at the base of odontocetes (Figure 4.21). This is consistent

with molecular analyses (Cassens *et al.*, 2000, Nikaido *et al.*, 2001; see Figure 4.20). The position of beaked whales, however, differs among morphological systematists. In one hypothesis, beaked whales are united in a clade with sperm whales (Fordyce, 2001; Geisler and Sanders, 2003). In an alternative arrangement, beaked whales are positioned with more crownward odontocetes (Delphinoidea and Platanistoidea) excluding sperm whales (Heyning, 1989; Heyning and Mead, 1990). The status of the Platanistoidea remains unresolved (see Messenger, 1994). The classic concept of Platanistoidea as including all extant river dolphins (i.e., Platanistidae, Pontoporiidae, Iniidae, and Lipotidae) is not supported by recent analyses of molecular data (Cassens *et al.*, 2000; Nikaido *et al.*, 2001) although the recent morphological analysis of Geisler and Sanders (2003) differs in supporting a monophyletic Platanistoidea. A third major odontocete clade, the Delphinoidea, although not identified by Geisler and Sanders (2003) has been traditionally recognized based on morphology (Heyning, 1997; Messenger and McGuire, 1998) and is strongly supported by molecular sequence data (Gatesy, 1998; Cassens *et al.*, 2000; Nikaido *et al.*, 2001).

Physeteroidea

Family Ziphiidae Beaked whales are a relatively poorly known but diverse group of toothed whales composed of at least 5 genera and 21 extant species. They are characterized by a snout that is frequently drawn out into a beak and from which the group obtains its common name, beaked whales. Ziphiids inhabit deep ocean basins and much of our information about them comes from strandings and whaling activities. One evolutionary trend in ziphiids is toward the loss of all teeth in the rostrum and most in the mandible, with the exception of one or two pairs of teeth at the anterior end of the jaw that become much enlarged (Figure 4.22). Phylogenetic analysis based on mtDNA data suggests species level taxonomic revisions (Dalebout *et al.*, 2002; Van Helden *et al.*, 2002). In addition to several features of the ear, premaxilla, and palatal region (e.g., see Fordyce,

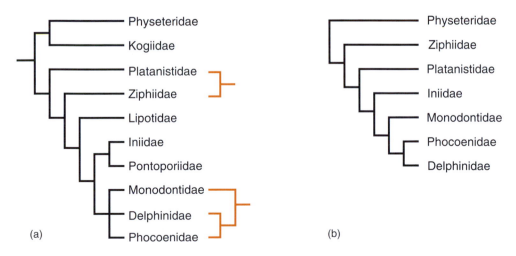

Figure 4.20. Alternative hypotheses for the phylogeny of extant odontocetes. **(a)** Cladogram based on retroposons and DNA sequence data (Nikaido *et al.*, 2001). **(b)** Cladogram based on morphologic data (Heyning, 1989, 1997; Heyning and Mead, 1990).

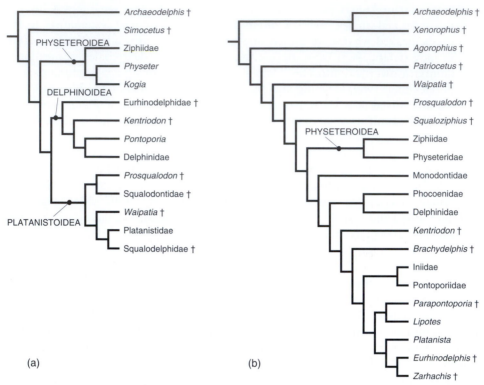

Figure 4.21.   Alternate phylogenies for fossil and recent odontocetes based on morphology. † = extinct taxa. **(a)** Fordyce, 2002, and **(b)** Geisler and Sanders, 2003.

1994), extant ziphiids can be distinguished from other odontocetes by possession of one pair of anteriorly converging throat grooves (see Figure 4.22).

Ziphiids have been classified either with sperm whales in the superfamily Physeteroidea or as a sister group to extant odontocetes other than physeterids. Ziphiids are known in the fossil record from the Miocene and Pliocene of Europe, North and South America, Japan, and Australia. A freshwater fossil ziphiid has been reported from the Miocene of Africa (Mead, 1975b).

Family Physeteridae The physeterids, or sperm whales, have an ancient and diverse fossil record, although only a single species, *Physeter macrocephalus,* survives. Derived characters of the skull that unite sperm whales include, among others, a large, deep, supracranial basin, which houses the spermaceti organ (Figure 4.23) and loss of one or both nasal bones (Fordyce, 1984). The terms sperm whale and spermaceti organ derive from the curious belief of those who named this whale that it carried its semen in its head. Sperm whales are the largest of the toothed whales, attaining a length of as much as 19 m and weighing 70 tons. They also are the longest and deepest diving vertebrates known (138 min and 3000 m; Clarke, 1976; Watkins *et al.*, 1985).

The fossil record of the Physeteridae goes back at least to the Miocene (late early Miocene 21.5–16.3 Ma) and earlier if *Ferecetotherium* from the late Oligocene (23+ Ma) of Azerbaidjan is included. By middle Miocene time, physeterids were moderately

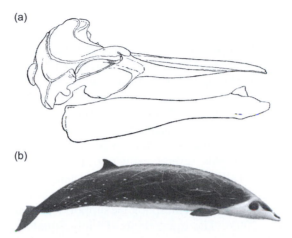

Figure 4.22. A representative of the Family Ziphiidae. **(a)** Lateral view of the skull and lower jaw of Gervais' beaked whale, *Mesoplodon europaeus*. Note the reduced dentition. (From Van Beneden and Gervais, 1880.) **(b)** Right side of the body of Stejneger's beaked whale, *Mesoplodon stejnegeri*. (Illustrated by P. Folkens.)

diverse and the family is fairly well documented from fossils found in South America, eastern North America, western Europe, the Mediterranean region, western North America, Australia, New Zealand, and Japan (Hirota and Barnes, 1995).

Family Kogiidae The pygmy sperm whale, *Kogia breviceps,* and the dwarf sperm whale, *Kogia simus,* are closely related to the sperm whale family, Physeteridae. The pygmy sperm whale is appropriately named, because males only attain a length of 4 m and females are no more than 3 m long. The dwarf pygmy sperm whale is even smaller, with adults ranging from 2.1 to 2.7 m. As in physeterids, there is a large anterior basin in the skull, but kogiids differ markedly in their small size, short rostrum, and other details of the skull (Fordyce and Barnes, 1994; Figure 4.24). The oldest kogiids are from the late Miocene (8.8–5.2 Ma) of South America and the early Pliocene (6.7–5 Ma) of Baja California.

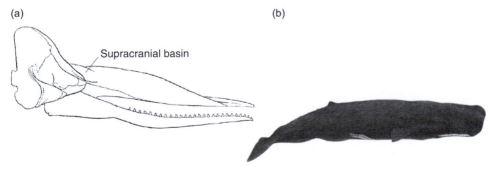

Figure 4.23. The Family Physeteridae (Sperm whale, *Physeter macrocephalus*). **(a)** Lateral view of the skull and lower jaw. Note the deep supracranial basin. (From Van Beneden and Gervais, 1880.) **(b)** Right side of the body. (Illustrated by P. Folkens.)

Platanistoidea

"River Dolphins" Living river dolphins include four families (Platanistidae, Lipotidae, Iniidae, and Pontoporiidae) that have invaded estuarine and freshwater habitats. According to Hamilton *et al.* (2001) and Cassens *et al.* (2000), river dolphins are a polyphyletic group of three lineages; the platanistids are sister to the remaining odontocetes and the remaining river dolphins are paraphyletically positioned at the base of the delphinoid clade (i.e., monodontids, delphinoids, and phocoenids). A once diverse radiation of platanistoids is apparent with inclusion of several extinct lineages. The superfamily Platanistoidea, a clade that according to Muizon (1987, 1988a, 1991, 1994) includes the Platanistidae plus several extinct groups (the Squalodontidae, the Squalodelphidae, and the Dalpiazinidae) and a closely related newly discovered lineage the Waipatiidae (Fordyce, 1994), has had a long and confusing history (Messenger, 1994; Cozzuol, 1996). There is some recent morphologic support for monophyly of the group (Geisler and Sanders, 2003). The squalodonts (Family Squalodontidae), or shark-toothed dolphins, named for the presence of many triangular, denticulate cheekteeth, are known from the late Oligocene to the late Miocene. They have been reported from North America, South America, Europe, Asia, New Zealand, and Australia. Squalodontids include a few species known from well-preserved skulls, complete dentitions, ear bones, and mandibles but many nominal species are based only on isolated teeth and probably belong in other families. Most squalodontids were relatively large animals with bodies 3 m or more in length. Their crania were almost fully telescoped, with the nares located on top of the head between the orbits. The dentition was polydont but still heterodont, with long pointed anterior teeth and wide, multiple-rooted cheekteeth (Figure 4.25). It is likely that the anterior teeth functioned in display rather than in feeding and the robust cheekteeth with worn tips may reflect feeding on prey such as penguins (Fordyce, 1996).

The Squalodelphidae include several early Miocene genera (*Notocetus, Medocinia,* and *Squalodelphis;* Muizon, 1981) with small, slightly asymmetrical skulls and moderately long rostra and near-homodont teeth (Muizon, 1987). The family Dalpiazinidae was established by Muizon (1988a) for *Dalpiazina ombonii,* an early Miocene species with a small symmetrical skull and a long rostrum armed with many near-homodont teeth (Fordyce and Barnes, 1994). Fordyce and Sampson (1992) reported an undescribed earliest Miocene species from the southwest Pacific.

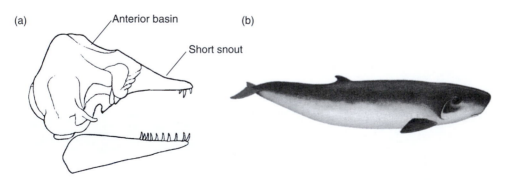

(a)   Anterior basin   (b)

Short snout

Figure 4.24.   The Family Kogiidae (Pygmy sperm whale, *Kogia breviceps*). **(a)** Lateral view of the skull and lower jaw. Note the short snout and anterior basin. (From Bobrinskii *et al.*, 1965, p. 197.) **(b)** Right side of the body. (Illustrated by P. Folkens.)

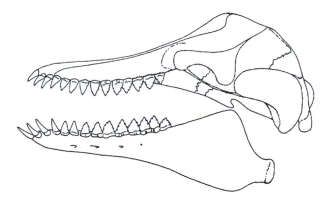

Figure 4.25. Skull and lower jaw of an archaic odontocete, *Prosqualodon davidsi,* from the early Miocene of Tasmania. (From Fordyce *et al.*, 1995.)

The family Waipatiidae was established by Fordyce (1994) for a single described species, *Waipatia maerewhenua,* characterized by a small slightly asymmetrical skull and long rostrum with small heterodont teeth.

Family Platanistidae The extant Asiatic river dolphins, *Platanista* spp. (the blind endangered Ganges and Indus River dolphins), comprise the family Platanistidae. They are characterized by a long narrow beak, numerous narrow pointed teeth, and broad paddle-like flippers. They have no known fossil record and the time of invasion into freshwater is unknown. Middle to late Miocene marine species of *Zarhachis* and *Pomatodelphis* are closely related to *Platanista,* although they differ in rostral profiles and cranial symmetry and in their development of pneumatized bony facial crests (Figure 4.26; Fordyce and Barnes, 1994).

Family Pontoporiidae The small, long-beaked franciscana, *Pontoporia blainvillei,* lives in coastal waters in the western South Atlantic and is the only extant pontoporiid. All pontoporiids except for the fossil *Parapontoporia* have virtually symmetrical cranial vertices and most have long rostra and many tiny teeth (Figure 4.27).

Fossil *Pontoporia*-like taxa include species of *Pliopontos* and *Parapontoporia* from temperate to subtropical marine settings in the east Pacific (Barnes, 1976, 1984b; Muizon, 1983, 1988b). Late Miocene *Pontistes* and *Pontoporia* came from marine

(a)   (b)

Bony facial crest

Narrow, elongated beak

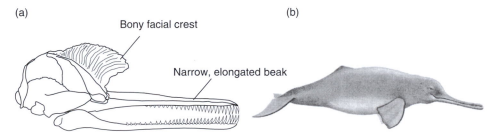

Figure 4.26. A representative of the Family Platanistidae (Ganges river dolphin, *Platanista gangetica*). **(a)** Lateral views of the skull and lower jaw. (From Duncan, 1877–1883: p. 248.) Note the development of bony facial crests. **(b)** Right side of the body. (Illustrated by P. Folkens.)

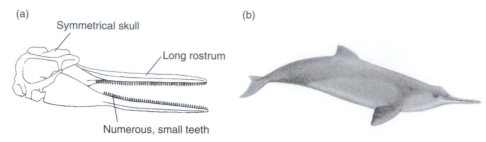

Figure 4.27.   The Family Pontoporiidae (franciscana, *Pontoporia blainvillei*). **(a)** Lateral view of skull and jaws. (From Watson, 1981.) **(b)** Right side of the body. (Illustrated by P. Folkens.) Note the symmetrical skull, long rostrum and numerous small teeth.

sediments of Argentina (Cozzuol, 1985, 1996) to colonize the nearshore coast of the La Plata estuary (Hamilton *et al.*, 2001).

Family Iniidae   The bouto, *Inia geoffrensis,* is a freshwater species with reduced eyes found only in Amazon River drainages. The name comes from the sound of its blow. According to Heyning (1989), the monotypic extant taxon *Inia* is diagnosed by having the premaxillae displaced laterally and not in contact with the nasals (Figure 4.28). Dentally they are diagnosed by conical front teeth and molariform posterior teeth. According to Cozzuol (1996), iniids (including fossil taxa) are characterized by an extremely elongated rostrum and mandible, very narrow supraoccipital, greatly reduced orbital region, and pneumatized maxillae forming a crest.

The fossil record of iniids goes back to the late Miocene of South America (Cozzuol, 1996) and the early Pliocene of North America (Muizon, 1988c; Morgan, 1994). The North American record of iniids is disputed by Cozzuol (1996, and references therein). The phylogenetic history and fossil record of iniids indicates that they originated in South America in the Amazonian basin, entering river systems along the Pacific coast (Cozzuol, 1996; Hamilton *et al.*, 2001).

Family Lipotidae   The endangered baiji, or Chinese river dolphin *(Lipotes vexillifer)*, lives in the Yangtze River, China. They are characterized by a long narrow upturned

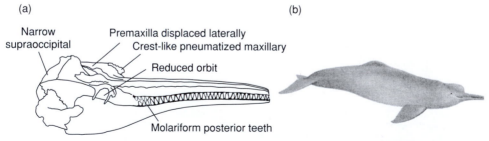

Figure 4.28.   The Family Iniidae (bouto, *Inia geoffrensis*). **(a)** Lateral view of the skull. (From Geibel, 1859: p. 498.) Note the premaxillae is displaced laterally and is not in contact with the nasals, narrow supraoccipital, reduced orbit, crest-like pneumatized maxillary, and molariform posterior teeth. **(b)** Right side of the body. (Illustrated by P. Folkens.)

(a)

(b)

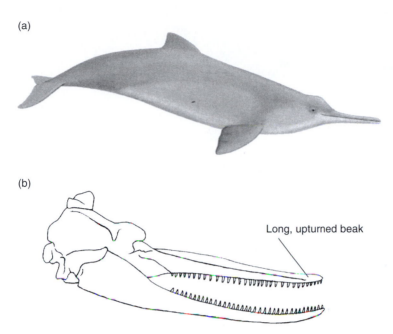

Long, upturned beak

Figure 4.29.   The Family Lipotidae (Chinese River dolphin, *Lipotes vexillifer*). **(a)** Right side of the body. (Illustrated by P. Folkens.) **(b)** Lateral view of skull and jaws. (From Watson, 1981.) Note the long, upturned beak.

beak, a low triangular dorsal fin, broad rounded flippers, and very small eyes (Zhou *et al.*, 1979; Figure 4.29).

The only fossil lipotid *Prolipotes,* based on a fragment of mandible from China (Zhou *et al.*, 1984) cannot be confirmed as belonging to this taxon (Hamilton *et al.*, 2001).

Archaic "Dolphins"   Archaic dolphins of the Miocene are grouped into one of three extinct families: the Kentriodontidae, the Albeirodontidae, and the Eurhinodelphidae. The earliest diverging lineage, the kentriodontids, were small animals approximately 2 m or less in length and with numerous teeth, elaborate basicranial sinuses, and symmetrical cranial vertices (Barnes, 1978; Dawson, 1996). This group's monophyly has been questioned (Cozzuol, 1996) because of relatively diverse species and widespread distribution ranging from the late Oligocene to late Miocene in both the Atlantic and Pacific Oceans (Ichishima *et al.*, 1995). Barnes (1984b) suggested that the Albeirodontidae, known by only one late Miocene species (Figure 4.30), was derived from kentriodontids, although Muizon (1988c) placed this taxon as sister group to phocoenids. The long beaked eurhinodelphids were widespread and moderately diverse during the early and middle Miocene and disappeared in the late Miocene (Figure 4.31). Eurhinodelphid relationships are contentious. Most recently they have been either included in a clade with kentriodontids and delphinids or allied with platanistoids (Fordyce, 2002; Geisler and Sanders, 2003).

Family Delphinidae   Delphinids are the most diverse of the cetacean families and include 17 genera and 36 extant species of dolphins, killer whales, and pilot whales. Most delphinids are small to medium sized, ranging from 1.5 to 4.5 m in length. The

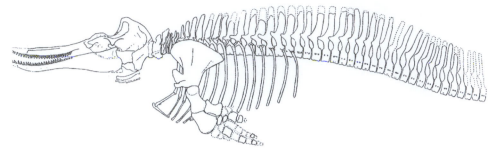

Figure 4.30.   Reconstruction of a fossil dolphin, *Albireo whistleri*. (From Fordyce *et al.*, 1995.)

giant among them, the killer whale, reaches 9.5 m in length. Although the Irrawaddy dolphin *(Orcaella brevirostris)* found only in the Indo-Pacific has been regarded as a monodontid by some (Kasuya, 1973; Barnes, 1984a), more recent morphologic and molecular work suggests that this species is a delphinid (Muizon, 1988c; Heyning, 1989; Árnason and Gullberg, 1996; Arnold and Heinsohn, 1996; Messenger and McGuire, 1998). Delphinids, including *Orcaella,* are united by the loss of the posterior sac of the nasal passage (Fordyce, 1994). Another distinguishing feature of delphinids is reduction of the posterior end of the left premaxilla so that it does not contact the nasal (Figure 4.32; Heyning, 1989). Le Duc *et al.* (1999) sequenced the cytochrome b gene for delphinids and found little resolution among subfamily groups and evidence for polyphyly in the genus *Lagenorhynchus*. The oldest delphinid is of latest Miocene age, possibly 11 Ma (Barnes, 1977).

Family Phocoenidae Porpoises include six small extant species. One of the most diagnostic features of phocoenids are premaxillae that do not extend posteriorly behind the anterior half of the nares. Phocoenids are further distinguished from other odontocetes by having spatulate-shaped rather than conical teeth (Figure 4.33; Heyning, 1989). Phocoenids and delphinids have been recognized by several workers (e.g., Barnes, 1990) as being more closely related to one another than either is to monodontids (see Figure 4.21). A recent comprehensive morphological study of cetaceans (Geisler and Sanders, 2003) rejected monophyly of the Delphinoidea and proposed that river dolphins are monophyletic and nested within that clade. Molecular data (Waddell *et al.*, 2000; Árnason *et al.*, 2004) supports an alliance between phocoenids and monodontids with delphinids as sister taxon to that clade.

Figure 4.31.   An archaic dolphin *(Eurhinodelphis cocheteuxi)* from the late Miocene of Belgium. (From Slijper, 1962.)

(a)   (b)

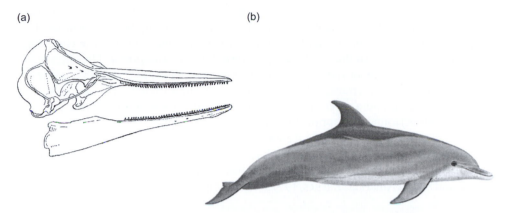

Figure 4.32. Representatives of the Family Delphinidae. **(a)** Lateral view of skull and lower jaw of common dolphin, *Delphinus delphis*. (From Van Beneden and Gervais, 1880.) **(b)** Right side of the body of bottlenose dolphin, *Tursiops trancatus*. (Illustrated by P. Folkens.)

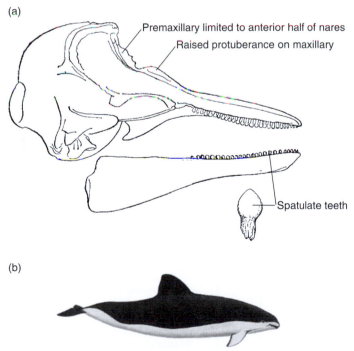

Figure 4.33. Representatives of the Family Phocoenidae (porpoises). **(a)** Lateral view of the skull and lower jaw of a phocoenid illustrating the raised rounded protuberances on the premaxillae (from Gervais 1855: 327) and spatulate-shaped teeth (from Flower and Lydekker, 1891: p. 263). **(b)** Right side of the body of spectacled porpoise, *Phocoena dioptrica*. (Illustrated by P. Folkens.)

Phylogenetic relationships among extant species based on cytochrome b sequence data (Rosel *et al.*, 1995; Figure 4.34) support a close relationship between Burmeister's porpoise, *Phocoena spinipinnis,* and the vaquita, *Phocoena sinus,* and also the association of these two species with the spectacled porpoise, *Phocoena dioptrica*. The latter result differs from a previous proposal based on morphology (Barnes, 1985) that groups *P. dioptrica* with Dall's porpoise, *Phocoenoides dalli,* in the subfamily Phocoeninae. The molecular analysis and a recent morphologic study of phocoenids (Fajardo, personal communication) found no support for this grouping. Morphologic and molecular data (Rosel *et al.*, 1995; Fajardo personal communication) indicate that the finless porpoise, *Neophocoena phocaenoides,* is the most basal member of the family. Like delphinids, phocoenids have a fossil record that extends back to the late Miocene and Pliocene in North and South America (Barnes, 1977, 1984b; Muizon, 1988a).

Family Monodontidae Monodontids include two extant species, the narwhal *(Monodon monoceros)* and the beluga *(Delphinapterus leucas)*. The narwhal is readily distinguished by the presence of a spiraled incisor tusk in males and occasionally in females (Figure 4.35). It has been suggested that the narwhal tusk may have been used in creating the legend of the unicorn, a horse with cloven hooves, a lion's tail, and a horn in the middle of its forehead that resembles the narwhal tusk (Slijper, 1962). The living beluga is characterized by its completely white coloration (see Figure 4.35).

The narwhal and beluga have a circumpolar distribution in the Arctic. During the late Miocene and Pliocene, monodontids occupied temperate waters as far south as Baja California (Barnes, 1973, 1977, 1984a; Muizon, 1988a).

An extinct relative of monodontids is the bizarre cetacean *Odobenocetops* convergent in its morphology and inferred feeding habits (see also Chapter 12) with the modern walrus (Muizon, 1993a, 1993b; Muizon *et al.*, 1999; Muizon *et al.*, 2001). *Odobenocetops* is known by two species from the early Pliocene of Peru.

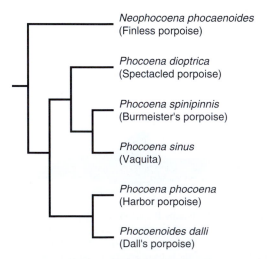

Figure 4.34. Species-level phylogeny of phocoenids (Rosel *et al.*, 1995).

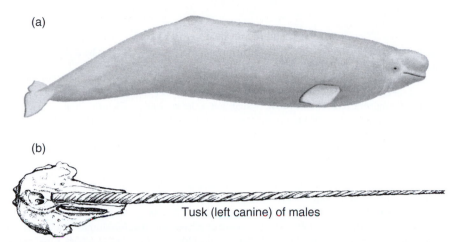

(a)

(b)

Tusk (left canine) of males

Figure 4.35. Representatives of the Family Monodontidae (narwhal, *Monodon monoceros* and beluga, *Delphinapterus leucas*). **(a)** Right side of the body of beluga. (Illustrated by P. Folkens.) **(b)** Dorsal view of the skull of the narwhal. Note the top of the nostrum has been removed to show the root of the large left tusk and the small, unerupted right tusk. (From Flower and Lydekker, 1891: p. 261).

4.3. Summary and Conclusions

Most morphologic and all molecular data are in general agreement that artiodactyls (specifically hippos) are the closest relatives of cetaceans. Odontocete monophyly is also widely accepted. The earliest archaeocete whales, a paraphyletic stem group that first appeared approximately 50 million years ago, are best known from India and Pakistan. A rapidly and continually expanding record provides evidence of considerable morphologic diversity among early whales, many with well-developed hind limbs and feet. Divergence estimates for baleen and toothed whales from a common archaeocete ancestor approximate 35 Ma based on molecular data that are in accord with the fossil record. There is evidence that some archaic mysticetes possessed both teeth and baleen. Later diverging mysticetes lost teeth but retained baleen. Relationships among modern families of baleen whales are unclear because of conflicting morphological results versus molecular data. Relationships among odontocetes are no less controversial. There is, however, general agreement of both molecular and morphological data that beaked whales and sperm whales are basal odontocetes. Relationships among other odontocete lineages will require comprehensive assessment of both fossil and recent taxa using both separate and combined analyses of morphological and molecular data.

4.4. Further Reading

The evolutionary history of fossil whales is summarized in Fordyce and Barnes (1994), Fordyce *et al.* (1995), and Fordyce and Muizon (2001). See Thewissen (1998) for an account of the early evolution of whales. For a popular treatment of the evolutionary significance of recent whale fossil discoveries see Gould (1994) and Zimmer (1998). The relationship of cetaceans to other ungulates based on morphologic and molecular data is reviewed in Geisler (2001) and O'Leary *et al.* (2003, 2004).

References

Árnason, U., and A. Gullberg (1994). "Relationship of Baleen Whales Established by Cytochrome b Gene Sequence Comparison." *Nature 367:* 726–727.

Árnason, U., and A. Gullberg (1996). "Cytochrome b Nucleotide Sequences and the Identification of Five Primary Lineages of Extant Cetaceans." *Mol. Biol. Evol. 13:* 407–417.

Árnason, U., A. Gullberg, S. Gretarsdottir, B. Ursing, and A. Janke (2000). "The Mitochondrial Genome of the Sperm Whale and a New Molecular Reference for Estimating Eutherian Divergence Dates." *J. Mol. Evol. 50:* 569–578.

Árnason, U., A. Gullberg, and A. Janke (2004). "Mitogenomic Analyses Provide New Insights into Cetacean Origin and Evolution." *Gene 333:* 27–34.

Árnason, U., A. Gullberg, and B. Widegren (1993). "The Complete Nucleotide Sequence of the Mitochondrial DNA of the Fin Whale, *Balaenoptera physalus." J. Mol. Evol. 33:* 556–568.

Arnold, P. W., and G. E. Heinsohn (1996). "Phylogenetic Status of the Irrawaddy Dolphin *(Orcaella brevirostris* Owen in Gray): A Cladistic Analysis." *Mem. Queensl. Mus. 39*(2): 141–204.

Bajpal, S., and J. G. M. Thewissen (1998). Middle Eocene cetaceans from the Harudi and Subathu formations of India, *In* "The Emergence of Whales: Patterns in the Origin of Cetacea" (J. G. M. Thewissen, ed.), pp. 213–233. Plenum Press, New York.

Barnes, L. G. (1973). "*Praekogia cedroensis,* A New Genus and Species of Fossil Pygmy Sperm Whale from Isla Cedros, Baja California, Mexico." *Nat. Hist. Mus., L. A. Cty. Sci. Bull. 247:* 1–20.

Barnes, L. G. (1977). "Outline of Eastern North Pacific Cetacean Assemblages." *Syst. Zool. 25:* 321–343.

Barnes, L. G. (1978). "A Review of *Lophocetus* and *Liolithax* and Their Relationships to the Delphinoid Family Kentriodontidae (Cetacea: Odontoceti)." *Nat. Hist. Mus. L. A. Cty. Sci. Bull. 28:* 1–35.

Barnes, L. G. (1984a). Whales, dolphins and porpoises: Origin and evolution of the Cetacea. *In* "Mammals Notes for a Short Course Organized by P. D. Gingerich and C. E. Badgley" (T. W. Broadhead, ed.), pp. 139–153. University of Tennessee, Department of Geological Science Knoxville, TS.

Barnes, L. G. (1984b). "Fossil Odontocetes (Mammalia: Cetacea) from the Almejas Formation, Isla Cedros, Mexico." *PaleoBios 42:* 1–46.

Barnes, L. G. (1985). "Evolution, Taxonomy and Antitropical Distributions of the Porpoises (Phocoenidae, Mammalia)." *Mar. Mamm. Sci. 1:* 149–165.

Barnes, L. G. (1990). The fossil record and evolutionary relationships of the genus *Tursiops. In* "The Bottlenose Dolphin" (S. Leatherwood and R. R. Reeves, eds.), pp. 3–26. Academic Press, San Diego, CA.

Barnes, L. G., D. P. Domning, and C. E. Ray (1985). "Status of Studies on Fossil Marine Mammals." *Mar. Mamm. Sci. 1*(1): 15–53.

Barnes, L. G., M. Kimura, H. Furusawa, and H. Sawamura (1995). "Classification and Distribution of Oligocene Aetiocetidae (Mammalia; Cetacea; Mysticeti) from Western North America and Japan." *Island Arc 3:* 392–431.

Barnes, L. G., and S. A. McLeod (1984). The fossil record and phyletic relationships of gray whales. *In* "The Gray Whale *Eschrichtius robustus"* (M. L. Jones, S. L. Swartz, and S. Leatherwood, eds.), pp. 3–32. Academic Press, New York.

Bérubé, M., and A. Aguilar (1998). "A New Hybrid Between a Blue Whale, *Balaenoptera musculus,* and a Fin Whale, *B. physalus:* Frequency and Implications of Hybridization." *Mar. Mamm. Sci. 14*(1): 82–98.

Bobrinskii, N. A., B. A. Kuznekov, and A. P. Kuzyakin (1965). *Key to the Mammals of the USSR,* 2nd ed. Moscow (In Russian).

Boisserie, J-R., F. Lihoreau, and M. Brunet (2005). "The Position of Hippopotamidae within Cetartiodactyla." *Proc. Nat. Acad. Sci. 102*: 1537–1541.

Bryant, P. J. (1995). "Dating Remains of Gray Whales from the Eastern North Atlantic." *J. Mammal. 76:* 857–861.

Cabrera, A. (1926). "Cetaceos Fosiles del Museo de La Plata." *Rev. Mus. La Plata 29:* 363–411.

Cassens, I., S. Vicario, V. Waddell, H. Balchowsky, D. Van Belle, W. Ding, C. Fan, R. S. Lal Mohan, P. C. Simoes-Lopes, R. Bastida, A. Meyer, M. Stanhope, and M. Milinkovitch (2000). "Independent Adaptation to Riverine Habitats Allowed Survival of Ancient Cetacean Lineages. *Proc. Nat. Acad. Sci. 97:* 11343–11347.

Clarke, M. R. (1976). "Observations on Sperm Whale Diving." *J. Mar Biol. Assoc. U.K. 56:* 809–810.

Cozzuol, M. A. (1985). "The Odontoceti of the `Mesopotamiense' of the Parana River Ravines. Systematic Review." *Invest. Cetacea 7:* 39–53.

Cozzuol, M. A. (1996). "The Record of the Aquatic Mammals in Southern South America. *Muench. Geowiss. Abh. 30:* 321–342.

Dalebout, M. L., J. G. Mead, C. S. Baker, A. N. Baker, and A. Van Helden (2002). "A New Species of Beaked Whale *Mesoplodon perrini* sp. n. (Cetacea: Ziphiidae) Discovered Through Phylogenetic Analyses of Mitochondrial DNA Sequences." *Mar. Mamm. Sci. 18:* 577–608.

Dawson, S. D. (1996). "A Description of the Skull of *Hadrodelphis calvertense* and Its Position Within the Kentriodontidae (Cetacea, Delphinoidea)" *J. Vert. Paleontol. 16*(1): 125–134.

Deméré, T. (1986). "The Fossil Whale *Balaenoptera davidsonii* (Cope 1872), with a Review of Other Neogene Species of *Balaenoptera* (Cetacea: Mysticeti)." *Mar. Mamm. Sci. 2*(4): 277–298.

Deméré, T. A., A. Berta, and M. R. McGowen (in press). "The Taxonomic and Evolutionary History of Fossil and Modern Balaenopteroid Mysticetes." *J. Mamm. Evol.*

Dooley, A. C., Jr., N. C. Fraser, and Z. Luo (2004). "The Earliest Known Member of the Rorqual-Gray Whale Clade (Mammalia, Cetacea)." *J. Vert. Paleontol. 24:* 453–463.

Duncan, P. M. (1877–1883). *Cassell's Natural History,* 3 vols. Cassell and Co., London.

Evans, P. G. H. (1987). *The Natural History of Whales and Dolphins.* Helm, London.

Flower, W. H. (1883). "On Whales, Past and Present, and Their Probable Origin." *Nat. Proc. R. Inst. G. B. 10:* 360–376.

Flower, W. H.. and R. Lydekker (1891). *An Introduction to the Study of Mammals Living and Extinct.* Adam and Charles Black, London.

Fordyce, R. E. (1980). "Whale Evolution and Oligocene Southern Ocean Environments." *Palaeogeogr. Palaeoclimatol. Palaeoecol. 31:* 319–336.

Fordyce, R. E. (1981). "Systematics of the Odontocete *Agorophius pygmeus* and the Family Agorophiidae (Mammalia: Cetacea)." *J. Paleontol. 55:* 1028–1045.

Fordyce, R. E. (1982). "A Review of Australian Fossil Cetacea." *Mem. Nat. Mus. Victoria 43:* 43–58.

Fordyce, R. E. (1984). Evolution and zoogeography of cetaceans in Australia. *In* "Vertebrate Zoogeography and Evolution in Australasia" (M. Archer and G. Clayton, eds.), pp. 929–948. Hesperian Press, Carlisle, Australia.

Fordyce, R. E. (1989). "Origins and Evolution of Antarctic Marine Mammals." *Spec. Publ. Geol. Soc. London 47:* 269–281.

Fordyce, R. E. (1992). Cetacean evolution and Eocene Oligocene environments. *In* "Eocene-Oligocene Climatic and Biotic Evolution" (D. Prothero and W. Berggren, eds.), pp. 368–381. Princeton University Press, Princeton, NJ.

Fordyce, R. E. (1994). "*Waipatia maerewhenua,* New Genus and Species (Waipatiidae, New Family), an Archaic Late Oligocene Dolphin (Cetacea: Platanistidae) from New Zealand." *Proc. San Diego Soc. Nat. Hist. 29:* 147–176.

Fordyce, R. E. (1996). New Zealand Oligocene fossils and the early radiation of platanistoid dolphins. *In* "Abstracts of the Conference on Secondary Adaptation to Life in the Water" (J.-M. Mazin, P. Vignaud, and V. de Buffrénil, eds.), p. 12. Poitiers, France.

Fordyce, R. E. (2001). "*Simocetus rayi* (Odontoceti: Simocetidae, New Family): A Bizarre New Archaic Oligocene Dolphin from the Eastern North Pacific." *Smithson. Contrib. Paleobiol. 93:* 185–222.

Fordyce, R. E., and L. G. Barnes (1994). "The Evolutionary History of Whales and Dolphins." *Annu. Rev. Earth Planet. Sci. 22:* 419–455.

Fordyce, R. E., L. G. Barnes, and N. Miyazaki (1995). "General Aspects of the Evolutionary History of Whales and Dolphins." *Island Arc 3:* 373–391.

Fordyce, E. and C. de Muizon. (2001). Evolutionary history of cetaceans: A review. *In* "Secondary Adaptation of Tetrapods to Life in Water" (J.-M. Mazin and V. de Buffrenil, eds.), pp. 169–233. Verlag Dr. Friedrich Pfeil, Munchen, Germany.

Fordyce, R. E., and C. R. Sampson (1992). "Late Oligocene Platanistoid and Delphinoid Dolphins from the Kokoamu Greensand-Otekaike Limestone, Waitaki Valley Region, New Zealand: An Expanding Record." *Geol. Soc. N. Z., Misc. Publ. 63a:* 66(abstr.).

Fraser, F. C., and P. E. Purves (1976). "Anatomy and Function of the Cetacean Ear." *Proc. R. Soc. London 152:* 62–78.

Gatesy, J. (1998). Molecular evidence for the phylogenetic affinities of Cetacea. *In* "The Emergence of Whales: Patterns in the Origin of Cetacea" (J. G. M. Thewissen, ed.), pp. 63–111. Plenum Press, New York.

Gatesy, J., C. Hayashi, M. A. Cronin, and P. Arctander (1996). "Evidence from Milk Casein Genes that Cetaceans are Close Relatives of Hippopotamid Artiodactyls." *Mol. Biol. Evol. 13:* 954–963.

Gatesy, J., C. Matthee, R. De Salle, and C. Hayashi (2002). "Resolution of a Supertree/Supermatrix Paradox." *Syst. Biol. 51:* 652–664.

Gatesy, J., M. Milinkovitch, V. Wassell, and M. Stanhope (1999a). "Stability of Cladistic Relationships Between Cetacea and Higher Level Artiodactyla Taxa." *Syst. Biol. 48:* 6–20.

Gatesy, J., P. O'Grady, and R. H. Baker (1999b). "Corroboration Among Data Sets in Simultaneous Analysis: Hidden Support for Phylogenetic Relationships Among Higher Level Artiodactyla Taxa." *Cladistics 15:* 271–314.

Gatesy, J., and M. A. O'Leary (2001). "Deciphering Whale Origins with Molecules and Fossils." *Trends Ecol. Evol. 16:* 562–570.

Geisler, J. H. (2001). "New Morphological Evidence for the Phylogeny of Artiodactyla, Cetacea and Mesonychidae." *Amer. Mus. Novitates 3344:* 1–53.

Geisler, J. H., and Z. Luo (1996). "The Petrosal and Inner Ear of *Herpetocetus* Sp. (Mammalia: Cetacea) and their Implications for the Phylogeny and Hearing of Archaic Mysticetes." *J. Paleontol. 70:* 1045–1066.

Geisler, J. H., and Z. Luo (1998). Relationships of Cetacea to terrestrial ungulates and the evolution of cranial vasculature in Cete. *In* "The Emergence of Whales: Patterns in the Origin of Cetacea" (J. G. M. Thewissen, ed.), pp. 163–212. Plenum Press, New York.

Geisler, J. H., and A. E. Sanders (2003). "Morphological Evidence for the Phylogeny of Cetacea." *J. Mamm. Evol. 10*(1/2): 23–129.

Geisler, J. H., and M. Uhen (2003). "Morphological Support for a Close Relationship Between Whales and Hippos." *J. Vert. Paleontol. 23:* 991–996.

Gervais, M. P. (1855). *Histoire naturelle de mammiferes*. L. Curtner, Paris.

Giebel, C. G. (1859). *Die Naturgeschichte des Thierreichs. Book 1: Die Saugethiere*. Verlag von Otto Wigand, Leipzig.

Gingerich, P. D. (1998). Paleobiological perspectives on Mesonychia, Archaeoceti, and the origin of whales. *In* "The Emergence of Whales: Patterns in the Origin of Cetacea" (J. G. M. Thewissen, ed.), pp. 423–449. Plenum Press, New York.

Gingerich, P. D, (2005). Cetacea. *In* "The Rise of Placental Mammals" (K. D. Rose and J. David Archibald, eds.), pp. 234–252. Johns Hopkins, New York.

Gingerich, P. D., M. Arif, M. Akram Bhatti, M. Anwar, and W. J. Sanders (1997). "*Basilosaurus drazindai* and *Basiloterus hussaini,* New Archaeoceti (Mammalia, Cetacea) from the Middle Eocene Drazinda Formation, with a Revised Interpretation of Ages of Whale-Bearing Strata in the Kirthar Group of the Sulaiman Range, Punjab (Pakistan)." *Contrib. Mus. Paleontol. Univ. Mich. 30*(2): 55–81.

Gingerich, P. D., M. Haq, I. Zalmout, I. Khan, M. Malkani (2001). "Origin of Whales from Early Artiodactyls: Hands and Feet of Eocene Protocetidae from Pakistan." *Science 239:* 2239–2242.

Gingerich, P. D., and D. E. Russell (1981). "*Pakicetus inachus*, a New Archaeocete (Mammalia, Cetacea) from the Early-Middle Eocene Kuldana Formation of Kohat (Pakistan)." *Contrib. Mus. Paleontol. Univ. Mich. 25:* 235–246.

Gingerich, P. D., and D. E. Russell (1990). "Dentition of Early Eocene *Pakicetus* (Mammalia, Cetacea)." *Contrib. Mus. Palaeontol. Univ. Mich. 28:* 1–20.

Gingerich, P. D., B. H. Smith, and E. L. Simons (1990). "Hind Limbs of Eocene *Basilosaurus:* Evidence of Feet in Whales." *Science 249:* 403–406.

Gingerich, P. D., and M. Uhen (1996). "*Ancalecetus simonsi,* a New Dorudontine Archaeocete (Mammalia, Cetacea) from the Early Late Eocene of Wadi Hitan, Egypt." *Contrib. Mus. Paleontol. Univ. Mich. 29:* 359–401.

Gingerich, P. D., N. A. Wells, D. E. Russell, and S. M. Shah (1983). "Origin of Whales in Epicontinental Remnant Seas: New Evidence from the Early Eocene of Pakistan." *Science 220:* 403–406.

Gould, S. J. (1994). "Hooking Leviathan by Its Past." *Nat. Hist. 5:* 8–15.

Hamilton, H. S., Caballero, A. G. Collins, and R. L. Brownell Jr. (2001). "Evolution of River Dolphins." *Proc. R. Soc. Lond. B 268:* 549–556.

Hasegawa, M., J. Adachi, and M. C. Milinkovitch (1997). "Novel Phylogeny of Whales Supported by Total Evidence." *J. Mol. Evol. 44*(Suppl. 1): S117–S120.

Heyning, J. E. (1989). "Comparative Facial Anatomy of Beaked Whales (Ziphiidae) and Systematic Revision Among the Family of Extant Odontoceti." *Nat. Hist. Mus. L. A. Cty. Contrib. Sci. 405:* 1–64.

Heyning, J. E. (1997). "Sperm Whale Phylogeny Revisited: Analysis of the Morphological Evidence." *Mar. Mamm. Sci. 13*: 596–613.

Heyning, J. E., and J. Mead (1990). Evolution of the nasal anatomy of cetaceans. *In* "Sensory Abilities of Cetaceans" (J. Thomas and R. Kastelain, eds.), pp. 67–79. Plenum Press, New York.

Hirota, K., and L. G. Barnes (1995). "A New Species of Middle Miocene Sperm Whale of the Genus *Scaldicetus* (Cetacea; Physeteridae) from Shiga-mura, Japan." *Island Arc 3:* 453–472.

Hulbert, R. C., Jr., R. M. Petkewich, G. A. Bishop, D. Bukry, and D. P. Aleshire (1998). "A New Middle Eocene Protocetid Whale (Mammalia: Cetacea: Archaeoceti) and Associated Biota from Georgia." *J. Paleontol. 72:* 907–927.

Ichishima, H., L. G. Barnes, R. E. Fordyce, M. Kimura, and D. J. Bohaska (1995). "A Review of Kentriodontine Dolphins (Cetacea; Detphinoidea; Kentriodontidae): Systematics and Biogeography." *Island Arc 3:* 486–492.

Irwin, D. M., and U. Árnason, (1994). "Cytochrome b Gene of Marine Mammals: Phylogeny and Evolution." *J. Mamm. Evol. 2:* 37–55.

Jefferson, T. A., S. Leatherwood, and M. A. Weber (1993). *FAO Species Identification Guide: Marine Mammals of the World.* Food and Agriculture Organization, Rome.

Kasuya, T. (1973). "Systematic Consideration of Recent Toothed Whales Based on the Morphology of Tympano-Periotic Bone." *Sci. Rep. Whales Res. Inst. 25:* 1–103.

Kellogg, R. (1936). "A Review of the Archaeoceti." *Carnegie Inst. Wash. Publ. 482:* 1–366.

Kellogg, R. (1968). "Fossil Mammals From the Miocene Calvert Formation of Maryland and Virginia." *U.S. Natl. Mus. Bull. 247:* 103–201.

Kimura, T., and T. Ozawa. (2001). "A New Cetothere (Cetacea: Mysticeti) from the Early Miocene of Japan." *J. Vert. Paleontol. 22*: 684–702.

Kumar, K., and A. Sahni (1986). "*Remingtonocetus harudiensis,* Nw Combination, a Middle Eocene Archaeocete (Mammalia: Cetacea) from Western Kutch, India." *J. Vert. Paleontol. 6:* 326–349.

Leatherwood, S., and R. R. Reeves (1983). *The Sierra Club Handbook of Whales and Dolphins.* Sierra Club Books, San Francisco.

LeDuc, R., W. F. Perrin, and A. E. Dizon (1999). "Phylogenetic Relationships Among the Delphinind Cetaceans Based on Full Cytochrome B Sequences." *Mar. Mamm. Sci. 15:* 619–648.

Luo, Z., and P. D. Gingerich (1999). "Terrestrial Mesonychia to Aquatic Cetacea: Transformation of the Basicranium and Evolution of Hearing in Whales." *Univ. Mich. Pap. Paleontol. 31:* 1–98.

MacFadden, B. J. (1992). *Fossil Horses: Systematics, Paleobiology and Evolution of the Family Equidae.* Cambridge University Press, Port Chester, NY.

McLeod, S. A., F. C. Whitmore, and L. G. Barnes (1993). Evolutionary relationships and classification. *In* "The Bowhead Whale" (J. J. Burns, J. J. Montague, and C. J. Cowles, eds.), Spec. Publ. No. 2, pp. 45–70. Soc. Mar. Mammal. Allen Press, Lawrence, KS.

Mead, J. (1975a). "Anatomy of the External Nasal Passages and Facial Complex in the Delphinidae (Mammalia: Cetacea)." *Smithson. Contrib. Zool. 207:* 1–72.

Mead, J. (1975b). "A Fossil Beaked Whale (Cetacea: Ziphiidae) from the Miocene of Kenya." *J. Paleontol. 49:* 745–751.

Mead, J., and R. Brownell (1993). Order Cetacea. *In* "Mammal Species of the World: A Taxonomic and Geographic Reference" (D. E. Wilson and D. M. Reeder, eds.), pp. 349–364. Smithsonian Institute Press, Washington, DC.

Messenger, S. L. (1994). "Phylogenetic Relationship of Platanistoid River Dolphins (Odontoceti, Cetacea): Assessing the Significance of Fossil Taxa." *Proc. San Diego Soc. Nat. Hist. 29:* 125–133.

Messenger, S. L. (1995). *Phylogenetic Systematics of Cetaceans: A Morphological Perspective.* Unpublished MS thesis, San Diego State University, San Diego, CA.

Messenger, S. L., and J. McGuire (1998). "Morphology, Molecules and the Phylogenetics of Cetaceans." *Syst. Biol. 47:* 90–124.

Milinkovitch, M. C. (1995). "Molecular Phylogeny of Cetaceans Prompts Revision of Morphological Transformations." *Trends Ecol. Evol. 10:* 328–334.

Milinkovitch, M. C., R. G. Le Duc, J. Adachi, F. Farnir, M. Georges, and M. Hasegawa (1996). "Effects of Character Weighting and Species Sampling on Phylogeny Reconstruction: A Case Study Based on DNA Sequence Data in Cetaceans." *Genetics 144:* 1817–1883.

Milinkovitch, M. C., G. Orti, and A. Meyer (1993). "Revised Phylogeny of Whales Suggested by Mitochondrial Ribosomal DNA Sequences." *Nature 361:* 346–348.

Milinkovitch, M. C., G. Orti, and A. Meyer (1994). "Phylogeny of All Major Groups of Cetaceans Based on DNA Sequences from Three Mitochondrial Genes." *Mol. Biol. Evol. 11:* 939–948.

Miller, G. S. (1923). "The Telescoping of the Cetacean Skull." *Smithson. Misc. Collec. 76:* 1–71.

Mitchell, E. D. (1989). "A New Cetacean from the Late Eocene La Meseta Formation, Seymour Island, Antarctic Peninsula." *Can. J. Fish. Aquat. Sci. 46:* 2219–2235.

Morgan, G. (1994). "Miocene and Pliocene Marine Mammal Faunas from the Bone Valley Formation of Central Florida." *Proc. San Diego Soc. Nat. Hist. 29:* 239–268.

Muizon, C, de (1981). *Les Vértebrés Fossiles de la Formation Pisco (Pérou). Part 1. Recherche sur les Grandes Civilisations,* Mém. No. 6. Institut Francais d'Etudes Andines, Paris.

Muizon, C. de (1983). "*Pliopontes littoralis* un nouveau Platanistidae Cetacea du Pliocene de la tote péruvienne." *C. R. Seances Acad. Sci. Ser 2 296:* 1101–1104.

Muizon, C. de (1987). The Affinities of *Notocetus vanbenedeni,* an Early Miocene Platanistoid (Cetacea, Mammalia) du Pliocene Inférieur de Sud-Sacaco, Mém. No. 50. Institut Francais d'Études Andines, Paris.

Muizon, C. de (1988a). *Les Vértebrés Fossiles de la Formation Pisco (Pérou). Part 3. Recherche sur les Grandes Civilisations,* Mém. No. 78. Institut Francais d'Études Andines, Paris.

Muizon, C. de (1988b). "Le polyphylétisme des Acrodelphidae, Odontocétes longirostres du Miocene européen." *Bull. Mus. Nat. Hist. Nat. 4 Serie 10C*(1): 31–88.

Muizon, C. de (1988c). "Les relations phylogénétiques des Delphinida (Cetacea, Mammalia)." *Ann. Paleontol. (Vértebr.-lnvértebr) 74*(4): 159–257.

Muizon, C. de (1991). "A New Ziphiidae from the Early Miocene of Washington State (USA) and Phylogenetic Analysis of the Major Groups of Odontocetes." *Bull. Mus. Nat. Hist. Nat. 4 Serie 12C*(3-4): 279–326.

Muizon, C. de (1993a). "Walrus-Feeding Adaptation in a New Cetacean from the Pliocene of Peru." *Nature 365:* 745–748.

Muizon, C. de (1993b). "*Odobenocetops peruvianus*: una remarcable convergencia de adaptacion alimentaria entre morsa y delphin." *Bull. de l'Institut Francais d'Etudes Andines 22:* 671–683.

Muizon, C. de (1994). "Are the Squalodontids Related to the Platanistoids?" *Proc. San Diego Soc. Nat. Hist. 29:* 135–146.

Muizon, C. de, D. P. Domning, and D. R. Ketten (2001). "*Odobenocetops peruvianus*, the Walrus Convergent Delphinoid (Cetacea, Mammalia) from the Lower Pliocene of Peru." *Smithson. Contrib. Paleobiol. 93:* 223–261.

Muizon, C de, D. P. Domning, and M. Parrish. (1999). "Dimorphic Tusks and Adaptive Strategies in the Odobenocetopsidar, Walrus-Like Dolphins from the Pliocene of Peru." *Comptes-Rendus de l'Academie des Sciences, Paris, Sciences de la Terre et des Planetes 329:* 449–455.

Murphy, W. J., E. Elizrik, W. E. Johnson, Y. P. Zhang, O. A, Ryder, and S. J. O'Brien (2001). "Molecular Phylogenetics and the Origins of Placental Mammals." *Nature 409:* 614–618.

Naylor, G. J. P., and D. C. Adams (2001). "Are the Fossil Data Really at Odds with the Molecular Data? Morphological Evidence for Cetartiodactyla Phylogeny Reexamined." *Syst. Biol. 50:* 444–453.

Naylor, G. J. P., and D. C. Adams (2003). "Total Evidence Versus Relevant Evidence: A Response to O'Leary *et al.* (2003)." *Syst. Biol. 52:* 864–865.

Nikaido, M., F. Matsuno, H. Hamilton, R. L. Brownell, Y. Cao, W. Ding, Z. Zuoyan, A. M. Shedlock, R. E. Fordyce, M. Hasegawa, and N. Okada. (2001). "Retroposon Analysis of Major Cetacean Lineages: The Monophyly of Toothed Whales and the Paraphyly of River Dolphins." *Proc. Nat. Acad. Sci. 98:* 7384–7389.

Nikaido, M., P. Rooney, and N. Okada (1999). "Phylogenetic Relationships Among Cetartiodactyls Based on Insertions of Short and Long Interspersed Elements: Hippopotamuses are the Closest Extant Relatives of Whales." *Proc. Nat. Acad. Sci. xx:* 10261–10266.

Novacek, M. J. (1992). "Mammalian Phylogeny: Shaking the Tree." *Nature 356:* 121–125.

Novacek, M. J. (1993). "Genes Tell a New Whale Tale." *Nature 361:* 298–299.

Oishi, M., and Y. Hasegawa (1995). "Diversity of Pliocene Mysticetes from Eastern Japan." *Island Arc 3:* 493–505.

O'Leary, M. (1998). Phylogenetic and morphometric reassessment of the dental evidence for a mesonychian and cetacean clade. *In* "The Emergence of Whales" (J. G. M. Thewissen, ed.), pp. 133–161. Plenum, New York.

O'Leary, M. A., M. Allard, M. J. Novacek, J. Meng, and J. Gatesy (2004). Building the mammalian sector of the Tree of Life. *In* "Assembling the Tree of Life" (J. Cracraft and M. J. Donoghue, eds.), pp. 490–516. Oxford University Press, New York.

O'Leary, M., J. Gatesy, and M. J. Novacek (2003). "Are the Dental Data Really at Odds with the Molecular Data? Morphological Evidence for Whale Phylogeny (Re)reexamined." *Syst. Biol. 52:* 853–863.

O'Leary, M., and J. Geisler (1999). "The Position of Cetacea Within Mammalia: Phylogenetic Analysis of Morphological Data from Extinct and Extant Taxa." *Syst. Biol. 48:* 455–490.

Perrin, W. F. (1997). "Development and Homologies of Head Stripes in the Delphinoid Cetaceans." *Mar. Mamm. Sci. 13:* 1–43.

Prothero, D. (1993). Ungulate phylogeny: molecular vs. morphological evidence. *In* "Mammal Phylogeny: Placentals" (F. S. Szaklay, M. J. Novacek, and M. C. McKenna, eds.), pp. 173–181. Springer-Verlag, New York.

Prothero, D. R., E. M. Manning, and M. Fischer (1988). The phylogeny of ungulates. *In* "The Phylogeny and Classification of the Tetrapods" (M. J. Benton, ed.), Vol. 2, pp. 201–234. Clarendon Press, Oxford.

Ray, J. (1693). *Synopsis methodica animalium quadrupedum et serpentine generis*. S. Smith and B. Walford, London.

Rice, D. W. (1998). *Marine Mammals of the World*. Spec. Publ. 4, Soc. Mar. Mammal., Allen Press, Lawrence, KS.

Rosel, P. E., M. G. Haygood, and W. F. Perrin (1995). "Phylogenetic Relationships Among the True Porpoises (Cetacea: Phocoenidae)." *Mol. Phylogenet. Evol. 4:* 463–474.

Rychel, A., T. Reeder, and A. Berta (2004). "Phylogeny of Mysticete Whales Based on Mitochondrial and Nuclear Data." *Mol. Phylogenet. Evol. 32:* 892–901.

Sanders, A. E., and L. G. Barnes (2002). "Paleontology of Late Oligocene Ashley and Chandler Bridge Formations of South Carolina, 3: Eomysticetidae, A New Family of Primitive Mysticetes (Mammalia: Cetacea)." *Smithson. Contrib. Paleobiol. 93:* 313–356.

Sasaki, T., M. Nikaido, H. Hamilton, M. Goto, H. Kato, N. Kanda, L.A. Pastene, Y. Cao, R. E. Fordyce, M. Hasegawa, and N. Okada (2005). "Mitochondrial Phylogenetics and Evolution of Mysticete Whales." *Syst. Biol. 54):* 77–90.

Scott, W. B. (1888). "On Some New and Little Known Creodonts." *J. Acad. Nat. Sci.* Philadelphia, *9:* 155–185.

Shedlock, A. M., M. C. Milinkovitch, and N. Okada. (2000). "SINE Evolution, Missing Data, and the Origin of Whales." *Syst. Biol. 49:* 808–817.

Shimamura, M., H. Abe, M. Nikaido, K. Oshima, and N. Okada. (1999). "Genealogy of Families of SINEs in Cetaceans and Artiodactyls: The Presence of a Huge Superfamily of tRNA ^{Glu}-Derived Families of SINEs." *Mol. Biol. Evol. 16:* 1046–1060.

Shimamura, M., H. Yasue, K. Ohshima, H. Abe, H. Kato, T. Kishiro, N. Goto, I. Munechika, and N. Okada (1997). "Molecular Evidence from Retroposons that Whales Form a Clade with Even-Toed Ungulates." *Nature 388:* 666–670.

Slijper, E. J. (1962). *Whales*. Basic Books, New York.

Szalay, F. (1969). "The Hapalodectinae and a Phylogeny of the Mesonychidae (Mammalia, Condylarthra)." *Am. Mus. Novit. 2361:* 1–26.

Szalay, F., and Gould, S. J. (1966)." Asiatic Mesonychidae (Mammalia, Condylarthra)." *Bull. Amer. Mus. Nat. Hist. 132:* 129–173.

Theodor, J. M., K. D. Rose, and J. Erfurt (2005). Artiodactyla. *In* "The Rise of Placental Mammals" (K. D. Rose and J. David Archibald, eds.), pp. 215–233. Johns Hopkins, New York.

Thewissen, J. G. M. (1994). "Phylogenetic Aspects of Cetacean Origins: A Morphological Perspective." *J. Mamm. Evol. 2*(3): 157–183.

Thewissen, J. G. M. (ed.) (1998). *The Emergence of Whales: Patterns in the Origin of Cetacea*. Plenum Press, New York.

Thewissen, J. G. M., and S. T. Hussain (1993). "Origin of Underwater Hearing in Whales." *Nature 361:* 444–445.

Thewissen, J. G. M., and S. T. Hussain (1998). "Systematic Review of the Pakicetidae, Early and Middle Eocene Cetacea (Mammalia) from Pakistan and India." *Bull. Carnegie Mus. Nat. Hist. 34:* 220–238.

Thewissen, J. G. M., and S. T. Hussain (2000). "*Attockicetus praecursor*, a New Remintonocetid Cetacean from Marine Eocene Sediments of Pakistan." *J. Mamm. Evol. 7:* 133–146.

Thewissen, J. G. M., S. T. Hussain, and M. Arif (1994). "*Ambulocetus natans,* the Walking Whale." *Science 263:* 210–212.

Thewissen, J. G. M., S. I. Madar, and S. T. Hussain (1996). "*Ambulocetus natans,* an Eocene Cetacean (Mammalia) from Pakistan." *CFS Cour Forschungsinst. Senckenberg 191:* 1–86.

Thewissen, J. G. M., S. I. Madar, and S. T. Hussain (1998). "Whale Ankles and Evolutionary Relationships." *Nature 395:* 452.

Thewissen, J. G. M., and E. M. Williams (2002). "The Early Radiations of Cetacea (Mammalia): Evolutionary Pattern and Developmental Correlations." *Annu. Rev. Ecol. Syst. 33:* 73–90.

Thewissen, J. G. M., E. S. Williams, L. J. Roe, and S. T. Hussain (2001). "Skeletons of Terrestrial Cetaceans and the Relationship of Whales to Artiodactyls." *Nature 413:* 277–281.

Uhen, M. D. (1998a). "New Protocetid (Mammalia: Cetacea) from the Late Middle Eocene Cook Mountain Formation of Louisiana." *J. Vert. Paleontol. 18:* 664–668.

Uhen, M. D. (1998b). Middle to late Eocene basilosaurines and dorudontines. *In* "The Emergence of Whales: Patterns in the Origin of Cetacea" (J. G. M. Thewissen, ed.), pp. 29–61. Plenum Press, New York.

Uhen, M. D. (1999). "New Species of Protocetid Archaeocete Whale, *Eocetus wardii* (Mammalia: Cetacea), from the Middle Eocene of North Carolina." *J. Paleontol.*

Uhen, M. D. (2002). Basilosaurids. *In* "Encyclopedia of Marine Mammals" (W. F. Perrin, B. Würsig, and J. G. M. Thewissen, eds.) pp.77–80. Academic Press, San Diego, CA.

Uhen, M. D. (2004). "Form, Function, and Anatomy of *Dorudon atrox* (Mammlia, Cetacea): An Archaeocete from the Middle to Late Eocene of Egypt." *Univ. Mich. Pap. Paleontol. 34:* 1–222.

Van Helden, A. L., A. N. Baker, M. L. Dalebout, J. C. Reyes, K. Van Waerebeek, and C. S. Baker (2002). "Resurrection of *Mesoplodon traversii* (Gray, 1874), Senior Synonym of *M. bahamondi* Reyes, Van Waerebeek, Cardenas and Yanex, 1995 (Cetacea: Ziiphidae)." *Mar. Mamm. Sci. 18:* 609–621.

Van Valen, L. (1966). "Deltatheridia, a New Order of Mammals." *Bull. Amer. Mus. Nat. Hist. 132:* 1–126.

Wada, S., M. Oishi, and T. K. Yamada (2003). "A Newly Discovered Species of Living Baleen Whale." *Nature 426:* 278–281.

Waddell, V. G., M. Milinkovitch, M. Berube, and M. J. Stanhope (2000). "Molecular Phylogenetic Examination of the Delphinoidea Trichotomy: Congruent Evidence of Three Nuclear Loci Indicates that Porpoises (Phocoenidae) Share a More Common Ancestry with White Whales (Monodontidae) than They Do with True Dolphins (Delphinidae)." *Mol. Phylogenet. Evol. 15:* 314–318.

Watkins, W. A., K. E. Moore, and P. Tyack (1985). "Investigations of Sperm Whale Acoustic Behaviors in the Southeast Caribbean." *Cetology 49:* 1–15.

Watson, L. (1981). *Whales of the World*. Hutchinson, London.

Westgate, J., and F. Whitmore (2002). "*Balaena ricei*, a New Species of Bowhead Whale from the Yorktown Formation (Pliocene) of Hampton, Virginia." *Smithson. Contrib. Paleobiol. 93:* 295–311.

Williams, E. (1998). Synopsis of the earliest cetaceans. *In* "The Emergence of Whales: Patterns in the Origin of Cetacea" (J. G. M. Thewissen, ed.), pp. 1–28. Plenum Press, New York.

Zhou K., W. Qian,. and Y. Li (1979). "The Osteology and Systematic of the Baji, *Lipotes vexillifer.*" *Acta Zool. Sin. 25:* 58–74.

Zhou K., M. Zhou, and Z. Zhao (1984). "First Discovery of a Tertiary Platanistoid Fossil from Asia." *Sci. Rep. Whales Res. Inst. 35:* 173–181.

Zimmer, C. (1998). *At the Water's Edge: Macroevolution and the Transformation of Life*. Free Press/Simon and Schuster, New York.

5

Sirenian and Other Marine Mammals: Evolution and Systematics

5.1. Introduction

The mammalian order Sirenia, or sea cows, includes two extant families, the Trichechidae (manatees) and the Dugongidae (the dugong). The name Sirenia comes from mermaids of Greek mythology known as sirens. Sirenians have a fossil record extending from the early Eocene (50 Ma) to the present (Figure 5.1). Manatees include three living species and are known from the early Miocene (15 Ma) to the Recent in the New World tropics. The dugong is represented by a single extant species, *Dugong dugon,* of the Indo-Pacific. Dugongs were considerably more diverse in the past, with 19 extinct genera described and a fossil record that extends back to the Eocene. A North Pacific lineage of dugongids survived into historic times and had successfully adapted to cold climates. Sirenians are unique among living marine mammals in having a strictly herbivorous diet, which is reflected in the morphology of their teeth and digestive system. The Desmostylia, the only extinct order of marine mammals, are relatives of sirenians and are discussed here, as is the extinct marine bear-like carnivoran, *Kolponomos*. Other marine mammals include members of two extant carnivore families, the Mustelidae (which includes the sea otter, *Enhydra lutris*), the Ursidae (containing the polar bear, *Ursus maritimus*), and the extinct sloth family Megalonychidae (which includes the aquatic sloth lineage *Thalassocnus*).

5.2. Origin and Evolution of Sirenians

5.2.1. Sirenians Defined

Sirenians possess relatively large stout bodies, downturned snouts, short rounded paddle-like flippers, and a horizontal tail fluke. Manatees can be readily distinguished from dugongs by their smaller size, a rounded rather than notched tail, and a less-pronounced

89

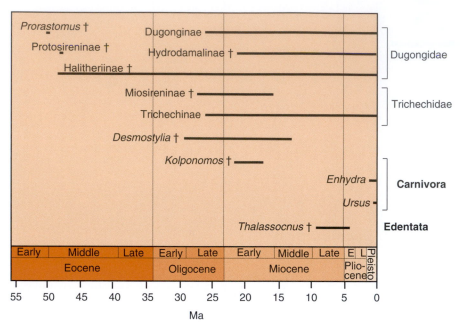

Figure 5.1. Chronologic ranges of living and extinct sirenians and other marine mammals. Ma = million years ago.

deflection of the snout. The latter feature enables manatees to feed at any level in the water column rather than being obligate bottom feeders, like the dugong with its strongly downturned snout.

The monophyly of sirenians is well established. Sirenians are united by possession of the following synapomorphies (Domning, 1994; Figures 5.2 and 5.3):

1. *External nares retracted and enlarged, reaching to or beyond the level of the anterior margin of the orbit*. In the primitive condition, the external nares (nostrils) are not retracted.

2. *Premaxilla contacts frontal*. All sirenians are characterized by a premaxilla-frontal contact. In the primitive condition, the premaxilla does not contact the frontal; instead it contacts the nasal posteriorly.

3. *Sagittal crest absent*. The skull of sirenians can be distinguished from other closely related mammals in lacking development of a sagittal crest.

4. *Five premolars, or secondarily reduced from this condition by loss of anterior premolars*. Early sirenians possess five premolars as did ancestral placental mammals (Archibald, 1996). This tendency was later reversed by post-Eocene sirenians, which often reduce the number of premolars. Ungulates show the primitive condition, possession of four premolars (Thewissen and Domning, 1992).

5. *Mastoid inflated and exposed through occipital fenestra*. In sirenians, the mastoid is inflated and fills a large fenestra (window-like opening) in the dorsal occiput. It does not extend around the base of the cranium to form a flange on the ventral occiput (Novacek and Wyss, 1987). In the primitive condition seen in most mammals, there is

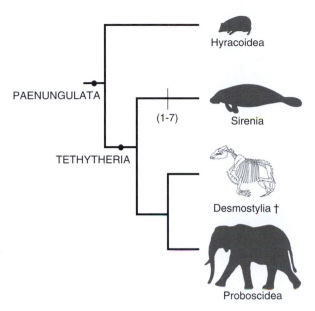

Figure 5.2. A cladogram depicting the relationship of sirenians and their close relatives. Numbers refer to sirenian synapomorphies, some of which are illustrated in Figure 5.3. † = extinct taxa.

continuous mastoid exposure between the horizontal basicranium and ventral (vertical) occiput.

6. *Ectotympanic inflated and drop-like*. Sirenians are distinguished by having an inflated ectotympanic (one of the bones forming the auditory bulla) that is drop-like in shape. In the primitive condition, the ectotympanic is uninflated (Tassy and Shoshani, 1988).

7. *Pachyostosis and osteosclerosis present in skeleton*. The skeleton of sirenians displays both pachyostosis and osteosclerosis, modifications involved in hydrostatic regulation (Domning and de Buffrénil, 1991).

5.2.2. Sirenian Affinities

Proboscideans (elephants) are usually considered the closest living relatives of sirenians (e.g., McKenna, 1975; Domning *et al*., 1986; Thewissen and Domning, 1992). Characters that unite proboscideans and sirenians include rostral displacement of the orbits with associated reorganization of the antorbital region, strongly laterally flared zygomatic process of the squamosal, and incipiently bilophodont (double crested) teeth (Savage *et al*., 1994). Sirenians, proboscideans, and the extinct desmostylians are recognized as a monophyletic clade, termed the **Tethytheria** (named because early members were thought to have inhabited the shores of the ancient Tethys Sea; McKenna, 1975; Figure 5.2). Morphological characters supporting an alliance between tethytheres, the perissodactyls (horses, rhinos, and tapirs), and the hyracoids (hyraxes), referred to as the Pantomesaxonia clade (Prothero *et al*., 1988), have been refuted (Savage *et al*., 1994). Molecular data remove perissodactyls from a relationship with tethytheres and hyracoids.

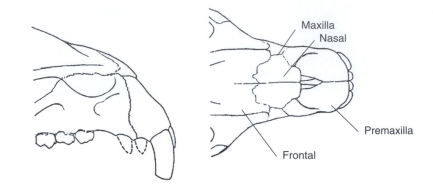

(a)

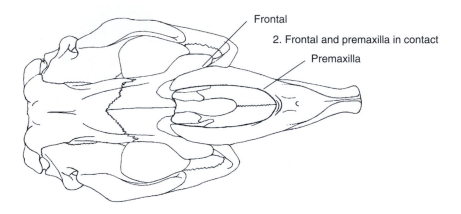

(b)

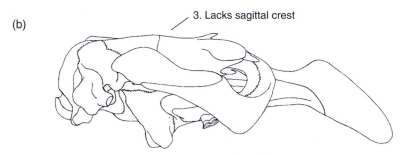

Figure 5.3.   Sirenian synapomorphies. **(a)** Snout of archaic elephant, *Moeritherium,* in dorsal and lateral views illustrating the lack of contact between the premaxilla and the nasals (primitive condition of character 2) (see text for further description). (Modified from Tassy and Shoshani, 1988.) **(b)** Skull of the sirenian *Dusisiren,* in dorsal and lateral views illustrating the derived condition of character 2, premaxilla lies in contact with nasals. (Modified from Domning, 1978.) Also visible are other sirenian synapomorphies, character: 1 = external nares retracted and enlarged, reaching to or beyond the anterior margin of the orbit and 3 = sagittal crest absent.

Recognition of another clade, the Paenungulata, composed of the Tethytheria and hyracoids (Novacek *et al.*, 1988; Shoshani, 1993), is more controversial. Fischer (1986, 1989) and Prothero (1993) maintained that morphological features supporting the Paenungulata can be disallowed as shared primitive characters and therefore are not indicative of relationship. These workers have argued for a closer relationship between hyracoids and perissodactyls. Molecular sequence data, however, strongly support the Paenungulata clade (sirenians, proboscideans, and hyracoids; Springer and Kirsch, 1993; Lavergne *et al.*, 1996; Stanhope *et al.*, 1998; Madsen *et al.*, 2001; Murphy *et al.*, 2001; Scally *et al.*, 2001).

An African clade of diverse mammals, named **Afrotheria,** that includes sirenians in addition to elephant shrews, tenrecs, golden moles, aardvarks, hyraxes, and elephants has received consistent and strong support from molecular data (e.g., Springer *et al.*, 1997; Stanhope *et al.*, 1998; Scally *et al.*, 2001; Murata *et al.*, 2003). Within Afrotheria, interrelationships are less clear although support was found for Tethytheria (i.e., sirenians + elephants), which is the sister taxon to hyraxes (Murphy *et al.*, 2001; Murata *et al.*, 2003). Discovery of a new family of retroposons among Afrotheria (AfroSINES) may help to resolve relationships among this group (Nikaido *et al.*, 2003).

5.2.3. Evolution of Early Sirenians

The earliest known sirenians are prorastomids *Prorastomus* and *Pezosiren* from early and middle Eocene age rocks (50 Ma) of Jamaica (Figures 5.4 and 5.5). The dense and swollen ribs of prorastomids point to a partially aquatic lifestyle, as does their occurrence in lagoonal deposits. The hip and knee joints of *Prorastomus* and *Protosiren* (Domning and Gingerich, 1994) and the nearly complete skeleton of *Pezosiren*

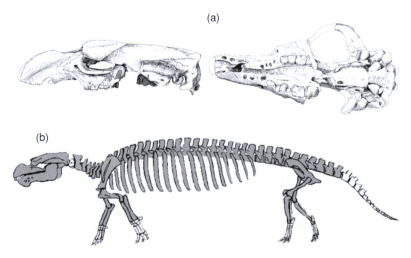

(a)

(b)

Figure 5.4. An early sirenian, *Prorastomus sirenoides,* from the late early Eocene of Jamaica. **(a)** Skull in lateral and ventral views. **(b)** Reconstructed composite skeleton of *Pezosiren portelli.* (From Domning, 2001.) (Unshaded areas are partly conjecture.)

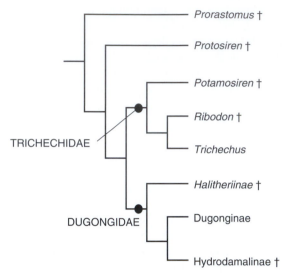

Figure 5.5.   Relationships among sirenians based on morphologic data. (Modified from Domning, 1994.)
——————   † = extinct taxa.

(Domning, 2001a) indicate that the earliest sirenians possessed well-developed legs (Figure 5.4).

Study of the type skull of *Protosiren fraasi* using CT scans (Gingerich *et al.*, 1994) reveals small olfactory bulbs, small optic tracts, and large maxillary nerves, consistent with the diminished importance of olfaction and vision in an aquatic environment and consistent with enhanced tactile sensitivity of the enlarged downturned snout of most Sirenia. *Prorastomus* and *Protosiren* were amphibious quadrupeds and not as fully aquatic as most later sirenians. The peculiar forceps-like snouts of *Prorastomus* and other early sea cows suggests a selective browsing habit by analogy with extant narrow-muzzled ungulates. Additional morphologic, ecologic, and **taphonomic** data support consideration of prorastomids as fluvatile (river) or estuarine semiaquatic herbivores (Savage *et al.*, 1994). Middle and late Eocene dugongids in need of taxonomic revision include *Eotheroides* and *Eosiren* from Egypt and *Prototherium* from Italy.

5.2.4. Modern Sirenians

5.2.4.1. Family Trichechidae

Some scientists as recently as the 19th century considered the manatee to be an unusual tropical form of walrus; in fact the walrus was once placed in the genus *Trichechus* along with the manatees (Reynolds and Odell, 1991). The family Trichechidae was expanded by Domning (1994) to include not only the manatees (Trichechinae) but also the Miosireninae, a northern European clade composed of two genera, *Anomotherium* and *Miosiren*. The trichechid clade as a whole appears to have been derived from late Eocene or early Oligocene dugongids or from protosirenids (see Gheerbrant *et al.*, 2005). The subfamily Trichechinae includes three living species: the West Indian manatee *(Trichechus manatus),* the West African manatee *(Trichechus senegalensis),* and the Amazon manatee *(Trichechus inunguis;* Figure 5.6). Two subspecies of the West Indian manatee can be distinguished on the basis of morphology and geography, the Antillean manatee, *T. m. manatus,* and the Florida manatee, *T. m. latirostris* (Domning and Hayek, 1986). Manatees are united as a monophyletic clade by features of the skull (e.g., ear region). Other derived characters include reduction of neural spines on the vertebrae, a possible tendency toward enlargement, and, at least in *Trichechus,* anteroposterior elongation of thoracic vertebral centra (Domning, 1994).

Morphologic data supports the West African manatee and the West Indian manatee sharing a more recent common ancestor than either does with the Amazon manatee (Domning, 1982; Domning and Hayek, 1986). Mitochondrial sequence data supports close divergence times for the three species (Parr and Duffield, 2002).

5.2.4.2. Family Dugongidae

The Family Dugongidae is paraphyletic as defined by Domning (1994). It includes two monophyletic subfamilies, the Dugonginae and extinct Hydrodamalinae, and the paraphyletic extinct "Halitheriinae."

The "Halitheriinae" includes the paraphyletic genera *Halitherium, Eotheroides, Prototherium, Eosiren, Caribosiren,* and *Metaxytherium.* The best known genus, *Metaxytherium,* was widely distributed in both the North Atlantic and Pacific during the Miocene. *Metaxytherium* had a strongly downturned snout and small upper incisor tusks. Members of this lineage were most likely generalized bottom-feeding animals that

(a)

(b)

(c)

Figure 5.6.   Modern manatee species. **(a)** West Indian manatee. **(b)** West African manatee. **(c)** Amazon manatee. (Illustrated by P. Folkens from Reeves *et al.,* 1992.)

probably consumed rhizomes (root-like stems) of small to moderate sized sea grasses and sea grass leaves (Domning and Furusawa, 1995). A Caribbean and West Atlantic origin for the genus, with subsequent dispersal to the North Pacific via the Central American Seaway and later dispersal to coastal Peru is suggested.

The extinct Hydrodamalinae includes the paraphyletic genus *Dusisiren* and the lineage that led to the recently extinct Steller's sea cow, *Hydrodamalis gigas* (Figure 5.7). *Dusisiren* evolved a very large body size, decreased snout deflection, and the loss of tusks, suggesting that these animals may have fed on kelp that grows higher in the water column than do sea grasses (Domning and Furusawa, 1995). Steller's sea cow, named for its discoverer, Georg W. Steller, a German naturalist, was a gigantic animal. It measured at least 7.6 m in length and was estimated to weigh between 4 and 10 tons. The sea cow was unusual in lacking teeth and finger bones and in possessing a thick, bark-like skin. Steller's sea cow lived in cold waters near islands in the Bering Sea, in contrast to the distribution of other sirenians in tropical or subtropical waters, and in prehistoric times from Japan to Baja California. The ancestry of this animal involves *Metaxytherium*

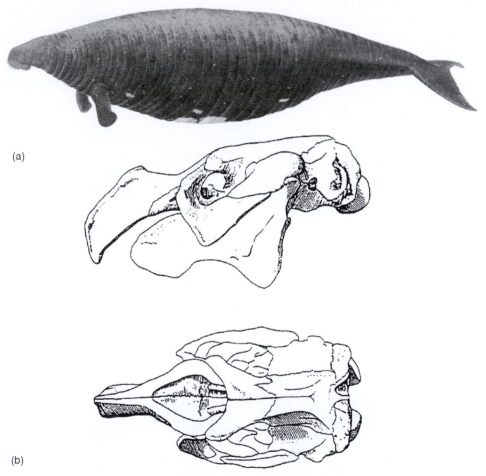

(a)

(b)

Figure 5.7.   Steller's sea cow. **(a)** Left side of the body. (Illustrated by P. Folkens from Reeves *et al.*, 1992.) **(b)** Lateral and dorsal views of the skull and mandible. (After Heptner, 1974.)

and *Dusisiren jordani* from the Miocene of California. *Dusisiren dewana,* described from 9-Ma rocks in Japan, makes a good structural intermediate between *D. jordani* and Steller's sea cow in showing a reduction of teeth and finger bones. A penultimate stage in the evolution of Steller's sea cow is represented by *Hydrodamalis cuestae* from 3- to 8-Ma deposits in California. *H. cuestae* lacked teeth, probably lacked finger bones, and was very large.

Steller unfortunately described the sea cow's blubber, 3–4 inches thick, as tasting something like almond oil. Steller's sea cow quickly became a major food resource for Russian hunters. By 1768, only 27 years after its discovery, the sea cow was extinct. Anderson (1995) proposed that the extinction of sea cows may also have been contributed to by a combination of predation, competition, and decline in food supplies that occurred when aboriginal human populations colonized mainland coastlines and islands along the North Pacific (further discussed in Chapter 12).

The subfamily to which the modern dugong belongs, the Dugonginae, includes in addition to *Dugong* the following extinct genera: *Bharatisiren, Corystosiren, Crenatosiren, Dioplotherium, Rytiodus,* and *Xenosiren.* Fossil remains of this dugongid clade have been found from 15-Ma rocks in the Mediterranean, western Europe, southeastern United States, Caribbean, Indian Ocean, South America, and the North Pacific. The most elaborate development of tusks in the Sirenia are found in later diverging dugongines such as *Rytiodus, Corystosiren, Xenosiren,* and *Dioplotherium.* These species possessed enlarged, blade-like, self-sharpening tusks that may have been used to dig up sea grasses (Figure 5.8). The modern dugong may have evolved large tusks for a similar reason, but now appears to use them chiefly for social interactions. The discovery of a fossil dugongine in the Indian Ocean (Bajpai and Domning, 1997) was not unexpected given the presence of living *Dugong* in that region today and it corroborates the earlier

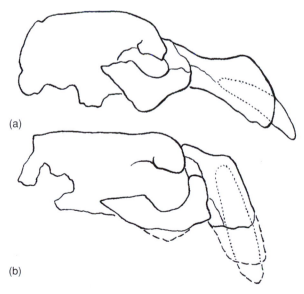

(a)

(b)

Figure 5.8.   Members of the dugong lineage illustrating differential development of the tusks. **(a)** *Dioplotherium manigaulti.* **(b)** *Rytiodus sp.* (From Domning, 1994.)

suggestion (Domning, 1994) that the discovery of additional fossils from that region would lend support for an Indo-Pacific origin for the genus.

The modern dugong, *Dugong dugon* (Figure 5.9), is distinguished by the following derived characters (Domning, 1994): nasals absent, constant presence in juveniles of a deciduous first incisor, frequent presence in adults of vestigial lower incisors, sexual dimorphism in size and eruption of permanent tusks (first incisor), and functional loss of enamel crowns on cheekteeth and persistently open roots of M2-3 and m2-3.

5.3. The Extinct Sirenian Relatives—Desmostylia

5.3.1. Origin and Evolution

First described on the basis of tooth fragments, the Desmostylia bear a name derived from the bundled columnar shape of the cusps of the molar teeth in some taxa (Figure 5.10). These bizarre animals constitute the only extinct order of marine mammals. They were confined to the North Pacific area (Japan, Kamchatka, and North America) dur-

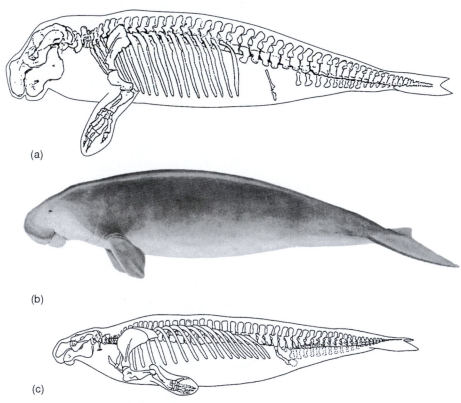

(a)

(b)

(c)

Figure 5.9.   Lateral view of skeleton of modern dugong and its fossil relative. **(a)** *Dugong dugon.* (Modified from Kingdon, 1971.) **(b)** Left side of body. (Illustrated by P. Folkens from Reeves *et al.*, 1992.) **(c)** *Dusisiren jordani* from the late Miocene-early Pliocene of California. (From Domning, 1978.)

ing the late Oligocene and middle Miocene epochs (approximately 33–10 Ma). Known
fossils represent at least 6 genera and 10 species, all hippo-sized amphibious quadrupeds
that probably fed on marine algae and sea grasses in subtropical to cool-temperate
waters (see Figure 5.10; Barnes *et al.*, 1985; Inuzuka *et al.*, 1995; Clementz *et al.*, 2003).

Basal desmostylians are represented by *Behemotops* from the middle or late Oligocene
of North America and Japan (Domning *et al.*, 1986; Ray *et al.*, 1994). *Cornwallius*, a later
diverging genus, is known from several eastern North Pacific late Oligocene localities.
Paleoparadoxia is a Miocene genus known on both sides of the Pacific. Sexual dimorphism in this species is suggested based on cranial and dental differences (Hasegawa

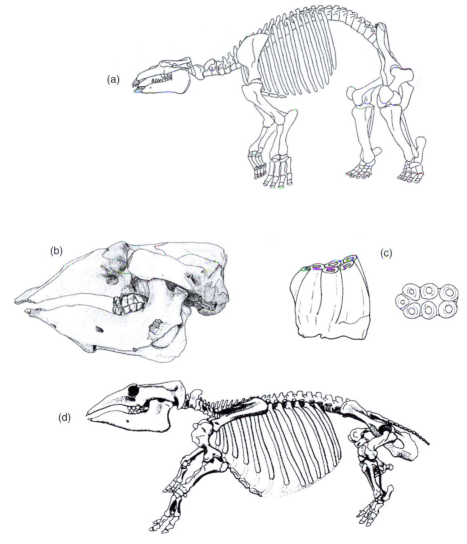

Figure 5.10.　　Representative desmostylans. **(a)** Restored skeleton of *Paleoparadoxia tabatai* (From
Domning, 2002). **(b)** Skull and mandible of *Desmostylus hesperus.* (From Domning, 2001b.)
(c) Lower molar of *Desmostylus* in lateral and occlusal aspect. (Modified from Vanderhoof,
1937.) **(d)** Restored skeleton of *Desmostylus.* (From Domning 2001.)

et al., 1995). A skeleton with skull from Point Arena, California was described as a new species, *Paleoparadoxia weltoni* (see Clark, 1991). Another new species of *Paleoparadoxia* has been reported from southern California and Mexico (Barnes and Aranda-Manteca, 1997). *Desmostylus,* the most specialized and best represented genus of the order, is found widely in Miocene coastal deposits of the North Pacific.

A phylogenetic analysis of desmostylians strongly supports a clade comprising *Desmostylus, Cornwallius, Paleoparadoxia,* and *Behemotops* as consecutive sister taxa (Clark, 1991; Ray *et al.*, 1994; Figure 5.11). Synapomorphies that unite desmostylians include lower incisors transversely aligned, the presence of an enlarged passage present through the squamosal from the external auditory meatus to roof of skull, roots of the lower first premolar fused, and paroccipital process elongated. Desmostylians are most closely related to proboscideans (elephants) on the basis of several characters of the lower molars and ear region, with sirenians forming the next closest sister group (Ray *et al.*, 1994).

Reconstructions of the skeleton and inferred locomotion of desmostylians have been controversial as recently reviewed by Domning (2002) and have included resemblances to sea lions, frogs, and crocodiles (e.g., Inuzuka, 1982, 1984, 1985; Halstead, 1985). Studies by Domning (2002) indicate that desmostylians had a more upright posture similar to that seen in some ground sloths and calicotheres. Locomotion in the water was by forelimb propulsion resembling polar bears. Dental morphology is varied, and later diverging species show adaptations for an abrasive diet, probably one that contained grit mixed with plant material scooped from the sea bottom or shore. A stable isotope study of tooth enamel from *Desmostylus* suggests that this taxon spent time in estuarine or freshwater environments rather than exclusively marine ecosystems and likely foraged on sea grasses as well as a wide range of aquatic vegetation (Clementz *et al.*, 2003).

5.4. The Extinct Marine Bear-Like Carnivoran, *Kolponomos*

5.4.1. Origin and Evolution

The large extinct carnivoran species *Kolponomos clallamensis* was originally described on the basis of an essentially toothless, incompletely preserved snout of middle Miocene age from Clallam Bay, Washington. Study of this specimen together with new material from coastal Oregon has resulted in the description of a second species, *K. newportensis* (Figure 5.12; Tedford *et al.*, 1994). *Kolponomos* had a massive skull with a markedly downturned snout and broad, crushing teeth.

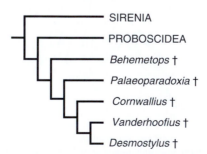

Figure 5.11.   Relationships among desmostylians and related taxa. (Modified from Domning, 2001b.) † = extinct taxa.

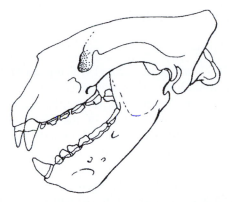

Figure 5.12.   Line drawing of the skull and lower jaw of *Kolponomos newportensis* from the early Miocene of Oregon. Original 25 cm long. (From Tedford *et al.*, 1994.)

The relationship of *Kolponomos* to other carnivores has been problematic. Originally this genus was questionably assigned to the Procyonidae, a family of terrestrial carnivores that includes raccoons and their allies. Study of additional specimens, including a nearly complete skull and jaw with some postcranial elements, has supported recognition of *Kolponomos* as an ursoid, most closely related to members of the extinct paraphyletic family Amphicynodontidae, which includes *Amphicynodon, Pachycynodon, Allocyon,* and *Kolponomos* (Tedford *et al.*, 1994). *Kolponomos* and *Allocyon* are hypothesized as the stem group from which the Pinnipedimorpha arose (Figure 5.13). Shared derived characters that link *Allocyon, Kolponomos,* and the pinnipedimorphs include details of the skull and teeth (Tedford *et al.*, 1994).

Kolponomos was probably coastal in distribution, because all specimens have been discovered in near-shore marine rocks. The crushing teeth would have been suited to a diet of hard-shelled marine invertebrates. *Kolponomos* probably fed on marine invertebrates living on rocky substrates, prying them off with the incisors and canines, crushing their shells, and consuming the soft parts as sea otters often do. *Kolponomos* represents a unique adaptation for marine carnivores; its mode of living and ecological niche are approached only by the sea otter (Tedford *et al.*, 1994).

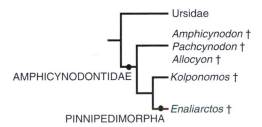

Figure 5.13.   Relationships among *Kolponomos* and related taxa. (Modified from Tedford *et al.*, 1994.)

5.5. The Extinct Aquatic Sloth, *Thalassocnus natans*

5.5.1. Origin and Evolution

In 1995, an aquatic sloth, *Thalassocnus natans* (Muizon and McDonald, 1995; Muizon 1996), represented by an abundance of associated complete and partial skeletons, was reported from early Pliocene marine rocks of the southern coast of Peru (Figure 5.14). Since that discovery four additional species of the *Thalassocnus* lineage have been described from the late Miocene-late Pliocene (McDonald and Muizon, 2002; Muizon *et al.*, 2003; Muizon *et al.*, 2004a). The aquatic sloth lineage spans over 4 Ma and was apparently endemic to Peru. *Thalassocnus* is a nothrotheriid ground sloth on the basis of a number of diagnostic cranial, dental, and postcranial features.

As previously known these sloths were medium to giant-sized herbivores with terrestrial or arboreal habits. As judged from its morphology and the paleoenvironment of the locality where these sloths have been recovered, *Thalassocnus* occupied an aquatic habit. The tail probably was used for swimming and a ventrally downturned premaxilla expanded at the apex suggests the presence of a well-developed lip for grazing. An increase in massiveness of the dentition and associated changes in the skull and mandible to permit crushing and grinding suggests that thalassocnines were grazers and fed primarily on sea grasses (Muizon *et al.*, 2004b). The morphological similarity of thalassocnines and desmostylians (i.e., elongate, spatulate rostra) raises the intriguing possibility that these animals were the ecologic homologues of desmostylians in the South Pacific (Domning, 2001b).

5.6. The Sea Otter, *Enhydra lutris*

Although sea otters (Figure 5.15) are the smallest marine mammals, they are the largest members of the Family Mustelidae, which includes 70 species of river otters, skunks, weasels, and badgers, among others. The generic name of the sea otter is from the Greek *enhydris,* for "otter," and the specific epithet is from the Latin *lutra,* for "otter." Three subspecies of sea otter are recognized based on differences in morphology as well as distribution: *Enhydra l. lutris* (Linnaeus, 1758) inhabits the Kuril Islands, the east coast of the Kamchatka Peninsula, and the Commander Islands; *Enhydra 1. kenyoni* (Wilson

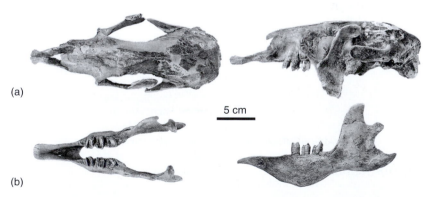

(a)

5 cm

(b)

Figure 5.14.   Aquatic sloth, *Thalassocnus natans* from the early Pliocene of Peru. **(a)** Skull. **(b)** Lower jaw in dorsal and lateral views. (From Muizon *et al.*, 2003.) (Courtesy of C. de Muizon.)

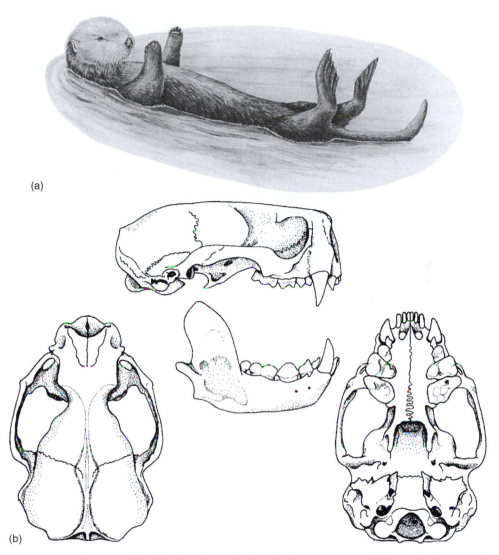

(a)

(b)

Figure 5.15. Sea otter, *Enhydra lutris*. **(a)** Ventral view of body. (Illustrated by P. Folkens in Reeves *et al.*, 1992.) **(b)** Skull in dorsal, lateral, and ventral views and lower jaw in lateral view. (From Lawlor, 1979.)

et al., 1991) ranges from the Aleutian Islands to Oregon; and *Enhydra l. nereis* (Merriam, 1904) had a historic range from northern California to approximately Punta Abrejos, Baja California.

Based on a cranial morphometric analysis, individuals of *E. l. lutris* are characterized by large wide skulls with short nasal bones. Specimens of *E. 1. nereis* have narrow skulls with a long rostrum and small teeth, and usually lack the characteristic notch in the postorbital region found in most specimens of the other two subspecies. Specimens of *E. 1. kenyoni* are intermediate to the other two but do not possess all characters and have longer mandibles than either of the other two subspecies (Wilson *et al.*, 1991).

5.6.1. Origin and Evolution

The modern sea otter *Enhydra* arose in the North Pacific at the beginning of the Pleistocene, about 1 to 3 Ma and has not dispersed since that time. There are records of *Enhydra* from the early Pleistocene of Oregon (Leffler, 1964) and California (Mitchell, 1966; Repenning, 1976). One extinct species, *Enhydra macrodonta* (Kilmer, 1972), has been described from the late Pleistocene of California.

The closest living relative of *Enhydra* are other lutine otters *Lutra* (Eurasian and spotted neck otters), *Aonyx* (short clawed otter), and *Amblonyx* (small clawed otter) based on separate and combined analysis of mitochondrial and nuclear sequence data (Koepfli and Wayne, 1998, 2003; Figure 5.16). The morphological analysis of extant mustelids by Bryant *et al*. (1993) differed in allying the giant otter *Pteronura* with other lutrines including *Enhydra* (see Figure 5.16). In a phylogenetic analysis that included both the living sea otter and related extinct taxa, Berta and Morgan (1985) proposed that there were two lineages of sea otters: an early diverging lineage that led to the extinct genus *Enhydriodon* and a later diverging lineage that led to the extinct giant otter *Enhydritherium* and the extant sea otter *Enhydra* (see Figure 5.16). *Enhydriodon* is known only from Africa and Eurasia, with three well-described species. In addition, there are several more poorly known specimens from Greece, England, and east Africa that have provisionally been assigned to the genus. All of this material is of late Miocene/Pliocene age. It is not known if *Enhydriodon* lived in marine or freshwater habitats or both. However, they were as large or larger than modern sea otters and had similarly well-developed molariform dentitions (Repenning, 1976). *Enhydritherium* is known from the late Miocene of Europe and the late Miocene/middle Pliocene of North America. Two species of *Enhydritherium* are described: *E. lluecai* from Spain and *E. terraenovae* from Florida and California. *Enhydritherium* is united with *Enhydra* based on dental synapomorphies.

An incomplete articulated skeleton of *Enhydritherium terraenovae* was described from northern Florida (Figure 5.17; Lambert, 1997). The depositional environment of this site, which is located a considerable distance from the coast, indicates that *E. terraenovae* frequented large inland rivers and lakes in addition to coastal marine environments. *Enhydritherium* was similar in size to *Enhydra,* with an estimated body mass of approximately 22 kg. The unspecialized distal hind limb elements and heavily developed humeral muscles of *Enhydritherium* strongly suggest that, contrary to *Enhydra,* this animal was primarily a forelimb swimmer. With its more equally proportioned forelimbs and hind limbs, *Enhydritherium* was almost certainly more effective at terrestrial loco-

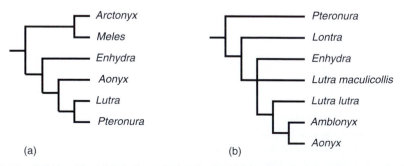

(a)   (b)

Figure 5.16.   Relationships of *Enhydra* and related taxa. **(a)** Morphological data (Bryant *et al*., 1993). **(b)** Molecular data (Koepfli and Wayne, 2003).

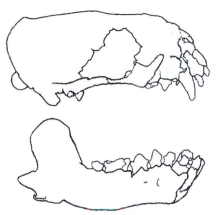

Figure 5.17.   Extinct giant otter, *Enhydritherium terranovae* from the late Miocene of Florida, skull and
lower jaw in lateral view. Original 16 cm long. (From Lambert, 1997.)

motion than *Enhydra*. The thickened cusps of the upper fourth premolars of *E. terra-novae* and their tendency to be heavily worn suggest that these otters, like *Enhydra,* consumed extremely hard food items such as molluscs (Lambert, 1997).

5.7. The Polar Bear, *Ursus maritimus*

5.7.1. Origin and Evolution

Polar bears are the only species of bear that spend a significant portion of their lives in the water. The generic name for the polar bear, *Ursus,* is the Latin word for bear, and its specific epithet, *maritimus,* refers to the maritime habitat of this species. The previous suggestion that the polar bear (Figure 5.18) might represent a separate genus, *Thalarctos,* because of its adaptation to aquatic conditions and its physical appearance is not supported. *Ursus maritimus* has a fossil record limited to the Pleistocene (Kurtén, 1964). Analysis of combined nuclear and mitochondrial sequence data (Yu *et al.*, 2004) corroborate a sister group relationship between brown and polar bears (Zhang and Ryder, 1994; Talbot and Shields, 1996; Waits *et al.*, 1999). Molecular data support divergence of polar bears from brown bears, *Ursus arctos,* 1–1.5 Ma (Yu *et al.*, 2004), which is approximately 10 times older than the fossil record (.07–.1 Ma; Kurtén, 1968).

5.8. Summary and Conclusions

The monophyly of sirenians is widely accepted and elephants are considered their closest living relatives. Sirenians, elephants, and extinct desmostylians form a monophyletic clade, the Tethytheria, that is part of a larger, diverse mammal clade, the Afrotheria. Sirenians are known in the fossil record from approximately 50 Ma. Early sirenians were fluvatile or estuarine semiaquatic herbivores with functional hind limbs. Manatees are likely derived from dugongids. An extinct lineage of dugongids led to the recently extinct Steller's sea cow that was cold adapted for life in the Bering Sea, in contrast to other members of this lineage distributed in tropical or subtropical waters. The hippopotamus-like desmostylians (33–10 Ma) have the distinction of composing the only extinct order of marine mammals. The large extinct bear-like

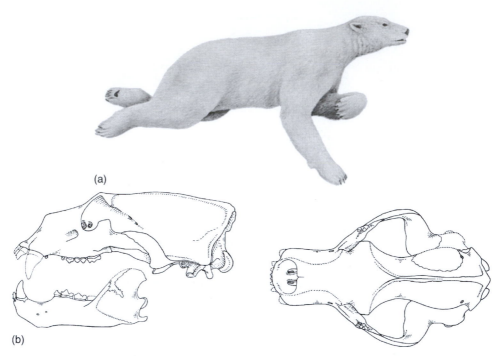

Figure 5.18.   Polar bear, *Ursus maritimus*. **(a)** Right side of body. **(b)** Lateral and dorsal views of skull and
                    lateral view of lower jaw. (From Hall and Kelson, 1959.)

carnivoran *Kolponomos* is now recognized as more closely related to amphicynodon-
tine ursids and pinnipedimorphs rather than its previous allocation to the raccoon
family. The range of adaptation of sloths, formerly known to have only terrestrial and
arboreal habits, was extended based on discovery of a diverse lineage of aquatic sloth
Thalassocnus. The modern sea otter appears to have evolved in the North Pacific 1–3
Ma. Among fossil sea otters is *Enhydritherium,* which likely frequented large rivers and
lakes as well as coastal marine environments. The most recently diverging lineage of
marine mammals, the polar bear, appears to have been derived from brown bears
between .5 and 1 Ma.

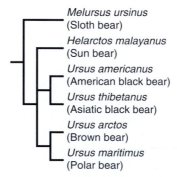

Figure 5.19.   Relationships of polar bears and their relatives. (From Yu *et al.*, 2004.)

5.9. Further Reading

Sirenian phylogeny is detailed in Domning (1994, 2001b) and a popular account of the evolution of manatees and dugongs can be found in Reynolds and Odell (1991). For a summary of the evolution and phylogeny of desmostylians see Domning (2001b, 2002). A description of the bear-like carnivoran *Kolponomos is* provided in Tedford *et al.* (1994). For descriptions of the aquatic sloth see Muizon and MacDonald (1995), McDonald and Muizon (2002), and Muizon *et al.* (2003). Sea otter evolution is reviewed by Berta and Morgan (1985) and Lambert (1997). Bear phylogeny, especially the molecular evidence, is discussed by Yu *et al.* (2004) (Figure 5.19).

References

Anderson, P. K. (1995). "Competition, Predation, and the Evolution and Extinction of Steller's Sea Cow." *Hydrodamalis gigas. Mar. Mamm. Sci. 11:* 391–394.

Archibald, J. D. (1996). "Fossil Evidence for a Late Cretaceous Origin of 'Hoofed Mammals.' " *Science 272:* 1150–1153.

Bajpai, S., and D. P. Domning (1997). "A New Dugongine Sirenian from the Early Miocene of India." *J. Vert. Paleontol. 17*(1): 219–228.

Barnes, L. G., and F. Aranda-Manteca (1997). New middle Miocene Paleoparadoxidae (Mammalia, Desmostylia) from Baja California and Mexico. *Abstr., 57th Ann. Meet. Soc. Vert. Paleo.,* Chicago, p. 30A.

Barnes, L. G., D. P. Domning, and C. E. Ray (1985). "Status of Studies on Fossil Marine Mammals." *Mar. Mamm. Sci. 1:* 15–53.

Berta, A., and G. S. Morgan (1985). "A New Sea Otter (Carnivora: Mustelidae) from the Late Miocene and Early Pliocene (Hemphillian) of North America." *J. Paleontol. 59:* 809–819.

Bryant, H. N., A. P. Russell, and W. D. Fitch (1993). "Phylogenetic Relationships Within the Extant Mustelidae (Carnivora): Appraisal of the Cladistic Status of the Simpsonian Subfamilies." *Zool. J. Linn. Soc. 108:* 301–334.

Clark, J. M. (1991). "A New Early Miocene Species of *Paleoparadoxia* (Mammalia: Desmostyha) from California." *J. Vert. Paleontol. 11:* 490–508.

Clementz, M. T., K. A. Hoppe, and P. L. Koch (2003). "A Paleoecological Paradox: The Habitat and Dietary Preferences of the Extinct Tethythere *Desmostylus*, Inferred from Stable Isotope Analysis." *Paleobiology 29:* 506–519.

Domning, D. P. (1978). "Sirenian Evolution in the North Pacific Ocean." *Univ. Calif. Publ. Geol. Sci. 118:* 1–176.

Domning, D. P. (1982). "Evolution of Manatees: A Speculative History." *J. Paleontol. 56:* 599–619.

Domning, D. P. (1994). "A Phylogenetic Analysis of the Sirenia." *Proc. San Diego Soc. Nat. Hist. 29:* 177–189.

Domning, D. P. (2001a). "The Earliest Known Fully Quadrupedal Sirenian." *Nature 413:* 625–627.

Domning, D. P. (2001b). Evolution of the Sirenia and Desmostylia. *In* "Secondary Adaptation to Life in Water" (J. M. Mazin and V. de Buffrenil, eds.), pp. 151–168. Verlag, Dr. Frederich Pfeil, Munchen Germany.

Domning, D. P. (2002) "The Terrestrial Posture of Desmostylians." *Smithson. Contrib. Paleobiol. 93:* 99–111.

Domning, D. P., and V. de Buffrénil. (1991). "Hydrostasis in the Sirenia: Quantitative Data and Functional Interpretations." *Mar. Mamm. Sci. 7:* 331–368.

Domning, D. P., and H. Furusawa (1995). "Summary of Taxa and Distribution of Sirenia in the North Pacific Ocean." *Island Arc 3:* 506–512.

Domning, D. P, and P. D. Gingerich (1994). "*Protosiren smithae,* New Species (Mammalia, Sirenia), from the Late Middle Eocene of Wadi Hitan, Egypt." *Contrib. Mus. Paleontol. Univ. Mich. 29*(3): 69–87.

Domning, D. P., and L. Hayek (1986). "Interspecific and Intraspecific Morphological Variation in Manatees (Sirenia: Trichechidae)." *Mar. Mamm. Sci. 2*(2): 87–141.

Domning, D. P., C. E. Ray, and M. C. McKenna (1986). "Two New Oligocene Desmostylians and a Discussion of Tethytherian Systematics." *Smithson. Contrib. Paleobiol. 59:* 1–56.

Fischer, M. S. (1986). "Die Stellung der Schliefer (Hyracoidea) im phylogenetischen System der Eutheria." *CFS Cour. Forschungsinst. Senckenberg 84:* 1–132.

Fischer, M. S. (1989). Hyracoids, the sister-group of perissodactyls. *In* "The Evolution of Perissodactyls" (D. R. Prothero, and R. M. Schoch, eds.), pp. 37–56. Clarendon Press, New York.

Gheerbrant, E., D. P. Domning, and P. Tassy (2005). Paenungulata (Sirenia, Proboscidea, Hyracoidea, and Relatives). *In* "The Rise of Placental Mammals" (K. D. Rose and J. David Archibald, eds.), pp. 84–105. Johns Hopkins, Baltimore.

Gingerich, P. D., D. P. Domning, C. E. Blane, and M. D. Uhen (1994). "Cranial Morphology of *Protosiren fraasi* (Mammalia, Sirenia) from the Middle Eocene of Egypt: A New Study Using Computed Tomography." *Contrib. Mus. Paleontol. Univ. Mich. 29*(2): 41–67.

Hall, E. R., and K. R. Kelson (1959). *The Mammals of North America.* Ronald Press, New York.

Halstead, L. B. (1985). "On the Posture of Desmostylians: A Discussion of Inuzuka's "Herpetiform Mammals." *Mem. Fac. Sci., Kyoto Univ., Ser. Biol. 10:* 137–144.

Hasegawa, Y., Y. Taketani, H. Taru, O. Sakamoto, and M. Manabe (1995). "On Sexual Dimorphism in *Paleoparadoxia tabatai.*" *Island Arc 3:* 513–521.

Heptner, V. G. (1974). Unterordnung Tricheciformes Hay, 1923, eigentliche Sirenen. *In* "Die Saugetier der Sowjetunion" (V. G. Heptner, and N. P. Naumov, eds.), pp. 30–51. Gustav Fischer Verlag, Jena., 2: 1-1006.

Inuzuka, N. (1982). *Published Reconstructions of Desmostylians. Saitama, (Japan).* Association for the Geological Collaboration in Japan, pp. 1–16. Ass. Geol. Col. Japan.

Inuzuka, N. (1984). "Skeletal Restoration of the Desmostylians: Herpetiform Mammals." *Mem. Fac. Sci., Kyoto Univ., Ser. Biol. 9:* 157–253.

Inuzuka, N. (1985). "Are 'Herpetiform Mammals' Really Impossible? A Reply to Halstead's Discussion." *Mem. Fac. Sci., Kyoto Univ., Ser. Biol. 10:* 145–150.

Inuzuka, N., D. P. Domning, and C. E. Ray (1995). "Summary of Taxa and Morphological Adaptations of Desmostylia." *Island Arc 3:* 522–537.

Kilmer, F. H. (1972). "A New Species of Sea Otter from the Late Pleistocene of Northwestern California." *Bull. South. Calif. Acad. Sci. 71*(3): 150–157.

Kingdon, J. (1971). *East African Mammals, an Atlas of Evolution in Africa.* Academic Press, London.

Koepfli,. K. P., and R. K. Wayne (1998). "Phylogenetic Relationships of Otters (Carnivora: Mustelidae) Based on Mitochondrial Cytochrome b Sequences." *J. Zool., Lond. 246:* 401–416.

Koepfli, K. P., and R. K. Wayne (2003). "Type I STS Markers are More Informative than Cytochrome b in Phylogenetic Reconstruction of the Mustelidae (Mammalia: Carnivora)." *Syst. Biol. 52:* 571–593.

Kurtén, B. J. (1964). "The Evolution of the Polar Bear. *Ursus maritimus* Phipps." *Acta Zool. Fenn. 108:* 1–26.

Kurtén, B. J. (1968). *Pleistocene Mammals of Europe.* Weidenfeld & Nicolson, London.

Lambert, W. D. (1997). "The Osteology and Paleoecology of the Giant Otter *Enhydritherium terranovae.*" *J. Vertebr. Paleontol. 17:* 738–749.

Lavergne, A., E. Douzery, T. Strichler, F. M. Catzefiis, and M. S. Springer (1996). "Interordinal Mammalian Relationships: Evidence for Paenungulate Monophyly is Provided by Complete Mitochondrial 12sRNA Sequences." *Mol. Phylogenet. Evol. 6:* 245–258.

Lawlor, T (1979). *Handbook to the Order and Families of Living Mammals.* Mad River Press, California.

Leffler, S. R. (1964). "Fossil Mammals from the Elk River Formation, Cape Blanco, Oregon." *J. Mammal. 45*(1): 53–61.

Linnaeus, C. (1758). *Systems naturae,* 10th ed., L. Salvii. Uppsala.

Madsen, O., M. Scally, C. J. Douady, D. J. Kao, R. W. DeBry, R. Adkins, H. M. Amrine, M. J. Stanhope, W. W. de Jong, and M. S. Springer (2001). "Parallel Adaptive Radiations in Two Major Clades of Placental Mammals. *Nature 409:* 610–614.

McDonald, H. G., and C. de Muizon (2002). "The Cranial Anatomy of *Thalassocnus* (Xenarthra, Mammalia) a Derived Nothrothere from the Neogene of the Pisco Formation (Peru)." *J. Vert. Paleontol. 22:* 349–365.

McKenna, M. C. (1975). Toward a phylogenetic classification of the Mammalia. *In* "Phylogeny of the Primates" (W. P. Luckett, and F. S. Szalay, eds.), pp. 21–45. Plenum Press, New York.

Merriam, C. H. (1904). "A New Sea Otter from Southern California." *Proc. Biol. Soc. Wash. 17:* 159–160.

Mitchell, E. D. (1966). "Northeastern Pacific Pleistocene Otters." *J. Fish. Res. Bd. Can. 23:* 1897–1911.

Muizon, C. de (1996). *Thalassocnus natans,* the marine sloth from Peru. *In* "Abstracts of the Conference on Secondary Adaptation to Life in the Water" (J.-M. Mazin, P. Vignaud, and V. de Buffrénil, eds.), p. 29. Poitiers, France.

Muizon, C. de, and H. G. McDonald (1995). "An Aquatic Sloth from the Pliocene of Peru." *Nature 375:* 224–227.

Muizon, C. de, H.G. McDonald, R. Salas, and M. Urbina (2003). "A New Early Species of the Aquatic Sloth *Thalassocnus* (Mammalia, Xenarthra) from the Late Miocene of Peru." *J. Vert. Paleontol. 23:* 886–894.

Muizon, C, de, H. G. McDonald, R. Salas, and M. Urbina (2004a). "The Youngest Species of the Aquatic Sloth and a Reassessment of the Relationships of Nothrothere Sloths (Mammalia: Xenarthra)." *J. Vert. Paleontol. 24:* 387–397.

Muizon, C. de, H. G. McDonald, R. Sala, and M. Urbina (2004b). "The Evolution of Feeding Adaptations of the Aquatic Sloth *Thalassocnus*." *J. Vert. Paleontol. 24:* 398–410.

Murata, Y., M. Nikaido, T. Sasaki, Y. Cao, Y. Fukumoto, M. Hasegawa, and N. Okada. (2003). "Afrotherian Phylogeny as Inferred from Complete Mitochondrial Genomes." *Mol. Phylogenet. Evol. 28:* 253–260.

Murphy, W. J., E. Elzirk, W.E. Johnson, Y.P. Zhang, O. A. Ryder, and S. J. O'Brien (2001). "Molecular Phylogenetics and the Origins of Placental Mammals." *Nature 409:* 614–618.

Nikaido, M. H. Nishihara, Y. Hukumotoa, and N. Okada (2003). "Ancient SINEs from African Endemic Mammals." *Mol. Biol. Evol. 20:* 522–527.

Novacek, M., and A. R. Wyss (1987). "Selected Features of the Desmostylian Skeleton and Their Phylogenetic Implications." *Am. Mus. Nov. 2870:* 1–8.

Novacek, M., A. R. Wyss, and M. C. McKenna (1988). The major groups of eutherian mammals. *In* "The Phylogeny and Classification of the Tetrapods" (M. J. Benton, ed.), Vol. 2, pp. 31–71. Clarendon Press, Oxford.

Parr, L., and D. Duffield (2002). Interspecific comparison of mitochondrial DNA among extant species of sirenians. *In* "Molecular and Cell Biology of Marine Mammals" (C. J. Pfeiffer, ed.), pp. 152–160. Krieger Publ, Malabar, FL.

Prothero, D. (1993). Ungulate phylogeny: Molecular vs. morphological evidence. *In* "Mammal Phylogeny: Placentals" (F. S. Szalay, M. J. Novacek, and M. C. McKenna, eds.), pp. 173–181. Springer-Verlag, New York.

Prothero, D. R., E. M. Manning, and M. Fischer (1988). "The Phylogeny of the Ungulates." *Syst. Assoc. Spec. Vol. 35B:* 201–234.

Ray, C. E., D. P. Domning, and M. C. McKenna (1994). "A New Specimen of *Behemotops proteus* (Order Desmostylia) from the Marine Oligocene of Washington." *Proc. San Diego Mus. Nat. Hist. 29:* 205–222.

Reeves, R. R., B. S. Stewart, and S. Leatherwood (1992). *The Sierra Club Handbook of Seals and Sirenians.* Sierra Club Books, San Francisco CA.

Repenning, C. A. (1965). [Drawing of *Paleoparadoxia* skeleton]. *Geotimes* 9(6): 1.3.

Repenning, C. A. (1976). "*Enhydra* and *Enhydriodon* from the Pacific Coast of North America." *J. Res. U.S. Geol. Surv. 4:* 305–315.

Reynolds, J. E., III, and D. K. Odell (1991). *Manatees and Dugongs.* Facts on File, New York.

Savage, R. G. J., D. P. Domning, and J. G. M. Thewissen (1994). "Fossil Sirenia of the West Atlantic and Caribbean Region. V. The Most Primitive Known Sirenian, *Prorastomus sirenoides* Owen, 1855." *J. Vert. Paleontol. 14:* 427–449.

Scally, M., O. Madsen, C. J. Douady, W. W. de Jong, M. J. Stanhope, and M. S. Springer (2001). "Molecular Evidence for the Major Clades of Placental Mammals." *J. Mamm. Evol. 8(4):* 239–277.

Shoshani, J. (1993). Hyracoidea-Tethytheria affinity based on myological data. *In* "Mammal Phylogeny: Placentals" (F. S. Szalay, M. J. Novacek, and M. C. McKenna, eds.), pp. 235–256. Springer-Verlag, New York.

Springer, M. S., G. C. Cleven, O. Madsen, W. W. de Jong, V. G. Waddell, H. M. Amrine, and M. J. Stanhope (1997). "Endemic African Mammals Shake the Phylogenetic Tree." *Nature 388:* 61–64.

Springer, M. S., and J. A. W. Kirsch (1993). "A Molecular Perspective on the Phylogeny of Placental Mammals Based on Mitochondrial 12s rDNA Sequences, with Special Reference to the Problem of the Paenungulata." *J. Mamm. Evol. 1(2):* 149–166.

Stanhope, M. O. Madsen, V. G. Waddell, G. C. Cleven, W. W. de Jong, and M. S. Springer (1998). "Highly Congruent Molecular Support for a Diverse Superordinal Clade of Endemic African Mammals." *Mol. Phylogenet. Evol. 9:* 501.

Talbot, S. L., and G. F. Shields (1996). "A Phylogeny of Bears (Ursidae) Inferred from Complete Sequences of Three Mitochondrial Genes." *Mol. Phylogenet. Evol. 5:* 567–575.

Tassy, P., and J. Shoshani (1988). "The Tethytheria: Elephants and Their Relatives." *Syst. Assoc. Spec. Vol. 35B:* 283–315.

Tedford, R. H., L. G. Barnes, and C. E. Ray (1994). "The Early Miocene Littoral Ursoid Carnivoran *Kolponomos:* Systematics and Mode of Life." *Proc. San Diego Mus. Nat. Hist. 29:* 11–32.

Thewissen, J. G. M., and D. P. Domning (1992). "The Role of Phenacodontids in the Origin of Modern Orders of Ungulate Mammals." *J. Vert. Paleontol. 12:* 494–504.

Vanderhoof, V. L. (1937). "A Study of the Miocene Sirenian *Desmostylus.*" *Univ. Calif. Publ. Bull. Dept. Geol. Sci. 24:* 169–262.

Waits, L. P., J. Sullivan, S. J. O'Brien, and R. H. Ward (1999). "Rapid Radiation Events in the Family Ursidae Indicated by Likelihood Estimation from Multiple Fragments of mtDNA." *Mol. Phylogenet. Evol. 13:* 82–92.

Wilson, D. E., M. A. Bogan, R. I. Brownell, Jr., A. M. Burdin, and M. K. Maminov (1991). "Geographic Variation in Sea Otters, *Enhydra lutris.*" *J. Mammal. 72*(1): 22–36.

Yu, L. Q-W Li, O. A. Ryder, and Y.-P. Zhang (2004). "Phylogeny of the Bears (Ursidae) Based on Nuclear and Mitochondrial Genes." *Mol. Phylogenet. Evol. 32*: 480–494.

Zhang, Y.-P., and O. A. Ryder (1994). "Phylogenetic Relationships of Bears (the Ursidae) Inferred from Mitochondrial DNA Sequences. *Mol. Phylogenet. Evol. 3:* 351–359.

6

Evolutionary Biogeography

6.1. Introduction—What Is Biogeography and Why Is It Important?

Biogeography involves the study of the geographic distributions of organisms, both past and present. It attempts to describe and understand patterns in the distributions of species and higher taxonomic groups and interprets aspects of both ecology and evolutionary biology. Understanding marine mammal distributions necessitates knowledge of a species' ecological requirements including both biotic and abiotic factors. Biotic factors such as food availability are discussed in the first part of the chapter. Among evolutionary questions of interest explored later in this chapter are: (1) How did a species come to occupy its present range?, (2) How have geologic events, such as the opening of the Bering Strait or the Central American Seaway, shaped this distribution?, and (3) Why are some closely related species confined to the same region, whereas others are widely separated and even found on opposite sides of the world?

6.2. Ecological Factors Affecting Distributions of Marine Mammals

Marine mammals have adapted to several properties of the ocean environment not experienced by their terrestrial ancestors. These include increased buoyancy derived from the relatively high density of sea water, frictional resistance to swimming created by viscous forces between water molecules, poor transmittance of light underwater, **osmotic** challenges created by **hyperosmotic** sea water, and substantial heat loss to the cold environment. Each of these is discussed in later chapters. Two additional features of the ecology of marine mammals that vary over oceanic distances appear to strongly influence the present and past distribution patterns of marine mammal species. These are the patterns of geographic and seasonal surface water temperature variations and the spatial and temporal patterns of **primary productivity** and the resulting distribution of food resources.

111

6.2.1. Water Temperature and Sea Ice

For most of their lives, marine mammals are in direct contact with seawater that is much colder than their core body temperatures. Even though large body sizes and streamlined body shapes characteristic of marine mammals serve to reduce thermal losses, the heat capacity of even temperate latitude sea water is about 25 times that of air of the same temperature. This means that marine mammals lose considerable heat to their aquatic environment, particularly when they are of small body size (discussed in Chapter 9). Geographically, surface ocean temperatures tend to be highest near the equator and decrease with increasing latitude toward both the north and south poles. This poleward gradient of surface ocean temperatures has been used to establish several latitudinal marine climatic zones shown (with approximate ranges of sea surface temperatures) in Plate 2a.

Sea ice forms only in polar and subpolar zones and reaches its maximum extent in winter. Seasonal cycles of freezing and melting of sea ice limit access to high latitudes from most species of marine mammals to the warmest summer months only. Two forms of ice are found near the poles: **fast ice** is ice attached to land; **pack ice** is free floating ice. Pinnipeds that inhabit pack ice are the walrus; harp, hooded, bearded, ribbon, ross, crabeater, leopard, largha, grey, and harbor seals; and the Steller's sea lion. Weddell, ringed, southern elephant, Caspian, Baikal, and grey seals occupy fast ice (Riedman, 1990). The importance of sea ice and oceanic islands as features critical for pinniped breeding and pupping is discussed in Chapter 13.

6.2.2. Distribution of Primary Productivity

The availability of food for marine mammals is established first by patterns of marine primary production and second by the number of **trophic levels** between primary production and the marine mammal consumer. **Phytoplankton** cell sizes are quite small, and numerous trophic levels link these extremely small primary producers with large animals that occupy high trophic levels (Steele, 1974). Sirenians are the only marine mammals to feed directly on primary producers (sea grasses), whereas some pinnipeds and odontocetes consume prey five or more trophic levels removed from the primary producers (Figure 6.1). Rates of marine primary production can vary by several orders of magnitude over geographic areas and also between seasons. Seasonal and spatial variations in primary production are related to differences in light intensity, water temperature, nutrient abundance, and grazing pressure. The underlying pulse for these temporal changes is the predictable seasonal variation in the intensity of sunlight reaching the sea surface. This in turn influences seasonal variations in water temperature, density, and the pattern of vertical mixing of water, with the magnitude of these variations between the summer and winter becoming more pronounced at higher latitudes (Figure 6.2).

In tropical and subtropical waters, sunlight is abundant all year long, but a strong permanent thermocline inhibits vertical mixing of nutrients from deeper waters. Low rates of nutrient return are partially compensated for by a year-round growing season and a deep **photic zone.** Even so, net primary production and standing crops are low, and seasonal variability in production is limited (see Figure 6.2).

A prominent feature in the production cycle of temperate seas is the **spring diatom bloom.** In general, bloom conditions in the open ocean occur as a broad band of primary production that sweeps poleward with the onset of spring and retraction of the seasonal ice cover (Brown *et al.*, 1985). The sweep moves north with the ice edge. At this time, the

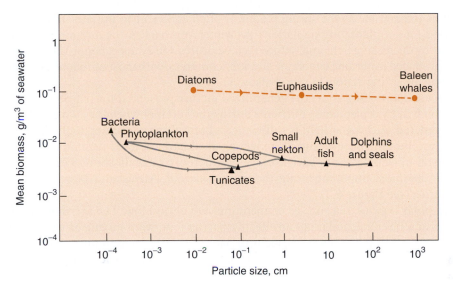

Figure 6.1. Relationship between food particle size and biomass in two pelagic food webs. Dashed line and closed circles are about 10x higher at all trophic levels than those in subtropical gyres (solid lines and triangles).

standing crop of diatoms increases quickly to the largest of the year and begins to deplete nutrient concentrations. Grazing zooplankton respond by increasing their numbers; the diatom population peaks, then declines and remains low throughout the summer (see Figure 6.2). Autumn air temperatures then begin to cool the water and allow the mixing of water of different temperatures; this renews the nutrient supply to the photic zone. Phytoplankton respond with another period of rapid growth, which, although typically

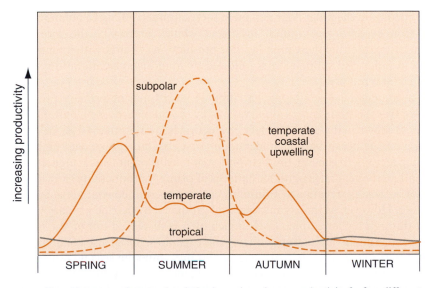

Figure 6.2. General patterns of seasonal variation in marine primary productivity for four different marine production systems.

not as remarkable as the spring bloom, is often sufficient to initiate another growth period for zooplankton populations. As winter approaches, the autumn bloom is cut short by decreasing light and reduced temperatures. On average, about 120 gC/m²/year are produced in oceanic temperate and subpolar areas, with most of that total occurring during the spring diatom bloom (Falkowski, 1980).

Coastal **upwelling** alters the generalized picture presented for temperate seas (see Figure 6.2) by replenishing nutrients during the summer when they would otherwise be depleted. As long as light is sufficient and upwelling continues, high phytoplankton production occurs and is reflected in abundant local animal populations. Coastal upwelling zones have average productivity rates of about 970 gC/m²/year. They are most apparent along the west coasts of Africa and North and South America (see Figure 6.2).

In polar regions, sea surface temperatures are always low. The thermocline, if one exists at all, is weak and is not an effective barrier to upward mixing of nutrients from deeper waters. Light, or more correctly the lack of it, is the major limiting factor for phytoplankton growth in polar seas. Sufficient light to sustain high primary production lasts for only a few months during the summer. During this time, photosynthesis can continue around the clock and thus produce huge phytoplankton populations quickly (see Figure 6.2). The short summer diatom bloom declines rapidly as light intensity and day length decline. Winter conditions resemble those of temperate regions except that in polar seas these conditions endure for much longer. The complete annual cycle of production consists of a single short period of phytoplankton growth, equivalent to a temperate climate spring bloom, immediately followed by an autumn bloom and then a decline that alternates with an extended winter of reduced net production. The annual average productivity rate for north polar seas is low (about 30 gC/m²/year) because so much of the year passes in darkness with almost no phytoplankton growth. In northern subpolar seas and around the Antarctic continent, upwelling of deep, nutrient-rich water supports very high summertime primary production rates and annual productivity rates of over 300 gC/m²/year (Pauly and Christensen, 1995). Although a few species of pinnipeds and the polar bear remain in these summer-intensive polar and subpolar production systems year-round, mysticete whales exemplify the more common approach of exploiting these high-latitude production systems, with intensive summer feeding in polar and subpolar seas followed by long-distance migrations to low latitudes in winter months.

The average annual geographic variation of global marine primary production, as compiled from several years of observations by the satellite-borne Coastal Zone Color Scanner, is shown in Plate 2b). Primary production is low (less than 60 gC/m²/year) in the central gyres of ocean basins, moderate in most coastal regions, and high in coastal upwelling regions. In general, the distribution of marine organisms at higher trophic levels resembles the general geographic patterns of primary productivity shown in Plate 2b, with the largest aggregations of animals concentrated in coastal areas and zones of upwelling. However, as the organic material produced by marine primary producers is moved through higher trophic levels, much of it is dispersed out of the near-surface photic zone. Animal populations utilizing this organic material also congregate at sharp density interfaces such as current eddies, water mass boundaries, and especially the sea bottom where some marine mammal species focus their feeding efforts. Zooplankton species such as krill (small crustaceans of the family Euphausiidae, Figure 6.3), and also squid, are concentrated in well-defined sound-reflecting layers during daylight at con-

Figure 6.3.   Lateral view of an adult *Euphausia superba,* actual size. (From Macintosh and Wheeler, 1929.)

siderable depth. These layers migrate upward to shallower water at night in temperate areas and during the warmer months in high latitudes. Krill are most efficiently captured by marine mammals in shallower water. Antarctic baleen whale and seal distributions largely mimic the distribution of krill, and both reflect the high phytoplankton biomass coinciding with the Antarctic Circumpolar Current (ACC; Tynan, 1998). The southern boundary was identified as an ecologically important oceanographic structure providing whales and other species a predictably profitable foraging area. Research has confirmed that fluctuations in the abundance of krill have taken place in the last 30 years and suggests lower abundances in recent years (Loeb *et al.*, 1997). Decreased krill availability may negatively affect their vertebrate predators, including pinnipeds and whales.

Water temperature has been shown to affect the diving behavior of the southern elephant seal (Boyd and Arnbom, 1991). The majority of dives of a southern elephant seal were spent at a relatively constant depth of 200–400 m, usually in association with transition to warm water. Given that there is little if any penetration of light to the normal depth of foraging, Boyd and Arnbom (1991) suggested that this species use other physical characteristics of its environment, such as the temperature discontinuity between water masses, to locate suitable foraging areas.

Water temperature and availability of food resources vary from one year to the next, in some years drastically, because of the effect of El Niño events. **El Niño-Southern Oscillation (ENSO)** is a meteorological and oceanographic phenomenon that occurs at irregular intervals of a few years. Its most obvious characteristic is the warming of surface waters in the eastern tropical Pacific (Figure 6.4), which

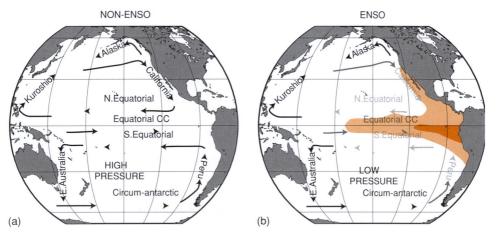

Figure 6.4.   Generalized Pacific Ocean surface currents during **(a)** non-ENSO and **(b)** ENSO condition. Note the marked relaxation of west-flowing equatorial currents during ENSO.

blocks the upward transport of deeper, nutrient-rich water from below. During a typical ENSO event, west-flowing equatorial currents in the tropical Pacific slow, blocking the flow of the cold California and Peru Currents in the eastern Pacific and halting their mechanism of upwelling. ENSO events are accompanied by large reductions in zooplankton. The 1982–1983 and 1997–1998 ENSO were the most intense in the last century, suggesting a trend toward more frequent and more intense ENSO events.

ENSO events, which heavily affect tropical marine environments where they originate, are propagated poleward. The severity of ENSO impacts on marine organism declines from low to high latitudes. Both the 1982–1983 and the 1997–1998 ENSO were times of food shortage and increased mortality, especially for temperate and tropical pinnipeds. Severe negative effects were observed among otariids on the Galapagos Islands, where 100% mortality of Galapagos fur seal pups and nearly 100% mortality of California sea lions occurred in 1982. An important long-term effect of the 1982–1983 ENSO was increased adult female mortality in these populations most likely due to decreased primary productivity and reduced availability of prey (Trillmich et al., 1991). Similar effects were observed in the aftermath of the 1997–1998 ENSO. Phocid seals, especially those living in tropical/subtropical areas, also have been affected by El Niño events. During the 1982–1983 ENSO for example, at the Farallon Islands (38° N) elephant seal pup mortality rates were normal, whereas at Ano Nuevo (37.5° N) pup mortality rates roughly doubled, and at San Nicholas Island (33° N) the pup mortality rate was five times above normal (Trillmich et al., 1991). Cetaceans have also been affected by El Niño events. High mortalities of dusky dolphins, bottlenose dolphins, common dolphins, and Burmeister's porpoises stranded on beaches around Pisco, Peru have been attributed to starvation due to El Niño in 1997–1998 (references cited in Domingo et al., 2002). In other cases, shifts in cetacean distributions have been described based on changes in prey resources, a result of El Niño events (examples cited in Wursig et al., 2002).

In addition to short-term climatic fluctuations such as ENSO events, long-term climate changes have the potential to profoundly affect marine mammals, particularly those living in the Arctic. Analyses continue to substantiate trends over the last several decades of decreasing sea ice extent in the Arctic Ocean. Such trends may be indicative of global climate warming. Direct effects of long-term climate change on marine mammals include the loss of ice-associated habitat and changes in prey availability (Tynan and DeMaster, 1997; Wursig et al., 2002). Ice seals (e.g., ringed seals, bearded seals, harp seals, and hooded seals), which rely on suitable ice substrate for resting, pupping, and molting, may be especially vulnerable to such changes. For example, declines in ringed seal density have been linked with ice conditions in the Beaufort Sea 1974–1975 and 1982–1985 (Stirling et al., 1977; Harwood and Stirling, 1992). For cetaceans inhabiting the arctic, climatic change is more likely to affect them via changes in their prey availability. For example, arctic cod are an important food source for belugas and narwhals, and lipid-rich copepods are important for bowhead whales. The distribution of these prey types vary with ice conditions (references cited in Tynan and DeMaster, 1997). Given the potential for climate-driven changes in the Arctic to affect habitat and prey availability, regional studies and monitoring of marine mammals are essential to document and interpret the effects of these long-term changes on the ecosystem (Loeng et al., 2005).

6.2.3. Ocean Temperatures and Productivity Fluctuations in the Past

Lipps and Mitchell (1976) proposed that radiations and declines of marine mammal species in the past were due to variations in the availability of resources in ocean environments. These trophic resources were related to upwelling processes. They proposed that increased upwelling intensity due to climatic or tectonic events permitted the initial invasions and radiations of pinnipeds and cetaceans. They further postulated that the apparent decline in marine mammal diversity reflects a decrease in thermal gradients, which led to decreased upwelling and primary productivity and thus limited dispersal and speciation in marine mammals.

This general hypothesis has been examined with respect to the evolution of cetaceans by numerous workers (Orr and Faulhaber, 1975; Gaskin, 1976; Fordyce, 1980, 1992, 2002). The late Eocene-early Oligocene (38–31.5 Ma) was marked by many environmental changes across the globe. Associated with the breakup of the southern continents was the opening of the Southern Ocean, restructuring of ocean circulation patterns, development of the Antarctic ice cap, and consequent changes in global climate patterns (Fordyce, 1992). Major continental glaciation occurred in Antarctica, which resulted in a marked drop in temperature. Concomitant with cooling was the development of bottom water currents, which increased the turnover and circulation of nutrient-rich water and increased abundance of small prey species. Fordyce (1980) proposed that the establishment of high-productivity areas in the southern hemisphere during the early Oligocene probably triggered odontocete and mysticete radiation. He further suggested that perhaps early odontocetes and mysticetes were restricted to the southern hemisphere because northern trophic resources were not affected enough by upwelling to have allowed exploitation by whales. During the middle Oligocene (31.5–28 Ma), the establishment of the Circum-Antarctic Current (also known as the ACC) and the effect of glaciations in Antarctica, which induced sea cooling, increased nutrient availability and hence productivity in shallow seas of the southwest Pacific. Fordyce (1980) suggested that the diversification of filter feeding mysticetes during the middle Oligocene is related to this increase in primary productivity. Because odontocetes consume prey (e.g., small fish and squid), which concentrate in areas of zooplankton abundance, the increase in odontocete diversity during this time likely also reflects these major water mass features. The persistence of archaic whales in the late Oligocene suggests an environment that supported several types of feeding adaptations, including the baleen filter feeding mechanism of mysticetes, the echolocation assisted feeding of odontocetes, and the less specialized feeding apparatus of toothed archaeocetes (Fordyce, 1980).

6.3. Present Patterns of Distribution

Two major patterns of modern marine mammal distribution can be identified: (1) **cosmopolitan** and (2) **disjunct.** Many species have wide, or cosmopolitan, distributions, inhabiting most of the world's oceans. Examples of cetaceans having cosmopolitan distributions are the common dolphin *(Delphinus delphis),* the false killer whale *(Pseudorca crassidens),* and Risso's dolphin *(Grampus griseus).* In addition, several widely distributed pinnipeds, such as the harbor seal *(Phoca vitulina),* may live in a wide range of environments including coastal areas, bays, and estuaries, as well as freshwater lakes.

Other marine mammals can be more or less widespread but are limited to a particular area and have either **endemic** or **circumpolar** distributions. Examples of marine mammals with occurrence restricted to cool-temperate Antarctic or Arctic waters (i.e., circumpolar distributions) include the narwhal *(Monodon monoceros)* and the beluga *(Delphinapterus leucas),* distributed in Arctic and subarctic waters. During the late Pliocene (4 Ma) warmer temperatures prevailed and monodontids occurred farther south in temperate to subtropical waters (e.g., Baja California). Occurrences naturally restricted to a particular area are also referred to as endemic distributions. The restricted distributions of "river dolphins" to river drainage systems in South America (Amazon River dolphin, *Inia geoffrensis*), China (Baiji, *Lipotes vexillifer*), and India and Pakistan (Indus and Ganges River susus *Platanista gangetica minor* and *P. gangetica gangetica*) are examples. Among pinnipeds, the Caspian seal *(Phoca caspica)* is endemic to the Caspian Sea and the Baikal seal *(Phoca siberica)* is restricted to Lake Baikal. Saimaa and Ladoga seals, both subspecies of the Arctic ringed seal *(Phoca hispida)*, are restricted to two additional inland bodies of water.

Other marine mammals occur in multiple regions that are separated from each other by a geographic barrier. These are **antitropical** or disjunct distributions. Antitropical distributions specifically involve different populations of the same species or sister species separated by the equator (Figure 6.5). For example, in the case of the porpoise genus *Phocoena* (see Figure 4.34), one member of a species pair, the vaquita *(P. sinus),* occupies the temperate/subtropical region of northern hemisphere and the other member Burmeister's porpoise *(P. spinipinnis)* occupies a similar southern hemisphere habitat. Another example is the northern right whale dolphin *(Lissodelphis borealis),* which occurs in the North Pacific, and the southern right whale dolphin *(Lissodelphis peronii),* which lives in the southern hemisphere. The ziphiid genus *Berardius* has a similar distribution pattern with Baird's beaked whale *(B. bairdii)* in the temperate North Pacific and Arnonx's beaked whale *(B. arnuxii)* in temperate and polar waters of the Southern Ocean. Among pinnipeds, antitropical distributions are exemplified by the northern and southern elephants seals *(Mirounga* spp.).

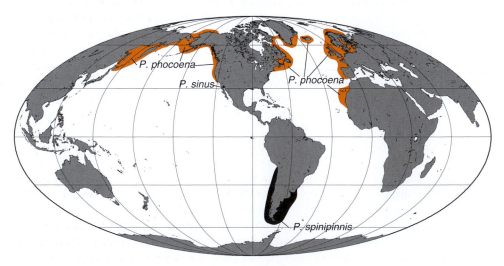

Figure 6.5.   Patterns of antitropical (black) and disjunct (color) distribution of three species of porpoises in the genus *Phocoena*. (Adapted from Gaskin, 1982.)

Antitropical distributions likely arose allopatrically when populations became isolated on either side of the tropics. Antitropical distribution patterns of marine organisms have been correlated with various geologic events. Jacobs *et al.* (2004) proposed that upwelling in the eastern North Pacific produced nutrient-rich waters that in turn led to the divergence of many marine lineages during the middle Miocene. Subsequent closure of the Panamanian seaway in the Pliocene altered oceanic temperature and circulation patterns and resulted in northern hemisphere species dispersing into the southern hemisphere (Lindberg, 1991). A cooling period in the Pleistocene allowed many species to cross the equatorial barrier and disperse into northern and southern hemispheres. This model has been used to explain phocoenid biogeography (Fajardo, personal communication). The fossil record and phylogeny of phocoenids suggests that this group originated in the eastern North Pacific in the Miocene. Closure of the Panamanian seaway resulted in dispersal of phocoenids into the southern hemisphere in the Pliocene and speciation of the southern species *Phocoena dioptrica* and *Phocoena spinipinnis*. Pleistocene cooling resulted in dispersal in the other direction, into the northern hemisphere (e.g., *Phocoena sinus*).

Examples of disjunct distributions include Pacific and Atlantic populations of the harbor porpoise *(Phocoena phocoena;* see Figure 6.5), subspecies of the walrus *(Odobenus rosmarus)*, fin whale *(Balaenoptera physalus)*, humpback whale *(Megaptera novaeangliae)*, western and eastern Pacific populations of the gray whale *(Eschrichtius robustus)*, and northern and southern populations of the sea otter *(Enhydra lutris)*.

6.4. Reconstructing Biogeographic Patterns

The fossil record coupled with reconstructions of the evolutionary relationships among marine mammal species provides evidence that can be used to explain these patterns. For example, the presence of beluga whales in fossil deposits of North Carolina (Whitmore, 1994) as well as Baja California (Barnes, 1984) indicates that this species historically has had a disjunct (Atlantic and Pacific) rather than a circumpolar distribution.

Traditionally, biogeographers, in attempting to interpret the distributions of organisms, have sought to discover a small area (or center) in which species originated and from which they dispersed, known as the **"center of origin/dispersalist explanation."** Rather than starting with an assumption about the center of origin, current workers prefer first to trace the geographic spread and fragmentation of the group through time and do not look for a specific locality for a group's origin. Dispersalist explanations are employed in situations in which organisms are assumed to have evolved elsewhere and dispersed into that area. For example, the lineage that led to the modern walrus, *Odobenus*, is hypothesized to have evolved in the North Atlantic and dispersed into the North Pacific across the Arctic corridor in the late Pleistocene. In this case, the modern pattern of distribution is explained by the merging of two faunal elements (North Pacific and North Atlantic).

An alternative hypothesis, the **vicariance explanation,** argues that organisms occur in an area because they evolved there and were later fragmented (by geographical, behavioral, or other means), with subsequent speciation. Accordingly, the formation of barriers (a process called vicariance) fragmented the ranges of once continuously distributed species. Thus if related species are found in different areas, it is not necessary to assume that there has been dispersal between these areas, only that a barrier (e.g., mountain range, river, or in the case of marine organisms, a land barrier separating two ocean basins or a temperature barrier or strong currents) has appeared between them.

It is likely that most species associations contain both vicariance and dispersalist elements. For example, consider the distribution of species 1–3 in areas A–C (Figure 6.6). This hypothesized biogeographic reconstruction is called an **area cladogram** because of its analogy to a phylogenetic cladogram. A vicariant hypothesis would have the common ancestor of species 1–3 occupying area A+B+C. In the case of marine organisms, the emergence of a land barrier would have first separated the oceans into A, occupied by species 1, and B+C, occupied by species 2–3. Subsequently, areas B+C occupied by the common ancestor of species 2 and 3 would have had another barrier divide their range.

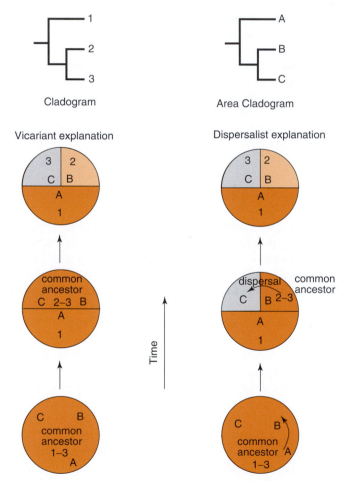

Figure 6.6.   Cladogram showing hypothesized phylogenetic relationships among species and corresponding dispersal and vicariant explanations to account for their distributions. Species are identified by numbers, area by letters. **(a)** In a vicariant explanation, an ancestor occupying A+B+C had a range split first into A and B+C and then the descendant in B+C had its range split. **(b)** In a dispersal explanation, an ancestral species with a center of origin in area A dispersed first to area B and a descendant from there then dispersed to area C. (Modified from Ridley, 1993.)

A dispersalist hypothesis would have the common ancestor of species 1–3 originating in area A. The common ancestor of species 2–3 would disperse to area B, followed by descendant species 3 dispersing into area C. These predictions can be compared with data on the geologic and climatic history of Earth and the distribution of other extinct and extant species that occur in the same area at the same time. If different species tend to speciate when their ranges are fragmented, they should all show similar relationships between phylogeny and biogeography and their area cladograms should match.

A real example of a vicariant explanation comes from a population study of Steller's sea lions (Bickham *et al.*, 1996). Analysis of mtDNA haplotypes revealed the presence of two genetically different populations of Steller's sea lions, a western population that included sea lions from rookeries in Russia, the Aleutian Islands, and the Gulf of Alaska (Beringia), and an eastern population that included sea lions from southeastern Alaska and Oregon (Pacific Northwest). Bickham *et al.* (1996) suggested that these two populations of Steller's sea lion were descended from populations isolated in glacial refugia in Beringia and the Pacific Northwest. In this case, the vicariant event was the onset of glacial periods and fluctuating sea levels, which fragmented the ranges of sea lion populations. They further noted that a similar pattern of population differentiation was observed in sockeye salmon and chinook salmon.

6.5. Past Patterns of Distribution

Past arrangements of continents and ocean basins have affected the distribution of marine mammals. The term **corridor** was proposed for a route that permits the spread of many from one region to another (Simpson, 1936, 1940). Some of the more important seaways that have been invoked as dispersal corridors and their affect on marine mammal distributions are reviewed later (Figures 6.7 to 6.9). In discussing the distributions of marine mammals it also is important to recognize the presence of barriers to dispersal, including physical barriers (continents or other landmasses), climatic barriers (the equator or cold temperatures), or biotic factors (low productivity).

The Tethys Sea (from Tethys of Greek mythology, a sea goddess who was the wife of Oceanus) was an equatorial sea that once divided the northern and southern continents (see Figure 6.8). The main body of the sea occupied an area now called the Mediterranean with a southern arm connected with the Indian Ocean. Restriction of the Tethys seaway occurred (40–45 Ma) when India became sutured to Eurasia. This was important in opening dispersal routes between North and South America from the Atlantic via the Caribbean into the Pacific and around the southern hemisphere via the Southern Ocean. The Paratethys Sea was a northern arm of the Tethys Sea stretching across the area now occupied by the Black, Caspian, and Aral Seas of Asia (see Figure 6.8). A large scale drying of the Mediterranean Sea (Messinian Salinity Crisis) occurred between 5 and 6 Ma, and both it and the Tethys seaway were reduced to series of lakes, including the hyposaline Black, Caspian, and Aral Seas. Marine mammals with Tethyan or Paratethyan distributions include the oldest whales and sirenians (see Figure 6.7) and possibly phocine pinnipeds (see Figure 6.7). The effect of the isolation of Paratethys on phocines has been disputed with some workers suggesting a Paratethyan origin of this lineage followed by dispersal into the Arctic and other workers proposing an Arctic origin followed by dispersal into Paratethys (Deméré *et al.*, 2003).

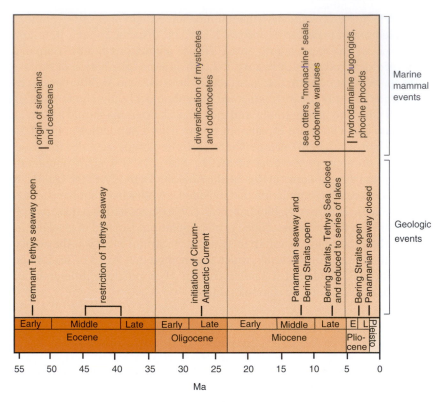

Figure 6.7.   Chronology of major geologic events affecting marine mammal distributions.

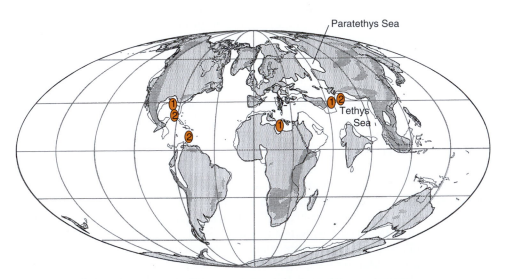

Figure 6.8.   Reconstruction of continents, ocean basins, and paleocoastlines in the early Eocene. (Base map from Smith *et al.*, 1994.) Note extent of Tethys and Paratethys seaways; 1 = early records of cetaceans; 2 = early records of sirenians.

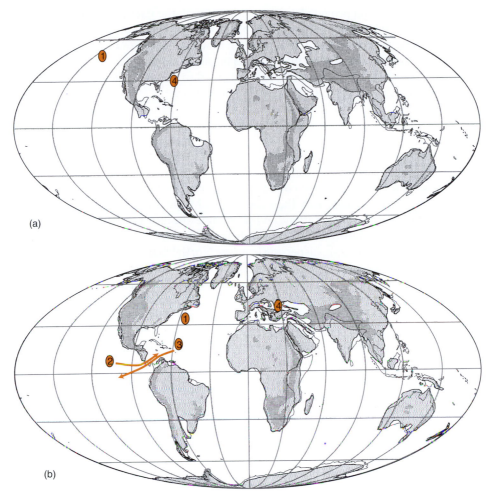

Figure 6.9.   Reconstruction of continents, ocean basins, and paleocoastlines in the **(a)** early Miocene
(20 Ma) (1 = early records of archaic pinnipeds, odobenids and desmatophocids), and **(b)** middle Miocene (12 Ma) (1 = early well documented phocids; 2 = dispersal of "monachines" and
odobenids to Atlantic; 3 = dispersal of phocines to South Pacific; 4 = isolation of phocines in
remnants of Parathethys Sea and in North Atlantic). (Base map from Smith *et al.*, 1994.)

The Central American, or Panamanian, Seaway separating North and South America
allowed faunal interchange between the Pacific and Atlantic oceans during much of the
Cenozoic. Transoceanic circulation was restricted 11 Ma owing to tectonic activity in the
region, then it was reestablished by 6.3 Ma until 5 Ma when it began to close. It was completely closed between 3.7 and 3.1 Ma with emergence of the Isthmus of Panama
(Duque-Caro, 1990; see Figure 6.9b). The Central American Seaway is the route most
likely followed by the lineage that led to the modern sea otter *Enhydra* (Berta and
Morgan, 1985), odobenine walruses, and "monachine" seals (see Figure 6.7).

The Bering Strait, a seaway between Alaska and Siberia, first opened as the result of
plate tectonic activity during the latest Miocene to earliest Pliocene (5.5–4.8 Ma) (but see

Marincovich, 2000) and provided a connection between the Pacific and Arctic Oceans. A land bridge established approximately 5.0 Ma disrupted the seaway. The seaway opened again between 4.0 and 3.0 Ma, at which time the Bering and Chukchi Seas were formed. The Arctic basin, during interglacial periods over the last 3 Ma, was an important dispersal route followed by the modern walrus, the bowhead whale, several phocine seals, hydrodamaline dugongids, and possibly the gray whale.

6.5.1. Pinnipeds

Walruses evolved in the North Pacific during the late early Miocene, approximately 18 Ma (Kohno *et al.*, 1995; Deméré and Berta, 2001; Deméré *et al.*, 2003; see Figure 6.9a). These basal walruses were confined to this region during the middle Miocene ranging as far south as northern Baja California, Mexico. During the late Miocene, odobenids diverged into dusignathine and odobenine lineages. Dusignathines remained endemic to the eastern North Pacific but odobenines (the modern walrus lineage) underwent dramatic diversification, dispersing from the North Pacific into the North Atlantic via the Central American Seaway (between 5 and 8 Ma; Repenning *et al.*, 1979). According to this hypothesis, odobenids became extinct in the Pacific in the Pliocene. Then <1 Ma, the modern genus *Odobenus* returned to the North Pacific through the Arctic Ocean (Repenning and Tedford, 1977; Repenning *et al.*, 1979). Alternatively, on the basis of new late Pliocene, 3-Ma fossil records of the modern walrus lineage from Japan, Kohno *et al.* (1998) have suggested that this lineage may not have become extinct in the North Pacific but may have instead continued to diversify in that region into the Pleistocene, while dispersing to the North Atlantic through the Arctic Ocean during the late Pliocene (Kohno *et al.*, 1995).

Otariids evolved in the eastern North Pacific during the middle Miocene (approximately 11–12 Ma) (Deméré *et al.*, 2003 and references cited therein; see Figure 6.9a). Basal otariids, including the extinct genus *Thalassoleon* and the modern Northern fur seal *Callorhinus*, remained in the North Pacific during the Miocene and Pliocene. The divergence of otariines (sea lions) from arctocephalines (fur seals) occurred by the late Pliocene/early Pleistocene. The timing of this divergence is difficult to determine because the fossil record of fur seals in the South Atlantic is poorly documented. Once in the southern hemisphere, speciation of the *Arctocephalus* clade likely occurred rapidly with most of the currently recognized species diversifying in high latitudes. The historical biogeography of sea lions also suffers from a limited fossil record. Basal sea lions, *Zalophus* and *Eumetopias*, are known from the Pleistocene in the North Pacific. They likely crossed the equator within the past 3 Ma coincident with the fur seal dispersal event. Speciation in the *Neophoca-Phocarctos-Otaria* clade most likely occurred during the Pleistocene in the southern hemisphere via the ACC and was likely synchronous with fur seal dispersal events. In support of this claim, Deméré *et al.* (2003) noted the sympatric distributions of several otariiines and arctocephalines including *Arctocephalus australis* and *Otaria byronia* in the eastern South Pacific and western South Atlantic.

Phocids likely evolved in the North Pacific with divergence of basal phocoids (phocids and extinct desmatophocids) occurring in the late Miocene sometime before 18 Ma (Deméré *et al.*, 2003). Both "monachine" and phocine lineages, were believed to have appeared during the middle Miocene (15 Ma) in the North Atlantic (North America and Europe; see Figure 6.9b). The discovery of a phocid in the South Atlantic in rocks that may be as old as 29 Ma (Koretsky and Sanders, 2002) suggests that the common ancestor of phocids had migrated to the North Atlantic, either northward through the Arctic basin or southward through the Central American Seaway. Costa (1993) suggested such a route (specifically via the southern route) based on the hypothesized sister group relationship of phocids and the extinct desmatophocids. A southern route of dispersal is more likely, as discussed by Costa (1993) and Bininda-Emonds and Russell (1996), given that the Bering land bridge blocked access to the Arctic through much of the late Oligocene and early Miocene. It may also be that the colder climate through the Arctic basin hindered phocid dispersal along the northern route. The Central American Seaway was open during this time and is in agreement with the warm water affinities of North Atlantic phocids at this time.

A North Atlantic origin for monachine seals is supported by the fossil record. The circum-Atlantic distribution (i.e., Paratethys and North Atlantic) of basal monachines suggests a trans-Atlantic dispersal event sometime in the middle Miocene (Deméré *et al.*, 2003). Fyler *et al.* (2005) confirmed an earlier proposal by Muizon (1982) that monk seals *(Monachus)* originated in the Mediterranean *(M. monachus)* with dispersal from east to west, first in the Caribbean *(M. tropicalis)* with subsequent dispersal to the central North Pacific (Hawaiian monk seal *M. schauinslandi*). The latter must have occurred prior to the mid-Pliocene closure of the Central American Seaway. An earlier previous hypothesis had speciation occurring as an interruption of the gene flow of a wide-ranging North Atlantic population (giving rise to a Mediterranean and Caribbean populations) followed by dispersal to the Hawaiian islands (Ray, 1976). Elephant seals *(Mirounga)* and lobodontine seals diversified subsequently in high southern latitudes in the cold waters of Antarctic region.

Phocine seals are also proposed to have a North Atlantic origin with considerable diversity in the Paratethys region (e.g., Koretsky, 2001; Koretsky and Holec, 2002). Modern phocines such as the bearded seals *(Erignathus)* and hooded seals *(Cystophora)* appear to have originated in the Arctic during the Pleistocene with *Erignathus* extending its range into the Bering Sea and North Atlantic and *Cystophora* dispersing into the North Atlantic (Deméré *et al.*, 2003). Ribbon seals *(Histriophoca)* evolved in the North Pacific with its sister taxon harp seals *(Pagophilus)* evolving in the North Atlantic and the *Phoca-Pusa* lineage diverging from *Halichoerus* in the Arctic. All likely represent Pleistocene events, the result of glacial/interglacial oscillations that would have cyclically expanded and fragmented species ranges, reduced gene flow, and promoted speciation.

6.5.2. Cetaceans

Cetacean historical biogeography is reviewed by Fordyce (2001) and Fordyce and Muizon (2001). The earliest known cetaceans, the archaeocetes, are from the early Eocene (50+ Ma). These basal cetaceans radiated rapidly, becoming diverse (see Thewissen 1998; Thewissen and Williams, 2002; Gingerich, 2003) in the warm, subtropical eastern Tethys (now India and Pakistan). Later diverging basal archaeocetes (protocetids) expanded their ranges into mid Tethys (Egypt, Nigeria) and the western Atlantic (southeastern United States). Basilosaurids were more widespread and able to occupy oceans around the Pacific. Closure of the Tethys Sea by collision of India with Asia by about 40–45 Ma concomitant with global climatic cooling and ocean restructuring, apparently resulted in a drop of species-level diversity of archaeocetes (Fordyce, 2002).

The earliest known mysticetes are toothed forms from the southern hemisphere, specifically from the Oligocene (30 Ma) southwest Pacific and proto-Southern Ocean (Antarctica and New Zealand). Diverse small toothed mysticetes (e.g., aetiocetids) are also reported from the eastern and western North Pacific during the late Oligocene. The explosive radiation of baleen-bearing mysticetes and the odontocetes in the Oligocene (marked by the advent of filter feeding in mysticetes) likely resulted from changes in zooplankton productivity of associated with restructuring of the Southern Ocean ecosystem (Fordyce, 1992). The early Miocene was marked by the high diversity of edentulous mysticetes (e.g., "cetotheres") known from the North Pacific, North Atlantic, and Paratethys and last recorded during the middle Pliocene.

The earliest record of modern mysticetes is the right whale (Balaenidae) from the early Miocene in the South Pacific (Patagonia). The later Miocene and Pliocene record of right whales document a modest diversity in the eastern North Atlantic. The origin of the pygmy right whale (Neobalaenidae) is unknown and there is no fossil record of this lineage. The earliest known balaenopterids are from the late Miocene of the North Pacific. Pliocene balaenopterids are diverse and known from the North and South Pacific and eastern North Atlantic Ocean basins. The gray whale lineage is known only as far back as the late Pliocene or Pleistocene.

The most basal odontocetes (i.e., *Agorophius*, *Archaeodelphis*, and *Xenorophus*) are from the Oligocene in the North Atlantic. Among living lineages, sperm whales (physeterids) are known from the late Oligocene in the Paratethys. Pygmy sperm whales (Kogiidae) have a record from the Miocene onward from the South Atlantic and eastern North Pacific. Beaked whales (Ziphiidae) are known beginning in the latest Oligocene-early Miocene in the eastern North Pacific with the majority of records from middle Miocene or younger deposits. Living platanistids are known only from rivers of Pakistan and India but fossil relatives are diverse. Related to platanistids are squalodelphids from the Mediterranean and Paratethys, South Atlantic, and South Pacific, and waipaittids from the South Pacific. The best known platanistoids from the eastern North Atlantic are the shark toothed dolphins (Squalodelphidae). The earliest true dolphins (Delphinoidea) are eurhinodelphinids from the early-middle Miocene in the North Atlantic. The earliest delphinids are from the late Miocene and they are important components of Pliocene assemblages. Porpoises (Phocoenidae) and white whales (Monodontidae) range back to the late Miocene. By the latest Miocene, phocoenids had dispersed to the western North Pacific and eastern South Pacific. Two extinct lineages of delphinoids from the late Miocene include the Albeirodontidae from the eastern North Pacific and two species of tusked "walrus" whales (Odobenocetopsidae) from the South Atlantic.

6.5.3. Sirenians

Much of this summary of sirenian biogeographic history is based on Domning (2001). The earliest known sirenians, the prorastomids, appeared in the early and middle Eocene (50 Ma) in the Caribbean (Jamaica). Before the end of the Eocene, protosirenids occupied warm waters from the western tropical Atlantic through the Tethys to the western Pacific.

The dugongids, the most diverse and successful sirenian lineage, appeared in the Mediterranean during the middle and late Eocene. In the Oligocene, halitheriine dugongids occupied the western North Atlantic with several lineages (e.g.,

Metaxytherium) dispersing into the eastern North Pacific through the Central American Seaway during the early Miocene. There was a major decline in the diversity of halitheriine sirenians in the late Miocene and they disappeared in the late Pliocene. Dugongines diverged from the halitheriines during the Oligocene and, for the most part, remained confined to the Atlantic. By the Pleistocene, a single genus, *Dugong,* had entered the Pacific and remained restricted to tropical latitudes in the Indo-Pacific Basin. Another lineage, hydrodamalines, radiated in the Pacific during the early and middle Miocene. This lineage culminated in Steller's sea cow *(Hydrodamalis gigas),* which occupied cold high water latitudes during the late Pleistocene until it was hunted to extinction by man.

Manatees appear to have diverged from dugongids in the late Eocene-middle Oligocene. By the middle Miocene (14 Ma) archaic manatees *(Potamosiren)* occupied coastal rivers and estuaries in South America. During the late Miocene and Pliocene (4–6 Ma), uplift of the Andes created erosion and runoff of dissolved nutrients into river systems, which greatly increased the abundance of abrasive grasses. Manatees adapted to this new food resource by evolving smaller, more numerous teeth that were continually replaced. In addition, enamel ridges that formed the chewing surfaces of the teeth became more complex. Some of these adaptations to freshwater grasses are seen in the Mio-Pliocene genus *Ribodon* from Argentina and North Carolina, the likely common ancestor for the living genus *Trichechus.* During the late Pliocene (3 Ma) manatees gained access to the Amazon basin, which at the time was temporarily isolated by the Andean uplift. Cut off from marine waters, the Amazonian manatee *(T. inunguis)* evolved. Manatees later dispersed to North America *(T. manatus)* and are known from the Pleistocene of the southeastern United States. There is no fossil record to reveal dispersal of *Trichechus* eastward to West Africa *(T. sengalensis;* Figure 6.10).

6.6. Summary and Conclusions

Understanding the geographic distribution of a species involves knowledge of both their ecology and evolutionary biology. The ecological requirements of a species limit where it can live, but past events can have a dramatic influence on where a species is actually

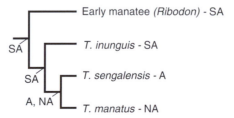

Figure 6.10. Biogeographic history of manatees. A = Africa; NA = North America; SA = South America (based on Domning, 1982). Timetable of Events: SA, 5 Ma—Manatees inhabit rivers and estuaries of SA. Uplift of Andes, increase in silt and grasses, corresponding changes in dentition: "conveyor belt" tooth replacement, increase in number of teeth, decrease in tooth size, increase in tooth (enamel) complexity. SA, 3 Ma—Manatees gain access to Amazon basin, localized speciation of *T. inunguis*. A, NA, 1–2 Ma—Dispersal of manatees: A *(T. sengalensis)* and NA *(T. manatus)*.

found. For marine mammals, ocean temperature patterns and the distribution of primary productivity influence their past and present distributions. Two major patterns of present day marine mammal distribution can be identified: widespread or cosmopolitan distributions and disjunct or antitropical distributions. Two major historical processes have influenced the geographic distribution of species: dispersal (movement of a species into an area) and vicariance (the formation of a barrier that splits the range of a species). An area cladogram shows the phylogenetic relationships between species inhabiting different areas. If geographic distributions are determined mainly by vicariant events, then the area cladograms of a taxon should match the geologic history of an area. Phylogenies have also been used to study the biogeography of host-parasite relationships, such as between tapeworms and phocine seals, and the evolution, biogeography, and feeding ecology of manatees.

The historical biogeography of pinnipeds suggests the following major patterns. The early evolution of walruses took place in the North Pacific approximately 18 Ma. The modern walrus lineage dispersed into the North Atlantic through an Arctic Ocean route. Otariid evolution largely took place in the North Pacific. Fur seals and sea lions crossed into the southern hemisphere by 6 Ma and rapidly diversified during the Pleistocene. The evolutionary history of phocids apparently began in the North Atlantic, although earlier the common ancestor of phocids likely migrated there from the North Pacific following a southern route through the Central American Seaway sometime before 18 Ma. Subsequent climatic cooling resulted in monachine seals retreating southward while the phocines adapted to colder climates in the north. Monachines evolved in the North Atlantic and apparently diversified later in the colder waters of the southern hemisphere to produce the lobodontine seal fauna present in the Antarctic today. The early biogeographic history of phocine seals centered in the Arctic and North Atlantic. Subsequent dispersal of phocines into Paratethys and the Pacific occurred during the Pleistocene; speciation in the Pacific and North Atlantic was affected by glacial events.

Both cetaceans and sirenians had a Tethyan origin. The earliest baleen and toothed whales are from the southern hemisphere. The evolution of filter feeding in mysticetes has been linked to initiation of the Circum-Antarctic Current and associated increases in zooplankton productivity. Recent diversification of cetaceans has been related to sea level changes that promoted isolation and speciation in some cases and extinction in other cases. Dugongids were considerably more diverse in the past but have been reduced to a single genus confined to a tropical distribution in the Indo-Pacific Basin. The lineage that led to Steller's sea cow had a North Pacific distribution and occupied the Bering Sea during the late Pleistocene until it was hunted to extinction by man.

6.7. Further Reading and Resources

A general introductory text on biogeography is Brown and Lomalino (1998) and see Crisci *et al.* (2003) for historical biogeography. Ecological influences on distribution are reviewed in Steele (1999) and Brown *et al.* (1985). The effect of El Niño on pinnipeds is summarized in Trillmich and Ono (1991). For vicariance biogeography see Humphries and Parenti (1999). A classic introduction to pinniped biogeography is Repenning *et al.* (1979); for other reviews of the topic see Muizon (1982), Bininda-Emonds and Russell (1996), Kohno *et al.* (1996), and Deméré *et al.* (2003). Cetacean biogeography is summarized by Gaskin (1976), Fordyce (2001), and Fordyce and Muizon (2001). For sirenian biogeography see Domning (1978, 1982, 2001), and Domning and Furusawa (1995).

An important website for software programs related to historical biogeography is one created by Joe Felsenstein (http://evolution.genetics.washington.edu).

References

Barnes, L. G. (1984). "Fossil Odontocetes from the Almejas Formation, Isla Cedros, Mexico." *PaleoBios 42:* 1–46.

Berta, A., and G. S. Morgan (1985). "A New Sea Otter (Carnivora: Mustelidae) from the Late Miocene and Early Pliocene (Hemphillian) of North America." *J. Paleontol. 59:* 809–819.

Bickham, J. W., J. C. Patton, and T. R. Loughlin (1996). "High Variability for Control-Region Sequences in a Marine Mammal: Implications for Conservation and Biogeography of Steller Sea Lions." *J. Mammal. 77:* 95–108.

Bininda-Emonds, O. R. P., and A. P. Russell (1996). "A Morphological Perspective on the Phylogenetic Relationships of the Extant Phocid Seals (Mammalia: Carnivora: Phocidae)." *Bonner Zool. Monogr. 41:* 1–256.

Boyd, I. L., and T. Arnbom (1991). "Diving Behavior in Relation to Water Temperature in the Southern Elephant Seal: Foraging Implications." *Polar Biol. 11:* 259–266.

Brown, J. H., and M. V. Lomalino (1998). *Biogeography*, 2nd ed. Sinauer Associates, Sunderland, MA.

Brown, O. B., R. H. Evans, J. W. Brown, H. R. Gordon, R. C. Smith, and K. S. Baker (1985). "Phytoplankton Blooming off the U.S. East Coast: A Satellite Description." *Science 229:* 163–167.

Costa, D. (1993). "The Relationship Between Reproductive and Foraging Energetics and the Evolution of the Pinnipedia." *Symp. Zool. Soc. London 66:* 293–314.

Crisci, J. V., L. Katinas, and P. Posadas. (2003). *Historical Biogeography*. Harvard University Press, Cambridge MA.

Deméré, T. A., and A. Berta (2001). "A Re-evaluation of *Proneotherium repenningi* from the Middle Miocene Astoria Formation of Oregon and Its Position as a Basal Odobenid (Pinnipedia: Mammalia)." *Vert. Paleontol. 21*(2): 279–310.

Deméré, T. A., A. Berta, and P. J. Adam (2003). "Pinnipedimorph Evolutionary Biogeography." *Bull. Amer. Mus. Nat. Hist. 279:* 32–76.

Domingo, M., S. Kennedy, and M. F. Van Bressum (2002). Marine mammal mass mortalities. *In* "Marine Mammals Biology and Conservation" (P. G. H. Evans and J. A. Raga, eds.), pp. 425–456. Kluwer Academic/Plenum, New York.

Domning, D. P. (1978). "Sirenian Evolution in the North Pacific Ocean." *Univ. Calif Publ. Geol. Sci. 118.*

Domning, D. P. (1982). "The Evolution of Manatees: A Speculative History." *J. Paleontol. 56:* 599–619.

Domning, D. P. (2001). "The Earliest Known Fully Quadrupedal Sirenian." *Nature 413:* 625–627.

Domning, D. P., and H. Furusawa (1995). "Summary of Taxa and Distribution of Sirenia in the North Pacific Ocean." *Island Arc 3:* 506–512.

Duque-Caro, H. (1990). "Neogene Stratigraphy, Paleoceanography, and Paleobiogeography in Northwestern South America and the Evolution of the Panama Seaway." *Paleogeogr Paleoclimatol. Paleoecol. 77:* 203–234.

Falkowski, P. G. (ed.) (1980). *Primary Productivity in the Sea*. Plenum, New York.

Fordyce, R. E. (1980). "Whale Evolution and Oligocene Southern Ocean Environments." *Palaeogeogr. Palaeoclimatol. Palaeoecol. 31:* 319–336.

Fordyce, R. E. (1992). Cetacean Evolution and Eocene/Oligocene Environments. *In* "Eocene-Oligocene Climatic and Biotic Evolution" (D. Prothero and W. A. Berggren, eds.), pp. 368–381. Princeton University Press, Princeton, NJ.

Fordyce, R. E. (2001). *"Simocetus rayi* (Odontoceti: Simocetidae, New Family): A Bizarre New Archaic Oligocene Dolphin from the Eastern North Pacific." *Smithson. Contrib. Paleobiol. 93:* 185–222.

Fordyce, R. E. (2002). Fossil record. *In* "Encyclopedia of Marine Mammals" (W. F. Perrin, B. Wursig, and J. G. M. Thewissen, eds.), pp. 453–471. Academic Press, San Diego, CA.

Fordyce, R. E., and C. de Muizon. (2001). Evolutionary history of cetaceans: a review. *In* "Secondary Adaptation of Tetrapods to Life in Water" (J.-M. Mazin and V. de Buffrenil, eds.), pp. 169–233. Verlag Dr. Friedrich Pfeil, Munchen, Germany.

Fyler, C. A., T. W. Reeder, A. Berta, G. Antonelis, A. Aguilar, and E. Androukaki (in review). "Molecular Phylogeny of Monachine Seals (Pinnipedia: Phocidae) with Implication for Their Origin and Diversification." *J. Biogeography. 32:* 1267–1279.

Gaskin, D. E. (1976). "The Evolution, Zoogeography and Ecology of Cetacea." *Oceanogr. Mar. Biol. 14:* 247–346.

Gaskin, D. E. (1982). *The Ecology of Whales and Dolphins.* Heinemann, London.

Gingerich, P. D. (2003). "Land-to-Sea Transition in Early Whales: Evolution of Eocene Archaeoceti (Cetacea) in Relation to Skeletal Proportions and Locomotion of Living Semiaquatic Mammals." *Paleobiology 29:* 429–454.

Harwood, L. A., and L. Stirling (1992). "Distribution of Ringed Seals in the Southeastern Beaufort Sea During the Late Summer." *Can. J. Zool. 70:* 891–900.

Humphries, C. J., and L. Parenti (1999). *Cladistic Biogeography.* 2nd edition, Oxford University Press, Oxford, UK.

Jacobs, D. K., T. A. Haney, and K. D. Louie (2004). "Genes, Diversity, and Geologic Process on the Pacific Coast." *Ann. Rev. Earth Planet. Sci. 32:* 601–652.

Kohno, N., L. G. Barnes, and K. Hirota (1995). "Miocene Fossil Pinnipeds of the Genera *Prototaria* and *Neotherium* (Carnivora: Otariidae; Imagotariinae) in the North Pacific Ocean: Evolution, Relationships and Distribution." *Island Arc 3:* 285–308.

Kohno, N., Y. Tomida, Y. Hasegawa, and H. Furusawa (1996). "Pliocene Tusked Odobenids (Mammalia: Carnivora) in the Western North Pacific, and Their Paleobiogeography." *Bull. Natal. Sci. Mus. Ser C: Geol. (Tokyo) 21:* 111–131.

Kohno, N., K. Narita, and H. Koike (1998). "An Early Pliocene Odobenid (Mammalia: Carnivora) from the Joshita Formation, Nagano Prefecture, Central Japan." *Shinshushinmachi Fossil Museum Research Report 1:* 1–7 [in Japanese with English abstract].

Koretsky, I. A. (2001). "Morphology and Systematics of Miocene Phocinae (Mammalia: Carnivora) from Paratethys and the North Atlantic Region." *Geol. Hung. Ser. Palaeontol. 54:* 1–109.

Koretsky, I., and P. Holec (2002). "A Primitive Seal (Mammalia: Phocidae) from the Badenian Stage (Early Middle Miocene) of Centralparatethys." *Smithson. Contr. Paleobiol. 93:* 163–178.

Koretsky, I., and A. E. Sanders (2002). "Paleontology of the Late Oligocene Ashley and Chandler Bridge Formation of South Carolina I: Paleogene Pinniped Remains; The Oldest Known Seal (Carnivora: Phocidea)." *Smithson. Contr. Paleobiol. 93:* 179–183.

Lindberg, D. (1991). "Marine Biotic Interchange Between the Northern and Southern Hemispheres." *Paleobiology 17:* 308–324.

Lipps, J., and E. D. Mitchell (1976). "Trophic Model for the Adaptive Radiations and Extinctions of Pelagic Marine Mammals." *Paleobiology 2:* 147–155.

Loeb, V., V. Siegel, O. Holm-Hansen, R. Hewitt, W. Fraser, W. Trivelpiece, and S. Trivelpiece (1997). "Effects of Sea-Ice Extent and Krill or Salp Dominance on the Antarctic Food Web." *Nature 387:* 897–898.

Loeng, H., K. Brander, E. Carmack, S. Denisenko, K. Drinkwater, B. Hansen, K.M. Kovacs, P. Livingston, F. McLaughlin, and E. Sakshaug (2005). Marine Systems *In* "Arctic Climate Impact Assessment."

Mackintosh, N. A., J. F. G. Wheeler, and A. J. Clowes (1929). "Southern Blue and Fin Whales." *Discovery Reports 1:* 257–540.

Marincovich, L., Jr. (2000). "Central American Paleogeography Controlled Pliocene Arctic Ocean Molluscan Migrations." *Geology 28:* 551–554.

Muizon, C. de (1982). "Phocid Phylogeny and Dispersal." *Ann. S. Afr. Mus. 89:* 175–213.

Orr, W. N., and J. Faulhaber (1975). "A Middle Tertiary Cetacean from Oregon." *Northwest Sci. 49:* 174–181.

Pauly, D., and V. Christensen. (1995). "Primary Production Required to Sustain Global Fisheries." *Nature 374:* 255–257.

Ray, C. E. (1976). "Geography of Phocid Evolution." *Syst. Zool. 25:* 391–406.

Repenning, C. A., and R. H. Tedford (1977). Otarioid seals of the Neogene. Professional Papers of the United States Geological Survey, 992: 93 pp.

Repenning, C. A., C. E. Ray, and D. Gigorescu (1979). Pinniped biogeography. *In* "Historical Biogeography, Plate Tectonics, and the Changing Environment" (J. Gray and A. J. Boucot, eds.), pp. 357–369. Oregon State University Press, Corvallis.

Ridley, M. (1993). *Evolution.* Blackwell Scientific, London.

Riedman, M. (1990). *The Pinnipeds. Seals, Sea Lions, and Walruses.* University of California Press, Berkeley.

Simpson, G. G. (1936). "Data on the Relationships of Local and Continental Mammalian Faunas." *J. Paleontol. 10:* 410–414.

Simpson, G. G. (1940). "Mammals and Land Bridges." *J. Wash. Acad. Sci. 30:* 137–163.

Smith, A. G., D. G. Smith, and B. M. Funnel (1994). *Atlas of Mesozoic and Cenozoic Coastlines*. Cambridge University Press, Cambridge.

Steele, J. H. (1974). *The Structure of Marine Ecosystems*. Harvard University Press, Cambridge, MA.

Stirling, I., H. Cleator, and T. G. Smith (1977). Marine mammals. *In* "Polynas in the Canadian Arctic. Canadian Wildlife Service Occasional Paper No. 45, pp. 45–58.

Thewissen, J. G. M. (ed.) (1998). *The Emergence of Whales: Evolutionary Patterns in the Origin of Cetacea*. Plenum Press, New York.

Thewissen, J. G. M., and E. M. Williams (2002). "The Early Radiations of Cetacea (Mammalia): Evolutionary Pattern and Developmental Correlations." *Annu. Rev. Ecol. Syst. 33:* 73–90.

Trillmich, F., and K. Ono (eds.) (1991). *Pinnipeds and El Nino*. Ecol. Stud. No. 88. Springer-Verlag, Berlin.

Trillmich, F., K. A. Ono, D. P. Costa, R. L. DeLong, S. D. Feldkamp, J. M. Francis, R. L. Gentry, C. B. Heath, B. J. LeBoeuf, P. Majluf, and A. E. York (1991). The effects of El Niño on pinniped populations in the eastern Pacific. *In* "Pinnipeds and El Niño" (F. Trillmich and K. A. Ono, eds.), pp. 247–288. Springer-Verlag, New York.

Tynan, C. T. (1998). "Ecological Importance of the Southern Boundary of the Antarctic Circumpolar Current." *Nature 392:* 708–710.

Tynan, C. T., and D. P. DeMaster (1997). "Observations and Predictions of Arctic Climate Change: Potential Effects on Marine Mammals." *Arctic 50:* 308–322.

Whitmore, F C., Jr. (1994). "Neogene Climatic Change and the Emergence of the Modern Whale Fauna of the North Atlantic Ocean." *Proc. San Diego Soc. Nat. Hist. 29:* 223–227.

Wursig, B. R. R. Reeves, and J. G. Ortega-Ortiz (2002). Global climate change and marine mammals. *In* "Marine Mammals Biology and Conservation" (P. G. H. Evans and J. A. Raga, eds.), pp. 589–608. Kluwer Academic/Plenum, New York.

7

Integumentary and Sensory Systems

7.1. Introduction

Before exploring the behavioral and ecological adaptations of marine mammals to their environment, it is necessary to discuss how the internal workings of these animals make it possible to live in the diverse ways exhibited by these groups. This chapter summarizes the functional anatomy and physiology of the integumentary and sensory systems of marine mammals. The review of the integumentary system includes a description of components of the system (i.e., skin, hair, glands, vibrissae, and claws). Discussion of the nervous system highlights how relative brain size is measured (and what it means) and how enlargements of different segments of the spinal cord are correlated with differing patterns of locomotion. Similarities and differences among marine mammals with respect to development of the visual, olfactory, and taste sensory systems are also described in the chapter. Sound production and reception are discussed in Chapter 11.

7.2. Integumentary System

7.2.1. Skin

As in all mammals, the skin of marine mammals consists of an outer layer (the epidermis), a middle layer (the dermis), and a deep layer (the hypodermis), which forms the blubber when present. The dermis contains hair follicles, sebaceous and sweat glands, and claws (in pinnipeds, the sea otter, and the polar bear). The skin of cetaceans is distinguished by the absence of glands and hair, except for bristle-like hairs that occur around the mouth. Sirenians also lack skin glands and have only sparse hair scattered over the dorsal surface.

   The structure of the epidermis in marine mammals is multilayered stratified squamous epithelium consisting of at least three of the five layers typical of mammals:

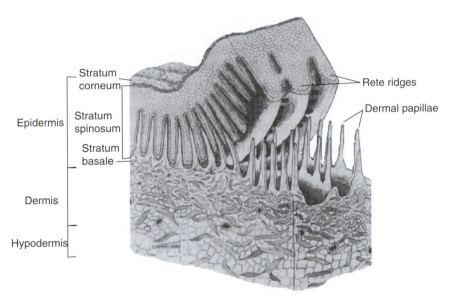

Epidermis
- Stratum corneum
- Stratum spinosum
- Stratum basale

Rete ridges

Dermal papillae

Dermis

Hypodermis

Figure 7.1. Skin of the bottlenose dolphin, *Tursiops truncatus,* illustrating the three layers of the epidermis (stratum basale, stratum spinosum, stratum corneum), the rete ridges between the epidermis and dermis, and the hypodermis. (From Geraci *et al.,* 1986.)

stratum basale, stratum spinosum, and stratum corneum (Figure 7.1). The outermost layer of the epidermis, the stratum corneum, consists of a layer of flattened, solid, keratinized cells. In pinnipeds, these keratinized cells are lubricated by lipids from sebaceous glands, which form a pliable waterproof layer. The thickness of the epidermis varies among pinnipeds; it is thickest in the walrus and thinnest in fur seals. The epidermis of phocids is heavily pigmented as discussed later. The epidermis of the walrus is characterized by the presence of transverse projections on the underside, similar to those seen in sirenians.

Cetacean skin is smooth with a rubbery feel. **Cutaneous ridges** have been described on the surface of the skin in many cetaceans. Although the function of these structures is unknown, it has been suggested that they may play a role in tactile sensing or in the hydrodynamic characteristics of the animal or both (Shoemaker and Ridgway, 1991). Another distinguishing feature of the cetacean integument is the development of **rete ridges** on the underside of the epidermis that are oriented parallel to the body axis (Parry, 1949). These ridges form slender flap-like projections between which dermal papillae are located (Figure 7.1). The epidermis of odontocetes is 10–20 times thicker than in terrestrial mammals (Geraci *et al.,* 1986). Thickness of the epidermis varies over the surface of the animal and changes with age (Harrison and Thurley, 1974). In a study of skin growth in the bottlenose dolphin, the average turnover rate of germinal cells was 70 days (Geraci *et al.,* 1986). However, the sloughing rate is high due to the large number of cells in the stratum basale. Combining this information with their growth studies, Geraci *et al.* (1986) calculated that the outermost layer of epidermis is renewed every 2 hours or 12 times/day, which is nine times faster than the sloughing rate in humans (Bergstresser and Taylor, 1977).

An accelerated turnover of superficial layers of the epidermis and an increase in production of germinal cells (stratum basale) in beluga whales provided evidence of a seasonal epidermal molt (St. Aubin *et al.*, 1990). This annual cycle of epidermal growth in
belugas is not seen in other whales. Belugas move into warm water estuaries during the
molt in some parts of their range; it is thought that they either save energy or molt more
rapidly than if they stayed in colder open waters (St. Aubin *et al.*, 1990; Boily, 1995).

The dermis of marine mammals is composed of dense irregular connective tissue
(Figure 7.1). It is well vascularized and contains an increasing number of fat cells with
depth as it becomes continuous with the hypodermis. Hair follicles are located in the dermis of pinnipeds, the sea otter, and the polar bear. The unusual thickness, strength, and
durability of walrus skin are functions mainly of the reticular layer of the dermis, in which
thick bundles of collagen fibrils form a particularly dense network (Sokolov, 1982). In the
walrus the skin attains its greatest thickness on the neck and shoulders of adult males,
where it is supplanted by rounded "bosses" or lumps about 1 cm thicker than the surrounding skin. These bosses are more pronounced in males than in females (Fay, 1982).

The most notable features of the cetacean dermis are the absence of hair follicles and
sebaceous and sweat glands, its thinness relative to the epidermis, and the elaboration of
finger-like projections (dermal papillae; Figure 7.1) that extend into epidermal ridges
(Ling, 1974). In sirenians, a very thick dermis is formed by dermal papillae and a sub-
papillary layer. Dermal papillae often penetrate the epidermis and may even extend into
the horny layer, so that they lie only a few cell rows below the surface. They may therefore
have a tactile function (Dosch, 1915).

The hypodermis, or blubber, is loose connective tissue composed of fat cells interlay-
ered with bundles of collagen (Figure 7.1). It is loosely connected to the underlying muscle layer. In the walrus, phocids, cetaceans, and sirenians, the fatty hypodermis functions
as insulation. Blubber is of variable thickness and lipid content depending on the age and
sex of the animal, as well as individual and seasonal variations. Blubber thickness varies
among pinnipeds and its pattern of distribution optimizes streamlining as well as insulation. In the harp seal, blubber is thickest and most variable dorsally along the posterior
region of the body, becoming gradually thinner through the neck region and around the
flippers (Beck and Smith, 1995). A similar distribution of blubber was reported for ringed
seals earlier (Ryg *et al.*, 1988). The blubber layer in cetaceans is more extensive than in
other marine mammals with some odontocetes reported to have as much as 80–90% of
the integument composed of blubber (Meyer *et al.*, 1995). A study of blubber growth in
the bottlenose dolphin revealed that blubber mass and depth increase proportionally with
body mass and length. Blubber lipid content increased with age of the animal. In terms of
lipid accumulation and depletion (e.g., during nutritional stress) the middle and the
deeper layers of blubber appear metabolically more active and important energy storage
sites that can be mobilized when necessary (e.g., during pregnancy or lactation), whereas
the superficial blubber layer likely serves a more structural role including streamlining
(Struntz *et al.*, 2004). Dolphins and porpoises inhabiting cold waters possess higher concentrations of an unusual fatty acid (isovaleric acid) in their outer blubber layers than
species from warmer regions suggesting a secondary role for isovaleric acid, the maintenance of blubber flexibility, in addition to its importance in echolocation (Koopman
et al., 2003; see Chapter 11). Mysticetes have the thickest blubber of any marine mammal,
and over some parts of the body of blue whales, it can attain a depth of 50 cm (Slijper,
1979). Blubber is thinner in sirenians and essentially absent in the sea otter (Ling, 1974).
Polar bears do not have "blubber," but they store large amounts of fat (Stirling, 1988).

7.2.1.1. Color

The outer epithelial portion of the skin or pelage of marine mammals may be pigmented. Color patterns in the pelage of pinnipeds occur almost exclusively among phocids (Figure 7.2). **Pagophilic,** or ice breeding, phocids (e.g., ribbon seal, harp seal, Ross seal, bearded seal, hooded seal, ringed seal, crabeater seal, Weddell seal, and leopard seal) show contrasting dark and light or disruptive color patterns. The unique, striking markings of the ribbon seal and harp seal are informative about age and gender because they develop with age and are most strongly expressed in males (Figure 7.3). The cryptic coloration of harbor seals and the more uniform coloration of grey seals may allow them to more readily blend into their coastal habitats.

Coloration patterns are determined by regional differences in the concentration of melanocytes in the epidermis. Numerous melanocytes are present throughout the epidermis of phocids, especially in the basal layers of dark regions. Otariid skin possesses only minor amounts of melanin. Walrus skin is dark in young individuals and becomes less so with age.

The color patterns of cetaceans have been discussed by Yablokov *et al.* (1972) and Mitchell (1970) and reviewed by Perrin (2001). Three basic color patterns are recognized: (1) uniform pattern (e.g., the completely white coloration of the beluga), (2) spotted or striped pattern with sharp colored areas on the head, side, belly, and flukes (e.g., the killer whale), and (3) saddled or counter-shading pattern, in which the animal is dark above and light below (e.g., most dolphins; Figure 7.4). Saddled and striped patterns primarily function in concealment and this is likely also true for spotted patterns.

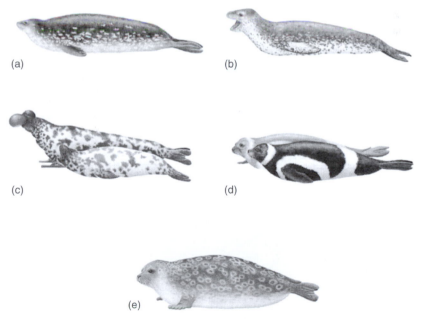

(a)   (b)

(c)   (d)

(e)

Figure 7.2.   Examples of phocid pelage patterns. **(a)** Weddell seal, *Leptonychotes weddellii*. **(b)** Leopard seal, *Hydrurga leptonyx*. **(c)** Hooded seal, *Cystophora cristata*. **(d)** Ribbon seal, *Histriophoca fasciata*. **(e)** Ringed seal, *Phoca hispida*. (Illustrated by P. Folkens from Reeves *et al.*, 1992.)

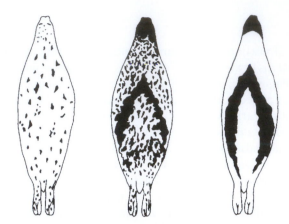

Figure 7.3.   Age variation in pelage markings of harp seals, *Pagophilus groenlandica*. Youngest animal is on the left and oldest animal is on the right. (From Lavigne and Kovacs, 1988.)

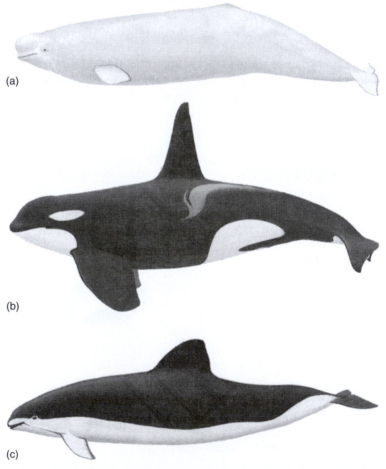

Figure 7.4.   Examples of cetacean pigmentation patterns. **(a)** Beluga, *Delphinapterus leucas*. **(b)** Killer whale, *Orcinus orca*. **(c)** Spectacled porpoise, *Phocoena dioptrica*. (Illustrated by P. Folkens from Leatherwood and Reeves, 1983.)

Spotted and striped patterns also are probably related to the search for food, protection from predators, and intraspecific communication. For example, signaling between individual Hawaiian spinner dolphins and whitesided dolphins is suggested as being accomplished by their body patterns which are a dark gray cape, lighter gray lateral field, and a white belly (Mitchell, 1970; Perrin, 1972). Norris *et al.* (1994) suggested that these patterns allow the animal to determine the degree of rotation of an adjacent neighbor. If rotation of the back is toward a neighbor, the dark cape increases in area at the expense of belly exposure. A larger area of the white belly is exposed if the animal rotates the other way (Figure 7.5). Such pattern relationship cues may be used as part of the overall system that allows synchrony of animals during dives and turns, especially when schooling.

Pigment granules and dendritic (branching) melanocytes of epidermal origin are present in sirenians. Numerous fibers occur both in and between the cells of the stratum spinosum layer and probably impart considerable elasticity to the epidermis. There are no major differences between the skin structure of manatees and the dugong except that the dugong skin may be less pigmented in the adult (Ling, 1974).

7.2.1.2. External Parasites

The distribution of parasites on marine mammals is reviewed in Aznar *et al.* (2002). Diatoms, **whale lice,** and barnacles are often found attached to the skin of the larger cetaceans and occasionally of the smaller ones. Many baleen whales have a thin yellow-green film of diatoms (e.g., *Cocconeis* on the blue whale) over their skin that is acquired during summer foraging in polar waters. Its presence has helped identify the migration patterns of some species (Nemoto, 1956; Nemoto *et al.*, 1980). Cyamid amphipod crustaceans (species of *Cyamus;* see Leung, 1967, for a review), commonly called whale lice (Figure 7.6), are found almost exclusively on the larger cetaceans, especially humpback, gray, and right whales.

There are anatomical and physiological explanations for the occurrence of whale lice on the surface of cetaceans. Whale lice require shelter where they will not be swept off the surface of the whale's skin. Cyamids aggregate in areas of reduced water flow such as the deep longitudinal grooves and prominent ridges covering the throat, chest, the rostrum, the margins of the lips, and on **callosities,** the raised patches of roughened skin found only on the head of right whales. Callosities that surround a single sensory hair may have evolved

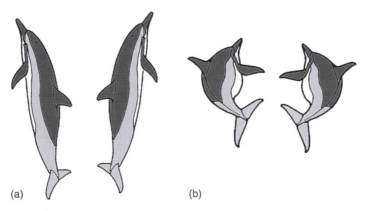

(a)   (b)

Figure 7.5.   Rotating pairs of Hawaiian spinner dolphins illustrating the use of pigment patterns in signaling **(a)** tilt away and **(b)** tilt-toward. (From Norris *et al.*, 1994.)

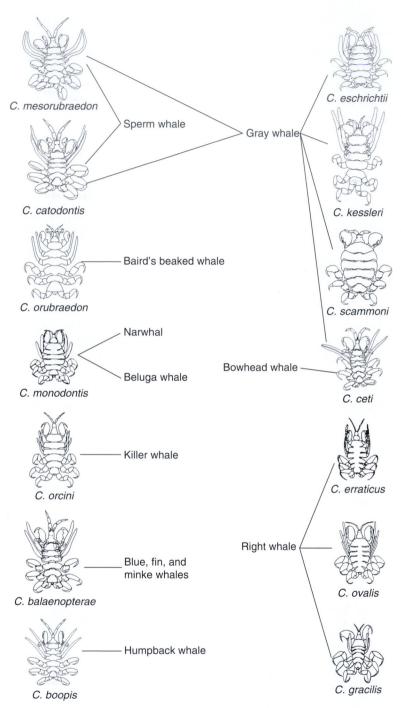

Figure 7.6.   Cetacean hosts of 14 species of cyamid whale "lice" in the genus *Cyamus*. Only males are illustrated. Actual sizes range from 5 mm *(C. balaenopterae)* to 27 mm *(C. scammoni)*. (Redrawn from Margolis *et al.*, 2000.)

to help the whale sense its environment, although their roughened surfaces also provide good attachment sites for cyamids. The cyamids that blanket a callosity could interfere with sensory reception of a whale by covering, breaking, or bending the sensory hairs.

Physiologically, *Cyamus* needs heat; hence these lice inhabit areas that are well vascularized. Observations indicate that cyamids feed on the epidermal layer of whale skin (Rowntree, 1996). In gray whales, cyamids also may actively undercut barnacles and increase their rates of detachment.

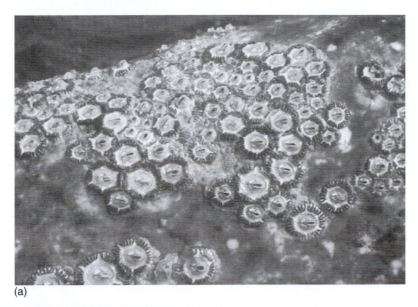

(a)

(b)

Figure 7.7. Acorn barnacles on the skin of **(a)** gray whales and stalked barnacles from **(b)** humpback whales. (Photo courtesy of P. Colla.)

Barnacles of three main types are found on the surfaces of slow-moving whales: mound-shaped acorn barnacles (*Coronula* and *Cryptolepas*), stalked or ship barnacles *(Conchoderma),* and pseudo-stalked barnacles (*Xenobalanus* and *Tubicinella;* Figure 7.7). These external parasites do not seem to cause any infection or inflammation.

The callosities on the head in front of the blowhole of right whales are infested with colonies of barnacles, parasitic worms, and whale lice. The largest patch, located on the snout, was called the "bonnet" by old time whalers and is a feature by which the species is easily recognized. The pattern of callosities is individually unique and can facilitate recognition of individuals (Figure 7.8).

External infestations of pinnipeds are most commonly sucking lice. Sucking lice of the family Echinophthiridae are obligate parasites exclusive to marine carnivores, mainly pinnipeds. Their claws may be modified for clinging to the fur or to the skin where they are found on the eyelids, nostrils, and anus (Kim, 1985). Pups are more prone to infestation than adults and may serve as hosts soon after birth. The evolution of sucking lice and their otariid pinniped hosts is discussed by Kim *et al.* (1975), and the evolution and historical biogeography of tapeworms (*Anophryocephalus* spp) and their pinniped hosts is discussed by Hoberg and Adams (1992) and Hoberg (1995).

Parasites are scarce on sirenians and are dominated by digenean flukes. Manatees are also often covered with algae and coronulid barnacles (Aznar *et al.*, 2002). Sea otters are infected by helminth parasites similar to those found in pinnipeds (Margolis *et al.*, 1997).

7.2.1.3. Hair

The hair or pelage of pinnipeds, sea otters, and the polar bear usually consists of two layers: outer protective **guard hairs** and inner soft **underfur hairs.** The longer, thicker guard hairs lie on top of the shorter, finer underfur hairs (Figure 7.9). Growth of the pelage is influenced by thyroid, adrenal, and gonadal hormones and by nutrition as well as indi-

Figure 7.8.   Bonnet of right whale with prominent callosities. (Photo courtesy of J. Goodyear.)

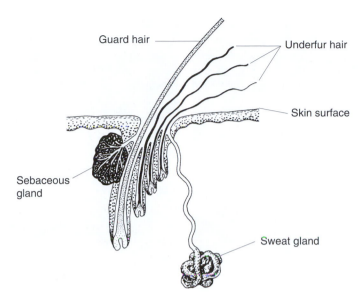

Figure 7.9.   Hair and related structures. (From Bonner, 1994.)

rectly by day length, thermal conditions, and reproduction (Ebling and Hale, 1970; Ling, 1970). Among the pinnipeds, phocids and the walrus lack underfur (Scheffer, 1964; Ling, 1974; Fay, 1982). The guard hairs have a pith, or medulla (central cavity), in otariids that is lacking in the walrus and phocids (Scheffer, 1964). Because medullated hair has been documented among other carnivores, the nonmedullated condition has been interpreted as a phylogenetically informative feature that unites phocids and the walrus (Wyss, 1987). Hair grows in distinct patterns over the body and among pinnipeds these patterns also are phylogenetically informative. In otariids, the hairs are uniformly arranged; in the walrus and in phocids and other carnivores, the hairs are arranged in clusters of two to four or in rows (Figure 7.10). As with sea otters, the hair of pinnipeds lack erector pili muscles, which serve to erect the hair. This may enhance the ability of the hairs to lie flat during submersion, further contributing to streamlining of the body.

   The lanugo or first coat of a pinniped differs from the adult pelage. Monk seals and elephant seals are born with black pelage (King, 1966; Ling and Button, 1975) whereas phocines and the walrus have white or gray lanugo (Ling and Button, 1975; Fay, 1982; King, 1983). Lanugo coats in otariids are dark brown to black (Ling and Button, 1975; King, 1983). Based on these patterns, Wyss (1988) interpreted the light-colored pelage of phocids (exclusive of elephant and monk seals) as a shared derived feature uniting phocines and lobodontines.

   The guard hairs of the sea otter, *Enhydra lutris,* are typically medullated; underfur hairs often are medullated at their base (Kenyon, 1969; Williams *et al.*, 1992; Figure 7.11). The sea otter has the most dense pelage of any mammal, an adaptation against heat loss. The mean hair density in the midback region is about 125,000 hairs/cm^2. This is more than twice that of the northern fur seal (Tarasoff, 1974). The guard hairs are sparse compared to underhairs and probably have little direct insulative value. Their main function may be one of protecting the integrity of the underhairs when the animal is immersed in water. Also, their relatively short shaft and springiness may assist in elevating the underhairs to renew trapped air when the animal emerges from the water

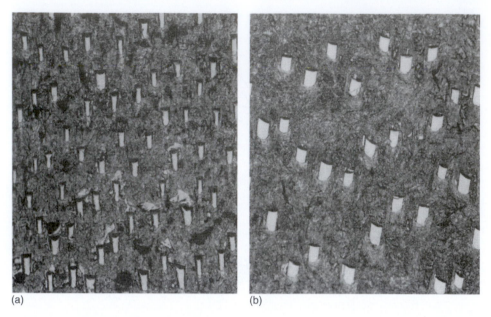

(a)   (b)

Figure 7.10.   Pinniped hair patterns. **(a)** Otariid pattern in which the hairs are uniformly arranged. **(b)** Pattern seen in the walrus and phocids in which the hairs are arranged in groups of two to four or in rows. (From Scheffer, 1964; reprinted courtesy of V. B. Scheffer.)

Figure 7.11.   Guard hairs and underfur of the sea otter, *Enhydra lutris.* (From Williams *et al.*, 1992.)

(Tarasoff, 1974). The wavy underhairs are shorter and of more uniform diameter than the guard hairs. Their waviness contributes to reducing the size of the air spaces in the longitudinal plane. Thus, when the animal is immersed, the underhairs overlap and interlock to help maintain trapped air.

Polar bear hair is medullated. The pelage of polar bears appears white, yellow, gray, or even brownish according to the season and lighting conditions. Immediately after molting, the fur is nearly pure white, matching the arctic winter landscape. In summer, it often takes on a yellowish cast, probably due to oxidation by the sun (Reeves *et al.*, 1992). A green coloration of the pelage that is sometimes seen in captivity is due to algal growth within the hair shaft (Lewin and Robinson, 1979). The claim that polar bear hairs act as optical fibers directing the sun's ultraviolet rays to the skin where it warms the animal has been refuted. Direct measurement of the transmission of light through polar bear hairs reveals a very small optical loss that is of little consequence to the thermal budgets of polar bears (Koon, 1998).

7.2.1.3.1. *Molt*

All phocid seals, the sea otter, and the beluga whale, are known to undergo an annual molt. Terrestrial mammals molt to generate a new pelt as part of an acclimation process to seasonal changes in the thermal environment. Because many phocids are pelagic for most of the year, the short molt period ashore likely serves a different purpose: an annual opportunity to repair and renew their pelt and epidermis (Boily, 1995).

The molt pattern varies among pinnipeds. The lanugo is molted in the uterus of hooded, harbor, and, to a degree, also in bearded seals before the pups are born (Kovacs *et al.*, 1996). McLaren (1966) proposed that prenatal shedding was related to selection of breeding site. According to McLaren, the natal coat of phocids is an adaptation to breeding on ice. The *in utero* molt of the lanugo of harbor seals reflects a secondary adaptation to breeding on land. More recently, it has been argued that prenatal molting, like prenatal blubber deposition, is instead an adaptation enabling newborn pups to enter cold water without adverse consequences (Bowen, 1991). This ability allows the use of pupping substrates that are unstable or regularly inundated with water, as illustrated by the breeding habitats of both hooded and harbor seals (Oftedal *et al.*, 1991). In all other pinnipeds the natal coat is molted at varying periods up to a few months of age. After this molt, the adult pelage is shed and replaced annually.

Molting takes place during the summer and autumn (December to April or May) in the southern hemisphere. In the northern hemisphere the timing of the molt is more variable, usually occurring during the spring (April to June). Molting in pinnipeds begins around the face and is followed by the flippers, abdomen, and finally the back (Stutz, 1967; Ashwell-Erikson *et al.*, 1986). The molt then spreads over the body surface, eventually replacing the entire pelage (Worthy and Lavigne, 1987). In the case of nutritionally stressed pups the molt occurs in a reversed order, with pups retaining the hair on the flippers and face because their lack of blubber means that they must mobilize all possible physiological responses to prevent heat loss (see Lydersen *et al.*, 2000).

The molt of phocids can be relatively fast (e.g., for elephant seals, about 25 days; Boyd *et al.*, 1993) or slow and gradual (e.g., for harbor seals, 26–43 days). However, the duration of the molt varies among different animals (e.g., spotted seals ranged from 7 to 154 days; Ashwell-Erikson *et al.*, 1986). Molting among the ice breeding phocids is shorter in years with ice available. Bearded seals lose significant amounts of hair in all months of the year but have a peak period of intense molting in June.

An unusual pattern of molting characterizes the elephant seals and the Hawaiian monk seal: the shed hairs are attached by their roots to larger sheets of molted epidermis (Figure 7.12; Kenyon and Rice, 1959; Ling and Thomas, 1967; Worthy *et al.*, 1992). In other pinnipeds, hairs are shed individually (Ling, 1970). Fur seals and sea lions do not molt annually but instead renew their pelt gradually all year long. Most of the guard hairs are shed every year, starting about a week after the underfur is shed, so at all times the animal retains a coat (Ling, 1970). Results of a heat flux study of marine mammals indicate that phocids must molt on land or ice to satisfy the thermal requirements of their epidermis. The model predicted that, although phocids should be able to produce enough heat in water to compensate for the heat loss resulting from molting in the water, it would be at a high energetic cost and only possible within a limited range of water temperature (Boily, 1995).

Molting in sea otters takes place gradually throughout the year, although a peak period of molting seems to take place in spring among captive Alaskan otters (Kenyon, 1969). Polar bears have an annual moult in the spring (Stirling, 1988).

7.2.1.4. Skin Glands

In pinnipeds, sebaceous glands are associated with each hair canal and their thick secretions primarily function to keep the epidermis pliable (Figure 7.9). The sweat glands of pinnipeds occur singly with guard hair follicles and their ducts open into the follicle (Figure 7.9). The sweat glands vary from large coiled structures in fur seals (related to their need for a large cooling surface, especially on land; Ling, 1965) to simple coiled tubules in phocids. The

Figure 7.12.   Molting pattern of elephant seals (and the Hawaiian monk seal) in which the hairs and upper layers of the skin are shed in large sheets. (From MacDonald, 1984.)

largest sweat glands of the walrus are located around the mouth and it has been suggested that their odiferous secretions may function in mother and pup recognition (Ling, 1974). Ring seal males secrete a very strong scent in the spring, during the reproductive period, from glands in the facial region ("tiggak" scent; Ryg *et al.*, 1992). The extremely strong smell is instantaneously recognizable in their snow dens, even if the actual seal is not present. The scent might be used to mark territorial boundaries. When in full production, the glandular tissue is swollen to the point that the faces of male ringed seals appear heavily creased with wrinkles and wet with the secretions produced by these facial glands (see Plate 4). Males of other phocid seals, such as harp and grey seals, also have this scent during the reproductive period but not nearly as intensely as in the ringed seal (Hardy *et al.*, 1991).

In the sea otter, sebaceous glands are associated with hair follicles and characteristically have a long, thin, tubular shape, but they often are dilated at their base. Sebaceous secretions in the sea otter are predominantly composed of lipids that differ from other mammals in having squalene as their major constituent (Williams *et al.*, 1992). Removal of lipids from these sebaceous secretions, especially in the sea otter, greatly lessen the water repellency of the fur, reducing its thermoregulatory capability. Apocrine sweat glands are more prominent than the other sebaceous glands (Williams *et al.*, 1992). There are no sebaceous or sweat glands in cetaceans or sirenians (with the exception of rudimentary sebaceous glands associated with snout hairs in sirenians).

7.2.1.5. Vibrissae

Vibrissae, or whiskers, are found to varying extents on the faces of all marine mammals. In pinnipeds, vibrissae are stiff hairs that occur only on the face (Figure 7.13). Differences between the vibrissae of other carnivores and those of pinnipeds include (1) enlargement of whiskers and their sites of innervation among pinnipeds, (2) stiffer hair in pinnipeds, and (3) follicles surrounded by three rather than the two blood sinuses found in terrestrial mammals. Pinnipeds possess three kinds of facial vibrissae: rhinal, superciliary, and mystacial. The most prominent and numerous vibrissae are mystacial whiskers.

Vibrissae appear early in the embryonic development of pinnipeds, long before pelage hairs appear (Ling, 1977). Vibrissae follicles are similar to hair follicles in development and structure but differ in being controlled by voluntary muscles. Pinniped vibrissae are heavily innervated, with blood vessels in the connective tissue sheath surrounding the

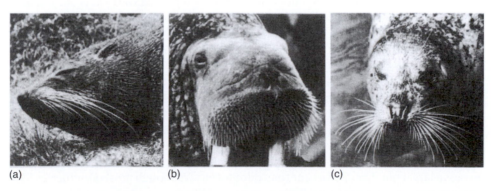

(a)   (b)   (c)

Figure 7.13.   Heads of various pinnipeds showing facial vibrissae. **(a)** New Zealand fur seal, *Arctocephalus forsteri*. **(b)** Walrus, *Odobenus rosmarus*. **(c)** Pacific harbor seal, *Phoca vitulina richardsi*. (From Ling, 1977; reprinted courtesy of J. K. Ling.)

follicle and prominent circular (ring) sinuses around the vibrissae. Walruses possess on average 600–700 vibrissae, more than other pinnipeds and most terrestrial carnivores.

Mystacial whiskers are arranged in rows along the sides of the snout and have either a smooth or a beaded outline. Beaded mystacial whiskers diagnose phocids (excluding monk, bearded, Ross, and leopard seals). The primitive smooth condition is seen in otariids and the walrus. The functional significance of these differences is not known. Mystacial whiskers range in size from the short stiff bristles of the walrus to the very long whiskers of fur seals. Walruses have the toughest mystacial vibrissae. The upper lip, in which the follicles are implanted, is characterized by small compartments with fibrous septa filled with adipose tissue. This gives the upper lip both strong rigidity and flexibility (Kastelein *et al.*, 1993). The walrus uses its mystacial vibrissae to locate prey in the substrate. As noted earlier by Murie (1871), the walrus can move its vibrissae together as a group or individually. Superciliary (supraorbital) whiskers, located above the eyes, are usually better developed in phocids than otariids. Rhinal whiskers, one to two on each side just posterior to each nostril, are found only in phocids (Ling, 1977).

Vibrissae function as tactile receptors and most research has been conducted on phocids. Renouf (1979) and Mills and Renouf (1986) found that vibrissal sensitivity to sound improved with increasing frequency of stimulation. The best sensitivity to vibrations was found at 100 Hz, with a range up to 2500 Hz, the upper limit of the test equipment. The Baltic ringed seal has exceptionally well-developed vibrissae that appear to help them find their way in the dark and often cloudy waters beneath the ice (Hyvärinen, 1989, 1995). A single vibrissae of the Baltic ringed seal contains 10 times the number of nerve fibers typically found in vibrissae of land mammals. Hyvärinen suggested that ringed seals can sense water-borne sound waves with their vibrissae (specifically by nerve fibers located within the ring sinus). Sound waves could be received by the blood sinuses and by tissue conduction through the vibrissae. In addition, the vibrissae may sense changes in swimming speed and direction, which could be important while navigating in darkness. Evidence of the heat conduction of vibrissae of harbor seals suggests that they also play a thermoregulatory role in maintaining high sensitivity in low ambient temperatures (Mauck *et al.*, 2000).

The possible use of vibrissae in prey detection also has been investigated. Cutting the vibrissae of a harbor seal disturbs its success rate in capturing fish (Renouf, 1980). Experimental observations of blindfolded captive ringed seals presented with acoustic cues (tapping on ice with a metal rod) showed that vibrissal sensation helped the seals position themselves within holes in ice-covered ponds but did not help in locating the hole (Wartzok *et al.*, 1992). In another experiment, blindfolded harbor seals were shown to use vibrations detected by their vibrissae to follow fish "trails" in the water (Denhardt *et al.*, 2001). Psychophysical studies of the Pacific walrus suggest that they employ their vibrissae to discriminate shape and size of food objects and the substrate (Kastelein and van Gaalen, 1988; Kastelein *et al.*, 1990). Similar studies have shown that the mystacial vibrissae of the California sea lion (Denhardt, 1994) and the Pacific harbor seal (Denhardt and Kaminski, 1995) allow excellent shape, size, and tactile discrimination and thus might be useful in prey detection.

Vibrissae occur only on the head in cetaceans where they are distributed along the margins of the upper and lower jaws. The structure and especially the innervation of vibrissae indicate a sensory role, and their location suggests a role in feeding. Baleen whales possess a larger number of vibrissae than odontocetes. Although most odontocetes have been reported as having vestigial follicles, work by Mauck *et al.* (2000) provides evidence that in the tuxuci (and perhaps other dolphins) the follicles possess well-developed sinus systems and may function as hydrodynamic receptors similarly to those of harbor seals.

The hair of sirenians is of the follicle sinus type, intermediate in form between pelage hairs and true vibrissae. The hairs are lightly scattered over the body, a novel development in marine mammals, but become denser and very robust on the muzzle and around the mouth. Study of body hairs in the Florida manatee revealed that they have the structural features of tactile hairs and may be analogous to the lateral line system of fish (Reep *et al.*, 2002). Unlike the body hairs, the facial hair surrounding the mouth are prominent bristles that exhibit characteristics of facial vibrissae with a dense connective tissue capsule, prominent blood sinus complex, and substantial innervation (Reep *et al.*, 2001).

In a study of bristle use and feeding in the Florida manatee, Marshall *et al.* (1998) reported that the bristles around the mouth are controlled by various facial muscles and their movements contribute significantly to changes in the shape of the mouth during feeding (see Chapter 12). The sea otter has mystacial, superciliary, and rhinal vibrissae, with mystacial whiskers being the most numerous (Ling, 1977). In the polar bear, vibrissae are few in number, and they are very stiff.

7.2.1.6. Claws/Nails

The fore and hind flippers of phocine seals are characterized by the presence of well-developed claws (Figure 7.14). In contrast, "monachine" seals claws tend to be poorly developed. Wyss (1988) interpreted the primitive condition in phocids to be the presence of reduced claws. The claws of the hand are reduced to small nodules in otariids and the walrus. The fore and hind flippers of otariids have cartilaginous extensions that project from the distal edge of the flipper border. Short cartilaginous extensions also are present on the flippers of the walrus (Figure 7.14).

Cetaceans have lost all trace of claws/nails on their fore flippers. Rudimentary nails are present on the second, third, and fourth digits in manatees, except for the Amazon manatee which lacks nails. Polar bears and sea otters possess claws on both fore and hind digits. The claws are not retractable in the polar bear, and in the sea otter only those on the forefeet are retractile.

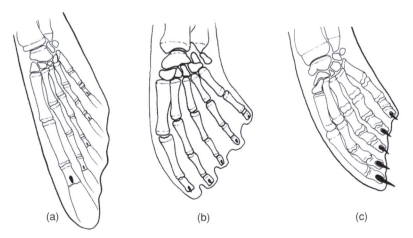

Figure 7.14.   Claws and cartilaginous extensions on the fore flippers of the following pinnipeds: **(a)** sea lion, **(b)** walrus, illustrating the otariid and walrus condition in which there are cartilaginous extensions to the digits and the nails are reduced to small nodules, and **(c)** harbor seal illustrating the phocine condition with well-developed claws. (Illustrated by P Adam; modified from Howell, 1930).

7.3. Nerves and Sense Organs

7.3.1. Nervous System

7.3.1.1. Brain

Descriptions of the anatomy of the pinniped brain are summarized in Harrison and Kooyman (1968), Harrison and Tomlinson (1963), and Flanigan (1972). Representative brain weights range from 320 g for the grey seal, 550 g for the Weddell seal, and 1020 g for the walrus (Table 7.1). The pinniped brain is more spherical than the brain of terrestrial carnivores and is more highly convoluted, especially in phocids. Compared with terrestrial carnivores, the olfactory area in pinnipeds is reduced (more so in phocids than in odobenids or otariids). The auditory area and visual system are well developed (King, 1983).

Comparative descriptions of cetacean brain anatomy are numerous, for example, see Morgane and Jacobs (1972), Morgane *et al.* (1986), and Ridgway (1986, 1990); for reviews see Glezer (2002) and Oelschlager and Oelschlager (2002). Computed tomography (CT) and magnetic resonance imaging (MRI) have recently been employed in studies of the neuroanatomy of cetacean brains (e.g., Marino *et al.*, 2001a, 2000b, 2003a, 2003b). These three-dimensional (3-D) visualization methods allow analysis of internal structures in their precise anatomical positions. A feature of the cetacean brain, similar to the human brain, is the large size of the cerebral hemispheres. The surface of the cetacean brain is highly convoluted (Figure 7.15), and for some odontocetes, it is even more convoluted than the human brain, although the cortex is thinner. In odontocetes there is also a trend toward increased relative size of the auditory processing region (see Chapter 11).

Table 7.1. Brain and Body Weights of Some Marine Mammals as Compared to Humans (Pinniped data from original sources listed in Spector, 1956; Bryden, 1972; Sacher and Staffeldt, 1974; Bryden and Erickson, 1976; Vaz-Ferreira, 1981; cetaceans from Bryden and Corkeron, 1988; sirenians from O'Shea and Reep, 1990)

Species	Brain Weight (gms)	Body Weight (kg)	(Brain Weight/Body Weight) × 100
Pinnipeds			
Otariids			
No. fur seal	355	250 (male)	.142
California sea lion	363	101	.359
Southern sea lion	550	260	.211
Phocids			
Bearded seal	460	281	.163
Gray seal	320	163	.196
Weddell seal	550	400	.138
Leopard seal	542	222	.244
Walrus	1020	600	.170
Cetaceans			
Odontocetes			
Bottlenose dolphin	1600	154	1.038
Common dolphin	840	100	.840
Pilot whale	2670	3178	.074
Killer whale	5620	5448	.103
Sperm whale	7820	33596	.023
Mysticetes			
Fin whale	6930	81720	.008
Sirenians			
Florida manatee	360	756	.047
Human	1500	64	2.344

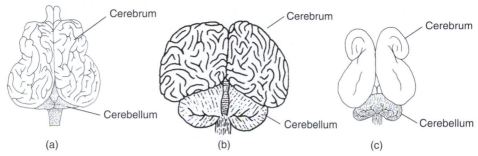

Figure 7.15. Comparison of the following marine mammal brains in dorsal view: **(a)** pinniped, *Otaria byronia,* **(b)** cetacean, *Tursiops truncatus,* and **(c)** sirenian, *Dugong dugon.* (Illustrated by P. Adam.)

The brain of sirenians is relatively small (Table 7.1) with few, shallow sulci (grooves in the cerebrum; Figure 7.15). A notable feature of sirenian brains is the general absence of convolutions on the surface of the cerebral cortex (Reep and O' Shea, 1990). The hemispheres of the manatee brain are divided by a deep, wide, longitudinal fissure and there is a deep Sylvian fissure. Although manatees have a relatively small brain with few surface convolutions, volume estimates of the major brain regions indicate that the manatee cerebral cortex is comparable to those seen among diverse taxa having large relative brain size, including primates (Reep and O'Shea, 1990). The dugong brain is more elongate than that of the manatee and the cerebral hemispheres are even more widely separated by the median longitudinal fissure. A shallow Sylvian fissure is present (see Figure 7.15). A detailed description of the dugong brain can be found in Dexler (1913).

The evolution of brain size is often considered with respect to **encephalization quotient** (EQ), defined by Jerison (1973) as a ratio of observed brain size relative to expected body size. Pinnipeds (Marino, 2002) as well as sea otters (Gittleman, 1986a, 1986b) and polar bears have brain sizes that are not significantly different than those of other terrestrial carnivores (Worthy and Hickie, 1986). Fossil whales apparently did not have large brains. Study of relative brain sizes in archaeocetes reveals that they had smaller brains than modern mammals (Gingerich, 1998) and low EQs (Marino *et al.*, 2000, 2003b). Mysticetes have small brains relative to their large absolute size (Table 7.1). EQs of modern mysticetes are below average (but EQ may not be an appropriate measure in this group owing to their disproportionate increase in body size (Marino, 2002).

Odontocetes, with the exception of the sperm whale, have relatively large brains. They are similar to the relative sizes of those found in anthropoid primates (gorillas, chimpanzees, and orangutans; Worthy and Hickie, 1986; Ridgway, 1986; Marino, 1998). An adult male bottlenose dolphin has an EQ almost 4.5 times larger than an average mammal of the same size. The closest rival in brain size of some delphinids is the gorilla, a similarity surpassed only by humans (Marino, 1998). Notwithstanding the debatable connection between large brain size and "intelligence" (e.g., Lilly, 1964, 1967; Worthy and Hickie, 1986), odontocetes demonstrate complex cognitive abilities (see also Chapter 11). For example, bottlenose dolphins share the extremely rare capacity for mirror self-recognition with great apes and humans (Reiss and Marino, 2001). By comparison, sirenians have relatively small brains, and the EQs for sirenians are among the lowest known for mammals (O'Shea and Reep, 1990).

EQ and related measures of relative brain size are also positively correlated with a number of life history patterns in terrestrial mammals, such as dietary strategy, social group size, and weaning age (Eisenberg, 1981; Marino, 1997 and references cited therein). The low EQ of sirenians is partly explained by large body size but is likely representative of other factors (i.e., prolonged postnatal growth or low metabolic rate;

O'Shea and Reep, 1990). This is further supported by the finding that the internal architecture of the manatee cerebrum is well defined, providing evidence that it is capable of extensive amount of information processing (e.g., control of tactile bristles of the upper lips; Reep *et al.*, 1998; Marshall and Reep, 1995).

Hypertrophy of the auditory region may be the primary reason for the large brain in dolphins. Considerable experimental work has involved the use of evoked potentials to study dolphin hearing and auditory processing (Morgane *et al.*, 1986; see also Chapter 11). These studies illustrate the comparatively well-developed auditory region of the cetacean brain.

Dolphins are also competent performers of complex tasks when the only information provided about the task is auditory. In a series of experiments (summarized in Herman, 1991; Herman *et al.*, 1994), simple artificial computer-generated languages have been created that dolphins have learned to understand. Rule-based sentences of these languages are created from artificial "words" referring to objects in the animal's tank, words referring to actions to take on those objects, and modifier words to express place or relative time. With these words and the rules for linking them, dolphins can, for example, distinguish the auditory instruction "Put the ball in the ring" from the instruction "Take the ring to the ball." Probably no other nonprimate mammal performs as competently on these types of tasks when the information about the task is solely auditory. Similar animal language research experiments conducted by Schusterman and colleagues (Schusterman and Kastak, 2002, and references cited therein) with California sea lions have shown that cognitive skills underlie auditory and visual task performance.

7.3.1.2. Spinal Cord

The spinal cord of pinnipeds is short; it terminates between the eighth and twelfth thoracic vertebrae (King, 1983). A functional relationship between neural canal size and locomotor type has been found among marine carnivores (Giffin, 1992). This relationship was interpreted to be the result of differential innervation and areas of musculature involved in generating movement of the limbs. Among otariids, lumbosacral (lower back to pelvis) enlargement of the neural canal is less than brachial (arm) enlargements. This pattern is correlated to their forelimb-dominated pattern of locomotion. The phocid neural canal is unique among marine carnivores. The lumbosacral enlargement is always higher than the forelimb peak. This is related to their mode of locomotion in which the hind limbs are principally employed. The walrus neural canal displays a hind limb enlargement, a pattern that is very similar to that of phocids. This is reflected in their locomotor pattern in which both the fore- and hind limb are used (see also Chapter 8).

The length of the cetacean spinal cord varies greatly depending on the body size of the species. The body length/spinal cord length ratio in whales is 4:1, approximately the same as that found in humans (Jansen and Jansen, 1969). Among characteristic features of the cetacean spinal cord is its nearly cylindrical form throughout. Associated with development of the flippers is a prominent cervical enlargement of the spinal cord. A lumbar enlargement is much less prominent, due no doubt to the vestigial character of the hind limbs. There are 40–44 pairs of spinal nerves (Flanigan, 1966). Dorsal sensory structures appear relatively small in comparison to ventral motor structures. Although this is not unique to cetaceans, the sensory component (dorsal roots and horns) is more reduced in cetaceans than in other mammals. This disparity has been attributed to a relative lack of peripheral sensory innervation and, especially in the lumbosacral region, the innervation of the powerful tail musculature.

The spinal cord of the manatee ends at the last lumbar or first caudal vertebra (Quiring and Harlan, 1953). In sea otters, the neural canal cross section increases in size posteriorly in the thoracic and lumbar region suggesting an extensive neural supply to the posterior torso and hind limb muscles. Polar bears differ from other marine mammals in showing enlargements of the neural canal at both the forelimb and hind limb levels (Giffin, 1992).

7.3.1.3. Endocrine Glands

The function of hormones produced by endocrine glands in marine mammals principally involve breeding and molting activities, metabolism, and reproduction (reviewed by St. Aubin, 2001a, 2001b). Many marine mammals occupy environments that are subject to seasonal variation in air temperature and light. Daylight, via the pineal gland of the brain, suppresses the production of melatonin. In the southern elephant seal, there is evidence for marked seasonality of the size of the pineal; it is largest with high circulating levels of melatonin in the winter darkness with low hormone levels during the spring breeding season (Griffths and Bryden, 1981). Although the pineal gland has been identified in cetaceans, little more is known. Future research of polar cetaceans (e.g., narwhal and beluga) is needed to determine if their melatonin levels respond similarly to seasonal changes in daylight.

The thyroid gland, located in the neck, and its principal hormone, thyroxin, is involved in the regulation of metabolism. The pinniped thyroid gland is very active within a few hours after birth in harp seals (Davydov and Makarova 1964) and elephant seals (Bryden 1994), suggesting the need for metabolic heat until the pup begins nursing and blubber reserves are established. The levels of thyroid hormones fluctuate throughout the year, especially in phocids, and molting is one event associated with these hormone level changes. Elevated thyroid hormone levels in terrestrial mammals stimulate hair growth and likely have the same effect in phocids. The seasonality of thyroid hormone activity has been reported for the beluga. with the summer increase linked to the epidermal molt (see p. 134). The thyroid gland of sirenians is exceptional in the relatively great abundance and density of its interfollicular connective tissue (Harrison, 1969). Colloid, a protein substance, is profuse, suggesting low thyroid activity. The pachyostotic bone, sluggish behavior, and relatively low rate of oxygen consumption of sirenians all have been linked to the unusual structure of the thyroid gland (Harrison and King, 1965).

The morphology of adrenal gland in marine mammals (located above the kidneys) is similar to other mammals in having two portions: an outer cortex and inner medulla. The adrenal gland, with its principal hormones epinephrine and norepinephrine, is one of the few tissues to which blood supply is maintained during a dive. This may be because epinephrine induces splenic contractions releasing stored red blood cells and enhancing aerobic dive capacity (see Chapter 10). In Weddell seals, epinephrine and norepinephrine increase during dives longer than a few minutes and rapidly return to normal levels after a dive suggesting that catecholamines are important in management of oxygen reserves (St. Aubin, 2001a,b).

7.3.2. Sense Organs

The sensory systems of marine mammals allow them to receive and process information from their surrounding environment. Sensory systems include mechanoreception (touch discussed earlier in this chapter), vision, chemoreception (olfaction and taste), and hearing (see Chapter 11). Excellent general reviews of the anatomy and physiology of sensory systems include Wartzok and Ketten (1999), Supin *et al.* (2001), and Denhardt (2002).

7.3.2.1. Visual System: Morphology of the Eye and Visual Acuity

Vision involves two important components: detection and acuity. The eyes of marine mammals must be able to discriminate objects in air as well as underwater. Most pinniped eyes are large relative to body size. Walruses are an exception and possess relatively small eyes in comparison to other pinnipeds. The exposed part of the eye is protected by a highly keratinized corneal epithelium and a thick sclera. The sclera prevents deformation of the eye as a result of the pressure variation (Pilleri and Wandeler, 1964) and provides protection during swimming (Jamieson and Fisher, 1972).

The observation that the sclera is thinner in the equatorial region of pinniped eyes suggests that it may function to flatten the eye when it is exposed to air, thereby reducing myopia (Pardue *et al.*, 1993). The thicker epithelial layer of the cornea in the bearded seal, in comparison to other phocids, may function in protecting the eye against debris and sand encountered when they feed on sedentary or burrowing animals as is their habit (Pardue *et al.*, 1993). The choroid, which lines much of the internal region of the sclera, is highly vascularized and contains a large cell layer, the **tapetum lucidum,** which makes the eye more light sensitive by acting as a mirror and reflecting the light so that it traverses the photoreceptive cells of the retina twice. The tapetum reaches its greatest development in seals (Jamieson and Fisher, 1972) and it is apparently lacking in the manatee (Walls, 1942).

The lenses of seals are nearly spherical, an adaptation that compensates for the loss of corneal refractive power when the eye is in water (Figure 7.16; Sivak, 1980). The iris is very muscular and well-vascularized and the contracted pupil is a vertical slit, except in the walrus in which it is a broad horizontal oval. In a study of the range of diameter of the pupil and lower limit of pupillary adjustment in northern elephant seals, harbor seals, and California sea lions, the northern elephant seal exhibited the largest range of pupillary area (Levenson and Schusterman, 1997). Furthermore, the elephant seal's visual system was found to function in dimmer conditions and to respond to greater changes in light levels than that of the other pinnipeds studied to date. The Harderian glands, located in the outer corners of the eye, produce an oily mucus that protects the eye and prevents the cornea from drying out. Lacrimal glands, which produce tears, are small and there is no nasolacrimal duct (King, 1983).When out of the water, the combination of the cornea and the bulbous lens probably makes the eyes strongly myopic, with the image focused in front of the retina (Figure 7.16). However, in most pinnipeds this is partly compensated for by a reduction in the size of the pupils. When the light level increases (the usual situation when going from the water into the air), the pupil of many pinnipeds becomes a narrow slit or pinhole. As such, pinnipeds probably have poor depth perception out of the water. The presence of a well-developed ciliary muscle in pinnipeds suggests accommodative movement of the lens that would reduce myopia when the eye is exposed to air (West *et al.*, 1991). The presence of circular muscle bundles indicates that the ciliary muscle is better developed in the elephant seal and California sea lion than in the walrus. Pinnipeds, sea otters, and polar bears, which make frequent transitions to land, require an accommodative mechanism. In contrast, in the com-

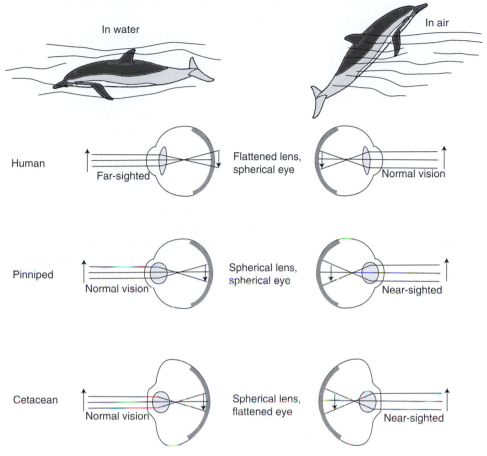

In water · In air

Human — Far-sighted / Flattened lens, spherical eye / Normal vision

Pinniped — Normal vision / Spherical lens, spherical eye / Near-sighted

Cetacean — Normal vision / Spherical lens, flattened eye / Near-sighted

Figure 7.16. Comparison of the eye and lens shape of a human, pinniped, and cetacean and optics in air and water.

pletely aquatic cetaceans and manatees, the ciliary muscle is rudimentary or lacking (West *et al.*, 1991).

The pinniped eye is well adapted for high visual acuity underwater because of the almost circular lens, the increased light sensitivity due to the thick retina, and the well-developed tapetum lucidum. The retina is dominated by rods, which give the animal increased sensitivity in dim light. Much of the interest in psychophysical research on pinniped visual acuity was initiated by Schusterman and his coworkers (e.g., for summaries, see Schusterman 1972, 1981). Visual acuity in the walrus is probably lower than in most other pinnipeds (most of which catch fast-moving prey) because the retina of the walrus eye is smaller than that of other pinnipeds and has relatively few rods per unit of surface area and many rods per optical nerve fiber (Kastelein *et al.*, 1993). Walrus eyes are placed laterally and provide limited binocular vision (Fay, 1985). Whether the optic nerves cross completely or partially in the walrus, as is necessary for stereoscopic vision, has not been investigated.

Demonstration of color discrimination by pinnipeds and cetaceans has been difficult due to differing results from behavioral data on the one hand and cellular, physiological, and genetic evidence on the other (Greibel and Peichl, 2003). Behavioral studies of the spotted seal (Wartzok and McCormick, 1978) and California sea lion (Griebel and Schmid, 1992) indicate some form of color discrimination. There is considerable evidence for the presence of both rods and cones in pinnipeds and cetaceans but their retinas are rod-dominated with sparse populations of cones (1% of photoreceptors). Recent immunohistochemical, physiological, and molecular genetic studies (reviewed in Greibel and Peichl, 2003) show that a number of pinnipeds as well as cetaceans generally lack one of the two cone receptors (S or short wavelength sensitive cones) present in mammals with dichromatic color vision. Although color vision is possible in cone monchromats, it appears to be limited to certain light conditions. Because loss of S-cone sensitivity would be disadvantageous in ocean waters (that are increasingly blue shifted with depth) it has been proposed that S-cone loss may be an earlier adaptation that evolved among ancestral pinnipeds and cetaceans inhabiting coastal waters (that are red-shifted; Peichl *et al.*, 2001). However, it is also possible that this change may have have been brought about by random mutation and that the loss of cone-based color vision is an inadvertent consequent loss of the S-cones. Although the precise origin of loss of S-cones is uncertain, a phylogenetic study of visual pigment genes (cone opsins) in odontocetes and mysticetes indicates that loss of the S-cone sensitivity occurred before divergence of these lineages (Levenson and Dizon, 2003).

In cetaceans, the eyes are located on the sides of the head but are directed forward. Rather than being spherical as in most mammals, the cetacean eye is markedly flattened anteriorly (Figure 7.16). The eyes of Indus river dolphins have no lens only, a flat membrane (Purves and Pilleri, 1973; Dral and Beumer, 1974). A reflective tapetum lucidum is also present behind the retina. The sclera of whales is very thick and may serve to hold the inner globe of the eye in an elliptical shape (Pilleri and Wandeler, 1964). Although cetaceans have no tear glands, the Harderian glands and glands in the conjunctiva of the eyelids regularly bathe the cornea in a viscous solution, thus protecting it from the effects of seawater (Dawson *et al.*, 1972, 1987). Although cellular evidence for loss of the S-cones includes several representative toothed whales (i.e., beaked, sperm, dolphins), data for baleen whales is limited to the gray whale (Griebel and Peichl, 2003).

Similar to pinnipeds, the retina of some cetaceans (notably the bottlenose dolphin) is duplex, meaning that it is composed of two classes of light sensitive cells. Duplicity is a necessity for color vision, although its presence is not a positive sign of the presence of color discrimination. Although rod-like and cone-like receptor cells are present in cetacean eyes, the inability of dolphins to discriminate colors is supported by several behavioral studies (Madsen and Herman, 1980; Griebel and Schmid, 2002). Large ganglion cells have been reported in dolphins and porpoises. Their specialized distribution in two high density areas has been used to suggest that one of these areas, the temporal, may function for frontal vision, with the rostral area functioning for side vision (Murayama *et al.*, 1995). The possibility of binocular vision has been suggested for some cetaceans based on behavioral observations (Caldwell and Caldwell, 1972). Similar kinds of evidence indicate that dolphins do not have binocular vision, although Ridgway (1990) suggested that bottlenose dolphins have small areas of overlap in their visual field. Behavioral data suitable to explore color vision for baleen whales is lacking.

The eyes of sirenians are small with a well-developed nictitating membrane (Harrison and King, 1965). The shape of the lens is nearly spherical and the sclera appears thicker posteriorly than elsewhere (Griebel and Schmid, 1992). As is true for cetaceans, secretions from glands under the eyelids create a copious flow of viscous tears (Kingdon, 1971). Absence of a well-developed ciliary muscle has been reported for both the dugong (Walls, 1942) and the manatee (West *et al.*, 1991) and is consistent with the poor visibility of their natural environment.

Behavioral observations of manatees in the wild indicate that they can see for considerable distances in clear water but that they may be farsighted inasmuch as they tend to bump into objects. Despite earlier reports of the presence of a tapetum lucidum in sirenians, there appears to be no tapetum lucidum in the manatee eye (Griebel and Peichl, 2003). The presence of binocular vision was suggested by an additional behavioral observation, their tendency to encounter objects head-on (Hartman, 1979). The retina of the West Indian manatee has been shown to have rod-like and cone-like receptors (Cohen *et al.*, 1982). Behavioral experiments support this conclusion and provide evidence that the West Indian manatee possesses dichromatic color vision (Griebel and Schmid, 1996).

The sea otter eye is characterized by extensively developed ciliary musculature associated with a well-developed accomodative mechanism that is able to change the refractive power of the lens (Sivak, 1980). In addition, the well-developed anterior epithelium may be an adaptation that helps the sea otter cope with salinity (Murphy *et al.*, 1990). Sea otters appear to have a well-developed tapetum lucidum (Riedman and Estes, 1990). No color vision studies are available for sea otters.

Visual acuity in the sea otter seems to be good both above and underwater. Murphy *et al.* (1990) found that sea otters are able to focus clearly on targets both underwater and in the air. Gentry and Peterson (1967) examined underwater visual discrimination in the sea otter. Results were compared with those of a similar study of the California sea lion and they suggested that the sea lion was better at making visual size discriminations than the sea otter. However, subsequent research has indicated that the visual size discrimination task is contaminated as a test for acuity by the brightness differences caused by differently sized targets (Schusterman, 1968; Forbes and Smock, 1981).

The eyes of polar bears are not well adapted for vision underwater. The retina contains rods and cones as indicated by separate light and dark adapted sensitivity (Ronald and Lee, 1981).

7.3.2.2. Olfactory System

The olfactory system of pinnipeds is less developed than in terrestrial mammals. Olfactory structures in phocids and the walrus are reduced compared to those of otariids (Harrison and Kooyman, 1968). There is extensive behavioral evidence (e.g., nose to nose nuzzlings) for the importance of olfaction in pinniped social interactions (e.g., Evans and Bastian, 1969; Ross, 1972; Miller, 1991).

Adult toothed whales typically do not possess olfactory bulbs or nerves. But there is evidence for development of the olfactory bulb in the early development of the harbor porpoise (Oelschlager and Buhl, 1985). In baleen whales the olfactory tract is present, but the olfactory bulb is present only in fetuses. The fact that mysticetes have retained some sense of smell is inferred primarily from the presence of ethmoid bones that have a cribriform plate perforated with small holes, allowing the terminal branches of the olfactory nerve to pass through (Slijper, 1936).

Manatees have poorly developed olfactory structures (Mackay-Sim *et al.*, 1985). Behaviorally they are often very selective in their food choice, but whether the selection is through smell, taste, or touch, or a combination of these, is not known. Olfaction is virtually unstudied in the sea otter. They have well-developed turbinates and acute olfactory sensitivity, typical of terrestrial carnivores. The social behavior of sea otters also suggests that scent production and acute olfactory sensitivity are important. Adult males may possibly locate and identify estrous females by olfactory cues (Riedman and Estes, 1990). Olfaction is important for polar bears both for hunting and for social interactions.

7.3.2.3. Taste

Although pinnipeds possess tastebuds (King, 1983) there have been few studies on their taste sensitivity; Friedl *et al.* (1990) demonstrated that the Steller sea lion detected three of the four primary tastes that humans perceive, sour, bitter, and salty, but were insensitive to sweet. Anatomical studies have revealed that cetaceans possess taste buds at the base of the tongue, and common and bottlenose dolphins and harbor porpoises have been shown to be able to distinguish different chemicals, even small concentrations of sour citric acid (Nachtigall, 1986; Kuznetsov, 1990).

In comparison to pinnipeds and cetaceans, taste buds are more numerous in sirenians. In dugongs, taste buds are located in a row of pits along the lateral margins of the tongue. Manatees possess a pair of swellings corresponding to the pits of the dugong tongue. Serous glands open into the pits and swellings and neighboring mucous glands open directly onto the surface of the tongue. The product of the glands may be used to rinse the taste buds, particularly in the case of the dugong, which feeds on sea grasses. It is also possible that some enzymes contained in the fluid convert the polysaccharides of sea grasses and algae into smaller molecules, which in turn stimulate the taste buds. Another characteristic feature of the dugong tongue is the presence of patches of fungiform papillae on the lateral surface of the tongue, which may serve as tactile organs (Yamasaki *et al.*, 1980).

7.4. Summary and Conclusions

The integumentary system of marine mammals functions in protection, thermoregulation, and communication. The skin of cetaceans and sirenians is distinguished from other marine mammals by the absence of glands and pelage hair. The outer epidermal layer of skin in beluga whales is unique among cetaceans in undergoing an annual molt. The inner layer of skin (hypodermis, or blubber) is of variable thickness and lipid content and is subject to age- and sex-related, individual, and seasonal changes. Color patterns of the skin and pelage of marine mammals primarily function in either concealment or communication. Barnacles and whale lice attached to the skin of cetaceans are probably best considered mutualists or commensals rather than parasites. The hair of pinnipeds, the sea otter, and the polar bear consists of longer, thicker guard hairs underlain by shorter, finer underfur (but phocids and walruses lack the latter). Unlike phocids, fur seals and sea lions do not molt annually but instead renew their pelt gradually all year long. Vibrissae function in tactile receptivity and may be particularly important in prey detection. One of the most discussed aspects of the nervous system of marine mammals is the large relative brain size of some odontocete cetaceans and its purported relation to intelligence. Comparisons of the EQ among marine mammals indicate that dolphins have the highest brain size to body size ratio and this is likely related to life history patterns, including foraging strategy and social group size and complexity. The EQ of sirenians is among the lowest known for mammals and has been related to their low metabolic rate and long period of postnatal growth. The differing degree of development of brachial and lumbosacral regions of the spinal cord have been related to the locomotor specializations of various marine mammals. The functioning of various endocrine glands has been related to the sluggish behavior of sirenians (thyroid gland) and the breeding cycle of some seals (pineal gland).

Of the three sensory systems reviewed in this chapter (vision, olfaction, and taste), the best developed and most studied is the visual system. The eye of marine mammals is characterized by a well-developed tapetum lucidum that functions to make the eye more light sensitive, especially under low light conditions, and by Harderian glands that produce an oily mucus to protect the eye. The accommodative mechanism (i.e., the ciliary musculature), which changes the refractive power of the lens, is especially well developed in pinnipeds and the sea otter in comparison to either cetaceans or sirenians. Demonstration of color discrimination in pinnipeds, cetaceans, and sirenians has been difficult, although behavioral experiments and the presence of both rods and cones in the retina have been documented for various species. Olfactory bulbs are small in pinnipeds and completely lacking in toothed whales. This is in contrast to both manatees and the sea otter, which have relatively large olfactory organs and presumably more acute olfactory sensitivity.

7.5. Further Reading

For reviews of the anatomy and physiology of pinnipeds see Ridgway (1972), Harrison (1972–1977), and King (1983). For summaries of cetacean anatomy see Slijper (1979), Norris (1966), and Yablokov *et al.* (1972). Accounts of sirenian anatomy can be found in Ridgway and Harrison (1985), Nishiwaki and Marsh (1985), and Caldwell and Caldwell (1985). The capabilities of various sensory systems of marine mammals have been reviewed (Popper, 1980; Dawson, 1980; Schusterman, 1981; Forbes and Smock, 1981) and results highlighted by Watkins and Wartzok (1985), Thomas and Kastelein (1990), and Kastelein *et al.* (1995).

References

Ashwell-Erikson, S., F. H. Fay, R. Elsner, and D. Wartzok (1986). "Metabolic and Hormonal Correlates of Molting and Regeneration of Pelage in Alaskan Harbour and Spotted Seal (*Phoca vitulina* and *Phoca largha*)." *Can. J. Zool. 64:* 1086–1094.

Aznar, F. J., J. A. Balbuena, M. Fernandez, and J. A. Raga (2002). Living together: The parasites of marine mammals. *In* "Marine Mammals: Biology and Conservation" (P. G. H. Evans and J. A. Raga, eds.), pp. 385–423. Kluwer/Academic/Plenum, New York.

Beck, G. G., and T. G. Smith (1995). "Distribution of Blubber in the Northwest Atlantic Harp Seal, *Phoca groenlandica*." *Can. J. Zool. 73:* 1991–1998.

Bergstresser, P. R., and J. R. Taylor, (1977). "Epidermal Turnover Time—New Examination." *Brit. J. Dermatology* 96(5): 503–509.

Boily, P. (1995). "Theoretical Heat Flux in Water and Habitat Selection of Phocid Seals and Beluga Whales During the Annual Molt." *J. Theor. Biol. 172:* 235–244.

Bonner, N. (1994). *Seals and Sea Lions of the World*. Facts on File, New York.

Bowen, W. D. (1991). Behavioural ecology of pinniped neonates. *In* "Behavior of Pinnipeds" (D. Renouf, ed.), pp. 66–127. Chapman & Hall, London.

Boyd, I., T. Arnbom, and M. Fedak (1993). "Water Flux, Body Composition and Metabolic Rate During Molt in Female Elephant Seals *(Mirounga leonina)*." *Physiol. Zool. 66:* 43–60.

Bryden, M. M. (1972). Growth in pinnipeds. *In* "Functional Anatomy of Marine Mammals" (R. J. Harrison, ed.), Vol. 1, pp. 2–79. Academic Press, London.

Bryden, M. M. (1994). Endocrine changes in newborn southern elephant seals. *In* "Elephant Seals: Population Ecology, Behavior, and Physiology" (B. J. Le Boeuf and R. M. Laws, eds.), pp. 387–397. University of California Press, Berkeley.

Bryden, M. M., and P. Corkeron (1988). Intelligence. *In* "Whales, Dolphins and Porpoises" (R. Harrison and M. M. Bryden, eds.), pp. 160–167. Facts on File, New York.

Bryden, M. M., and A. W. Erickson (1976). "Body Size and Composition of Crabeater Seals *(Lobodon carcinophagus),* with Observations on Tissue and Organ Size in Ross Seals *(Ommatophoca rossi)*." *J. Zool. London 179:* 235–247.

Caldwell, D. K., and M. C. Caldwell (1972). Senses and communication. *In* "Mammals of the Sea: Biology and Medicine" (S. H. Ridgway, ed.), pp. 419–465. Charles C Thomas, Springfield, IL.

Cohen, J. L., G. S. Tucker, and D. K. Odell (1982). "The Photoreceptors of the West Indian Manatee." *J. Morphol. 173:* 197–202.

Davydov, A. F., and A. R. Makarova (1964). "Changes in Heat Regulation and Circulation in Newborn Seals on Transition to Aquatic Form of Life." *Fizol. Zhur SSSR 50:* 894–897.

Dawson, W. W. (1980). The cetacean eye. *In* "Cetacean Behavior" (L. Herman, ed.), pp. 53–100. Wiley-Interscience, New York.

Dawson, W. W., L. A. Birndorf, and J. M. Perez (1972). "Gross Anatomy and Optics of the Dolphin Eye." *Cetology 11:* 1–11.

Dawson, W. W., J. P. Schroeder, and S. N. Sharpe (1987). "Corneal Surface Properties of Two Marine Mammal Species." *Mar. Mamm. Sci. 3:* 186–197.

Denhardt, G. (1994). "Tactile Size Discrimination by a California Sea Lion *(Zalophus californianus)* Using its Mystacial Vibrissae." *J. Comp. Physiol. A 175:* 791–800.

Denhardt, G. (2002). Sensory Systems. *In* "Marine Mammal Biology" (R. Hoelzel, ed.), pp. 116–141. Blackwell, Oxford.

Denhardt, G., and A. Kaminski (1995). "Sensitivity of the Mystical Vibrissae of Harbour Seals *(Phoca vitulina)* for Size Differences of Actively Touched Objects." *J. Exp. Biol. 198:* 2317–2323.

Denhardt, G., B. Mauck, W. Hanke, and H. Bleckmann (2001). "Hydrodynamic Trail-following in Harbor Seals *(Phoca vitulina)*." *Science 293:* 102–104.

Dexler, H. (1913). "Das Him von Halicore dugong Erxl." *Morphol. Jahrb. 45:* 95–195.

Dosch, F. (1915). *Structure and Development of the Integument of the Sirenia* (D. A. Sinclair, transl.), Tech. Trans. No. 1624. NRC, Ottawa, Canada.

Dral, A. D. G., and L. Beumer (1974). "The Anatomy of the Eye of the Ganges River Dolphin *Platanista gangetica* (Roxburgh, 1801)." *Z f. Saugetierk 9:* 143–167.

Ebling, F. J., and P. A. Hale (1970). "The Control of the Mammalian Molt." *Mem. Soc. Endocrinol. 18:* 215–235.

Eisenberg, J. F. (1981). *The Mammalian Radiations*. University Chicago Press, Chicago.

Evans, W. E., and J. Bastian (1969). Marine mammal communication: Social and ecological factors. *In* "The Biology of Marine Mammals" (H. T. Andersen, ed.), pp. 425–476. Academic Press, New York.

Fay, F. (1982). Ecology and biology of the Pacific walrus, *Odobenus rosmarus divergens* Illiger. *North Am. Fauna Ser., U.S. Dep. Int., Fish Wildl. Serv.*, No. 74, Washington, D.C.

Fay, E (1985). *"Odobenus rosmarus." Mamm. Species 238:* 1–7.

Fernand, V. S. V. (1951). "The Histology of the Pituitary and the Adrenal Glands of the Dugong *(Dugong dugong)*, Ceylon." *J. Sci. 8:* 57–62.

Flanigan, N. J. (1966). The anatomy of the spinal cord of the Pacific striped dolphin, *Lagenorhynchus obliquidens*. *In* "Whales, Dolphins and Porpoises" (K. S. Norris, ed.). pp. 207–231. University of California Press, Berkeley, CA

Flanigan, N. J. (1972). The central nervous system. *In* "Mammals of the Sea: Biology and Medicine" (S. Ridgway, ed.), pp. 215–246. Thomas, Springfield, IL.

Forbes, J. L., and C. C. Smock (1981). "Sensory Capabilities of Marine Mammals." *Psychol. Bull. 89:* 288–307.

Gentry, R. L., and R. S. Peterson (1967). "Underwater Vision of the Sea Otter." *Nature 216:* 435–436.

Geraci, J. R., D. J. St. Aubin, and B. D. Hicks (1986). The epidermis of odontocetes: A view from within. *In* "Research on Dolphins" (M. M. Bryden and R. Harrison, eds.). pp. 3–31. Oxford Science Publ., Oxford.

Giffin, E. (1992). "Functional Implications of Neural Canal Anatomy in Recent and Fossil Marine Carnivores." *J. Morphol. 214:* 357–374.

Gingerich, P. D. (1998). Paleobiological perspectives on Mesonychia, Archaeoceti, and the origin of whales. *In* "The Emergence of Whales: Patterns in the Origin of Cetacea" (J. G. M. Thewissen, ed.), pp. 423–449. Plenum Press, New York.

Gittleman, J. L. (1986a). "Carnivore Brain Size, Behavioral Ecology and Phylogeny." *J. Mammal. 67:* 23–36.

Gittleman, J. L. (1986b). "Carnivore Life History Patterns: Allometric, Ecological and Phylogenetic Associations." *Am. Nat. 127:* 744–771.

Glezer, I. I. (2002). Neural morphology. In "Marine Mammal Biology" (A. R. Hoelzel, ed.), pp. 98–115. Blackwell, Oxford.

Griebel, U., and L. Peichl (2003). "Colour Vision in Aquatic Mammals-Facts and Open Questions." *Aquat. Mamm.* 29(1): 18–30.

Griebel, U., and A. Schmid (1992). "Color Vision in the California Sea Lion *(Zalophus californianus)*." *Vision Res. 32:* 477–482.

Griebel, U., and A. Schmid (1996). "Color Vision in the Manatee *(Trichechus manatus)*." *Vision Res. 36:* 2747–2757.

Griebel, U., and A. Schmid (2002). "Spectral Sensitivity and Color Vision in the Bottlenose Dolphin *(Tursiops truncatus)*." *Mar. Freshwater Behav. Physiol. 35:* 129–137.

Griffiths, D. J., and M. M. Bryden (1981). The annual cycle of the pineal gland of the elephant seal *(Mirounga leonina)*. *In* "Pineal Function" (C. Matthews and R. Seamark, eds.), pp. 57–66. Elsevier, Amsterdam.

Hardy, M. H., E. Roff, T. G. Smith, and M. Ryg (1991). "Facial Skin Glands of Ringed and Gray Seals, and Their Possible Function as Odoriferous Organs." *Can. J. Zool. 69:* 189–200.

Harrison, R. J. (1969). Endocrine organs: Hypophysis, thyroid, and adrenal. *In* "The Biology of Marine Mammals" (H. T. Andersen, ed.), pp. 349–390. Academic Press, New York.

Harrison, R. J. (1972-1977). *Functional Anatomy of Marine Mammals,* Vols. 1–3. Academic Press, London.

Harrison, R. J., and J. E. King (1965). *Marine Mammals.* Hutchinson, London.

Harrison, R. J., and G. L. Kooyman (1968). General physiology of the Pinnipedia. *In* "The Behavior and Physiology of the Pinnipeds" (R. J. Harrison, R. C. Hubbard, R. S. Peterson, C. E. Rice, and R.J. Schusterman, eds.), pp. 211–296. Appleton-Century-Crofts, New York.

Harrison, R. J., and Thurley (1974). Structure of the epidermis in *Tursiops, Delphinus, Orcinus* and *Phocoena. In* "Functional Anatomy of Marine Mammals," Vol. 1 (R. J. Harrison, ed.), pp. 45–71. Academic Press, New York.

Harrison, R. J., and J. D. W. Tomlinson (1963). Anatomical and physiological adaptations in diving mammals. *In* "Viewpoints in Biology" *2*: 115–162. Butterworth, London.

Hartman, D. S. (1979). "Ecology and Behavior of the Manatee *(Trichechus manatus)* in Florida." *Am. Soc. Mammal. Spec. Publ. 3:* 1–153.

Herman, L. M. (1991). What the dolphin knows, or might know, in its natural world. *In* "Dolphin Societies," (K. Pryor, and K. S. Norris, eds.) pp. 349–363. University of California Press, Berkeley, CA.

Herman, L. M., A. A. Pack, and A. M. Wood (1994). "Bottlenose Dolphins Can Generalize Rules and Develop Abstract Concepts." *Mar. Mamm. Sci. 10:* 70–80.

Hoberg, E. P. (1995). "Historical Biogeography and Modes of Speciation Across High Latitude Seas of the Holarctic: Concepts for Host-Parasite Co-Evolution Among the Phocini (Phocidae) and Tetrabothriidae (Eucestoda)." *Can. J. Zool. 73:* 45–57.

Hoberg, E. P., and A. M. Adams (1992). "Phylogeny, Historical Biogeography, and Ecology of *Anophryocephalus* spp. (Eucestoda: Tetrabothrii*dae*) Among Pinnipeds of the Holarctic During the Late Tertiary and Pleistocene." *Can. J. Zool. 70:* 703–719.

Howell, A. B. (1930). *Aquatic Mammals.* Charles C. Thomas, Springfield, IL.

Hyvärinen, H. (1989). "Diving in Darkness: Whiskers as Sense Organs of the Ringed Seal *(Phoca hispida saimensis)*." *J. Zool. Soc. London* 218: 663–678.

Hyvärinen, H. (1995). Structure and function of the vibrissae of the ringed seal *(Phoca hispida L.)*. *In* "Sensory Systems of Aquatic Mammals" (R. A. Kastelein, J. A. Thomas, and P. E. Nachtigall, eds.), pp. 429–445. De Spit Publishers, Woerden, The Netherlands.

Jamieson, G. S., and H. D. Fisher (1972). The pinniped eye: A review. *In* "Functional Anatomy of Marine Mammals," Vol. 1 (R. J. Harrison, ed.), pp. xx–xx. Academic Press, London.

Jansen, J., and J. K. S. Jansen (1969). The nervous system of Cetacea. *In* "The Biology of Marine Mammals" (H. T. Andersen, ed.), pp. 175–252. Academic Press, New York.

Jerison, H. J. (1973). *Evolution of the Brain and Intelligence.* Academic Press, New York.

Kastelein, R. A. S. Stevens, and P. Mosterd (1990). "The Tactile Sensitivity of the Mystacial Vibrissae of a Pacific Walrus *(Odobenus rosmarus divergens)*. Part 2: Masking." *Aquat. Mamm. 16:* 78–87.

Kastelein, R. A., J. A. Thomas, and P. E. Nachtigall (eds.) (1995). *Sensory Systems of Aquatic Mammals*. De Spit Publishers, Woerden, The Netherlands.

Kastelein, R. A., and M. A. van Gaalen (1988). "The Sensitivity of the Vibrissae of a Pacific Walrus *(Odobenus rosmarus divergens)*. Part 1." *Aquat. Mamm. 14:* 123–133.

Kastelein, R. A., R. C. V. J. Zweypfenning, H. Spekreijse, J. L. Dubbeldam, and E. W. Born (1993). "The Anatomy of the Walrus Head *(Odobenus rosmarus)*. Part 3: The Eyes and Their Function in Walrus Ecology." *Aquat. Mamm. 19:* 61–92.

Kenyon, K. W. (1969). The sea otter in the eastern Pacific Ocean. *North Am. Fauna Ser., U.S. Dep. Int., Fish Wildl. Serv.* No. 68.

Kenyon, K. W., and D. W. Rice (1959). "Life History of the Hawaiian Monk Seal." *Pac. Discovery 13:* 215–252.

Kim, K. C. (1985). Evolution and host associations of Anoplura. *In* "Coevolution of Parasitic Arthropods and Mammals" (K. C. Kim, ed.), pp. 197–232. Wiley, New York.

Kim, K. C., C. A. Repenning, and G. V. Morejohn. (1975). "Specific Antiquity of the Suckling Lice and Evolution of Otariid Seals." *Rapp. P-v. Reun. Cons. Int. Explor. Mer. 169:* 544–549.

King, J. E. (1966). "Relationships of the Hooded and Elephant Seals (Genera *Cystophora* and *Mirounga*)." *J. Zool. London 148:* 385–398.

King, J. E. (1983). *Seals of the World,* 2nd ed. Cornell University Press, Ithaca, NY.

Kingdon, J. (1971). *East African Mammals, an Atlas of Evolution in Africa,* Vol. 1. Academic Press, London.

Koon, D. W. (1998). "Is Polar Bear Hair Fiber Optic?" *Appl. Optics 37:* 3198–3200.

Koopman, H. N., S. J. Iverson, and A. J. Read (2003). "High Concentrations of Isovaleric Acid in the Fats of Odontocetes: Variation and Patterns of Accumulation in Blubber vs. Stability in the Melon." *J. Comp. Physiol. B 173:* 247–261.

Kovacs, K. M., C. Lydersen, and I. Gjertz (1996). "Birth Site Characteristics and Prenatal Molting in Bearded Seals *(Erignathus barbatus)*." *J. Mammal. 77:* 1085–1091.

Kuznetsov, V. B. (1990). Chemical senses of dolphins: quasiolfaction. *In* "Sensory Abilities of Cetaceans" (J. A. Thomas and R. A. Kastelein, eds.), pp. 481–503. Plenum, New York.

Lavigne, D. M., and K. M. Kovacs. (1988). *Harps and Hoods*. University of Waterloo Press, Ontario, Canada.

Leatherwood, S., and R R. Reeves (1983). *The Sierra Club Handbook of Whales and Dolphins*. Sierra Club Book, San Francisco.

Leung, Y. M. (1967). "An Illustrated Key to the Species of Whale Lice." *Crustaceana 12:* 279–291.

Levenson, D. H., and A. Dizon (2003). "Genetic Evidence for the Ancestral Loss of Short-Wavelength-Sensitive Cone Pigments in Mysticete and Odontocete Cetaceans." *Proc. Royal Soc. London B. 270:* 673–679.

Levenson, D. H., and R. J. Schusterman (1997). "Pupillometry in Seals and Sea Lions: Ecological Implications." *Can. J. Zool. 75:* 2050–2057.

Lewin, R. A., and P. T. Robinson (1979). "The Greening of Polar Bears in Zoos." *Nature 278:* 445–447.

Lilly, J. C. (1964). Animals in aquatic environments: Adaptation of mammals to the ocean. *In* "Handbook of Physiology-Environment," pp. 741–747. American Physiological Society, Washington, D. C.

Lilly, J. C. (1967). *The Mind of the Dolphin*. Doubleday, Garden City, NY.

Ling, J. K. (1965). "Functional Significance of Sweat Glands and Sebaceous Glands in Seals." *Nature 208:* 560–562.

Ling, J. K. (1970). "Pelage and Molting in Wild Mammals with Special Reference to Aquatic Forms." *Q. Rev. Biol. 45:* 16–54.

Ling, J. K. (1974). The integument of marine mammals. *In* "Functional Anatomy of Marine Mammals," Vol. 2 (R. J. Harrison, ed.), pp. 1–44. Academic Press, London.

Ling, J. K. (1977). Vibrissae of marine mammals. *In* "Functional Anatomy of Marine Mammals," Vol. 3 (R. J. Harrison, ed.), pp. 387–415. Academic Press, London.

Ling, J. K., and C. E. Button (1975). "The Skin and Pelage of Gray Seal Pups (*Halichoerus grypus* Fabricus): With a Comparative Study of Foetal and Neonatal Moulting in the Pinnipedia." *Rapp. R.-v. Réun. Cons. Int. Explor. Mer. 169:* 112–132.

Ling, J. K., and C. D. B. Thomas (1967). "The Skin and Hair of the Southern Elephant Seal, *Mirounga leonina (L.)* 2. Prenatal and Early Post-Natal Development and Moulting." *Aust. J. Zool. 15:* 349–365.

Lydersen, C., K. M. Kovacs, and M. O. Hammill (2000). "Reversed Molting Pattern in Starvling Gray *(Halichoerus grypus)* and Harp *(Phoca groenlandica)* Seal Pups." *Mar. Mamm. Sci. 16:* 489–493.

MacDonald, D. (ed.) (1984). *Sea Mammals*. Torstar Books, New York.

MacKay-Sim, A., D. Duvall, and B. M. Graves (1985). "The West Indian Manatee, *Trichechus manatus,* Lacks a Vomeronasal Organ." *Brain Behav. Evol. 27:* 186–194.

Madsen, C. J., and L. M. Herman (1980). Social and ecological correlates of cetacean vision and visual appearance. *In* "Cetacean Behavior" (L. Herman, ed.). Wiley-Interscience, New York.

Margolis, L. J. M. Groff, S. C. Johnson, T. E. McDonald, M. L. Kent, and R. B. Blaylock (1997). "Helminth Parasites of Sea Otters *(Enhydra lutris)* from Prince William Sound, Alaska: Comparisons with Other Populations of Sea Otters and Comments on the Origin of Their Parasites." *J. Helminthol. Soc. Wash.* 64(2): 161–168.

Margolis, L., T. E. McDonald, and E. L. Bousfield. (2000). "The Whale Lice (Amphipoda: Cyamidae) of the Northeastern Pacific Region." *Amphipacifica 2:* 63–119.

Marino, L. (1997). "The Relationship Between Gestation Length, Encephalization, and Body Weight in Odontocetes." *Mar Mamm. Sci. 13:* 133–138.

Marino, L. (1998). "A Comparison of Encephalization Between Odontocete Cetaceans and Anthropoid Primates." *Brain Behav. Evol. 51:* 230–238.

Marino, L. (2002). "Brain Size Evolution. *In* "Encyclopedia of Marine Mammals" (W. F. Perrin, B. Würsig, and J. G. M. Thewissen, eds.), pp. 158–162. Academic Press, San Diego, CA.

Marino, L., T. L. Murphy, L. Gozal, and J. L. Johnson (2001a). "Magnetic Resonance Imaging and Three-Dimensional Reconstructions of the Brain of the Fetal Common Dolphin, *Delphinus delphis.*" *Anat. Embryol.* 203: 393–402.

Marino, L., K. Sudheimer, T. L. Murphy, K. K. Davis, D. A. Pabst, W. McClellan, J. K. Rilling, and J. I. Johnson. (2001b). "Anatomy and Three-Dimensional Reconstructions of the Brain of Bottlenose Dolphin *(Tursiops truncatus)* from Magnetic Resonance Images." *Anat. Rec. 264:* 397–414.

Marino, L., K. Sudheimer, D. Sarko, G. Sirpenski, and J. I. Johnson. (2003a). "Neuroanatomy of the Harbor Porpoises *(Phocoena phocoena)* from Magnetic Resonance Images." *J. Morph. 257:* 308–347.

Marino, L., M. Uhen, B. Froelich, J. M. Aldag, C. Blane, D. Bohaska, and F. Whitmore (2000). "Endocranial Volume of Mid-Late Eocene Archaeocetes (Order Cetacea) Revealed by Computed Tomography: Implications for Cetacean Brain Evolution." *J. Mamm. Evol. 7:* 81–94.

Marino, L., M. D. Uhen, N. D. Peyenson, and B. Froelich (2003b). "Reconstructing Cetacean Brain Evolution Using Computed Tomography." *Anat. Rec. 272b:* 107–117.

Marshall, C. D., L. A. Clark, and R. L. Reep (1998). "The Muscular Hydrostat of the Florida Manatee *(Trichechus manatus latirostris):* Functional Morphological Model of Perioral Bristle Use." *Mar Mamm. Sci. 14:* 290–305.

Marshall, C. D., and R. L. Reep (1995). "Manatee Cerebral Cortex: Cytoarchitecture of the Caudal Region in *Trichechus manatus latirostris.*" *Brain Behav. Evol. 45:* 1–18.

Mauck, B., U. Eysel, and G. Dehnhardt (2000). "Selective Heating of Vibrissal Follicles in Seals *(Phoca vitulina)* and Dolphins *(Sotalia fluviatus guianensis).*" *J. Exp. Biol. 203:* 2125–2131.

McLaren, I. A. (1966). "Taxonomy of Harbor Seals of the Western North Pacific and the Evolution of Certain Other Hair Seals." *J. Mammal. 47:* 466–473.

Miller, E. H. (1991). Communication in pinnipeds, with special reference to non-acoustic signaling. *In* "Behaviour of Pinnipeds" (D. Renouf, ed.), pp. 128–235. Chapman and Hall, London.

Mills, F., and D. Renouf (1986). "Determination of the Vibrational Sensitivity of the Harbour Seal, *Phoca vitulina* Vibrissae." *J. Exp. Mar. Biol. Ecol. 100:* 3–9.

Mitchell, E. D. (1970). "Pigmentation Pattern Evolution in Delphinid Coloration: An Essay in Adaptive Coloration." *Can. J. Zool. 48:* 717–740.

Morgane, P. J., and M. Jacobs (1972). Comparative anatomy of the cetacean nervous system. *In* "Functional Anatomy of Marine Mammals," Vol. X (R. J. Harrison, ed.), pp. 117–244. Academic Press, London.

Morgane, P. J., M. S. Jacobs, and A. Galaburda (1986). Evolutionary morphology of the dolphin brain. *In* "Dolphin Cognition and Behavior: A Comparative Approach" (R. J. Schusterman, J. A. Thomas, and F. G. Wood, eds.), pp. 5–28. Erlbaum, Hillsdale, NJ.

Murayama, T., H. Somiya, I. Aoki, and T. Ishii (1995). "Retinal Ganglion Cell Size and Distribution Predict Visual Capabilities of Dall's Porpoise." *Mar. Mamm. Sci. 11:* 136–149.

Murie, J. (1871). "Researchers Upon the Anatomy of the Pinnipedia. Part I. On the Walrus (*Trichechus rosmarus* Linn.)." *Trans. Zool. Soc. London 7:* 411–464.

Murphy, C. J., R. W. Bellhorn, R. W. Williams, T. Burns, M. S. Schaeffel, and H. C. Howland (1990). "Refractive State, Ocular Anatomy, and Accommodative Range of the Sea Otter." *Vision Res. 30:* 23–32.

Nachtigall, P. E. (1986). Vision, audition, and chemoreception in dolphins and other marine mammals. *In* "Dolphin Cognition and Behavior: A Comparative Approach" (R. J. Schusterman, J. A. Thomas, and F. G. Wood, eds.), pp. 79–113. Erlbaum, Hillsdale, NJ.

Nemoto, T. (1956). "On the Diatoms of the Skin Film of Whales in the Northern Pacific." *Sci. Rep. Whales Res. Inst. Tokyo 11:* 99–132.

Nemoto, T., P. B. Best, K. Ishimaru, and H. Takano (1980). "Diatom Films on Whales in South African Waters." *Sci. Rep. Whales Res. Inst. 32:* 97–103.

Nishiwaki, M., and H. Marsh (1985). *Dugong dugon* (Muller, 1776). *In* "Handbook of Marine Mammals," Vol. 3 (S. H. Ridgway and R. Harrison, eds.), pp. 1–31. Academic Press, New York.

Norris, K. S. (ed.) (1966). *Whales, Dolphins and Porpoises.* University of California Press, Berkeley, CA.

Norris, K. S., R. S. Wells, and C. M. Johnson (1994). The visual domain. *In* "The Hawaiian Spinner Dolphin" (K. S. Norris, B. Wursig, R. S. Wells, and M. Wursig, eds.), pp. 141–160. University of California Press, Berkeley, CA.

Oelschlager, H. A., and E. H. Buhl (1985). "Development and Rudimentation of the Peripheral Olfactory System in the Harbor Porpoise *Phocoena phocoena* (Mammalia: Cetacea)." *J. Morphol. 184:* 351–360.

Oelschlager, H. A., and J. S. Oelschlager (2002). Brain. *In* "Encyclopedia of Marine Mammals" (W. F. Perrin, B. Wursig, and J. G. M. Thewissen, eds.), pp. 133–158. Academic Press, San Diego, CA.

Oftedal, O. T., W. D. Bowen, E. M. Widdowson, and D. J. Boness (1991). "The Prenatal Molt and Its Ecological Significance in Hooded and Harbor Seals." *Can. J. Zool. 69:* 2489–2493.

O'Shea, T. J., and R. L. Reep (1990). "Encephalization Quotients and Life-History Traits in the Sirenia." *J. Mammal. 71:* 534–543.

Pardue, M. T., J. G. Sivak, and K. M. Kovacs (1993). "Corneal Anatomy of Marine Mammals." *Can. J. Zool. 71:* 2282–2290.

Parry, D. A. (1949). "The Structure of Whale Blubber, and a Discussion of Its Thermal Properties." *Q. J. Microsc. Sci. 90:* 13–25.

Peichl, L., G. Behrmann, and R. H. H. Kroger (2001). "For Whales and Seals the Ocean Is Not Blue: A Visual Pigment Loss in Marine Mammals." *Eur. J. Neurosci. 13:* 1520–1528.

Perrin, W. F. (1972). "Color Patterns of Spinner Porpoises (*Stenella* cf. *S. longirostris*) of the Eastern Pacific and Hawaii, with Comments on Delphinid Pigmentation." *Fish. Bull. 70:* 983–1003.

Perrin, W. F. (2001). Coloration. *In* "Encyclopedia of Marine Mammals" (W. F. Perrin, B. Würsig, and J. G. M. Thewissen, eds.), pp. 236–245. Academic Press, San Diego, CA.

Pilleri, G., and A. Wandeler (1964). "Ontogeny and Functional Morphology of the Eye of the Fin Whale (*Balaenoptera physalus*)." *Acta Anat. Suppl. 50:* 179–245.

Popper, A. N. (1980). Sound emission and detection by delphinids. *In* "Cetacean Behavior" (L. M. Herman, ed.), pp. 1–52. Wiley-Interscience, New York.

Purves, P. E., and G. Pilleri (1973). "Observations on the Ear, Nose, Throat and Eye of *(Platanista indi)*." *Invest. Cetacea 5:* 13–57.

Quiring, D. P., and C. F. Harlan (1953). "On the Anatomy of the Manatee." *J. Mammal. 34:* 192–203.

Reep, R. L., C. D. Marshall, and M. L. Stoll (2002). "Tactile Hairs on the Postcranial Body in Florida Manatees: A Mammalian Lateral Line?" *Brain Behav. Evol. 59:* 141–154.

Reep, R. L., C. D. Marshall, M. L. Stoll, and D. M. Whitaker (1998). "Distribution and Innervation of Facial Bristles and Hairs in the Florida Manatee *(Trichechus manatus latirostris)*." *Mar Mamm. Sci. 14:* 257–273.

Reep, R. L., and T. J. O'Shea (1990). "Regional Brain Morphometry and Lissencephaly in the Sirenia." *Brain Behav. Evol. 35:* 185–194.

Reep, R. L., M. L. Stoll, C. D. Marshall, B. L. Homer, and D. A. Samuelson (2001). "Microanatomy of Facial Vibrissae in the Florida Manatee: The Basis for Specialized Sensory Function and Oripulation." *Brain Behav. Evol. 58:* 1–14.

Reeves, R. R., B. S. Stewart, and S. Leatherwood (1992). *The Sierra Club Handbook of Seals and Sirenians.* Sierra Club Books, San Francisco, CA.

Reiss, D., and L. Marino (2001). "Mirror Self-Recognition in the Bottlenose Dolphin: A Case of Cognitive Convergence." *Proc. Natl. Acad. Sci. 98:* 5937–5942.

Renouf, D. (1979). "Preliminary Measurements of the Sensitivity of the Vibrissae of Harbour Seals *(Phoca vitulina)* to Low Frequency Vibrations." *J. Zool. Soc. London 188:* 443–450.

Renouf, D. (1980). "Fishing in Captive Harbour Seals *(Phoca vitulina concolor):* A Possible Role for Vibrissae." *Neth. J. Zool. 30:* 504–509.

Ridgway, S. H. (1972). Homeostasis in the aquatic environment. *In* "Mammals of the Sea: Biology and Medicine" (S. H. Ridgway, ed.), pp. 590–747. Charles C. Thomas, Springfield, IL.

Ridgway, S. H. (1986). Dolphin brain size. *In* "Research on Dolphins" (M. M. Bryden and R. Harrison, eds.), pp. 59–70. Oxford Science Publications, Oxford.

Ridgway, S. H. (1990). The central nervous system of the bottlenose dolphin. *In* "The Bottlenose Dolphin" (S. Leatherwood and R. R. Reeves, eds.), pp. 69–97. Academic Press, San Diego, CA.

Ridgway, S. H., and R. J. Harrison (eds.) (1985). *Handbook of Marine Mammals,* Vol. 3. Academic Press, New York.

Riedman, M. L., and J. A. Estes (1990). "The Sea Otter (*Ehhydra lutris*): Behavior, Ecology, and Natural History. *U.S. Fish Wildl. Serv., Biol Rep. 90* (14).

Ronald, K., and J. Lee (1981). "The Spectral Sensitivity of a Polar Bear." *Comp. Biochem. Physiol. A 70:* 595–598.

Ross, G. J. B. (1972). "Nuzzling Behavior in Captive Cape Fur Seals." *Int. Zool. Yearbook 12:* 183–184.

Rowntree, V. J. (1983). "Cyamids: The Louse that Moored." *Whalewatcher. J. Am. Cetacean Soc. 17:* 14–17.

Rowntree, V. J. (1996). "Feeding, Distribution, and Reproductive Behavior of Cyamids (Crustacea: Amphipoda) Living on Humpback and Right Whales." *Can. J. Zool. 74:* 103–109.

Ryg, M., T. G. Smith, and N. A. Øritsland (1988). "Thermal Significance of the Topographical Distribution of Blubber in Ringed Seals *(Phoca hispida)*." *Can. J. Fish. Aquat. Sci. 45:* 985–992.

Ryg, M., Y. Solberg, C. Lydersen, and T. G. Smith (1992). "The Scent of Rutting Male Ringed Seals *(Phoca hispida)*." *J. Zool. London 226:* 681–689.

Sacher, G. A., and E. F. Staffeldt (1974). "Relation of Gestation Time to Brain Weight for Placental Mammals: Implications for the Theory of Vertebrate Growth." *Am. Naturalist 108:* 593–615.

Scheffer, V. (1964). "Hair Patterns in Seals (Pinnipedia)." *J. Morphol. 115:* 291–304.

Schusterman, R. J. (1968). Experimental laboratory studies of pinniped behavior. *In* "The Behavior and Physiology of Pinnipeds" (R. J. Harrison, R. C. Hubbard, R. S. Peterson, C. E. Rice, and R.J. Schusterman, eds.), pp. 87–171. Appleton-Century-Crofts, New York.

Schusterman, R. J. (1972). Visual acuity in pinnipeds. *In* "Behavior and Marine Animals" (H. E. Winn, and B. L. Olla, eds.), pp. 469–491. Plenum Publishers, New York.

Schusterman, R. J. (1981). "Behavioral Capabilities of Seals and Sea Lions: A Review of Their Hearing, Visual, Learning and Diving Skills." *Psychol. Rec. 31:* 125–143.

Schusterman, R. J., and D. Kastak (2002). Problem solving and memory. *In* "Marine Mammal Biology" (A. Rus Hoelzel, ed.), pp. 371–387. Blackwell Science, Oxford.

Shoemaker, P. A., and S. H. Ridgway (1991). "Cutaneous Ridges in Odontocetes. *Mar. Mamm. Sci. 7:* 66–74.

Sivak, J. G. (1980). "Accommodation in Vertebrates: A Contemporary Survey." *Curr Topics. Eye Res. 3:* 281–330.

Slijper, E. J. (1936). "Die Cetaceen, vergleichend anatomisch and systematisch, Diss. Utrecht." *Capita Zoologica 7.*

Slijper, E. J. (1979). *Whales,* 2nd ed. Hutchinson, University Press, London.

Sokolov, V. E. (1982). *Mammal Skin.* University of California Press, Berkeley, CA.

Spector, W. S. (1956). *Handbook of Biological Data.* Saunders, Philadelphia, PA.

St. Aubin, D. J. (2001a). Endocrinology. *In* "CRC Handbook of Marine Mammal Medicine," 2nd ed. (L. A. Dierauf, and F. M. D. Gulland, eds.), pp. 165–192. CRC Press, Boca Raton, FL.

St. Aubin, D. J. (2001b). Endocrine systems. "Encyclopedia of Marine Mammals" (W. F. Perrin, B. Würsig, and J. G. M. Thewissen. eds.), pp. 382–387. Academic Press, San Diego, CA.

St. Aubin, D. J., T. G. Smith, and J. R. Geraci (1990). "Seasonal Epidermal Molt in Beluga Whales, *Delphinapterus leucas*." *Can. J. Zool. 68:* 359–367.

Stirling. I. (1988). *Polar Bears.* University of Michigan Press, Ann Arbor.

Struntz, D. J., W. A. McLellan, R. M. Dillaman, J. E. Blum, J. R. Kucklick, and D. A. Pabst (2004). "Blubber Development in Bottlenose Dolphins *(Tursiops truncatus)*." *J. Morphol. 259:* 7–20.

Stutz, S. S. (1967). "Moult in the Pacific Harbour Seal, *Phoca vitulina richardsi*." *J. Fish. Res. Board Can. 24:* 435–441.

Supin, A. Ya., V. V. Popov, and A. M. Mass (2001). *The Sensory Physiology of Aquatic Mammals.* Kluwer Academic Publishers, .

Tarasoff, F. J. (1974). Anatomical adaptations in the river otter, sea otter and harp seal with reference to thermal regulation. *In* "Functional Anatomy of Marine Mammals," Vol. 2 (R. J. Harrison, ed.), pp. 111–142. Academic Press, London.

Thomas, J. A., and R. A. Kastelein (eds.) (1990). *Sensory Abilities of Cetaceans: Laboratory and Field Evidence.* Plenum Press, New York.

Vaz-Ferreira, R. (1981). South American sea *lion-Otaria flavescens. In* "Handbook of Marine Mammals," Vol. 1 (S. H. Ridgway and R. J. Harrison, eds.), pp. 39–66. Academic Press, London.

Walls, G. L. (1942). *The Vertebrate Eye and Its Adaptive Radiation.* Cranbrook Institute of Science, Bloomfield Hills, MI.

Wartzok, D., R. Elsner, H. Stone, B. P. Kelly, and R W. Davis (1992). "Under-Ice Movements and Sensory Basis of Hole Finding by Ringed and Weddell Seals." *Can. J. Zool. 70:* 1712–1722.

Wartzok, D., and D. Ketten (1999). Marine mammal sensory systems. *In* "Biology of Marine Mammals" (J. E. Reynolds, and S. Rommel, eds.), pp. 117–175. Smithsonian Press, Washington, D. C.

Wartzok, D., and M. G. McCormick (1978). "Color Discrimination by a Bering Sea Spotted Seal, *Phoca larga.*" *Vision Res. 18:* 781–784.

Watkins, W. A., and D. Wartzok (1985). "Sensory Biophysics of Marine Mammals." *Mar Mamm. Sci. 1:* 219–260.

West, J. A., J. G. Sivak, C. J. Murphy, and K. M. Kovacs (1991). "A Comparative Study of the Anatomy of the Iris and Ciliary Body in Aquatic Mammals." *Can. J. Zool. 69:* 2594–2607.

Williams, T. M., J. E. Haun, W. A. Friedl, and L. W. Bivens (1992). "Assessing the Thermal Limits of Bottlenose Dolphins: A Cooperative Study Between Trainers, Scientists, and Animals." *IMATA Strandings 4:* 16–17.

Worthy, G. A. J., and J. P. Hickie (1986). "Relative Brain Size in Marine Mammals." *Am. Nat. 128:* 445–449.

Worthy, G. A. J., and D. M. Lavigne (1987). "Mass-Loss, Metabolic-Rate, and Energy-Utilization by Harp and Gray Seal Pups During the Postweaning Fast." *Physiol. Zool. 60:* 352–364.

Worthy, G. A. J., P. A. Morris, D. P. Costa, and B. J. Le Boeuf (1992). "Moult Energetics of the Northern Elephant Seal *(Mirounga angustirostris)*." *J. Zool. London 227:* 257–265.

Wyss, A. R. (1987). "The Walrus Auditory Region and the Monophyly of Pinnipeds." *Am. Mus. Novit. 2871:* 1–31.

Wyss, A. R. (1988). "On "Retrogression" in the Evolution of the Phocinae and Phylogenetic Affinities of the Monk Seals." *Am. Mus. Novit. 2924:* 1–38.

Yablokov, A. V., V. M. Bel'kovich, and V. I. Borisov (1972). *Whales and Dolphins.* Israel Programs for Scientific Translations, Jerusalem.

Yamasaki, F., S. Komatsu, and T. Kamiya (1980). "A Comparative Morphological Study on the Tongues of Manatee and Dugong (Sirenia)." *Sci. Rep. Whales Res. Inst. Tokyo 32:* 127–144.

8

Musculoskeletal System and Locomotion

8.1. Introduction

Rather than providing detailed anatomical descriptions for individual marine mammal taxa, the focus of this chapter is one of comparing and considering variation in the musculoskeletal anatomy of the major marine mammal groups, especially as it relates to locomotion. Propulsion for swimming by marine mammals is derived from paired flipper movements (pinnipeds and sea otters) or vertical movements of caudal flukes (cetaceans and sirenians). Paired flipper propulsion is more efficient at low speeds when maneuverability is critical. The evolution of locomotion in each of the major groups of marine mammals also is reviewed. Anatomical specializations that specifically relate to sound production are covered in Chapter 11 and those related to feeding are further discussed in Chapter 12. The nomenclature and topographical and directional terms used are from the *Nomina Anatomica Veterinaria* (1983) and *Illustrated Veterinary Anatomical Nomenclature* (Schaller, 1992).

8.2. Pinnipeds

This discussion of pinniped musculoskeletal anatomy is based on Howell (1929) and King (1983).

8.2.1. Skull and Mandible

The pinniped skull is similar to that of terrestrial mammals and is characterized by large eye orbits, a relatively short snout, a constricted interorbital region, and large orbital vacuities (unossified spaces in the ventromedial wall of the eye orbit; see Figures 3.3 and 8.1). The skull of otariids is readily distinguishable from those of phocids and the walrus in the development of large, shelf-like supraorbital processes of the frontal bones (Figure 8.1). In addition, the relationship of the nasals to the frontals is unique among otariids; the frontals extend anteriorly between the nasals forming a W-shape nasofrontal contact (see Figures 3.11 and 8.1).

(a)   (b)   (c)

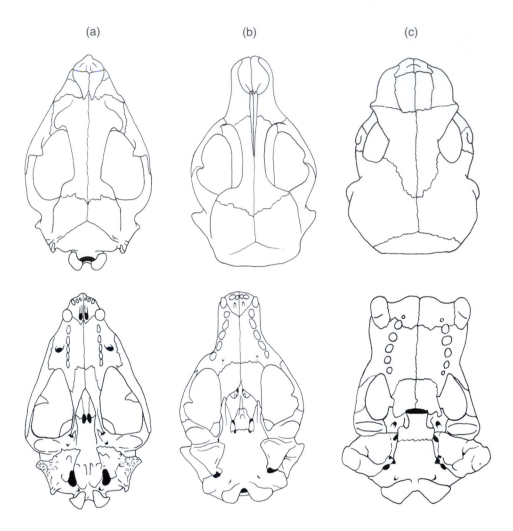

Figure 8.1.   Dorsal and ventral views of skulls of representative pinnipeds. **(a)** Otariid, *Callorhinus ursinus*. **(b)** Phocid, *Phoca hispida*. (Modified from Howell, 1929.) **(c)** Walrus, *Odobenus rosmarus*. (Illustrated by P. Adam.)

The modern walrus skull is distinctive; the large maxilla accommodates the upper canine tusks. The entire skull is heavily ossified and shortened anteriorly (see Figure 8.1). The modern walrus lacks development of supraorbital processes. Prominent antorbital processes formed of the maxilla and frontal bones and enlarged infraorbital foramina are diagnostic features of the walrus. One of the principal eye muscles, the mm. orbicularis oculi, which closes the upper and lower eyelids, is attached to the antorbital process. The short supercilaris muscle, which functions to lift the upper eyelid, also is attached to this process. The modern walrus has a highly vaulted palate. This vaulting occurs in both the transverse and longitudinal planes, with the degree of vaulting greatest between the anterior teeth (incisors) and the end of the postcanine toothrow (Deméré, 1994). The external nares are elevated above the toothrow.

The phocid skull is characterized by the lack of supraorbital processes, an inflated tympanic bulla, and nasals that narrow greatly posteriorly and terminate posterior to the frontal-maxillary contact in a V-shaped suture with the frontals (see Figures 3.11 and 8.1).

In otariids, the angular (pterygoid) process is well developed and positioned near the base of the ascending ramus of the mandible. This process is the point of attachment of the pterygoideus medialis muscle, which elevates the mandible. In walruses and phocids this process is reduced and elevated above the base of the ascending ramus (Figure 8.2). The phocid lower jaw is characterized by a thinning and ventral extension of the posterior end to form a thin bony flange (see Figure 8.2). The modern walrus is characterized by a solid joint between the anterior ends of the lower jaws, referred to as a fused mandibular symphysis.

8.2.2. Hyoid Apparatus

In pinnipeds, the hyoid bone, which serves for the attachment of tongue muscles, is composed of the same elements seen in other carnivores: the basihyal, thryohyal, epihyal, and ceratohyal. In two phocids, the Ross seal and the leopard seal, only the ceratohyal and epihyal are present. In addition, these taxa have proximal unossified ends of the epihyal that lie freely inside a fibrous tube (King, 1969).

The prehyoid (mylohyoid and stylohyoid) and pharyngeal and tongue musculature are well developed in the Ross seal (King, 1969; Bryden and Felts, 1974). This musculature may be involved in feeding by allowing grasping and swallowing of large cephalopods. This musculature, and particularly the pharyngeal musculature, also may be involved in sound production by the Ross seal because the throat is expanded considerably before sounds are produced. The muscles of the walrus tongue are involved in suction feeding and are discussed further in Chapter 12.

8.2.3. Vertebral Column and Axial Musculature

The typical pinniped vertebral formula is C7, T15, L5, S3, C10-12. The cervical (neck) vertebrae of otariids are large, with well-developed transverse processes and neural spines associated with muscles for movement of the neck and head (Figure 8.3).

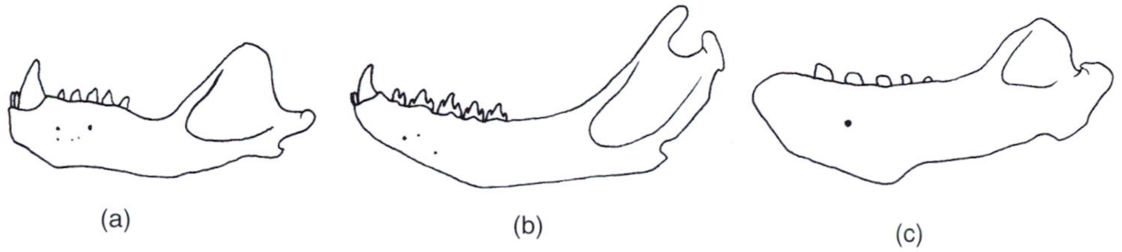

(a)   (b)   (c)

Figure 8.2. Lateral views of lower jaw of representative pinnipeds. **(a)** Otariid, *Callorhinus ursinus*. **(b)** Phocid, *Phoca hispida*. (Modified from Howell, 1929.) **(c)** Walrus, *Odobenus rosmarus*. (Illustrated by P. Adam.)

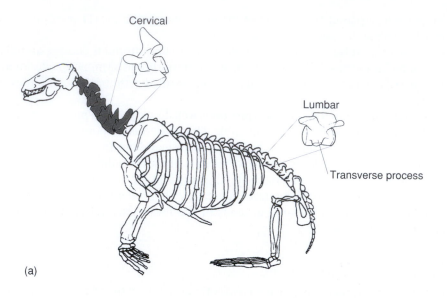

(a)

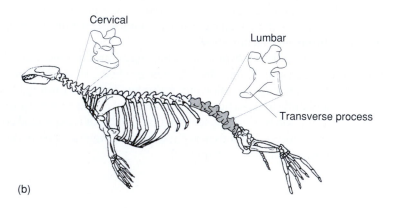

(b)

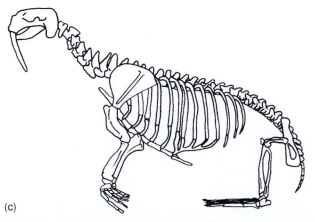

(c)

Figure 8.3. Lateral view of generalized skeleton of representative pinnipeds. **(a)** Otariid. (Modified from Macdonald, 1984.) **(b)** Phocid. (Modified from Macdonald, 1984.) **(c)** Walrus. (Illustrated by P. Adam.)

Extensive movements of the head and neck in otariids occur during terrestrial locomotion and help maintain balance by lifting the forelimbs from the ground (English, 1976). The walrus and phocids have cervical vertebrae that are smaller than the thoracic and lumbar vertebrae with small transverse processes and neural spines (Fay, 1981). In contrast to the low neural spines of phocids and the walrus, the thoracic vertebrae of otariids possess large neural spines. Increased height of neural spines provides larger attachment points for epaxial (dorsal to the spine) musculature (mm. multifidus lumborum and longissimus thoracics).

The lumbar vertebrae of otariids have small transverse processes and closely set zygopophyses (one of the processes by which a vertebra articulates with another), whereas those of phocids have larger transverse processes and more loosely fitting zygopophyses (King, 1983). In the walrus and in phocids the transverse processes are two or three times as long as they are wide (Fay, 1981), whereas in otariids these processes are about as long as they are wide. In all pinnipeds except the walrus, the number of lumbar vertebrae is five. Walruses typically have six lumbar vertebrae (Fay, 1981). In phocids, the elongated transverse processes provide larger attachment points for the hypaxial (ventral to the spine) musculature (mm. quadratus lumborum, longissimus thoracics, and iliocaudalis), which is correlated with horizontal movements in the posterior end of the body (see Figure 8.3). Because the tail is not used in swimming, the caudal vertebrae are small, cylindrical, and without strong processes.

8.2.4. Sternum and Ribs

Pinnipeds are characterized by elongated manubria (one of the bones that makes up the sternum). In otariids, the length of the manubrium is increased by a bony anterior extension at the point of attachment of the first pair of ribs. In phocids and the walrus, the length of the manubrium is increased by cartilage (King, 1983). Ribs have a well-developed capitulum and head and are firmly articulated to the thoracic vertebrae. There are typically eight true ribs, four false ribs, and three floating ribs.

8.2.5. Flippers and Locomotion

8.2.5.1. Pectoral Girdle and Forelimb

Front and hind limb bones are relatively short and lie partially within the body outline. The axilla (arm pit) in otariids and the walrus falls at about the middle of the forearm, and in phocids at the wrist. The hind flipper is free distal to the ankle.

The otariid scapula is unique among pinnipeds in possessing at least one ridge (called accessory spines) subdividing the supraspinous fossa (King, 1983; see Figures 3.11 and 8.4). A large supraspinous fossa is a consistent feature of otariids and the walrus. Enlargement of the supraspinous fossa and the development of a scapular ridge in otariids are correlated with strong development of the supraspinatus muscle, which possesses pinnated heads divided by an aponeurosis (fibrous or membranous sheets). In relation to the size of the infraspinous fossa, the supraspinous fossa tends to be substantially reduced in phocids, particularly among phocines (Figure 8.4; Wyss, 1988).

The humerus is short and robust. In pinnipeds the greater and lesser tubercles are prominent relative to those in terrestrial carnivores. In otariids the greater tubercle is elevated above the head of the humerus; in phocids it is the lesser tubercle that is ele-

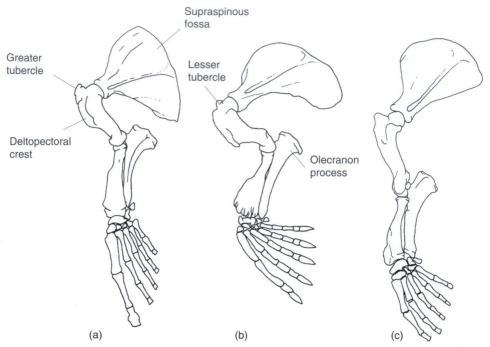

Figure 8.4.   Forelimb of representative pinnipeds in anatomical position. **(a)** Otariid, *Zalophus californi-anus*. (Modified from Howell, 1930a.) **(b)** Phocid, *Phoca hispida*. (Modified from Howell, 1930a.) **(c)** Walrus, *Odobenus rosmarus*. (Illustrated by P. Adam.)

vated above the head of the humerus. These enlarged humeral tubercles serve as expanded areas of insertion for the large rotator cuff musculature (mm. deltoideus, infraspinatus, subscapularis, supraspinatus), and, by lying above the proximal joint surface of the head, the moment arm of these muscles as protactors and humeral rotators is greater than in animals without enlarged humeral tubercles (English, 1974). In "monachines," otariids, and odobenids the deltopectoral crest is elongated, extending two-thirds to three-quarters the length of the shaft where it ends abruptly, forming an acute angle with the shaft in lateral view. The extraordinary development of the deltopectoral crest is associated with enlargement of the sites of attachment of the deltoid and pectoralis muscles and the pectoral portion of the latissimus dorsi muscle (English, 1977). Enlargement of both the deltoid and triceps muscles in pinnipeds increases thrust. Among phocids, the harbor seal and southern elephant seal share a unique component of the pectoralis muscle, an ascending pectoral muscle that extends over the humerus (Rommel and Lowenstine, 2001). A supracondylar (entepicondylar) foramen is usually found in phocines but not in "monachines" (some fossil "monachines" are exceptions; Wyss, 1988).

There are several unique structural features of the elbow joint of otariids. One is the position of the annular ligament, which in terrestrial carnivores forms a ring around the neck of the radius attaching to the sides of the articular surface of the ulna, and the semilunar notch, to help secure the proximal radioulnar articular surfaces. In fur seals and seal lions, the annular ligament is not continuous; it terminates on the articular capsule and lateral epicondyle. This arrangement of the annular ligament inhibits or stops rotary forearm movements in otariids. A second major structural feature of the elbow joint of otariids is the shape of the articular surface of the ulna. In fur seals and sea lions, the lateral half of the semilunar notch is poorly developed, unlike the condition in terrestrial carnivores. A prominent coronoid process forms the medial border of an expanded trough for articulation with the radius (English, 1977, Figure 15). This modification of joint structure may be advantageous in supporting the load that the joint must carry during locomotion.

The radius and ulna are short and flattened anteroposteriorly. The olecranon process of the ulna is laterally flattened and very enlarged in all pinnipeds (see Figure 8.4). The olecranon process of the ulna of otariids forms the sole source of origin of the mm. flexor and extensor carpi ulnaris, both of which display some humeral origin in terrestrial carnivores (English, 1977). The large flattened olecranon allows insertion on the ulna by muscles of the triceps complex, which extend the elbow.

Pinnipeds (except phocines) are characterized by having the first metacarpal in the hand greatly elongated and thicker in comparison to metacarpal II (King, 1966; Wyss, 1988). Strong reduction of the fifth intermediate phalanx of the hand occurs in all pinnipeds. Pinnipeds possess cartilaginous rods distal to each digit that support an extension of the flipper border. Long cartilaginous extensions occur on both the fore and hind flipper of otariids, short extensions occur in walruses, and shorter extensions occur in at least some phocids (see Figure 7.14; King, 1969, 1983).

8.2.5.2. Pelvic Girdle and Hind Limb

The hip (or innominate bone) of pinnipeds has a short ilium and elongated ischium and pelvis. The connection of the hip bones anteriorly, at the pubic symphysis, is unfused but has a ligament binding adjoining bones. This differs from the condition in terrestrial carnivores in which the symphysis is bony and fused. Phocines, except the bearded seal, are characterized by lateral eversion of the ilium accompanied by a deep lateral excavation of the iliac wing (King, 1983). The great eversion of the phocid ilium means that the medial surface of the bone now faces almost anteriorly. This presumably gives a much greater attachment area to the strong iliocostalis lumborum muscle that is, to a large extent, responsible for much of the lateral body movement used in swimming (King, 1983). This flaring also results in an increased area of origin of the mm. gluteus medius, minimis, and pyriformis. These muscles attach to the greater trochanter of the femur and on contraction cause the femur to be adducted and rotate. As noted by King (1983), a strongly developed, dorsally directed ischiatic spine is present only in phocids, where the deep head of m. biceps femoris and the muscles attached to it helps in elevating the hind flippers to produce the characteristic phocid posture (Figure 8.5).

The femur is short, broad, and flattened anteroposteriorly (see Figures 3.5 and 8.6). The position of the fovea capitis is barely visible on the head of the femur, and the round ligament between the head of the femur and the acetabulum does not occur in pinnipeds. This ligament fixes the femoral head within the acetabulum to provide a more secure joint when the weight of the animal is borne on the hind limb. As pinnipeds spend a lot of time in the water, the need of the round ligament has been lost (Tarasoff, 1972). The lesser trochanter (one of several bony processes at the proximal end of the femur) is present only as a small knob distal to the head in otariids and is reduced or absent in phocids. In the walrus, the lesser trochanter is marked by a slightly raised area (see Figure 3.5). In

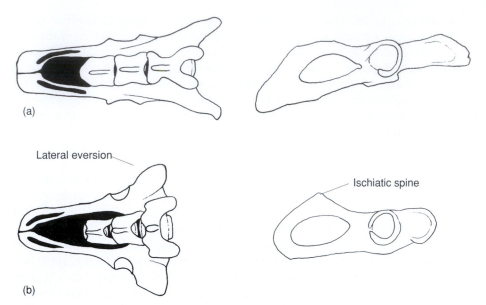

Figure 8.5.   Pelvis of representative pinnipeds in dorsal and lateral views. **(a)** Otariid, *Callorhinus ursinus*.
(b) Phocid, *Monachus schauinslandi*. (Illustrated by P. Adam.)

phocid seals (that do not bring their hind limbs forward), decrease in size or loss of the lesser trochanter represent a decrease in the area of insertion of muscles that rotate the femur posteriorly (mm. iliacus and psoas major). In phocid seals, insertion is onto the iliac wing (m. psoas major) or distal to the medial femur (m. iliacus). The new points of attachment for these muscles aid in increasing the strength of lateral undulation of the lumbosacral region of the spine and flexion of the leg.

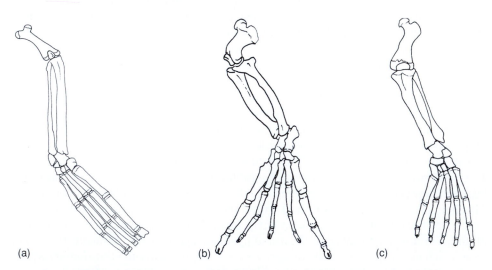

Figure 8.6.   Hind limb of representative pinnipeds. **(a)** Otariid, *Callorhinus ursinus*. **(b)** Phocid, *Monachus tropicalis*. **(c)** Walrus, *Odobenus rosmarus*. (Illustrated by P. Adam.)

The morphology of the fore flipper is correlated with that of the foot. Pinnipeds (except phocines) are characterized by relatively long, flattened metatarsal shafts with flattened heads associated with smooth, hinge-like articulations (Wyss, 1988). They also are characterized by having elongated digits I and V (metatarsal I and proximal phalanx) in the foot (see Figure 8.6). In "monachines" and the hooded seal, the third metatarsal is considerably shorter than the others. The phocid ankle bone (astragalus) is character- ized by a strong caudally directed process (calcaneal process) over which passes the ten- don of the flexor hallucis longus (see Figure 3.17). Tension on this tendon prevents the foot from being dorsiflexed in phocids such that the foot lies at right angles to the leg, as in an otariid. In walruses there is a slight posterior extension of this process; in otariids there is not. There are also differences between the heel bone (calcaneum) of phocids and otariids. Among the more obvious of these are the presence of a groove for the Achilles tendon on the tuberosity and the presence of a medially directed process (the sustentac- ulum) in the otariid heel bone. Both of these characters are lacking in phocids (King, 1983).

8.2.5.3. Mechanics of Locomotion

Among modern pinnipeds, terrestrial and aquatic locomotion are achieved differently. Three distinct patterns of pinniped swimming are recognized, yet all create thrust with the hydrofoil surfaces of their flippers. When swimming, these hydrofoils are oriented at an angle (the angle of attack) to their direction of travel, producing thrust parallel to the direction of travel and generating lift perpendicular to that direction (Figure 8.7). One of these patterns, pectoral oscillation (forelimb swimming), is seen in otariids. Sea lions and fur seals move their forelimbs to produce thrust in a manner similar to flap- ping birds in flight. Observations indicate that the hind limbs are essential in providing maneuverability and directional control but play little role in propulsion (Godfrey, 1985). The larger pectoral flippers, with nearly twice the surface area of the pelvic flip- pers move in unison, acting as oscillatory hydrofoils in a stroke that includes power, paddle, and recovery phases (Figure 8.8; Feldkamp, 1987; English, 1976). The power stroke is generated by medial rotation and adduction and retraction of the forelimbs, in contrast to nearly pure limb retraction seen in walking terrestrial carnivores. Combined with an extremely flexible body, the large pectoral flippers of otariids can turn in tighter circles than can the less flexible phocids or cetaceans of similar body size (e.g., Fish *et al.,* 2003). However, this increased maneuverability is accomplished at some expense to trajectory stability.

The otariid terrestrial posture allows weight to be borne on all four limbs, with the hind flippers facing forward. On land, locomotion is limb based, although extensive movements of the head and neck contribute more to propulsion than hind limb

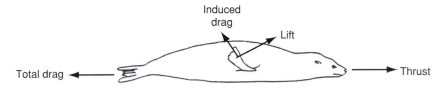

Figure 8.7.   Lateral view of drag components encountered by a swimming sea lion.

Figure 8.8.   Oblique view of aquatic locomotion in the otariid, *Zalophus califomianus*. Tracings of limb and body movements. (From English, 1976.)

movements (Figure 8.9). Beentjes (1990) documented walk and gallop gaits in the New Zealand sea lion and the New Zealand fur seal. He observed gaits in which the limbs are moved in sequence alternately and independently, as seen in the sea lion, typical of otariids that live on relatively sandy substrates. New Zealand fur seals, as well as other fur seals, move with a bounding gait that displaces their center of gravity vertically. This type of gait, in which the hind limbs are moved in unison, is more often used by otariids that live on rocky substrates.

A second method of aquatic locomotion in pinnipeds is pelvic oscillation (hind limb swimming) of seals. Among phocids, the hind limbs are the major source of propulsion in the water and the forelimbs function principally for steering (Figure 8.10). During swimming, seals also laterally undulate the lumbosacral region of their bodies enhancing propulsive forces produced by the hind limbs. Phocids are incapable of turning the hind limbs forward, and consequently the hind limbs are not used in terrestrial locomotion. Movement on land by phocids is accomplished generally by vertical undulations of the trunk. Arching of the lumbar region allows the pelvis to be brought forward while the weight of the body is carried by the sternum. This action is followed by extension of the anterior end of the body while the weight is carried by the pelvis (Figure 8.11). The fore flippers may facilitate this movement by lifting or thrusting the anterior body off of the substrate; hind limbs are typically held up and do not contribute to this shuffling movement. The exceptional role of forelimbs in undulatory terrestrial locomotion of the grey seal was noted by Backhouse (1961). A second, apparently more rapid, mode of terrestrial locomotion has been noted in a few species. Crabeater, ribbon, harp, ringed, grey, and leopard seals have been observed moving in a sinuous fashion with lateral undulations of the body, particularly when on an icy substrate (O'Gorman, 1963; Burns, 1981; Kooyman, 1981). This pattern involves backward strokes by the fore flippers and lateral movement by the posterior torso with the hind flippers lifted above the substrate. The unusually rapid aquatic and terrestrial locomotion of the leopard seal has been noted (O'Gorman, 1963).

A variant of pelvic oscillation, similar to that described for true seals, is exhibited by the walrus (Figure 8.12). The hind limbs of the walrus generate the dominant propulsive force; forelimbs are used either as rudders or as paddles at slower speeds. The fore-

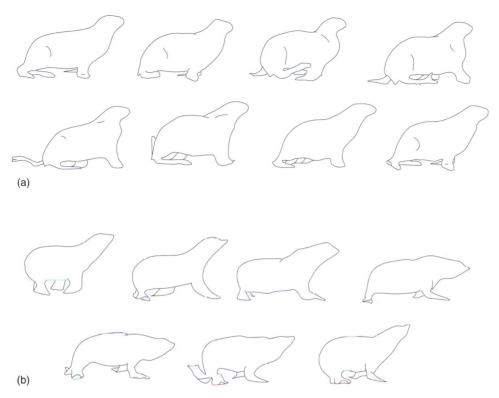

Figure 8.9. Terrestrial locomotion in otariids. Tracings of limb and body movements. **(a)** New Zealand sea lion, *Phocarctos hookeri*. **(b)** New Zealand fur seal, *Arctocephalus forsteri*. (From Beentjes, 1990.)

limb stroke cycle is bilateral and contains both power and recovery strokes. The power stroke consists of adduction and retraction of the medially rotated limbs. The recovery stroke is accomplished by lateral rotation followed by abduction and protraction of the forelimb. The hind limb stroke is unilateral and consists also of a power and recovery stroke. The power stroke is a rearward translation of the flipper followed by a medial flexion of the flipper and lower leg. The recovery stroke returns the flipper to its protracted position but represents the power stroke of the opposing leg. In walruses, as in otariids, the hind limbs can be rotated forward in terrestrial locomotion. Terrestrial locomotion is unusual in that the body of the animal is supported in large part by the belly not by the limbs (Figure 8.13). The movement of the feet is a lateral sequence walk that alternates with a lunge that is responsible for forward progression. In the lunge, the chest is raised off the ground by the forelimbs while the lumbar and posterior thoracic regions of the torso are flexed. The hind limbs and torso then extend, pushing the body forward. Young (small) walruses are able to execute a walk similar to that of sea lions without the lunge (Gordon, 1981). Fay (1981) reported observing terrestrial locomotion of a walrus on ice using only the forelimbs, with the hindquarters and limbs being dragged passively.

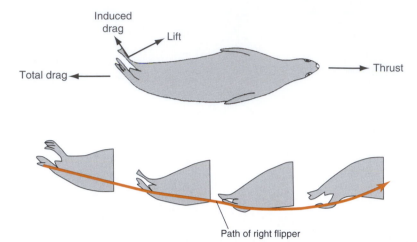

Path of right flipper

Figure 8.10.   Aquatic locomotion in phocids, tracings in dorsal view of the hind flipper. (Adapted from Fish
                    et al., 1988.)

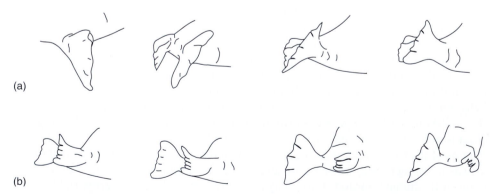

Figure 8.11.   Terrestrial locomotion in the harbor seal, *Phoca vitulina*. (Courtesy of T. Berta.)

(a)

(b)

Figure 8.12.   Aquatic locomotion in the walrus, *Odobenus rosmarus*. Tracings in dorsal view of the hind
                    flipper. **(a)** Power stroke. **(b)** Recovery stroke. (Modified from Gordon, 1981.)

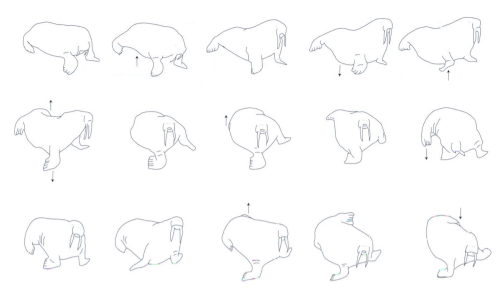

Figure 8.13.   Terrestrial locomotion in the walrus, *Odobenus rosmarus*. Tracings of limb and body movements. (Modified from Gordon, 1981.)

Swimming speeds for a variety of pinnipeds have been determined in both controlled laboratory and unrestrained natural situations. California sea lions swimming in tanks exhibited velocities ranging from 2.7 to 3.5 m/s (Feldkamp, 1987; Godfrey, 1985), and Ponganis *et al.* (1990) reported swimming velocities between 0.6 and 1.9 m/s for surface swimming and between 0.9 and 1.9 m/s for submerged swimming in four species of otariids. Le Boeuf *et al.* (1992) obtained similar velocities (0.9–1.7 m/s) for an unrestrained adult female elephant seal foraging at sea. Swimming velocities of unrestrained foraging Weddell seals ranged from 1.3 to 1.9 m/s (Sato *et al.,* 2003) with fatter animals performing regular stroke and glide swimming, while thinner (and presumably less buoyant) seals descended with significantly longer glide periods between power strokes. Surface recovery times suggested that prolonged-glide swimming was the more energy efficient mode.

8.2.5.3.1. Case Study: Integration of Phylogeny and Functional Morphology
The evolution of pinniped locomotor patterns was investigated by mapping skeletal characters associated with a particular locomotor mode (ambulation or undulation for terrestrial locomotion and forelimb or hind limb for aquatic locomotion) onto a phylogeny for pinnipeds (Figure 8.14). The distribution of skeletal characters suggests the following transformations. The undulatory movements of phocids on land seems to have evolved only once in pinnipeds and morphologic features of this locomotion characterize the phocid clade, including desmatophocids and basal phocids (e.g., *Acrophoca* and *Piscophoca*). Forelimb swimming appears to be primitive for the group (*Enaliarctos* shows features consistent with forelimb and hind limb swimming, but seems more specialized for forelimb swimming) but was lost to hind limb swimming once within the phocid clade (and once within the Odobenidae, *Imagotaria* and all later diverging walruses). The basal phocoid *Allodesmus* retains several features consistent with forelimb propulsion but also displays adaptations for hind limb swimming. Forelimb swimming

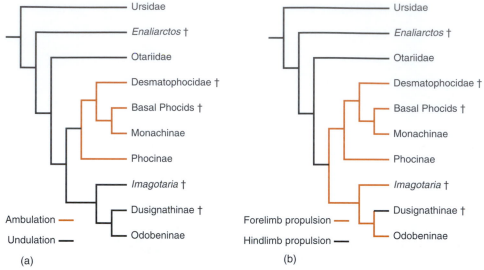

Figure 8.14. Evolution of **(a)** terrestrial and **(b)** aquatic locomotion in pinnipeds, † = extinct taxa. (Berta and Adam, 2001.)Terrestrial (ambulation) characters: relatively short ischiopubic region, mortise tibio-astragalar joint, unmodified tarsals; Terrestrial (undulation) characters: elongation of ischiopubic region, spheroid-like tibio-astragalar joint, posterior extension of the astragalus; Forelimb swimming characters: enlarged supraspinous scapula with accessory spines, round scapular glenoid fossa, cartilaginous extensions on manus digits; Hind limb swimming characters: enlargement of lumbar vertebrae and their processes, mediolaterally narrow scapular glenoid fossa and hind flippers with a lunate trailing edge (see text for more explanation).

was independently regained in the fossil walrus *(Gomphotaria)*. Although this case study focused on aquatic locomotion, researchers are incorporating terrestrial locomotion into a more general framework of locomotor evolution (e.g., Berta and Adam, 2001).

8.3. Cetaceans

The following discussion is based on the broad comparative treatments of cetacean anatomy found in Slijper (1936) and Yablokov *et al.* (1972).

8.3.1. Skull

In comparison to the typical mammalian skull, the cetacean skull is telescoped (as described in Chapter 4); the braincase, or the portion of the skull behind the rostrum, has been shortened (see Figure 4.11). This telescoping has altered the size, shape, and relationship of many of the skull bones. The external narial opening (blowhole[s]) has migrated posteriorly to a more dorsal position on the head, and the nasals have become small, tabular bones situated immediately behind it. Telescoping of the cranial bones in the mysticete skull involves the maxilla extending posteriorly underneath the frontal. In odontocetes, the premaxilla and maxilla extend posteriorly and laterally so as to override the frontals and crowd the parietals laterally.

Typically, the odontocete skull displays cranial and facial asymmetry in which those bones and soft anatomical structures on the right side are larger than those on the left side (Figure 8.15; see also Chapter 4). It has been suggested that asymmetry has evolved with the right side being more specialized for producing sound and the left side adapted more for respiration (Norris, 1964; Wood, 1964; Mead, 1975).

The skulls of several odontocetes are variously ornamented. A cranial feature of phocoenids and some river dolphins is the presence of small rounded protuberances on the premaxillae (Heyning, 1989; see Figure 4.33). The facial region of the beluga is unique among odontocetes because the entire surface is slightly convex rather than concave. The susu has unique maxillary crests that overhang the facial region anteriorly (Heyning, 1989). The ventral surface of these crests is covered by a thin, flat, complicated air sac derived from the pterygoid air sinus system (Fraser and Purves, 1960; Purves and Pilleri, 1973; also see Figure 4.26).

The rostra of many toothed whales differ from those of baleen whales in being made of dense bone. This is especially true of beaked whales, particularly Blainville's beaked whale (de Buffrénil and Casinos, 1995; Zioupos *et al.*, 1997). Several hypotheses have been

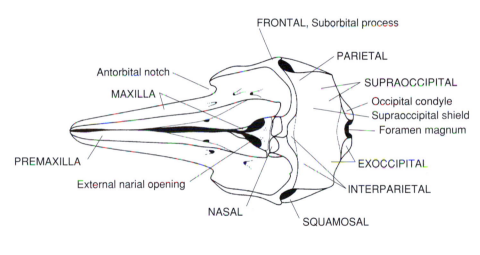

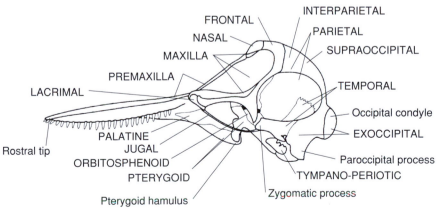

Figure 8.15. The skull of a representative odontocete, *Tursiops truncatus*. **(a)** Dorsal and **(b)** lateral views. (Modified from Rommel, 1990.)

proposed to explain the functional significance of a dense rostrum. Heyning (1984) suggested that the compaction of the rostrum in various beaked whales increases the strength of the rostrum and reduces the risk of fracture during intraspecific fights between adult males. However, De Buffrénil and Casinos (1995) and Zioupos *et al.* (1997) presented data demonstrating that the rostrum, in addition to being highly mineralized, was a stiff, brittle, and fragile structure. They suggested that similarity in composition and mechanical properties between the rostrum and the tympanic bulla indicated an acoustics-related function for the rostrum.

In the mysticete skull, the entire facial region is expanded and the rostrum is arched to accommodate the baleen plates that hang from the upper jaw (Figure 8.16). The degree of rostral arching varies; it is slightly arched in balaenopterids (less than 5% between the basicranium and the base of the rostrum), moderately arched in *Eschrichtius* and *Caperea* (10 and 17%, respectively), and greatly arched (over 20%) in the balaenids (Barnes and McLeod, 1984). The high rostral arch of balaenids accommodates their exceptionally long baleen plates. In mysticetes, unlike odontocetes, there is strong development of several facial and skull bones (i.e., vomer, temporal, and palatine bones).

All mysticetes have two unbranched nasal passages leading to paired blowholes. In odontocetes (with the exception of physeterids), the single nasal passage (vestibule) extends vertically to the bony nares. A series of blind-ended sacs branch off this nasal passage. The nasal air spaces are variable, both within a single species and between

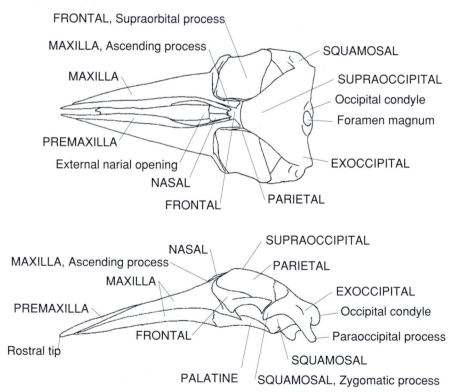

Figure 8.16. The skull of a representative mysticete, minke whale, *Balaenoptera acutorostrata*. **(a)** Dorsal and **(b)** lateral views. (After True, 1904.)

species. Detailed descriptions of these nasal sacs are provided by Heyning (1989). The musculature associated with the nasal passages is extremely complex (see Lawrence and Schevill, 1956; Mead, 1975; Heyning, 1989). Sperm whales are unique among odonto-cetes in the retention of two nasal passages (Figure 8.17). The left nasal passage extends anteriorly in nearly a straight line from the enlarged left bony naris to the blowhole and lacks air sacs. Just superficial to the bony naris is the frontal air sac off the right nasal passage. The right nasal passage extends anteriorly to the distal vestibular air sac and then connects vertically to the blowhole. On the posterior wall of this sac is a muscular ridge with a narrow slit-like orifice that opens to the right nasal passage. The gross appearance of this valve-like structure suggested a monkey's muzzle, or **museau de singe** in the original French (Pouchet and Beauregard, 1892). Cranford (1999) has proposed the more descriptive term "phonic lips" for this structure.

Several functions have been attributed to the complex facial structure of odontocetes, including buoyancy adjustments during dives (Clarke, 1970, 1979) and ramming structures in competitive encounters between sperm whale males (Carrier *et al.*, 2002). However, there is no question that the principal functions of these facial specializations are related to feeding, respiration, and sound production and reception (discussed in greater detail in Chapter 11). A pair of large fleshy masses of tissue, the **nasal plugs,** occlude the bony nares of cetaceans. These nasal plugs consist of connective tissue, muscle, and fat. Relative to odontocetes, the nasal plugs of mysticetes contain less fatty tissue but are otherwise similar in general morphology. The nasal plugs are retracted by paired nasal plug muscles that originate from the premaxillae. Contraction of the nasal plug muscles pulls the nasal plugs anterolaterally. When the nasal plugs are retracted during inhalation and exhalation, the nasal passage remains fully open. In addition to this respiratory function the nasal plugs play a major role in sound production (see Chapter 11).

A large ovoid melon is located in the facial region of odontocetes. The melon typically is asymmetrically positioned, slightly off to the right side. It rests on a pad of dense connective tissue on top of the bony rostrum of the skull. The melon and mandibular lipid

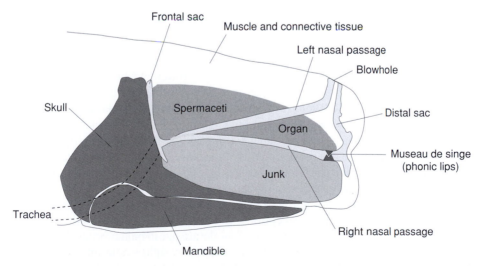

Figure 8.17.   Diagram of skull of sperm whale, *Physeter macrocephalus* with nasal passages. (Modified from Evans, 1987.)

tissues, which are involved in sound production and reception (see Chapter 11), are composed of unique fatty acids rich in isovaleric acid (Varanasi and Malins, 1970; Gardner and Varanasi, 2003). Differences in the biochemical composition of these lipids in several different aged odontocetes (extrapolated from size differences among specimens) suggest that echolocation is not fully developed at birth (Gardner and Varanasi, 2003).

As discussed in Chapter 4, a vestigial melon has been described in baleen whales by Heyning and Mead (1990). They suggest that the original function of the melon in cetaceans was to allow the free movement of the nasal plugs as they are drawn posteriorly over the premaxillae during contraction of the nasal plug muscles. Alternatively, Milinkovitch (1995) suggests that the presence of a melon in mysticetes indicates that the ancestor of all whales may have possessed a well-developed melon and correspondingly well-developed echolocation abilities. Comparative study of melon morphology using computed tomography (CT) data indicates that the melon evolved early in odontocete evolution and is more likely a synapomorphy for this lineage (McKenna, personal communication).

In the forehead of sperm whales is a large **spermaceti organ,** which may occupy more than 30% of the whale's total length and 20% of its weight (see Figure 8.17). The fatty tissue in the facial region of physeterids is highly modified and very different in structure than the melon of all other odontocetes. Because it is situated posterodorsal to the nasal passages, the spermaceti organ is not thought to be homologous with the melon but rather is an extremely hypertrophied structure homologous to the posterior bursa of other odontocetes (Cranford *et al.,* 1996). In adult sperm whales there is an elongate connective tissue sac, or case, that is filled with viscous waxy fluid called **spermaceti.** This is the sperm whale oil that was most sought after by whalers for candlemaking and for burning in lanterns. The spermaceti organ is contained within a thick case of muscles and ligaments. Below the spermaceti organ is a region of connective tissue alternating with spaces filled with spermaceti oil. This region was called the **junk** by whalers because it contains an oil of poorer quality (Clarke, 1978; see Figure 8.17). The spermaceti "junk" is probably homologous with the melon of other odontocetes. The function of these structures in sound production is discussed in more detail in Chapter 11.

8.3.2. Mandible

The lower jaws, or mandibles, of odontocetes appear straight when viewed dorsally. The posterior non-tooth bearing part of the jaw has thin walls that form the fat filled **pan bone** (Figure 8.18). Norris (1964, 1968, 1969) proposed that this region is the primary site of sound reception in odontocetes (discussed in Chapter 11). In mysticetes the mandible curves laterally and there is no pan bone. The mandibular symphysis, a fibrocartilage articulation, connects the tapered distal ends of the paired dentary bones. It is analogous in structure to the intervertebral joints; its center, filled with a gelatinous substance, is surrounded by a dense fibrocartilaginous capsule.

The coronoid process (see Figure 8.18) for attachment of the temporalis muscles is reduced in most odontocetes (an exception is the susu, which retains a distinct process). Among mysticetes, the coronoid process is of moderate size in balaenopterids and developed as a slightly upraised area in the gray whale, pygmy right whale, and the balaenids (Barnes and McLeod, 1984). The function of the coronoid process in mysticete feeding is discussed in Chapter 12.

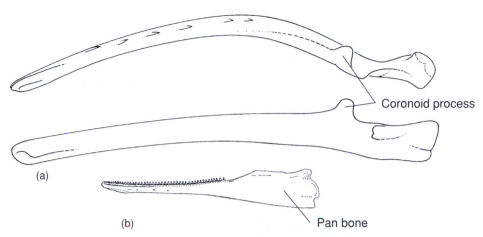

Figure 8.18.   Lateral views of lower jaw of representative cetaceans. **(a)** Mysticete, minke whale, *Balaenoptera acutorostrata.* (From Deméré, 1986.) **(b)** Odontocete, spotted dolphin, *Stenella* sp.

8.3.3. Hyoid Apparatus

The hyoid bones are well developed in all cetaceans. In odontocetes the hyoids are divisible into a basal portion (basihyal, paired thyrohyals) and a suspensory portion (paired ceratohyals, epihyals, stylohyals, and tympanohyals; Reidenberg and Laitman, 1994). Muscles that retract the hyoid apparatus (e.g., sternohyoid) or control the tongue (e.g., styloglossus, hyoglossus) are enlarged and it is suggested that they may be important in suction feeding in some species of odontocetes and mysticetes (see Chapter 12).

8.3.4. Vertebral Column and Axial Musculature

In cetaceans, the vertebral column does not contain a sacral region because the pelvic girdle is absent. The boundaries between cervical, thoracic, and lumbar regions are established according to the presence of ribs, and the boundary between the lumbar and caudal segments is determined by the presence of **chevron bones** (Figure 8.19). The typical cetacean vertebral formula is C7, TI 1–12, L usually 9–24 (range 2–30), C usually 15–45 (range 15–49) (Yablokov *et al.,* 1972).

All cetaceans have seven cervical vertebrae (Figure 8.20). The vertebral bodies of cervical vertebrae differ from those in other mammals in being extremely flat and occasionally consisting of only thin, osseous plates that have lost the main characteristics of vertebrae. Most cetaceans, including balaenids, neobalaenids, and odontocetes (e.g., bottlenose whale, pygmy sperm whale, bottlenose dolphin) have two or more of the cervicals fused (Rommel, 1990; see Figure 8.20a). In the sperm whale, the last six cervical vertebrae are fused (DeSmet, 1977). The resulting short, rigid neck adds to the streamlining of the body and stabilizes the head (Slijper, 1962). The unfused cervical vertebrae (see Figure 8.19b) in balaenopterids, eschrichtiids, platanistids, iniids, and pontoporiids (Barnes and McLeod, 1984) and in the beluga (Yablokov *et al.,* 1972) allows considerable neck mobility.

The thoracic vertebrae are flanked by ribs. Typically, 11 or 12 thoracic vertebrae are present. The common feature of thoracic vertebrae is relatively poor development of the articular surfaces on the vertebral bodies; usually only thoracic vertebrae 1–4 or 5 have

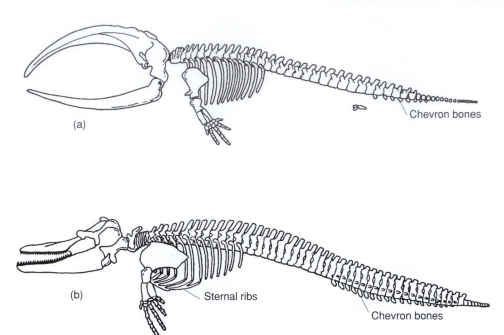

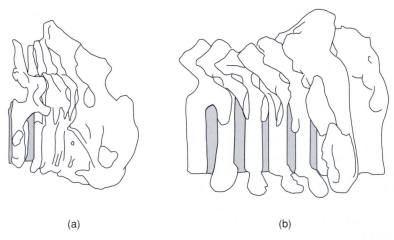

Figure 8.19.   Vertebral columns of representative cetaceans. **(a)** Mysticete. **(b)** Odontocete. (Modified from Harrison and Bryden, 1988.)

Figure 8.20.   Lateral view of the cervical vertebrae of **(a)** a pilot whale, *Globicephala* sp. and **(b)** a blue whale, *Balaenoptera musculus*. The first six vertebrae of the pilot whale are fused. The vertebrae of the blue whale are unfused and are separated by intervertebral disks (gray). (Redrawn from Slijper, 1962.)

these articulations (Yablokov *et al.,* 1972). In mysticetes, the relative size of the vertebral body is considerably larger than the size in odontocetes; usually in the latter the vertebral body comprises more than 50% of vertebral height. Some thoracic vertebrae possess ventral projections or hypophyses that are not structurally part of the vertebrae. Hypophyses may also occur among the posterior thoracic and anterior lumber vertebrae in some species (e.g., pygmy and dwarf sperm whales). These intervertebral structures increase the mechanical advantage of the hypaxial muscles (Rommel and Reynolds, 2001).

Posterior to the thoracic vertebrae, the vertebral series continues as far as the notch of the tail flukes. Lumbar vertebrae lack ribs. The lumbar vertebrae possess the largest vertebral bodies and have the best-developed transverse and spinous processes. There is considerable variation in the number of lumbar vertebrae. The maximum number (29–30) is reported in Dall's porpoise; the fewest are reported from the pygmy sperm whale (2), the Platanistidae (3–5), and the Monodontidae (6; Yablokov *et al.,* 1972). There are 16–18 lumbar vertebrae in the bottlenose dolphin (Rommel, 1990).

There are no defined sacral vertebrae in cetaceans. Early whale evolution is characterized by reduction in the number of sacrals, loss of fusion between sacral vertebrae, and the loss of articulation of sacral vertebrae with the pelvis (Buchholtz, 1998). The tail or caudal vertebrae of cetaceans are defined as the first vertebra with a chevron bone immediately posterior to its caudal epiphysis and all vertebrae posterior to this (see Figure 8.19). Chevron bones are paired ventral intervertebral ossifications found in the caudal region of many vertebrates. They articulate via paired vertebral facets of the vertebra in front of them and are held in position by ligaments. When seen from the front they have a Y or a V shape. Pairs of chevrons form arches, creating a hemal canal that serves to protect blood vessels that supply the tail (Rommel and Reynolds, 2001). Variation also exists in the number of caudal vertebrae, ranging from a minimum of 13 in the pygmy right whale to a maximum of 49 in the finless porpoise, the Cuvier's beaked whale, and the pygmy sperm whale (Yablokov *et al.,* 1972; Rommel and Reynolds, 2001). There are 25–27 caudal vertebrae in the bottlenose dolphin (Rommel, 1990).

In a study of vertebral morphology among extant delphinids, Buchholtz and Schur (2004) described significant intrafamilial variation (i.e., differences in vertebral count, shape, and neural spine orientation). Most living delphinids differ from their ancestors in having localized flexibility to anterior (synclinal) and posterior (fluke base) sites, identified as a key innovation signaling the evolution of a bimodal torso. Bimodal torsos are associated with other vertebral changes that result in an increase in the flexibility of the tailstock.

During normal swimming the thoracic and lumbar vertebrae of cetaceans are restrained by a strong collagenous **subdermal connective tissue sheath** that gives rigidity to the thorax and provides an enlarged surface to anchor the flexor and extensor muscles of the tail (Pabst, 1993; Figure 8.21). Among the dorsal axial muscles, the position of the semispinalis muscle suggests that its action is to change the position of the skull relative to the vertebral column (providing extension and lateral flexion of the skull). This muscle also helps tense the anterior superficial and deep tendon fibers. The action of another major epaxial muscle group, the multifidus, is to stiffen its deep tendon of insertion, thereby forming a stable platform for the m. longissimus, another epaxial muscle. The longissimus muscle transmits the majority of its force to the caudal spine by way of a novel interaction between its insertional tendons and the subdermal connective tissue sheath (Pabst, 1993).

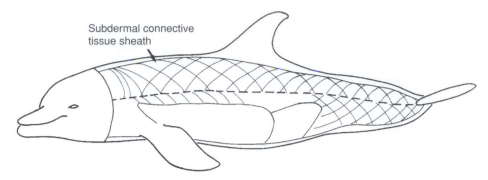

Figure 8.21.　Subdermal connective tissue sheath (SDS) of the bottlenose dolphin, *Tursiops truncatus* (blubber and m. cutaneous trunci removed). (From Pabst, 1990.)

The vertebral column is the primary structure used by dolphins to generate the dorsoventral bending characteristic of cetacean swimming (discussed later in this chapter). In dorsoventral bending, the intervertebral joints of the saddleback dolphin are stiff near the middle of the body and become more flexible toward the head and tail (Long *et al.,* 1997). The pattern of intervertebral joint stiffness is consistent with the hypothesis that axial muscles anchored in the lumbar spine cause extension of the tail (Pabst, 1993). Results of a mechanical study of the bending dynamics of the saddleback dolphin suggest that the intervertebral joint at the base of the flukes acts as a "low resistance hinge, permitting subtle and continuous alterations of the angle of attack of the flukes" on the water (Long *et al.,* 1997, p. 75). These workers propose a novel structural mechanism that functions to stiffen the joint in both tail extension and flexion. This mechanism works by placing ligaments in tension during both of these movements. As the articular processes of a caudal vertebra shear past the neural spine of the cranial vertebrae, mediolaterally oriented ligaments are lengthened (Figure 8.22). Another conclusion from this study was that the dolphin vertebral column has the capacity to

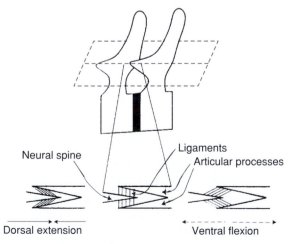

Figure 8.22.　Articular process stiffening mechanism in the saddleback dolphin. Left lateral view of a vertebral pair; the dashed and obliquely oriented plane is the frontal section through the articular processes shown below. (From Long *et al.,* 1997.)

store elastic energy and to dampen oscillations as well as to control the pattern of body deformation during swimming.

8.3.5. Sternum and Ribs

The sternum of odontocetes differs from that of mysticetes in several features. In odontocetes, 5–7 pairs of ribs usually are attached to the sternum (sternal ribs), which consists of a long, flat bone, usually segmented and widened anteriorly, with flared cup-like recesses at the site of attachments for ribs (see Figure 8.18). In mysticetes, there is only one pair of ribs attached to the sternum and the sternum is shorter and broader than in odontocetes (Yablokov *et al.,* 1972). Also unlike most other mammals, odontocetes have bony rather than cartilaginous sternal ribs (Rommel and Reynolds, 2001).

Cetaceans are unique in having single-headed ribs (but see later) that attach to their respective vertebrae only to the transverse processes instead of having two vertebral attachments points, one to the vertebral body and the other to the transverse processes, as in most mammals. There are 12–14 vertebral ribs in the bottlenose dolphin (Rommel, 1990). The anterior-most 4–5 vertebral ribs are double headed, with a proximal capitulum and distal tuberculum. Double headed ribs increase the number of joints between the ribs and vertebrae and allow considerable mobility of the rib cage. In the sperm and beaked whales, true sternal ribs, of which there are only 3–5 pairs, are cartilaginous. True sternal ribs are lacking in mysticetes and only the first pair of ribs is attached by a ligament to the extremely small sternum. The first pair of ribs does not have a head in mysticetes (the gray whale is an exception; Yablokov *et al.,* 1972). Cetaceans have poorly developed intercostal muscles and depend on the diaphragm for forceful inhalation.

8.3.6. Flippers and Locomotion

In cetaceans, the forelimb proportions are so altered from those of terrestrial mammals that the elbow is at approximately the body contour and the visible extremity consists almost entirely of the forearm and hand. Cetacean flippers are variable in size and shape. The flippers of sperm, killer, and beluga whales are almost round (in the beluga, the posterior edge may even be curved upward). The flippers are triangular in the La Plata, Amazonian, and Ganges river dolphins.

The humpback whale has the longest flippers of any cetacean, with their length varying from 25 to 33% of the total body length (Fish and Battle, 1995). The flipper shape is long, narrow, and thin. The humpback whale flipper is also unique because of the presence of large protuberances, or tubercles, located on the leading (anterior) edge, which give the flippers a scalloped appearance (Figure 8.23). Typically, barnacles are found on the upper leading edge of the tubercles. The flipper has a cross-sectional design typical of low-drag hydrofoils for lift generation and maneuverability. The position and number of tubercles on the flipper suggest that they function as enhanced lift devices to control flow over the flipper and to maintain lift at high angles of attack. The morphology of the humpback whale flipper further suggests that it is adapted for high maneuverability associated with the whale's unique "bubble-cloud" feeding behavior (Fish and Battle, 1995; Miklosovic *et al.,* 2004 discussed in more detail in Chapter 12).

The ratio of radius-ulna length to that of the humerus varies slightly among odontocetes; generally the forearm skeleton is longer, and in mysticetes, this is particularly true of the long and slender radius and ulna. The interphalangeal (finger) joints have nearly

Figure 8.23.   Flipper of humpback whale showing tubercles and cross-section illustrating hydrodynamic
                    design. (From Fish and Battle, 1995; photo courtesy Felipe Vallejo.)

flat surfaces and, as in other joints distal to the shoulder, capsules and dorsal and ventral
ligamentous arrangements provide almost complete immobilization between bones
(Felts, 1966).

The reduced density of cetacean limb bones (de Buffrénil *et al.,* 1986) is accompanied
by extensive development of a dense connective tissue matrix that may help to maintain
flipper strength. A study of the allometric scaling relationships of the cetacean forelimb
suggests that either the bones of large cetacean are underbuilt and less robust than
expected or that limb bones of small cetaceans are more robust than expected. A possible explanation for this negative allometry is the greater relative swimming speed of
small delphinids and phocoenids, as well as the greater stresses incurred by high speed
swimming (Dawson, 1994).

8.3.6.1. Pectoral Girdle and Forelimb

The scapula is typically wide, flat, and fan-shaped (Figure 8.24). In odontocetes, the
coracoid and acromial processes are well developed. In some mysticetes (e.g., the humpback whale), the acromion is reduced (Howell, 1930a). The infraspinous fossa occupies
practically the entire lateral aspect of the bone (although the muscle covers only 1/2 to
2/3 of the area), whereas the bony area of the supraspinous fossa is insignificant. The
supraspinatus muscle is therefore of lessened importance (Howell, 1930a). The scapula
of the franciscana differs from those of described delphinoids (dolphins, porpoises, narwhal, beluga) and physeteroids (sperm and dwarf sperm whales) in the relatively large
size of the supraspinous fossa (Strickler, 1978). Cetaceans lack the trapezius muscle.
Among odontocetes, the serratus ventralis muscle occurs only in the franciscana,
although it exists in rorquals among the mysticetes. The pectoralis abdominalis and
three rhomboideus divisions are found in the franciscana and the pygmy sperm whale
but in relatively few dolphins. Strickler (1978) suggested that these characteristics are
associated with a more generalized use of the forelimb in the franciscana.

The humerus, radius, and ulna are relatively short and flattened in cetaceans (see Figure
8.24). The radius and ulna exceed the humerus in length and have dorsoventrally compressed shafts. The elbow joint is immobile due to flattened articular facets. The lesser
tubercle of the humerus is medially positioned and reduced in size, indicated only by a
pronounced rugosity. The lack of development of the lesser tubercle indicates that either

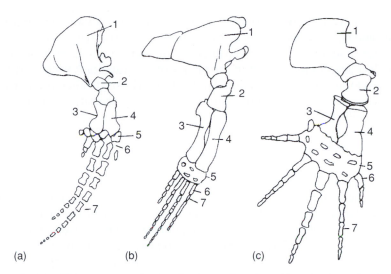

1=Scapula, 2=Humerus, 3=Ulna, 4=Radius, 5=Carpals, 6=Metacarpals, 7=Phalanges

Figure 8.24. Forelimb of representative cetaceans. **(a)** Pilot whale, *Globicephala* sp. **(b)** Blue whale, *Balaenoptera musculus.* **(c)** Right whale, *Eubalaena* sp. (From Evans, 1987.)

the subscapularis is unusually weak, which its extent belies, or else that it has a somewhat altered function (i.e., to effect rotation of the humerus when the arm is considerably flexed). The latter is regarded as more likely (Howell, 1930a). In mysticetes, the greater tubercle is well developed and may be nearly as high as the head. Although this process appears homologous with the greater tubercle it is hardly so, except in position, and is actually a deltoid process. In odontocetes, the proximal humerus differs. Instead of the head being located toward the rear of the shaft it is situated to the lateral side. Because of the shift of the head, the greater tubercle is located so as to allow the subscapularis (which inserts on it) to act as an efficient adductor (Howell, 1930a). Some fossil archaeocete whales retained a moveable elbow joint, a long humerus, and a flipper that is not as streamlined as in extant cetaceans. Among the most significant events in the evolution of the cetacean flipper is immobilization of the elbow and manus joints; among mysticetes (i.e., aetiocetids) this occurred in the early Oligocene (Cooper, personal communication).

The cetacean hand, unique among mammals, exhibits **hyperphalangy** (they have an unusually large number of finger bones, phalanges; see Figure 8.24). The development of hyperphalangy also occurs in two extinct lineages of aquatic reptiles, ichthyosaurs (Sedmera *et al.,* 1997a) and mosasaurs (Caldwell, 2002). Among cetaceans, hyperphalangy is limited to the central digits (second and third). The maximum number of phalanges found in an odontocete is the pilot whale (as many as 4, 14, and 11 in the first, second, and third digits, respectively: Fedak and Hall, 2004). Among mysticetes, balaenids have five digits, and all other mysticetes with the exception of the gray whale have four digits. When the distribution of hyperphalangy is optimized onto cetacean phylogeny it appears to have evolved several times and is further hypothesized to have evolved separately among mysticetes and odontocetes (Fedak and Hall, 2004). Although the phylogenetic distribution and developmental basis for hyperphalangy is unclear it is now under investigation (e.g., Richardson and Oelschlager, 2002; Cooper, personal communication). Watson (1994) reported a rare case of the congenital malformation **polydactyly,** or the presence of additional digits, in the bottlenose dolphin. In this case, one flipper had duplicated the fourth digit making the number of digits total six. Polydactyly has also been reported in the manus of the vaquita (Ortega-Ortiz *et al.,* 2000). The development of polydactylous limbs has been linked to altered activity of the homeobox genes during limb bud development caused by genetic, developmental, or environmental influences.

8.3.6.2. Pelvic Girdle and Hind Limb

Only a few altered and reduced pelvic bones remain in cetaceans. They have no direct connection to the vertebral column and are imbedded in the visceral musculature. In some cases the femur or tibia remains; in even rarer cases, elements of the foot exist (Yablokov *et al.,* 1972). Hind limb buds reported in the embryos of different cetacean species regress in early fetal life (for a review, see Sedmera *et al.,* 1997b). Various mechanisms are likely involved in the loss of the hind limb, including the nonexpression of certain homeobox genes. There is an intriguing similarity between the developmental pattern in snakes (i.e., progressive limb loss coincides with an increase in number of vertebrae) and the evolutionary pattern of cetaceans (i.e., decrease in number of digits in archaic cetaceans to loss of hind limb among modern taxa) raising the question of whether the same genes control development in the regions in both animals (Thewissen and Williams, 2002).

Some (possibly all) archaeocete cetaceans had external hind limbs. The pelvic facets on sacral vertebrae of *Protocetus* suggest a well-developed pelvis. Hind limbs have been discovered on *Prozeuglodon, Ambulocetus natans* has a large hindleg and foot, *Rodhocetus kasrani* has a large pelvis, and a protocetid from Georgia also has a pelvis (Hulbert, 1998). Archaeocetes may have been able to haul out on beaches as do pinnipeds. It has been suggested that the small functional hind limbs of *Basilosaurus isis* perhaps aided copulation (Gingerich *et al.,* 1990) or locomotion in shallow waters (Fordyce and Barnes, 1994; see also Chapter 4). However, it could just as reasonably be interpreted as vestigial structures without a function (Berta, 1994).

8.3.6.3. Tail (Fluke)

The tail, or **fluke,** of cetaceans (Figure 8.25) has the following basic components: (1) a cutaneous layer not significantly different from that described for other regions of the body; (2) a blubber layer far thinner than that the blubber layer over the rest of the body; (3) a ligamentous layer extending from the caudal keels and sides of the tail; and (4) a core of extremely tough, dense, fibrous tissue within the ligamentous envelope, forming the bulk of the fluke (Felts, 1966). Penetrating the fibrous core are numerous blood vessels arranged as heat-retaining countercurrent systems (Figure 8.26).

The flukes are an outgrowth of the lateral caudal region (similar to the development of limb buds from the lateral body wall) and are supported centrally by dorsoventrally compressed caudal vertebrae that extend almost to the fluke notch (Rommel, 1990). The shape of the flukes differs among cetaceans (see Figure 8.25) in response to varying hydrodynamic parameters (Fish, 1998). The trailing edges of most are slightly convex, but some are almost straight (sperm whale), and others are conspicuously curved (humpback), falcate (sickle-shaped; rorquals), or even biconvex (narwhal).

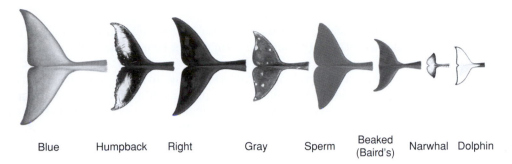

| Blue | Humpback | Right | Gray | Sperm | Beaked (Baird's) | Narwhal | Dolphin |

Figure 8.25. Different shaped flukes of various cetaceans. (Illustrated by P. Folkens.)

The evolution of cetacean flukes remains speculative although it seems likely that they evolved by the late Eocene as judged from the vertebral morphology of dorudontines and basilosaurids (Buchholtz, 1998; Bajpai and Thewissen, 2001; Gingerich *et al.*, 2001).

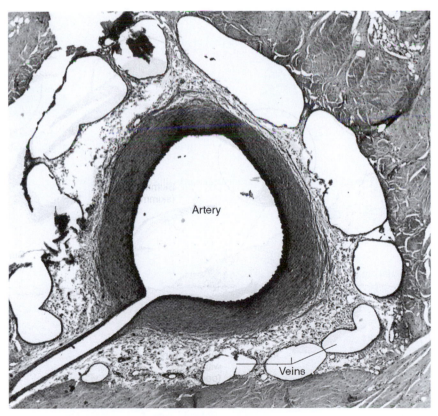

Figure 8.26. Cross-section of a vascular countercurrent artery (center) surrounded by several thin-walled veins embedded within the fluke connective tissue of a bottlenose dolphin. (Courtesy of S. Ridgway.)

8.3.6.4. Dorsal Fin

There is generally a prominent dorsal fin on the back of most cetaceans (Figure 8.27). Some cetaceans lack a dorsal fin (balaenids, sperm whales, beluga, narwhal, and some porpoises such as *Neophocoena*). In others it is poorly developed and too small to function as a mechanism preventing rolling during swimming. In most odontocetes it is larger, culminating in the adult male killer whales in which it may attain a length of almost 2 m. The fin is supported not by bone but by tough fibrous tissue similar in structure to flukes. Dorsal fins provide additional planing surfaces to assist in the maintenance of balance and maneuverability, for thermoregulation, and possibly for individual or conspecific recognition.

8.3.6.5. Mechanics of Locomotion

Modern cetaceans are caudal oscillators; they swim by vertical movements of the flukes (Figure 8.28) by the alternate actions of the epaxial and hypaxial muscles. The flukes act as paired dynamic wings that generate lift-derived thrust (Fish, 1998). Except for their different planes of tail motion, cetacean swimming is quite similar to that of tunas and billfishes. This derived mode of locomotion differs from that of all other marine

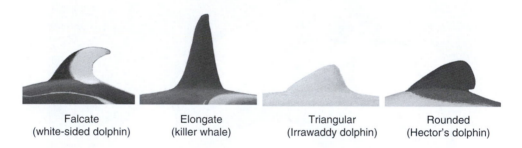

Falcate	Elongate	Triangular	Rounded
(white-sided dolphin)	(killer whale)	(Irrawaddy dolphin)	(Hector's dolphin)

Figure 8.27.   Representative dorsal fins of various cetaceans. (Illustrated by P. Folkens.)

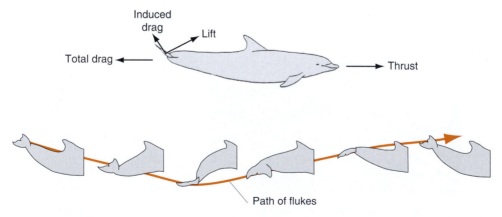

Figure 8.28.   Cetacean propulsion, tracings of flukes of bottlenose dolphin. (Modified from Coffey, 1977.)

mammals with the exception of sirenians. Caudal propulsion using high aspect ratio flukes that are relatively narrow and pointed at the tips provides improved efficiency at sustained high velocities (Fish, 1997). Differences between the sizes of the epaxial and hypaxial muscle masses led to the early contention that the up and down movements of the tail were not equivalent; the upstroke was considered to produce thrust while the downstroke functioned only as a recovery stroke. Newer data indicate thrust is produced during both phases, although the magnitude of thrust from the downstroke is greater than that produced by the upstroke (Fish and Hui, 1991). Maintaining the flukes in a positive angle of attack throughout the greater part of the stroke cycle ensures nearly continuous thrust generation (Fish, 1998).

The upstroke, which involves muscles that extend the tail, is powered primarily by two epaxial muscles: (1) the m. multifidus with its caudal extension (the m. extensor caudae medialis) and (2) the m. longissimus with its caudal extension (the m. extensor caudae lateralis; Pabst, 1990, 1993). The caudal extension of the m. longissimus is the only epaxial muscle that acts to control the angle of attack of the flukes. The down-stroke, which involves muscles that flex the tail and depress the flukes, is powered by the mm. flexor caudae lateralis and flexor caudae medialis; both are extensions of the m. hypaxalis lumborum. Another major tail flexor is the m. ischiocaudalis, which also provides some lateral torque. Strong elastic connective tissues in the flukes of cetaceans have been found to help in transmitting propulsive forces. Energy is tem-porarily stored as elastic strain energy in connective tissue (e.g., tendons) in the fluke and subdermal connective tissue sheath, and then is recovered in as elastic recoil instead of being dissipated and subsequently being replaced (at metabolic cost) by muscular work (Blickhan and Cheng, 1994). The overall effect is to restrict propulsive flexing to the posterior third of the body and concentrate thrust production on the flukes.

For individual animals, the frequency of fluke stroke cycles varies directly with swim-ming velocities. Trained captive bottlenose and oceanic spinner dolphins have been clocked in controlled test situations at greater than 40 km/hr (Lang and Pryor, 1966; Rohr et al., 2002), and peak velocities of killer whales are estimated to exceed 50 km/hr. Other maximum swimming velocities by both captive and free-ranging dolphins are summarized by Rohr et al. (2002). At maximum velocities, fine maneuvering becomes difficult. However, most marine mammals seldom approach their peak swimming veloc-ities because these velocities are energetically very expensive.

8.3.6.5.1. *Evolution of Cetacean Locomotor Patterns*

The caudal oscillation mode of cetacean swimming evolved from an initial quadrupedal locomotor stage. This was followed by a pelvic phase *(Ambulocetus),* a caudal undula-tion phase *(Kutchicetus),* and the final adoption of caudal oscillation, the swimming mode seen in dorudontids and all modern cetaceans (Figure 8.29; Fish, 1996; Thewissen and Fish, 1997; Thewissen and Williams, 2002; Gingerich, 2003). Fossil discoveries in Eocene rocks of Pakistan have revealed several of the critical evolutionary steps involved in the transition from land to the sea (Thewissen et al., 1996; Thewissen and Fish, 1997). *Ambulocetus natans,* the "walking whale," walked on land but could also swim using a pelvic paddling provided by the hind limbs. Its locomotion in the water has been compared to that of the modern otter *Lutra.* The hand of *Ambulocetus* was small relative to the foot and probably did not provide any propulsion. The thumb was mobile at the wrist but the arm was rigid. Forelimb motions were limited and the shoulder and

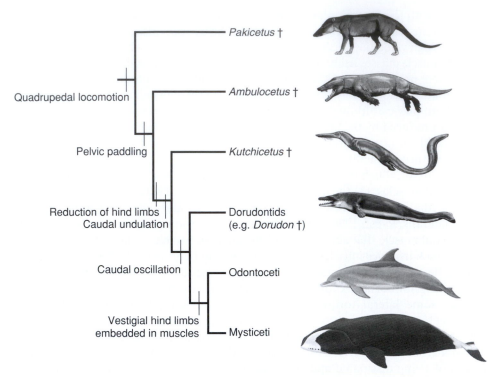

Figure 8.29. Evolution of locomotion in cetaceans. (Based on Thewissen and Williams, 2002.) †=extint
taxa. Illustrations by Carl Buell (fossil reconstructions) and Pieter Folkens (modern whales).

thumb were the primary means of modifying the position of the forelimb during swimming. Joints of the remaining digits were mobile and were likely used for locomotion on
land. The large feet with elongated, flattened phalanges, similar to those of pinnipeds,
suggest that *Ambulocetus* may have had webbed feet.

A later diverging archaic whale, *Kutchicetus* from Pakistan, spent most of its time in
the water. *Kutchicetus* was small with a long and muscular back and flat tail. Swimming
in *Kutchicetus* may have been similar to the South American giant freshwater otter
(*Pteronura;* Bajpai and Thewissen, 2000).

Another stage in the evolution of locomotion in cetaceans is represented by discoveries of two still later diverging archaic whales from the southeastern United States
(Hulbert *et al.,* 1998; Uhen, 1999). The protocetid, *Georgiacetus vogtlensis,* has been
described as having a pelvis that did not articulate with the hind limb, suggesting a loss
of any significant locomotor function of the hind limb in terrestrial locomotion.
Aquatic locomotion in *Georgiacetus* is inferred to have been primarily caudal oscillation
with a secondary contribution from the hind limb (Hulbert *et al.,* 1998). A final stage in
the evolution of swimming in cetaceans is seen in dorudontids, late Eocene whales that
have vertebral specializations (i.e., ball vertebra; Uhen, 1998) that indicate that they had
a fluke. They probably swam by dorsoventral oscillations of their tail flukes like modern
whales.

8.4. Sirenians

8.4.1. Skull and Mandible

The dugong skull has remarkably enlarged downturned premaxillae (Figure 8.30). Nasal bones are absent. Several features of the skull show sexual dimorphism, the most obvious of which is a difference in thickness of the premaxillae; they are more robust in the male, presumably because of the difference in tusk eruption (Nishiwaki and Marsh, 1985).

The manatee skull is broad with a relatively short snout and an expanded nasal basin. The premaxillae are only slightly downturned, relatively small, and lack tusks (Figure 8.30b). The vomer is short, extending anteriorly only to the level of the middle of the orbit, in the West African manatee, whereas in the West Indian manatee the vomer extends to the posterior edge of the incisive foramina or beyond. The lower jaw is massive. Manatees possess a specialized, lipid-filled structure (the zygomatic process of the squamosal) that has been suggested to function in sound reception in a similar manner to the fatty-filled mandibular canal of odontocetes (further discussed in Chapter 11).

Domning (1978), in his description of the musculature of the Amazonian manatee, made comparisons with other manatees and the dugong. Several notable differences exist. In the Amazonian manatee, the rectus capitis muscle has been modified from its function of bending the head laterally and strengthened to serve also as an extensor of the atlantooccipital joint in conjunction with the semispinalis muscle.

8.4.2. Vertebral Column and Axial Musculature

Sirenians have experienced a lengthening of the thorax, which has resulted in a shortening of the lumbar region. There are 57–60 vertebrae in the dugong (C7, T17-19, L4, S3, C28-29). Vertebrae number is 43–54 in manatees (C6, T15-19, and LSC23 to 29; Husar, 1977). In sirenians, the articulations of the vertebrae in the lumbocaudal region are reduced, more markedly in the manatee (Howell, 1930a). The presence of six rather than seven cervical vertebrae in manatees (also seen in two sloth lineages) appears to have evolved independently in mammals at least three times (Giffin and Gillett, 1996). Manatees differ from all other marine mammals is having indistinct or no vertebral epiphyses (Rommel and Reynolds, 2001).

(a)   (b)

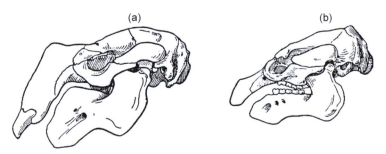

Figure 8.30.   Lateral views of **(a)** dugong and **(b)** manatee skulls. (From Gregory, 1951.)

8.4.3. Sternum and Ribs

The sternum in both the dugong and manatee is a broad, flat, single bone (the manubrium), with no indication of additional separate or fused sternal elements as is the normal mammalian condition. The manubrium of the West Indian manatee has a deep median notch in the anterior border, whereas that of the West African manatee lacks a deep notch (Husar, 1977, 1978). Only the first three pairs of ribs join the sternum; the others are free of distal articulation.

8.4.4. Flippers and Locomotion

8.4.4.1. Pectoral Girdle and Forelimb

The axilla in sirenians is situated just proximal to the elbow. Because sea cows, like all other marine mammals, lack a clavicle, the shoulder girdle is composed only of the scapula. The scapula of the dugong has a short acromion and a well-developed coracoid. The acromion is better developed in manatees. The humerus has prominent tubercles (Figure 8.31). The humeral head in both the manatee and dugong is situated fairly posterior to the main shaft axis. This suggests that the chief direction of movement is in the

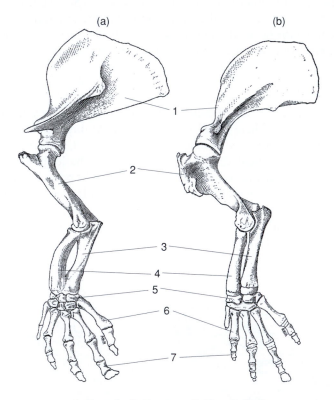

1=Scapula, 2=Humerus, 3=Ulna, 4=Radius
5=Carpals, 6=Metacarpals, 7=Phalanges

Figure 8.31.   Forelimb of **(a)** manatee and **(b)** dugong. (From Howell, 1930a.)

sagittal plane. This is also supported by the fact that the lesser tubercle is not medially positioned. Instead it is continuous with the greater tubercle and the two are conjoined to form a broad, transverse ridge anterior to the head. There is no teres minor muscle in the dugong so the scapular muscles inserting on this ridge are the subscapularis on the medial part, the infraspinatus on the lateral portion, and the supraspinatus between them. In the manatee, the deltoid is inserted on a rugosity on the middle of the shaft (Howell, 1930a). In the dugong, the humeral head is also posterior to the shaft axis, indicating that flexion and extension are the chief movements. The greater tubercle is higher than the lesser tubercle, and both are distinct instead of continuous as in manatees. In addition, there is a heavy and high deltoid crest that continues distally from the greater tubercle (Howell, 1930a).

The humerus, radius, and ulna are well developed in the dugong. This is especially true with regard to the proximal portion of the humerus, with its much stouter processes. The radius and ulna of both the manatee and the dugong are proximally fused and the dugong also exhibits fusion distally (see Figure 8.31). The olecranon process of the ulna is also much better developed in the dugong. The elbow joint is movable in both (Kaiser, 1974). The trochlea spiral of the humeroulnar joint of sirenians is in the opposite direction from that of terrestrial mammals (i.e., it is expanded posteriorly and medially and anteriorly and laterally). Domning (1978) suggests that this may be because the principal propulsive force is transmitted from the shoulder muscles through the humerus and elbow to the forearm, with negligible force exerted in the opposite direction; that is, the forearm flexors play little role in propulsion.

The forelimbs of Steller's sea cow are described as short, blunt, and hook-like and therefore would have been of little use in paddling. This animal probably remained in shallow water and employed the forelimbs to pull the body forward along the bottom. The wrist joint of sirenians is moveable. The pisiform (one of the wrist bones) is absent. The carpals of the dugong show a tendency for fusion (Harrison and King, 1965). The fifth digit is poorly developed with a single phalanx present (Kaiser, 1974). In the dugong, the carpal elements are reduced to three but in manatees there are six carpals (five carpals according to Quiring and Harlan, 1953, with fusion of the radiale and intermedium). The metacarpals and phalanges are flattened, especially in manatees. Two phalanges occur in four of the five digits; the first digit has one phalanx (Quiring and Harlan, 1953). The ungual phalanges (equivalent to finger tips) are of irregular shape and are particularly flat, and the thumb is reduced. The fourth digit is the longest and represents the tip of the flipper (Howell, 1930a).

The shoulder musculature in the dugong and manatees differs and Domning (1978) suggested that this may reflect the requirements of maneuvering in different habitats; the marine habitat of the dugong is more open than that of the manatee (Nishiwaki and Marsh, 1985).

8.4.4.2. Pelvic Girdle and Hind Limb

The pelvic girdle in sirenians is vestigial; pubic bones are absent and the ischium and ilium, which are both rod-like, are fused in adults into an innominate bone. Sexual dimorphism in the innominate bone of dugongs was described by Domning (1991). The pelvic girdle is absent in the West African manatee; innominate bones are reduced in both the West Indian manatee and West African manatee (Husar, 1977). Distal elements of the hind limbs are absent.

8.4.4.3. Hydrostatic Adaptations

The skeleton of sirenians is both pachyostotic and osteosclerotic. Heavy bones and horizontal lungs are adaptations involved in maintaining neutral buoyancy (Domning and de Buffrénil, 1991). A hydrostatic function for the diaphragm of sirenians has been also suggested (Domning, 1977). The diaphragm of the West Indian manatee differs from that of other marine mammals in lying in a horizontal plane and extending the length of the body cavity. Additionally, it does not attach to the sternum; it attaches medially at a central tendon forming two distinct hemidiaphragms. This unique orientation and extreme muscularity of the diaphragm has been related to buoyancy control. Accordingly, contractions of the diaphragm and abdominal muscles may change the volume in the surrounding pleural cavities to affect buoyancy, roll, and pitch (Rommel and Reynolds, 2000).

8.4.4.4. Mechanics of Locomotion

Sirenians, like cetaceans, use caudal oscillation to create propulsion. Nearly all of our information on sirenian swimming comes from Hartman's (1979) work on the West Indian manatee. Compared to cetaceans, manatees are poor swimmers and are unable to reach or sustain high speeds. According to Hartman (1979) movement is initiated from a stationary position by an upswing of the tail followed by a downswing, repeated until undulatory movement is established. Each stroke of the tail displaces the body vertically, the degree of pitching increasing with the power stroke (Figure 8.32). The tail also serves as a rudder. Cruising animals can bank, steer, and roll by means of the tail alone. The use of the flippers in locomotion differs somewhat from cetaceans. While cruising, the flippers of adult manatees are held motionless at the sides. Juvenile manatees have been reported to swim exclusively with their flippers (Moore, 1956, 1957). Flippers (either independently or simultaneously) are normally used only for precise maneuvering and for corrective movements to stabilize and orient the animal while it is feeding. The primary locomotory use of flippers is to turn an animal to the right or left. This is in contrast to dugongs, which may employ the flippers when cruising; dugongs also employ the flippers to turn and maintain balance.

Dugongs and manatees typically swim in a leisurely manner (Hartman, 1979; Nishiwaki and Marsh, 1985). As herbivores, sirenians do not require speed and rapid

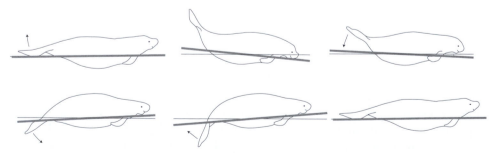

Figure 8.32.   Lateral view of a manatee swimming, tracings of body, limb, and tail movements. Arrows indicate direction of movement. (From Hartman, 1979.)

acceleration to catch prey. Low speed swimming allows for precise maneuverability. Swimming velocities of manatees vary greatly depending on activity; idling speed is 0.5–1 m/s and cruising speed is 1–2 m/s (Hartman, 1979). During flight, sprint speeds of over 6 m/s (22 km/h) aid in escape from predators such as sharks and crocodiles (Hartman, 1979; Nishiwaki and Marsh, 1985; Reynolds and Odell, 1991).

8.4.4.4.1. *Evolution of Sirenian Locomotor Patterns*

According to Domning (1996; Figure 8.33), sirenians passed through the following stages in their adaptation to water: (1) as mostly terrestrial quadrupeds, swimming by alternate thrusts of the limbs; followed by (2) amphibious quadrupeds swimming by dorsoventral spinal undulation and bilateral thrusts of the hind limbs; and (3) completely aquatic animals swimming with the tail only. The earliest stage in the evolution of sirenian locomotor patterns is represented by prorastomids from the early and middle Eocene of Jamaica (see Figure 8.33). Prorastomids had well-developed pelvic and hind limb bones. The development of large and expanded neural processes of the posterior vertebrae suggests unusual development of the longissimus dorsi muscles for spinal extension. Because the caudal vertebrae lack transverse processes, it is likely that the hind limbs rather than the tail were the major propulsive organs. Also apparent at this early stage of locomotor evolution were pachyostotic and osteosclerotic bones suggesting the need for heavy ballast when swimming. The next stage is represented by *Protosiren* from the middle Eocene of Egypt. Although this animal retained a complete hind limb, the pelvis was modified and the sacrum was only weakly connected to the

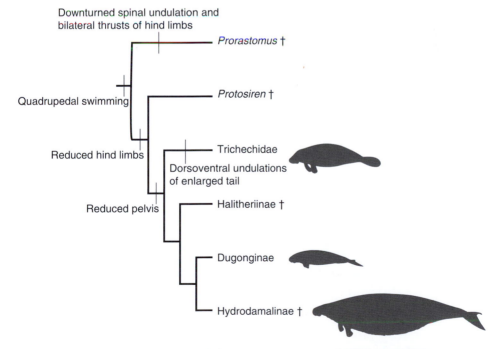

Figure 8.33.   Evolution of sirenian locomotion, † = extinct taxa. (Based on Domning, 1996.)

pelvis. The caudal vertebrae had broad transverse processes typical of later sirenians, suggesting that the tail had become the major propulsive organ, although the hind limbs probably played some role in swimming. Shortening of the neck occurred simultaneously with a reduction of the hind limbs and is apparent in *Protosiren*. Archaic dugongids living contemporaneously with some species of *Protosiren* represent a more advanced stage of evolution. The pubis was very reduced and it is clear that these animals could no longer support their weight on the hind limbs. Late Eocene and Oligocene dugongids show considerable variation in the pelvic and hind limb bones. The recent *Dugong* and *Hydrodamalis* lack femora and elements of the pelvis (including the acetabulum) entirely. Trichechids have completely lost the pelvis and, with it, connection to the vertebral column.

8.5. Sea Otter

8.5.1. Skull and Mandible

The sea otter skull is short and robust and is characterized by a blunt rostrum, prominent zygomatic arches, and well-developed sagittal and occipital crests. Auditory bullae are large and inflated. The lower jaw exhibits a large coronoid process and weakly developed angular process (see Figure 5.15). Skull length data indicate significant sexual dimorphism (Estes, 1989). Asymmetry has been reported in the sea otter skull (Barabash-Nikiforov *et al.,* 1947, 1968) in which the left side of the cranium tends to be larger in most individuals. More detailed study of this phenomenon indicates that the asymmetry varies both between individuals and between populations (Roest, 1993).

8.5.2. Vertebral Column and Axial Musculature

Sea otters possess 50–51 vertebrae with a typical vertebral formula, C7, T14, L6, S3, C20-21. The vertebrae of the sea otter differ from those of the river otter in having larger intervertebral foraminae, especially posteriorly, and very reduced vertebral processes. The neck of the sea otter is shorter relative to the length of the trunk in comparison to the river otter (Taylor, 1914). This shortness of the neck is associated with streamlining of the body and the development of the thoracolumbar and caudal regions used for propulsion.

   During rapid aquatic locomotion in the sea otter (as well as in the river otter), the lumbosacral region is moved vertically, paralleling the activity of the tail and hind foot movements. As one would expect, there is an increased development of these regions displayed by increased muscle mass of the epaxial musculature (mm. multifidis lumborum and longissimus thoracis; Gambarajan and Karapetjan, 1961) and an increased height of the neural spines and transverse processes, which provide attachment points for those muscles.

8.5.3. Sternum and Ribs

Sea otters possess 14 pairs of ribs. The first 10 pairs of ribs articulate loosely with the sternum and contribute to the mobility of the thoracic region.

8.5.4. Pectoral Girdle and Forelimb

The sea otter scapula is relatively smaller than in the river otter and not so long antero-posteriorly. This has been related to the lack of dependence on forelimbs for support (Taylor, 1914). There is no clavicle, allowing extreme mobility of the pectoral girdle. The forelimbs are proportionally smaller than in the river otter. The forelimb in general and the wrist in particular are highly mobile. The forefeet are small with reduced metacarpals and phalanges (Figure 8.34).

8.5.5. Pelvic Girdle and Hind Limb

The sea otter pelvis is elevated, lying more nearly parallel to the vertebral column than is found in the river otter. The ilia are markedly turned outward anteriorly. The femur, tibia, and fibula are relatively short. There is no round ligament in the sea otter, which confers greater mobility to the femur (see Figure 8.34). Both the metatarsals and phalanges are elongated (see Figure 8.34b) and a large web of skin exists between the digits of the hind feet, so that the foot becomes twice as wide when the digits are spread (Taylor,

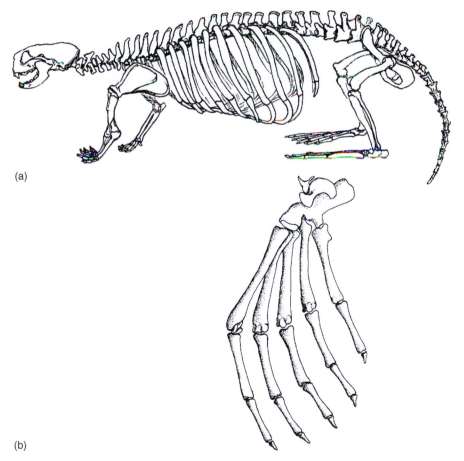

(a)

(b)

Figure 8.34. Sea otter, *Enhydra lutris*. **(a)** Skeleton. (From Chanin, 1985.) **(b)** Hind foot. (From Taylor, 1989.)

1989). The fourth and fifth digits are closely bound to give rigidity to the hind flipper for propulsion. In the sea otter, the proportion of the hind limb protruding from the body contours is reduced.

8.5.6. Locomotion

The hind limbs of sea otters are so much larger than the forelimbs that terrestrial locomotion is clumsy and slow (Tarasoff *et al.,* 1972; Kenyon, 1981). The sea otter on land exhibits two patterns of locomotion: walking and bounding. Unlike the river otter, there are no running movements. Instead, for rapid forward locomotion, bounding occurs. For walking, the general pattern of movement is one of forward movement of alternate limbs.

Aquatic locomotion in modern sea otters is achieved by pelvic paddling when at the surface and pelvic undulation when submerged (Williams, 1989). Pelvic paddling involves hind limb propulsion and pelvic undulation is provided by vertical flexing of the vertebral column (Figure 8.35). This is in contrast to a forelimb-specialized swimming that has been suggested for the extinct giant otter *Enhydritherium* (Lambert, 1997; further discussed in Chapter 5; Thewissen and Fish, 1997). Specializations for pelvic paddling in modern sea otters include a robust femur and elongated distal hind limb elements (Tarasoff *et al.* 1972). Three primary modes of swimming were observed by Williams (1989): (1) ventral surface up swimming; (2) ventral surface down swimming; and (3) alternate ventral up and down swimming. Ventral surface up swimming was used during periods of food manipulation and ingestion and in the initial stages of an escape response from a disturbance. In this position, the animal's body is partially submerged with the head and chest held above the water surface. The forefeet are folded close to the chest above the water line while the hind feet provided propulsion. Both alternate and simultaneous strokes of the hind feet were observed. Occasionally otters maneuver slowly by making lateral undulations of the tail while the hind feet are raised above the water surface. During ventral surface down swimming, the head and scapular region of the back remain above the water surface. Propulsion is provided by either alternate or simultaneous strokes of the hind feet (see Figure 8.35). The forepaws are held against the submerged chest. Neither the forefeet nor the tail appear to play a role in propulsion. This position is often used during intermediate speed travel between areas and prior to a dive and high speed submerged swimming. An intermediate form of surface swimming also is observed that is associated with grooming behavior. Rather than maintaining a fixed position, the animals alternately swam with their ventral surface up or down. Such rolling along the long axis of the body is superimposed on the forward progression. As with the other forms of surface swimming, the hind paws provided the propulsion.

In the same study, sea otters were found to have two distinct speed ranges that varied with swimming mode (Williams, 1989). Sustained surface swimming, including ventral

Figure 8.35.   Aquatic locomotion in the sea otter, tracings of body, limb, and tail movements. (From Tarasoff *et al.*, 1972.)

surface up, ventral surface down, and rolling body positions, occurred at speeds less than 0.8 m/s. Generally, ventral up swimming was used at the lower range of preferred speeds (0.1–0.5 m/s). With increases in surface swimming speed, this position was replaced by ventral surface down and rolling positions. Often ventral down swimming preceded submerged swimming; consequently, submerged swimming occurred over a higher range of speeds. Steady swimming by submerged sea otters ranged from 0.6 to 1.4 m/s. Based on these results, crossover speeds between surface and submerged swimming ranged from 0.6 to 0.8 m/s.

8.6. Polar Bear

In comparison to other marine mammals, considerably less information is available on the anatomy of the polar bear. Modern polar bears are the largest bear species, although they are smaller than their Pleistocene ancestors. Males are larger than females and there is considerable sexual dimorphism in the skull as well as overall differences in body dimensions.

Polar bears (Figure 8.36) possess 39–45 vertebrae distributed to the formula C7, T14-15, L5-6, S4-6, and C9-11. The muscles of the polar bear's long neck are especially well developed (Uspenskii, 1977).

Polar bears, like most other terrestrial mammals, have few morphological adaptations for efficient swimming. They have large robust limbs **plantigrade** feet, which form flat plates oriented perpendicular to the direction of motion, creating dragbased thrust to move the animal forward. Polar bears swim with a stroke like a crawl, pulling through the water with the forelimbs while the hind legs trail behind (Flyger and Townsend, 1968). This is reflected in the development of a wide flange on the posterior margin of the scapula called the postscapular fossa (Davis, 1949; see Figure 8.36). The subscapularis muscle arises in part from this fossa and the unusual position of this muscle in bears is correlated with their method of climbing, or in the case of polar bears, with swimming, which involves pulling up the heavy body by the forelimbs.

When traveling on land, the polar bear's huge paws help distribute their body weight, whereas their foot pads, which are covered with small soft papillae, increase friction between the feet and the ice (Stirling, 1988). The claws of polar bears are relatively large and robust. They are used to grip the ice when speed is required and they also assist them in pulling seals out of their breathing holes. The polar bear uses a terrestrial walk (Dagg, 1979) similar to that of other large carnivores; lateral legs are used to a large extent, whereas diagonal legs are seldom used (Taylor, 1989). Terrestrial locomotion in polar bears has been measured at speeds up to 11 m/s (Hurst *et al.*, 1982).

8.7. Summary and Conclusions

The pinniped skull is characterized by large orbits, a relatively short snout, a constricted interorbital region, and large orbital vacuities. The modern walrus skull is easily distinguished from that of other pinnipeds by being foreshortened and having large maxillae for accommodation of the upper canine tusks. The axial skeleton is differently developed among otariids, phocids, and the walrus. The neck vertebrae of otariids are large with well-developed processes associated with muscles for movements of the head and neck.

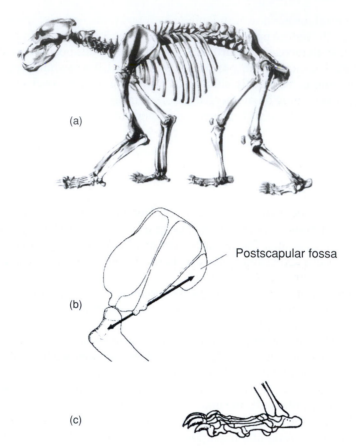

Figure 8.36.   Polar bear, *Ursus maritimus*. **(a)** Skeleton. **(b)** Scapula with arrow showing line of action of
subscapularis muscle. **(c)** Plantigrade foot. (From Ewer, 1973.)

In the walrus and in phocids the enlarged processes of the lumbar vertebrae provide
attachment surfaces for hypaxial muscles associated with horizontal movements of the
posterior end of the body. The bones of the pinniped fore and hind limb are short, flattened, and modified as flippers. Propulsion for swimming by marine mammals is derived
from paired flipper movements (pinnipeds and the sea otter) or vertical movements of
caudal flukes (cetaceans and sirenians). Among modern pinnipeds, aquatic and terrestrial locomotion are achieved differently. Otariids use pectoral oscillation and phocids
and the walrus use pelvic oscillation for propulsion in water. Terrestrial locomotion is
likewise achieved differently: it is ambulatory in walruses and otariids, and phocids
employ sagittal undulation.

The cetacean skull differs profoundly from the typical mammalian skull because it is
telescoped, the result of migration of the external narial opening to a dorsal position on
the skull. In cetaceans, the vertebral column does not contain a sacral region because the
pelvic girdle is absent. The subdermal connective tissue sheath provides an enlarged surface to anchor flexor and extensor muscles of the fluke. The elbow joint of modern
cetaceans is uniquely immobile. The cetacean fore flipper is used primarily for steering
rather than for propulsion. The cetacean flipper is unique among mammals in exhibiting

hyperphalangy. The fossil record has revealed several critical steps in the transition of whales from land to sea. Archaic whales possessed a well-developed pelvis and hind limbs and walked on land. Later diverging whales reduced the hind limbs and developed large processes on the sacral vertebra indicating that caudal undulation of the body was well developed, like in modern whales.

The sirenian skull is distinguished from other marine mammals by its downturned premaxilla. It is more sharply inclined in the dugong than in manatees. The heavy dense bones of sirenians in addition to the unique morphology of the lungs and diaphragm are adaptations involved in hydrostatic regulation. As in cetaceans, the forelimb flipper is primarily used for steering. The pelvic girdle is vestigial.

The sea otter skull is short and robust and is characterized by a blunt rostrum, prominent cheeks, and well-developed bony crests for the attachment of powerful jaw-closing muscles. Because the hind limbs are larger than the forelimbs, locomotion on land is slow and clumsy. Swimming involves undulations of the vertebral column, tail, and hind foot. Correlated with this is an increased development of vertebral processes in the lumbo-sacral region for attachment of the epaxial musculature involved in swimming. Polar bears, the largest bears, use their huge plantigrade forefeet to generate propulsion in swimming. Terrestrial locomotion involves walking and running with their large paws serving for weight distribution.

8.8. Further Reading

For excellent recent comparative works on the anatomy of marine mammals consult Pabst *et al.* (1999), Rommel and Reynolds (2000, 2001), and Rommel and Lowenstine (2001). The classical comparative treatment of the functional anatomy of pinnipeds, cetaceans, sirenians, and the sea otter is found in Howell (1930a); a summary focused on the energetics of aquatic locomotion is provided by Fish (1992). A good general introduction to pinniped musculoskeletal anatomy is King (1983). The osteology and myology of fossil and extant phocids is described by Muizon (1981). The most thorough comparative survey of the morphology of skulls of southern fur seals is provided by Repenning *et al.* (1971). Descriptions of pinniped osteology and musculature in varying degrees of detail are given for the bearded seal (Mikker, 1888), Ross seal (King, 1969; Pierard and Bisaillon, 1979), Weddell seal (Howell, 1929; Pierard, 1971), hooded seal, Antarctic fur seal (Miller, 1888), California sea lion (Howell, 1929; Mori, 1958), Steller sea lion (Murie, 1872, 1874), Pacific walrus (Murie, 1871; Bisaillon and Pierard, 1981; Kastelein *et al.*, 1991), harbor seal (Miller, 1888), baikal seal (Koster *et al.*, 1990), ringed seal (Miller, 1888; Howell, 1929), and the southern elephant seal (Bryden, 1971). A classic treatment of North American pinnipeds is Allen (1880).

Accounts of the cetacean skeleton are provided for the right whale (Eschricht and Reinhardt, 1866), minke whale (Omura, 1975), pygmy blue whale (Omura *et al.*, 1970), Cuvier's beaked whale (Omura, 1972), and the bottlenose dolphin (Rommel, 1990). Detailed descriptions of cetacean musculature are given for the pygmy sperm whale (Schulte and Smith, 1918), sperm whale (Berzin, 1972), porpoises (Boenninghaus, 1902; Schulte, 1916; Howell, 1927; Moris, 1969; Sokolov and Rodinov, 1974; Mead, 1975; Smith *et al.*, 1976; Kastelein *et al.*, 1997), Sei whale (Carte and McAlister, 1868; Schulte, 1916), Risso's dolphin (Murie, 1871), pilot whale (Murie, 1874), narwhal (Hein, 1914; Howell, 1930b), bottlenose dolphin (Huber, 1934), and susu (Pilleri *et al.*, 1976).

Principal references for descriptions of the facial anatomy of odontocetes include Lawrence and Schevill (1956), Mead (1975), Purves and Pilleri (1973), Schenkkan (1973), Heyning (1989), Curry (1992), and Cranford *et al.* (1996).

An atlas of the osteology of sirenians is provided by Kaiser (1974). Detailed descriptions of the myology of the dugong are given by Domning (1977) and for the Amazonian manatee see Domning (1978). The osteology of the West Indian manatee is described by Quiring and Harlan (1953). Brief reviews of the musculoskeletal anatomy and locomotion in the manatee are in Caldwell and Caldwell (1985); for the dugong see Nishiwaki and Marsh (1985) and Reynolds and Odell (1991).

Principal contributors to the osteology and myology of the sea otter include Taylor (1914), Gambarajan and Karapetjan (1961), and Howard (1973, 1975). For an account of the limb anatomy of the sea otter with reference to locomotion see Tarasoff *et al.* (1972).

References

Allen, J. A. (1880). *History of North American Pinnipeds: A Monograph of the Walruses, Sea-Lions, Sea-Bears and Seals of North America*. Government Printing Office, Washington, D. C.

Backhouse, K. M. (1961). "Locomotion of Seals with Particular Reference to the Forelimb." *Symp. Zool. Soc. London 5:* 59–75.

Bajpai and J. G. M. Thewissen (2000). "NEW Dimension Eocene Whale from Kachchh (Gujarat, India) and its Implications for Locomotor Evolution of Cetaceans." *Curr. Sci. 79:* 1478–1482

Barabash-Nikiforov, I. I. (1947). *"Kalan" (The Sea Otter)*. Soviet Ministrov RSFSR. (Translated from Russian, Israel Program for Scientific Translations, Jerusalem, 1962).

Barabash-Nikiforov, I. I., S. V. Marakov, and A. M. Nikolaev (1968). *Otters (Sea Otters)*. Izd-vo Nauka, Leningrad [in Russian].

Barnes, L. G., and S. A. McLeod (1984). The fossil record and phyletic relationships of gray whales. *In* "The Gray Whale" (M. L. Jones and S. L. Swartz, eds.), pp. 3–28. Academic Press, New York.

Beentjes, M. P. (1990). "Comparative Terrestrial Locomotion of the Hooker's Sea Lion *(Phocarctos hookeri)* and the New Zealand Fur Seal *(Arctocephalus forsteri):* Evolutionary and Ecological Implications." *Zool. J. Linn. Soc. 98:* 307–325.

Berta, A. (1994). "What Is a Whale?" *Science 263:* 181–182.

Berta, A., and P. Adam (2001). Evolutionary biology of pinnipeds. *In* "Secondary Adaptation of Tetrapods to Life in Water" (J-M. Mazin and V. de Buffrenil, eds.), pp. 235–260. Verlag Dr. Friedrich Pfeil, Munchen, Germany.

Berzin, A. A. (1972). *The Sperm Whale* (A. V. Yablokov, ed.). (Translated from Russian, Israel Program for Scientific Translations, Jerusalem).

Bisaillon, A., and J. Pierard (1981). "Oseologie du Morse de l'Atlantique *(Odobenus rosmarus, L.,* 1758) Ceintures et membres." *Zentralbl. Veterinaer med. Reihe C. 10:* 310–327.

Blickhan, R., and J.-Y. Cheng (1994). "Energy Storage by Elastic Mechanisms in the Tail of Large Swimmers-A Re-Revaluation." *J. Theor. Biol. 168:* 315–321.

Boenninghaus, G. (1902). "Der Rachen von *Phocaena communis." Zool. Jahrb. 17:* 1–98.

Bryden, M. M. (1971). "Myology of the Southern Elephant Seal *Mirounga leonina (L.)." Antarc. Res. Ser. 18:* 109–140.

Bryden, M. M., and W. J. L. Felts (1974). "Quantitative Anatomical Observations on the Skeletal and Muscular System of Four Species of Antarctic Seals." *J. Anat. 118:* 589–600.

Buchholtz, E. A. (1998). Implications of vertebral morphology for locomotor evolution in early Cetacea. *In* "The Emergence of Whales: Patterns in the Origin of Cetacea" (J. G. M. Thewissen, ed.), pp. 325–351. Plenum Press, New York.

Buchholtz, E. A., and S. A. Schur (2004). "Vertebral Osteology in Delphinidae (Cetacea)." *Zool. Jour. Linnean Soc. 140:* 383–401.

Burns, J. J. (1981). Ribbon seal-*Phoca fasciata* Zimmerman, 1783. *In* "Handbook of Marine Mammals," (S. H. Ridgway and R. J. Harrison, eds.), pp. 89–109. Academic Press, London.

Caldwell, M.W. (2002). "From Fins to Limbs: Limb Evolution in Fossil Marine Reptiles." *Amer. Jour. Med. Genetics 112:* 236–249.

Caldwell, D. K., and M. C. Caldwell (1985). Manatees. *In* "Handbook of Marine Mammals," Vol. 3 (S. H. Ridgway and R. Harrison, eds.), pp. 33–66. Academic Press, New York.

Carrier, D. R., S. M. Deban, and J. Otterstrom. (2002). "The Face That Sank the Essex: Potential Function of the Spermaceti Organ in Aggression." *J. Exp. Biol. 205:* 1755–1763.

Carte, A., and A. McAlister (1868). "On the Anatomy of *Balaenoptera rostrata*." *Phil. Trans. Royal Soc. London 158:* 201–261.

Chanin, P. (1985). *The Natural History of Otters*. Croom Helm, London.

Clarke, M. R. (1970). "Function of the Spermaceti Organ of the Sperm Whale." *Nature 228:* 873–974.

Clarke, M. R. (1978). "Buoyancy Control as a Function of the Spermaceti Organ in the Sperm Whale." *J. Mar. Biol. Assoc. U.K. 58:* 27–51.

Clarke, M. R. (1979). "The Head of the Sperm Whale." *Sci. Am. 240:* 106–117.

Coffey, D. L. (1977). *Dolphins, Whales and Porpoises*. MacMillan Press, New York.

Cranford, T. W. (1999). "The Sperm Whale's Nose: Sexual Selection on a Grand Scale?" *Mar. Mamm. Sci. 15:* 1133–1157.

Cranford, T. W., M. Amundin, and K. S. Norris (1996). "Functional Morphology and Homology in the Odontocete Nasal Complex: Implications for Sound Generation." *J. Morphol. 228:* 223–285.

Curry, B. E. (1992). "Facial Anatomy and Potential Function of Facial Structures for Sound Production in the Harbor Porpoise *(Phocoena phocoena)* and Dall's Porpoise *(Phocoenoides dalli)*." *Can. J. Zool. 70:* 2103–2114.

Dagg, A. L. (1979). "The Walk of the Large Quadrupedal Mammals." *Can. J. Zool. 57:* 1157–1163.

Davis, D. D. (1949). "The Shoulder Architecture of Bears and Other Carnivores." *Fieldiana. Zool. 31:* 285–305.

Dawson, S. D. (1994). "Allometry of Cetacean Forelimb Bones." *J. Morphol. 222:* 215–221.

de Buffrénil, V., and A. Casinos (1995). "Observations on the Microstructure of the Rostrum of *Mesoplodon densirostris* (Mammalia, Cetacea, Ziphiidae): The Highest Density Bone Known." *Ann. Sci. Nat., Zool. Biol. Anim. [14] 16:* 21–32.

de Buffrénil, V., J. Y. Sire, and D. Schoevaert (1986). "Comparison of the Skeletal Structure and Volume Between a Delphinid *(Delphinus delphis* L.) and a Terrestrial Mammal *(Panthera leo L.)*." *Can. J. Zool. 64:* 1750–1756.

Deméré, T. A. (1986). "The Fossil Whale *Balaenoptera davidsonii* (Cope, 1872), with a Review of Other Neogene Species of *Balaenoptera* (Cetacea: Mysticeti)." *Mar Mamm. Sci. 2:* 277–298.

Deméré, T. A. (1994). "The Family Odobenidae: A Phylogenetic Analysis of Fossil and Living Taxa." *Proc. San Diego Soc. Nat. Hist. 29:* 99–123.

DeSmet, W. M. A. (1977). The regions of the cetacean vertebral column. *In* "Functional Anatomy of Marine Mammals," Vol. 3 (R. J. Harrison, ed.), pp. 59–79. Academic Press, London.

Domning, D. P. (1977). "Observations on the Myology of *Dugong dugong* (Muller)." *Smithson. Contrib. Zool. 226:* 1–56.

Domning, D. P. (1978). "The Myology of the Amazonian Manatee, *Trichechus inunguis* (Natterer) (Mammalia: Sirenia)." *Acta Amazonica, 8:* Suppl. 1.

Domning, D. P. (1991). "Sexual and Ontogenetic Variation in the Pelvic Bones of *Dugong dugon* (Sirenia)." *Mar. Mamm. Sci. 7:* 311–316.

Domning, D. P. (1996). The readaptation of Eocene sirenians to life in the water. *In* "Abstracts of the Conference on Secondary Adaptation to Life in the Water" (J.-M. Mazin, P. Vignaud, and V. de Buffrénil, eds.), p. 5. Poitiers, France.

Domning, D. P., and V. de Buffrénil (1991). "Hydrostasis in the Sirenia: Quantitative Data and Functional Interpretations." *Mar. Mamm. Sci. 7:* 331–368.

English, A. W. (1974). *Functional Anatomy of the Forelimb in Pinnipeds*. PhD thesis, University of Illinois MedCenter, Chicago.

English, A. W. (1976). "Limb Movements and Locomotor Function in the California Sea Lion *(Zalophus californianus)*." *J. Zool. Soc. London 178:* 341–364.

English, A. W. (1977). "Structural Correlates of Forelimb Function in Fur Seals and Sea Lions." *J. Morphol. 151:* 325–352.

Eschricht, D. F., and J. Reinhardt (1866). On the Greenland right-whale *(Balaena mysticetus)*. *In* "Recent Memoirs on the Cetacea" (W. H. Flower, ed.), pp. 1–45. Ray Society, London.

Estes, J. (1989). Adaptations for aquatic living by carnivores. *In* "Carnivore Behavior, Ecology and Evolution" (J. L. Gittleman, ed.), pp. 242–282. Cornell University Press, Ithaca, NY.

Evans, P. G. H. (1987). *The Natural History of Whales and Dolphins*. Facts on File, New York.

Ewer, R. F. (1973). *The Carnivores*. Cornell University Press, Ithaca, NY.

Fay, F. H. (1981). Walrus-*Odobenus rosmarus*. *In* "Handbook of Marine Mammals," Vol. 1 (S. H. Ridgway, and R. J. Harrison, eds.), pp. 1–23. Academic Press, London.

Fedak, T. J., and B. K. Hall (2004). "Perspectives on Hyperphalangy: Patterns and Processes." *J. Anat. 204:* 151–163.

Feldkamp, S. D. (1987). "Forelimb Propulsion in the California Sea Lion *Zalophus californianus*." *J. Zool. Soc. London 212:* 4333–4357.

Felts, W. J. L. (1966). Some functional and structural characteristics of cetacean flippers and flukes. *In* "Whales, Dolphins and Porpoises" (K. S. Norris, ed.), pp. 255–276. University of California Press, Berkeley.

Fish, F. E. (1992). Aquatic locomotion. *In* "Mammalian Energetics: Interdisciplinary Views of Metabolism and Reproduction"(T. E. Tomasci and T. H. Horton, eds.), pp. 34–63. Cornell University Press, Ithaca, NY.

Fish, F. E. (1996). "Transition from Drag-Based to Lift-Based Propulsion in Mammalian Swimming." *Am. Zool. 36:* 628–641.

Fish, F. E. (1997). Biological designs for enhanced maneuverability: Analysis of marine mammal perform-ance. *Tenth International Symposium on Unmanned Untethered Submersible Technology*. 10 pp.

Fish, F. E. (1998). Biomechanical perspective on the origin of cetacean flukes. *In* "The Emergence of Whales: Patterns in the Origin of Cetacea"(J. G. M. Thewissen, ed.), pp. 303–324. Plenum Press, New York.

Fish, F. E., and J. M. Battle (1995). "Hydrodynamic Design of the Humpback Whale Flipper." *J. Morphol. 225:* 51–60.

Fish, F. E., and C. A. Hui (1991). "Dolphin Swimming—A Review." *Mamm. Rev. 21:* 181–195.

Fish, F. E., S. Innes, and K. Ronald (1988). "Kinematics and Estimated Thrust Production of Swimming Harp and Ringed Seals." *J. Exp. Biol. 137:* 157–173.

Fish, F. E., J. E. Peacock, and J. J. Rohr. (2003). "Stabilization Mechanism in Swimming Odontocete Cetaceans by Phased Movements." *Mar. Mamm. Sci. 19:* 515–528.

Flyger, V, and M. R. Townsend (1968). "The Migration of Polar Bears." *Sci. Am. 218:* 108–116.

Fordyce, E., and L. G. Barnes (1994). "The Evolutionary History of Whales and Dolphins." *Ann. Rev. Earth Planet. Sci. 22:* 419–455.

Fraser, F C., and P. E. Purves (1960). "Hearing in Cetaceans." *Bull. Brit. Mus. (Nat. Hist.) Zool. 7:* 1–140.

Gambarajan, P. P., and W. S. Karpetjan (1961). "Besonderheiten im Bau des Seelöwen *(Eumetopias californi-anus), der* Baikalrobbe *(Phoca sibirica)* and des Seotters *(Enhydra lutris)* in anpassung an die Fortbewegung im Wasser." *Zool. Jahrb. Anat. 79:* 123–148.

Gardner, S. C., and U. Varanasi (2003). "Isovaleric Acid Accumulation in Odotocete Melon During Development." *Naturwiss. 90:* 528–531.

Giffin, E. B., and M. Gillett (1996). "Neurological and Osteological Definition of Cervical Vertebrae in Mammals." *Brain Behav. Evol. 47:* 214–218.

Gingerich, P. D. (2003). "Land-to-Sea Transition in Early Whales." *Paleobiology 29:* 429–454.

Gingerich, P. D., B. H. Smith, and E. L. Simons (1990). "Hind Limbs of *Basilosaurus:* Evidence of Feet in Whales." *Science 249:* 403–406.

Gingerich, P. D., M. ul Haq, I. S. Zalmout, I. H. Khan, and M. S. Malkani (2001). "Origin of Whales from Early Artiodactyls: Hands and Feet of Eocene Protocetidae from Pakistan." *Science 293:* 2239–2242.

Godfrey, S. J. (1985). "Additional Observations of Subaqueous Locomotion in the California Sea Lion *(Zalophus californianus)*." *Aquat. Mamm. 11*(2): 53–57.

Gordon, K. (1981). "Locomotor Behaviour of the Walrus *(Odobenus)*." *J. Zool. Soc. London 195:* 349–367.

Gregory, W. K. (1951). *Evolution Emerging,* Vol. 2. Maximillan, New York.

Harrison, R. H., and M. M. Bryden (1988). Whales, Dolphins and Porpoises. Facts on File, New York.

Harrison, R. H., and J. E. King (1965). *Marine Mammals,* 2nd ed. Hutchinson, London.

Hartman, D. S. (1979). Ecology and behavior of the manatee *(Trichechus manatus)* in Florida. *Am. Soc. Mammal. Spec. Publ.*, No. 5.

Hein, S. A. A. (1914). "The Larynx and Its Surrounding in *Monodon*." *Verhand. Kor Adkad. Wetensch.* 2-18??: 4–54.

Heyning, J. E. (1984). "Functional Morphology Involved in Intraspecific Fighting of the Beaked Whale *Mesoplodon carlhubbsi*." *Can. J. Zool. 62:* 1645–1654.

Heyning, J. E. (1989). "Comparative Facial Anatomy of Beaked Whales (Ziphiidae) and a Systematic Revision Among the Families of Extant Odontoceti." *Nat. Hist. Mus. L. A. Cty. 405:* 1–64.

Heyning, J. E., and J. G. Mead (1990). Function of the nasal anatomy of cetaceans. *In* "Sensory Abilities of Cetaceans"(J. Thomas and R. Kastelein, eds.), pp. 67–79. Plenum, New York.

Howard, L. D. (1973). "Muscular Anatomy of the Forelimb of the Sea Otter *(Enhydra lutris)*." *Proc. Calif. Acad. Sci. 39:* 411–500.

Howard, L. D. (1975). "Muscular Anatomy of the Hind Limb of the Sea Otter *(Enhydra lutris)*." *Proc. Calif. Acad. Sci. 40:* 335–416.

Howell, A. B. (1927). "Contribution to the Anatomy of the Chinese Finless Porpoise, *Neomeris phocaenoides.*" *Proc. U. S. Natl. Mus. 70*(Artic. 13): 1–43.

Howell, A. B. (1929). "Contribution to the Comparative Anatomy of the Eared and Earless Seals (Genera *Zalophus* and *Phoca*)." *Proc. U. S. Natl. Mus. 73*(Artic. 15): 1–142.

Howell, A. B. (1930a). *Aquatic Mammals.* Charles C. Thomas, Springfield, IL.

Howell, A. B. (1930b). "Myology of the Narwhal *(Monodon)*." *Am. J. Sci. 46:* 187–215.

Huber, E. (1934). "Anatomical Notes on Pinnipedia and Cetacea." *Carnegie Inst. Washington Publ. 447:* 105–136.

Hulbert, R.C., Jr. (1998). Postcranial osteology of the North American Middle Eocene protocetid *Geogiacetus.* In "The Emergence of Whales" (J. G. M. Thewissen, ed.), pp. 235–268. Plenum, New York.

Hulbert, R. C. Jr., R. M. Petkewich, G. A. Bishop, D. Bukry, and D. P. Aleshire (1998). "A New Middle Eocene Protocetid Whale (Mammalia: Cetacea: Archaeoceti) and Associated Biota from Georgia." *J. Paleontol. 72:* 907–927.

Hurst, R. J., M. L. Leonard, P. Beckerton, and N. A. Oritsland (1982). "Polar Bear Locomotion: Body Temperature and Energetic Cost." *Can. J. Zool. 60:* 222–228.

Husar, S. (1977). "The West Indian Manatee *(Trichechus manatus)*." *Res. Rep. U. S. Fish Wildl. Serv. 7:* 1–22.

Husar, S. (1978). *"Trichechus sengalensis."* *Mamm. Species 89:* 1–3.

Kaiser, H. E. (1974). *Morphology of the Sirenia: A Macroscopic and X-ray Atlas of the Osteology of Recent Species.* Karger, Basel.

Kastelein, R. A., J. L. Dubbledam, J. Luksenburg, C. Staal, and A. A. H. van Immerseel (1997). Anatomical atlas of an adult female harbour porpoise *(Phocoena phocoena).* In "The Biology of the Harbour Porpoise" (A. J. Read, P. R. Wiepkema, and P. E. Nachtigall, eds.), pp. 87–178. De Spil Publishers, Woerden, The Netherlands.

Kastelein, R. A. N. M. Gerrits, and J. L. Dubbledam (1991). "The Anatomy of the Walrus Head *(Odobenus rosmarus)*. Part 2: Description of the Muscles and of Their Role in Feeding and Haul-Out Behavior." *Aquat. Mamm. 17:* 156–180.

Kellogg, R. (1928). "The History of Whales—Their Adaptation to Life in the Water." *Q. Rev. Biol. 3:* 29–76.

Kenyon, K. (1981). Sea Otter—*Enhydra lutris.* In "Handbook of Marine Mammals" (S. H. Ridgway and R. J. Harrison, eds.), pp. 209–223. Academic Press, New York.

King, J. E. (1966). "Relationships of the Hooded and Elelphant Seals (genera *Cystophora* and *Mirounga*)." *J. Zool. Soc. London 148:* 385–398.

King, J. E. (1969). "Some Aspects of the Anatomy of the Ross Seal, *Ommatophoca rossi* (Pinnipedia: Phocidae)." *Brit. Antarct. Surv. Sci. Rep. 63:* 1–54.

King, J. E. (1983). *Seals of the World.* Oxford University Press, London.

Kooyman, G. (1981). Leopard seal, *Hydrurga leptonyx.* In "Handbook of Marine Mammals, Vol. 2, Seals" (S. H. Ridgway and R. J. Harrison, eds.), pp. 261–274. Academic Press, London.

Koster, M. D., K. Ronald, and P. Van Bree (1990). "Thoracic Anatomy of the Baikal Seal Compared with Some Phocid Seals." *Can. J. Zool. 68:* 168–182.

Lambert, W. D. (1997). "The Osteology and Paleoecology of the Giant Otter *Enhydritherium terranovae*." *J. Vertebr. Paleontol. 17:* 738–749.

Lang, T. G., and K. Pryor (1966). "Hydrodynamic Performance of Porpoises *(Stenella attenuata)*." *Science 152:* 531–533.

Lawrence, B., and W. E. Schevill (1956). "The Functional Anatomy of the Delphinid Nose." *Bull. Mus. Comp. Zool. 114:* 103–151.

Le Boeuf, B. J., Y. Naito, T. Asaga, D. Crocker, and D. P. Costa (1992). "Swim Speeds in a Female Northern Elephant Seal: Metabolic and Foraging Implications." *Can. J. Zool. 70:* 786–795.

Long, J. H., D. A. Pabst, W. R. Shepard, and W. A. McLellan (1997). "Locomotor Design of Dolphin Vertebral Columns: Bending Mechanics and Morphology of *Delphinus delphis*." *J. Exp. Biol. 200:* 65–81.

Macdonald, D. (ed.) (1984). *All the World's Animals Sea Mammals.* Torstar, New York.

Mead, J. (1975). "Anatomy of the External Nasal Passages and Facial Complex in the Delphinidae (Mammalia:Cetacea)." *Smithson. Contrib. Zool. 207:* 1–72.

Mikker 1888 cited in the text.

Miklosovic, D. S., M. M. Murray, L. E. Howle, and F. E. Fish (2004). "Leading-Edge Tubercles Delay Stall on Humpback Whale *(Megaptera novaeangliae)* Flippers." *Phys. Fluids 16*(5): 39–42.

Milinkovitch, M. (1995). "Molecular Phylogeny of Cetaceans Prompts Revision of Morphological Transformations." *Trends Ecol. Evol. 10:* 328–334.

Miller, W. C. G. (1888). The myology of the Pinnipedia. *In* "Report on the Scientific Results of the Voyage of the H.M.S. Challenger," Vol. 26, (2):139–240. Challenger Office, 1880–1895, Edinburgh.

Moore, J. C. (1956). "Observations of Manatees in Aggregations." *Am. Mus. Novit. 1811:* 1–24.

Moore, J. C. (1957). "Newborn Young of a Captive Manatee." *J. Mamm. 38:* 137–138.

Mori, M. (1958). "The Skeleton and Musculature of *Zalophus*." *Okajimas Folia Anat. Jpn. 31:* 203–284.

Moris, F. (1969). "Étude anatomique de la region cephalique du marouin, *Phocoena phocoena* L. (cétacé odontocete)." *Mammalia 33:* 666–726.

Muizon, C. de (1981). *Les Vertébrés Fossiles de la Formation Piscou (Pérou) Part 1. Recherches sur les Grandes Civilisations,* Mem. No. 6. Institut Francais d'Etudes Andines, Paris.

Murie, J. (1871). "On Risso's *Grampus*." *J. Anat. Physiol. Norm. Pathol. Homme Anim. 5:* 118.

Murie, J. (1872). "Researches Upon the Anatomy of the Pinnipedia. Part 2. Descriptive Anatomy of the Sea Lion *(Otaria jubata)*." *Trans. Zool. Soc. London 7:* 527–596.

Murie, J. (1874). "Researches Upon the Anatomy of the Pinnipedia. Part 3. Descriptive Anatomy of the Sea Lion *(Otaria jubata)*." *Trans. Zool. Soc. London 8:* 501–582.

Nishiwaki, M., and H. Marsh (1985). Dugong *Dugong dugon* (Muller, 1776). *In* "Handbook of Marine Mammals," Vol. 33 (S. H. Ridgway and R. Harrison, eds.), pp. 1–31. Academic Press, New York.

Nomina Anatomica Veterinaria (1983). 5th ed. Williams and Wilkens, Baltimore, MD.

Norris, K. S. (1964). Some problems of echolocation in cetaceans. *In* "Marine Bio-acoustics" (W. N. Tavolga, ed.), pp. 317–336. Pergamon Press, Oxford.

Norris, K. S. (1968). The evolution of acoustic mechanisms in odontocete cetaceans. *In* "Evolution and Environment" (E. T. Drake, ed.), pp. 297–324. Yale University Press, New Haven, CT.

Norris, K. S. (1969). The echolocation of marine mammals. *In* "The Biology of Marine Mammals" (H. T. Anderson, ed.), pp. 391–423. Academic Press, New York.

Norris, K. S., and C. M. Johnson (1994). Locomotion. *In* "The Hawaiian Spinner Dolphin" (K. S. Norris, B. Wursig, R. S. Wells, and M. Wursig, eds.), pp. 201–205. University of California Press, Berkeley, CA.

O'Gorman, F. (1963). "Observations on Terrestrial Locomotion in Antarctic Seals." *Proc. Zool. Soc. London 141:* 837–850.

Omura, H. (1972). "An Osteological Study of the Cuvier's Beaked Whale *Ziphius cavirostris,* in the Northwest Pacific." *Sci. Rep. Whales Res. Inst. 24:* 1–34.

Omura, H. (1975). "Osteological Study of the Minke Whale from the Antarctic." *Sci. Rep. Whales Res. Inst. 27:* 1–36.

Omura, H., T. Ichihara, and T. Kasuya (1970). "Osteology of Pygmy Blue Whale with Additional Information on External and Other Characteristics." *Sci. Rep. Whales Res. Inst. 22:* 1–27.

Ortega-Ortiz, J. G, B. Villa-Ramirez, and J. R. Gersenowies (2000). "Polydactyly and Other Features of the Manus of the Vaquita, *Phocoena sinus*." *Mar. Mamm. Sci. 16:* 277–286.

Pabst, D. A. (1990). Axial muscles and connective tissues of the bottlenose dolphin. *In* "The Bottlenose Dolphin" (S. Leatherwood and R. R. Reeves, eds.), pp. 51–67. Academic Press, San Diego, CA.

Pabst, D. A. (1993). "Intramuscular Morphology and Tendon Geometry of the Epaxial Swimming Muscles of Dolphins." *J. Zool. Soc. London 230:* 159–176.

Pabst, D. A. S. A. Rommel, and W. A. McLellan (1999). The functional morphology of marine mammals. *In* "Biology of Marine Mammals" (eds. J. E. Reynolds and S. A Rommel), pp. 15–72. Smithsonian University Press, Washington D. C.

Pierard, J. (1971). "Osteology and Myology of the Weddell Seal *Leptonychotes weddelli* (Lesson, 1826)." *Antarct. Res. Ser. 18:* 53–108.

Pierard, J., and A. Bisaillon (1979). "Osteology of the Ross Seal, *Ommatophoca rossi* Gray, 1844." *Antarct. Res. Ser. 31:* 1–24.

Ponganis, P. J., E. P. Ponganis, K. V. Ponganis, G. L. Kooyman, G. L. Gentry, and F. Trillmich (1990). "Swimming Velocities in Otariids." *Can. J. Zool. 68:* 2105–2112.

Pouchet, M. G., and H. Beauregard (1892). "Sur "l'organe des spermaceti."" *Compt. Rend. Soc. Biol. 11*: 343–344.

Purves, P. E., and G. Pilleri (1973). "Observations on the Ear, Nose, Throat and Eye of *Platanista indi*." *Invest. Cetacea 5:* 13–57.

Quiring, D. P., and C. F. Harlan (1953). "On the Anatomy of the Manatee." *J. Mammal. 34:* 192–203.

Reidenberg, J. S., and J. T. Laitman (1994). "Anatomy of the Hyoid Apparatus in Odontoceti (Toothed whales): Specializations of Their Skeleton and Musculature Compared with Those of Terrestrial Mammals." *Anat. Rec. 240:* 598–624.

Repenning, C. E., R. S. Peterson, and C. L. Hubbs (1971). Contributions to the systematics of the southern fur seals, with particular reference to the Juan Fernandez and Guadalupe species. *In* "Antarctic Pinnipedia" (W. H. Burt, ed.), pp. 1–34. *Antarct. Res. Ser.* 18, American Geophysical Union, Washington, D. C.

Reynolds, J., and D. Odell (1991). *Manatees and Dugongs.* Facts on File, New York.

Richardson, M. K., and H. H. A. Oelschlager (2002). "Time Pattern, and Heterochrony: A Study of Hyperphalangy in the Dolphin Embryo Flipper." *Evol. Develop. 4:* 435–444.

Roest, A. I. (1993). "Asymmetry in the Skulls of California Sea Otters *(Enhydra lutris nereis)*." *Mar. Mamm. Sci. 9:* 190–194.

Rohr, J. J., F. E. Fish, and J. W. Gilpatrick (2002). "Maximum Swim Speed of Captive and Free-Ranging Delphinids: Critical Analysis of Extraordinary Performance." *Mar. Mamm. Sci. 18:* 1–19.

Rommel, S. A. (1990). Osteology of the bottlenose dolphin. *In* "The Bottlenose Dolphin" (S. Leatherwood and R. R. Reeves, eds.), pp. 29–49. Academic Press, San Diego, CA.

Rommel, S. A., and L. J. Lowenstine (2001). Gross and microscopic anatomy. *In* "CRC Handbook of Marine Mammal Medicine" (L. A. Dierauf, and F. M. D. Gulland, eds.), pp. 129–164. CRC Press, Boca Raton, FL.

Rommel, S. A., and J. E. Reynolds III (2000). "Diaphragm Structure and Function in the Florida Manatee *(Trichechus manatus latriostris)*." *Anat. Rec., 259:* 41–51.

Rommel, S. A., and J. E. Reynolds III (2001). Skeletal anatomy. *In* "Encyclopedia of Marine Mammals" (W. F. Perrin, B. Wursig, and J. G. M. Thewissen, eds.), pp. 1089-1103. Academic Press, San Diego, CA.

Sato, K., Y. Mitani, M. F. Cameron, D. B. Siniff, and Y. Naito (2003). "Factors Affecting Stroking Patterns and Body Angle in Diving Weddell Seals Under Natural Conditions." *J. Exp. Biol. 206:* 1461–1470.

Schaller, O. (1992). *Illustrated Veterinary Anatomical Nomenclature.* Fedinan Enke Verlag, Stuttgart.

Schenkkan, E. J. (1973). "On the Comparative Anatomy and Function of the Nasal Tract in Odontocetes (Mammalia, Cetacea)." *Bijdra. Dierkd. 43:* 127–159.

Schulte, H. von W. (1916). "Anatomy of a Fetus of *Balaenoptera borealis*." *Mem. Am. Mus. Nat. Hist. 1:* 389–502.

Schulte, H. von W., and M. De F. Smith (1918). "The External Characters, Skeletal Muscles and Peripheral Nerves of *Kogia breviceps*." *Bull. Am. Mus. Nat. Hist. 38:* 7–72.

Sedmera, D., I. Misek, and M. Klima (1997a). "On the Development of Cetacean Extremities: II. Morphogenesis and Histogenesis of the Flippers in the Spotted Dolphin *(Stenella attenuata)*." *Eur. J. Morphol. 35*(2): 117–123.

Sedmera, D., I. Misek, and M. Klima (1997b). "On the Development of Cetacean Extremities: 1. Hind Limb Rudimentation in the Spotted Dolphin *(Stenella attenuata)*." *Eur. J. Morphol. 35*(1): 25–30.

Slijper, E. J. (1936). "Die Cetaceen, vergleichendantomisch and systematisch." *Capita Zool. 6/7:* 1–600.

Slijper, E. J. (1962). *Whales.* Hutchinson, London.

Smith, G. J. D., K. W. Browne, and D. E. Gaskin (1976). "Functional Myology of the Harbour Porpoise, *Phocoena phocoena (L.)*." *Can. J. Zool. 54:* 716–729.

Stirling, I. (1988). *Polar Bears.* University of Michigan Press, Ann Arbor.

Strickler, T. L. (1978). "Myology of the Shoulder of *Pontoporia blainvillei,* Including a Review of the Literature on Shoulder Morphology in the Cetacea." *Am. J. Anat. 152:* 419-432.

Tarasoff, F. J. (1972). Comparative aspects of the hind limbs of the river otter, sea otter and seals. *In* "Functional Anatomy of Marine Mammals," Vol. 1 (R. J. Harrison, ed.), pp. 333–359. Academic Press, London.

Tarasoff, F. J., A. Bisaillon, J. Pierard, and A. P. Whitt (1972). "Locomotory Patterns and External Morphology of the River Otter, Sea Otter, and Harp Seal (Mammalia)." *Can. J. Zool. 50:* 915–929.

Taylor, M. E. (1989). Locomotor adaptations by carnivores. *In* "Carnivore Behavior, Ecology and Evolution" (J. L. Gittleman, ed.), pp. 382–409. Cornell University Press, Ithaca, NY.

Taylor, W. P. (1914). "The Problem of Aquatic Adaptation in the Carnivora, as Illustrated in the Osteology and Evolution of the Sea-Otter." *Univ. Calif., Berkeley, Publ. Dept. Geol. 7*(25): 465–495.

Thewissen, J. G. M., and F. E. Fish (1997). "Locomotor Evolution in the Earliest Cetaceans: Functional Model, Modern Analogues, and Paleontologic Evidence." *Paleobiology 23:* 482–490.

Thewissen, J. G. M., S. I. Madar, and S. T. Hussain (1996). "*Ambulocetus natans,* an Eocene Cetacean (Mammalia) from Pakistan." *CFS Cour Forschungsinst. Senckenberg 191:* 1–86.

Thewissen, J. G. M., and E. M. Williams (2002). "The Early Radiations of Cetacea (Mammalia): Evolutionary Pattern and Developmental Correlations." *Annu. Rev. Ecol. Syst. 33:* 73–90.

True, F. W. (1904). "The Whalebone whales of the Western North Atlantic Compared with those Occurring in European Waters with Some Observations on the Species of the North Pacific." *Smithson. Contrib. to Knowledge 33:* 1–332.

Uhen, M. D. (1999). "New Species of Protocetid Archaeocete Whale, *Eocetus wardii* (Mammalia: Cetacea), from the Middle Eocene of North Carolina." *J. Paleontol. 73:* 512–528.

Uspenskii, S. M. (1977). *Belyi Medved (The Polar Bear)*. Navka, Moscow. Unedited translation by Government of Canada Translation Bureau, No. 1541321, June, 1978.

Varanasi, U., and D. C. Malins (1970). "Ester and Ether-Linked Lipids in Mandibular Canal of a Porpoise *(Phocoena phocoena)*—Occurrence of Isovaleric Acid in Glycerolipids." *Biochemistry 9:* 4576–4579.

Watson, A. (1994). "Polydactyly in a Bottlenose Dolphin, *Tursiops truncatus*." *Mar Mamm. Sci. 10:* 93–100.

Williams, T. M. (1989). "Swimming by Sea Otters: Adaptations for Low Energetic Cost Locomotion." *J. Comp. Physiol., A 164:* 815–824.

Wood, F. G. (1964). General discussion. *In* "Marine Bio-Acoustics," Vol 1. (W. N. Tavolga, ed.), pp. 395–398. Pergamon Press, New York.

Wyss, A. R. (1988). "On "Retrogression" in the Evolution of the Phocinae and Phylogenetic Affinities of the Monk Seals." *Am. Mus. Novit. 2924:* 1–38.

Yablokov, A. V., V. M. Bel'kovich, and V. I. Borisov (1972). *Whales and Dolphins*. Israel Programs for Scientific Translations, Jerusalem.

Zioupos, P., J. D. Currey, A. Casinos, and V. de Buffrénil (1997). "Mechanical Properties of the Rostrum of the Whale *Mesoplodon densirostris,* a Remarkably Dense Bony Tissue." *J. Zool. 241:* 725–737.

9

Energetics

9.1. Introduction

The field of marine mammal **energetics** employs a variety of methods to evaluate the energetic costs and benefits of life processes. These processes include the energetic costs of acquiring needed resources (including energy resources) and the manner in which those resources are allocated. Studies of energy allocation by marine mammals help to elucidate the evolution of physiological adaptations of mammals to the challenges imposed by existence in seawater, with its high density, viscosity, and thermal conductivity. These adaptations are expressed in the ways marine mammals allocate energy for daily or seasonal activities, ranging from their choices about when and where to migrate or reproduce to their choices of prey items.

The patterns of cost-benefit interactions are often portrayed as energy flow models at scales ranging from communities to individuals (Figure 9.1). The need to integrate the energetics of individuals into a dynamic picture of population or community energetics is addressed in Chapter 14. In this chapter, we focus on major patterns of energy expenditures by individuals, including basal metabolism, thermoregulation, swimming, and osmoregulation. Benefits are typically expressed as improved performance or as additional tissue building, either through growth or reproduction. Diving physiology is considered separately in Chapter 10 as are maternal costs of reproduction in Chapter 13. Energy acquisition, including diet and foraging energetics, is discussed in Chapter 12.

9.2. Metabolic Rates

The cellular mechanisms used by marine mammals to process energy-rich substrates (primarily lipids) are no different than those of other mammals. When submerged, however, marine mammals are isolated from their access to O_2. Consequently, many species must ration their stored O_2 during dives and must deal with higher concentrations of lactate produced in anaerobic glycolysis than terrestrial mammals.

Whole-animal metabolic rates represent the total of all the individual organ and tissue metabolic rates of an individual. Metabolic rates of individuals can be measured in a variety of ways, although some widely employed methods are not appropriate for larger

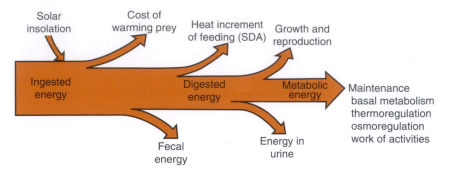

Figure 9.1.   Idealized model of energy flow through a marine mammal. (Adapted from Costa, 2002.)

or noncaptive marine mammals (Table 9.1); a detailed discussion of these methods is beyond the scope of this book (see Boyd, 2002, and Costa, 2002 for more details).

The units used to measure metabolic rates are often a consequence of the method of measurement. For example, O_2 respirometry yield results in units of O_2 volume used per unit time, whereas heat loss measurements results in calorie units. Both O_2 consumption and heat production measurements only estimate metabolic rates indirectly because metabolic rates represent units of work per units of time (joules/s or watts). Most measures of metabolic rates can be converted approximately to watts (W). Units of O_2 used per unit time derived from direct or indirect respirometry methods are presented as $\dot{V}_{O2}$ (ml $O_2 \cdot s^{-1}$, pronounced "vee-dot-O-two." Although $\dot{V}_{O2}$ does not measure work, it can be converted to equivalent units of work with knowledge of an animal's respiratory quotient (RQ = CO_2 produced/O_2 used). RQ values range from 0.7 when lipids are oxidized for energy to 1.0 for carbohydrates. With the exception of herbivorous sirenians, RQ values for marine mammals typically range from 0.74 to 0.77, reflecting a mixed diet of lipid and protein.

Metabolic rate comparisons rely on accepted standardized methods of not only the units but also the biological state of an animal that influence metabolic rates. **Basal metabolic rate** (BMR) is a measure of the metabolic rate of sexually and physically mature, postabsorptive individuals when at rest in a **thermoneutral** environment (Kleiber, 1932, 1947, 1975). BMR is often the preferred measure for comparison across species or larger taxonomic units because it is not complicated by activity levels, feeding or reproductive behaviors, or variable environmental conditions. Kleiber (1947) published an analysis of BMRs for a variety of domesticated mammals and derived a

Table 9.1. Examples of Methods Used for Measuring Metabolic Rates of Marine Mammals (Adapted from Boyd, 2002)

Method	Species	Published example
Min. heat loss	*Balaenoptera acutorostrata*	Blix and Folkow, 1995
Ventilation rate	*Eschrichtius robustus*	Sumich, 1983
O_2 respirometry	*Tursiops truncatus*	Williams *et al.*, 1993
Heart rate	*Zalophus californianus*	Boyd *et al.*, 1995
Mass balance	*Phoca vitulina*	Markussen *et al.*, 1990
Labelled water	*Zalophus californianus*	Boyd *et al.*, 1995

regression equation to describe the relationship between BMR and body mass. This regression curve, often referred to as the "Kleiber curve" (Figure 9.2), scales BMR to body mass to the 0.75 power (BMR = $aM^{0.75}$), or to the –0.25 power for mass-specific BMR. The Kleiber curve is commonly used as a benchmark for comparing metabolic rates derived from animals in nonbasal conditions; yet its applicability for comparisons with marine mammals is compromised by the fact that metabolic rate (MR) measurements performed on marine mammals rarely satisfy the conditions for basal metabolic measurements. Also, their metabolic processes do not necessarily compare well to the species, many of which were domestic animals genetically selected for high rates of production and metabolism, that were the basis of Kleiber's curve.

Attempts to estimate metabolic rates of marine mammals are often complicated by the large size of many species their habit of staying in an environment that is not thermally neutral, their large seasonal variation in body mass as a result of seasonal bouts of feeding and fasting, and the lack of agreement on how allometric relationships developed from studies of small cetaceans, pinnipeds, and other captive mammals might serve as appropriate metabolic models for much larger animals (Lavigne *et al.*, 1986; McNab, 1988; Boyd, 2002). More practical measures of marine mammal metabolic rates include resting metabolic rates (RMR, measured at rest but not satisfying other basal conditions), active metabolic rates (AMR, measured during specific activities such as swimming or diving), and field metabolic rates (FMR, measured in field situations with unrestrained animals and sometimes reported as mean daily metabolic rates, MDMR).

The issue of whether marine mammals have higher metabolic rates than terrestrial mammals of comparable size continues to be debated. Boyd (2002) suggested that

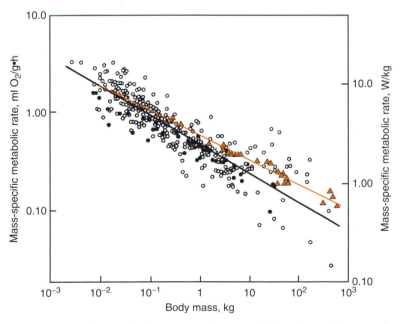

Figure 9.2.   Relationships between body masses and mass-specific basal metabolic rates of mammals. Colored points and regression line based on domestic mammals (Kleiber, 1947); circles (open = placental mammals; filled = marsupial mammals) and black regression line based on wild mammals (McNab, 1988).

measurements of FMR tend to converge on Kleiber's curve at larger body sizes (Figure 9.3). Several studies of pinniped metabolism (as measured by O_2 respirometry) have shown that RMRs of pinnipeds are about 1.5–3 times greater than those of terrestrial mammals of similar size (e.g., Kooyman, 1981). Similarly it has been argued that cetaceans have a higher metabolic rate than average terrestrial mammals of similar size (Irving et al., 1941; Pierce, 1970; Hampton et al., 1971; Ridgway and Patton, 1971; Kasting et al., 1989), although the range of body masses available for comparison is limited. Lavigne et al. (1986) and Huntley et al. (1987) have argued that metabolic rates of pinnipeds are not significantly different from those of terrestrial mammals and suggested that previous studies did not always make comparisons under standardized conditions (Figure 9.4). Opposing evidence has also been presented for cetaceans (Karandeeva et al., 1973; Lavigne et al., 1986; Innes and Lavigne, 1991) using similar data to reject claims of higher metabolic rates.

Metabolic rates of manatees are 20–35% of that predicted by Kleiber's regression (Gallivan and Best, 1980; Irvine, 1983), and dugongs are also presumed to have low metabolic rates (Marsh et al., 1978). It has been suggested that the low metabolic rates reported for sirenians (Figure 9.4) are a result of their relatively poor quality diet, as discussed further in Chapter 12. A lifestyle based on foraging aquatic plants does not demand complex foraging strategies or elaborate sensory systems. Additionally, as suggested by O'Shea and Reep (1990), it is likely that the low metabolic rate of sirenians placed constraints on potential selection for larger relative brain size. Thus, sirenians evolved large body sizes that were advantageous for energy conservation and for processing low-quality forage, but not larger relative brain size (see also Chapter 7).

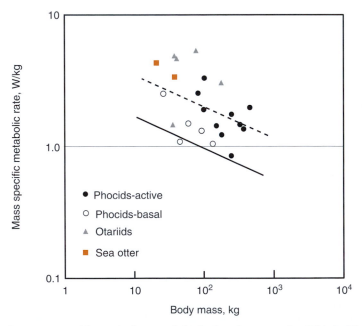

Figure 9.3.   Some mass-specific metabolic rates of pinnipeds and sea otters (see Table 1 of Boyd, 2002, for sources). Lower line is the Kleiber curve for BMR; upper line is 2× Kleiber.

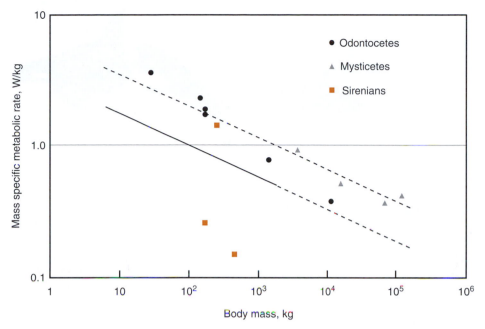

Figure 9.4. Some mass-specific metabolic rates of cetaceans and sirenians (see Table 1 of Boyd, 2002, for sources). Lower line is the Kleiber curve for BMR; upper line is 2× Kleiber.

9.3. Thermoregulation

Marine mammals live in a cold, highly conductive medium. The heat capacity of water is 25 times that of air. Even in tropical waters a 10°C temperature differential exists between the core temperature of a mammal and its aquatic environment, and this difference can be 35–40°C in polar waters. Marine mammals employ several methods to reduce heat losses in cold environments: large body size and reduced surface-to-volume ratios, increased insulation, and heat-conserving vascular countercurrent systems. These structural adaptations provide many marine mammals with very broad thermal neutral zones and only rarely are they required to elevate their metabolism for purely thermoregulatory reasons.

9.3.1. Body Size and Surface-to-Volume Ratios

Marine mammals span a body mass range of approximately four orders of magnitude, from sea otter neonates (~5 kg) to adult blue whales (10×10^4 kg). The surface area of a body increases proportional to the square of its length while its volume (approximately equal to its mass) increases proportional to its cube. It follows that for bodies of the same general shape, the larger ones will have low surface area-to-volume ratios (Figure 9.5). Most marine mammals are large and they are thus capable of producing considerable heat with relatively little loss at the surface. Sea otters, although small in comparison to most marine mammals, are more than twice as large as the largest terrestrial mustelid, the European badger, and about the same mass as the giant South American river otter.

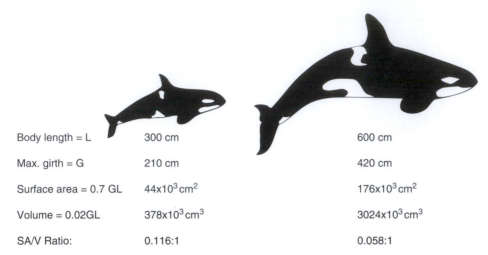

Body length = L	300 cm	600 cm
Max. girth = G	210 cm	420 cm
Surface area = 0.7 GL	$44 \times 10^3 \, cm^2$	$176 \times 10^3 \, cm^2$
Volume = 0.02GL	$378 \times 10^3 \, cm^3$	$3024 \times 10^3 \, cm^3$
SA/V Ratio:	0.116:1	0.058:1

Figure 9.5.   Surface-to-volume relationship. As a body increases in size (left to right), its surface area and volume also increase; however, the ratio of its surface area to its volume decreases.

Sea otters depend on heat generated from grooming and swimming activities, as well as the heat increment of feeding, to offset some of their thermoregulatory costs. In sea otters, the heat increment of feeding represents a metabolic rate increase of more than 50% for 4–5 hours after feeding (Costa, 2002).

   Marine mammals also reduce the amount of surface area across which heat can be lost with streamlined body forms and few projecting appendages. Streamlining approximates the minimum possible surface area for a given volume while still maintaining a shape that is appropriate for efficient locomotion, foraging, and reproduction. The surface areas of pinnipeds, cetaceans, and sea otters are, on average, about 23% smaller than those of terrestrial mammals of similar body mass (Innes *et al.*, 1990).

9.3.2. Insulation

The dense fur or blubber covering marine mammals minimizes heat loss in the water. Blubber has many functions including thermoregulation, energy storage, buoyancy control, and streamlining (see references in Struntz *et al.*, 2004). In contrast to the fat reservoirs of adult seals and whales, most pinniped pups are born with very little blubber and must develop a covering of fatty insulation quickly to survive. In Weddell seals, the lanugo is several times longer than the fur of the adult, which helps to conserve heat (Elsner *et al.*, 1977). Harp seal pups born on open pack ice make good use of sunlight, which is reflected by their translucent (white) hairs onto their dark skin where the heat is absorbed. The pup's coat traps this warmth between hair and skin, producing a "greenhouse effect" that prevents heat loss (Øritsland, 1970, 1971). In addition, although the newborns lack a layer of blubber, they do possess subcutaneous **brown fat** in a layer along the back and at several internal sites, around the neck, heart, kidneys, and abdominal walls (Blix *et al.*, 1975). This specialized brown fat, also found in human babies and many hibernating mammals, helps keep the pups warm by means of **nonshivering thermogenesis** (Grav *et al.*, 1974; Blix *et al.*, 1975; Grav and Blix, 1976). Instead of shivering, seals metabolize this high energy fat to produce considerable heat. After a few days the thermogenic fat is used up, but by then the pups have accumulated some insulating blubber (Blix *et al.*, 1979).

The thermal conductivity of blubber is the inverse of its insulative value and is a function of its thickness, its lipid content, and its peripheral blood flow. For cetaceans, the thermal conductivity of blubber is mostly a function of its lipid content. Worthy and Edwards (1990) found both higher lipid content and greater thickness in the blubber of temperate-water harbor porpoises when compared to blubber of tropical spotted dolphins, giving the former a four-fold increase in overall insulative value (Figure 9.6). Williams and Friedl (1990) showed that in bottlenose dolphins the blubber layer thickness adapts to water temperature by changing as quickly as 2 mm/month. Bryden (1964) has shown that seal blubber is a better insulator than whale blubber, probably because there is very little fibrous tissue present in seal blubber. In addition to blubber conductivity, another important variable with respect to heat loss models is the distribution of blubber around the body core. Kvadsheim *et al.* (1997) demonstrated that some heat loss models, which are based on calculations of heat flux across the blubber layer, overestimate heat loss from the body of seals because they fail to take into account that the blubber is distributed asymmetrically (as previously noted in Chapter 7).

Research on thermoregulation in harbor seals and grey seals (Hansen *et al.*, 1995; Hansen and Lavigne, 1997a, 1997b) provides evidence that ambient air temperature plays an important role in limiting the distribution of seals. For grey seals, cold air temperatures seem to affect distribution mainly through the breeding season, primarily through their thermoregulatory effects on small, fasting pups before they enter the water. In spite of this, these seals give birth in January on the east coast of Canada, the coldest month of the year. Presumably, the adaptive advantage of gaining access to extensive areas of winter pack ice or the late winter/early spring availability of food for

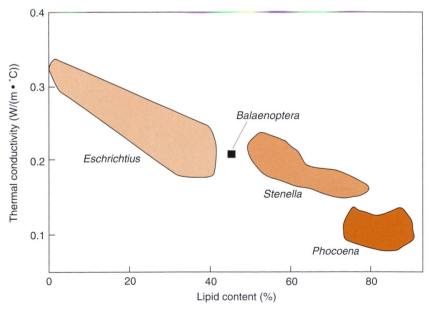

Figure 9.6.   The relationship between lipid content and thermal conductivity of the blubber in four genera of cetaceans. (Adapted from Worthy and Edwards, 1990.)

pups is worth incurring these higher thermoregulatory costs imposed by low winter air temperatures (see Chapter 14).

The distribution of pinnipeds toward the equator is limited by their inability to thermoregulate at elevated temperatures, especially during those periods when they are on land (i.e., during parturition, nursing, and the post-weaning fast and for some species during the annual molt) (Hansen *et al.*, 1995). In a study of the effect of water temperature on the thermoregulatory costs of bottlenose dolphins, Williams *et al.*, (1992) found that changes in water temperature altered energetic costs. Animals acclimated to 15°C had a resting metabolic rate that was 1.4 times the value measured for dolphins acclimated to 25°C. The lower critical temperature was less than 6°C for the cold acclimated dolphins and 11–16°C for the animals living in warm water.

It is important to remember that these same anatomical adaptations (i.e., an insulative layer of blubber or fur) that conserve heat while in water function to inhibit heat dissipation when out of water. For those marine mammals living in warm environments, overheating can pose a problem. When temperatures are high, fur seals, sea lions, and seals enter the water or rest in tidepools. Some pinnipeds, such as northern elephant seals, New Zealand sea lions, and Southern sea lions, cool off by flipping sand onto their backs. The characteristic "sand flipping" behavior during exposure to direct sunlight affords the animal a layer of cool damp sand that decreases heat gain and enhances conductive and evaporative heat losses (references cited in Hansen *et al.*, 1995). The northern fur seal actually pants like a dog to dissipate heat (Gentry, 1981). The monk seal, the only phocid living in a tropical environment, cools off on warm days by resting near the water in damp sand, often digging holes to expose cooler layers (Riedman, 1990). When on land, Galapagos fur seal pups too young to cool off in the sea seek shelter in the shade of boulders and curtail their activities during the heat of the day (Linberger *et al.*, 1986). Adult Galapagos and Guadalupe fur seals and Galapagos sea lions also use large boulders for shade when they are available and sometimes enter caves to seek shade (Figure 9.7).

Figure 9.7.   Guadalupe fur seals seeking shade in and near caves on Isla San Benito, Mexico.

9.3.3. Countercurrent Vascular Heat Exchange Systems

In addition to reduced surface areas, marine mammals can reduce heat losses to the environment by controlling peripheral blood circulation through vascular heat exchange systems in appendages, blubber, nasal mucosa, and reproductive organs. Whales and seals rely on controlled peripheral blood flow from the body core to the skin and appendages for heat conservation or loss. Unlike terrestrial mammals, whose insulation typically overlies vascular circulation to the skin, the insulating blanket of whale and seal blubber is penetrated by vascular beds to the base of the epidermis (Parry, 1949). Consequently, minimum heat flux is achieved when peripheral vasoconstriction is greatest and is independent of maximum heat flux achieved during periods of heat stress.

Countercurrent heat exchangers (CCHEs; Figure 9.8) conserve heat by maintaining a heat differential between oppositely directed flows of blood, thereby increasing the amount of heat transferred (compared to a concurrent flow). So that flippers, fins, and flukes remain functional and avoid freezing, it is necessary to maintain some blood circulation through them. In these tissues, the blood supply is arranged so that the main arteries carrying blood to cooler extremities are closely surrounded by veins, bringing the blood back to the body core (see Figure 8.26). Cold blood from the veins absorbs heat from the arterial blood coming from the body core so that the blood returning to the body is warmed and heat loss is minimized.

The CCHEs of sirenians are in the form of vascular bundles scattered throughout the body, including the body walls, face and jaw, tail and, spinal cord (Husar, 1977), but they are best developed in the flippers (Caldwell and Caldwell, 1985) and tail. The vigorous swimming activity of manatees in warm water necessitates a mechanism that prevents elevated core temperatures. In the manatee tail, deep caudal veins provide a collateral venous return. This return allows arterial expansion and increased arterial supply to the skin, thus preventing elevated core temperatures (Rommel and Caplan, 2003).

CCHEs also are associated with reproductive tracts where they prevent overheating in these heat-sensitive organs. Phocid seals and dolphins possess a vascular countercurrent exchanger that functions to cool their testes (Rommel *et al.*, 1995; Pabst *et al.*, 1995, 1998). In phocids, venous plexuses located between the veins of the distal hind limbs and the pelvis suggest that the testes are cooled directly. Cooled blood from the hind flipper flowing through the inguinal plexus located in the groin would introduce a thermal gradient

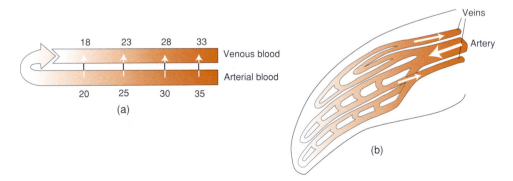

Figure 9.8.   General pattern of heat exchange in **(a)** an ideal countercurrent system and **(b)** a simplified vascular exchange network of a dolphin flipper.

that could allow local and direct heat transfer from the testes and adjacent muscles. This explanation is supported by measurements of testicular temperatures, which can be lower than core body temperatures by as much as 6–7°C in elephant seals (Bryden, 1967).

In dolphins, spermatic arteries in the posterior abdomen are juxtaposed to veins returning cooled blood from the surfaces of the dorsal fin and tail flukes (Figure 9.9). Immediately after vigorous swimming, temperatures at the CCHE decreased relative to resting and preswim values; postswim temperatures at the CCHE were maximally 0.5°C cooler than preswim temperatures (Rommel *et al.*, 1992). These data suggest that the CCHE has an increased ability to cool the arterial blood supply to the testes when the dolphin is swimming. Subsequently, Rommel *et al.* (1994) reported deep body temperature measurements in male dolphins, which supported this thermoregulatory hypothesis. Low temperatures were present in the region of the pelvic CCHE in males. Female dolphins and phocid seals use a similar countercurrent exchange system with cool blood flowing directly to the fetus from the flukes a of dolphins or from the hind flippers of seals. This prevents the fetus (which has a metabolic rate about twice that of its mother) from overheating and experiencing thermal distress, developmental disorders, or death (Rommel *et al.*, 1993, 1995). Similar vascular structures for reproductive cooling are found in both male and female manatees (Rommel *et al.*, 2001) indicating convergent adaptations that evolved among three clades of marine mammals.

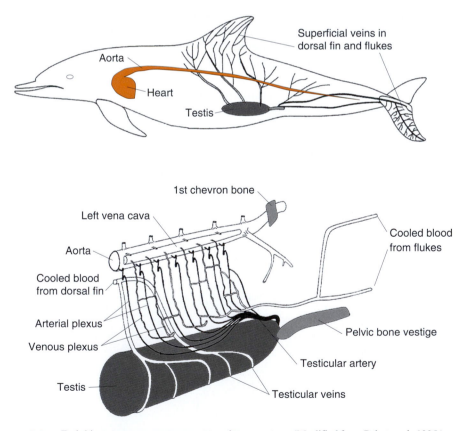

Figure 9.9.   Dolphin testes as a countercurrent exchange system. (Modified from Pabst *et al.*, 1995.)

Right and gray whales possess CCHEs in their mouths, enabling them to reduce heat losses when feeding in cold waters (Ford and Kraus, 1992; Heyning and Mead, 1997; Figure 9.10). Another example of a countercurrent exchange system is located the nasal passages of elephant seals, hooded seals and other phocids, enabling them to reduce their respiratory water losses (e.g., Huntley *et al.*, 1984; Folkow and Blix, 1987). Their nasal structure is similar to that of desert rodents for which water conservation is of great importance. Countercurrent blood flow in the nose creates temperature gradients that allow retention and recycling of moisture in the nasal passages. This moisture would normally be lost with exhalation, but because of the temperature gradient and the increased surface area of the nasal passages, much of the moisture in the exhaled air condenses on the cooled epithelium of the nasal passage. This moisture is then used during inhalation to humidify inspired air on its way to the lungs.

9.4. Energetics of Locomotion

Swimming and diving represent major energetic expenditures in marine mammals. In wild bottlenose dolphins, locomotor activities can represent more than 80% of the animal's daytime activity budget (Hansen and Defran, 1993). The cost of swimming is affected by water flow patterns around the swimmer as dictated by physical properties of

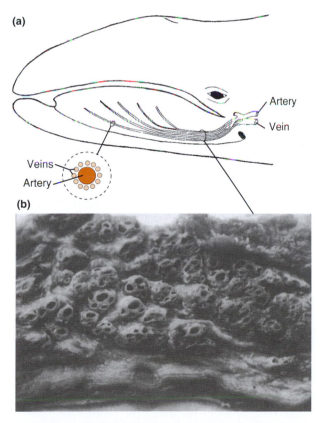

Figure 9.10.   Countercurrent exchange system in gray whale mouth **(a)** schematic diagram and **(b)** cross section through lingual rete. (From Heyning and Mead, 1997; courtesy of John Heyning.)

the aquatic medium. Important among these properties are density, viscosity, and kinematic viscosity (the ratio of density to viscosity). Water is over 800 times more dense than air and at least 30 times more viscous. Thus, the frictional resistance to moving through water is considerably greater than in air. As a consequence, movement through water imposes severe limitations on speed and energetic performance for swimming mammals. There are, however, advantages to moving in water. For example, propulsive forces are easier to generate in water than in air. Because the density of mammalian bodies is similar to that of water, most marine mammals are approximately neutrally buoyant. Consequently, swimming mammals can maintain their vertical position in the water with little energy expenditure because they do not need to support their weight during locomotion, as do terrestrial mammals and flyers (Rayner, 1986; Withers, 1992; Fish, 1993).

Marine mammals must overcome several different components of hydrodynamic drag in order to maintain high swimming velocities or to swim in an energetically efficient manner, each of which is influenced by the swimmer's body size and shape. The body shape of a fast swimming marine mammal, such as a dolphin, is a compromise between different hypothetical body forms, each of which reduces some component of the total drag and allows the animal to slip through the water with as little resistance as possible. **Frictional drag** (D_f) is a result of the animal's wetted surface area, the kinematic viscosity of the water, and the character of the water flow in the boundary layer near the swimmer's body surface. D_f is low if the flow in the boundary layer is laminar, and high if it is turbulent. **Pressure drag** (D_p), sometimes called form drag, is the consequence of displacing an amount of water equal to the swimmer's largest cross-sectional area (from a head-on view). When swimming horizontally at or near the sea surface, a swimming animal creates an additional **wave drag** (D_w) from the production of surface waves. As all marine mammals must surface to breathe, D_w can contribute substantially to the total drag that must be overcome to swim. A fourth drag component is **induced drag** (D_i) created by the hydrofoils (flukes or flippers) that marine mammals use to produce thrust (Fish, 1993).

Frictional and pressure drag of a fast swimmer are reduced with a streamline body form that is roundly blunt at the front end, tapered to a point in the rear, and round in cross section. A measure of streamlining, the **fineness ratio** (the ratio of maximum body length to maximum body diameter) for efficient swimmers ranges from 3 to 7, with the ideal near 4.5 (Webb, 1975). Most cetaceans exhibit fineness ratios near 6–7, with killer whales, right whales, and some phocids having nearly ideal ratios of 4.0–5.0.

For submerged swimming marine mammals, both inertial and viscous forces are important. Although the inertial forces dominate, the viscosity of the fluid is responsible for the formation of the boundary layer, which is important in the creation of skin friction due to the shear forces near the surface of the animal. A widely used comparative indicator of the forces acting on such submerged bodies, the **Reynolds number** (R), provides an order of magnitude approximation of these forces (Webb, 1975):

$$R = \text{body length} \times \text{swimming velocity/kinematic viscosity of water, or}$$

$$R = LV/\omega$$

Because the kinematic viscosity changes little over typical ranges of seawater temperatures, calculated R values are influenced primarily by body lengths and swimming velocities. For submerged swimming animals with streamlined bodies and R values $< 5 \times 10^5$, water flow over the body remains laminar and stable, with low total drag characteristics. At R values $> 5 \times 10^6$, the pattern of water flow becomes unstable and turbulent, resulting in greatly increased drag characteristics (when $5 \times 10^5 < R < 5 \times 10^6$, flow is transitional between laminar and turbulent conditions; Webb, 1975). Thus, low R values and patterns of laminar water flow are associated with small or slow-moving swimmers.

Models of hydrodynamic performance indicate that power requirements for sub-merged swimming by streamlined animals are proportional to the cube of the swimming velocity. Predicted power requirements (P) for streamlined submerged swimmers can be approximated by the following equation (Webb, 1975):

$$P = 0.5\rho C_t SWV^3$$

where ρ = water density

$$C_t = \text{coefficient of total drag}$$

$$SW = \text{wetted surface area of the swimmer}$$

$$V = \text{swimming velocity}$$

The magnitude of C_t is estimated to be at least an order of magnitude greater for animals experiencing turbulent boundary conditions than it is for animals swimming with lami-nar flow characteristics. Empirical determinations of C_t are available for only a few species of small odontocetes ($C_t = 0.03$–0.04; Lang, 1975) and one mysticete ($C_t = 0.06$; Sumich, 1983). The lower C_t value for the small odontocetes is attributed to laminar water flow characteristics over at least a substantial part of the body. Completely turbu-lent patterns characterize the water flow over the much larger bodies of mysticetes, and the resulting calculated value of C_t is correspondingly greater. Model-based esti-mates of C_t are summarized by Fish (1993).

Some of the complexities involved in estimating swimming power requirements from drag factors and hydrodynamic equations can be side-stepped by determining the **cost of transport** (COT) instead (Sumich, 1983; Williams *et al.*, 1993; Rosen and Trites, 2002; Williams *et al.*, 2004). COT is a useful measure for comparing the locomotory efficien-cies between different modes of locomotion or between different species employing the same mode of locomotion. COT for a swimming animal is defined as the power (P) required to move a given body mass (M) at some velocity (V). Thus, in appropriate units, COT = P/MV and is inversely proportional to energetic efficiency of swimming. Measures of power output can be estimated from direct or indirect O_2 respirometry methods in controlled or free-swim situations. The curve for most swimming animals describing the relationship between total power and resulting swimming velocity is U-shaped, with the power requirements reaching a minimum at some intermediate optimum velocity. The COT at that velocity is the COT_{min}, the velocity at which the ener-getic efficiency (but not necessarily the time efficiency) of swimming is highest. Tucker (1975) summarized the known COT_{min} values for a variety of swimming, flying, and running animals spanning several orders of magnitude in body size (see Figure 9.11).

It is apparent from Figure 9.11 that, within any one mode of locomotion (running, fly-ing, or swimming), the COT_{min} decreases with increasing body size and is essentially independent of taxonomic affiliation. Of the three general modes of transport, swim-ming is the least costly because swimmers need not support their body weight against the constant pull of gravity. Conspicuously absent from Tucker's summary were estimates of the COT for cetaceans or other marine mammals. The general relationship between body size and COT_{min} suggests that large swimming animals should have exceedingly low COT_{min}, but experimental evidence to test that prediction is still sparse.

To measure power output rates, the metabolic rates of a subject animal must be measured, and that is difficult to do with large, unrestrained, swimming cetaceans.

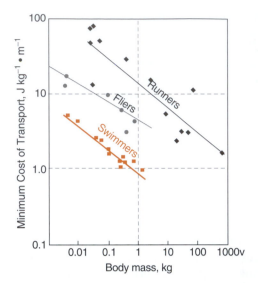

Figure 9.11. Relationships between COT_{min} and body weight for swimmers, fliers, and runners, regardless of taxonomic affiliation. (Adapted from Tucker, 1975.)

T. M. Williams *et al.* (1992, 1993) trained two Atlantic bottlenose dolphins to swim in open water beside a pace boat and to match its speed. Heart and respiratory rates, previously calibrated to oxygen consumption rates, were monitored and recorded continuously during each 20- to 25-minute test session to estimate metabolic rate. Blood samples for lactate analysis (a product of anaerobic respiration) were collected immediately after each session.

The results of this study indicated that the COT is minimum for these swimmers at speeds of 2.1 m/s and that the COT is doubled at 2.9 m/s. These results, at least for COT_{min}, have been confirmed for dolphin swimming speed measurements made in the wild using a multidirectional video sonar (Ridoux *et al.*, 1997). However, when speeds were increased above approximately 3 m/s, the dolphins in the T. M. Williams *et al.* (1992) study invariably switched to **wave riding,** a behavior that is best described as surfing the stern wake of the pace boat. When wave riding at 3.8 m/s, the COT was only 13% higher than the minimum value at 2.1 m/s. The large energetic saving that accompanies wave riding at higher speed explains the common practice of dolphins riding the bow or stern waves of ships and even other large whales, apparently with little effort (Figure 9.12).

Interestingly, the swimming velocity at which COT was minimum is nearly identical, 2 m/s, for both dolphins and gray whales, and is within 10% of the mean migrating speeds of southbound gray whales (Sumich, 1983). The energetic implications of swimming at speeds that minimize their COT and maximize their range, as gray whales do, are likely critical factors in successfully covering the exceedingly long migratory distances that they travel. Observations of Ridoux *et al.* (1997) on the swimming speed and activities of wild dolphins indicate that they continue their foraging activities during rising tides and that the additional cost of such a foraging strategy must be balanced by an increased prey density, availability, or catchability during rising tides. In another study that linked COT to the cost of thermoregulation, COT_{min}, and the optimal swim speed were found to be a function of water temperature and to vary between species (Hind and Gurney, 1997).

How does the COT of cetaceans compare to the COT of other swimmers? Williams (1999) compared calculated values for three species of cetaceans and three pinnipeds

Figure 9.12.   Common dolphins riding the stern wave of a vessel.

(Figure 9.13). It is apparent that cetaceans are efficient swimmers, with COT_{min} about an order of magnitude lower than that of humans or other surface swimmers. However, the regression line for cetacean COT_{min} is still nearly 10 times higher than that of a hypothetical ectothermic fish scaled to comparable body size. The additional costs incurred by cetaceans are presumably associated with the additional costs of

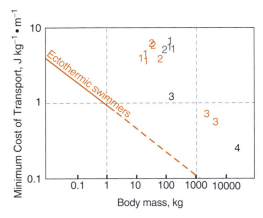

Figure 9.13.   Measured COT_{min} as a function of body mass for (1) otariids, (2) phocids, (3) odontocetes, and (4) a mysticete. Colors indicate different species. The regression line for ectothermic swimmers from Figure 9.11 is extrapolated to larger body masses. (Data from Williams, 1999, and Rosen and Trites, 2002.)

endothermy. In essence, this is the overhead cost of keeping the motor warmed up and running regardless of whether the animal is moving. The extended regression of cetacean COT_{min} is generally lower than that of pinnipeds, presumably reflecting the somewhat higher efficiency of cetacean flukes as propulsive organs.

When submerged swimming mammals approach the sea surface to breathe, they encounter additional wave drag due to gravitational forces associated with creating waves at the air-sea interface. At depths below 2.5 times an animal's greatest body diameter, these gravitational forces are negligible, but as a swimming mammal ascends, its total drag rapidly increases to a maximum of five-fold at the sea surface due to energy lost in the formation of surface waves. This pattern of increasing drag nearer the sea surface is shown in Figure 9.14. Any mammal can reduce its COT by swimming at depths below 2.5 body diameters for a period approaching the duration of its aerobic breathhold capacity (defined in Chapter 10), then surfacing to breathe only as frequently as O_2 demands dictate. This is reflected in the apneustic breathing patterns of many swimming mammals (Fig. 9.14), featuring extended submerged breathhold swims punctuated by surface bouts of several closely spaced ventilatory cycles. Thus short breathholds apparently serve to achieve efficient rates of oxygen assimilation, whereas long breathholds permit more efficient locomotion over extended or even migratory distances.

Marine mammals must surface to breathe more frequently at higher swimming velocities. Small species must surface to breathe more frequently than large ones, which makes it more difficult for them to escape beneath the high drag associated with the sea surface. At high speeds, the high wave drag associated with surfacing can be partly avoided by leaping above the water-air interface and gliding airborne for a few body lengths (Figure 9.15). The aerial phase of this type of **porpoising** or leaping locomotion removes the animal from the high drag environment of the water surface while providing an opportunity to breathe (Norris and Johnson, 1994). The velocity at which it becomes more efficient to leap rather than to remain submerged is known as the **crossover speed** (Fish and Hui, 1991). The crossover speed is estimated to be about 5 m/s for spotted dolphins and increases with increasing body size until leaping becomes a prohibitively expensive mode of locomotion in cetaceans longer than about 10 m.

Diving dolphins also use a swim-and-coast mode of swimming to reduce locomotor costs. At low to moderate swim speeds, swim-and-coast modes without aerial leaping pro-

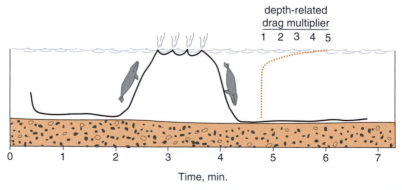

Figure 9.14.   Apneustic breathing pattern of a migrating gray whale. The graph at the right illustrates the relative increase in drag as the whale approaches the surface.

Figure 9.15. Porpoising white-sided dolphins, *Lagenorhynchus.* (Courtesy of NOAA.)

vide a definite energetic advantage over constant velocity swimming (Weihs, 2002). In an experiment in which dolphins were trained to dive in a straight path to submerged targets at different depths, Williams *et al.* (1996) reported that rather than swim at constant speeds, the dolphins consistently switched to swim-and-coast modes of swimming, thus incurring lower energy costs during long dives. However, when small cetaceans must move horizontally at high speeds, they experience high metabolic demands and high breathing rates. To accommodate this need to remain in the high drag environment near the surface, porpoising and swim-and-coast modes are often combined into a three-phase leap-coast-burst swim mode. Weihs (2002) has proposed that substantial energy savings occurs with this mode while maintaining high average swim velocities and breathing rates.

Another energy savings occurs when calves position themselves alongside their mothers in a "drafting" position. The displacement effect that results from the motion of the mother's body causes water in front to move forward and outward and the water behind the body to move forward to replace the animal's mass. The net result of this and other hydrodynamic effects of drafting is that the calf can gain up to 90% of the thrust needed to move alongside its mother at speeds up to 2.4m/sec. A comparison with observations of eastern spinner dolphins indicates that a savings of up to 60% in the thrust is achieved by calves swimming alongside their mothers (Weihs, 2004).

9.5. Osmoregulation

Osmoregulation includes those processes that balance water intake with excretion and loss. For general reviews of osmoregulation in marine mammals, see Ortiz (2001), Elsner (1999), and Williams and Worthy (2002). Most marine mammals are **hypoos-**

motic; their body fluids have a lower ionic content than their surrounding seawater environment, and they are constantly losing some water to the hyperosmotic seawater in which they live.

Marine mammals obtain the water they need from the food they eat: preformed water in their diet and subsequent metabolically derived water. Most fish and invertebrate prey consists of 60–80% water and the metabolism of fat, protein, and carbohydrates provide metabolic water during the digestion of food. It has been shown experimentally that seals can obtain all the water they need from the food they eat. If seawater is given to seals, the stomach becomes upset and the excess salts have to be eliminated using body water. Despite this, seals occupying warm climates have been observed drinking seawater (King, 1983), a practice called **mariposia.** It has been suggested that seals intermittently consume small amounts of seawater at intervals that would not be enough to cause digestive problems but would be sufficient for facilitating nitrogen excretion. Mariposia is especially common among adult male otariids (Riedman, 1990). Why do pinnipeds drink seawater? Gentry (1981) noted that most of the otariids observed ingesting seawater live in warmer climates and lose water by urination, panting, and sweating. He suggested that such water loss, along with prolonged fasting by territorial males, may be severe enough to promote the drinking of seawater. The behavior may play a role in nitrogen excretion by supplementing water produced oxidatively from metabolized fat reserves. Mariposia also has been reported for Atlantic bottlenose dolphins, common dolphins, and harbor porpoises.

Not all marine mammals live in a highly saline environments. A study comparing the Baikal seal, an inhabitant of freshwater, and in the ringed seal, its marine counterpart, revealed no major differences in renal function. It was concluded that the Baikal seal, isolated from seawater and living in freshwater for 0.5 Ma, retained the renal function they possessed at the time of isolation (Hong et al., 1982).

The isotopic concentration of mammalian tooth phosphate reflects the type of water ingested (i.e., marine vs freshwater), and it is possible to determine when cetaceans adapted to the excess salt load associated with ingesting seawater. From studies of isotopic concentrations of fossil and living cetacean teeth (and bone) it was determined that the earliest whales (e.g., the pakicetids) were tied to terrestrial sources of freshwater and food. By the middle Eocene the first fully marine cetaceans (i.e., protocetids and remingtonocetids) appeared (Thewissen et al., 1996; Roe et al., 1998). The transition of cetaceans from terrestrial to marine habitats is reflected by changes in osmoregulatory function and diet that allowed them to leave the coast and rapidly disperse across oceans.

Costa (1978) demonstrated that sea otters actively drink seawater. Because sea otters primarily consume invertebrates (which possess higher electrolyte concentrations than bony fish or mammals), they must process large amounts of electrolytes, nitrogen, and water. Ingestion of seawater promotes urea elimination by increasing the urinary osmotic space without increasing the electrolyte concentration in the urine (Costa, 1982; Riedman and Estes, 1990).

Marine mammals reduce their water losses by excreting concentrated urine. In general, marine mammals can produce urine with an osmolality slightly greater than that of seawater, although their capability is far exceeded by the desert dwelling kangaroo rat, which can excrete urine with a concentration 14–17 times that of their plasma (Elsner, 1999). Other water retention mechanisms include CCHEs located in the nasal passages of pinnipeds (described in the previous section).

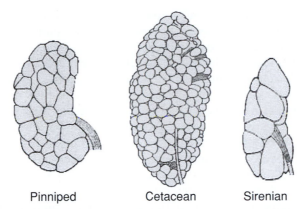

Pinniped   Cetacean   Sirenian

Figure 9.16.   Comparison of generalized kidney appearance of marine mammals emphasizing extensive reniculi in both pinnipeds and cetaceans but not in sirenians. (From Slijper, 1979.)

The kidneys of marine mammals are generally larger than those found in terrestrial mammals of similar body mass (Beuchat, 1996). All marine mammals except sirenians possess a **reniculate** kidney (Figure 9.16). In these animals each kidney is made up of small discrete lobes, the reniculi. Each reniculus is like a small individual kidney with its own cortex, medulla, and calyx, and the duct from it joins with others to form the ureter.

The number of reniculi in cetaceans ranges from hundreds to thousands per kidney (Ommanney, 1932; Gihr and Kraus, 1970). One structure in the reniculus that appears unique to cetaceans and some pinnipeds is the sporta perimedullaris musculosa, a layer of smooth muscle elastic fibers and collagen that penetrates the reniculus and surrounds the medulla (Vardy and Bryden, 1981). This feature together with other features of the cetacean reniculus, such as large reservoirs of glycogen and unique medullary blood vessels, are adaptations that may facilitate diving by providing local tissue stores of glycogen when blood perfusion of the kidneys is reduced during dives (Pfeiffer, 1997; and see Chapter 10). Dugong kidneys are elongate (Batrawi, 1957) unlike the lobulate kidneys of cetaceans, pinnipeds, and manatees; the latter is superficially lobulate (see Figure 9.16).

9.6.   Summary and Conclusions

Major energetic costs for marine mammals include metabolism, temperature regulation, locomotion, and osmoregulation. The issue of whether marine mammals have higher than expected metabolic rates remains contentious although it is generally acknowledged that different methods of measurement and the lack of standardized conditions have confounded comparisons. Although newborn seals of many species lack blubber, a layer of brown fat under the skin helps keep them warm by nonshivering thermogenesis until this fat and the milk they ingest is converted into insulative blubber. Because the thermal conductivity of blubber is partly a function of its thickness, temperate water porpoises have thicker blubber than tropical spotted dolphins. Research on the thermoregulation of seals suggests that ambient air temperatures play an important role in limiting their geographic distribution. Further work is necessary to test whether this is a more general phenomenon among pinnipeds. CCHEs, which regulate temperature, occur throughout the body but especially in the flippers, fins, and flukes as well as the reproductive organs of marine mammals. Marine mammals also reduce heat loss to the environment by reducing their surface areas. Even the smallest marine mammals are relatively larger compared to their terrestrial counterparts and thus obtain favorable surface-to-volume ratios. Because swimming is a major energy expenditure in marine mammals they employ various methods to reduce the COT (e.g., wave riding, porpoising, and swim and coast swimming in dolphins). The reniculate kidney of marine mammals allows the production of large volumes of concentrated urine. Marine mammals obtain the water they need from the food they eat. The practice of drinking seawater, mariposia, by seals, dolphins, and porpoises, has been related to nitrogen excretion and may provide a supplement to water produced oxidatively from metabolizing fat reserves. Among marine mammals the oldest whales were freshwater animals indicating that their adaptation to a marine environment occurred later.

9.7.   Further Reading

For comprehensive recent summaries of marine mammal energetics consult Costa and Williams (1999) and Boyd (2002).

References

Batrawi, A. (1957). "The Structure of the Dugong Kidney." *Publ. Mar. Biol. Sm., Ghardaqa, Red Sea 9:* 51–68.

Beuchat, C.A. (1996). "Structure and Concentrating Ability of the Mammalian Kidney: Correlations with Habitat." *Am. J. Physiol. 271:* R157–179.

Blix, A. S., and L. P. Folkow (1995). "Daily Energy Expenditure of Free-Living Minke Whales." *Acta Physiol. Scand. 153:* 61–66.

Blix, A. S., H. J. Grav, and K. Ronald (1975). "Brown Adipose Tissue and the Significance of the Venous Plexes in Pinnipeds." *Acta Physiol. Scand. 94:* 133–135.

Blix, A. S., L. K. Miller, M. C. Keyes, H. J. Grav, and R. Elsner (1979). "Newborn Northern Fur Seals—Do They Suffer from Cold?" *Am. J. Physiol. 236:* 322–327.

Boyd, I. L. (2002). Energetics: Consequences for fitness. *In* "Marine Mammal Biology, An Evolutionary Approach" (A. R. Hoelzel, ed.), p. 247–277. Blackwell Science, Oxford, UK.

Boyd, I. L., A. J. Woakes, P. J. Butler, R. W. Davis, and T. M. Williams (1995). "Validation of Heart Rate and Doubly Labelled Water as Measures of Metabolic Rate During Swimming in California Sea Lions." *Functional Ecol. 9:* 151–160.

Bryden, M. M. (1964). "Insulating Capacity of the Subcutaneous Fat of the Southern Elephant Seal." *Nature 203:* 1299–1300.

Bryden, M. M. (1967). "Testicular Temperature in the Southern Elephant Seal, *Mirounga leonina (L.)*." *J. Reprod. Fertil. 13:* 583–584.

Caldwell, D. K., and M. C. Caldwell (1985). Manatees-*Trichechus manatus, Trichechus sengalensis,* and *Trichechus inuguis. In* "Handbook of Marine Mammals, Vol. 3, The Sirenians and Baleen Whales" (S.H. Ridgway and R. J. Harrison, eds.), pp. 33–36. Academic Press, London.

Costa, D. P. (1978). "The Sea Otter: Its Interaction with Man." *Oceanus 21:* 24–30.

Costa, D. P. (1982). "Energy, Nitrogen, and Sea-Water Drinking in the Sea Otter, *Enhydra lutris*." *Physiol. Zool.* **55:** 35–44.

Costa, D. P. (2002). Energetics. *In* "Encyclopedia of Marine Mammals" (W. F. Perrin, B. Wursig and J. G. M. Thewissen, eds.), pp. 387–394. Academic Press, San Diego, CA.

Costa, D. P., and T. M. Williams (1999). Marine mammal energetics *In* "Biology of Marine Mammals"(J. E. Reynolds III and S. A. Rommel, eds.), pp. 176–217. Smithsonian Institution Press, Washington, D. C.

Elsner, R. (1999). Living in water: Solution to physiological problems. *In* "Biology of Marine Mammals"(J. E. Reynolds and S. Rommel, eds.), pp. 73–116. Smithsonian Institution Press, Washington, D. C.

Elsner, R., D. D. Hammond, D. M. Dension, and R. Wyburn (1977). Temperature regulation in the newborn Weddell seal *Leptonychotes weddelli*. *In* "Adaptation within Antarctic Ecosystems" (G. A. Llano, ed.), pp. 531–540. Proceedings of the Third SCAR Symposium on Antarctic Biology, Aug. 26–30, 1974, Smithsonian Institution, Washington D. C.

Fish, F. E. (1993). "Influence of Hydrodynamic Design and Propulsive Mode on Mammalian Swimming Energetics." *Aust. J. Zool. 42:* 79–101.

Fish, F. E., and C. A. Hui (1991). "Dolphin Swimming—A Review." *Mamm. Rev. 21:* 181–195.

Folkow, L. R., and A. S. Blix (1987). "Nasal Heat and Water Exchange in Gray Seals." *Am. J. Physiol. 253:* R883–R888.

Ford, T. J., and S. D. Kraus (1992). "A Rete in the Right Whale." *Nature 359:* 680.

Gallivan, G. J., and R. C. Best (1980). "Metabolism and Respiration of the Amazonian Manatee *(Trichechus inunguis)*." *Physiol. Zool. 59:* 552–557.

Gentry, R. L. (1981). "Sea Water Drinking in Eared Seals." *Comp. Biochem. Physiol. 68A:* 81–86.

Gihr, M., and C. Kraus (1970). Quantitative investigation on the cetacean kidney. *In* "Investigations on Cetacea," Vol. 2 (G. Pilleri, ed.), pp. 168–176. Berne, Switzerland.

Grav, H. H., and A. S. Blix (1976) "Brown Adipose Tissue—A Factor in the Survival of Harp Seal Pups." *Can. J. Physiol. Pharmacol. 54:* 409–412.

Grav, H. H., A. S. Blix, and A. Pasche (1974). "How Do Seal Pups Survive Birth in Arctic Winter?" *Actas Physiol. Scand. 92:* 427–429.

Hampton, I. F. G., G. C. Whitlow, J. Szekercezes, and S. Rutherford (1971). "Heat Transfer and Body Temperature in the Atlantic Bottlenosed Dolphin, *Tursiops truncatus*." *Int. J. Biometeorol. 15:* 247–253.

Hansen, M. T., and R. H. Defran (1993). "The Behavior and Feeding Ecology of the Pacific Coast Bottlenose Dolphin, *Tursiops truncatus*." *Aquat. Mamm. 19:* 127–142.

Hansen, S., and D. M. Lavigne (1997a). "Ontogeny of the Thermal Limits in the Harbor Seal *(Phoca vitulina)*." *Physiol. Zool. 70*(1): 85–92.

Hansen, S., and D. M. Lavigne (1997b). "Temperature Effects on the Breeding Distribution of Gray Seals *(Halichoerus grypus)*." *Physiol. Zool. 70:* 436–443.

Hansen, S., D. M. Lavigne, and S. Innes (1995). "Energy Metabolism and Thermoregulation in Juvenile Harbor Seals *(Phoca vitulina)*." *Physiol. Zool. 68:* 290–315.

Heyning, J., and J. Mead (1997). "Thermoregulation in the Mouths of Feeding Gray Whales." *Science 278:* 1138–1139.

Hind, A. T., and W. S. C. Gurney (1997). "The Metabolic Cost of Swimming in Marine Homeotherms." *J. Exp. Biol. 200:* 531–542.

Hong, S. K., R. Elsner, J. R. Claybaugh, and K. Ronald (1982). "Renal Functions of the Baikal Seal *Pusa sibirica* and Ringed Seal *Pusa hispida*." *Physiol. Zool. 55:* 289–299.

Huntley, A. C., D. P. Costa, and R. D. Rubin (1984). "The Contribution of Nasal Countercurrent Heat-Exchange to Water-Balance in the Northern Elephant Seal, *Mirounga angustirostris*." *J. Exp. Biol. 113:* 447–454.

Huntley, A. C., D. P. Costa, G. A. J. Worthy, and M. A. Castellini (1987). *Approaches to Marine Mammal Energetics*. Allen Press, Lawrence, KS.

Husar, S. (1977). *The West Indian Manatee: Trichechus manatus*. Wildl. Res. Rep. No. 7. U.S. Department of the Interior, Fish and Wildlife Service, Washington, D. C.

Innes, S., and D. M. Lavigne (1991). "Do Cetaceans Really Have Elevated Metabolic Rates?" *Physiol. Zool. 64:* 1130–1134.

Innes, S., G. A. J. Worthy, D. M. Lavigne, and K. Ronald (1990). "Surface Area of Phocid Seals." *Can. J. Zool. 68:* 2531–2538.

Irvine, A. B. (1983). "Manatee Metabolism and Its Influence on Distribution in Florida." *Biol. Conservation 25:* 315–334.

Irving, L., S. P. F., and S. W. Grinnell. (1941). "The Respiration of the Porpoise, *Tursiops truncatus*." *J. Cell Comp. Physiol. 17:* 145.

Karandeeva, O. G., S. K. Matisheva, and V. M. Shapunov (1973). Features of external respiration in the delphinidae. *In* "Morphology and Ecology of Marine Mammals: Seals, Dolphins, and Porpoises" (K. K. Chapskii and V. E. Soklov, eds.) pp. 196–206. Wiley, New York.

Kasting, N. W., S. A. L. Adderley, T. Safford, and K. G. Hewlett (1989). Thermoregulation of whales and dolphins. *In* "Whales, Dolphins and Porpoises" (K. Norris, ed.), pp. 397–407. University of California Press. Berkeley.

King, J. E. (1983). *Seals of the World,* 2nd ed. Cornell University Press, Ithaca, NY.

Kleiber, M. (1932). "Body Size and Metabolism." *Hilgardia 6:* 315–353.

Kleiber, M. (1947). "Body Size and Metabolism." *Physiol. Rev. 27:* 511–541.

Kleiber, M. (1975). *The Fire of Life*. Robert E. Krieger Publishing Co, Huntington, NY.

Kooyman, G. (1981). Leopard seal, *Hydrurga leptonyx, In* "Handbook of Marine Mammals," Vol. 2 (S. H. Ridgway and R. J. Harrison, eds.), pp. 261–274. Academic Press, London.

Kvadsheim, P. V., A. R. L. Gotaas, L. P. Folkow, and A. S. Blix (1997). "An Experimental Validation of Heat Loss Models for Marine Mammals." *J. Theor Biol. 184:* 15–23.

Lang, T. G. (1975). Speed, power, and drag measurement of dolphins and porpoises. *In* "Swimming and Flying in Nature," Vol 2. (T.Y.-T. Wu, C. J. Brokaw, and C. Brennen, eds.), pp. 553–621. Plenum Press, New York.

Lavigne, D. M., S. Innes, G. A. J. Worthy, K. M. Kovacs, O. J. Schmitz, and J. P. Hickie. (1986). Metabolic rates of seals and whales. *Can. Jour. Zool*. 64:279-284

Linberger, D., F. Trillmich, H. Biebach, and R. D. Stevenson (1986). "Temperature Regulation and Microhabitat Choice by Free-Ranging Galapagos Fur Seals *(Arctocephalus galapagoensis)*." *Oecologia 69:* 53–59.

Markussen, N. H., M. Ryg, and N. A. Øritsland. (1990). "Energy Requirements for Maintenance and Growth of Captive Harbour Seals, *Phoca vitulina*." *Can. J. Zool. 68:* 423–426.

Marsh, H., A. V. Spain, and G. E. Heinsohn. (1978). "Physiology of the Dugong." *Comp. Biochem. Physiol. Pt. A 225:* 711–715.

McNab, K. (1988). "Complications Inherent in Scaling the Basal Rate of Metabolism in Mammals." *Q. Rev. Biol. 63:* 25–54.

Norris, K. S., and C. M. Johnson (1994). Locomotion. *In* "The Hawaiian Spinner Dolphin" (K. S. Norris, B. Wursig, R. S. Wells, and M. Wursig, eds.), pp. 201–205. University of California Press, Berkeley.

Ommanney, F. C. (1932). "The Urogenital System of the Fin Whale *(Balaenoptera physalus)*. With Appendix: The Dimensions and Growth of the Kidney of Blue and Fin Whales." *Discovery Rep. 5:* 363–466.

Øritsland, N. A. (1971). "Wavelength-Dependent Solar Heating of Harp Seals *(Pagophilus groenlandicus)*." *Comp. Biochem. Physiol. 40A:* 359–361.

Øritsland, T. (1970). Energetic significance of absorption of solar radiation in polar homeotherms. *In* "Antarctic Ecology," Vol. 1 (M. W. Holgate, ed.), pp. 464–470. Academic Press, London.

Ortiz, R. M. (2001). "Osmoregulation in Marine Mammals." *J. Exp. Biol. 204:* 1831–1844.

O'Shea, T. J., and R. L. Reep (1990). "Encephalization Quotients and Life-History Traits in the Sirenia." *J. Mamm. 71:* 534–543.

Pabst, D. A., S. A. Rommel, and W. A. McLellan (1998). Evolution of the thermoregulatory function in the cetacean reproductive systems. *In* "The Emergence of Whales: Patterns in the Origin of Cetacea" (J. G. M. Thewissen, ed.), pp. 379–397. Plenum Press, New York.

Pabst, D. A., S. A. Rommel, W. A. McLellan, T M. Williams, and T. K. Rowles (1995). "Thermoregulation of the Intra-Abdominal *testes* of the Bottlenose Dolphin *(Tursiops truncatus)* During Exercise." *J. Exp. Biol. 198:* 221–226.

Parry, D. A. (1949). "The Structure of Whale Blubber, and a Discussion of Its Thermal Properties." *Q.J. Microsc. Sci. 90:* 13–25.

Pfeiffer, C. J. (1997). "Renal Cellular and Tissue Specializations in the Bottlenose Dolphin *(Tursiops truncates)* and Beluga Whale *(Delphinapterus leucas)*." *Aquat. Mamm. 23:* 75–84.

Pierce, W. H. (1970). *Design and Operation of a Metabolic Chamber for Marine Mammals*. Ph.D. thesis, University of California, Berkeley.

Rayner, J. M. V. (1986). "Pleuston: Animal Which Moves in Water and Air." *Endeavour 10:* 58–64.

Ridgway, S. H., and G. S. Patton (1971). "Dolphin Thyroid: Some Anatomical and Physiological Findings." *Z. Verg. Physiol. 71:* 129–141.

Ridoux, V., C. Guinent, C. Liret, P. Creton, R. Steenstrup, and G. Beauplet (1997). "A Video Sonar as a New Tool to Study Marine Mammals in the Wild: Measurements of Dolphin Swimming Speed." *Mar Mamm. Sci. 13:* 196–206.

Riedman, M. L. (1990). *The Pinnipeds: Seals, Sea Lions and Walruses*. University of California Press, Berkeley.

Riedman, M. L., and J. A. Estes (1990). The sea otter *(Enhydra lutris):* Behavior, ecology, and natural history. *U.S. Fish Wildl. Sers:, Biol. Rep*. 90(14).

Roe, L. J., J. G. M. Thewissen, L. Quade, J. R. O'Neill, S. Bajpal, A. Sahni, and S. T. Hussain (1998). Isotopic approaches to understanding the terrestrial-to-marine transition of the earliest cetaceans. *In* "The Emergence of Whales: Patterns in the Origin of Cetacea" (J. G. M. Thewissen, ed.), pp. 399–422. Plenum Press, New York.

Rommel, S. A., and H. Caplan (2003). "Vascular Adaptations for Heat Conservation in the Tail of Florida Manatees." *J. Anat. 202:* 343–353.

Rommel, S. A., G. A. Early, K. A. Matassa, D. A. Pabst, and W. A. McLellan (1995). "Venous Structures Associated with Thermoregulation of Phocid Seal Reproductive Organs." *Anat. Rec. 243:* 390–402.

Rommel, S. A., D. A. Pabst, and W. A. McLellan (1993). "Functional Morphology of the Vascular Plexuses Associated with the Cetacean Uterus." *Anat. Rec. 237:* 538–546.

Rommel, S. A. , D. A. Pabst, and W. A. McLellan (2001). "Functional Morphology of Venous Structures Associated with the Male and Female Reproductive Systems in Florida Manatees *(Trichechus manatus latirostris)*." *Anat. Rec. 264:* 339–347.

Rommel, S. A., D. A. Pabst, W. A. McLellan, J. G. Mead, and C. W. Potter (1992). "Anatomical Evidence for a Countercurrent Heat Exchanger Associated with Dolphin Testes." *Anat. Rec. 232:* 150–156.

Rommel, S. A., D. A. Pabst, W. A. McLellan, T. M. Williams, and W. A. Friedl (1994). "Temperature Regulation of the Testes of the Bottlenose Dolphin *(Tursiops truncatus):* Evidence from Colonic Temperatures." *J. Comp. Physiol. B 164:* 130–134.

Rosen, D. A. S., and A. W. Trites (2002). "Cost of Transport in Steller Sea Lions, *Eumetopias jubatus*." *Mar. Mamm. Sci. 18*(2): 513–524.

Struntz, D. J. W. A. McLellan, R. M. Dillaman, J. E. Blum, J. R. Kucklick, and D. A. Pabst (2004). "Blubber Development in Bottlenose Dolphins *(Tursiops truncatus)*." *J. Morphol. 259:* 7–20.

Sumich, J. (1983). "Swimming Velocities, Breathing Patterns, and Estimated Costs of Locomotion in Migrating Gray Whales, *Eschrichtius robustus*." *Can. J. Zool. 61:* 647–652.

Thewissen, J. G. M., L. J. Roe, J. R. O'Neill, S. T. Hussain, A. Sahni, and S. Bajpal (1996). "Evolution of Cetacean Osmoregulation." *Nature 381:* 379–380.

Tucker, V. A. (1975). "The Energetic Cost of Moving About." *Am. Sci. 63:* 413–419.

Vardy, P. H., and M. M. Bryden (1981). "The Kidney of *Leptonychotes weddelli* (Pinnipedia: Phocidae) with Some Observations on the Kidneys of Two Other Southern Phocid Seals." *J. Morphol. 167:* 13–34.

Webb, P. W. (1975). "Hydrodynamics and Energetics of Fish Propulsion." *Fisheries Res. Board. Can. Bull. 190:* 1–170.

Weihs, D. (2002). "Dynamics of Dolphin Porpoising Revisited." *Integr. Comp. Biol. 42:* 1071–1078.

Weihs, D. (2004). "The Hydrodynamics of Dolphin Drafting." *J. Biol. 3:* 8–21.

Williams, T. D., D. D. Allen, J. M. Groff, and R. L. Glass (1992). "An Analysis of California Sea Otter *(Enhydra lutris)* Pelage and Integument." *Mar. Mamm. Sci. 8:* 1–18.

Williams, T. M. (1999). "The Evolution of Cost Efficient Swimming in Marine Mammals: Limits to Energetic Optimization." *Phil. Trans. Royal Soc. London B 354:* 193–201.

Williams, T. M., and W. A. Friedl (1990). "Heat Flow Properties of Dolphin Blubber—Insulating Warm Bodies in Cold Water." *Am Zool. 30:* A33–A33(publ. abstract).

Williams, T. M., W. A. Friedl, M. L. Fong, R. M. Yamada, P. Sedivy, and J. E. Haun (1992). "Travel at Low Energetic Cost by Swimming and Wave-Riding Bottlenose Dolphins." *Nature 355:* 821–823.

Williams, T. M., W. A. Friedl, and J. E. Haun (1993). "The Physiology of Bottlenose Dolphins (*Tursiops truncatus*): Heart Rate, Metabolic Rate and Plasma Lactate Concentrations During Exercise." *J. Exp. Biol. 179:* 31–46.

Williams, T. M., L. A. Fuiman, M. Horning, and R. W. Davis.(2004). "The Cost of Foraging by a Marine Predator, the Weddell Seal *Leptonychotes weddellii*: Pricing by the Stroke." *J. Exp. Biol. 207:* 973–982.

Williams, T. M., S. F. Shippee, and M. J. Rothe (1996). Strategies for reducing foraging costs in dolphins. *In* "Aquatic Predators and Their Prey" (S. P. R. Greenstreet and M. L. Tasker, eds.), pp. 4–9. Fishing News (Blackwell Science), London.

Williams. T. M., and G. A. J. Worthy (2002). Anatomy and physiology: The challenge of aquatic living. *In* "Marine Mammal Biology: An Evolutionary Approach" (A. R. Hoelzel, ed.), pp. 73–97. Blackwell Publ., Oxford, UK.

Withers, P. C. (1992). *Comparative Animal Physiology*. Saunders, Fort Worth, TX.

Worthy, G. A. J., and E. F. Edwards (1990). "Morphometric and Biochemical Factors Affecting Heat Loss in a Small Temperate Cetacean *(Phocoena phocoena)* and a Small Tropical Cetacean *(Stenella attenuata)*." *Physiol. Zool. 63:* 432–442.

10

Respiration and Diving Physiology

10.1. Introduction

A prominent feature of the normal behavior of marine mammals is diving. The details of diving behavior often are difficult to observe and interpret, for they usually occur well below the sea surface. Whether diving below the surface to forage for food, to increase swimming efficiency by diving below the high drag conditions found at the surface, to save energy by reducing metabolic costs, or to sleep while minimizing the risk of predation, most marine mammals spend a large portion of their lives below the surface of the water (e.g., Le Boeuf, 1994; Thorson and Le Boeuf, 1994; Le Boeuf and Crocker, 1996; Andrews *et al.*, 1997).

Measured either as maximal achieved depth or maximal duration of a dive, the diving capabilities of marine mammals vary immensely. Some species of marine mammals are little better than the best human free-divers, whereas some whales and pinnipeds are capable of astounding feats that include diving for periods of hours to depths of kilometers (Table 10.1). At either end of the scale of ability, the diving capabilities of marine mammals allow them to explore and exploit the oceans of the world. Although considerable progress has been made in documenting the diving behavior of marine mammals in recent years, knowledge regarding diving ability is limited to very few species and our understanding regarding the physiology of diving is still elementary. Much of the information gathered to date regarding how marine mammals dive has come from studies of Weddell seals and elephant seals, two of the most extreme pinniped divers. Additional experimentation with other seals and small cetaceans has lead to the conclusion that there is no generic marine mammal and that extrapolations from one species to another must be done with caution. However, some basic patterns are seen among marine mammals with respect to diving. These are the focus of much of the following discussion.

10.2. Problems of Deep and Prolonged Dives for Breath-Holders

Aristotle recognized over 20 centuries ago that dolphins are air-breathing mammals (see Chapter 1). However, it was not until early in the 20th century that the physiological basis

Table 10.1. Maximum Dive Depths and Breath-Holding Capabilities of Marine Mammals (depths and durations do not necessary correspond to same dive or the same animal). Data from original sources listed in Schreer and Kovacs (1997) unless otherwise indicated.

Species	Maximal depth (m)	Maximal duration of breath-hold (min.)	Source
Pinnipeds			
Phocids			
Northern elephant seal			
Male	1530	77	
Female	1273	62	
Southern elephant seal			
Female	1430	120	Slip *et al.* (1994)
Male	1282	88.5	Slip *et al.* (1994)
Weddell seal	626	82	Castellini *et al.* (1992)
Crabeater seal	528	10.8	
Harbor seal	508	7	
Harp seal	370	16	
Grey seal	268	32	
Hawaiian monk seal		12	Hochachka and Mottishaw (1998)
Ross seal		9.8	Hochachka and Mottishaw (1998)
Otariids			
California sea lion	482	15	
Hooker's sea lion	474	12	
Steller sea lion	424	6	
New Zealand fur seal	474	11	Costa *et al.* (1998)
Northern fur seal	207	8	
South African fur seal	204	8	
Antarctic fur seal	181	10	
South American fur seal	170	7	
Galapagos fur seal	115	8	
Southern sea lion	112	6	
Australian sea lion	105	6	Costa and Gales (2003)
Guadalupe fur seal	82	18	
Walrus	300	12.7	Lydersen, personal communication
Cetaceans			
Odontocetes			
Sperm whale	3000	138	
Bottlenose whale	1453	120	Hooker and Baird (1999)
Narwhal	1400	20	Laidre *et al.* (2003)
Blainsville beaked whale	890	23	Baird *et al.* (2004)
Beluga	647	20	
Pilot whale	610	20	
Bottlenose dolphin	535	12	
Killer whale	260	15	
Pacific white sided dolphin	214	6	
Dall's porpoise	180	7	
Mysticetes			
Fin whale	500	30	
Bowhead whale	352	80	Krutikowsky and Mate (2000)
Right whale	184	50	
Gray whale	170	26	
Humpback	148	21	
Blue whale	153	50	Lagerquist *et al.* (2000)
Sirenians			
West Indian manatee	600	6	
Dugong	400	8	
Sea otter	100		Bodkin *et al.* (2004)

for the deep and prolonged dives of marine mammals was explored. Unlike human SCUBA divers who carry air for breathing underwater, marine mammals must cease breathing during dives, and that leads to several worsening, or at least conflicting, physiological conditions during those **apneic** conditions (Castellini, 1985a, 1991; Castellini *et al.*, 1985). First, oxygen stores begin to deplete when the intensity of activity is increasing and the demand for oxygen is highest. Second, without ventilation, CO_2 and lactate (a metabolic end product produced in vertebrates when oxygen stores are depleted) increase in both blood and muscle tissues, making the blood serum and cell fluid more acidic. During these periods of **hypoxia,** continued muscle activity is maintained anaerobically, which results in an even greater accumulation of lactate. Decades ago, when relatively little was known about the extent of the diving abilities of marine mammals, the anaerobic aspect of marine mammal physiology received considerable research attention. However, the focus more recently has turned to how marine mammals manage to remain within their aerobic limits so much of the time while performing remarkable diving feats. Documentation of at-sea behavior in a variety of marine mammal species has shown that they routinely dive in sequence over many hours; aerobic metabolism is the only practical option for this behavior.

In addition to dealing with the challenges of limited available oxygen, when mammals dive below the sea surface, they must also tolerate an increase in water pressure; increasing pressures of 1 atmosphere (atm) for each 10 m of water depth must be dealt with at depth and the consequence of having been under pressure must be dealt with on surfacing (300 atm for a sperm whale at 3000 m, Table 10.1). Increasing water pressure on the outside of an animal squeezes air-filled spaces inside the animal, causing the spaces to distort or collapse as the air they contain is compressed, which can damage membranes or rupture tissues. Absorbing gases from air at high pressures poses some potentially serious problems for diving mammals, because oxygen can be toxic at high concentrations, nitrogen can have a narcotic effect on the central nervous system, and both can form damaging bubbles in tissues and blood vessels during and immediately following ascent (Moon *et al.*, 1995). The nervous systems of mammals also are generally sensitive to exposure to high pressure.

10.3. Pulmonary and Circulatory Adaptations to Diving

10.3.1. Anatomy and Physiology of the Cardiovascular System

10.3.1.1. Heart and Blood Vessels

Marine mammals undergo important circulatory changes during diving. In general, the structure of the heart of cetaceans and pinnipeds closely resembles that of other mammals (Figure 10.1). At the ultrastructural level, several distinctive differences are observed in the hearts of ringed and harp seals, such as enlarged stores of glycogen, which strongly suggest that the cardiac tissues of theses animals are capable of a greater anaerobic capacity that those of terrestrial mammals (Pfeiffer, 1990; Pfeiffer and Viers, 1995). Ridgway and Johnston (1966) suggested that the relatively larger heart size of the Dall's porpoise, when compared to the less active inshore Atlantic bottlenose dolphin was related to its pelagic deep-diving habit. However, neither is heart size fundamentally different in marine versus terrestrial mammals nor would it be expected to be as marine mammal hearts do not work particularly hard to support circulation when diving (see later).

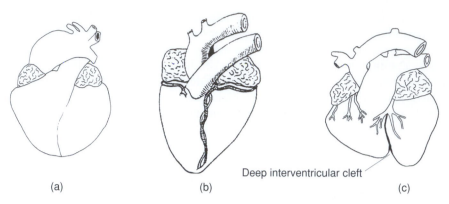

(a)   (b)   (c)

Deep interventricular cleft

Figure 10.1.   Ventral view of heart of **(a)** pinniped (Weddell seal) (modified from Drabek, 1977), **(b)** cetacean (porpoise) (modified from Slijper, 1979), and **(c)** sirenian (dugong) (modified from Rowlatt and Marsh, 1985).

The ascending aorta in pinnipeds increases in diameter by 30–40% immediately outside the heart to form an expanded elastic aortic bulb (aortic arch). After all the great vessels (i.e., brachiocephalic, left common carotid and left subclavian arteries) have branched from the aorta, at about the level of the ductus arteriosus, there is a sudden decrease in the diameter of the aorta by about 50%, and it then continues posteriorly as a relatively slender abdominal aorta. The aortic bulb and smaller arteries at the base of the brain serve to dampen blood pressure pulses at each heart beat.

Although the enlarged aortic bulb is characteristic of all pinnipeds (Drabek, 1975, 1977; King, 1977), there is thought to be some correlation between the size of the bulb and the diving habits of the species. The shallow water leopard seal, for example, has a smaller bulb than the deep-diving Weddell seal. The proportionately larger heart and aortic bulb of deep-diving hooded seals likely function to increase lung perfusion during surface recovery and to assist in maintaining high blood pressures throughout the cardiac cycle during dives. The absence of age-related differences in heart morphology of hooded seals suggests that these adaptations are important in the development of diving behavior (Drabek and Burns, 2002).

A bulbous expansion of the aortic arch similar to that seen in pinnipeds has also been described in some cetaceans (Shadwick and Gosline, 1994; Gosline and Shadwick, 1996; Melnikov, 1997). Further examination of the structural properties of the aorta and associated vessels (Gosline and Shadwick, 1996) indicates that most of the arterial compliance in a whale's circulation results from volumetric expansion in the aortic arch and differences in the mechanical properties of its walls such as the thickness and organization of elastic tissues.

In addition to having a heart with a deep interventricular cleft (double ventricular apex) extending almost the full length of the ventricles, sirenians are distinguished from other marine mammals in having a dorsally located left atrium (see Figure 10.1). In most respects the manatee heart is very similar to the dugong heart. Among the differences are the quadrate shape of the manatee left ventricle, as opposed to its conical shape in the dugong, and the bulbous swelling of the aortic arch in the manatee, similar to that seen in pinnipeds (Drabek, 1977) and fin whales (Shadwick and Gosline, 1994). It seems unlikely that expansion of the aortic arch in manatees can be attributed to differences in diving performance because all sirenians are modest divers compared to most other marine mammals (Rowlatt and Marsh, 1985, and see Table 10.1).

The circulatory systems of marine mammals are characterized by groups of blood vessels (**retia mirabilia**). Retia mirabilia are tissue masses containing extensive contorted spirals of blood vessels, mainly arteries but with thin-walled veins among them, that usually form blocks of tissue on the inner dorsal wall of the thoracic cavity and extremities or periphery of the body (Figure 10.2). The sperm whale has been described as having the most extensively developed thoracic retia among cetaceans (Melnikov, 1997). These structures serve as blood reservoirs to increase oxygen stores for use during diving (e.g., Pfeiffer and Kinkead, 1990).

The main changes in the venous system of pinnipeds are the enlargement and increased complexity of veins to enhance their capacity. Most of our knowledge of these adaptations comes from work done on the venous system of phocid seals (Harrison and Kooyman, 1968; Ronald *et al.*, 1977; Figure 10.3). In phocids, the posterior vena cava is frequently is developed as a pair of vessels each capable of considerable distension of their thin, elastic walls. Each branch of the posterior vena cava drains extensive plexi from the veins in the flippers, pelvis, and lateral abdominal wall. Each branch also receives several tributaries of varying sizes from the stellate plexus that encloses the kidneys (see Figure 10.3). Just posterior to the diaphragm and covered by the lobes of the liver, lies the hepatic sinus, which is formed from enlarged hepatic veins. It also receives blood from the posterior vena cava and conveys it to the heart. Immediately anterior to the diaphragm, the vena cava has a muscular caval sphincter surrounding it (see Figure 10.3). Anterior to the sphincter the veins from the pericardial plexus enter the vena cava. This convoluted mass of interconnected veins (which are often encased in brown adipose tissue) forms a ring around the base of the pericardium and sends out leaf-like projections into the pleural cavities containing the lungs.

(a)   (b)

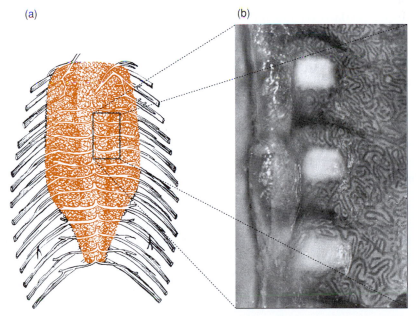

Figure 10.2.   Retia mirabilia, their anatomical position relative to the ribs (**a**) (adapted from Slijper, 1979), and the right thoracic retia of a spotted dolphin (**b**).

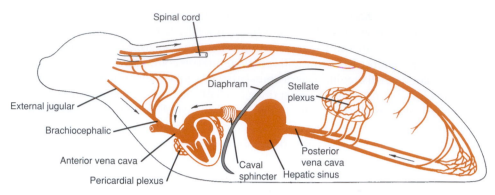

Figure 10.3.   Diagram of the venous circulation in a phocid with the major vessels labeled. (From King, 1983.)

The vein walls in this region are relatively thick and contain coiled collagenous elastic and smooth muscle fibers, which suggest considerable capacity for expansion.

In phocids, the greatest number of these venous modifications is combined, including the pericardial plexus, renal stellate plexus, large extradural vein receiving the main cranial drainage, hepatic sinus, and caval sphincter (King, 1983; Munkasci and Newstead, 1985). Walruses resemble phocids in having a large hepatic sinus and a well-developed caval sphincter but resemble otariids in having a single azygous vein, no well-developed pericardial plexus, and no prominent stellate plexus (Fay, 1981).

Relative to body size, the veins of cetaceans are not as enlarged as those of pinnipeds, although the posterior vena cava is enlarged in the hepatic region in some species. The vena cava shows marked variation in its anatomy not only in different species but even within a species (Harrison and Tomlinson, 1956). As in pinnipeds, the posterior vena cava frequently occurs as paired vessels. A vascular distinction of most cetaceans is the development of a pair of large veins ventral to the spinal cord that have been suggested to be related to their diving ability (Slijper, 1979). There is no caval sphincter or hepatic sinus in cetaceans (Harrison and King, 1980).

10.3.1.2. Total Body Oxygen Stores

The largest and most important stores of oxygen in diving mammals are maintained in chemical combination with **hemoglobin** within the red blood cells of the blood and with **myoglobin,** the oxygen binding molecule found in muscle cells that gives them a dark red appearance. By packaging hemoglobin in red blood cells, mammals can maintain high hemoglobin concentrations without the associated increase in plasma osmotic pressure. Each molecule of hemoglobin can bind with up to four O_2 molecules, thereby increasing the O_2 capacity of blood nearly 100-fold, from 0.3 ml O_2/100 ml for plasma alone to 25 ml O_2/100 ml for whole blood (Berne and Levy, 1988). In contrast to the anatomically fixed nature of myoglobin, hemoglobin circulates, and arterial regulation can direct the delivery of hemoglobin-bound O_2 to organ systems most in need of it.

Red blood cells (which contain hemoglobin) are about the same size in diving and non-diving mammals; however, some species of diving mammals have higher relative blood volumes that appear to be at least roughly correlated with the species diving capabilities and more red blood cells per unit volume of blood. These last two features together effectively increase the **hematocrit,** or packed red blood cell volume, and consequently the hemoglobin volume of the blood. Hematocrit values range from 40 to 45% in gray whales and 48% in California sea lions to 50 to 60% in elephant seals (Table 10.2). Hedrick and Duffield (1991) suggested that species of marine mammals with the greatest adaptation for increased oxygen storage (with higher hematocrit levels) actually experience a decreased capacity for oxygen transport and limited ability to sustain fast swimming speeds because of increased blood viscosity and decreased blood flow.

Much of the total store of oxygen available for use during a dive is bound to the myoglobin of the skeletal muscles (Figure 10.4). The high concentrations of this pigment in all mammals that dive to depths greater than about 100 m strongly suggests that myoglobin is a key adaptation for diving (Noren and Williams, 2000; Kooyman, 2002). The skeletal muscles are very tolerant of the hypoxic conditions experienced during dives; when the muscles deplete their myoglobin-oxygen store, they and other peripheral organs (such as the kidneys and digestive organs) may be largely deprived of the circulating hemoglobin-bound oxygen stored in the blood. This is essential because the higher oxygen binding efficiency of myoglobin compared to hemoglobin would strip the oxygen from the blood and hence deprive vital organs such as the brain that are less tolerant of hypoxia. However, recent evidence suggests that occasional, brief reperfusions of skeletal muscle likely do take place during diving (Guyton *et al.*, 1995) so that the oxidative integrity of the tissue may be for the most part maintained.

10.3.1.3. Splenic Oxygen Stores

It has been noted that the spleen of seals and sea lions is large (4.5% of body weight). The elephant seal has the largest spleen, its relative weight being more than three times that of terrestrial mammals. The spleen provides reserve storage for oxygenated red blood cells. Diving capacity in phocids (but not in otariids) is strongly correlated with spleen size (Hochachka and Mottishaw, 1998).

The spleen of cetaceans is very small (0.02% of body weight) in comparison with most terrestrial mammals (Bryden, 1972). No correlation has yet been described between diving ability and spleen size in cetaceans.

Table 10.2. Blood Values That Relate to Oxygen Capacity in Selected Marine Mammals

Species	Blood vol. (mL/kg)	Hematocrit (%)	Source
Pinnipeds			
Phocids			
N. elephant seal	100–175	50–62	Thorson and Le Boeuf (1994)
S. elephant seal		46–62	Lewis *et al.* (2001)
Otariids			
California sea lion		48	Ridgway (1972)
Cetaceans			
Odontocetes			
Sperm whale	204	52	Sleet *et al.* (1981)
Bottlenose dolphin	71–95	42–52	Ridgway and Johnston (1966)
Mysticetes			
Gray whale	61–81	40–45	Gilmartin *et al.* (1974)
		38–46	T. Reidarson, personal communication

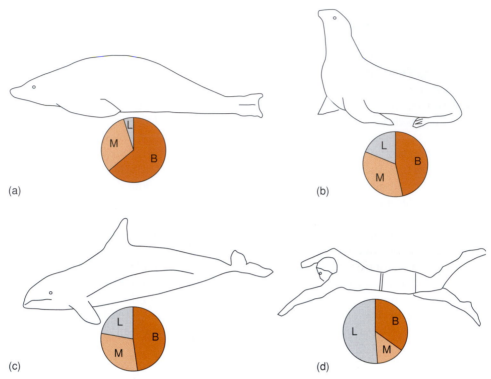

Figure 10.4.   Generalized comparison of relative blood (B), muscle (M), and lung (L) oxygen stores for
(a) phocids, (b) otariids, (c) odontocetes, and (d) humans. (After Kooyman, 1985.)

10.3.2. Anatomy and Physiology of the Respiratory System

The respiratory tract of marine mammals begins with the external nares of pinnipeds
and sirenians and blowhole(s) of cetaceans and ends with the lungs. Opening of the nares
in all of these marine mammal groups is accomplished by contraction of skeletal muscle,
whereas closure is a passive process. The active opening of the nostrils or blowholes is an
energy-conserving adaptation to spending life in an aquatic medium, avoiding the need
for continuous muscle contraction to keep water out of the respiratory tract.

The position of the nares in pinnipeds is not exceptional for mammals and no valve is
present in the external nares of pinnipeds. Instead, the flat medial surfaces of two throat
(arytenoid) cartilages lie in close proximity, which also abut the posterior surface of the
epiglottis, providing a tight seal against the entry of water into the trachea. The powerful muscles of the larynx also help to keep this entrance closed (King, 1983).

Cetaceans breathe through the blowholes (external nares), which have migrated to a
position high on the top of the head. The sperm whale, in which the blowhole is on the
anterior top end of the head, slightly left of center, is the exception. Mysticetes possess
two blowholes in contrast to the single blowhole of odontocetes.

Manatees breathe through two-valved nostrils situated at the tip of the rostrum. The
nostrils of the dugong lie close together and are situated anterodorsally, thereby allowing the dugong to breathe with most of its body submerged. The nostrils are closed during diving by anteriorly hinged valves (Nishiwaki and Marsh, 1985).

Deep-diving marine mammals have flexible chest walls and other structures that are capable of sufficient collapse to render the lungs virtually airless. The trachea in most pinnipeds is supported by cartilaginous rings that either form complete circles or are incomplete and overlap dorsally. This morphology permits the alveoli to collapse while the proximal airways are still open, ensuring that pulmonary gas exchange is eliminated and pulmonary entrapment of air is avoided when the animal is at depth.

The tracheal rings are reduced to ventral bars in Ross, Weddell, and leopard seals (King, 1983). In the ribbon seal, a longitudinal slit occurs between the ends of the incomplete cartilage rings on the dorsal surface of the trachea just before it bifurcates to form the bronchi. A membranous sac of unknown function extends both anteriorly and posteriorly from the trachea on the right side of the body. Male ribbon seals possess this sac, and it increases in size with age suggesting that it may have a role in sound production associated with mating behavior. The tracheal slit is present in females, but the sac is small or absent (King, 1983). In phocids and the walrus, the trachea divides into two primary bronchi immediately outside the substance of the lung. In otariids this bifurcation is located more anteriorly, at approximately the level of the first rib, and the two elongated bronchi run parallel until they diverge to enter the lungs dorsal to the heart (see Figure 3.11).

The cetacean larynx is composed of a cartilaginous framework held together by a series of muscles. To keep inhaled air separate from food, the larynx of odontocetes (but not mysticetes) has two elongate cartilages (Figure 10.5; Lawrence and Schevill, 1965; Slijper, 1979; Reidenberg and Laitman, 1987). This provides a more direct connection between the trachea and blowhole than is found in mysticetes.

In cetaceans the trachea is short and broad and consists of several cartilaginous rings (varying from 5 to 7 rings in the beluga and sperm whale to 13 to 15 in the fin whale) that are interconnected with each other. Unlike baleen whales, the tracheal rings of toothed whales are closed and form a noncollapsing tube (Yablokov et al., 1972).

The dugong trachea is short (only four cartilage rings) and it is deeply divided by a medial septum (Hill, 1945; Harrison and King, 1980). The manatee trachea is longer and is supported by 8–12 tracheal rings (Harrison and King, 1980).

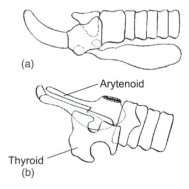

Figure 10.5. Lateral view of cetacean larynx in **(a)** mysticete (fin whale) and **(b)** odontocete (narwhal). Note the elongated arytenoid and thyroid cartilages in the odontocete. (From Slijper, 1979.)

10.3.2.1. Lungs

The lungs of marine mammals are not larger than those of terrestrial mammals, but some important differences exist between the lungs of marine mammals and those of terrestrial mammals. The left and right lungs of pinnipeds are approximately equal in size and are lobulated as in terrestrial carnivores; both lungs have three main lobes, but the right lung has an additional small intermediate lobe. There appears to be a tendency to reduce lobulation in the lungs of some pinnipeds, such as the walrus and ribbon, harp, and spotted seals. The latter species has been reported as having little or no lobulation (King, 1983). The bronchi subdivide within the lungs to form bronchioles and eventually end in alveoli, but the details of this transition vary among the three families (Figure 10.6; see King, 1983).

The lungs of cetaceans are distinct from those of all other mammals in their overall sacculate shape and lack of lobes (Figure 10.7). Only occasionally is the apical part of the right lung somewhat prominent, resembling the apical lobe of the lungs of other mammals. The right lung is usually larger, longer, and heavier than the left. Such asymmetry of lung size, which is related in whales as in other mammals to the somewhat asymmetric position of the heart in the chest cavity, is seen in virtually all studied cetaceans: rorquals, various dolphins, and sperm and beluga whales.

The lungs of cetaceans, compared to those of terrestrial mammals, show greater rigidity and elasticity because of increased cartilaginous support. In baleen, sperm, and bottlenose whales, the septa projecting into the proximal portion of the air sacs contain heavy myoelastic bundles. In the smaller toothed whales these bundles are atrophied, but there are a series of myoelastic sphincters in the smallest bronchioles. In both the bundles and the sphincters, the muscular part may act to close the air sacs, whereas the elastic part may facilitate rapid expiration. Airway closure should delay alveolar collapse during dives and allow additional pulmonary gas exchange to occur for those odontocete families known to possess myoelastic terminal airway sphincters (Drabek and Kooyman, 1983).

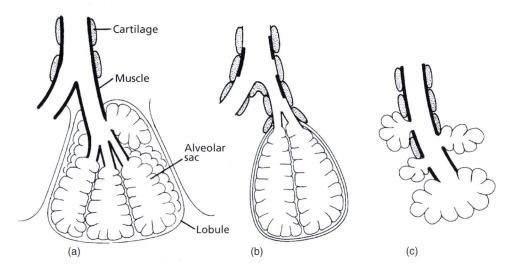

Figure 10.6.   Diagram of the structure of alveoli and associated cartilage and muscle in pinnipeds. **(a)** Phocid. **(b)** Otariid. **(c)** Odobenid. (From King, 1983.)

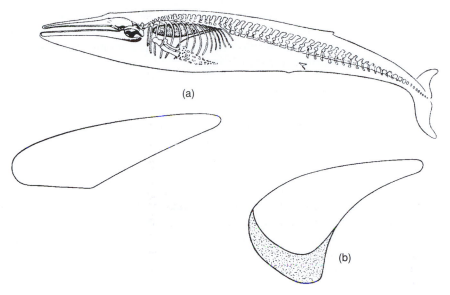

Figure 10.7. Diagram showing position of the cetacean lung and comparison of **(a)** mysticete (fin whale), **(b)** and odontocete (bottlenose dolphin) lung. Stippled area indicates particularly thin section of lung. (From Slijper, 1979.)

Relative lung volume is lower for cetaceans than terrestrial mammals. The relatively small lung volume of deep-diving species is a logical consequence of the inability of the respiratory tract to store gas because of the risks of an **embolism** and other difficulties that divers that breathe gas under pressure experience when surfacing rapidly (Kooyman and Andersen, 1969; Yablokov *et al.*, 1972). The unusual oblique position of the diaphragm permits the abdominal contents to occupy part of the thorax when the animal is under pressure. Deep divers also have a very small residual lung volumes, which means that the lung empties more completely and gas exchange can occur more fully in a respiratory cycle (e.g., Olsen *et al.*, 1969; Denison *et al.*, 1971)

The lungs of sirenians are long and extend posteriorly almost as far back as the kidneys. They are separated from the abdominal viscera by a large obliquely sloped diaphragm, bronchial tree, and respiratory tissue (Engel, 1959a, 1959b, 1962). The primary bronchi run almost the length of the lungs with only a few smaller side branches or secondary bronchi. The secondary bronchi pass into smaller tubes, which in turn give rise to minute tubules supplying the respiratory vesicles. Another unique sirenian feature is that these vesicles arise laterally along the length of the bronchioles rather than from their ends, as is the typical mammalian condition (Engel, 1959a). The bronchioles are very muscular and may function to close off respiratory vesicles when desired. For example, the dugong may use this technique to compress the volume and density of air in the lungs, thus enabling it to surface or sink without the use of flippers or tail and without expelling air (Engel, 1962). Cartilage occurs throughout the length of the air passages (Nishiwaki and Marsh, 1985).

The thoracic cavity of the sea otter is large and the diaphragm is positioned obliquely (Barabash-Nikiforov, 1947). The right lung has four lobes and the left has two lobes (Tarasoff and Kooyman, 1973a, 1973b). The lungs are large in relation to body size, nearly 2.5 times that found in other mammals of similar size. Large lungs serve more to regulate buoyancy than to store oxygen (Lenfant *et al.*, 1970; Kooyman, 1973; Leith, 1976; Costa and Kooyman, 1982). The polar bear respiratory system is not unlike that of other bears. Although they are powerful swimmers they are not known to have any special physiological adaptations specific to diving.

10.3.2.2. Breathing

Breathing patterns of marine mammals vary. Pinnipeds breathe vigorously and frequently during the recovery phase after prolonged diving, but many species are typically periodic breathers under other circumstances. Particularly when resting or sleeping it is normal for seals to perform quite long apneas with short periods of rapid respiration between these nonbreathing periods (e.g., Huntley *et al.*, 1984). Although pinnipeds commonly exhale prior to diving, Hooker *et al.* (2005) found that Antarctic fur seals consistently dive with full lungs and exhale during the latter stage of the ascent portion of a dive.

Cetaceans exhale and inhale singly but very rapidly on surfacing. Whale **blows** represent the rapid emptying or expiration of whales' lungs through their blowholes in preparation for the next inspiration. A blow is one of the most visible behaviors of whales when they are observed at the sea surface. A particularly large amount of water may be spouted in baleen whales, whose blowholes are located in rather deep folds. The visibility of a blow is due to a mixture of vapor and seawater entrained into the exhaled column of air at the sea surface (Figure 10.8). When a blow occurs below the sea surface,

Figure 10.8.   Towering blow of a blue whale. (Courtesy of P. Colla.)

as it sometimes does as bubble-blasts in gray whales or bubble trains in humpback whales, it may be intended as an audible or visual signal to other nearby whales. The size, shape, and orientation of a blow can help to identify some species of whales from a distance.

Vapor formed by contact between air warmed in the lungs and the cold external air sometimes enhances the visibility of the blow. The expired air of a blow also contains **surfactant** from the lungs. Surfactant is a complex a mixture of lipoproteins that reduce surface tension in pulmonary fluids and facilitate easy reinflation of collapsed lungs on surfacing. Pulmonary surfactant is secreted by alveolar type II cells and is necessary for normal mammalian lung function (Miller *et al.*, 2004). Spragg *et al.* (2004) have found differences in surfactant of nondiving and diving mammals and suggest that these differences are associated with the repetitive collapse and reinflation of the lungs of divers.

A complete breathing cycle typically consists of a very rapid expiration (the blow) immediately followed by a slightly longer and much less obvious inspiration, then an extended yet variable period of breath-holding, or apnea. The rapid expiration of a typical whale blow provides more time to complete the next inspiration as the blowholes of a swimming animal breaks through the sea surface and results in little delay before submerging again (Kooyman and Cornell, 1981).

The rapidity of the blow is accomplished by maintaining high flow rates throughout almost the entire expiration (Figure 10.9), which is in strong contrast to humans and other land mammals. The high expiratory flow rates of cetaceans are enhanced by very flexible chest walls and by cartilage reinforcement of the smallest terminal air passages of the lungs to prevent them from collapsing until the lungs are almost completely emptied. Small dolphins, for instance, expire and inspire in about 0.1 s, then hold their breath for 20 to 30 s before taking another breath. Even adult blue whales can empty their lungs of 1500 liters of air and refill them in as little as 2 s.

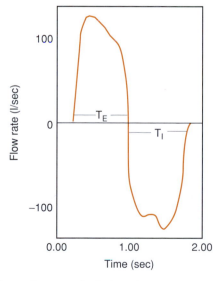

Figure 10.9. Typical ventilatory flow rates of a single expiratory (T_E)/inspiratory (T_I) event from a rehabilitating gray whale calf.

Blow patterns of whales vary, depending on their behaviors. In small cetaceans, swimming at low speeds, blowhole exposure during a blow is minimal and gradually changes to porpoising above the sea surface at higher swimming speeds (see Chapter 9). When migrating or feeding, larger baleen whales typically surface to blow several times in rapid succession, then make an extended dive of several minutes duration.

During inspiration, extensive elastic tissue in the lungs and diaphragm (Figure 10.10) is stretched by diaphragm and intercostal musculature. These fibers recoil during expiration to rapidly and nearly completely empty the lungs. Oxygen uptake within the alveoli of the lungs may be enhanced as lung air is moved into contact with the walls of the alveoli by the action of small myoelastic bundles scattered throughout the lungs. In some species, the alveoli are highly vascularized to promote rapid uptake of oxygen. Bottlenose dolphins, for example, can remove nearly 90% of the oxygen available in each breath (Ridgway *et al.*, 1969). In comparison, humans and most other terrestrial mammals use only about 20% of the inspired oxygen. It has been proposed that foreign particulate matter in inspired air or seawater may result in the formation of biomineral concretions, or calculi, discovered in the nasal sacs of some delphinids (Curry *et al.*, 1994).

Manatees exhale after surfacing and, like cetaceans, can renew about 90% of the air in the lungs in a single breath. By comparison, humans at rest renew about 10% of the air in the lungs in a single breath (Reynolds and Odell, 1991).

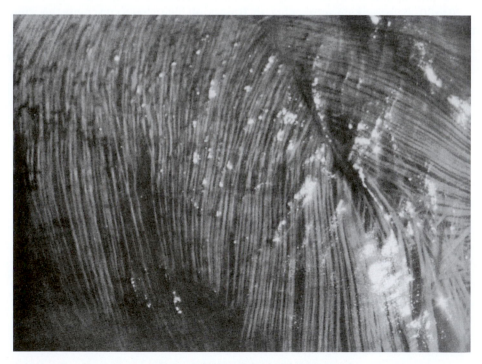

Figure 10.10.   Extensive elastic fibers of diaphragm tissue of dolphin *Stenella*.

10.3.2.3. Respiratory Systems and Diving

Cetaceans typically dive with full lungs, whereas pinnipeds often exhale prior to diving. These differences support the contention that the volume of lung air at the beginning of a dive is of little importance in supplying oxygen during a dive, but it may be adjusted to achieve neutral buoyancy during some types of diving. Moreover, the lungs and their protective rib cage are modified to allow the lungs to collapse as the water pressure increases with depth (Figure 10.11). Complete lung collapse occurs at depths of 25–50 m for Weddell seals (Falke *et al.*, 1985), 70 m for the bottlenose dolphin (Ridgway and Howard, 1979), and probably occurs in the first 50–100 m for most marine mammals. Any air remaining in the lungs below that depth is squeezed out of the alveoli and into the bronchi and trachea of the lungs.

By tolerating complete lung collapse, these animals avoid the need for respiratory structures capable of resisting the extreme water pressure experienced during deep dives. They also receive an additional bonus. As the air is forced out of the collapsing alveoli during a dive, the compressed air still within the larger air passages is blocked from contact with the thin, gas-exchanging walls of the alveoli. Consequently, little of this compressed gas is absorbed by the blood during dives and marine mammals avoid the potentially serious diving problems of decompression sickness (also called the **bends**) and **nitrogen narcosis** that sometimes plague human divers breathing compressed air (Moon *et al.*, 1995).

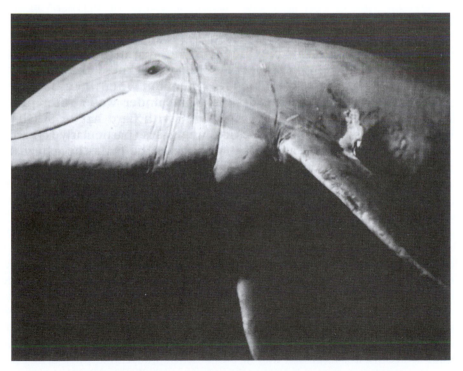

Figure 10.11. Bottlenose dolphin at a depth of 300 m, experiencing obvious thoracic collapse visible behind the left flipper. (Courtesy of S. Ridgway and used by permission.)

10.4. Diving Response

When a marine mammal leaves the surface its on-board oxygen stores (described previously) must satisfy its needs throughout submergence. As a dive proceeds there is a steady decline in the amount of available oxygen (hypoxia) and an increase in carbon dioxide **(hypercapnia),** which together create a condition known as **asphyxia.** Eventually, if the dive continues beyond a time that can be serviced by aerobic metabolism, byproducts of anaerobic metabolism such as lactic acid and hydrogen ions also begin to accumulate. However, marine mammals and other animals that have become adapted to dealing with periods of asphyxia have a complex array of physiological responses that extend the time that a given oxygen supply can service their bodies. These responses include a pronounced decline in heart rate **(bradycardia)** accompanied by regional vasoconstriction (selective **ischemia**) that entails a preferential distribution of circulating blood to oxygen-sensitive organs as well as a drop in core body temperature and likely metabolic rate in regions that receive reduced blood supplies.

Bradycardia and its implied reduction in metabolic costs have been recognized as being a diving response since the late 1800s (Bert, 1870; Richet, 1894, 1899). However, it was not until the 1930s that Irving's (e.g., Irving et al., 1935; Irving, 1939) experimental laboratory studies provided evidence that oxygen was conserved during diving by selective circulatory adjustments during periods of bradycardia. A series of elaborate physiological experiments conducted by Irving and Scholander in the laboratory with forced-dived marine mammals, demonstrated the fundamental factors employed by diving animals to conserve oxygen and deal with the products of anaerobic metabolism postdiving (e.g., Irving and Scholander, 1941a, 1941b; Irving et al., 1942; Scholander, 1940, 1960, 1964; Scholander et al., 1942a, 1942b). These experiments attracted criticism for a time, because of the unnatural conditions under which the animals were forced to dive. However, as technology has advanced and studies have been performed in the wild on unrestrained animals, it has become clear that the early experiments did evoke natural dive responses although they tended to be extreme, presumably because the animals did not know when they were going to be able to surface.

The range of heart-rate responses to diving is variable between species and within a species under different circumstances. However, in general, short dives evoke only slight responses, whereas long dives promote more intense levels of bradycardia. The most extreme levels of heart rate reduction, down to 5% of predive levels, have been recorded during free diving by phocid seals (e.g., Jones et al., 1973; Elsner et al., 1989; Thompson and Fedak, 1993; Andrews et al., 1995). Variable responses are at least in part due to what appears to be a remarkable level of voluntary control over the cardiovascular system in at least some species. Experiments with free-diving seals of several species have shown that seals have heart rates at the start of dives that are correlated with the subsequent duration of the dive, strongly suggesting that the animal prepares for a dive of a certain duration on leaving the surface, and anticipatory elevations in heart rate take place prior to the end of dives as well (e.g., Fedak, 1986; Hill et al., 1987; Elsner et al., 1989; Wartzok et al., 1992). There is increasing evidence that fine adjustments can be made during a dive in some species (Andrews et al., 1997). Although much less data are available for cetaceans, it is clear that they also experience bradycardia when diving (e.g., Elsner et al., 1966; Spencer et al., 1967). Diving bradycardia in manatees and dugongs is modest, as is their diving ability (Elsner, 1999).

Despite the marked decline in cardiac output that accompanies the drop in heart rate during diving, core body arterial blood pressure remains relatively constant so that perfusion of vital organs (brain, heart, placenta, etc.) is maintained. This is achieved in part by the elastic recoil of the aortic bulb in the hearts of marine mammals but primarily through ischemia that restricts blood flow to the visceral organs, skin, and muscles. Tissues such as those in the liver and kidney that regularly experience drastic reductions in blood flow during diving show extreme tolerance of these conditions and their consequences. Deprivation of arterial blood flow to selected organs produces a gradual reduction in body temperature (Scholander *et al.*, 1942b; Hammel *et al.*, 1977; Hill *et al.*, 1987; Andrews *et al.*, 1995); in the extreme, perhaps even brain cooling occurs (Odden *et al.*, 1997). There is evidence that even normally sensitive tissues such as the brain and heart are adapted to dealing with low oxygen conditions in some marine mammal species (Ridgway *et al.*, 1969; Kjukshus *et al.*, 1982; Elsner and Gooden, 1983; White *et al.*, 1990).

Hypometabolism almost certainly occurs during diving, because the metabolic cost of diving is so low (Kooyman *et al.*, 1973; Castellini *et al.*, 1992; Costa, 1993; Andrews *et al.*, 1995), but direct evidence is difficult to obtain in the wild. Depressed metabolic rates have been documented directly during voluntary diving in captive grey seals (Sparling and Fedak, 2004). In addition to reduction in temperature, another mechanism that might result in metabolic inhibition during diving is increasing tissue acidity (Harken, 1976). Although the details of how an animal that is hypometabolic is able to retain the ability to actively swim remains unclear, it is clear that compromises must be made between diving time, exertion, and oxygen economy (Castellini, 1985a, 1985b). Marine mammals have relatively low aerobic scope for activity, but their anaerobic capabilities are well beyond those of terrestrial mammals (e.g., Elsner, 1987; Ponganis *et al.*, 1990; Williams *et al.*, 1991, 1993).

The biochemical manifestations of cellular resistance to conditions of apnea remain elusive and continue to be the subject of some debate (Blix, 1976; Castellini *et al.*, 1981; Kooyman *et al.*, 1981; Hochachka *et al.*, 1988; Hochachka, 1992), but it is conclusively established that marine mammals can tolerate anaerobic cellular conditions with elevated lactic acid and declining pH that terrestrial mammals would find disruptive or even lethal. In an extensive comparison of diving and nondiving mammals, the most significant differences in tissue biochemistry were found in the levels of myoglobin and muscle buffering capacity (Castellini and Somero, 1981; Castellini *et al.*, 1981). Buffering capacity is the ability to hold tissue pH constant in the face of an increasing amount of acidic end products created by anaerobic metabolism. Marine mammals have a greater buffering ability than terrestrial mammals, and phocids have higher non-carbonate plasma buffering capacities than either otariids or most cetaceans (Boutilier *et al.*, 1993). This probably reflects their general patterns of breath-hold diving and relative potentials for tolerating low oxygen and metabolic acidosis. Several aspects of the biochemistry of marine mammals suggest that their lifestyles generally do not require sustained endurance of intense exercise, but rather they seem to support burst activity levels that can switch over to anaerobic sources of energy for short periods (Costa and Williams, 1999).

The ability to dive in all species of marine mammals studied to date shows an ontogenetic pattern of development; one that is very rapid in some species (e.g., Le Boeuf *et al.*, 1996; Horning and Trillmich, 1997; Baker and Donohue, 1999; Lydersen and Kovacs, 1999; Jørgensen *et al.*, 2001). Of course, marine mammals are "experienced" divers when they are born; fetuses show heart rate declines when their mothers dive (Elsner *et al.*, 1970; Liggins *et al.*, 1980; Hill *et al.*, 1987), although hypoxia is likely first experienced postbirth because uterine arterial blood flow is maintained during dives undertaken during pregnancy (Elsner *et al.*, 1970). Although neonatal dolphins and seals lack the myoglobin concentrations required for prolonged dive durations, myoglobin content in skeletal muscles increases significantly during subsequent development, (Noren and Williams, 2000). Noren *et al.* (2001) also demonstrated an age-related capability of *Tursiops* to decrease heart rate during dives.

Many questions remain to be resolved regarding how marine mammals are protected against the adverse effects of their frequent exposures to high pressure in deep dives, as well as questions such as how those species without sonar manage to find food at great depths (see Chapter 7, Sensory Systems). However, general patterns about their diving performances and patterns are beginning to emerge.

10.5. Diving Behavior and Phylogenetic Patterns

Simple observations and incidental catches of marine mammals in gear set at depth provided evidence that some marine mammals dive for extended period of time to great depths. However, it was not until the development of time-depth recorders (TDRs) of various types that systematic data on the diving behavior of marine mammals started to accumulate. Pioneering studies beginning in the late 1960s with mechanical devices developed by Kooyman and his colleagues that were deployed on Weddell seals in the Antarctic (e.g., Kooyman, 1966, 1985; Kooyman and Campbell, 1972). The exciting results stimulated rapid technological advances in the development of smaller electronic instruments with increasing complex sensors. TDRs, which had to be recovered to download data, and later independently reporting satellite-linked platform transmitter terminals (PTTs) provided new opportunities for deployments of instruments on a wide variety of marine mammal species. Although data on many species are still lacking, or available for only some age or sex classes during part of their annual cycle, some exciting data sets are available for several species (most notably Weddell seals and elephant seals) and at least fragmented patterns are emerging among marine mammals regarding their diving behavior.

Diving ability is of course intimately linked with physiological capabilities, but body size, ecological niche, and life-history strategy also play a role in the type of diving that dominates a species repertoire (Boyd and Croxall, 1996; Boyd, 1997; Schreer and Kovacs, 1997; Schreer *et al.*, 1998; Costa and Williams, 1999; Costa *et al.*, 2001). For their body size, phocid seals are the most capable divers among all of the marine mammals. They utilize a strategy that has been referred to as "energy conserving" (Hochachka *et al.*, 1997; Mottishaw *et al.*, 1999), and all phocids exhibit deep and long dives compared to otariids or whales of similar size. Phocid seals tend to be large which means that they have a low mass specific metabolic rate and they can carry a large amount of oxygen in their tissues. They also have higher blood oxygen storage capacities in their blood because they have elevated hematocrits (Lenfant *et al.*, 1970). Phocids also perform the most profound bradycardia responses and the greatest degree of vasoconstriction among pinnipeds resulting in extremely low metabolism during diving (e.g., Castellini *et al.*, 1992; Costa, 1993). They have slow swimming speeds that minimizes the cost of locomotion.

The Weddell seal has been the subject of the most extensive and comprehensive examinations of diving physiology in the wild (Figure 10.12). Kooyman and others have found that Weddell seals perform dives up to about 20 minutes in length without adjusting their heart rate or circulatory patterns. These results suggest that Weddell seals have sufficient stored oxygen to last about 20 minutes (Figure 10.13). Only during dives lasting longer

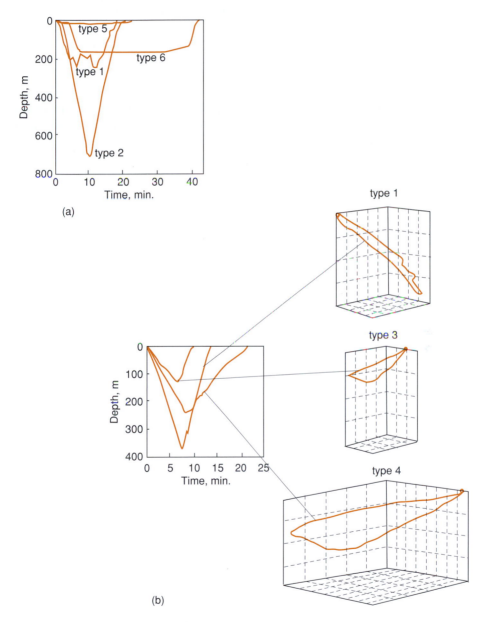

Figure 10.12. (a) Time-depth profiles of four Weddell seal dive types characterized by Schreer and Testa (1996). (b) Three Weddell seal dive profiles, although similar in shape to type 2 dives in (a), their 3-D maps emphasize large differences in horizontal as well as vertical movements (from Davis *et al.*, 2003). Three-D maps scaled in 100-m increments.

than this duration do they show signs of a dive response. This appears to be the **aerobic dive limit** (ADL) of this species, defined specifically by Kooyman (1985) as the longest dive that does not lead to an increase in blood lactate concentration during the dive. If an animal dives within its ADL, there is no lactate accumulated to metabolize after the dive, and a subsequent dive can be made as soon as the depleted body oxygen stores are replenished, which occurs very rapidly in diving animals. ADL can be calculated by measuring available oxygen stores and dividing by either a measured or an estimated metabolic rate (measured as oxygen consumption). Only Weddell seals and Baikal seals have had their ADLs verified by the measurement of blood lactic acid concentrations following free diving (Kooyman *et al.*, 1983; Ponganis *et al.*, 1997). The actual duration of an animal's ADL is dependent on its level of activity and is very difficult to calculate accurately because myoglobin levels vary in different skeletal muscles and circulatory splenic reservoirs of blood complicate calculations. Even so, ADL remains a useful concept for identifying a demarcation between oxidative and anaerobic metabolic processes during diving.

If animals exceed their ADL and accumulate lactate, a surface recovery period is required. After very long dives, Weddell seals are exhausted and sleep for several hours (Kooyman *et al.*, 1980; Castellini *et al.*, 1988). During dives of long duration (see Table 10.1), Weddell seals exhibit full dive responses; both peripheral vasoconstriction and bradycardia are initiated maximally from the onset of the dive and remain throughout the dive. On extended dives approaching one hour in duration, core body temperature can be depressed to 35°C, kept depressed between dives, and then rapidly elevated after the last dive of a dive series. Although Weddell seals are capable of remaining submerged for more than an hour, they seldom do. Approximately 85% of the dives observed by Kooyman and Campbell (1972) were within the presumed 20-minute ADL of this species (see Figure 10.13). Although the actual duration varies markedly between species, this general pattern is observed in all species studied to date. Within a species, the vast majority of dives occur within quite a narrow duration limit, extending beyond this duration only during a very small proportion of their dives. These values likely approximate their species-specific ADL.

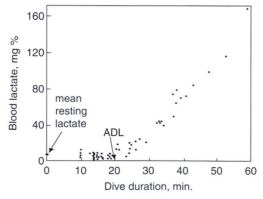

Figure 10.13.   Peak arterial lactate concentrations obtained following dives of varying durations for three Weddell seals. Commencement of postdive increases in arterial lactate define ADL. (Redrawn from Kooyman *et al.*, 1981.)

The most extreme depths and durations for diving among the phocid seals have been recorded for the largest members of this group, the northern and southern elephant seals (see Le Boeuf and Laws, 1994). Both northern and southern elephant seals exhibit extraordinary capacities for diving for extended periods at sea during long migratory/foraging phases of the year (see Chapter 11). During two feeding migrations each year, they remain at sea for months. During such trips the dives of northern elephant seals are typically 20–30 minutes long, with females usually going to depths of about 400 m (e.g., Le Boeuf *et al.*, 1989) and males dive to average depths of 750–800 m (DeLong and Stewart, 1991; Stewart and DeLong 1995; see also Chapter 11). Both sexes dive night and day for days at a time without prolonged rest periods at the sea surface. After each dive, they typically spend only a few minutes at the surface before the next dive (Figure 10.14). The short duration of these surface bouts suggests that the 20- to 30-minute dives are within the ADL of this species. These seals probably adjust their swimming speeds and metabolic rates to sustain almost all dives aerobically, so that little surface recovery time is required before commencing the next dive. However, some dives do extend well beyond the norm, with maximal depths reaching beyond 1500 m with maximal durations of 77 minutes. Similar diving patterns are seen among southern elephant seals following breeding (Hindell *et al.*, 1992). However, females of this species extend their diving times such that about 44% of their dives exceed the calculated ADL. Most of these extended dives are thought to be foraging dives, and they often occurred in bouts. The longest recorded bout consisted of 63 consecutive dives over a 2-day period, with little extended postdive recovery time required between dives. Similarly, free-ranging grey seals do not seem to require extended surface times to recover from unusually long dives (Thompson and Fedak, 1993). This suggests that a reduction in overall metabolic rate, rather than a switch to anaerobic metabolism, is the most likely mechanism for extending dive time in these species. It also appears that these animals do not have a set heart rate or metabolic rate throughout a dive, but rather it appears that these responses may be adjusted during a dive (Andrews *et al.*, 1997). At some locales,

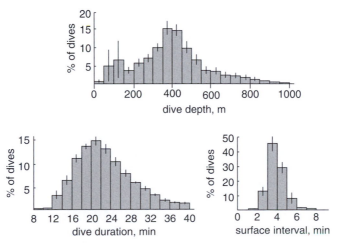

Figure 10.14. Summaries of depths, durations, and postdive intervals of over 36,000 dives made by six adult male northern elephant seals (error bars = ±1 s.e.). (Redrawn from DeLong and Stewart, 1991.)

hooded seals, another large, deep-diving phocid seal, seem to specialize on deep dwelling fish and can dive to depths of 1000 m for periods of 52 minutes or more. However, most dives in this species are much shorter and occur to depths of 100–600 m (Folkow and Blix, 1995).

Phocid seals of intermediate size tend to display diving behavior that is concomitant with their body sizes. Species such as grey seals, harp seals, Ross seals, and crabeater seals do most of their diving to depths of about 100 m for less than 10 minutes, although maximal values can be significantly longer and deeper (e.g., Lydersen and Kovacs, 1993; Bengston and Stewart, 1992, 1997; Lydersen et al., 1994; Folkow et al., 2004). Small phocids tend to be the most conservative divers in this group with Baikal, Saimaa, ringed, and harbor seals usually diving for periods of only a few minutes to relatively shallow depths. However, even these small species do on occasion display deep and long dives. Harbor seals have been documented to dive to over 450 m during dives that last more than 30 minutes (Bowen et al., 1999; Gjertz et al., 2001), and juvenile ringed seals weighing less than 40 kg dive to over 500 m and remain submerged for longer than 30 minutes (Lydersen, Kovacs, and Fedak, unpublished data).

Hawaiian monk seals and bearded seals are both relatively large phocids, but both species display quite shallow and short duration diving patterns, because these two species usually forage in shallow coastal waters. However, adult monk seals do sometimes dive outside lagoon areas, where adult males have been recorded as deep as 550 m and bearded seals are also clearly capable of more extreme diving than is performed in coastal areas. Bearded seals pups of only a few months of age hold the dive records for this species. During their early wanderings, 7 of 7 postmolting pups dove to depths of over 400 m (Gjertz et al., 2000), whereas adults of this species normally dive for only a few minutes (2–4) to depths of 20 m (Gjertz et al., 2000; Krafft et al., 2000). These two species illustrate the strong influence of foraging preference on behavioral patterns in diving.

Otariids do not spend as much time diving as phocids and they usually dive for only a few minutes to relatively shallow depths. They are also at sea for relatively short periods of time compared to many phocids that spend a lot of time pelagically. The otariid strategy has been described as being energy dissipative (Hochachka et al., 1997; Mottishaw et al., 1999; Table 10.3). Otariids have a relatively small and hydrodynamically sleek body consistent with a high speed predator lifestyle. They appear to sacrifice extended foraging time for energy needed for higher speed swimming (Costa, 1991). This pinniped group shows the same basic patterns with respect to ADL as do the phocids, in that they remain within their aerobic limits during most of their diving, although patterns are perhaps somewhat more variable. Data from Antarctic fur seals indicate that less than 6% of their dives exceed the estimated ADL but some dive bouts were of significantly longer duration and are followed by longer surface intervals (Boyd and Croxall, 1996). Although some individual dives exceeded their ADL, these specific dives did not appear to have any immediate effect on their subsequent diving behavior. Relatively deep foraging dives (>75 m) are common for northern fur seals (Gentry et al., 1986; Figure 10.15), and in one study 92% of dives exceeded the calculated ADL for this species, whereas only 8% of shallow foraging dives (<40 m) exceeded the ADL (Ponganis et al., 1992). This suggests that at least some otariids may also employ energy conservation/metabolic strategies in addition to anaerobic metabolism for extended diving duration.

Among otariid species, allometric patterns are apparent intra- and interspecifically with males of sexually dimorphic species tending to dive deeper than females and small

Table 10.3. Comparison of Phocid and Otariid Diving Strategies (Based on Mottishaw *et al.*, 1999)

Phocid Characteristics
1. Apnea, with exhalation on initiation of diving
2. Bradycardia in 1:1 proportion with changes in cardiac output
3. Peripheral vasoconstriction and hypoperfusion (to conserve O_2 for central nervous system and the heart)
4. Hypometabolism of (vasoconstricted) ischemic tissues
5. Enhanced O_2 carrying capacity (enlarged blood volume, expanded red blood cell [RBC] mass within the blood volume–i.e., higher hematocrit, higher hemoglobin concentration, possibly higher myoglobin concentration in muscles and heart)
6. Enlarged spleen (for regulating the hematocrit so that very high RBCs need not be circulated under all physiological conditions)

Otariid Characteristics
1. Apnea, initiated on inhalation and a gas exchange system that does not completely collapse
2. Bradycardia
3. Peripheral vasoconstriction, with propulsive muscle microvasculature presumably more relaxed than in phocids
4. Hypometabolism of ischemic tissues
5. O_2 carrying capacity intermediate between large phocids and terrestrial mammals
6. Spleen not much larger as a percentage of body weight than in terrestrial mammals

species tending to dive for shorter periods to lesser depths than large species (Schreer and Kovacs, 1997). However dive depths for otariids generally reflect the vertical distribution of prey in the water column as opposed to the physiological limits of the divers (Gentry, 2001; Costa *et al.*, 2001). Dives of more than 5–7 minutes are uncommon. However, there are exceptions. The New Zealand sea lion is a relatively deep diver, performing deeper dives than many midsized and small phocids. Lactating females of this species

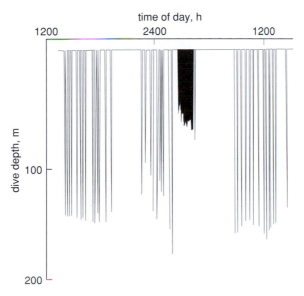

Figure 10.15. Deep dive bouts during foraging by an adult female northern fur seal. (Redrawn from Gentry and Kooyman, 1986.)

dive to average depths of 123 m during their 4- to 6-minute average dives, with maximal dive records of 474 m and 11 minutes. This otariid has the highest blood volume yet reported for an otariid, one that falls within the range of phocids (Costa *et al.*, 1998).

Although walruses are among the largest pinnipeds they usually dive for short periods to very shallow depths (Fay and Burns, 1988). This reflects the fact that, similar to bearded seals, they feed in shallow waters on benthic prey. However, walruses are capable of diving to much greater depths than most studies report; walruses moving between summering areas in Svalbard and winter breeding areas in Frans Joseph Land dive to over 300 m where the bathymetry permits (Lydersen *et al.*, unpublished data). The physiological limits of walruses are not known, but they almost certainly exceed diving records documented to date on this species.

On the basis of body size it would be expected that cetaceans should be able to dive longer and deeper than other marine mammals because they can store more oxygen and have lower mass-specific metabolic rates. However, this is not the case and cetaceans (with the exception of the sperm whale and the northern bottlenose whale; Hooker and Baird, 1999; Watkins *et al.*, 2002; Amano and Yoshioka, 2003) are surpassed in average diving capacity by considerably smaller phocids (see Schreer and Kovacs, 1997; and Costa and Williams, 1999 for reviews). There are several reasons for this. First, the feeding ecology of many cetaceans involves exploitation of prey that is located at shallow depths and therefore they may not need to dive as deep or for as long as species feeding on more diverse or generally deeper dwelling prey. It can also be argued that the body sizes of mysticete whales are allometrically distorted by their extensive blubber volumes and enlarged mouths (see Chapter 12). Approximately one third of the body length of balaenopterids is dedicated just to the jaw and its support. Among odontocetes, allometric patterns are apparent with the largest species, sperm whales and northern bottlenose whales, performing the longest and deepest dives among all marine mammals (see Table 10.1).

Small social groups of sperm whales spend hours together at the surface each day, then initiate a series of long and deep foraging dives. Adult Caribbean sperm whales tracked with acoustic transponders exhibited dive profiles like those in Figure 10.16, with extended periods of resting and socializing at the surface, especially during afternoon hours (Watkins *et al.*, 1993). Diving was most common at night and during early morning hours. Dives typically lasted more than 30 minutes to depths usually below 400 m. Like the dives of foraging elephant seals and migrating gray whales, little surface recovery time was apparent between dives. Sperm whales, like the other species of marine mammals discussed in this chapter, presumably dive within their ADL or are compensating by other mechanisms.

Some midsized odontocetes display considerable diving abilities. Narwhals and white whales in some parts of their range routinely dive deeply (>500 m) and they can dive to over 1000 m (e.g., Heide-Jørgensen and Dietz, 1995; Martin and Smith, 1999). Their dive durations are relatively modest with most dives lasting only 5–15 minutes. Killer whales are remarkably shallow divers for their size, whereas the long-finned pilot whales and bottlenose dolphins dive to significant depths (max. 500–600 m, 16 minutes) although durations still tend to remain under 10 minutes. Other, smaller dolphins such as the short-beaked common dolphin and pantropical spotted dolphin routinely dive to about 100 m although they can go to approximately twice this depth. All of the smaller dolphins studied to date usually dive for periods of only a few minutes (e.g., Ridgway, 1986).

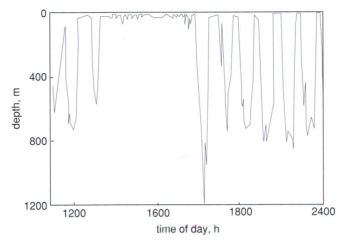

Figure 10.16.   Continuous dive profile of a single sperm whale, generated by connecting depth points recorded by a digital depth monitor tag. (Redrawn from Watkins *et al.*, 1993.)

Mysticete whales as a group can be characterized as shallow divers that do not remain submerged for long periods between periods at the surface (Schreer and Kovacs, 1997). They do not exhibit clear allometric patterns within the group and are anomalous within marine mammals as having very modest diving abilities for their size. Even benthic feeding gray whales typically dive for very short durations (average 4–5 minutes) over the continental shelf of the Bering Sea. Large blue whales can dive to 150–200 m for periods up to 50 minutes, but most of their recorded dives have been very short and occur in the top 100 m (Langerquist *et al.*, 2000). Fin whales often target deep-dwelling krill, so tend to be deeper divers; they have been documented to dive repeatedly to 180 m for an average of 10 minutes (max. 20 minutes) in the Ligurian Sea, although elsewhere they do not show these long, deep dives. Humpback whales studied in Alaska had dive durations, surface times, and the numbers of blows per surfacing that were correlated with the depths of foraging dives (Dolphin, 1987). Most of their dives were less than 3 minutes and shallow (84.6% were to depths of less than 60 m). Dolphin (1987) postulated that dives to depths of 40–60 m and 4–6 minutes in duration likely represented the ADL of humpback whales because the percentage of time spent at the surface increased with progressively for dives greater than 60 m.

Manatees are slow moving animals that feed on floating and shallowly submerged vegetation in coastal areas, so it is unlikely that they ever dive very deeply. Observations of animals in the wild suggest that most dives are for less than 5 minutes. Some longer dives up to 20 minutes have been observed, but it is thought that these longer submersions may occur when the animals sink to the bottom to sleep. Dugongs are faster swimmers and they do dive down to 20 m to feed on offshore seagrass beds. Most dives in this species are 2–5 minutes long, but foraging dives have lasted more than 12 minutes (Chilvers *et al.*, 2004).

Sea otters live along coastlines, feeding in shallow near-shore waters. They tend to dive in bouts, resting in floating vegetation for significant periods at the surface between bouts, grooming their fur and resting. The diving capacity of polar bears is quite limited. They do dive beneath the surface to approach seals that are resting on the edges of ice floes for periods of a minute or two following a stealthful swimming approach at the surface. They have also been seen diving in attempts to hunt white whales, diving in from the ice edge and remaining submerged for a minute or so, but polar bears spend most of their aquatic time swimming on the surface.

10.6. Summary and Conclusions

The effects of pressure on the diving animal involve circulatory and respiratory adaptations. Among circulatory changes in pinnipeds and cetaceans are enlargement and increased complexity of blood vessels, including the development of retia mirabilia throughout the body that serve as oxygen reservoirs during deep dives. The muscles, blood, and spleen are important for oxygen stores in marine mammals. Respiratory adjustments that occur during diving involve modifications in the structure of the lungs, especially the bronchioles.

Bradycardia, peripheral vasoconstriction, and other circulatory adjustments are important components of an integrated set of diving responses. Research on the Weddell seal and other pinnipeds revealed that the majority of dives are aerobic and that only dives exceeding an animal's ADL elicited one or more of the diving responses. The phylogenetic implications of the diving patterns indicate two strategies among pinnipeds, an otariid "energy dissipative" strategy and a phocid "energy conserving" strategy. It is suggested that the walrus is capable of deep dives but has little reason to do so because of the availability of their prey in shallow water. In addition to their function in energy conservation and foraging, the diving patterns of pinnipeds may also play a role in the avoidance of predators. The diving and breath-hold capacities of most whales (with the exception of the sperm whale) are exceeded by considerably smaller phocid seals. Possible explanations for this include less accurate measurements of cetacean diving behavior and their exploitation of prey located at shallower depths.

10.7. Further Reading

For a general introduction to diving physiology and behavior in marine mammals see Kooyman (1989); popular accounts of diving in the Weddell seal can be found in Kooyman (1981) and Williams (2004). A comprehensive account of elephant seal diving behavior is summarized in an edited volume of their biology (Le Boeuf and Laws, 1994) and see Gentry and Kooyman (1986) for diving behaviors of various fur seals. Summary accounts of cetacean diving include Ridgway (1986) and Ridgway and Harrison (1986).

References

Amano, M., and M. Yoshioka (2003). "Sperm Whale Diving Behavior Monitored Using a Suction-Cup-Attached TDR Tag." *Mar. Ecol. Progr. Ser. 258:* 291–295.

Andrews, R. D., D. R. Jones, J. D. Williams, D. E. Crocker, D. P. Costa, and B. J. Le Boeuf (1995). "Metabolic and Cardiovascular Adjustments to Diving in Northern Elephant Seals *(Mirounga angustirostris)*." *Physiol. Zool. 68:* 105.

Andrews, R. D., D. R. Jones, J. D. Williams, P. H. Thorson, G. W. Oliver, D. P. Costa, and B. J. Le Boeuf (1997). "Heart Rates of Northern Elephant Seals Diving at Sea and Resting on the Beach." *J. Exp. Biol. 200:* 2083–2095.

Baird, R. W., D. J. McSweeney, A. D. Ligon, and D. L. Webster (2004). Tagging Feasibility and Diving of Cuvier's Beaked Whales *(Ziphius cavirostris)* and Blainville's Beaked Whales *(Mesoplodon densirostris)* in Hawaii. Report prepared under Order No. AB133F-03-SE-0986 to the Hawai'i Wildlife Fund, Volcano, HI.

Baker, J. D., and M. J. Donohue (1999). "Ontogeny of Swimming and Diving in Northern Fur Seal *(Callorhinus ursinus)* Pups." *Can. J. Zool. 78:* 100–109.

Barabash-Nikiforov, I. I. (1947). *"Kalan" (The Sea Otter).* Soviet Ministrov RSFSR. (Translated from Russian, Israel Program for Scientific Translations, Jerusalem, 1962).

Bengtson, J. L., and B. S. Stewart (1992). "Diving and Haulout Behavior of Crabeater Seals in the Weddell Sea, Antarctic During March 1996." *Polar Biol. 12:* 635–644.

Bengtson, J. L., and B. S. Stewart (1997). "Diving Patterns of a Ross Seal *(Ommatophoca rossii)* near the Eastern Coast of the Antarctic Peninsula." *Polar Biol. 18:* 214–218.

Berne, R. M., and M. N. Levy (1988). *Physiology.* Mosby, St. Louis, MO.

Bert, P. (1870). *In* "Leçons sure la Physiologie Comparéde la Respiration," pp. 526–553. Baillière, Paris.

Blix, A. S. (1976). "Metabolic Consequences of Submersion Asphyxia in Mammals and Birds." *Biochem. Soc. Symp. 41:* 169–178.

Bodkin, J. L., G. G. Esslinger, and D. H. Monson. (2004). "Foraging Depths of Sea Otters and Implications to Coastal Marine Communities." *Mar. Mamm. Sci. 20:* 305–321.

Boutilier, R. G., M. Nikinmaa, and B. L. Tufts (1993). "Relationship Between Blood Buffering Properties, Erythrocyte pH and Water Content, in Gray Seals *(Halichoerus grypus)*." *Acta. Physiol. Scand. 147:* 241–247.

Bowen, W. D., D. J. Boness, and S. J. Iverson (1999). "Diving Behaviour of Lactating Harbour Seals and Their Pups During Maternal Foraging Trips." *Can. J. Zool. 77:* 978–988.

Boyd, I. L. (1997). "The Behavioural and Physiological Ecology of Diving." *Trends Ecol. Evol. 12:* 213–217.

Boyd, I. L., and J. P. Croxall (1996). "Dive Durations in Pinnipeds and Seabirds." *Can. J. Zool. 74:* 1696–1705.

Bryden, M. M. (1972). Growth and development of marine mammals. *In* "Functional Anatomy of Marine Mammals," Vol. 1 (R. J. Harrison, ed.), pp. 1–79. Academic Press, New York.

Castellini, M. A. (1985a). "Metabolic Depression in Tissues and Organs of Marine Mammals During Diving: Living with Less Oxygen." *Mol. Physiol. 8:* 427–437.

Castellini, M. A. (1985b). Closed systems: resolving potentially conflicting demands of diving and exercise in marine mammals. *In* "Circulation, Respiration and Metabolism" (R. Gilles, ed.), pp. 220–226. Springer-Verlag, Berlin.

Castellini, M. A. (1991). "The Biology of Diving Mammals: Behavioral, Physiological, and Biochemical Limits." *Adv. Comp. Physiol. 8:* 105–134.

Castellini, M. A., Davis, R. W., and G. L. Kooyman (1988). "Blood Chemistry Regulation During Repetitive Diving in Weddell Seals." *Physiol. Zool. 61:* 379–386.

Castellini, M. A., G. L. Kooyman, and P. J. Ponganis. (1992). "Metabolic Rates of Freely Diving Weddell Seals: Correlations with Oxygen Stores, Swim Velocity and Diving Duration." *J. Exp. Biol. 165:* 181–194.

Castellini, M. A., B. J. Murphy, M. Fedak, K. Ronald, N. Gofton, and P. W. Hochachka. (1985). "Potentially Conflicting Metabolic Demands of Diving and Exercise in Seals." *J. Appl. Physiol. 58:* 392–399.

Castellini, M. A., and G. N. Somero (1981). "Buffering Capacity of Vertebrate Muscle: Correlations with Potentials for Anaerobic Function." *J. Comp. Physiol. 143:* 191–198.

Castellini, M. A., G. N. Somero, and G. L. Kooyman (1981). "Glycolytic Enzyme Activities in Tissues of Marine and Terrestrial Mammals." *Physiol. Zool. 54:* 242–252.

Chilvers, B. L., S. Delean, N. J. Gales, D. K. Holley, I. R. Lawler, H. Marsh, and A. R. Preen (2004). "Diving Behaviour of Dugongs, *Dugong dugon*." *J. Exp. Mar. Biol. Ecol. 304:* 203–224.

Costa, D. P. (1991). Reproductive and foraging energetics of pinnipeds: Implications for life history patterns. *In* "The Behavior of Pinnipeds" (D. Renouf, ed.), pp. 300–344. Chapman & Hall, London.

Costa, D. P. (1993). The relationship between reproduction and foraging energetics and the evolution of the Pinnipedia. *In* "Recent Advances in Marine Mammal Science" (I. Boyd, ed.), pp. 293–314. Symp. Zool. Soc., Lond. No. 66, Oxford University Press, London.

Costa, D. P., and N. J. Gales (2003) "Energetics of a Benthic Diver: Seasonal Foraging Ecology of the Australian Sea Lion, *Neophoca cinera*." *Ecol. Monogr. 73:* 27–43.

Costa, D. P, N. J. Gales, and D. E. Crocker (1998). "Blood Volume and Diving Ability of the New Zealand Sea Lion, *Phocarctos hookeri*." *Physiol. Zool. 71:* 208–213.

Costa, D. P., N. J. Gales, and M. E. Goebel (2001). "Aerobic Dive Limit: How Often Does It Occur in Nature?" *Comp. Biochem. Physiol. A 129:* 771–783.

Costa, D. P., and G. L. Kooyman (1982). "Oxygen Consumption, Thermoregulation, and the Effect of Fur Oiling and Washing on the Sea Otter, *Enhydra lutris*." *Can. J. Zool. 60:* 2761–2767.

Costa, D. P., and T. M. Williams (1999). Marine mammal energetics. *In* "Biology of Marine Mammals" (R. E. Reynolds and S. A. Rommel, eds.), pp. 176–217. Smithsonian Institute Press, Washington, D. C.

Curry, B. E., J. Mead, and A. P. Purgue (1994). "The Occurrence of Calculi in the Nasal Diverticula of Porpoises (Phocoenidae)." *Mar Mamm. Sci. 10:* 81–86.

Davis, R. W., L. A. Fuiman, T. M. Williams, M. Horning, and W. Hagey (2003). "Classification of Weddell Seal Dives Based on 3 Dimensional Movements and Video-Recorded Observations." *Mar. Ecol. Prog. Ser. 264:* 109–122.

DeLong, R. L., and B. S. Stewart (1991). "Diving Patterns of Northern Elephant Seal Bulls." *Mar. Mamm. Sci. 7:* 369–384.

Denison, D. M., D. A. Warrell, and J. B. West (1971). "Airway Structure and Alveolar Emptying in the Lungs of Sea Lions and Dogs." *Respir. Physiol. 13:* 253–260.

Dolphin, W. F (1987). "Dive Behavior and Estimated Energy Expenditure of Foraging Humpback Whales in South-East Alaska." *Can. J. Zool. 65:* 354–362.

Drabek, C. M. (1975). "Some Anatomical Aspects of the Cardiovascular System of Antarctic Seals and Their Possible Functional Significance in Diving. *J. Morphol. 145*(1): 85–106.

Drabek, C. M. (1977). Some anatomical and functional aspects of seal hearts and aortae. *In* "Functional Anatomy of Marine Mammals," Vol. 3 (R. J. Harrison, ed.), pp. 217–234. Academic Press, London.

Drabek, C. M., and J. M. Burns (2002). "Heart and Aorta Morphology of the Deep-Diving Hooded Seal *(Cystophora cristata)*." *Can. J. Zool. 80:* 2030–2036.

Drabek, C. M., and G. L. Kooyman (1983). "Terminal Airway Embryology of the Delphinid Porpoises, *Stenella attenuata* and *S. longirostris*." *J. Morphol. 175:* 65–72.

Elsner, R. (1987). The contribution of anaerobic metabolism to maximum exercise in seals. *In* "Marine Mammal Energetics" (A. C. Huntley, D. P. Costa, G. A. J. Worthy, and M. A. Castellini, eds.), pp. 109–114. Soc. Mar. Mammal., Spec. Publ. No 1, Lawrence, KS.

Elsner, R. (1999). Living in water: Solutions to physiological problems. *In* "Biology of Marine Mammals" (R. E. Reynolds and S. A. Rommel, eds.), pp. 73–116. Smithsonian Institute Press, Washington, D. C.

Elsner, R., and B. Gooden (1983). *Diving and Asphyxia: A Comparative Study of Animals and Men.* Cambridge University Press, New York.

Elsner, R., D. D. Hammond, and H. R. Parker (1970). "Circulatory Responses to Asphyxia in Pregnant and Fetal Animals: A Comparative Study of Weddell Seals and Sheep." *Yale J. Biol. Med. 42:* 202–217.

Elsner, R., D. W. Kenney, and K. Burgess (1966). "Diving Bradycardia in the Trained Dolphin." *Nature 212:* 407–408.

Elsner, R., D. Wartzok, N. B. Sonafrank, and B. P. Kelly (1989). "Behavioural and Physiological Reactions of Arctic Seals During Under-Ice Pilotage." *Can. J. Zool. 67:* 2506–2513.

Engel, S. (1959a). "The Respiratory Tissue of Dugong *(Halicore dugong)*." *Anat. Anz. 106:* 90–100.

Engel, S. (1959b). "Rudimentary Mammalian Lungs." *Gegenbaurs Morphol. Jahrb. 106:* 95–114.

Engel, S. (1962). "The Air Passages of the Dugong Lung." *Acta Anat. 48:* 95–107.

Falke, K. J., R. D. Hill, J. Qvist, R. C. Schneider, M. Guppy, G. C. Liggins, P. W. Hochachka, R. E. Elliott, and W. M. Zapol (1985). "Seal Lungs Collapse During Free Diving: Evidence From Arterial Nitrogen Tensions." *Science 229:* 556–557.

Fay, F. (1981). Walrus-*Odobenus rosmarus. In* "Handbook of Marine Mammals" (S. H. Ridgway and R. J. Harrison, eds.), pp. 1–23. Academic Press, London.

Fay, F H., and J. J. Burns (1988). "Maximum Feeding Depth of Walruses." *Arctic 41:* 239–240.

Fedak, M. A. (1986). Diving and exercise in seals a benthic perspective. *In* "Diving in Animals and Man" (A. Brubakk, J. W. Kanwisher, and G. Sundnes, eds.), pp. 11–32. Tapir Publ., Trondheim, Norway.

Folkow, L. P., and A. S. Blix (1995). Distribution and diving behaviour of hooded seals. *In* "Whales, Seals, Fish and Man" (A. S. Blix, L. Walløe, and Ø. Ulltang, eds.), pp. 193–202. Elsevier, Amsterdam.

Folkow, L. P., E. S. Nordøy, and A. S. Blix (2004). "Distribution and Diving Behaviour of Harp Seals *(Pagophilus groenlandicus)* from the Greenland Sea Stock." *Polar Biol. 27:* 281–298.

Gentry, R. L. (2001). Eared seals. *In* "Encyclopedia of Marine Mammals" (W. F. Perrin, B. Würsig, and J. G. M. Thewissen, eds.), pp. 348–351. Academic Press, San Diego, CA.

Gentry, R. L., and G. L. Kooyman (1986). *Fur Seals: Maternal Strategies on Land and at Sea.* Princeton University Press, Princeton, NJ.

Gentry, R. L., G. L. Kooyman, and M. E. Goebel (1986). Feeding and diving behavior of northern fur seals. *In* "Fur Seals: Maternal Strategies on Land and at Sea" (R. L. Gentry, and G. L. Kooyman, eds.), pp. 61–78. Princeton University Press, Princeton, NJ.

Gilmartin, W. G., R. W. Pierce, and G. A. Antonelis (1974). "Some Physiological Parameters of the Blood of the California Gray Whale." *Marine Fisheries Rev. 36:* 28–31.

Gjertz, I., K. M. Kovacs, C. Lydersen, and Ø. Wiig (2000). "Movements and Diving of Bearded Seal *(Erignathus barbatus)* Mother and Pups During Lactation and Post-Weaning." *Polar Biol. 23:* 559–566.

Gjertz, I., C. Lydersen, and Ø. Wiig (2001). "Distribution and Diving of Harbour Seals *(Phoca vitulina)* in Svalbard." *Polar Biol. 24:* 209–214.

Gosline, J. M., and R. E. Shadwick (1996). "The Mechanical Properties of Fin Whale Arteries are Explained by Novel Connective Tissue Designs." *J. Exp. Biol. 199:* 985–997.

Guyton, G. P., K. S. Stanek, R. C. Schneider, P. W. Hochachka, W. E. Hurford, D. G. Zapol, G. C. Liggins, and W. M. Zapol (1995). "Myoglobin Saturation in Free-Diving Weddell Seals." *J. Appl. Physiol. 79:* 1148–1155.

Hammmel, H. T., R. Elsner, H. C. Heller, J. A. Maggert, and C. R. Bainton (1977). "Thermoregulatory Responses to Alternating Hypothalamic Temperature in the Harbor Seal." *Am. J. Physiol. 232:* R18–R26.

Harken, A. H. (1976). "Hydrogen Ion Concentration and Oxygen Uptake in an Isolated Canine Limb." *J. Appl. Physiol. 40:* 1–5.

Harrison, R. J., and J. E. King (1980). *Marine Mammals,* 2nd ed. Hutchinson, London.

Harrison, R. J., and G. L. Kooyman (1968). General physiology of the pinnipedia. *In* "The Behavior and Physiology of the Pinnipeds" (R. J. Harrison, R. C. Hubbard, R. S. Peterson, R. E. Rice, and R. J. Schusterman, eds.), pp. 211–296. Appleton-Century-Crofts, New York.

Harrison, R. J., and J. D. W. Tomlinson (1956). "Observations on the Venous System in Certain Pinnipedia and Cetacea." *Trans. Zool. Soc. London 126*(part 2): 205–233.

Hedrick, M. S., and D. A. Duffield (1991). "Haematological and Theological Characteristics of Blood in Seven Marine Mammal Species: Physiological Implications for Diving Behaviour." *J. Zool. London 225:* 273–283.

Heide-Jorgensen, M. P., and R. Dietz (1995). "Some Characteristics of Narwhal, *Monodon monoceros,* Diving Behaviour in Baffin Bay." *Can. J. Zool. 73:* 2120–2132.

Hill, R. D., R. C. Schneider, G. C. Liggins, A. H. Schuette, R. L. Elliott, M. Guppy, P. W. Hochachka, J. Qvist, K. J. Falke, and W. M. Zapol (1987). "Heart Rate and Body Temperature During Free Diving of Weddell Seals." *Am. J. Physiol. 253:* R344–R351.

Hill, W. C. O. (1945). "Notes on the Dissection of Two Dugongs." *J. Mammal. 26:* 153–175.

Hindell, M. A., D. J. Slip, H. R. Burton, and M. M. Bryden (1992). "Physiological Implications of Continuous, Prolonged, and Deep Dives of the Southern Elephant Seal *(Mirounga leonina)*." *Can. J. Zool. 70:* 370–379.

Hochachka, P. W. (1992). "Metabolic Biochemistry and the Making of a Mesopelagic Mammal." *Experentia 48:* 570–575.

Hochachka, P. W., J. M. Castellini, R. D. Hill, R. C. Schneider, J. L. Bengtson, S. E. Hill, G. C. Liggins, and W. M. Zapo (1988). "Protective Metabolic Mechanisms During Liver Ischemia: Transferable Lessons from Long-Diving Animals." *Mol. Cell. Biochem. 23:* 12–20.

Hochachka, P. W., S. C. Land, and L. T. Buck (1997). "Oxygen Sensing and Signal Transduction in Metabolic Defense Against Hypoxia: Lessons from Vertebrate Facultative Anaerobes." *Comp. Biochem. Physiol. A 118:* 23–29.

Hochachka, P. W., and P. D. Mottishaw (1998). Evolution and adaptation of the diving response: Phocids and otariids. *In* "Cold Ocean Symposia" (H. O. Portwer and R. C. Playle, eds.), pp. 391–431. Cambridge University Press, Cambridge, MA.

Hooker, S. K., and R. W. Baird (1999). "Deep-Diving Behaviour of the Northern Bottlenose Whale, *Hyperoodon ampullatus* (Cetacea: Ziphiidae)." *Proc. R. Soc. London B 266:* 671–676.

Hooker, S. K., P. J. O. Miller, M. P. Johnson, O. P. Cox, and I. L. Boyd. (2005). "Ascent Exhalations of Antarctic Fur Seals: A Behavioural Adaptation for Breath-Hold Diving? *Proc. Royal Soc. London B 272:* 355–363.

Horning, M., and F. Trillmich (1997). "Ontogeny of Diving Behaviour in the Galapagos Fur Seal." *Behaviour 134:* 1211–1257.

Huntley, A. C., D. P. Costa, and R. D. Rubin (1984). "The Contribution of Nasal Countercurrent Heat Exchange to Water Balance in the Northern Elephant Seal, *Mirounga angustirostris*." *J. Exp. Biol. 113:* 447–454.

Irving, L. (1939). "Respiration in Diving Mammals." *Physiol. Rev. 19:* 112–134.

Irving, L., P. F. Scholander, and S. W. Grinnell (1941a). "The Respiration of the Porpoises, *Tursiops truncatus*." *J. Cell. Comp. Physiol. 17:* 1–45.

Irving, L., P. F. Scholander, and S. W. Grinnell. (1941b). "Significance of the Heart Rate to the Diving Ability of Seals." *J. Cell. Comp. Physiol. 18:* 283–297.

Irving, L., P. F. Scholander, and S. W. Grinnell. (1942). "The Regulation of Arterial Blood Pressure in the Seal During Diving." *Am. J. Physiol. 135:* 557–566.

Irving, L., O. M. Solandt, D. Y. Solandt, and K. C. Fisher (1935). "Respiratory Characteristics of the Blood of the Seal." *J. Cell. Comp. Physiol. 6:* 393–403.

Jones, D. R., H. D. Fisher, S. McTaggart, and N. H. West (1973). "Heart Rate During Breath-Holding and Diving in the Unrestrained Harbor Seal, *Phoca vitulina richardsi.*" *Can. J. Zool. 51:* 671–680.

Jørgensen, C., C. Lydersen, O. Brix, and K. M. Kovacs (2001). "Diving Development in Nursing Harbour Seals." *J. Exp. Biol. 204:* 3993–4004.

King, J. E. (1977). "Comparative Anatomy of the Blood Vessels of the Sea Lions *Neophoca* and *Phocarctos;* with Comments on the Differences Between the Otariid and Phocid Vascular Systems." *J. Zool. London 181:* 69–94.

King, J. E. (1983). *Seals of the World,* 2nd ed. Comstock, Ithaca, NY.

Kjekshus, J. K., A. S. Blix, R. Elsner, R. Hol, and E. Amundsen. (1982). "Myocardial Blood Flow and Metabolism in the Diving Seal." *Am. J. Physiol. 242:* R79–R104.

Kooyman, G. L. (1966). "Maximum Diving Capacities of the Weddell Seal *(Leptonychotes weddelli)*." *Science 151:* 1553–1554.

Kooyman, G. L. (1973). "Respiratory Adaptations in Marine Mammals." *Am. Zool. 13:* 457–468.

Kooyman, G. L. (1981). *Weddell Seal: Consummate Diver.* Cambridge University Press, Cambridge.

Kooyman, G. L. (1985). "Physiology Without Restraint in Diving Mammals." *Mar. Mamm. Sci. 1:* 166–178.

Kooyman, G. L. (1989). *Diverse Divers: Physiology and Behavior.* Springer-Verlag, New York.

Kooyman, G. L. (2002). Diving physiology *In* "Encyclopedia of Marine Mammals" (W. F. Perrin, B. Würsig, and J. G. M. Thewissen, eds.), pp. 339–344. Academic Press, San Diego, CA.

Kooyman, G. L., and H. T. Anderson (1969). Deep diving. *In* "Biology of Marine Mammals" (H. T. Andersen, ed.), pp. 65–94. Academic Press, New York.

Kooyman, G. L., and W. B. Campbell (1972). "Heart Rates in Freely Diving Weddell Seals, *Leptonychotes weddellii.*" *Comp. Biochem. Physiol. A 43:* 31–36.

Kooyman, G. L., M. A. Castellini, and R. W. Davis (1981). "Physiology of Diving in Marine Mammals." *Annu. Rev. Physiol. 43:* 343–356.

Kooyman, G. L., M. A. Castellini, R. W. Davis, and R. A. Mauae. (1983). "Aerobic Diving Limits of Immature Weddell Seals." *J. Comp. Physiol. B 151:* 171–174.

Kooyman, G. L., and L. H. Cornell (1981). "Flow Properties of Expiration and Inspiration in a Trained Bottle-Nosed Porpoise." *Physiol. Zool. 54:* 55–61.

Kooyman, G. L., D. H. Kerem, W. B. Campbell, and J. J. Wright. (1973). "Pulmonary Gas Exchange in Freely Diving Weddell Seals *(Leptonychotes weddelli)*." *Resp. Physiol. 17:* 283–290.

Kooyman, G. L., E. A. Wahrenbrock, M. A. Castellini, R. W. Davis, and E. E. Sinnett (1980). "Aerobic and Anaerobic Metabolism During Voluntary Diving in Weddell Seals: Evidence of Preferred Pathways from Blood Chemistry and Behavior." *J. Comp. Physiol. 138:* 335–346.

Krafft, B. A., C. Lydersen, K. M. Kovacs, I. Gjertz, and T. Haug (2000). "Diving Behaviour of Lactating Bearded Seals *(Erignathus barbatus)* in the Svalbard Area." *Can. J. Zool. 78:* 1408–1418.

Krutikowsky, G. K., and B. R. Mate, (2000). "Dive and Surfacing Characteristics of Bowhead Whales *(Balaena mysticetus)* in the Beaufort and Chukchi Seas." *Can. J. Zool. 78:* 11:82–1198.

Lagerquist, B. A., K. M. Stafford, and B. R. Mate. (2000). "Dive Characteristics of Satellite-Monitored Blue Whales *(Balaenoptera musculus)* off the Central California Coast." *Mar. Mammal Sci. 16:* 375–391.

Laidre, K. L., M. P. Heide-Jorgensen, R. Dietz, R. C. Hobbs, and O. A. Jorgensen (2003). "Deep-Diving by Narwhals *Monodon monoceros*: Differences in Foraging Behavior Between Wintering Areas?" *Mar. Ecol. Prog. Ser. 261:* 269–281.

Langerquist, B. A., K. M. Stafford, and B. R. Mate (2000). "Dive Characteristics of Satellite-Monitored Blue Whales *(Balaenoptera musculus)* off the California Coast." *Mar. Mamm. Sci. 16:* 375–391.

Lawrence, B., and W. E. Schevill (1965). "Gular Musculature in Delphinids. *Bull. Mus. Comp. Zool. 133:* 1–65.

Le Boeuf, B. J. (1994). Variation in the diving pattern of Northern elephant seals with age, mass, sex and reproductive condition. *In* "Elephant Seals" (B. J. Le Boeuf and R. M. Laws, eds.), pp. 237–252. University of California Press, Berkeley.

Le Boeuf, B. J., and D. E. Crocker (1996). Diving behavior of elephant seals: Implications for predator avoidance. *In* "Great White Sharks, The Biology of *Carcharodon carchias*" (A. P. Klimley and D. G. Ainley, eds.), pp. 193–206. Academic Press, San Diego, CA.

Le Boeuf, B. J., and R. M. Laws, eds. (1994). *Elephant Seals.* University of California Press, Berkeley.

Le Boeuf, B. J., P. A. Morris, S. B. Blackwell, D. E. Crocker, and D. P. Costa (1996). Diving behavior of juvenile northern elephant seals. *Can. J. Zool. 74:* 1632-1644.

Le Boeuf, B. J., Y. Naito, A. C. Huntley, and T. Asaga (1989). "Prolonged, Continuous, Deep Diving by Northern Elephant Seals." *Can. J. Zool. 67:* 2514–2519.

Leith, D. (1976). Physiological properties of blood of marine mammals. *In* "The Biology of Marine Mammals" (H. T. Anderson, ed.), pp. 95–116. Academic Press, New York.

Lenfant, C., K. Johansen, and J. D. Torrance (1970). "Gas Transport and Oxygen Storage Capacity in Some Pinnipeds and the Sea Otter." *Respir. Physiol. 9:* 227–286.

Lewis, M., C. Campagna, M. Uhart, and C. L. Ortiz (2001). "Ontogenetic and Seasonal Variation in Blood Parameters in Southern Elephant Seals." *Mar. Mammal Sci. 17:* 862–872.

Liggins, G. C., J. Qvist, P. W. Hochachka, B. J. Murphy, R. K. Creasy, R. C. Schneider, M. T. Snider, and W. M. Zapol (1980). "Fetal Cardiovascular and Metabolic Responses to Simulated Diving in the Weddell Seal." *J. Appl. Physiol. 49:* 424–430.

Lydersen, C., and K. M. Kovacs (1993). "Diving Behaviour of Lactating Harp Seal, *Phoca groenlandica,* Females from the Gulf of St. Lawrence, Canada." *Anim. Behav. 46:* 1213–1221.

Lydersen, C., and K. M. Kovacs (1999). "Behaviour and Energetics of Ice-Breeding, North Atlantic Phocid Seals During the Lactation Period." *Mar. Ecol. Progr. Ser. 187:* 265–281.

Lydersen, C., M. O. Hammill, and K. M. Kovacs (1994). "Activity of Lactating Ice-Breeding Grey Seals *(Halichoerus grypus)* from the Gulf of St Lawrence, Canada." *Anim. Behav. 48:* 1417–1425.

Martin, A. R., and T. G. Smith (1999). "Strategy and Capability of Wild Belugas, *Delphinapterus leucas,* During Deep, Benthic Diving." *Can. J. Zool. 77:* 1783–1793.

Melnikov, V. V. (1997). "The Arterial System of the Sperm Whale *(Physeter macrocephalus).*" *J. Morphol.* 234: 37–50.

Miller, N. J., C. B. Daniels, D. P. Costa, and S. Orgeig (2004). "Control of Pulmonary Surfactant Secretion in Adult California Sea Lions." *Biochem. Biophy. Res. Comm. 313:* 727–732.

Moon, R. E., R. D. Vann, and P. B. Bennett (1995). "The Physiology of Decompression Illness." *Sci. Am. 273:* 70–77.

Mottishaw, P. D., S. J. Thornton, and P. W. Hochachka (1999). "The Diving Response Mechanism and Its Surprising Evolutionary Path in Seals and Sea Lions." *Am. Zool. 39:* 434–450.

Munkacsi, I. M., and J. D. Newstead (1985). "The Intrarenal and Pericapsular Venous Systems of Kidneys of the Ringed Seal, *Phoca hispida.*" *J. Morphol. 184:* 361–373.

Nishiwaki, M., and H. Marsh (1985). Dugong *Dugong dugon* (Muller, 1776). *In* "Handbook of Marine Mammals," Vol. 3 (S. H. Ridgway, and R. Harrison, eds.), pp. 1–31. Academic Press, New York.

Noren, S. R., and T. M. Williams (2000). "Body Size and Skeletal Muscle Myoglobin of Cetaceans: Adaptations for Maximizing Dive Duration." *Comp. Biochem. Physiol. A 126:* 181–191.

Noren, S. R., T. M. Williams, D. A. Pabst, W. A. McLellan, and J. L. Dearolf (2001). "The Development of Diving in Marine Endotherms: Preparing the Skeletal Muscles of Dolphins, Penguins, and Seals for Activity During Submergence." *Jour. Comp. Physiol. B 171:* 127–134.

Olsen, C. R., F. C. Hale, and R. Elsner (1969). "Mechanics of Ventilation in the Pilot Whale." *Resp. Physiol. 7:* 137–149.

Pfeiffer, C. J. (1990). "Observations on the Ultrastructural Morphology of the Bowhead Whale *(Balaena mysticetus)* Heart." *J. Zool. Wild. Med. 21:* 48–55.

Pfeiffer, C. J., and T. P. Kinkead (1990). "Microanatomy of Retia Mirabilia of Bowhead Whale Foramen Magnum and Mandibular Foramen." *Acta Anat. 139:* 141–150.

Pfeiffer, C. J., and V S. Viers (1995). "Cardiac Ultrastructure in the Ringed Seal, *Phoca hispida* and Harp Seal, *Phoca groenlandica.*" *Aquat. Mamm. 21:* 109–119.

Ponganis, P. J., R. L. Gentry, E. P. Ponganis, and K. V. Ponganis (1992). "Analysis of Swim Velocities During Deep and Shallow Dives of Two Northern Fur Seals, *Callorhinus ursinus.*" *Mar Mamm. Sci. 8:* 69–75.

Ponganis, P. J., G. L. Kooyman, E. A. Baranov, P. H. Thorson, and B. S. Stewart (1997). "The Aerobic Submersion Limit of Baikal Seals, *Phoca sibirica.*" *Can. J. Zool. 75:* 1323–1327.

Ponganis, P. J., E. P. Ponganis, K. V. Ponganis, G. L. Kooyman, R. L. Gentry, and F. Trillmich (1990). "Swimming Velocities in Otariids." *Can. J. Zool. 68:* 2105–2115.

Reidenberg, J. S., and J. T. Laitman (1987). "Position of the Larynx in Odontoceti." *Anat. Rec. 218:* 98–106.

Reynolds, J., and D. Odell (1991). *Manatees and Dugongs.* Facts on File, New York.

Richet, C. (1894). "La résistance des canards á l'asphyxie." *J. Physiol. Path. Gen. 1:* 244–245.

Richet, C. (1899). "De la résistance des canards á l'asphyxie." *J. Physiol. Path. Gen. 5:* 641–650.

Ridgway, S. H. (1972). Homeostasis in the aquatic environment. *In* "Mammals of the Sea" (S. H. Ridgway, ed.), pp. 590–747. Thomas Publ., Springfield, IL.

Ridgway, S. H. (1986). Diving by cetaceans. *In* "Diving in Animals and Man" (A. OBrubakk, J. W. Kanwisher, and G. Sundness, eds.), pp. 33–62. The Royal Norwegian Society of Science and Letters, Trinheim, Norway.

Ridgway, S. H., and R. J. Harrison (1986). Diving dolphins. *In* "Research on Dolphins" (M. M. Bryden and R. H. Harrison, eds.), pp. 33–58. Oxford University Press, Oxford, UK.

Ridgway, S. H., and R. Howard (1979). "Dolphin Lung Collapse and Intramuscular Circulation During Free Diving: Evidence from Nitrogen Washout." *Science 206:* 1182–1183.

Ridgway, S. H., and D. G. Johnston (1966). "Blood Oxygen and Ecology of Porpoises of Three Genera." *Science 151:* 456–458.

Ridgway, S. H., B. L. Scronce, and J. Kaniwisher (1969). "Respiration and Deep Diving in the Bottlenose Porpoise." *Science 166:* 1651–1653.

Ronald, K., R. McCarter, and L. J. Selley (1977). Venous circulation in the harp seal *(Pagophilus groenlandicus)*. *In* "Functional Anatomy of Marine Mammals," Vol. 3 (R. J. Harrison, ed.), pp. 235–270. Academic Press, New York.

Rowlatt, U., and H. Marsh (1985). "The Heart of the Dugong *(Dugong dugon)* and the West Indian Manatee *(Trichechus manatus)* Sirenia." *J. Morphol. 186:* 95–105.

Scholander, P. F. (1940). "Experimental Investigations on the Respiratory Function in Diving Mammals and Birds." *Hvalradets Skr. 22:* 1–131.

Scholander, P. F. (1960). "Oxygen Transport Through Hemoglobin Solutions." *Science 131:* 585–590.

Scholander, P. F. (1964). Animals in aquatic environments: diving mammals and birds. *In* "Handbook of Physiology, Section 4: Adaptations to the Environment" (D. B. Dill, E. . Adolph, and C. G. Wiber, eds.), pp. 729–739. American Physiology Society, Washington, D. C.

Scholander, P. F., L. Irving, and S. W. Grinell (1942a). "Aerobic and Anaerobic Changes in Seal Muscles During Diving." *J. Biol. Chem. 142:* 431–440.

Scholander, P. F., L. Irving, and S. W. Grinnell. (1942b). "On the Temperature and Metabolism of the Seal During Diving." *J. Cellular Comp. Physiol. 21:* 53–63.

Schreer, J. F., and K. M. Kovacs (1997). "Allometry of Diving Capacity in Air-Breathing Vertebrates." *Can. J. Zool. 75:* 339–358.

Schreer, J. F., R. J. O'Hara Hines, and K. M. Kovacs. (1998). "Classification of Multivariate Diving Data from Air-Breathing Vertebrates: A Comparison of Traditional Statistical Clustering and Unsupervised Neural Networks." *J. Agric. Biol. Environ. Stat. 3:* 383–404.

Schreer, J. F., and J. W. Testa (1996). "Classification of Weddell Seal Diving Behavior." *Mar. Mamm. Sci. 12:* 227–250.

Shadwick, R. E., and J. M. Gosline (1994). "Arterial Mechanics in the Fin Whale Suggest a Unique Hemodynamic Design." *Am. J Physiol. 267:* R805–R818.

Sleet, R. B., J. L. Sumich, and L. J. Weber (1981). "Estimates of Total Blood Volume and Total Body Weight of a Sperm Whale *(Physeter catadon)*." *Can. J. Zool. 59:* 567–570.

Slijper, E. J. (1979). *Whales,* 2nd ed. Hutchinson, London.

Slip, D. J., M. A. Hindell, and H. R. Burton (1994). Diving behavior of southern elephant seals from Macquarie Island. *In* "Elephant Seals" (B. J. Le Boeuf, and R. M. Laws, eds.), pp. 253–270. University of California Press, Berkeley.

Sparling, C. E., and M. A. Fedak (2004). "Metabolic Rates of Captive Grey Seals During Voluntary Diving. *J. Exp. Biol.* 201: 1615–1624.

Spencer, M. P., T. A. Gornall, and T. C. Poulter (1967). "Respiratory and Cardiac Activity of Killer Whales." *J. Appl. Physiol. 22:* 974–981.

Spragg, R. G., P. J. Ponganis, J. J Marsh, G. A. Rau, and W. Bernhard (2004). "Surfactant from Diving Aquatic Mammals." *J. Appl. Physiol. 96:* 1626–1632.

Stewart, B. S., and R. L. DeLong (1995). "Double Migrations of the Northern Elephant Seal, *Mirounga angustirostris.*" *J. Mammal. 76:* 196–205.

Tarasoff, F. J., and G. L. Kooyman (1973a). "Observations on the Anatomy of the Respiratory System of the River Otter, Sea Otter, and Harp Seal I. The topography, Weight, and Measurements of the Lungs." *Can. J. Zool. 51:* 163–170.

Tarasoff, F. J., and G. L. Kooyman (1973b). "Observations on the Anatomy of the Respiratory System of the River Otter, Sea Otter, and Harp Seal II. The Trachea and Bronchial Tree." *Can. J. Zool. 51:* 171–177.

Thompson, D., and M. A. Fedak (1993). "Cardiac Responses of Gray Seals During Diving at Sea." *J. Exp. Biol. 174:* 139–164.

Thorson, P. H., and B. J. Le Boeuf (1994). Developmental aspects of diving in northern elephant seal pups. *In* "Elephant Seals" (B. J. Le Boeuf and R. M. Laws, eds.), pp. 271–289. University of California Press, Berkeley.

Wartzok, D., R. Elsner, H. Stone, B. P. Kelly, and R. W. Davies. (1992). "Under-Ice Movements and the Sensory Basis of Hole Finding by Ringed and Weddell Seals." *Can. J. Zool. 70:* 1712–1722.

Watkins, W. A., M. A. Daher, N. A. DiMarzio, A. Samuels, D. Wartzok, K. M. Fristrup, P. W. Howey, and R. R. Maiefski (2002). "Sperm Whale Dives Tracked by Radio Tag Telemetry." *Mar. Mamm. Sci. 18:* 55–68.

Watkins, W. A., M. A. Daher, K. M. Fristrup, T. J. Howard, and G. N. Sciara (1993). "Sperm Whales Tagged with Transponders and Tracked Under Water by Sonar." *Mar Mamm. Sci. 9:* 55–67.

White, F. C., R. Elsner, D. Willford, E. Hill, and E. Merhoff (1990). "Responses of Harbor Seal and Pig Heart to Progressive and Acute Hypoxia." *Am. J. Physiol. 259:* R849–R856.

Williams, T. M. (2004). *The Hunter's Breath*. M. Evans and Company, New York.

Williams, T. M., W. A. Friedl, and J. E. Haun. (1993). "The Physiology of Bottlenose Dolphins *(Tursiops truncates)*: Heart Rate, Metabolic Rate and Plasma Lactate Concentration During Exercise." *J. Exp. Biol. 179:* 31–46

Williams, T. M., G. L. Kooyman, and D. A. Croll (1991). "The Effect of Submergence on Heart Rate and Oxygen Consumption of Swimming Seals and Sea Lions." *J. Comp. Physiol. B 160:* 637–644.

Yablokov, A. V, V. M. Bel'kovich, and V. I. Borisov (1972). *Whales and Dolphins*. Israel Programs for Scientific Translations, Jerusalem.

11

Sound Production for Communication, Echolocation, and Prey Capture

11.1. Introduction

This chapter deals with the production, transmission, and reception of sounds produced by vocalizing marine mammals in air and underwater. The manner in which vocalizations are produced and received differs between marine mammal taxa and also according to the medium in which they are produced (i.e. airborne or waterborne sounds). The purpose of vocalizations ranges from communicating with individuals of the same species to locating unseen targets with echolocation.

11.2. Sound Propagation in Air and Water

Acoustic energy can be characterized by its velocity (dependent on the density of the transmitting medium), its frequency, its wavelength, and its amplitude. The frequency and wavelength of sound are related to velocity by the following equation:

$$\text{Velocity (m/s)} = \text{Frequency (vibrations/s)} \times \text{wavelength (m)}$$

The human ear is an extremely sensitive instrument for analyzing and comparing airborne auditory signals of other animals and for characterizing their qualitative features. The frequencies detectable by most people ranges from about 18 vibrations/s, or hertz (Hz) to 15,000 Hz (or 15 kHz). Marine mammal vocalizations often extend both above and below the range of human hearing. For our convenience, we have labeled sounds with frequencies lower than 18 Hz as **infrasonic** and those higher than 20 kHz as **ultrasonic.**

Sound travels in water about five times faster than in air. Sound velocity in air is approximately 340 m/s and in water between 1450 and 1550 m/s depending on tempera-

270

ture and salinity (which vary with depth; Sverdrup *et al.*, 1970). Some marine mammals have co-opted the increased velocity of sound underwater to compensate for diminished transmission of light and consequent poor vision in water.

Soon after the first microphones were lowered into the sea in the early part of the 20th century, it became apparent that the ocean is a very noisy environment. Nonbiological sounds from waves and surf, anthropogenic noise, as well as sounds from biological sources, contribute to the symphony of underwater sounds. Fish and crustaceans, as well as whales and pinnipeds, generate a tremendous repertoire of underwater sounds. Although the emphasis of this chapter is on the active uses of phonation, many animals with acute hearing may be able to obtain substantial information about their immediate acoustic environment without giving away similar information about themselves just by passively listening to the sounds made by others.

Several distinct functions of intentionally produced sounds have been demonstrated or suggested. Dolphins produce a large variety of whistle-like sounds (Popper, 1980), and captive individuals have been shown to understand complex linguistic subtleties (Herman, 1991). Many of the moans, squeals, and wails are used for communication. Several species of whales produce unique signature sounds for individual identification, including whistles of dolphins, click **codas** of sperm whales, and very low frequency tones of blue and fin whales. Other sounds, especially those of the humpback whale, have a fascinating musical quality and are thought to be produced primarily by adult males during courtship displays. Loud impulse sounds, so far recorded from *Tursiops, Orcinus,* and *Physeter,* have been suggested as possible acoustic mechanisms to overload the sensory systems in other individuals, for debilitation of prey, self-defense, or intimidation of conspecifics (Norris and Mohl, 1983; Herzing, 2004). Finally, the most studied function of underwater vocalization is **echolocation,** the active detection and identification of targets with sound.

The basic characterizing features of any acoustic signal, its frequency, duration, and energy level, are conventionally portrayed graphically as a **spectogram** (frequency with time; Figure 11.1a), a **power spectrum** (sound pressure levels with time; Figure 11.1b), and a **frequency spectrum** (sound pressure levels with frequency; Figure 11.1c). These representations are used in this chapter to assist in visualizing the sound characteristics being discussed. The time scales of spectograms are represented in appropriate units (varying from milliseconds to minutes), frequency is measured in Hz or kHz, and sound pressure level is measured in a logarithmic decibel scale (Au, 1993).

With the velocity of sound in water nearly constant, the wavelength of any underwater sound varies with its frequency. Low-frequency sounds attenuate more slowly and are good for long distance communication, not echolocation. If they are used for echolocation, their ability to resolve target size cannot be finer than the wavelength of the sound (about 15 m at 100 Hz and 1.5 cm at 100 kHz). Higher frequency sounds attenuate more quickly, but they have the potential to provide more information on target resolution because of their shorter wavelengths. The spatial resolution of a sound (sonar) depends directly on the wavelength used. The shorter the wavelength (i.e., the higher the frequency), the better the spatial resolution, and vice versa (for more see Supin *et al.*, 2001).

The conventional measure of propagated sound energy in water is sound pressure rather than intensity (amplitude). It is defined in terms of sound pressure level (SPL) in units of decibels (dB):

$$\text{SPL in decibels} = 20 \log (p/p_o),$$

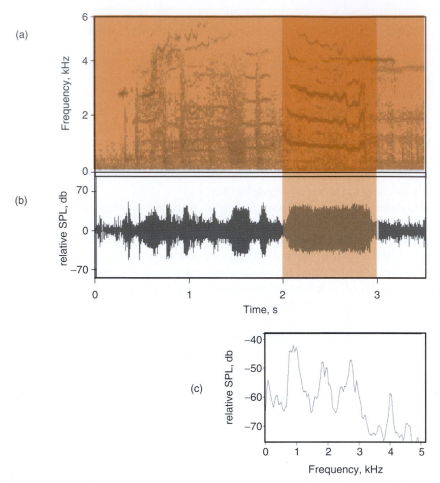

Figure 11.1.   A complex whistle vocalization of a beluga whale, displayed as **(a)** a spectrogram, **(b)** power spectrum, and **(c)** frequency spectrum of the portion of **a** and **b** shaded in color. In spectograms, the relative SPL of the sound is represented by variations in signal intensity.

where p_o is a standardized reference pressure (typically 1 μPascal of pressure underwater). Decibels are used in acoustics as a convenient measure of the ratio of the measured sound pressure relative to the reference sound pressure. These units provide a convenient logarithmic scale by which to compare vastly different sound pressure levels.

11.3. Anatomy and Physiology of Sound Production and Reception

11.3.1. The Mammalian Ear

The mammalian ear evolved for the detection of sound vibrations in air. The typical mammalian ear includes an outer ear or pinna that collects sound waves and funnels them into an auditory canal (Figure 11.2a) to the tympanic membrane, or eardrum, which separates the outer and middle ear (Figure 11.2b). The middle ear is an air-filled

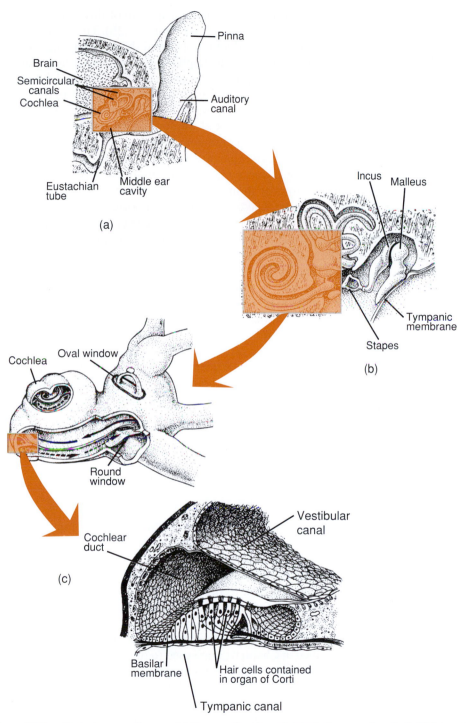

Figure 11.2. Typical mammalian ear. **(a)** Cross section through a typical mammalian skull. **(b)** Internal structure of the cochlea. **(c)** Section through the organ of Corti. (From Kardong, 1995.)

chamber inside the tympanic bone or bulla, containing a chain of three small bony elements, or ossicles (the malleus, incus, and stapes; see Figure 11.2b). These bones form a continuous bony bridge to conduct sound vibrations from the inside of the tympanic membrane to the oval window of the inner ear, amplifying them considerably along the chain. The organ of Corti for sound reception is located in the cochlea, which is the auditory part of the inner ear. The cochlea is a coiled organ that is divided lengthwise into three parallel tubular canals that become progressively narrower toward the apex (Figure 11.2c). The stapes is located at the oval window, which is the opening to the vestibular canal. The tympanic canal is continuous with the vestibular canal and is closed by the round window. These two canals are filled with perilymph, Between these two parallel canals lies the cochlear duct containing the organ of Corti (Figure 11.3c). Within the organ of Corti are rows of thousands of sensory hair cells, each connected with neurons of the auditory nerve. These hair cells are supported on the basilar membrane with the tectorial membrane directly over them. The entire organ of Corti is bathed in the endolymph of the cochlear duct. The mammalian inner ear houses two organs of equilibrium, the vestibule and the semicircular canals. As an animal's head changes position, moving fluid in the semicircular canals and vestibule puts shearing forces on the hair cells. Changes in shearing forces are transmitted into neuronal impulses that pass this information to the brain.

The energy of airborne sound waves striking the tympanic membrane is conducted and amplified through the bones of the middle ear to the oval window, where its oscillations are transmitted to the fluids of the vestibular and tympanic canals. These oscillating fluids simultaneously cause the basilar membrane supporting the hair cells to vibrate. Different portions of the basilar membrane respond to different frequencies of sound depending on the membrane's width and stiffness. The basilar membrane is narrow and thick at the base, where high frequencies are detected, and wide and thin at the apex, where low frequencies are detected. The amplitude of sound, or loudness, is determined by the number of hair cells stimulated, and its frequency, or pitch, depends on the distribution pattern of stimulated hair cells.

11.3.2. Sound Production and Reception in Pinnipeds

Airborne sounds produced by pinnipeds usually are within the range of human hearing and are often described as grunts, snorts, or barks or are identified with their presumed social function, such as "threat calls" of breeding males or "pup-attraction calls" of mothers. Most pinniped vocalizations are produced in the larynx, although male walruses also make clacking noises with their teeth and produce distinctive bell-like sounds in air and underwater with their inflated pharyngeal (throat) pouches. These pouches are only present on males, and the bell-like sounds are produced almost exclusively by adult males during the breeding season as part of a courting display. The hood and nasal septum of hooded seals (described in Chapter 13) are used to produce sounds both underwater and in air. These sounds are emitted by adult males in either courtship or combat (Terhune and Ronald, 1973; Ballard and Kovacs, 1995).

With the exception of the reduction in size (in otariids) or complete absence (in phocids and walruses) of the pinnae (Figure 11.3), the system for in-air sound reception in pinnipeds is not markedly different from the typical mammalian ear. Pinnipeds have relatively large tympanic bulla, and thus large middle ear cavities, enabling better low-frequency hearing (in air). In air, pinnipeds hear like terrestrial mammals; sound is

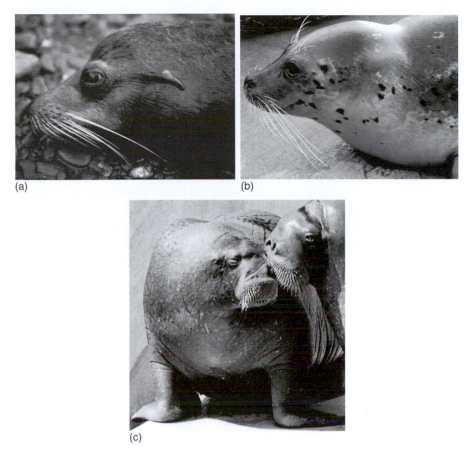

(a)   (b)

(c)

Figure 11.3.   Lateral views of ears of pinnipeds. **(a)** Sea lion. **(b)** Harbor seal. **(c)** Walrus.

conducted through the external auditory meatus to the tympanic membrane and through the ossicles to the inner ear. The pinniped ear shows several modifications for hearing underwater. These modifications amplify sound reception. The outer and middle ear contain cavernous tissue capable of being engorged with blood when the animal is submerged. In addition to helping in pressure equalization when diving, these cavernous tissues may enhance the transmission of sound to the inner ear in particular, making the ear more sensitive to high frequencies (Repenning, 1972; Kastelein *et al.*, 1996). The middle ear is specialized for bone conducted hearing in water (Mohl, 1968; Nummela, 1995). Phocid middle ear bones show a number of modifications (e.g., extreme expansion of the incus to form a head and, in some, extra articulations on the malleus) not seen in otariids, the walrus, or other carnivores. The middle ear bones are enlarged in both the walrus and phocids and they also share specializations of the malleus (Wyss, 1987). The enlarged ossicles bring extra mass to the vibrating ossicular chain, and this shifts the rotational axis of the chain and enables bone conducted hearing. The increase in ossicular mass shifts the hearing frequency range in air toward lower frequencies, as is the case for phocids and the walrus; otariid hearing is at slightly higher frequencies (Nummela, 1995; Hemila *et al.*, 1995).

11.3.3. Sound Production and Reception in Cetaceans

11.3.3.1. Sound Production

After decades of study, there is little debate about the anatomical source of sound production in odontocetes. The results of many studies (Cranford *et al.*, 1996, 1997) conclusively support the region of the nasal sac system just inside the blowhole as the whistle and echolocation click producing structures in small odontocetes (Figure 11.4).

The basic odontocete sound-production system (see Figure 11.4) consists of a structural complex associated with the upper nasal passages termed the **"monkey lips"/dorsal bursa (MLDB complex).** The term "monkey lips" derived from their appearance in sperm whales (Figure 11.5a) although they appear very differently in smaller odontocetes. Consequently, the less colorful but more descriptive term "phonic lips" is preferred for this structure, although the MLDB label remains. All odontocetes except sperm whales possess two bilaterally placed MLDB complexes. Each MLDB complex is located just below the ventral floor of the vestibular air sac and is composed of a pair of fat-filled anterior and posterior dorsal bursae in which a pair of slit-like muscular phonic lips are embedded, a resilient cartilaginous blade (the bursal cartilage), and a stout blowhole ligament, all suspended within a complex array of muscles and air spaces (Cranford *et al.*, 1996). Cranford (1988, 1992) and Cranford *et al.* (1996) have proposed that, in spite of the obvious structural differences between the heads of dolphins and sperm whales, the mechanism for click production is homologous between sperm whales (physeterids) and other odontocetes. For example, they suggest that the junk of the sperm whales (Figure 11.5b) is homologous to the melon in other odontocetes and that the spermaceti organ of sperm whales is homologous to the right posterior bursa of other odontocetes.

Comparison of these homologous structures suggests that all odontocetes make their pulsed echolocation signals by a similar mechanism. These clicks are produced by pneumatic pressurization within intranarial spaces. Cranford *et al.* (1996) hypothesized that sounds are generated as air is forced between the phonic lips, setting the MLDB complex

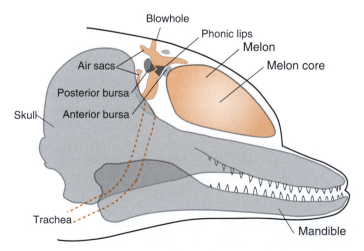

Figure 11.4.   Semitransparent illustration of a dolphin head showing the position of the melon and associated sound-producing structures. The variation in lipid density is indicated with shading. (Modified from Cranford *et al.*, 1996.)

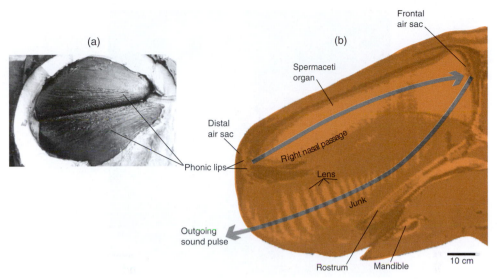

Figure 11.5. **(a)** Phonic lips from a sperm whale with the distal air sac partially removed. (Courtesy of T. Cranford.) **(b)** Sagittal CAT section of a neonate sperm whale head, with major structures implicated in phonation labeled. Arrows indicate putative path of sound pulse from the phonic lips through the spermaceti organ to the frontal air sac where it is reflected anteriorly and focused through the fatty lenses of the junk. (CAT section from Cranford, 1999.)

into vibration. The periodic opening and closing of the phonic lips breaks up the air flow and determines the click repetition rate of the train. When the phonic lips snap together during click production, vibrations in the bursae are likely produced. The nasal plugs and their nodes along with the blowhole ligament and other membranes are most likely involved in regulating air movement in the passages dorsal to the nares and perhaps are used in whistle production. Direct observations by Cranford *et al.* (1997) of vocalizing bottlenose dolphins using a high-speed video endoscope have confirmed the MLDB as the only structure responsible for echolocation signal generation.

The sound generation system of small odontocetes is coupled to the sound-propagation structure, the melon, to focus and direct emitted sounds forward into the water. The melon sits atop the skull anterior to the MLDB (Figure 11.6), and consists of low density lipids which serve as an acoustic lens to create focused directional beams in front of the melon (Figure 11.7). The larger and structurally more complex sound production system of sperm whales exhibits strong homologies with those of smaller odontocetes, yet there are important differences. In addition to the obvious spermaceti organ, the phonic lips (i.e., monkey muzzle) are large, are cornified, and are located at the anterior end of the junk and spermaceti organ (see Figure 11.5a).

Sperm whale vocalizations consist of reverberant pulses that are repeated more slowly and at lower frequencies than those of delphinids (Figure 11.8a). Each click contains a series of uniformly spaced pulses, each lasting about 24 ms, that gradually decay in amplitude (Figure 11.8b). The mechanism for the multipulsed nature of sperm whale clicks was first proposed in 1972 by Norris and Harvey. They suggested that a sound pulse is produced by the phonic lips and is transmitted forward into the water from the whale's head. A portion of this sound energy is reflected posteriorly by the distal air sac

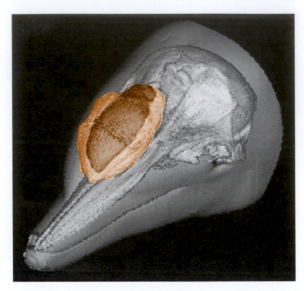

Figure 11.6.   Oblique CAT scan of a bottlenose dolphin head showing position of melon (color). (Courtesy of Megan McKenna and Ted W. Cranford.)

through the spermaceti organ, then again forward from the frontal air sac. With each successive reflection, some of the acoustic energy is transmitted into the water forward of the whale, and some is reflected again back through the spermaceti organ. With less energy, the SPL of each successive pulse within a click decreases, although the interpulse interval remains constant (Figures 11.8b and 11.9a). The interpulse interval is interpreted as the two-way travel time of the sound pulse between the reflecting distal and frontal air sacs, and is constant for individual whales.

However, recent work by Mohl *et al.* (2003) suggests a different and more complex picture of sperm whale echolocation capabilities. Using a star array of hydrophones to determine the directionality of vocalizations from foraging sperm whales, Mohl *et al.* found that when one hydrophone of the array is "on-axis" relative to the whale (pre-

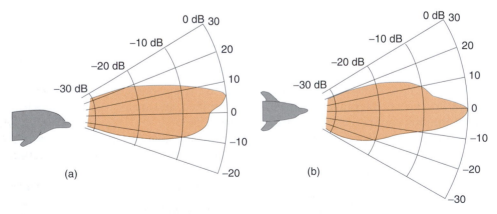

Figure 11.7.   Focused transmission beam patterns of bottlenose dolphins in the **(a)** vertical and **(b)** horizontal planes. (Redrawn from Au, 1993.)

sumed to be aligned with the whale's echolocation beam), the multipulsed nature of the click disappears (Figure 11.9b). The mono-pulsed click is very directional and has a high source level (approx. 235 dB re 1 μPa rms; likely the loudest sound known to be produced by a nonhuman animal). Mohl *et al.* (2003) suggest that the sperm whale nose is an acoustical horn doubled back on itself. Almost all the energy of a sound pulse produced by the phonic lips (p0 in Figure 11.9b) is transmitted backward through the spermaceti organ to the frontal air sac then reflected forward through the junk rather than the spermaceti (see Figure 11.5) to be emitted as the p1 pulse of Figure 11.9b. It is the fatty lenses of the junk that focuses the sound into a forward-directed beam, as the melon does in smaller odontocetes. The "off-axis" hydrophones, however, record a lower intensity, nondirectional, multipulsed click described by Figures 11.8b and 11.9a. As only on-axis hydrophones can record the directional mono-pulsed click characteristics, most recordings of sperm whale clicks include only the off-axis click characteristics that were used to describe the nature of sperm whales clicks for three decades.

No anatomical studies have shown equivalent structural specializations for sound generation or transmission in mysticetes. Mysticetes have a larynx but lack vocal cords. The cranial sinuses of mysticetes are thought to be involved in phonation although no precise or general mechanism has so far been demonstrated.

11.3.3.2. Sound Reception

Behavioral studies of both wild and captive subjects suggest that all species of cetaceans have good hearing, with odontocetes having high sensitivity across a broad range of

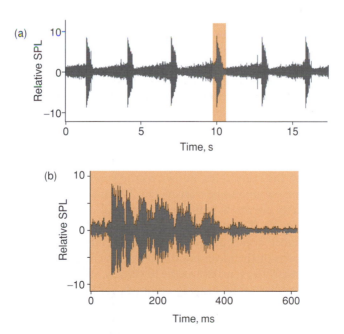

Figure 11.8. **(a)** Relative SPL of a series of sperm whale echolocation clicks about 2.5 s apart. **(b)** Decaying SPL of sequential pulses of a single click from **(a)** (shown in color). Note the change in time scale. (Recording Courtesy of J. Fish.)

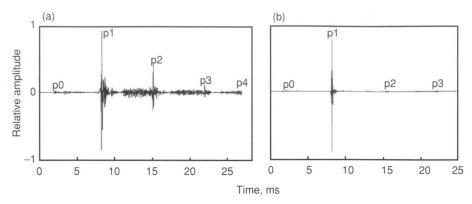

Figure 11.9.   **(a)** Relative SPL of a single sperm whale click recorded off-axis, showing several reverbera-
tions of diminishing intensity **(b)** on-axis monopulse showing the almost complete absence of
reverberating pulses shown in **(a)**. p0 represents the time of the actual phonic lips pulse
(Adapted from Mohl *et al.*, 2003).

frequencies (Figure 11.10). Experimental evidence, however, is again largely restricted to
studies of captive small-toothed whales. Their sound-detection systems must be attuned
to very faint echoes of their own clicks but must simultaneously withstand the intense
power of outgoing clicks generated in adjacent regions of the head.

The external auditory canal is the typical mammalian sound-conducting channel connecting the external and middle ears. This structure is extremely narrow in odontocetes
or completely plugged in mysticetes, and there is debate regarding whether it is functional. In mysticetes, an extension of the eardrum pushes into the ear canal. This **glove
finger** ends in a horn-like plug (composed of dead cells from the canal lining) that may be
a meter long (Figure 11.11).

Mapping of acoustically sensitive areas of dolphins' heads has shown the external
auditory canal to be about six times less sensitive to sound than the lower jaw.
Additionally, in experimental discrimination tests, the echolocating performance of
a dolphin was significantly reduced when a sound-attenuating rubber hood was worn
over the lower jaw (Brill *et al.*, 1988). These results support the hypothesis first
proposed by Norris (1964) of a unique sound reception pathway in odontocetes. The
posterior portions of the mandibles, known as the **pan bones**, are flared toward the
rear and often are thin enough to be translucent. Within each half of the lower jaw is
a fat body that directly connects with the lateral wall of the auditory bulla of the
middle ear (Figure 11.12). These fat bodies, like the lipid of a dolphin's melon or a
sperm whale's spermaceti organ, act as low density sound channels to conduct sounds
from the flared portions of the lower jaw directly to the middle ear. An area on either
side of the melon is nearly as sensitive as the lower jaw, suggesting that dolphins may
possess two other very sensitive hearing channels for sound reception. It is likely
that mysticetes differ in their sound reception mechanism and receive sound from the
ear canal rather than from the jaw, although this has not yet been demonstrated.

The ear of odontocetes has two distinct components, the tympanic and the periotic
bones **(tympano-periotic complex),** both of which are constructed from very dense or
pachyostotic bone (see Oelschlager, 1986a, 1986b, for discussion of the evolution of

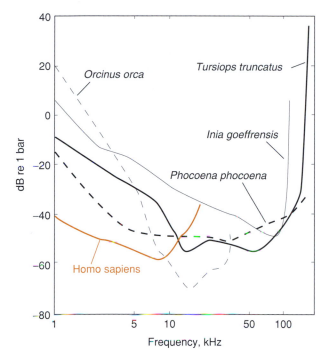

Figure 11.10.   Auditory sensitivity curves (audiograms) for several species of odontocetes. (Redrawn from Au, 1993.)

this region in toothed whales). Cetacean tympanic and periotic bones differ from those of other mammals in appearance, construction, and cranial associations. Mysticete and odontocete ear complexes differ in size, in shape, and in the relative volumes of the tympanic and periotic bones, but several structures scale with each other in both lineages (Nummela *et al.*, 1999a, 1999b). Bullar dimensions are strongly correlated with animal size; mysticete bullae are two to three times larger than those of most odontocetes. In toothed whales, the tympano-periotic complex is

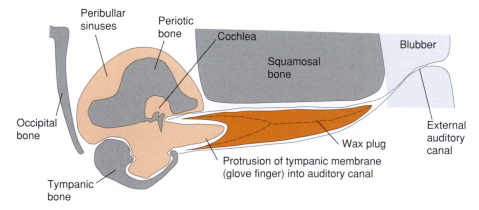

Figure 11.11.   Simplified cross-sections of the ear region of a mysticete. (Adapted from Reysenbach de Haan, 1956.)

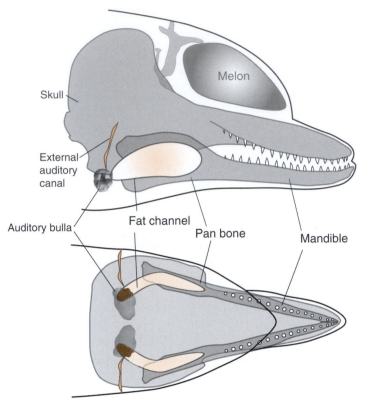

Figure 11.12. Semitransparent illustration of a dolphin head showing the position of the mandibular and lateral fat channels. The variation in lipid density is indicated with shading. The approximate positions of the middle ear and cochlea are also indicated.

separated from adjacent bones of the skull by peribullar sinuses filled with an insulating emulsion of mucus, oil, and air. The ear complex is suspended in this emulsion by a sparse network of connective tissue (Figure 11.13a). Thus, the ears are isolated from the skull and from each other and function as independent sound receivers able to better localize the directional characteristics of sound sources or of received echoes.

In both odontocetes and mysticetes the tympanic membrane is reduced to a calcified ligament (often called tympanic ligament). The tip of this ligament is attached to the malleus (Figure 11.13b).

The actual mechanism of sound transmission in the middle ear is controversial but the best functional middle ear model for odontocetes has been proposed by Hemila *et al.* (1999, 2001). According to this model, sound brings the tympanic bone (especially its thin ventrolateral wall or tympanic plate) into vibration. The malleus is ossified to the tympanic plate through a thin processus gracilis, and so the vibrations of the tympanic plate are transmitted to the oval window and the inner ear fluid through the ossicular chain. This bony mechanism contains two levers, one created by the tympanic plate and the other by the ossicles, that help match the sound vibrations to enter the inner ear fluid. This model correlates well with behavioral audiogram data for the hearing of several odontocete species throughout their hearing frequency range and can be used for predicting theoretical hearing limits for animals for which sufficient middle ear anatomical data are available.

In the inner ear of dolphins, the cochlea is similar to that of humans, with about the same number of hair cells, but the ganglion cell-to-hair cell ratio (2:1 in humans) is 5:1 in *Tursiops*. In addition, the basilar membrane is thicker and stiffer, again presumably to

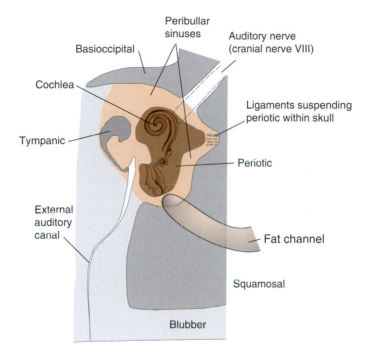

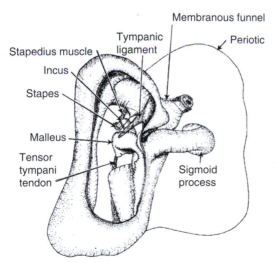

Figure 11.13.   **(a)** Schematic drawing of odontocete ear region in ventral view (without tympanic bone) illustrating periotic, peribullar sinuses, mastoid, occipital, and paroccipital process (from Oelschlager, 1986a). **(b)** Right auditory bulla of bottlenose dolphin opened to show middle ear. Tympanic bone is shown in detail with the periotic shown only in outline (from McCormick *et al.*, 1970).

enhance higher frequency sensitivity. Other cochlear differences seen in odontocetes are the number of cochlear turns and the distribution of membrane support structures (see Ketten *et al.*, 1992, Spoor et al., 2002).

11.3.4. Sound Production and Reception in Other Marine Mammals

The vocal and hearing systems of sea otters and polar bears resemble those of other terrestrial carnivores, with no apparent specializations for underwater vocalization or hearing.

The external pinnae of manatees are absent and the external meatus is reduced to a tiny opening that leads to a narrow external auditory canal. The ear complex consists of a large bilobed periotic and a smaller tympanic. The tympano-periotic complex is composed of exceptionally dense bone, similar to that of cetaceans. Unlike the cetacean tympano-periotic complexes, which are external to the skull, manatee tympano-periotics are attached to the inner wall of the cranium and are attached to bone (Figure 11.14). The intracranial position of the periotic and its fusion with the squamosal has important implications for hearing. The periotic connects via a bony bridge to an enlarged zygomatic process of the squamosal bone. The zygomatic process is an inflated, oil-filled, bony sponge that is analogous to the fatty filling of the mandibular canal of odontocetes and may have a significant role in manatee sound reception as a low-frequency sound channel (Bullock *et al.*, 1980; Reynolds and Odell, 1991; Ketten *et al.*, 1992).

Comparison of the lipid composition of the zygomatic process in the Florida manatee to the pan bone fat body of the bottlenose dolphin revealed that the manatee samples did not contain isovaleric acid found in the bottlenose dolphin and some other odontocetes and thought to be related to sound conduction (see also Chapter 8). These results suggest that a different complex of lipids may be involved in sound conduction in manatees (Ames *et al.*, 2002). The middle ear structures of manatees imply that they lack sensitivity and directionality compared to most mammals. There is no indication that any species of manatee has ultrasonic hearing. The combined effects of poor directionality and lack of high-frequency sensitivity of the manatee ear may explain the absence of avoidance maneuvers that result in large numbers of manatee deaths each year from boat collisions (Nowacek *et al.*, 2004).

The tympano-periotic complexes of extinct sirenians are very similar to those of the modern Florida manatee and are consistent with fully aquatic animals. They imply that few functional changes have occurred in the sirenian auditory system since the appearance of the group 50 Ma. Thus, manatees appear to represent an exception to the convention that hearing is the most significant and developed of marine mammal senses.

11.4. Functions of Intentionally Produced Sounds

11.4.1. Pinnipeds

In-air vocalizations of pinnipeds can be grouped by species, age, and sex class and whether individuals are in breeding or nonbreeding groups. Some phocid species are

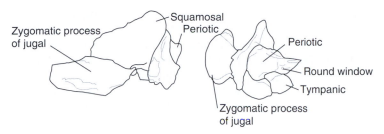

Figure 11.14. Diagram of manatee tympano-periotic complex in **(a)** lateral and **(b)** posterior views. (From Ketten, 1992b.)

virtually silent when on land, whereas most otariid colonies are a cacophony of noise. Male California sea lions produce a loud directional bark that advertises dominance as well as threatens other males. Fur seal males exhibit a more complex repertoire of vocalizations, including a nondirectional "trumpeted roar" threat call that is sufficiently distinctive to allow males occupying neighboring breeding territories to recognize one another and respond less often to vocalizations of immediate neighbors than to those of strange males encroaching upon their territory (Roux and Jouventin, 1987). Dominant male elephant seals also produce loud and repetitive vocalizations in their crowded breeding rookeries, presumably to communicate their relative breeding status to other nearby males over the cacophony of background rookery noises. These threat vocalizations differ sufficiently from one rookery to another to be considered distinctive regional **dialects** (Le Boeuf and Petrinovich, 1975). Such geographic variation of a species' vocal calls also has been found in high-latitude Weddell and bearded seals and is discussed later in this chapter.

Walruses utter sounds in most social interactions on land or on ice. Among the airborne sound classes produced by walruses are roars, grunts, and guttural sounds used in threat displays, sometimes in combination with tusk presentation. They also produce barks, distinctive loud calls with a variety of functions that range from a submissive bark given only by adults to the bark of young calves when they are distressed (Miller, 1985). Walruses also produce what is considered a rutting whistle (Verboom and Kastelein, 1995a).

A final class of above-water pinniped vocalizations includes mother-pup calls. Mothers and pups of most species of pinnipeds have specific vocalizations to assist mother-pup pairs in recognizing and locating each other. For pinnipeds such as elephant seals, whose mothers and pups remain together throughout the nursing period, these calls help a pair to maintain contact in crowded breeding rookeries. Sea lion and fur seal mothers produce distinctive calls to attract their pups when returning from foraging at sea. Even though several pups may respond to the call of a single female, her own pup is capable of recognizing her mother's vocalizations, and individual pups are identified by their mothers by a combination of its vocalizations as well as visual, olfactory, and spatial cues (Roux and Jouventin, 1987; Hanggi, 1992; Reiman and Terhune, 1993; Kovacs, 1995). Harp seals have very complex pup vocalizations. Despite having a very short period of maternal care it has been suggested that these vocalizations may represent a developmental step toward the complex system of adult underwater communication (Miller and Murray, 1995).

Pinnipeds produce a variety of underwater sounds that appear most often related to breeding activities (Stirling and Thomas, 2003) and social interactions. Whistles, chirps, trills, and low pitched buzzes characterize Weddell seals and are used as territorial declarations (Thomas and Kuechle, 1982; Figure 11.15). Moors and Terhune (2004) suggest that rhythmic repetitions of these calls may enhance the likelihood that they will be detected by conspecifics. Trills similar to those of Weddell seals have been recorded for hooded seals (Ballard and Kovacs, 1995) and bearded seals (e.g., Cleator *et al.*, 1989) and it has been suggested that they may be used in establishing and maintaining aquatic territories as well as attracting female mating partners (Van Parijs *et al.*, 2001, 2003a, 2004). Male bearded seals display significant individual variation in their trill-calls (Figure 11.16), consisting of oscillating warbles that change frequencies and are punctuated by brief unmodulated low-frequency moans (Ray *et al.*, 1969). Some males show site fidelity in their calling location over a period of years (Van Parijs *et al.*, 2001, 2003a). Leopard seal vocalizations are described as soft, lyrical calls rather than the aggressive sounding grunts, barks, and groans found in most other phocids and may be related to their solitary social system that does not require calls for territorial defense or inter-animal disputes (Rogers *et al.*, 1995; Thomas and Golladay, 1995). The variation in the call repertoires of leopard seals on opposite sides of the Antarctic suggests that there is geographic variation between repertoires (Thomas and Golladay, 1995). Similar research has suggested that this is also true for populations of Weddell seals around Antarctica (Pahl *et al.*, 1997, and references cited therein) and for male harbor seals at the oceanic, regional, population and subpopulation level (Van Parijs *et al.*, 2003b). Study of the behavioral context of leopard seal vocalizations revealed that vocalizations are used by mature males to advertise their sexual readiness (Rogers *et al.*, 1996). Similarly, evidence that male harbor seals are vocal underwater during the breeding season was reported by Hangii and Schusterman (1994). Results of this study suggested that these vocalizations are used in male-male competition and/or advertisement displays to attract females.

Among the most distinctive underwater pinniped sounds are those made by male walruses during and outside the breeding season (Figure 11.17). Males produce a series of knocking sounds (including bells, bell-knocks, double knocks, and double-knock bell phonations) often described as "ringing bells" that are produced both in air and underwater (Schevill *et al.*, 1966; Fay *et al.*, 1984; Miller, 1985; Stirling *et al.*, 1987). The loud, repetitive underwater vocal displays (i.e., intense, slower repetition "knock" and less intense, quick "tap"), best studied in Atlantic walruses, have been described as songs (Sjare *et al.*, 2003 and references cited therein) in the same sense as humpback whale songs (see section 11.4.2.5). Walrus songs are of shorter duration and exhibit less variation in sound composition. The singing behavior of male walruses appears to reinforce dominance status in the absence of physical interactions. Underwater knocking sounds have also been recorded from hooded seals (Ballard and Kovacs, 1995), Weddell seals (Thomas and Kuechle, 1982), and grey seals (Asselin *et al.*, 1993).

Underwater clicks have also been recorded from several phocids, including harbor, ringed, harp, grey, and hooded seals (Ballard and Kovacs, 1995, and references cited therein) and several species of Antarctic ice seals. Renouf and Davis (1982) have speculated that these clicks are used for echolocation, although this is still controversial, and is challenged by Schusterman *et al.* (2000) and Holt *et al.* (2004), who argue that

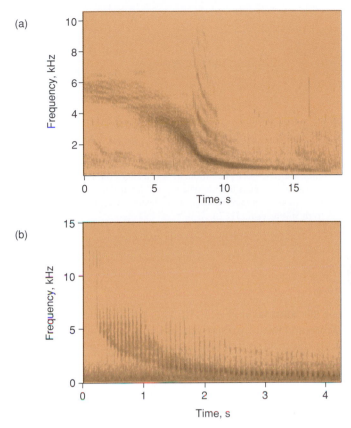

Figure 11.15.   Spectograms of a **(a)** descending trill and **(b)** descending buzz vocalizations of a Weddell seal.

the amphibious lifestyle of pinnipeds has precluded the sensory specializations needed for effective echolocation. Although some pinnipeds have the acoustic repertoire to echolocate, currently there is no confirming evidence that they do so (Evans *et al.*, 2004)

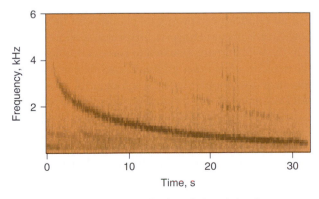

Figure 11.16.   Spectogram of a descending vocalization of a bearded seal.

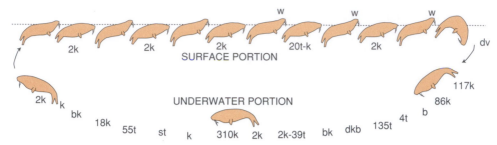

Figure 11.17.   Spatial and vocal structure of a single courting display of a male walrus, with eight discrete phonations: b = bell phonation; bk = bell knock phonation; dk = double knock phonation; dkb = double knock bell phonation; dv = diving phonation; k = knock phonation; st = strum phonation; t = tap phonation; w = whistle; numerals = number of repetitions. (Redrawn from Stirling *et al.*, 1987.)

11.4.2. Cetaceans

11.4.2.1. Echolocation

About 20% of all mammalian species (mostly bats) have overcome the problem of orienting themselves and locating objects in darkness or where vision is otherwise limited by producing short-duration sounds and listening for reflected echoes as the sounds bounce off objects. Essentially, echolocation is a specialized type of acoustic communication in which an animal sends information to itself. Echolocation has evolved independently in at least five mammalian lineages. Microchiropteran bats are well-known echolocators and so too are some shrews, golden hamsters, flying lemurs, and some marine mammals, notably odontocete cetaceans. By 1938, the echolocating abilities of bats were clearly demonstrated (Pierce and Griffin, 1938); yet another 25 years passed before echolocation in dolphins was reported. Kellogg *et al.* (1953) reported that captive dolphins could hear sounds up to 80 kHz, and McBride (1956) presented some of the first evidence that bottlenose dolphins could use echolocation to detect underwater objects. For a more detailed account of the early dolphin echolocation studies, see Au (1993).

Since those first reports, the echolocating abilities of odontocetes (especially captive dolphins) have been the subject of intense research. As echolocation can only be confirmed for captive animals that are deprived of access to their other senses, fewer than a dozen species of small odontocetes (mostly delphinids) are the only marine mammals unequivocally known to echolocate. However, all species of odontocetes produce click- or pulse-like sounds in the wild and are assumed to possess echolocation capabilities, and echolocation is suspected in some other groups of marine mammals. Several new investigative techniques, such as computer assisted tomography (CAT) and magnetic resonance imaging (MRI) scanning of the structures involved in the production and reception of these complex sounds have helped elucidate the functions of these structures. However, these techniques can only be used to study captive or freshly dead animals.

Behavioral studies of the echolocation sensitivities of captive animals typically have involved a training regimen to establish stimulus control of the behavior of an animal (Au, 1993). Thus, when any stimulus is changed, such as the size or shape of a target object presented to a dolphin wearing opaque eyecups, a resulting change in behavioral performance can be measured. From such behavioral evidence, captive odontocetes have uniformly performed well on target discrimination tests and sound frequency and intensity tests. They also can localize sound sources to within a couple of degrees, comparable to humans, and resolve time differences of a few millionths of a second, an order of magnitude finer than humans.

Small dolphins have been the principal subjects of most captive echolocation studies, although other species are being studied in field conditions that permit simultaneous identification of the echolocator and its associated behavior (Au and Herzing, 2003). The sound production system of all odontocetes generates trains or pulses of broad-frequency **clicks** of very short duration (Figure 11.18). As each click strikes a target, a portion of its sound energy is reflected back to the source (Figure 11.19). Each click lasts from 10 to 100 μs and may be repeated as many as 600 times each second. *Tursiops* uses clicks that are composed of a wide range of frequencies often exceeding 150 kHz, with most of the acoustic energy between 30 and 150 kHz (Figure 11.20a). Acoustic signals produced by harbor porpoises cover a very broad frequency range, from 40 Hz to at least 150 kHz (Verboom and Kastelein, 1995b, 1997, 2004). White-beaked, spotted, and dusky dolphins, as well as killer whales, also employ echolocation signals with bimodal frequency patterns (Au, 2004). Individual signals consist of low-frequency (80–10 kHz), midfrequency (10 kHz), and high-frequency (100–160 kHz) components (see Figure 11.20b). The low-frequency components of high amplitude sounds probably are used for detection. The midfrequency components have such a low energy level that they may not be of much practical function. The high-frequency components are used for bearing detection and classification of objects such as prey items. Even more different are Commerson's dolphins (see Figure 11.20b), which produce a narrow band frequency with nearly all of the energy in a high-frequency band between 100 and 200 kHz (Evans and Awbrey, 1988).

While these trains of rapidly repeated clicks are being produced, their rate of repetition is adjusted to allow the click echo to return to the animal during the time between outgoing clicks. The time required for a click to travel from its producer to the reflecting

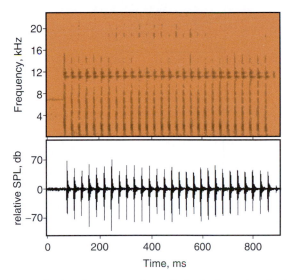

Figure 11.18. Spectogram (*top*) and power spectrum (*bottom*) of a series of echolocation clicks of a Commerson's dolphin.

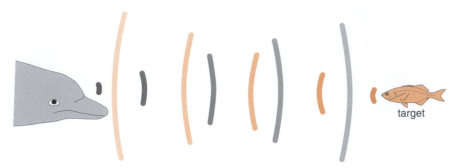

Figure 11.19.   Pattern of click train production and echo return for an echolocating dolphin. Outgoing clicks occur between returning echoes to reduce interference.

target and back again is a measure of the distance to the target. As that distance varies so will the time necessary for the echo to return. Continued evaluation of returning echoes from a moving target can indicate the target's speed and direction of travel. As a dolphin closes in on a target, its interclick interval (ICI) decreases corresponding to the distance to the target, and each click's SPL decreases so that the intensity of the returning echo remains nearly constant (Au and Benoit-Bird, 2003). Altes *et al.* (2003) suggest that dolphins use many successive echolocation clicks to interrogate a target, then use multiclick

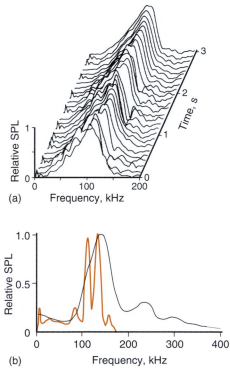

Figure 11.20.   **(a)** Frequency spectra of a sequence of dolphin echolocation clicks (Redrawn from Au, 1993). **(b)** Frequency (power) spectra of harbor porpoise (*fine line*) and Commerson's dolphin (*heavy line*), with most acoustic power in a high-frequency component at 100–200kHz. (Redrawn from Verboom and Kastelein, 2004, and Evans and Awbrey, 1988.)

processing to combine the resulting echoes to obtain more refined information about the target.

To make this acoustic picture more complicated, at the same time a dolphin is producing a train of click pulses, it can simultaneously produce frequency-modulated tonal whistle signals (Figure 11.21) that vary from 2 to 30 kHz and direct those emitted sound signals forward from the melon in different beam patterns. It can do this while continually varying the frequency content of the clicks to adjust to changing background noises or to the acoustic characteristics of the target.

The echolocation clicks of most odontocetes consist of broadband frequency spectra and short duration pulses, with the actual frequency used being continually adjusted to avoid competing background noises and to maximize the return of information about the target. The maximum range exhibited in target discrimination tests by *Tursiops* is about 100 m (Au and Snyder, 1980), although Ivanov (2004) provides evidence of target detection ranges exceeding 650 m. These clicks are of high intensity, with repetition rates adjusted to changing animal-target distance. Slight changes in signal characteristics from click to click may be due to the interaction of the two sets of phonic lips. The echolocation abilities of captive dolphins wearing eyecup blinders are sufficiently refined to discriminate target diameter ratios as small as 1:1.25 for metal targets of the same shape, and for different thicknesses of the same metals of the same size. By extension, it is presumed that these discriminatory abilities are sufficient to acoustically identify preferred prey and other similar items in the dolphin's natural habitat. Harley *et al.* (2003) present evidence that *Tursiops* extracts information about target objects directly from returning echoes rather than by storing whole-object mental "sound templates" and matching them to particular echo patterns. For more details on target shape and size discrimination capabilities, see also Roitblat (2004) and Pack *et al.* (2004).

The highly directional, intense clicks of foraging sperm whales occur in different patterns. "Usual clicks" have long and variable ICIs, whereas "creaks" have long durations

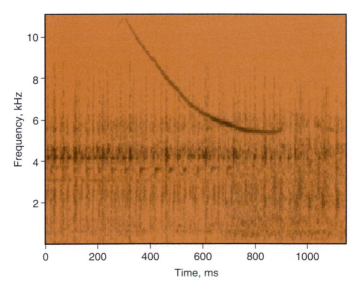

Figure 11.21.   Spectogram of a common dolphin whistle call superimposed on a series of its echolocation clicks.

(5–6 s), very short ICIs (0.03–0.04 s), and fast repetition rates (Madsen *et al.*, 2002). When a foraging whale dives from the surface, usual clicks are produced initially. Thode *et al.* (2002) found that within 6–12 minutes of the start of a dive, usual click ICIs matched the two-way travel time of the click from the whale's position to the sea floor. Usual clicks likely serve as long-range echolocators of the sea floor and the individual prey items above it.

Sperm whale creaks typically are produced only at the bottoms of foraging dives following a train of usual clicks, presumably during the terminal phase of a whale's approach to prey items at depth. Miller *et al.* (2004) used suction-cup attached digital acoustic recording tags (DTAGs) that record sound, pitch and roll, heading, and depth to demonstrate that creaks are associated with body rolls and other rapid changes in body positions. These data support the contention by Clarke and Paliza (2003) that sperm whales are upside down when they attack their prey. Theoretical calculations indicate that, in deep water, usual clicks have sufficient power and directionality to detect targets (prey or the sea floor) from distances up to 16 km and creaks at distances to 6 km (Madsen *et al.*, 2002).

11.4.2.2. Evidence for Echolocation in Mysticetes

How common is echolocation in cetaceans? Presently, it is uncertain because it is difficult to establish whether wild populations are indeed using echolocation-like clicks for the purposes of orientation and location. If judgments can be made from the types of sounds produced, then echolocation probably occurs in all toothed whales.

Broad-spectrum trains of clicks or short pulses have been recorded in the presence of gray whales in the North Pacific (Figure 11.22) and blue, fin, and minke whales in the Atlantic and Pacific Oceans. The fact that the majority of sounds produced by gray whales along their migration route are at frequencies below 200 Hz and that they have a pattern of repetition interspersed with long periods of silence suggest their use is in communication rather than echolocation (Crane and Lashkari, 1996).

In 1992, the U.S. Navy initiated a test program to make available to marine mammal scientists the North Atlantic Ocean undersea listening capabilities of the Integrated Undersea Surveillance System (IUSS). IUSS is part of the U.S. submarine defense system developed over 4 decades ago to acoustically detect and track Soviet submarines. The system consists of networks of hydrophone arrays, some towed by ships and others fixed to the sea bottom; it is also sufficiently sensitive to locate and track individual whales over hundreds of km for weeks.

Prior to 1992, the Navy made no systematic effort to record or archive any of the whale sounds they detected. The IUSS study has resulted in a wealth of acoustical data on large baleen whales, especially blue, fin, and minke whales. The vocalizations of these whales are typically very loud, low-frequency pulses of varying spectral complexity. The pulses of blue whales are between 15 to 20 Hz, mostly below the range of human hearing (Figure 11.23), whereas those of fin whales are only slightly higher at 20–30 Hz. Their function is not known, but two plausible explanations have been put forward. It is reasonable to conclude that if we can detect these sounds at long distances, other whales should be able to as well. They may therefore function in long-distance communication. These loud, low-frequency, patterned sequences of tones propagate through water with much less attenuation than do the higher frequency whistles or echolocation clicks of small toothed whales. Mellinger and Clark (2003) found that the frequency, duration,

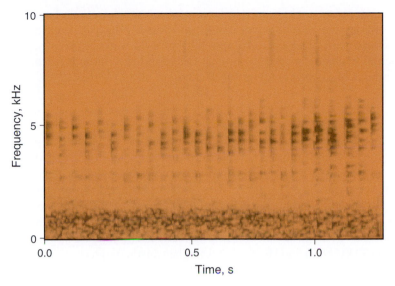

Figure 11.22. Spectogram of click series from a feeding gray whale. (From Fish *et al.*, 1974.)

and repetition patterns of blue whale calls in the North Atlantic differ from blue whale calls in other oceans, supporting the hypothesis that distinctive acoustic displays are characteristic of geographically separate blue whale populations.

In addition to signature or identity calls, it has been proposed that the low-frequency, short duration tone pulses may serve an echolocation function, although a very different one than that described for small toothed whales. A typical single blue whale call like that shown in Figure 11.23 lasts for 20 s and, in water, extends for approximately 30 km. The low frequency of blue and fin whale tones have very long wavelengths, from 50 m at 30 Hz to 100 m at 15 Hz, and, if used for echolocation, cannot resolve target features smaller than those wavelengths. Clark (1994) and Clark and Ellison (2004) have speculated

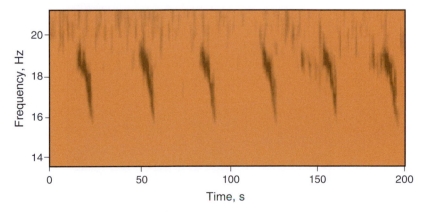

Figure 11.23. Spectogram of a series of calls from a blue whale. (From Clark, 1994.)

that these tone pulses might be used by large mysticetes to detect very large scale oceanic features, such as continental shelves, islands, and possibly sharp differences in water density associated with divergences or upwelling of cold water. Currently, however, the experimental evidence needed to support the sound reception capability needed for echolocation in mysticetes is completely lacking.

11.4.2.3. Signature Whistles of Dolphins

In addition to echolocation clicks and loud impulse sounds, dolphins also produce another type of vocalization, a narrow band frequency modulated (FM) sound often with a harmonic structure, usually described as a whistle or squeal. Typically, the frequency of these pure-tone emissions rises and falls between about 7 and 15 kHz and averages less than 1 s in duration (Figure 11.24). However, recordings of wild spotted dolphins in social settings demonstrate fundamental whistle frequencies regularly extend above 20 kHz (ultrasonic to us), with many harmonics above 50 kHz and occasionally to 100 kHz (Au and Herzing, 2003). The whistle frequencies of Hawaiian spinner dolphins span most of their range of hearing sensitivity (Lammers *et al.*, 2003).

Since 1965, the Caldwells (see Caldwell *et al.*, 1990 for a detailed review) have studied the acoustic characteristics of over 100 captive *Tursiops* of all ages and both sexes. They observed that each individual dolphin in a group produces an individual whistle contour so distinct that each animal can be identified from the whistle contour on a spectogram. These sounds came to be called **signature whistles.** The Caldwells hypothesized that the distinctiveness of an individual's whistle served to broadcast the identity of the animal producing the whistle and possibly to communicate other information, such as their state of arousal or fear, to group mates.

In addition to individual identification, a broader social function of signature whistles is suggested by a growing body of evidence. Dolphins often whistle when separated from other group members or in response to the whistles of group members. Captive dolphins have been trained to mimic electronically generated dolphin-like whistles and may imitate each other's signature whistle up to 20% of the time (Tyack, 1991). Tyack proposed

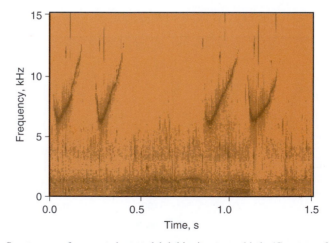

Figure 11.24.   Spectogram of a repeated spotted dolphin signature whistle. (Courtesy of D. Herzing.)

that a dolphin in a large group will mimic the signature whistle of another group member to initiate a social interaction. Dolphin species that are more social also are known to whistle more. There are, however, suggestions that the signature whistle hypothesis needs to be reevaluated. The results of McCowan and Reiss (1995, 2001) indicate that captive bottlenose dolphins share several different whistle types and that signature whistles may play a less predominant role than previously suspected. Other studies have investigated the whistle repertoires of wild dolphins. Comparisons of the whistles of several bottlenose dolphin populations suggest that although there may be differences between whistles from different individuals within the same population, there are still some characteristics that are unique for each population (Ding *et al.*, 1995a). In another study, the fact that common dolphin whistle repertoires were not individual-specific nor context-specific led Moore and Ridgway (1995) to suggest that they may represent a portion of a regional dialect, similar to the pod-specific dialects proposed for killer whales (see later). Additional work analyzing dolphin whistle repertoires with respect to behavioral contexts and social relationships is needed.

In a study comparing whistle structure among various odontocetes, some of the observed differences were correlated with taxonomic relationships, habitats, and body lengths (Ding *et al.*, 1995b). For example, whistle differences of the freshwater river dolphin *Inia geoffrensis* and whistles of oceanic delphinid species were related to habitat differences. The low and narrow frequency signals of *Inia* have better refractive capabilities, important to species whose habitats are rivers, which have higher noise levels than pelagic environments and carry more suspended material (Evans and Awbrey, 1988). Finally, a limitation of sound production capability related to body length is suggested because, in general, larger bodies lower the maximum whistle frequency range that can be produced.

11.4.2.4. Vocal Clans of Killer Whales and Sperm Whales

Killer whales have been found to produce repetitious calls that are now considered to be group **dialects** (Ford, 1991). Repertoires consisting of a small number of discrete calls (averaging about 10, such as the 2 shown in Figure 11.25) are shared by individuals within a pod and appear to persist unchanged for several decades. These pod-specific repertoires seem to serve as vocal indicators of pod affiliation and help to make vocal communication within a pod more efficient. Sixteen pods of resident killer whales studied by Ford in British Columbia coastal waters formed four distinct acoustic associations, or clans. All the pods within a clan shared several but not all calls. No sharing of calls occurred between different clans. This hierarchy of call associations, with individuals within a pod sharing a repertoire of pod calls, pods within a clan sharing some of those calls, and different clans sharing no calls, led Ford (1991) to propose that each group's vocal tradition had evolved over generations, with related pods in a clan having descended from a common ancestral group through growth and division of the group along matrilineal lines (Figure 11.26). Paralleling these group divergences were divergences of the group's vocal traditions as innovations of new calls and loss of old calls accumulated over time. These results have been confirmed by other studies of call repertoires in killer whale pods (Strager, 1995). In contrast to other delphinid species, killer whale whistles seem not to be used as individual "signature" calls to maintain acoustic contact with each other. Instead, these calls are most commonly associated with close-proximity social interactions and may play an important role in close-range communication among pod members (Thomsen *et al.*, 2002).

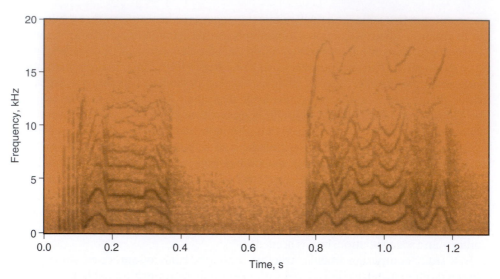

Figure 11.25.   Spectogram of a killer whale vocal call. (Courtesy Hubbs/Sea World Research Institute.)

The same type of clicks used by sperm whales for echolocation also serve as a means of communication. In female social units, rhythmic patterns of clicks, known as codas, last up to 1–2 s and consist of 3–30 clicks (Figure 11.27). By localizing and recording sperm whale sound sources with arrays of hydrophones, Watkins and Schevill (1977) found that individual whales repeatedly produce unique codas, and they suggest that these codas may serve as recognition codes for individual whales. These identity codas may allow pod members to keep track of each other when they disperse over several square kilometers during foraging dives (see Chapter 12). Other sperm whale codas are shared by several whales in local groups, suggesting that some communication function

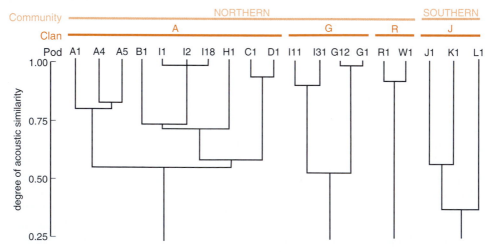

Figure 11.26.   Clan association diagram, showing the likely genealogies for the resident pods of killer whales in the Puget Sound/Vancouver Island region. (Redrawn from Ford, 1991.)

other than individual identity is served (Moore *et al.*, 1993; Weilgart and Whitehead, 1993). Coda repertoires distinctive to groups seem to persist over several years and show significant, but weaker, geographical differences (Weilgart and Whitehead, 1997). Rendell and Whitehead (2001) reported that all coda repertoires of 18 known social units could be assigned to one of 6 acoustic clans. Each of these clans are sympatric, range over thousands of km of ocean, and use coda patterns most likely transmitted culturally between individuals and units within each clan (Whitehead, 2003; Whitehead *et al.*, 2004). Some criticism has been directed at the temporary nature of these social aggregations in terms of maintenance of stable cultural characteristics (Mesnick, 2001; Tyack, 2001; Rendell and Whitehead, 2003).

11.4.2.5. Humpback Whale Sounds

Humpback whales are the most sonorous of the mysticetes. While on their winter breeding grounds, they sing long and acoustically complex songs that are shared by all singing whales occupying the same breeding ground (Payne and McVay, 1971). Consequently, the songs of North Atlantic humpback whales are identifiably different from those in the North Pacific Ocean. Each song, although often repeated for hours, lasts 10–15 minutes and is composed of repeated themes, phrases, and subphrases (Figure 11.28). Individual units that make up subphrases are typically a few seconds in duration with frequencies generally below 1.5 kHz (Payne *et al.*, 1983). In general terms, the songs of southern hemisphere humpback whales resemble those published for northern hemisphere humpback whales (Jenkins *et al.*, 1995).

The structure of humpback whale songs changes progressively over time. Most of the changes occur during winter months on breeding grounds and typically include changes in the frequency and duration of individual units as well as deletion of old and insertion

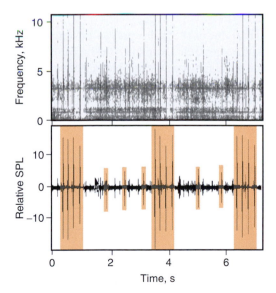

Figure 11.27. Spectogram (*top*) and power spectrum (*bottom*) of a series of three regular four-click codas (*large color boxes*) interspersed between several "usual" clicks (*small color boxes*) recorded in the presence of a female/young unit of sperm whales near the Azores. (Recording courtesy of P. Colla.)

of new phrases or themes (Payne *et al.*, 1983). The composition and sequence of themes within a song also change with time so that the song structure at the beginning of a winter breeding season is modified substantially by the end of that season a few months later, although all whales sing the same song at any point in time. Individual whales express the same changes at the same rate as the rest of the group of singing whales (Guinee *et al.*, 1983). Thus, singing whales seem to actively learn songs from each other; forgetfulness or changing membership in a singing population does not seem to account for the changes in song structure over time.

Underwater observations of singing humpbacks in the clear tropical waters of the Hawaiian breeding grounds (Chapter 13) have established that only adult males sing. The identification of the sex of the singers has also been confirmed by DNA analysis of humpbacks from the breeding grounds of the Mexican Pacific Ocean (Medrano *et al.*, 1994). These observations support the notion that humpback songs play a reproductive role similar to that of bird songs, communicating a singer's species, sex, location, readiness to mate with females, and readiness to engage competitively with other whales (Tyack, 1981). Additionally, the simultaneous singing by many males may serve as a communal display to synchronize the ovulation of females (Mobley and Herman, 1985). These songs also are thought to function as acoustic signals to mark underwater territories of adult males (Tyack, 1981; Darling *et al.*, 1983). Occasionally, humpback whales have been heard singing in pelagic waters during the spring migration well away from known breeding areas (Barlow, personal communication), as well as in summer feeding grounds in southeast Alaska (McSweeney *et al.*, 1989). These songs were essentially abbreviated composites of the breeding ground songs recorded in Hawaii in the previous and following winters. The gender of the Alaska singers was not determined, and the function of singing during migration or feeding is not known.

11.4.2.5. Prey Stunning Sounds

It has been discovered that one of the major prey of odontocetes, clupeid fish (e.g., herring, shad), can detect ultrasound clicks up to 180 kHz (Mann *et al.*, 1997). The ability of clupeids to detect ultrasound may be an example of convergent evolution such as that seen in moths and other insects capable of detecting ultrasounds of their predators. Shad readily detect echolocating pulses of dolphins, and like moths, they respond to detection with escape behavior. Because fossil clupeids are known from the early Cretaceous

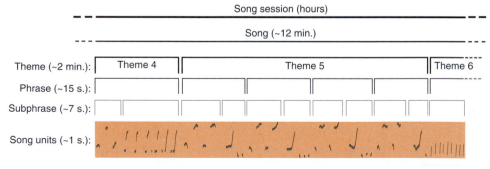

Figure 11.28. Hierarchical structure of humpback whale songs, each illustrated with spectogram tracings. (From Payne, 1983.)

(130 Ma), long before the first odontocetes in the Oligocene (25–38 Ma), it is likely that the ultrasound detection capability of clupeids was present before there were echolocating predators.

To counter the possibility of acoustic detection and avoidance by prey species, some species of toothed whales may debilitate or temporarily disorient small prey with loud blasts of sound called "bangs." Presumably the same sound production system employed for echolocation is used to produce these bangs. The concept of stunning prey with acoustic energy was first proposed for sperm whales by Bel'kovich and Yablokov (1963), and was supported by the work of Berzin (1971). The concept has been evaluated in the context of other odontocete species by Norris and coworkers (i.e., Norris and Mohl, 1983; Marten *et al.*, 1988). In addition, it has been suggested that jaw claps, loud long-duration multipulsed sounds associated with rapid jaw closure in odontocetes, may be similar to sounds implicated in debilitation of prey (Johnson and Norris, 1994). The general arguments used to support the concept of prey debilitation are based on anatomical, behavioral, and acoustic evidence as well as anecdotal information. Among the anatomical arguments used to support the concept is the serious mismatch between successful prey capture (based on examinations of stomach contents) and the feeding structures exhibited by several predaceous odontocete species. This mismatch is especially noticeable in the absence of functional teeth in most beaked whales, the narwhal, and in sexually immature sperm whales. That these whales successfully capture fast, slippery prey without functional teeth suggests that they must be able to approach very closely before engulfing their prey with a piston-like action of the tongue (see also Chapter 12).

Acoustic evidence for prey debilitation is difficult to collect, for sound bursts of sufficient pressure to damage or disorient small fish (estimated at 240–250 dB by Zagaeski, 1987) are very difficult to record in natural conditions, and captive animals are unlikely to emit such loud sounds in reverberant concrete tanks. Norris and Mohl (1983) calculated that sperm whales may emit click pulses at 265 dB. However, recording such an emission in a natural setting usually saturates the electronics of standard recording systems, and, unless the orientation angle and precise location of the whale are known, evaluating the SPL of bangs is difficult.

Some recordings in natural settings have been made that are suggestive of prey debilitation (Figure 11.29). Similar loud, low-frequency stunning sounds have been recorded from wild bottlenose dolphins feeding in coastal Australia and California waters, from killer whales in the North Atlantic and Northeast Pacific Oceans, and from sperm whales in the Indian Ocean (Marten *et al.*, 1988). Although the SPL of these sounds could not be measured, they were typically much higher than the SPL of the presumed echolocation clicks immediately preceding the bang. In summary, despite several decades of discussion and research, it remains unclear whether odontocetes actually use these sounds to debilitate prey and, if they do, which species produce these sounds and at what frequencies and sound pressure levels.

11.4.2.5. The Evolution of Cetacean Hearing

We are now able to link specific auditory structures in whales with entry into the water (Nummela *et al.*, 2004). The earliest cetaceans, pakicetids, used the same sound transmission mechanisms as land mammals in air (external auditory meatus, tympanic membrane, and ossicles), and in water they used bone conduction hearing mechanism. Their

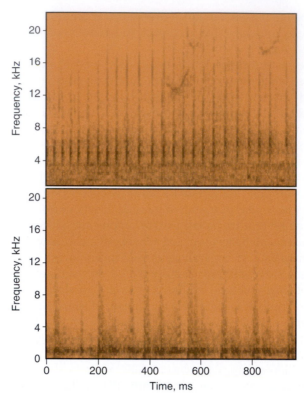

Figure 11.29. Comparison of spectograms of typical echolocation clicks (*top*) and a series of popping sonations (*bottom*) produced by bottlenose dolphin(s), with higher amplitude, longer durations, and lower frequencies than typical echolocation clicks. (Courtesy of V. Dudley.)

hearing sensitivity in water was apparently poor, as well as their directional hearing. The ear complex was not isolated from the skull, and the ear morphology was of the land mammal type. Later diverging remingtonocetids and protocetids retained the land mammal system but also developed a new sound transmission system similar to that of modern whales. This "key" innovation, the presence of a large mandibular foramen, heralded development of a fat-filled pad that directs sound to the earbones best developed in toothed whales for reception of high-frequency sound. In air, remingtonocetids heard like land mammals, but with low sensitivity, and in water they used the generalized cetacean hearing mechanism, in which sound arrives to the middle ear through the mandibular fat pad. Directional hearing was possible to some degree. The land mammal ear disappeared in the totally marine basilosaurids and the modern cetacean ear with its acoustic isolation (air filled sinuses) was further developed.

The mammalian inner ear contains two organs of equilibrium, the vestibule and the semicircular canals. Cetaceans are exceptional in having semicircular canals that are significantly smaller than the cochlear canal. If the semicircular canals are vestigial, these animals may not have any rotational or three-dimensional positioning sense, which may permit the flying turns of spinner dolphins without the side effects of motion sickness (Ketten, 1992a). Examination of the semicircular canals in archaic whales revealed that the first appearance of small semicircular canals appeared early in cetacean evolution (i.e., in middle Eocene remingtonocetids). This "key" event in the evolution of aquatic behavior is hypothesized to have led to a fundamental shift from the land to the marine environment (Spoor *et al.*, 2002).

Fossil evidence suggests that the ability to use high frequencies may have originated early in cetacean history. The recent discovery of the ability of clupeid fish (herring and shad) to detect ultrasound pulses from odontocetes indicates that echolocation may have evolved prior to the appearance of odontocetes (Mann *et al.*, 1997). The presence of related osteological changes (telescoping of the skull, concave mandible, separate bullae, and enlarged peribullar spaces) in the earliest odontocetes, the agorophids, indicates the development of echolocation because in modern odontocetes these features are associated with soft tissue developments principally related to high-frequency sound reception. The presence of high-frequency hearing structures (i.e., numerous foramina for the ganglion cells of the auditory nerve) and the highly specialized vestibule and semicircular canals in squalodontoids (including agorophids) suggests that high-frequency hearing and other specializations of the inner ear among cetaceans occurred before the early Oligocene (Luo and Eastman, 1995). However, the lack of complete isolation of the ear complex from the skull and limited telescoping of the skull among Eocene whales suggests further study is necessary before concluding that echolocation evolved prior to the appearance of odontocetes.

Because the earliest cetaceans, common ancestors of odontocetes and mysticetes, probably used high, but not ultrasonic, frequencies, it is likely that the low-frequency hearing of mysticetes evolved subsequently. Why did low-frequency hearing evolve? The appearance of mysticetes coincides with the opening of new oceanic regions in the southern hemisphere. In addition to an increase in primary productivity (see Chapters 6 and 12), there was a substantial reduction in surface temperatures at higher latitudes. In colder regions, an increase in body size would offer a substantial metabolic advantage. As cochlea measurements scale isometrically to body size, Ketten (1992a) proposes that a lower frequency cochlea may have resulted as a consequence of the increased body size of mysticetes. She further suggested that with less pressure to use echolocation as a foraging strategy in more productive waters, a decrease in the reception of higher frequency sounds may not have been a significant disadvantage. Therefore, as larger mysticetes evolved, increased size of cochlear structures may have constrained the resonance characteristics of the ear to progressively lower frequencies.

11.4.3. Other Marine Mammals

Sounds described as chirp-squeaks, identified as short, FM signals, have been reported for both manatees and dugongs (2.5–8 kHz range for manatees, Evans and Herald, 1970; Sousa-Lima *et al.*, 2002; 3–18 kHz range for dugongs, Anderson and Barclay, 1995; Figure 11.30). Hartman (1979) described chirp-squeaks, squeals, and screams for the West Indian manatee. Additional categories of sound were reported for this same species and were further distinguished by gender and age (Steel, 1982). Analysis of West Indian manatee recordings indicates that vocalizations are stereotyped and show little geographic variation (Nowacek *et al.*, 2003). Observations in the field indicate that manatee vocalizations play a key role in keeping mothers and calves together. In addition to chirp-squeaks, other vocalizations of dugongs have been described as barks and trills (Anderson and Barclay, 1995). According to Anderson and Barclay, low pitched whistles previously attributed to dugongs are more likely an abnormality in the respiratory system rather than a means of communication, given their production during breathing.

On the basis of behavioral observations, it appears that the chirp-squeaks and other sounds of the dugong, originate in the frontal region of the head, suggesting a mechanism similar to that in whales, rather than in the larynx as previously suggested. Chirp-squeaks are emitted when male dugongs feed at the bottom or patrol territories. Barks have physical characteristics appropriate for aggressive behavior and have been recorded in situations suggesting a role in territorial defense. Trills have characteristics more appropriate in advertisement of a territory or readiness for mating (Anderson and Barclay, 1995). The vocal cords of sirenians are absent and are replaced by fleshy, prominent cushions (Harrison and King, 1980).

The vocal repertoire of sea otters consists of above-water, low-frequency, low intensity signals that are similar in complexity to those of certain pinnipeds, notably the California sea lion and the northern elephant seal (McShane *et al.*, 1995). One moderately long distance (<1 km) signal is likely important for mother-pup recognition. Of particular interest is the presence of graded signals (i.e., those that vary over a continuum of frequencies rather than forming discrete units) that are most appropriate for short range communication between familiar individuals (McShane *et al.*, 1995).

Field observations suggest that polar bears lack most of the distinctive vocalizations that characterize other carnivores (Stirling and Derocher, 1990). Females produce mother-cub recognition calls and defensive roars, growls, and snarls and males snort and chuff. The latter is a low intensity repetitive call during interactions with one another (Wemmer *et al.*, 1976).

11.5. Acoustic Thermometry of Ocean Climate and Low-Frequency Military Sonars

In 1991, oceanographers from the United States and other countries initiated an experiment to determine if measuring the transmission time of low-frequency sounds across ocean basins could be used to detect changes in ocean temperature to monitor global warming. This experiment, referred to as the Heard Island Feasibility Test, was successful and a follow-up study was authorized. This study, the **Acoustic Thermometry of Ocean Climate (ATOC)** Program, called for installing 260-watt, low-frequency sound

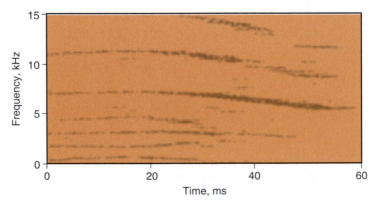

Figure 11.30.   Spectogram of a dugong call. (From Anderson and Barclay, 1995.)

generators in deep water off the coast of Hawaii and central California. A research program was later included to ensure that the sounds produced by ATOC would not have a negative impact on marine mammals and other ocean animals. The location of the planned California sound source was moved further north to a location off the San Francisco coast and testing was begun late in 1995. Analysis of the data collected indicates that the ATOC transmissions have had few detectable effects on marine mammals (Marine Mammal Commission, 1998), although these conclusions remain inconclusive and controversial.

The U.S. Navy has deployed a new generation of active sonar systems known as SUR-TASS LFAs (for Surveillance Towed Array Sensor System, Low Frequency Active) sonar. This LFA system has both passive listening and active broadcast components, and it is the active component of this and other military sonar systems that are of serious concern to marine mammal researchers. LFA systems project high-intensity (source levels up to 235 dB), low-frequency (100–500 Hz) sound from underwater arrays of up to 18 acoustic projectors towed below a surface ship. At a distance of 750 km from the source, the sound pressure level is estimated to be 120 dB, well below the generally accepted threshold level of 180 dB needed to cause trauma to marine mammals. The acoustic signal produced is discontinuous, with individual signals lasting from 6 to 100 s (*Federal Register*, Anon., 2002).

In the past decade, three major and several smaller multiple species stranding events have occurred in close proximity to large-scale U.S. or NATO naval exercises. In 1996, 13 beaked whales stranded during a NATO exercise; in 2000 in the Bahamas Islands, 17 beaked whales in three species and 2 minke whales; and in 2002 in the Canary Islands, 14 beaked whales in three species. Most of the stranded animals that were examined while fresh presented symptoms of acoustic trauma. Postmortem conditions of examined animals included hemorrhage and edema and major-organ blood vessels congested with small gas bubbles. This type of bubble formation within blood vessels is characteristic of decompression sickness in humans. These symptoms were new to marine mammal pathologists and were consistent with exposure to intense acoustical activity characteristic of military sonar or seismic blasting (Jepson *et al.*, 2003). The damage patterns suggest that beaked whales may have a particular sensitivity to these sonar sounds, either because they have airspace and tissue resonance frequencies that are sensitive to sonar frequencies or because their deep foraging habits make them particularly vulnerable to *in vivo* gas-bubble formation in the presence of intense sound pressures. All marine mammals live in an acoustic world, and further research on the physical and behavioral effects of exposure to military sonar is needed. Yet it is becoming clear that the 180-dB threshold for protecting marine mammals from sound-induced trauma is much too high, and immediate efforts should be made to make these human-generated underwater noises less harmful without waiting until we completely understand the mechanisms of acoustically induced trauma on marine mammals.

11.6. Summary and Conclusions

Airborne and waterborne sounds are produced by marine mammals for echolocation, communication, and prey capture. The majority of airborne sounds of pinnipeds are produced in the larynx. Other airborne pinniped vocalizations include those of the walrus, produced by the teeth and throat pouches. Modifications of the pinniped ear for underwater hearing include enlarged ossicles, especially the incus, which enhances bones conduction hearing, and the development of cavernous tissue in the outer and middle ear, which functions to equalize pressure when diving as well as to enhance the transmission of high-frequency sound to the inner ear. Echolocation involves the production of short duration sounds (clicks) and listening for reflected echoes as sounds bounce off objects. The echolocation abilities of toothed whales have been the subject of intense research. The source of production of echolocation sounds has been determined to be a structural complex associated with the upper nasal passages (MLDB complex). The periodic opening and closing of the monkey lips break up the air flow between the lips and determines the click repetition rate. This sound generation system is coupled to a sound propagation structure, the melon, which serves as an acoustic lens to focus sound. The sound reception system of odontocetes involves a number of anatomical specializations of the ear (e.g., development of air sinuses, increase in number of receptors for hearing (hair cells) and lower jaw (e.g., fat pad that acts as sound channel to conduct sound to middle ear). There is no experimental evidence to support the claim that some mysticetes and pinnipeds echolocate. Echolocation likely evolved among odontocetes as judged by transformations of the ear and jaw. Earlier diverging whales developed some characters related to echolocation including directional hearing. Sounds produced by sirenians likely originate in the frontal region of the head, indicating a mechanism similar to whales. Sound reception may occur via a low-frequency sound channel in an oil-filled bony sponge in the cheek region of the skull. Sea otter vocalizations are low intensity, low-frequency signals similar to those of pinnipeds.

In air vocalizations of pinnipeds include mother-pup calls and those used to communicate breeding status. Underwater sounds of pinnipeds range from the soft lyrical sounds of the leopard seal to the aggressive sounding grunts, barks, and groans of most other phocids. Social sounds produced by whales for communication include the signature whistles of dolphins, dialects of killer whales, and the songs of humpback whales. Loud blasts of sound, "bangs," produced by toothed whales that may be used to debilitate or temporarily disorient small prey. Sirenian vocalizations described as chirps and squeaks, which are short-frequency modulated signals, have been identified. The issue of how anthropogenic sound affects marine mammals is contentious and additional research is necessary to address gaps in our current understanding of this issue.

11.7. Further Reading

A general review of echolocation in dolphins can be found in Au (1993) and Cranford *et al.* (1996) provide a review of recent developments in the mechanics of sound production in dolphins and other odontocetes. Summaries of the function of acoustic communication in marine mammals can be found in a number of edited volumes (Bushel and Fish, 1980; Nachtigall and Moore, 1988; Thomas and Kastelein, 1990; Renouf, 2001; Thomas *et al.*, 1992; Webster *et al.*, 1992; Kastelein *et al.*, 1995; Hoelzel, 2002; Au *et al.*, 2000; Supin *et al.*, 2001; Tyak and Miller, 2002). More information about ATOC and its effect on marine mammals can be found at the following Internet address: http://atoc.ucsd.edu; and see the Acoustical Society of America (http://asa.aip.org), a professional organization with an animal bioacoustics group interested in cetacean sonar and vocalizations among other topics.

References

Altes, R. A., L. A. Dankiewicz, P. W. Moore, and D. A. Helweg (2003). "Multiecho Processing by an Echolocating Dolphin." *J. Acoust. Soc. Am. 114:* 1155–1166.

Ames, A. L., E. S. Van Vleet, and J. E. Reynolds (2002). "Comparison of Lipids in Selected Tissues of the Florida Manatee (Order Sirenia) and Bottlenose Dolphin (Order Cetacea: Suborder Odontoceti)." *Comp. Biochem. Physiol. B 132:* 625–634.

Anderson, P. K., and R. M. R. Barclay (1995). "Acoustic Signals of Solitary Dugongs: Physical Characteristics and Behavioral Correlates." *J. Mammal. 76:* 1226–1237.

Anon. (2002). "Taking and Importing Marine Mammals; Taking Marine Mammals Incidental to Navy Operations of Surveillance Towed Array Sensor System Low Frequency Active Sonar; Final Rule." *Fed. Reg. 67* (136): 46711–46789.

Asselin, S., M. O. Hammill, and C. Barette (1993). "Underwater Vocalizsations of Ice Breeding Gray Seals." *Can. J. Zool. 71:* 2211–2219.

Au, W. W. L. (1993). *The Sonar of Dolphins*. Springer-Verlag, New York.

Au, W. W. L. (2004). "Echolocation Signals of Wild Dolphins." *Acoust. Phys. 50:* 454–462.

Au, W. W. L., and K. J. Benoit-Bird (2003). "Automatic Gain Control in the Echolocation System of Dolphins." *Nature 423:* 861–863.

Au, W. W. L., and D. L. Herzing (2003). "Echolocation Signals of Wild Atlantic Spotted Dolphin *(Stenella frontalis)*." *J. Acoust. Soc. Am. 113:* 598–604.

Au, W. W. L., A. N. Popper, and R. R. Fay (eds.) (2000). *Hearing in Whales and Dolphins*. Springer Handbook of Auditory Research, vol. 12. Springer, New York.

Au, W. W. L., and K. J. Snyder (1980). Long-range target detection in open waters by an echolocating Atlantic Bottlenose Dolphin (*Tursiops truncatus*). *J. Acoust. Soc. Am.* 68: 1077-1084.

Ballard, K. A., and K. M. Kovacs (1995). "The Acoustic Repertoire of Hooded Seals *(Cystophora cristata)*." *Can. J. Zool. 73:* 1362–1374.

Bel'kovich, V. M., and A. V. Yablokov (1963). "Marine Mammals Share Experience with Designers." *Nauka-Zhizn. 30:* 61–64.

Berzin, A. A. (1971). "The Sperm Whale." *Pac. Sci. Res. Inst. Fish. Oceanogr. Sp. Publ.* 1–394.

Brill, R. L., M. S. Sevenich, T. J. Sullivan, J. D. Sustman, and R. E. Witt (1988). "Behavioural Evidence for Hearing Through the Lower Jaw by an Echolocating Dolphin *(Tursiops truncatus)*." *Mar. Mamm. Sci. 4:* 223–230.

Bullock, T. H., D. P. Domning, and R. C. Best (1980). "Evoked Brain Potentials Demonstrate Hearing in a Manatee *(Trichechus inunguis)*." *J. Mammal. 61:* 130–133.

Bushel, R. G., and J. F. Fish (eds.). (1980). *Animal Sonar Systems*. Plenum Press, New York.

Caldwell, M. C., D. K. Caldwell, and P. L. Tyack (1990). Review of the signature-whistle hypothesis for the Atlantic Bottlenose Dolphin. *In* "The Bottlenose Dolphin" (S. R. Leatherwood, and R. R. Reeves, eds.), pp. 199–234. Academic Press, San Diego, CA.

Clark, C. W. (1994). "Blue Deep Voices: Insights from the Navy's Whales '93 Program." *Whalewatcher 28:* 6–11.

Clark, C. W., and W. T. Ellison (2004). "Potential Use of Low-Frequency Sound by Baleen Whales for Probing the Environment: Evidence from Models and Empirical Measurements." *In* "Echolocation in Bats and Dolphins" (J. Thomas, C. Moss, and M. Water, eds.) pp. 564–581. Univ. Chicago Press, Chicago, IL.

Clarke, R., and O. Paliza (2003). "When Attacking Their Prey Sperm Whales are Upside Down." *Mar. Mamm. Sci. 19:* 607–608.

Cleator, H. J., I. Stirling, and T. G. Smith (1989). "Underwater Vocalizations of the Bearded Seal *(Erignathus barbatus)*." *Can. J. Zool. 67:* 1900–1910.

Crane, N. L., and K. Lashkari (1996). "Sound Production of Gray Whales, *Eschrichtius robustus,* Along Their Migration Route: A New Approach to Signal Analysis." *J. Acoust. Soc. Am. 100:* 1878–1886.

Cranford, T. W. (1988). The anatomy of acoustic structures in the spinner dolphin forehead as shown by X-ray computed tomography and computer graphics. *In* "Animal Sonar: Processes and Performance" (P. E. Nachtigall, and P. W. Moore, eds.), pp. 67–77. Plenum Press, New York.

Cranford, T. W. (1992). *Functional Morphology of the Odontocete Forehead: Implications for Sound Generation*. Ph.D. Thesis, University of California, Santa Cruz.

Cranford, T. W. (1999). "The Sperm Whale's Nose: Sexual Selection on a Grand Scale?" *Mar. Mamm. Sci. 15*: 1133–1157.

Cranford, T. W., M. Amundin, and K. S. Norris (1996). "Functional Morphology and Homology in the Odontocete Nasal Complex: Implications for Sound Generation." *J. Morphol. 228:* 223–285.

Cranford, T. W., W. G. Van Bonn, S. H. Ridgway, M. S. Chaplin, and J. R. Carr (1997). "Functional Morphology of the Dolphins Biosonar Signal Generator Studied by High-Speed Video Endoscopy." *J. Morphol. 232:* 243-????.

Darling, J. D., K. M. Gibson, and G. K. Silber (1983). Observations on the abundance and behavior of hump-back whales *(Megaptera novaeangliae)* off West Maui, Hawaii, 1977–1979. *In* "Communication and Behavior of Whales" (R. Payne, ed.), pp. 201–222. Westview Press, Boulder, CO.

Ding, W., B. Wursig, and W. Evans (1995b). Comparisons of whistles among seven odontocete species. *In* "Sensory Systems of Aquatic Mammals" (R. A. Kastelein, J. A. Thomas, and P. E. Nachtigall, eds.), pp. 299–323. De Spil Publishers, Woerden, The Netherlands.

Ding, W., B. Wursig, and W. Evans (1995a). "Whistles of Bottlenose Dolphins: Comparisons Among Populations." *Aquat. Mamm. 21:* 65–77.

Evans, W. E., and F. Awbrey (1988). "High Frequency Pulses Produced by Free-Ranging Commerson's Dolphin *(Cephalorhynchus commersonii)* Compared to Those of Phocoenids." *IWC Spec. Iss. 9:* 173–181.

Evans, W. E., and E. S. Herald (1970). "Underwater Calls of a Captive Amazon Manatee, *Trichechus inunguis.*" *J. Mammal. 51:* 820–823.

Evans, W. E., J. A. Thomas, and R. W. Davis (2004). Vocalizations from Weddell seals *(Leptonychotes weddelli)* during diving and foraging. *In* "Echolocation in Bats and Dolphins" (J. A. Thomas, C. F. Moss, and M. Vater, eds.), pp. 541–546. University of Chicago Press, Chicago.

Fay, F. H., G. C. Ray, and A. A. Kibal'chich (1984). Time and location of mating and associated behavior of the Pacific walrus, *Odobenus rosmarus divergens* Illiger. *In* "Soviet-American Cooperative Research on Marine Mammals: vol. 1. Pinnipeds." NOAA Tech. Rep. NMFS 12.

Fish, J. F., J. L. Sumich, and G. L. Lingle (1974). "Sounds Produced by the Gray Whale *Eschrichtius robustus.*" *Mar Fish. Rev. 36:* 38–49.

Ford, J. B. (1991). "Vocal Traditions Among Resident Killer Whales *(Orcinus orca)* in Coastal Waters of British Columbia." *Can. J. Zool. 69:* 1454–1483.

Guinee, L. N., K. Chu, and E. M. Dorsey (1983). Changes over time in the songs of known individual Humpback whales *(Megaptera novaeangliae)*. *In* "Communication and Behavior of Whales" (R. Payne, ed.), pp. 59–80. Westview Press, Boulder, CO.

Hanggi, E. (1992). "Importance of Vocal Cues in Other-Pup Recognition in a California Sea Lion." *Mar. Mamm. Sci. 8:* 430–432.

Hanggi, E., and R. J. Schusterman (1994). "Underwater Acoustic Displays and Individual Variation in Male Harbor Seals, *Phoca vitulina.*" *Anim. Behav. 48:* 1275–1283.

Harley, H. E., Erika A. Putman, and H. L. Roitblat (2003). "Bottlenose Dolphins Perceive Object Features Through Echolocation." *Nature 424:* 667–669.

Harrison, R. H., and J. E. King (1980). *Marine Mammals,* 2nd. ed., Hutchinson, London.

Hartman, D. S. (1979). "Ecology and Behavior of the Manatee *(Trichechus manatus)* in Florida." *Am. Soc. Mamm. Spec. Publ. 5:* 1–153.

Hemilä, S., S. Nummela, and T. Reuter (1995). "What Middle Ear Parameters Tell About Impedance Matching and High Frequency Hearing." *Hear. Res. 85:* 31–44.

Hemilä, S., S. Nummela, and T. Reuter (1999). "A Model of the Odontocete Middle Ear." *Hear. Res. 133:* 82–97.

Hemilä, S., S. Nummela, and T. Reuter (2001). "Modeling Whale Audiograms: Effects of Bone Mass on High-Frequency Hearing." *Hear. Res. 151:* 221–226.

Herman, L. M. (1991). What the dolphin knows, or might know, in its natural world. *In* "Dolphin Societies" (K. Pryor and K. S. Norris, eds.), pp. 349–363. University of California Press, Berkeley.

Herzing, D. L. (2004). Social and Nonsocial Uses of Echolocation in Free-Ranging *Stenella frontalis* and *Tursiops truncatus. In* "Echolocation in Bats and Dolphins" (J. A. Thomas, C. F. Moss, and M. Vater, eds.), pp. 404–413. Chicago University Press, Chicago.

Hoelzel, A. Rus (ed.) (2002). *Marine Mammal Biology: An Evolutionary Approach.* Blackwell Publishing, Oxford, UK.

Holt, M. M., R. J. Schusterman, B. L. Southhall, and D. Kastak (2004). "Localization of Aerial Broadband Noise by Pinnipeds." *J. Acoust. Soc. Am. 115*(5): 2339–2345.

Ivanov, M. P. (2004). "Dolphin's Echolocation Signals in a Complicated Acoustic Environment." *Acoust. Phys. 50:* 469–479.

Jenkins, P. F., D. A. Helweg, and D. H. Cato (1995). Humpback whale song in Tonga: Initial findings. *In* "Sensory Systems of Aquatic Mammals" (R. A. Kastelein, J. A. Thomas, and P. E. Nachtigall, eds.), pp. 335–348. De Spil Publishers, Woerden, The Netherlands.

Jepson, P. D., M. Arbelo, R. Deaville, I. A. P. Patterson, P. Castro, J. R. Baker, E. Degollada, H. M. Ross, P. Herraez, A. M. Pocknell, F. Rodriguez, F. E. Howie, A. Expinosa, R. J. Reid, J. R. Jaber, V. Martin, A. A. Cunningham, and A. Fernandez. (2003). "Gas-Bubble Lesions in Stranded Cetaceans: Was Sonar Responsible for a Spate of Whale Deaths After an Atlantic Military Exercise?" *Nature 425:* 575–576.

Johnson, C. H., and K. S. Norris (1994). Social behavior. *In* "The Hawaiian Spinner Dolphin" (K. S. Norris, B. Wursig, R. S. Wells, and M. Wursig, eds.), pp. 243–286. University of California Press, Berkeley.

Kardong, K. V. (1995). *Vertebrates*. William C. Brown, Dubuque, IA.

Kastelein, R. A., J. L. Dubbledam, M. A. G. de Bakker, and N. M. Gerrits (1996). "The Anatomy of the Walrus Head *(Odobenus rosmarus)*. Part 4: The Ears and Their Function in Aerial and Underwater Hearing." *Aquat. Mamm. 22:* 95–125.

Kastelein, R. A., J. A. Thomas, and P. E. Nachtigall (eds.) (1995). *Sensory Systems of Aquatic Mammals*. De Spil, Netherlands.

Kellogg, W. N., R. Kohler, and H. N. Morris (1953). "Porpoise Sounds as Sonar Signals." *Science 117:* 239–243.

Ketten, D. M. (1992a). The marine mammal ear: Specializations for aquatic audition and echolocation. *In* "The Evolutionary Biology of Hearing" (D. B. Webster, R. R. Fay, and A. N. Popper, eds.), pp. 717–750. Springer-Verlag, New York.

Ketten, D. M. (1992b). The cetacean ear: Form, frequency, and evolution. *In* "Marine Mammal Sensory Systems" (J. Thomas, R. A. Kastelein, and A. Ya. Supin, eds.), pp. 53–75. Plenum Press, New York.

Ketten, D. M., D. K. Odell, and D. P. Domning (1992). Structure, function, and adaptation of the manatee ear. *In* "Marine Mammals Sensory Systems" (J. Thomas, R. A. Kastelein, and A. Ya. Supin, eds.), pp. 77–95. Plenum Press, New York.

Kovacs, K. M. (1995). "Mother-Pup Reunions in Harp Seals, *Phoca groenlandica*: Cues for the Relocation of Pups." *Can. J. Zool. 73:* 843–849.

Lammers, M. O., W. W. L. Au, and D. L. Herzing (2003). "The Broadband Social Acoustic Signaling Behavior of Spinner and Spotted Dolphins." *J. Acoust. Soc. Am. 114:* 1629–1639.

Le Boeuf, B. J., and L. F. Petrinovich (1975). "Elephant Seal Dialects: Are They Reliable?" *Rapp. P.-V. Reun. Cons. Int. Exp. MerCons. 169:* 213–218.

Luo, Z., and E. R. Eastman (1995). "Petrosal and Inner Ear of a Squalodontoid Whale: Implications for Evolution of Hearing in Odontocetes." *J. Vertebr. Paleontol. 15:* 431–442.

Madsen, P. T., Wahlberg, M., and Mohl, B. (2002). "Male Sperm Whale *(Physeter macrocephalus)* Acoustics in a High-Latitude Habitat: Implications for Echolocation and Communication." *Behav. Ecol. Sociobiol. 53:* 31–41.

Mann, D. A., Z. Lee, and A. N. Popper (1997). "A Clupeid Fish Can Detect Ultrasound." *Nature 389:* 341.

Marine Mammal Commission (1998). *Annual Report to Congress 1997*. Washington, D. C.

Marten, K., K. S. Norris, P. W. B. Moore, and K. Englund (1988). Loud impulse sounds in odontocete predation and social behavior. *In* "Animal Sonar: Processes and Performance" (P. E. Nachtigall, and P. W. B. Moore, eds.), pp. 281–285. Plenum Press, New York.

McBride, A. F. (1956). "Evidence for Echolocation by Cetaceans." *Deep-Sea Res. 3:* 153–154.

McCormick, J. G., E. G. Wever, G. Palm, and S. H. Ridgway (1970). "Sound Conduction in the Dolphin Ear." *J. Acoust. Soc. Am. 48:* 1418–1428.

McCowan, B., and D. Reiss (1995). "Quantitative Comparison of Whistle Repertoires from Captive Adult Bottlenose Dolphins *(*Delphinidae, *Tursiops truncatus)*: A Re-Evaluation of the Signature Whistle Hypothesis." *Ethology 100:* 194–209.

McCowan, B., and D. Reiss (2001). "The Fallacy of `Signature Whistles' in Bottlenose Dolphins: A Comparative Perspective of "Signature Information" in Animal Vocalizations." *Anim. Behav. 62:* 1151–1162.

McShane, L. J., J. A. Estes, M. L. Riedman, and M. M. Staedler (1995). "Repertoire, Structure, and Individual Variation of Vocalizations of the Sea Otter." *J. Mammal. 76:* 414–427.

McSweeney, D. J., K. C. Chu, W. F Dolphin, and L. N. Guinee (1989). "North Pacific Humpback Whale Songs: A Comparison of Southeast Alaskan Feeding Ground Songs with Hawaiian Wintering Ground Songs." *Mar. Mamm. Sci. 5:* 139–148.

Medrano, L., M. Salinas, I. Salas, P. Ladronde Guevara, and A. Aguayo (1994). "Sex Identification of Humpback Whales, *Megaptera novaeangliae,* on the Wintering Grounds of the Mexican Pacific Ocean." *Can. J. Zool. 72:* 1771–1774.

Mellinger, D. K., and C. W. Clark (2003). "Blue Whale *(Balaenoptera musculus)* Sounds from the North Atlantic." *J. Acoust. Soc. Am. 114:* 1108–1119.

Mesnick, S. L. (2001). "Genetic Relatedness in Sperm Whales: Evidence and Cultural Implications." *Behav. Brain Sci. 24:* 346–347.

Miller, E. H. (1985). "Airborne Acoustic Communication in the Walrus *Odobenus rosmarus.*" *Nat. Geogr. Res. 1:* 124–145.

Miller, P. J. O., M. Johnson, and P. L. Tyack (2004). "Sperm Whale Behaviour Indicates the Use of Echolocation Click Buzzes 'Creaks' in Prey Capture." *Proc. R. Soc. Lond. B 271:* 2239–2247.

Miller, E. H., and A. V. Murray (1995). Structure, complexity, and organization of vocalizations in harp seal *(Phoca groenlandica)* pups. *In* "Sensory Systems of Aquatic Mammals" (R. A. Kastelein, J. A. Thomas, and P. E. Nachtigall, eds.), pp. 237–264. De Spil Publishers, Woerden, The Netherlands.

Mobley, J. R. J., and L. M. Herman (1985). "Transcience of Social Affiliations Among Humpback Whales *(Megaptera novaeangliae)* on the Hawaiian Wintering Grounds." *Can. J. Zool. 63:* 762–772.

Mohl, B. (1968). Hearing in seals. *In* "The Behavior and Physiology of Pinnipeds" (R. J. Harrison, R. C. Hubbard, R. S. Peterson, C. E. Rice, and R. J. Schusterman, eds.), pp. 172–195. Appleton-Century-Crofts, New York.

Mohl, B., M. Wahlberg, P. T. Madsen, A. Heerfordt, and A. Lund (2003). "The Monopulsed Nature of Sperm Whale Clicks." *J. Acoust. Soc. Am. 114:* 1143–1154.

Moore, S. E., and S. H. Ridgway (1995). "Whistles Produced by Common Dolphins from the Southern California Bight." *Aquat. Mamm. 21:* 55–63.

Moore, K. E., W. A. Watkins, and P. L. Tyack (1993). "Pattern Similarity in Shared Codas from Sperm Whales *(Physeter catodon)*." *Mar. Mamm. Sci. 9:* 1–9.

Moors, H. B., and J. M. Terhune. (2004). "Repetition Patterns in Weddell Seal *(Leptonychotes weddellii)* Underwater Multiple Element Calls." *J. Acoust. Soc. Am. 116:* 1261–1270.

Nachtigall, P. E., and P. W. B. Moore (eds.) (1988). *Animal Sonar: Processes and Performance.* Plenum Press, New York.

Norris, K. (1964). Some problems of echolocation in cetaceans. *In* "Marine Bio-Acoustics" (W. N. Tavolga, ed.), pp. 317–336. Pergamon Press, New York.

Norris, K. S., and G. W. Harvey. (1972). A theory for the function of the sperm whale *(Physeter catodon)*. *In* "Animal Orientation and Navigation" (S. R. Galler, K. Schmidt-Koenig, G. J. Jacobs, and R. Belleville, eds.), pp. 397–417. National Aeronautics and Space Administration. Washington, D. C.

Norris, K. S., and B. Mohl (1983). "Can Odontocetes Debilitate Prey with Sound?" *Am. Nat. 122:* 85–104.

Nowacek, D. P., B. M. Casper, R. S. Wells, S. M. Nowacek, and D. A. Mann (2003). "Intraspecific and Geographic Variation of West Indian Manatee *(Trichechus manatus* spp.*)* Vocalizations (L)." *J. Acoust. Soc. Am. 114:* 66–69.

Nowacek, S. M., R. S. Wells, E. C. G. Owen, T. R. Speakman, R. O. Flamm, and D. P. Nowacek. (2004). "Florida Manatees, *Trichechus manatus latirostris,* Respond to Approaching Vessels." *Biol. Conserv. 119:* 517–523.

Nummela, S. (1995). "Scaling of the Mammalian Middle Ear." *Hear. Res. 85:* 18–30.

Nummela, S., T. Reuter, S. Hemilä, P. Holmberg, and P. Paukku (1999a). "The Anatomy of the Killer Whale Middle Ear *(Orcinus orca)*." *Hear. Res. 133:* 61–70.

Nummela, S., J. G. M. Thewissen, S. Bajpai, S. T. Hussain, and K. Kumar (2004). "Eocene Evolution of Whale Hearing." *Nature 430:* 776–778.

Nummela, S., T. Wägar, S. Hemilä, and T. Reuter (1999b). "Scaling of the Cetacean Middle Ear." *Hear. Res. 133:* 71–81.

Oelschlager, H. A. (1986a). "Comparative Morphology and Evolution of the Otic Region in Toothed Whales (Cetacea, Mammalia)." *Am. J. Anat. 177:* 353–368.

Oelschlager, H. A. (1986b). "Tympanohyal Bone in Toothed Whales and the Formation of the Tympano-Periotic Complex (Mammalia: Cetacea)." *J. Morphol. 188:* 157–165.

Pack, A. A., L. M. Herman, and M. Hoffmann-Kuhnt (2004). Dolphin echolocation shape perception: from sound to object. *In* "Echolocation in Bats and Dolphins" (J. A. Thomas, C. F. Moss, and M. Vater, eds.), pp. 288–298. Chicago University Press, Chicago.

Pahl, B. C., J. M. Terhune, and H. R. Burton (1997). "Repertoire and Geographic Variation in Underwater Vocalisations of Weddell Seals *(Leptonychotes weddellii,* Pinnipedia: Phocidae*)* at the Vestfold Hill, Antarctica." *Aust. J. Zool. 45:* 171–187.

Payne, K., P. Tyack, and R. Payne (1983). Progressive changes in the songs of humpback whales *(Megaptera novaeangliae):* A detailed analysis of two seasons in Hawaii. *In* "Communication and Behavior of Whales" (R. Payne, ed.), pp. 9–57. Westview Press, Boulder, CO.

Payne, R. (ed.) (1983). *Communication and Behavior of Whales,* pp. 59–80. Westview Press, Boulder, CO.

Payne, R., and S. McVay (1971). "Songs of the Humpback Whales." *Science 173:* 587–597.

Pierce, G. W., and D. R. Griffin (1938). "Experimental Determination of Supersonic Notes Emitted by Bats." *J. Mammal. 19:* 454– 455.

Popper, A. N. (1980). Sound emission and detection by delphinids. *In* "Cetacean Behavior: Mechanisms and Functions" (L. M. Herman, ed.), pp. 1–49. Wiley, New York.

Ray, C., W. A. Watkins, and J. Burns (1969). "The Underwater Song of *Erignathus barbatus* (Bearded Seal)." *Zoologica 54:* 79–83.

Reiman, A. J., and J. M. Terhune (1993). "The Maximum Range of Vocal Communication in Air Between a Harbor Seal *(Phoca vitulina)* Pup and Its Mother." *Mar. Mamm. Sci. 9:* 182–189.

Rendell, L., and H. Whitehead (2001). "Culture in Whales and Dolphins." *Behav. Brain. Sci. 24:* 309–382.

Rendell, L. E., and H. Whitehead. (2003). "Vocal Clans in Sperm Whales *(Physeter macrocephalus)*." *Proc. R. Soc. Lond. B 270:* 225–231.

Rendell, L., and H. Whitehead (2004). "Do Sperm Whales Share Coda Vocalizations? Insights into Coda Useage from Acoustic Size Measurement." *Anim. Behav. 67:* 865–874.

Renouf, D. (ed.) (2001). *Behaviour of pinnipeds.* Chapman & Hall, New York.

Renouf, D., and M. B. Davis (1982). "Evidence That Seals May Use Echolocation." *Nature 300:* 635–637.

Repenning, C. A. (1972). "Adaptive Evolution of Sea Lions and Walruses." *Syst. Zool. 25:* 375–390.

Reynolds, J. E., and D. Odell (1991). *Manatees and Dugongs.* Facts on File, New York.

Reysenbach de Haan, F. W. (1956). "Hearing in Whales." *Acta Otolaryngol. Suppl. 134:* 1–114.

Rogers, T., D. Cato, and M. M. Bryden (1995). Underwater vocal repertoire of the leopard seal *(Hydrurga leptonyx)* in Prydz Bay, Antarctica. *In* "Sensory Systems of Aquatic Mammals" (R. A. Kastelein, J. A. Thomas, and P. E. Nachtigall, eds.), pp. 223–236. De Spil Publishers, Woerden, The Netherlands.

Rogers, T., D. Cato, and M. Bryden (1996). "Behavioral Significance of Underwater Vocalizations of Captive Leopard Seals, *Hydrurga leptonyx*." *Mar. Mamm. Sci. 12:* 414–427.

Roitblat, H. L. (2004). Object recognition by dolphins. *In* "Echolocation in Bats and Dolphins" (J. A. Thomas, C. F. Moss, and M. Vater, eds.), pp. 278–282. Chicago University Press., Chicago.

Roux, J. P., and P. Jouventin (1987). Behavioral cues to individual recognition in the subantarctic fur seal, *Arctocephalus tropicalis. In* "Status, Biology and Ecology of Fur Seals" (J. R. Croxall, and R. L. Gentry, eds.), pp. 95–102. NOAA Tech Rep., National Oceanic Atmospheric Administration, Washington, D. C.

Schevill, W. E., W. A. Watkins, and G. C. Ray (1966). "Analysis of Underwater *Odobenus* Calls with Remarks on the Development and Function of the Pharyngeal Pouches." *Zoologica 51:* 103–106.

Schusterman, R. J., D. Kastak, D. H. Levenson, C. Reichmuth, and B. L. Southall (2000). "Why Pinnipeds Don't Echolocate." *J. Acoust. Soc. Am. 107:* 2256–2264.

Sjare, B., I. Stirling, and C. Spencer (2003). "Structural Variation in the Songs of Atlantic Walruses Breeding in the Canadian High Arctic." *Aquat. Mamm. 29:* 297–318.

Sousa-Lima R. S., A. P. Paglia, and G. A. B. Da Fonesca (2002). "Signature Information and Individual Recognition in the Isolation Calls of Amazonian Manatees, *Trichechus inunguis* (Mammalia: Sirenia)." *Anim. Behav. 63:* 301–310.

Spoor, F., S. Bajpai, S. T. Hussain, K. Kumar, and J. G. M. Thewissen (2002). "Vestibular Evidence for the Evolution of Aquatic Behavior in Early Cetaceans." *Nature 417:* 163–165.

Steel, C. (1982). *Vocalization Patterns and Corresponding Behavior of the West Indian Manatee (Trichechus manatus).* Ph.D. dissertation thesis, Florida Institute of Technology, Melbourne, FL.

Stirling, I., and A. E. Derocher (1990). "Factors Affecting the Evolution and Behavioral Ecology of the Modern Bears." *Int. Cont. Bear Res. Manage. 8:* 189–204.

Stirling, I., W. Calvert, and C. Spencer (1987). "Evidence of Stereotyped Underwater Vocalizations of Male Atlantic Walruses *(Odobenus rosmarus rosmarus)*." *Can. J. Zool. 65:* 2311–2321.

Stirling, I., and J. A. Thomas (2003). "Relationships Between Underwater Vocalizations and Mating Systems in Phocid Seals." *Aquat. Mamm. 29:* 227–246.

Strager, H. (1995). "Pod Specific Call Repertoires and Compound Calls of Killer Whales, *Orcinus orca* Linnaeus, 1758 in the Waters of Northern Norway." *Can. J. Zool. 73:* 1037–1047.

Supin, A. Ya., V. V. Popov, and A. M. Mass (2001). *The Sensory Physiology of Aquatic Mammals.* Kluwer Academic Publishers, Boston.

Sverdup, H. U., M. W. Johnson, and R. H. Fleming (1970). *The Oceans: Their Physics, Chemistry, and Biology.* Prentice-Hall, Englewood Cliffs, NJ.

Terhune, J. M., and K. Ronald (1973). "Some Hooded Seal *(Cystophora cristata)* Sounds in March." *Can. J. Zool. 51:* 319–321.

Thewissen, J. G. M., and S. T. Hussain (1993). "Origin of Underwater Hearing in Whales." *Nature 361:* 444–445.

Thewissen, J. G. M., S. I. Madar, and S. T. Hussain (1996). "*Ambulocetus natans,* an Eocene Cetacean (Mammalia) from Pakistan." *CFS Cour. Forschungsinsr. Senckenberg 191:* 1–86.

Thode, A. M., D. K. Mellinger, S. Stienessen, A. Martiez, and K. Mullin (2002). "Depth-Dependent Acoustic Features of Diving Sperm Whales *(Physeter macrocephalus)* in the Gulf of Mexico." *J. Acoust. Soc. Am. 112:* 308–321.

Thomas, J., R.A. Kastelein, and A. Ya. Supin (1992). *Marine Mammal Sensory Systems,* Plenum Press, New York.

Thomas, J. A., and C. L. Golladay (1995). Geographic variation in leopard seal *(Hydrurga leptonyx)* underwater vocalizations. *In* "Sensory Systems of Aquatic Mammals" (R. A. Kastelein, J. A. Thomas, and P. E. Nachtigall, eds.), pp. 201–221. De Spil Publishers, Woerden, The Netherlands.

Thomas, J.A., and R. A. Kastelein (eds.) (1990). *Sensory Abilities of Cetaceans: Laboratory and Field Evidence.* Plenum Press, New York.

Thomas, J. A., and V. B. Kuechle (1982). "Quantitative Analysis of the Weddell Seal *(Leptonychotes weddellii)* Underwater Vocalizations at McMurdo Sound, Antarctica." *J. Acoust. Soc. Am. 72:* 1730–1738.

Thomsen, F., D. Franck, and J. K. B. Ford (2002). "On the Communicative Significance of Whistles in Wild Killer Whales *(Orcinus orca)*." *Naturwissenschaften 89:* 404–407.

Tyack, P. (1981). "Interactions Between Singing Hawaiian Humpback Whales and Conspecifics Nearby." *Behav. Ecol. Sociobiol. 8:* 105–116.

Tyack, P. (1991). Use of a telemetry device to identify which dolphin produces a sound. *In* "Dolphin Societies" (K. Pryor and K. S. Norris, eds.), pp. 319–344. University of California Press, Berkeley.

Tyack, P. (2001). "Cetacean Culture: Humans of the Sea." *Behav. Brain Sci. 24:* 358–359.

Tyack, P. L., and E. H. Miller (2002). Vocal anatomy, acoustic communication and echolocation. *In* "Marine Mammal Biology: An Evolutionary Approach" (A. R. Hoelzel, ed.), pp. 143–184. Blackwell Publ., Oxford, UK.

Van Parijs, S. M., P. J. Corkeron, J. Harvey, S. A. Hayes, D. K. Mellinger, P. A. Rouget, P. M. Thompson, M. Wahlberg, and K. M. Kovacs (2003b). "Patterns in the Vocalizations of Male Harbor Seals." *J. Acoust. Soc. Am. 113:* 3403–3410.

Van Parijs, S. M., K. M. Kovacs, and C. Lydersen (2001). "Spatial and Temporal Distribution of Vocalizing Male Bearded Seals—Implications for Male Mating Strategies." *Behaviour 138:* 905–922.

Van Parijs, S. M., C. Lydersen, and K. M. Kovacs (2003a). "Vocalizations and Movements Suggest Alternative Mating Tactics in Male Bearded Seals." *Anim. Behav. 65:* 273–283.

Van Parijs, S. M., C. Lydersen, and K. M. Kovacs (2004). "Effects of Ice Cover on the Behavioural Patterns of Aquatic-Mating Male Bearded Seals." *Anim. Behav. 68:* 89–96.

Verboom, W. C., and R. A. Kastelein (1995a). Rutting whistles of a male Pacific walrus *(Odobenus rosmarus divergens)*. *In* "Sensory Systems of Aquatic Mammals" (R. A. Kastelein, J. A. Thomas, and P. E. Nachtigall, eds.), pp. 287–298. De Spill Publishers, Woerden, The Netherlands.

Verboom, W. C., and R. A. Kastelein (1995b). Acoustic signals by harbour porpoises *(Phocoena phocoena)*. *In* "Harbour Porpoises, Laboratory Studies to Reduce Bycatch" (P. E. Nachtigall, J. Lien, W. W. Au, and A. J. Read, eds.), pp. 1–40. De Spill Publishers, Woerden, The Netherlands.

Verboom, W. C. and R. A. Kastelein (1997). Structure of harbour porpoises *(Phocoena phocoena)* click train signals. *In* "The Biology of the Harbour Porpoise" (A. J. Read, P. R. Wiepkema, and P. E. Nachtigall, eds.), pp. 343–363. De Spill Publishers, Woerden, The Netherlands.

Verboom, W. C., and R.A. Kastelein (2004). Structure of harbor porpoise *(Phocoena phocoena)* acoustic signals with high repetition rates. *In* "Echolocation in Bats and Dolphins" (J. A. Thomas, C. F. Moss, and M. Vater, eds.), pp. 40–42, Chicago University Press, Chicago.

Watkins, W. A., and W. E. Schevill (1977). "Spatial Distribution of *Physeter catodon* (Sperm Whales) Underwater." *Deep-Sea Res. 24:* 693–699.

Webster, D. B., R. R. Fay, and A. N. Popper (eds.) (1992). *The Evolutionary Biology of Hearing.* Springer-Verlag, New York.

Weilgart, L., and H. Whitehead (1993). "Coda Communication by Sperm Whales *(Physeter macrocephalus)* Off the Galapagos Islands." *Can. J. Zool. 71:* 744–752.

Weilgart, L., and H. Whitehead (1997). "Group-Specific Dialects and Geographical Variation in Coda Repertoire in South Pacific Sperm Whales." *Behav. Ecol. Sociobiol. 40:* 277–285.

Wemmer C., M. von Ebers, and K. Scaw (1976). "An Analysis of the Chuffing Vocalization of the Polar Bear." *J. Zool. Soc. London 180:* 425–439.

Whitehead, H. (2003). *Sperm Whales: Social Evolution in the Ocean*. University of Chicago Press, Chicago.

Whitehead, H., L. Rendell, R. W. Osborne, and B. Würsig (2004). "Culture and Conservation of Non-Humans with Reference to Whales and Dolphins: Review and New Directions." *Biol. Conserv. 120:* 427–437.

Wyss, A. R. (1987). "The Walrus Auditory Region and the Monophyly of Pinnipeds." *Am. Mus. Novit. 2871:* 1–31.

Zagaeski, M. (1987). "Some Observations on the Prey Stunning Hypothesis." *Mar. Mamm. Sci. 3:* 275–279.

12

Diet, Foraging Structures, and Strategies

12.1. Introduction

Marine mammals occupy the large end of the range of body sizes displayed by animals that live in the sea. One consequence of their large size is that most marine productivity (phytoplankton and small zooplankton) is unavailable to marine mammals simply because of the size disparity between predator and prey. Mysticetes and some pinnipeds (i.e., crabeater seals) have anatomical specializations that permit exploitation of smaller (one to several centimeters long) prey from lower trophic levels, but no marine mammals can harvest phytoplankton or the smallest zooplankton. Most pinnipeds and odontocetes prey on larger and less abundant animals several trophic levels removed from the primary producers. Only the sirenians graze directly on the plants, the primary trophic level, and these herbivorous animals are restricted to shallow coastal waters where larger rooted vegetation occurs. In this chapter, the diets of marine mammals are examined as are several of the more obvious structural and behavioral specializations employed by marine mammals to capture their prey. We also briefly examine the foraging energetics and the anatomy and physiology of digestion.

Research on the foraging activities of marine mammals is difficult, because feeding typically occurs below the sea surface. Standard approaches to analyzing foraging behavior, in addition to direct observations of predators pursuing and capturing prey, include stomach content analysis of dead, stranded, or net-entangled animals, stomach lavage of healthy restrainable animals, studies of fecal remains, and dietary patterns derived from stable isotope signatures in various tissues (see Hobson and Welch, 1992; Hobson and Sease, 1998). Serological methods have shown some potential to identify prey of marine mammals (Pierce *et al.*, 1993). Molecular genetic techniques also contribute to diet analyses. Mitochondrial and nuclear DNA markers have been used to assign individual, sex, and species identities of prey in pinniped fecal samples and have enabled assessments of dietary preferences (Reed *et al.*, 1997). Tracer fatty acids can provide very specific information about the diet of some marine animals (Hooper *et al.*, 1973) and fatty acid signature analyses may be useful in tracking major dietary shifts and likely also provide at least qualitative information about marine mammal diets (see Iverson, 1993; Smith *et al.*, 1997; Kirsch *et al.*, 2000). New technology such as the "crittercam," an animal-borne video system and data recorders that document marine mammal hunting behavior, permit detailed examination of the foraging behavior and success rates of individual animals (e.g., Davis *et al.*, 1999; Parrish *et al.*, 2000; Davis *et al.*, 2003).

12.2. Seasonal and Geographical Patterns of Prey Abundance

Marine mammal diets and foraging behaviors are affected by demographic factors including age, sex, and reproductive status, anatomical and physiological constraints, risk of predation, competitive interactions, and the distribution and abundance of potential prey. The latter factor is a direct consequence of both the spatial and temporal patterns of marine primary productivity (see Chapter 6). In general terms, primary productivity in marine systems is highest over continental shelves and other shallow banks, in regions of upwelling, and in areas that cool appreciably in winter months. Most tropical and subtropical areas of the open sea are characterized by very low rates of primary production and little seasonal variability. Only extremely small phytoplankton cell sizes occur at the primary trophic level in warm water regions, and several trophic levels exist between these phytoplankton and the sparse populations of animals sufficiently large enough to be prey for pinnipeds or odontocetes (see Figure 6.1).

At higher latitudes, small phytoplankton are also the dominant primary producers, but the high degree of seasonality they exhibit has resulted in forms of zooplankton and other animals whose life-history strategies are adapted to surviving periods without food. These small animals such as cananoid copepods and krill, store lipids for their own biological maintenance and hence represent energy-rich food sources for secondary consumers and thus are essential links to the higher trophic levels. Higher latitude marine systems generally have shorter food chains, exhibit high seasonal variability, and support aggregations of larger animals (see Figure 6.1). Marine mammals in these high latitude waters must have mechanisms to cope with the extreme seasonal variability in their food supplies, including the ability to fast for extended periods, exhibit migratory excursions to lower latitude waters, or exhibit some combination of these two patterns.

Most of the animal species commonly preyed on by marine mammals tend to concentrate near their own food supplies in the near surface photic zone. These include sharp density interfaces, such as the sea surface, the sea floor, and the thermocline near the bottom of the photic zone, areas of upwellings associated with open-ocean eddies and fronts between differing water masses or ocean current boundaries, or fronts of glaciers in Arctic or Antarctic regions. The presence of marine mammal populations is generally a good indicator of high productivity marine systems.

The feeding ecology of marine mammals is complex and our current knowledge is limited. Considerable information exists on the energetic benefits of selective prey consumption in whales (Gaskin, 1982; Lockyer, 1981a, 1981b), pinnipeds (e.g., Gentry *et al.*, 1986; Costa, 1993), sea otters (Riedman and Estes, 1990), and polar bears (Stirling, 1988). Although many species might be considered generalist opportunistic predators, they do exhibit selection in the food they consume when choices are available. Because of the potential economic implications of marine mammal feeding to fisheries, considerable research attention has been dedicated to bioenergetics studies of marine mammals and the construction of energy budgets for some marine mammal species (e.g., Lavigne *et al.*, 1982, 1985; Lockyer, 1981a, 1981b). These models are useful for identifying gaps in our knowledge regarding population demographics and essential information about foraging behavior as well as for highlighting essential aspects of energetic balance such as the ratios of various species in the diet, the nutritional content of prey of different age, sex, size, etc., and the digestibility of various types of prey the energetic cost of different sorts of foraging (discussed later in this chapter).

12.3. Adaptations for Foraging in Pinnipeds

12.3.1. Teeth

Pinnipeds have heterodont dentitions, similar to most other mammals; i.e., they have different types of teeth that are specialized for different tasks. Following the typical mammalian convention, the teeth of pinnipeds are named according to their type and position in the tooth-row. Premolars and molars are typically similar in size and shape and are often collectively called postcanines. Abbreviations used for each tooth type are as follows: I = incisors (the small gripping teeth at the front of the mouth), C = canines (the often powerful puncturing, holding teeth that occur in pairs), P = premolars, M = molars, PC = postcanines (the teeth toward the back of the mouth that do the shearing and crushing when such actions take place during feeding). Numbers affixed to these letters refer to specific teeth of each type; superscript numbers refer to teeth in the upper jaw (e.g., I^3 = upper third incisor), whereas subscript numbers refer to teeth in the lower jaw (e.g., PC_2 = lower second postcanine). Small letters indicate milk dentition. The number of teeth of each type (i.e., incisors, canines, premolars, and molars) is referred to as the **dental formula.** It is written as the number of each type of tooth in one side of the upper jaw over the corresponding number occurring in one side of the lower jaw. For humans, the dental formula is written as incisors 2/2, canines 1/1, premolars 3/3, and molars 2/2. Because the tooth types are always presented in the same order, this formula may be abbreviated as I 2/2, C1/1, P3/3, M2/2 × 2 = 32 (total). There is a reduction in the number of teeth in pinnipeds compared to most carnivores. The typical mammalian number is 44, whereas seals have between 22 and 38 teeth. In all pinnipeds, a double rooted condition of the postcanine teeth is the ancestral condition, apart from the first premolar that is single rooted, but King (1983) suggests that there is a trend in all lineages toward evolution of single rooted teeth.

   In pinnipeds the deciduous, or milk, teeth are very small and simple in shape. The permanent incisors and canines have deciduous precursors, which represent the typical mammalian condition, but only the postcanines P2-4 are preceded by milk teeth. The first permanent teeth to erupt are the upper and lower incisors. In phocids the milk teeth are usually resorbed before birth or shed from the gums shortly after birth. They are relatively small compared to those of otariids (King, 1983).

   The deciduous dentition of the walrus consists of three incisors, a canine, and three postcanines in each side of both upper and lower jaws, the first permanent postcanine not being preceded by a milk tooth. The formula for the deciduous dentition is therefore i3/3, c1/1, p3/3, m0/0 × 2 = 28. The milk teeth are shed shortly after birth but not all of them are replaced. Apart from the canines, the four teeth usually found in one half of the upper jaw of an adult walrus are the third incisor and three postcanines. In the lower jaw there are no permanent incisors and the four teeth are the molariform canine and three postcanines. A small tooth sometimes erupts behind the third milk postcanine. This usually belongs to the permanent dentition (a vestigial p4/4) but it is usually shed quite early, although it may persist into adult life. Because the teeth that erupt after the milk teeth do not all remain into adult life, the functional dentition is I1/0, C1/1, P3/3×2 = 18, although considerable variation may be observed between individuals (King, 1983).

The most noticeable aspect of the dentition of walruses is a pair of long tusks, present in both sexes but more slender in females. The tusks are modified upper canines, and their large size has resulted in changes to the shape of the anterior end of the skull (Figure 12.1). The permanent upper canines erupt at about 4 months of age. They have persistently open pulp cavities and continue to grow throughout the life of the animal, although the tips are abraded constantly during feeding such that there is a limit to the lengths they reach. The granular nature of the tooth material filling the pulp cavity (globular dentine) is structurally unique and characteristic of walrus ivory. The tusks of males are larger, thicker, and spaced farther apart at the base. A single tusk of an adult male may reach 1 m in length and weigh as much as 5.4 kg (King, 1983; Fay, 1982, 1985). Male walruses use their tusks mainly in dominance displays and females use them to defend themselves and their young. They are occasionally used by both sexes to pull themselves onto ice floes. Tusks are not usually used in feeding, although rarely they are used to stab seal prey. The abrasion to the tips, noted previously, is the result of the tusks being abraded by sediment when an animal is searching for benthic dwelling molluscs, primary prey of walruses.

Otariids have more teeth (34–38) than phocids (22–36), and both have more than the walrus (18–24). The permanent teeth of otariids are less diverse in morphology than are those of phocids. The typical otariid dental formula is I 3/2, C1/1, PC 5-6/5 with some interspecific differences and individual variation. Otariids have pigmented teeth, which is uncommon in mammals (some shrews also show similar dark pigmentation of their enamel).

Because there is little or no need for slicing or chewing in pinnipeds while processing food (prey are swallowed whole), the cheekteeth of both phocids and otariids are typically homodont with a single pointed cusp (Figure 12.2) for gripping slippery prey. Additional cusps can be present and are more frequent on the posterior cheekteeth. The cheekteeth of crabeater seals are highly modified with complex elongated cusps

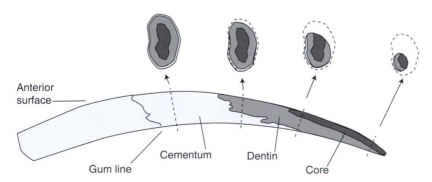

Figure 12.1. Lateral view of a mature male walrus tusk with several cross-sections indicating the extent of tusk abrasion. Dashed line = original tusk surface. (Redrawn from Fay, 1982.)

to trap and strain krill (see Figure 12.2). Ringed, harp, and leopard seals also show some complexity in terms of the number of cusps on the cheekteeth. Ringed and harp seals also eat some invertebrate prey, particularly when they are subadults. Both crabeater and leopard seals are thought to suck krill into their open mouths by retracting their tongues, then forcing excess water out through the sieving cheekteeth (Bonner, 1982). Leopard seals also possess well-developed canines for preying on penguins and other sea birds and occasionally on fur seal pups.

Among phocids, the phocines have incisors 3/2 (except for the hooded seal, which has 2/1) and the monachines have incisors 2/2 (except for the elephant seal, which has 2/1). In all other phocids the rest of the dental formula is C1/1, PC5/5. There is considerable variation in tooth shape and degree of cusping (see Figure 12.2).

12.3.2. Anatomy and Physiology of Digestion

Pinniped salivary glands are relatively small. They produce mucus to help in swallowing, but as the food is usually swallowed whole with no mastication, the saliva has no digestive enzymes (e.g., salivary amylase). The salivary glands are slightly better developed in otariids and the walrus than in phocids. The largest salivary gland is usually the submandibular gland. The sublingual gland is variably present and the parotid gland is very

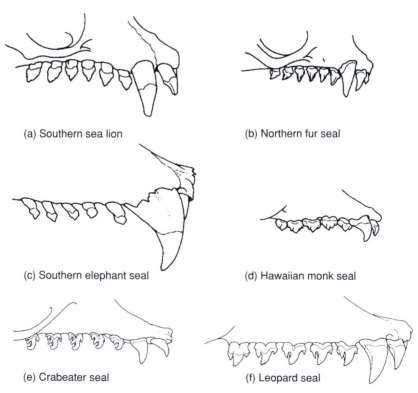

(a) Southern sea lion   (b) Northern fur seal

(c) Southern elephant seal   (d) Hawaiian monk seal

(e) Crabeater seal   (f) Leopard seal

Figure 12.2.   Representative otariid (**a, b**) and phocid (**c, d, e, f**) upper dentitions. (From King, 1983.)

small or may be absent. Palatine tonsils have been reported in embryonic Weddell seals and harbor seals but have not been observed in adult Weddell seals; a pharyngeal tonsil is present in Weddell seals (Eastman and Coalson, 1974).

The esophagus of pinnipeds frequently has deep longitudinal folds and is well supplied with mucous glands. The esophagus appears particularly muscular and dilatable in the Ross seal (Bryden and Erikson, 1976). In adult Weddell seals, the large size and extensive distribution of the esophageal glands indicates that they probably augment salivary gland secretions to facilitate the passage of food through the esophagus to the stomach (Eastman and Coalson, 1974).

The stomach of all pinnipeds is simple and similar in structure to terrestrial carnivores. A characteristic feature is that the pyloric region is sharply bent back on the body of the stomach. The inner mucosal layer is arranged with well-developed rugae. Gastric glands are the most widely distributed gland in the pinniped stomach.

In most species of pinnipeds there is no cecum (first part of the large intestine) and no clear division between small and large intestines. The first segment of the small intestine, the duodenum, is only indistinctly recognizable from the much longer segments of the small intestine, the jejunum, and the ileum. The small intestine of pinnipeds is between 8 (California sea lion) and 25 times (elephant seal) their body length (Helm, 1983). It has been suggested that intestine length is related to diet and prey digestibility. Those animals that have a diet composed of large quantities of hard to digest squid (e.g., elephant seals) have relatively long intestines, whereas those animals that eat more readily digested warm-blooded prey or fish (e.g., leopard seals) have much shorter intestines (Bryden, 1972). It also has been suggested that the great length of the pinniped small intestine may provide greater surface area for food and perhaps also water absorption. However, a strictly dietary explanation for the long intestine of pinnipeds now seems unlikely because some squid-eating pinnipeds such as the Ross seal do not have an unusually long intestine. A more likely explanation, according to Krockenberger and Bryden (1994), is that it is an adaptation to frequent, deep, and long-term diving. The long intestine increases the volume of the digestive tract, which serves as an extended storage compartment when the animal is submerged. Most absorption occurs during brief times when the animal is at the surface and the gut is perfused with blood. In another study, the different dietary habits of crabeater and Weddell seals were compared with respect to the anatomy of their gut and only subtle rather than major morphological differences were seen despite the major dietary differences between these two species (Schumacher *et al.*, 1995; see also Martenssen *et al.*, 1998).

Pinnipeds lack an appendix. The large intestine of pinnipeds is relatively short and only slightly larger in diameter than the small intestine. It has been suggested that the shorter large intestine of the elephant seal in comparison to that of the California sea lion and harbor seal may be related to their highly efficient ability to conserve water.

The liver is relatively large in pinnipeds compared to those of terrestrial carnivores. It is deeply divided into five to eight long, and weakly pointed, lobes that surround the hepatic sinus. A gallbladder is present in pinnipeds and serves to store and concentrate bile, which is involved in emulsifying fats during digestion. The pancreas is an elongated organ extending transversely across the dorsal abdominal wall from the duodenum to the spleen and posteriorly to the stomach. The relative size and growth of the pancreas in pinnipeds is thought to be related to diet to some extent. For example, in the southern elephant seal the relative growth of the pancreas changes during different growth phases associated with changes in diet (Bryden, 1971).

12.3.3. Diet and Feeding Strategies

Pinniped foraging activities include a complex suite of energetically costly behaviors, especially the additional costs of locomotion and the heat increment of digestion. Williams *et al.* (2004), using video-data loggers attached to free-ranging Weddell seals, found that costs of locomotion increased linearly with the number of swimming strokes taken during a foraging dive. Additionally, foraging dives required 45% more postdive oxygen recovery than did nonforaging dives of similar duration.

Fish and cephalopods are the most common prey eaten by pinnipeds. However, other invertebrates are important to several pinniped species. Euphausids (krill) are an important dietary component of three Antarctic species. Crabeater seals and leopard seals both eat significant quantities of krill (Laws, 1984), as do lactating Antarctic fur seals (Doidge and Croxall, 1985). In the Arctic, harp seal diets often contain significant quantities of crustaceans, particularly in pups and immature animals (Lawson *et al.*, 1994; Nilssen *et al.*, 1995). Walruses depend heavily on another invertebrate group, bivalve mollusks (Fay, 1982). Some walruses and other pinnipeds also occasionally eat birds and other pinnipeds (Riedman, 1990). Although leopard seals are perhaps the best known predator of other warm-blooded animals, other pinnipeds including the New Zealand, Antarctic, sub-Antarctic, and South African fur seals, as well as New Zealand, Australian, and southern sea lions and grey seals, also eat birds or other seals. The northern fur seal also sometimes eats birds, although they represent an insignificant part of the diet. This large, warm-blooded prey aside, pinnipeds typically consume animals that are small enough to be swallowed whole. Grey seals, harbor seals, ribbon seals and harp seals all target fish prey that fall within a 10–35 cm size range (see Bowen and Siniff, 1999 for a review), despite the large range in body size that is covered within this group. Even squid-eating elephant seals, the largest of the pinnipeds, appear to target relatively small prey (e.g., Rodhouse *et al.*, 1992; Slip, 1995). Adults of a given species do seem to target larger prey than immature animals (Frost and Lowry, 1986) and in species in which there is marked sexual dimorphism the diets of adult males and females may differ in size selectivity as well. Pinniped diets are characterized by a large number of species being consumed (e.g., Benoit and Bowen, 1990; Lowry *et al.*, 1990; Hjelseth *et al.*, 1999; Wallace and Lavigne, 1992), but in virtually all cases only a few species accounts for most of the energy ingested by pinnipeds in any one season at a particular geographic area.

There are both marked geographic and seasonal variations in the diet of many pinnipeds but these differences have not been explored in detail. Otariids living in temperate and tropical areas certainly experience considerable interannual variability in potentially available prey related to El Niño-Southern Oscillation (ENSO) events (discussed in Chapter 6). Diet also varies with age, with juvenile diets differing from those of adult animals. For example, adult harp seals typically feed on fish and some crustaceans whereas pups eat mainly zooplankton. Riedman (1990) suggested that one reason for this difference might be that juveniles require prey that is easier to capture. A difference in foraging strategies by young and adult female Steller sea lions has been implicated as a factor in the recent decline in their Alaskan populations (Merrick and Loughlin, 1997, and see Chapter 15). It was suggested that, because the young have more limited diving abilities than older animals, they are more affected by changes in prey distribution. Contrary to expectations, a summer study comparing maternal attendance and foraging patterns of Steller sea lions from stable and declining populations found no evidence that lactating females from a declining population had difficulty obtaining prey (Milette and Trites, 2003).

Pinnipeds forage both individually and cooperatively in groups. Resources most efficiently captured by an individual animal include nonschooling fish, slow moving or sessile invertebrates, or relatively small, warm-blooded prey. Many phocids, including elephant and harbor seals, are solitary foragers but many of these species target schooling pelagic fishes as well as more sessile species (Riedman, 1990). Cooperative foraging occurs most frequently among pinnipeds that exploit large, patchily distributed schools of fish or squid. Many fur seal and sea lions cooperate and forage in groups. Foraging strategies exhibit considerable plasticity depending on the type and distribution of the food resource. For example, Steller sea lions forage together in large groups when feeding on schooling fish or squid but feed singly or in small groups otherwise (Fiscus and Baines, 1966). Lactating northern fur seals from the same breeding sites also tend to forage together and avoid areas and hydrographic domains preferred by fur seals from different breeding sites (Robson et al., 2004). The foraging strategies of pinnipeds in relation to maternal care strategies are discussed in Chapter 13 and diving patterns are reviewed in Chapter 10. Although most pinniped species prey on small fish, they display specializations and sometimes quite distinct niche partitioning when there is potential for intraspecific competition. Ringed seal and harp seal studies in the Barents Sea suggested a high degree of spatial and prey species overlap between these species. However, detailed studies of prey size selection and depth of foraging showed distinct specializations that result in these two seal species exploiting different fractions of the same resource base (Wathne et al., 2000).

Walruses forage for clams, their typical prey, by swimming along the sea bottom in a head-down position, rooting with their snout, sensitive vibrissae, and flippers (Figure 12.3). Levermann et al. (2003) used underwater video and examinations of museum specimens to demonstrate a strong preference for using the right flipper to remove sediments around prey items. When a clam is located, it is either sucked into the mouth directly or is first excavated with jets of water squirted from the mouth. The soft tissue of the clam is sucked out of the shell by the tongue, which acts as a piston creating very low pressure in the mouth cavity. Pressure changes are caused by retraction of the tongue by the large styloglossus muscle and its depression by the large genioglossus muscles (see illustrations in Kastelein et al., 1991). Fay (1982) reported that the walrus tongue can create a vacuum with a pressure of −76 cm Hg in the mouth cavity. The tip of the walrus tongue has a well-developed tactile sense as evidenced by the presence of a large number of mechanoreceptors (i.e., lamellated corpuscles). Compared to many terrestrial mammals, the walrus has relatively few, but large, taste buds (Kastelein et al., 1997a).

Because the walrus consumes a large quantity of clams per day (up to 6000 individual prey per meal as reported by Fay, 1985) the feeding activities described previously have to be performed very fast and very often. Oliver et al. (1983) recorded a feeding rate of six clams per minute for the walrus. This would require 16–17 hours foraging time in order to capture enough prey for a full meal. In their foraging activities, walruses produce long furrows and pits on the sea bottom and can be important agents of disturbance in shallow soft-bottom Arctic benthic communities.

Based on the shape of the buccal cavity of bearded seals and their benthic feeding habits Burns (1981) and Adam and Berta (2002) suggested that this species might also use suction feeding. These suggestions have been confirmed in laboratory experiments that have demonstrated that bearded seals use considerable amounts of suction to extract prey

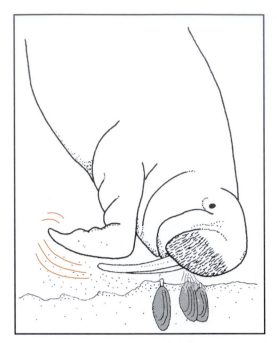

Figure 12.3.   Walrus excavating benthic mollusks using both flipper fanning and water jetting (*color*) to expose prey. (Modified from Werth, 2000a.)

from tight spaces and also employ hydraulic jetting to loosen prey that do not respond readily to suction forces (Marshall, Lydersen, and Kovacs, unpublished data).

12.3.4. Foraging Movements

Seasonal migratory patterns are known for several pinniped species. Harp seals, hooded seals, both northern and southern elephant seals, and possibly also Weddell seals change their locations in predictable, seasonal patterns that are likely at least in part linked to availability of prey. Inshore and offshore migratory patterns are seen in other species such as the spotted seals. Even quite sedentary species such as grey and harbor seals display quite marked patterns of seasonal movements even though they remain in roughly the same geographic areas. These movements are likely linked to local seasonal changes in prey abundance.

One of the best studied foraging patterns among pinnipeds is that of the elephant seal. Adults of both species of elephant seals stay at sea for 8–9 months each year to forage. During this time they make two round-trip migrations between near-shore island breeding rookeries and offshore foraging regions. With the development of microprocessor-based geographic location dive time and depth recorders (GLTDRs; also see Chapter 10 and section 14.3.2 of Chapter 14), details of the foraging behavior of these deep-diving seals have been documented. By attaching GLTDRs to adults as they leave their island rookeries, dive behaviors of individual animals can be recorded continuously for several months until they return to their rookeries where their GLTDRs and the data they contain can be recovered (Le Boeuf *et al.*, 1989; DeLong and Stewart, 1991; Stewart and

DeLong, 1993, 1994, 1995). The following discussion focuses on northern elephant seals, although most generalizations apply to southern elephant seals as well. For a summary of the movements of southern elephant seals monitored by satellite telemetry, see McConnell and Fedak (1996).

Although many vertebrates migrate long distances between breeding seasons, elephant seals are the first reported to make a double migration each year. Their individual annual movements of 18,000–21,000 km rival gray and humpback whales in terms of the greatest distance traveled. When ashore for breeding or for molting, adult elephant seals fast (see Chapter 13), and each fasting period is followed by a prolonged period of foraging in offshore waters of the North Pacific Ocean. The postbreeding migration begins in February or early March, when seals return to sea after spending 1–3 months ashore fasting during the breeding season. This migration averages 2.5 months for females and 4 months for males (Stewart and DeLong, 1993). After returning to their breeding rookery for a short molting period of 3–4 weeks, they again migrate north and west into deeper water. During this postmolt migration, females forage for about 7 months, whereas males feed at sea for about 4 months (Figure 12.4). Both males and females then return again to their rookeries for breeding. Individuals generally are thought to return to the same foraging areas during postbreeding and postmolt migrations.

While at sea, both males and females dive almost continuously, remaining submerged for about 90% of the total time. Dives for both sexes average 23 minutes, with

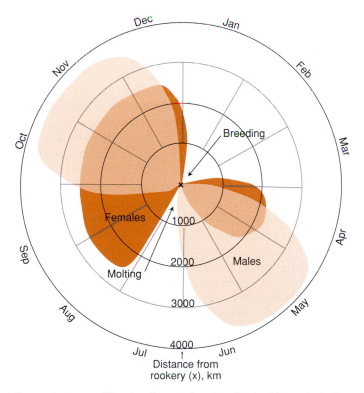

Figure 12.4. Seasonal patterns of foraging distances from breeding/molting rookeries for male and female northern elephant seals. (From Stewart and DeLong, 1993.)

males having slightly longer maximum dive durations. These dive durations probably are near the aerobic dive limit for this species (see Chapter 10; DeLong and Stewart, 1991). Elephant seals of both sexes spend about 35% of their dive time near maximum depth, between 200 and 800 m. In general, daytime dives are 100–200 m deeper than nighttime dives.

Antonelis *et al.* (1994) used stomach lavage and identification of hard parts of prey to show that the principal prey of elephant seals are mesopelagic squid and fish that spend daylight hours below 400 m and ascend nearer the surface at night. This upward night-time movement reflects the general daily behavior of many common deep scattering layer species and is likely the basis for the marked difference in maximum depths between day and night dives of elephant seals. The preferred squid prey of elephant seals are slow-swimming, neutrally buoyant animals of relatively low nutritional quality. Stewart and DeLong (1993) suggest that, in selecting neutrally buoyant squid of lower nutritional quality instead of more muscular, faster-swimming, and more nutritious squid, elephant seals are trading nutritional content of their prey for ease of capture and shortened pursuit times.

Foraging elephant seals are widely dispersed over the northeast Pacific Ocean and exhibit strong gender differences for preferred foraging locations. Females tend to remain south of 50° N. latitude, particularly in the subarctic Frontal Zone (Figure 12.5). Adult males travel farther north, through the feeding areas of females, to subarctic waters of the Gulf of Alaska and the offshore boundary of the Alaska Current that flows along the south side of the Aleutian Islands. These areas of aggregation are regions of high primary productivity that contain abundant fish and squid.

The geographical segregation of foraging areas by male and female northern elephant seals may reflect the preferences of males for larger and more oil-rich species of squid found in higher latitude sub-arctic waters, even though they must travel farther to get to these high latitude foraging grounds (Stewart and DeLong, 1993). The considerably smaller females might have different energetic requirements that encourage them to undertake shorter migratory distances at the expense of access to energy-rich prey at higher latitudes. Sperm whales also exhibit patterns of latitudinal segregation by gender in North Pacific foraging areas. Their prey is also primarily mesopelagic squid, and their seasonal migrations and diving patterns may also reflect the geographic and vertical distribution of their preferred squid prey (Jaquet and Whitehead, 1999).

In a further study investigating the cause of the different migrations of the sexes in northern elephant seals, Stewart (1997) reported that the sex-specific patterns appear to develop during puberty, when growth rates of males are substantially greater than those of females. These patterns are well established by the time males are 4.5–5 years old. Stewart's (1997) results suggest that the development of sexual segregation results from different metabolic needs of males during the period of sexual maturation and accelerated growth rather than to sexual differences in the energy requirements of adults. High-latitude segregation apparently confers survival benefits to pubescent males that more than compensate for the higher mortality generally associated with rapid growth during puberty.

12.3.5. Evolution of Feeding Strategies in Pinnipeds

Most otariids and phocids employ a generalist or pierce feeding strategy and feed on a variety of fish and squid. A few invertebrate specialists (e.g., crabeater seals) use a form

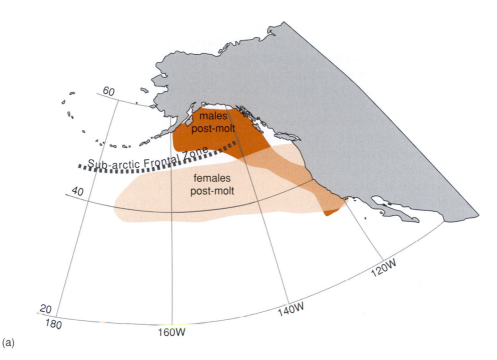

(a)

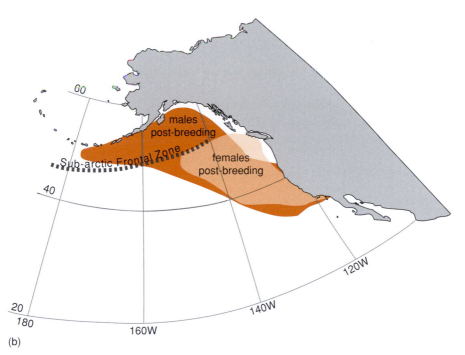

(b)

Figure 12.5. Geographical distribution of male and female northern elephant seals during **(a)** postbreeding and **(b)** postmolt foraging migrations. (From Stewart and DeLong, 1993.)

of filter feeding. Walruses, southern sea lions, and bearded seals employ a specialized suction feeding strategy using their powerful tongues and suction forces created in their small mouths to suck meat out of the shells of mollusks. A study of the structure of the palate, jaws, and teeth of pinnipeds indicates a number of features that can be used to distinguish pierce feeding from specialist feeding strategies (e.g., palate shape and length, tooth reduction; Adam and Berta, 2002). In an effort to understand the evolution of these feeding patterns in pinnipeds, morphologic evidence for feeding pattern was mapped onto the phylogenetic framework for pinnipeds (Figure 12.6). A generalist (pierce) strategy, presumed to be an adaptation for a piscivorous diet, appears to be the ancestral feeding condition in archaic pinnipeds, otariids, phocids, and archaic walruses (i.e., *Neotherium* and *Imagotaria*). A specialized suction feeding strategy evolved secondarily in both dusignathine and odobenine walruses, and an otariine seal and the bearded seal. The shift in feeding strategy among walruses took place at least 11 Ma.

12.4. Feeding Specializations of Cetaceans

Cetaceans prey on an extensive range of food particle sizes, from millimeter-sized copepods to large squids and other cetaceans exceeding several meters in length (Figure 12.7). Odontocetes tend to specialize on prey at the larger end of this range of sizes and mysticetes on the smaller end, although there is considerable overlap in the middle. Due to

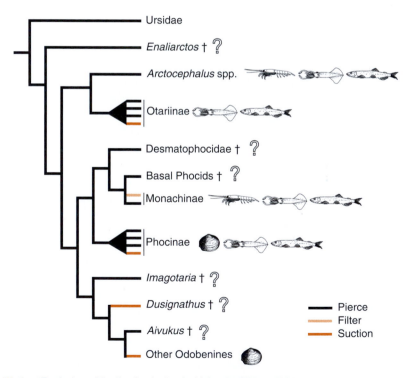

Figure 12.6.   Evolution of feeding in pinnipeds. (Adam and Berta, 2002.)

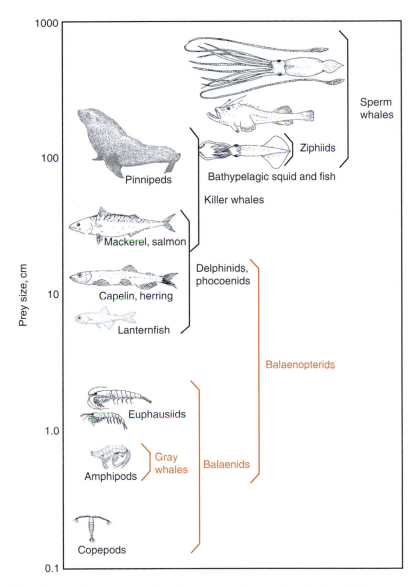

Figure 12.7.   Range of preferred prey sizes for cetaceans. (Redrawn from Gaskin, 1982.)

the fundamental differences in feeding specializations of odontocetes and mysticetes, they are examined separately here.

12.4.1. Mysticetes

Mysticetes feed primarily on planktonic or micronectonic crustaceans (e.g., krill) or small pelagic fish found at relatively shallow depths. Mysticete feeding dives are not known to exceed 500 m (Panigada *et al.*, 1999). Mysticetes are characterized by a complete absence of teeth (although teeth may be found in fetal baleen whales). The feeding apparatus consists of two rows or racks of **baleen** plates (called **whalebone** by whalers) that project ventrally from the outer edges of the roof of the mouth. The base of each plate is deeply embedded into the epidermal pad of the palate. The plates grow continuously from the base but are also continually eroded at the edges by movements of the tongue and by abrasion of prey items. The baleen plates grow from a foundation layer of cells that sends small projections (papillae) into the underlying layer. The papillae are covered by sheaths of keratin-filled cells (the same material that makes up hair, claws and fingernails) that form a single fringe on the border of the baleen plate. An intermediate layer of tissue produces a sheet of keratin on each side of the row of fringes, which serve to bind them into plates (Figure 12.8). The sheets thicken into a solid mass, anchoring the adjacent plates to the soft tissue of the palate. As the outer fringes are gradually worn away through use, the inner fringes become frayed and tangled. Collectively these tangled fibrous strands overlap each other to form an extended filtering surface on the inner sides of each of row of baleen plates. The density of the resulting baleen fringe (number

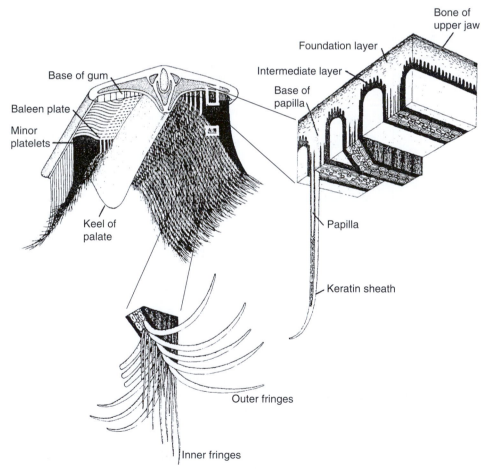

Figure 12.8.   Generalized structure of baleen. (Redrawn from Pivorunas, 1979.)

of baleen fibers/cm^2) varies at least tenfold among species. The number and length of individual baleen plates on each side of the mouth ranges from about 155 (the longest 0.4–0.5 m long) in gray whales (Figure 12.9; Sumich, 2001) to more than 350 in right whales (some exceeding 3 m in length). The flexible, light, yet strong characteristics of baleen plates of most mysticete species made them commercially valuable in the 19th century, utilized in many of the roles that plastics do today.

Mysticetes have evolved three different strategies for feeding efficiently on small prey, skimming, engulfment feeding (including lunging), and suction feeding (Figure 12.10). These strategies are reflected in characters of the baleen (see Kawamura, 1980), mouth, tongue, and the method by which the mouth is expanded (Heyning and Mead, 1996; Werth, 2000a). Right and bowhead whales (Balaenidae) and pygmy right whales (Neobalaenidae) are slow swimmers, with a highly arched rostrum and very long and finely textured baleen that is housed within a tall, blunt mouth. These whales feed at the surface and at depth by skimming the surface of the water collecting copepod prey on the baleen as water flows into the mouth through a frontal gape between the baleen plate rows. Water is subsequently expelled by the muscular tongue (Figure 12.10a). Anatomical observations suggest that during feeding rather than trap prey in the baleen bristles the tongue may control the flow of water in the mouth to sweep prey posteriorly where prey compaction occurs between the base of the tongue and the oropharyngeal wall (Lambertsen *et al.*, 2005). In contrast to bowheads and right whales, rorquals (Balaenopteridae) are fast, streamlined swimmers with shorter and coarser baleen that employ engulfment feeding, a method of prey capture, in which large volumes of water and prey are taken in. The mouths of all rorqual species are enormous, occupying most of the anterior half of the body (Figure 12.11). All members of this family have 70–80 external grooves (furrows) in the ventral wall of the mouth and throat, collectively referred to as **throat grooves** (Figure 12.12). During feeding the grooved mouth floor is opened and the pleats extend to increase vastly the volume that can be contained within the mouth; in blue and fin whales the volume of water that can be contained is equivalent

Figure 12.9. Baleen of a captive rehabilitating gray whale. (Courtesy of R. Botten.)

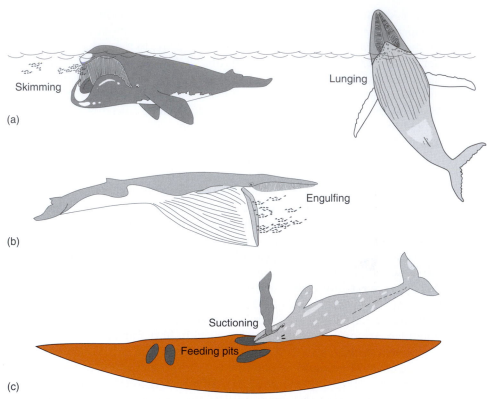

Figure 12.10.   Mysticete feeding types. **(a)** Skim feeding by a right whale. **(b)** Engulfment feeding by a fin whale and lunge feeding by a humpback whale. **(c)** Suction feeding by a gray whale. (Adapted from Pivorunas, 1979; Werth, 2000b.)

to 70% of the animal's body weight. A mature blue whale can engulf as much as 70 tons of water at one time (Pivorunas, 1979). Alternating longitudinal strips of muscle and blubber, both with large amounts of the protein elastin, facilitate the extension that takes place in the pleats of the grooves (Pivorunas, 1979; Orton and Brodie, 1987). During extension, capillary networks within the tissue give the throat a reddish color. It is this feature that has conferred the name "rorqual" or red throat to the balaenopterids.

Water and the small prey it contains enter the open mouth by negative pressure that is produced by the backward and downward movement of the tongue and by the forward swimming motion of the feeding animal. After engulfing shoals of euphausiids, sand lances, capelin, or other small prey in this manner, the lower jaw is slowly closed around the mass of water and prey. Then the muscular tongue acts in concert with contraction of the ventral wall muscles of the mouth (and sometimes with vertical surfacing behavior) to force the water out through the baleen and to assist in swallowing trapped prey (Kawamura, 1980). Prey is compacted against the inside of the baleen racks differing from the process in balaenids (Lambertsen *et al.*, 2005). During feeding the elastic tongue is capable of invaginating to form a hollow, sac-like structure, the **cavum ventrale,** which lines the ventral pouch of the body. In this way the everted, elastic tongue further enlarges the capacity of the mouth (Lambersten, 1983). Examination of the morphology of the grooved ventral pouch reveals that, in addition to fat tissue, thick layers of elastic

(a)

(b)

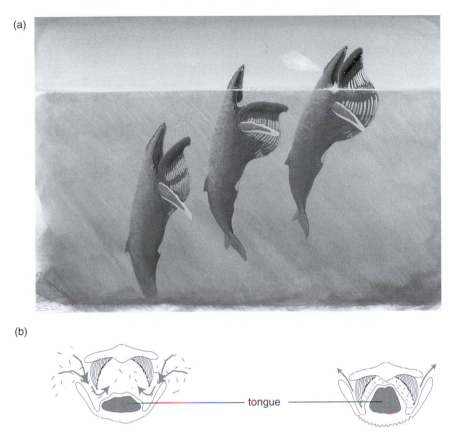

tongue

Figure 12.11.   **(a)** Lunge-feeding sequence of a humpback whale. (Illustrated by P. Folkens.) **(b)** Cross-sectional views of mouth, illustrating the action of the tongue and pattern of water flow within the baleen basket. (Redrawn from Pivorunas, 1979.)

connective tissue and muscle with layered corpuscles are found closely associated with the grooves. It has been suggested that these corpuscles may have a sensory function that helps let the whale control the appropriate timing of mouth closure during feeding (Bakker *et al.*, 1996).

A functionally important component of the rorqual engulfment feeding mechanism is the well-developed coronoid process of the lower jaw (see Figure 8.18). It serves principally as the site of insertion of the large temporalis muscle. By comparison, balaenid, neobalaenid, and eschrichtiid species, which are not specialized for engulfment feeding, lack a distinctive coronoid process. A tendinous part of the temporalis muscle, the **frontomandibular stay,** provides a continuous, mechanical linkage between the skull and lower jaw and serves to optimize the gape of the mouth opening during engulfment feeding. In addition, the frontomandibular stay may act to amplify the mechanical advantage of the temporalis during mouth closure (Figure 12.13; Lambertsen *et al.*, 1995). Another important, novel anatomical structure involved in balaenopterid feeding is the maxillo-mandibular cam articulation. This structure, formed of the suborbital plate of the maxilla, provides a thin shelf of bone at the back of the mouth for the tendinous origin of the superficial masseter muscle which, when contracted, creates a cam (the

Figure 12.12.   Lateral view of expanded throat pleats of a feeding blue whale. (Courtesy Richmond Productions, Inc.)

maxilla) and follower (the mandible) mechanism against which the tongue works in feeding (Lambertsen and Hintz, 2001, 2004).

Humpback whales display a wide variety of engulfment feeding strategies. Bottom feeding has been observed for this species at several locations off the North American Atlantic coast (Swingle *et al.*, 1993; Hain *et al.*, 1995). Freshly abraded areas on the right side of the rostrum and tips of the flukes also suggest that at least some individual humpbacks feed on or near the sea floor. When feeding on herring and other small fishes, humpback whales often produce a large (about 10 m diameter) cloud of ascending bubbles of expired air, a behavior referred to as **bubble-cloud feeding** (Figure 12.14). This net of bubbles is produced by one member of a group of several cooperating whales that

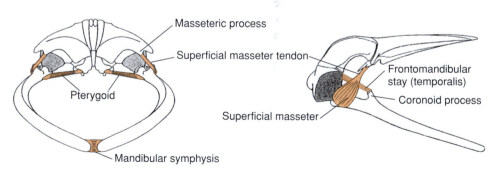

Figure 12.13.   Frontomandibular stay of balaenopterids. (From Lambertsen *et al.*, 1995, courtesy of R. Lambertsen.)

Figure 12.14. Surface view of bubble curtain produced by humpback whales foraging on herring in southeastern Alaska waters. (Courtesy of F. Sharpe.)

form long-term foraging associations. These bubble clouds seem to confuse prey and cause them to clump into tight food balls, which allows for more efficient feeding by the whales (Weinrich *et al.*, 1992). As one whale produces the bubble curtain, other members of the group dive below the target herring school and force it into the curtain, then lunge through the confused school from below (see Figure 12.14). Low-frequency calls are sometimes produced as the whales lunge through the school of prey. These calls and may aid in orienting whales during this complex maneuver (A. Sharpe and L. M. Dill, personal communication).

Observations of humpbacks feeding in New England waters shows that bubble-cloud feeding was followed by a tail slap, or lobtail, referred to as **lobtail feeding** (Weinrich *et al.*, 1992). It is hypothesized that this behavior is initiated if the whales switch from feeding on herring to feeding on the smaller sand lance, which may react to a lobtail by clumping more densely, making the bubble cloud more effective. The release of bubbles during foraging activities has also been reported for spotted dolphins and killer, gray, fin, and Bryde's whales (Sharpe and Dill, 1997, and references cited therein). A series of laboratory experiments conducted to determine the effects of bubbles on schooling prey concluded that herring have a strong aversion to bubbles and can readily be manipulated or contained within bubble nets by predators. Some further experimental evidence suggests that herring may be responding to acoustic and visual aspects of the bubbles (Sharpe and Dill, 1997).

Differences in rorqual feeding patterns are best observed in the northern hemisphere, where competition has encouraged partitioning of available food resources (Nemoto, 1959). In high latitudes of the southern Hemisphere, all species of large baleen whales exploit a single abundant krill species, *Euphausia superba* (see Figure 6.3), of the Antarctic upwelling region. In both hemispheres, extreme seasonal variations in the abundance of high-latitude prey forces large mysticetes to meet their total annual food requirements in just 4–6 months of intensive feeding, then abandon their feeding areas during the dark, cold months of winter when prey are scarce or not available at all in the near surface layers. The resulting annual migrations are discussed later in this chapter.

Gray whales (Eschrichtiidae) have shorter and coarser baleen and fewer throat grooves than rorquals and employ suction feeding in which the animal rolls to one side and uses the tongue to draw water and sediment into the mouth (Figure 12.15a), sieving the sediment through the baleen while retaining prey. Gray whales forage principally on bottom invertebrates, especially small amphipod crustaceans, in their shallow summer feeding grounds of the Bering and Arctic Seas (Nerini, 1984). Gray whales possess 2–5 throat grooves that may function to expand the floor of the mouth somewhat during feeding. Direct observations on the feeding behavior of gray whales has demonstrated that after several minutes working bottom sediments, feeding gray whales surface to replenish their blood oxygen stores while flushing residual mud from their mouths (Figure 12.15b). Gray whales leave behind mouth-sized depressions in the bottom sediments following their feeding activities. Similar feeding behaviors are employed by some gray whales that do not make the complete northward migration; rather, they forage on a wide variety of benthic invertebrates and shoaling mysid crustaceans in shallow coastal waters throughout the gray whale's migratory range south of the Aleutian Islands (Darling, 1984; Nerini, 1984; Darling *et al.*, 1998; Dunham and Duffus, 2002).

12.4.2. Odontocetes

Fish, squid, large crustaceans, birds, and, more rarely, other marine mammals are the principal prey of modern odontocetes. The shape of the mouth and jaw, and the number, size, and shape of teeth vary considerably between odontocete families, and most exhibit extensive modifications in tooth numbers and shapes typical of placental mammals. Most often the teeth of odontocetes are simple and peg-shaped with single, open roots with all of the teeth in the mouth being similar. This condition is referred to as **homodont dentition.** There is no milk or deciduous dentition. In odontocetes most of the tooth consists of dentin covered by a thin layer of cementum on the tooth root and a very thin layer

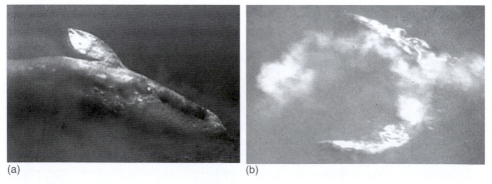

(a)   (b)

Figure 12.15.   Feeding behavior of gray whales. **(a)** Foraging on the sea floor (from video). **(b)** Flushing collected sediments from food at the surface.

of enamel on the crown. In many toothed whale species there is considerable individual variation in tooth number. For example, in the sperm whale, pilot whale, and common dolphin the total number of teeth can range from a 1.5- to 4.5-fold difference among individuals. Other odontocete species (e.g., beaked whales) possess a relatively constant number of teeth. The teeth are often only loosely socketed in irregularly shaped alveoli.

In many odontocetes, because of the nearly homodont condition of the teeth (Figure 12.16), it is impossible to make distinctions between tooth types (e.g., premolars vs molars). Thus, the total tooth number for each side of the jaw is simply combined as a single number, without indication of tooth type (e.g., 15/18 indicates 15 teeth per side in the upper jaw and 18 per side in the mandible). In some odontocetes (i.e., beaked whales, pygmy and true sperm whales, and Dall's porpoises) there is drastic reduction in the number of teeth in the jaws (see Figure 12.16); for some there are few or no teeth in the upper jaw but numerous teeth in the lower jaw (e.g., 0/25 for the sperm whale and 0/8–16 for the pygmy sperm whale).

Among beaked whales, the strap-toothed whale *(Mesoplodon layardii)* with its very reduced dentition (only 1 tooth is present in each side of the lower jaw) and small mouth has raised questions about how adult males feed. Their erupted mandibular teeth curve medially above the rostrum, thereby limiting the extent to which they can open their mouths. It has been suggested that the erupted teeth may function to guide prey in the mouth (Leatherwood and Reeves, 1983). However, because females and immature males

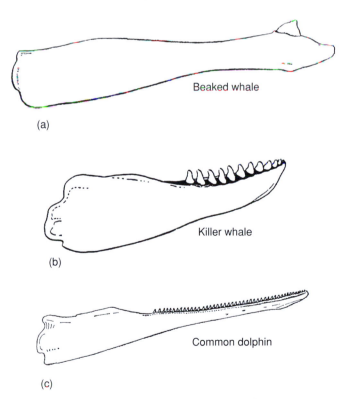

(a)

(b)

(c)

Figure 12.16.   Representative lower dentitions of odontocetes. **(a)** Beaked whale. **(b)** Killer whale. **(c)** Common dolphin. (Redrawn from Slijper, 1979.)

lack these teeth, Heyning (1984) suggested that it was more likely these teeth function in intraspecific fighting.

Finally, there is a large group of cetaceans with an equal number of teeth in the upper and lower jaws (e.g., tooth numbers in the Phocoenidae vary from 15/15 to 30/30). Among members of the Delphinidae (see Figure 12.16) the number varies from 0/2 in the short-snouted Risso's dolphin to 65/58 in the long-snouted spotted dolphin. Tooth morphology also differs between the dolphins and the porpoises. Phocoenids have spatulate teeth and delphinids have conical teeth.

Much attention has been given to the presence of the remarkable tusk in the narwhal, which may exceed a length of 3 m (Tomilin, 1957). The tusk extends anteriorly from the head, and is likely responsible for the myth of the unicorn (see Figure 4.35). As a general rule, the left canine erupts as a tusk only in males. The outer surface of the tusk is marked by a series of prominent spiral ridges that may number five or six 360-degree turns on a large specimen. The spiral of the tusk is always left-handed when viewed from the root. Male narwhals occasionally develop two tusks or may lack an erupted tusk, and females occasionally develop a left tusk or both tusks. When present, the tusks of females are shorter and less robust than those of males.

The large disparity in size between male and female narwhals and the later attainment of sexual maturity in males strongly suggest that this species is polygynous, with males competing for mating opportunities. Hence, it is likely that their tusks are used in sexual displays between males. It has been suggested that the tusk is used for nonviolent assessment of hierarchical status on the basis of relative tusk size during frontal encounters (Best, 1981), but evidence from head scarring and broken tusks indicates that adult males also indulge in aggressive displays that involve violent fighting using the tusk (see Figure 13.21). Ford and Ford (1986) describe the only known instance of a narwhal being speared in the abdomen by the tusk of another whale. The injured animal subsequently died, almost certainly as a result of this massive injury. The attacked animal was not a male, as one might expect, but a female that had recently given birth (Hay and Mansfield, 1989).

12.4.3. Anatomy and Physiology of Digestion

Cetacean digestive systems, like those of pinnipeds, are striking in their extreme length. This may be related to the increased metabolic demands of a larger body size or the need to have efficient water handling in the body and may not be directly related to dietary demands. The esophagus of cetaceans consists of a long, thick-walled tube that is lubricated by mucoserous glands (Tarpley, 1985). Its length depends on the size of the animal, constituting approximately one quarter of the total body length in toothed whales (Yablokov et al., 1972). The importance of the esophagus in prey compaction for balaenids is discussed by Lambertsen et al. (2005).

All cetaceans are characterized by a complex stomach with multiple divisions that resemble those of ruminants (e.g., cattle, goats, and sheep). The four main compartments are (1) a **forestomach,** which is devoid of glands and lined by keratinized stratified squamous epithelium; (2) a **fundic chamber** or **main stomach** with a folded mucosa and gastric glands; (3) a **connecting stomach,** between the main stomach and pyloric stomach; and (4) a relatively smooth **pyloric stomach,** which may be bent on itself or show several dilations and which has some glands in its walls (Figure 12.17). The organization of the cetacean stomach differs from that of ruminants. In whales, only the initial

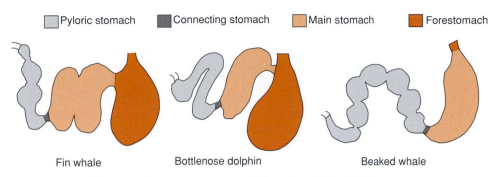

Pyloric stomach   Connecting stomach   Main stomach   Forestomach

Fin whale   Bottlenose dolphin   Beaked whale

Figure 12.17.   Representative cetacean stomachs. (Redrawn from Slijper, 1979.)

chamber is nonglandular (vs the first three chambers of the ruminant stomach) (Olsen *et al.*, 1994).

The cetacean forestomach is homologous with the rumen, reticulum, and omasum of ruminants and is where digestion is initiated. The forestomach of baleen whales often contains stones and pebbles of various sizes, which, with the powerful contractions of the muscular walls, help to grind fish bones and crustaceans. Additionally, high concentrations of volatile fatty acids and anaerobic bacteria have been observed in the forestomach of baleen whales as well as small toothed whales, indicating that microbial fermentation occurs here as it does in the rumen (Olsen *et al.*, 1994). The forestomach is usually the largest chamber in odontocetes but it is lacking in beaked whales and the franciscana (see references cited in Langer, 1996). The forestomach of dolphins, with volumes of several liters, has a muscular sphincter between the esophagus and the forestomach (Harrison *et al.*, 1970). Harrison *et al.* (1970) observed dolphins ejecting water from the mouth in volumes large enough to suggest that some of it came from the forestomach. One function of the dolphin forestomach may be to hold swallowed water while they secure prey against the palate and then expel the water past the secured prey as was suggested by Heyning and Mead (1996). The fundic chamber is equivalent to the fundic portion of the simple mammalian stomach or to the abomasum of ruminants. Its gastric glands contain parietal and chief cells, which produce a gastric juice containing pepsin, hydrochloric acid, and some lipase. The forestomach and main stomach of a large blue whale can hold up to 1000 kg of krill. The connecting stomach acts as a third chamber, separated from the main and pyloric chamber by sphincters. It is lined with a glandular mucus-secreting mucosa that is thought to protect the epithelium from mechanical damage and to facilitate transport of digested food through the channel.

The pyloric chamber is equivalent to the pyloric region of the abomasum of ruminants. It is lined by columnar mucous cells. Gastric digestion continues in this chamber and much of the solid food is broken down before it passes through the connecting channel. A narrow pylorus leads into the duodenum, the first part of which is sometimes markedly dilated and called the duodenal ampulla. This allows further digestion of food before it reaches the duodenum proper, indicating that enzymatic digestion may have a greater significance in cetaceans compared to ruminants (Olsen *et al.*, 1994). It has been suggested that differences in the stomachs of odontocetes (i.e., number of forechambers, number of main chambers, shape of connecting channel, and number and shape of pyloric chambers) may be phylogenetically informative (Rice and Wolman, 1990; Langer, 1996).

The basic difference between the intestine of baleen and toothed whales is that there is a cecum in baleen whales and none (with the exception of the Ganges River dolphin) in toothed whales (Yablokov *et al.*, 1972; Takahashi and Yamasaki, 1972). High concentrations of volatile fatty acids and anaerobic bacteria in the colon of fin and minke whales indicate the occurrence of bacterial fermentation in the colon of these animals (Herwing and Staley, 1986). Because they possess relatively shorter small intestines compared to those of pinnipeds, which have a similar diet, the multichambered stomach of minke whales is essential for optimal utilization of prey (Olsen *et al.*, 1994). According to Cowan and Smith (1995), cetaceans lack an appendix. However, a complex of lymphoepithelial organs called "anal tonsils," is a consistent structure of the anal canal of the bottlenose dolphin, and is likely to occur in most, if not all, cetaceans (Cowan and Smith, 1999). It is possible that the anal tonsil fulfills some of the functions of the appendix.

Cetacean livers are usually bilobed, but a third lobe is sometimes present (Harrison and King, 1980). Whales do not have a gallbladder. The pancreas is firm, elongated, and connected to the intestine through a special duct. It is larger in adult females than in males. Data (Kamiya, 1962; Bryden, 1972) indicate that there are two main ducts leading from the pancreas to the common bile duct in mysticetes, whereas in the toothed whales examined, there is only a single main duct (Yablokov *et al.*, 1972).

12.4.4. Seasonal Feeding Migrations of Mysticetes

Intensive high latitude summer feeding followed by long distance migrations to low latitudes in winter months is typical behavior of large mysticetes of both hemispheres (Figure 12.18). When away from their feeding grounds, these whales reduce their intake

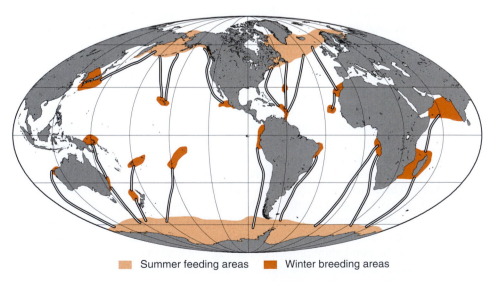

■ Summer feeding areas   ■ Winter breeding areas

Figure 12.18.   Distribution of principal humpback whale feeding and breeding areas and the migration routes linking them. (Partly from Slijper, 1979.)

of food or completely cease feeding. The seasonal variations of feeding intensity impose annual cycles of fattening and fasting on these animals that are as extreme as those of lactating phocids or small hibernating terrestrial mammals. Although neither the migratory routes nor the patterns of energy apportionment for these migrations are known in detail for most mysticete species, a few generalizations can be made.

The function of the annual migrations of mysticete whales is unknown. Some hypotheses attempting to explain baleen whale migrations have focused on the thermoregulatory benefits to the calf by being born in subtropical waters. Others (Sumich, 1986) suggest that these migrations are an evolutionary holdover evolved in a different set of foraging and reproductive circumstances. Corkeron and Connor (1999) have proposed that these winter migrations to low latitudes reduce the risk of killer whale attacks on the young, vulnerable calves. As killer whale abundance is lower in low latitudes and most killer whales do not appear to migrate with baleen whales, the migrations of baleen whales may avoid the risks of killer whale predation until the calves achieve sufficient body mass and swimming capability to improve their likelihood of evading predation attempts on their initial migration to high latitudes.

Successful winter migrations to low latitudes where fasting occurs require that sufficient energy be stored as lipids before summer feeding ceases. Additionally, the fasting migratory animals must be large enough to have sufficient absolute body volumes to accommodate these lipid stores. The threshold body size for successful winter fasting migrations is unclear. Small juvenile animals of all mysticete species tend to feed throughout the year. So too do the adults of the smallest species of mysticetes, the minke and pygmy right whales (Sekiguchi and Best, 1992), although only the former makes annual latitudinal migrations. Hinga (1978) argued that the much larger fin whale must also feed, although at a reduced rate, during winter months. However, adult gray whales fast for most or all of the 5–6 months that they are away from their high-latitude feeding grounds (Nerini, 1984; Sumich, 1986).

Gray whales exhibit evolutionarily conservative feeding structures, especially the relatively small head with few extensible throat pleats, in addition to short, coarse baleen, which suggest that they are poor competitors with other mysticete whales for pelagic food resources at high latitudes. However, the abundance of shallow water benthic prey, which is not exploited by any other whale species, permits gray whales to avoid competition. To maintain access to these food resources, Northeast Pacific gray whales accomplish one of the longest annual migrations of any mammal, covering 15,000–20,000 km. This migration spans as much as 50° of latitude and links summer feeding areas in the Bering, Chukchi, and Okhotsk Seas with warmer breeding, calving, and assembling grounds along subtropical Pacific coastlines in winter (Figure 12.19). Almost all of this migration occurs within 10 km of shore, and the details of this migration have been intensively studied (Gilmore, 1960; Pike, 1962; Rice and Wolman, 1971; Reilly, 1981; Herzing and Mate, 1984; Poole, 1984; Perryman et al., 1999). As with other mysticete species, the highest energy costs during migration are incurred by pregnant or lactating females. Small lactating gray whales are estimated to require about 3.2 metric tons of lipids (more than 70% of their estimated total lipid reserve) for feeding their young and for their own maintenance costs through the winter lagoon season (Sumich, 1986). These small females must resume feeding sometime during their northbound migration, whereas whales unburdened with the costs of lactation, as well as older, larger pregnant females, likely have the necessary lipid reserves to complete the north migration without feeding.

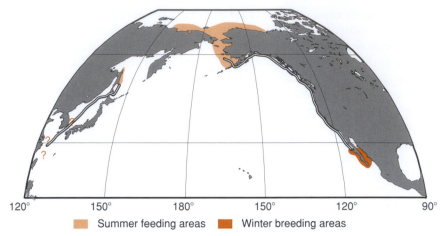

Summer feeding areas ■ Winter breeding areas

Figure 12.19.   Distribution of principal gray whale feeding and breeding areas and the migration routes linking them.

During the Last Glacial Maximum (LGM, 18,000 years before present) global sea level was about 150 m below its present stand (Imbrie *et al.*, 1983). With the shift of polar and subpolar waters southward, all of the shallow sea floor areas presently used by gray whales were inaccessible during the LGM. The shallow portions of the Bering, Chukchi, and Okhotsk Seas now used intensively for summer foraging were either drained of water during the LGM or they were completely icebound. Thus the extensive prey assemblages presently exploited by gray whales either did not exist or were unavailable to gray whales during the LGM. Most of the present continental shelves in both the North Pacific and the North Atlantic Ocean were also exposed above sea level, leaving much narrower continental shelves than exist at present.

The strong contrast between present and LGM oceanographic conditions in both the North Pacific and North Atlantic Oceans leads to the conjecture that gray whales may be thriving now because a particularly fortuitous set of environmental circumstances exists in polar seas during this (and possibly earlier) interglacial periods, but which disappears during glacial maxima. The present foraging flexibility expressed by gray whales feeding in their migratory corridor suggests that they always have been food generalists but that they have specialized temporarily into a narrow ecological niche to capitalize on benthic amphipods, which are presently abundantly available. During glacial advances and consequent sea level lowering and exclusion by polar sea ice, access to this food resource might disappear.

12.4.5. Odontocete Prey Capture Strategies

Odontocetes differ from mysticetes in feeding principally on fish or squid that live at greater depths than mysticete prey. The diets of odontocetes is reflected in the morphology of the jaw and type and number of teeth and depends on aspects of the distribution and ecology of their prey. For a summary of the diets of various odontocete species see Evans (1987). Among species that subsist largely or entirely on squid are the sperm and beaked whales. These squid feeders exhibit a number of adaptations for employing the tongue as a piston and sucking squid into the mouth. This suction feeding strategy is characterized by a reduced number of teeth and a small gape to allow unobstructed entry of small prey, a ribbed palate to help to hold slippery-bodied squid, and throat grooves to allow for expansion of the throat region (Heyning and Mead, 1996). The suction feeding mechanism in beaked whales, described by Heyning and Mead (1996), involves distension of the floor of the mouth provided by the throat grooves and retraction of the tongue by the styloglossus and hyoglossus muscles (Figure 12.20). The extensive sliding movement of the tongue is facilitated by smooth connective tissue between the tongue and the genioglossus muscles, similar to the cavum ventrale of rorquals. In the latter, however, there is significant lateral movement of the throat floor as the tongue inverts during feeding. It is hypothesized that the throat region returns to its resting position after distension primarily as the result of elastic recoil of surrounding tissues, acting in a

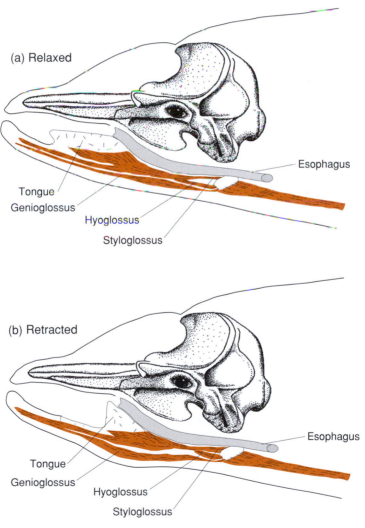

Figure 12.20. Mechanics of suction feeding in beaked whales with the tongue in the relaxed **(a)** and retracted **(b)** positions. (After Heyning and Mead, 1996.)

similar manner to the cavum ventrale in rorquals (see Figure 12.11). Video-documentation of suction feeding in captive long-finned pilot whales *(Globicephala melas)* indicates that a typical 90 msec ingestion sequence is initiated with a partial gape of the jaw, followed by jaw opening and rapid depression of the hyoid to suck in prey from mean distances of 14 cm. The piston-like tongue is depressed and retracted to generate negative intraoral pressures for prey capture without using their teeth (Werth, 2000a).

Other odontocetes (i.e., species of delphinids, monodontids, phocoenids) also use suction to obtain prey (Werth, 2000b). Pelagic dolphins, for example *Stenella* and *Delphinus,* are wide-ranging in the subtropical and tropical regions of the world and are opportunistic feeders. Their diets can vary seasonally and from year to year. The pincer-like action of the jaws of these long-snouted forms allows fish to be trapped by the interlocking tips of the teeth when the jaw closes and water is forced out (Norris and Mohl, 1983). Short-snouted forms that capture prey by suction feeding, such as beaked whales, the narwhal, and immature sperm whales, may be more effective if the prey is first debilitated by high-intensity sounds before engulfment (discussed in Chapter 11). The harbor porpoise also employs a suction feeding strategy in which the tongue tip creates a watertight seal around a fish, while the base of the tongue moves down to create suction toward the throat (Kastelein *et al.*, 1997b).

Bottlenose dolphins exhibit a broad range of foraging behaviors, reflecting their wide array of prey species. In addition to typical capture of pelagic fish prey (Barros *et al.*, 2000), geographically localized groups bottlenose dolphins express other prey-capture techniques, including benthic crater-feeding on Bahama sand banks (Rossbach and Herzing, 1997) and mud-plume feeding in very shallow waters of the Florida Keys (Lewis and Schroeder, 2003). An intriguing behavior that might be a foraging specialization has been observed among a small number of female bottlenose dolphins in Shark Bay, Australia. These animals have been observed consistently over a number of years carrying sponges (Smolker *et al.*, 1997). Suggestions for how the sponges might be used range from their employment as "tools" to disturb bottom sediments and flush out prey species to protection for the dolphin's face from the spines and stingers of stingrays, stone fish, and scorpion fish. Future work is needed to address such questions as how the behavior is acquired and the cost and benefits of sponge carrying as a foraging specialization (Smolker *et al.*, 1997).

The broad ecological and latitudinal range of killer whales corresponds to a wide variety of known prey species, including fish, cetaceans, pinnipeds, birds, cephalopods, sea turtles, and sea otters (Hoyt, 1984; Estes *et al.*, 1998; Pyle *et al.*, 1999). Although killer whales have been described as feeding opportunists or generalists, Felleman *et al.* (1991) suggested that they can also behave as feeding specialists, with the necessary flexibility to respond to variations in the type and abundance of their preferred prey.

In Icelandic coastal waters, some pods of killer whales sometimes feed on marine mammals, then switch to fish prey at other times of the year. Farther east, enormous schools of herring congregate in the deep waters (160–370 m) of Norwegian fjords each winter (Nøttestad and Similä, 2001), where they are preyed upon by other fish-eating killer whales. To more efficiently forage on these small but numerous fish, killer whales cooperate in groups to swim below these herring schools and to split them into small dense schools, then force them over 150 m upward to near-surface waters where they are easier to corral against the sea surface. This lifting of entire schools of small prey is often accompanied by intense tail-slapping. Scaling of the unsteady locomotor performances involved in prey capture suggests that whole-body attacks on individual prey items becomes inefficient at predator:prey body size ratios of about 10:2 (Domenici, 2001). Stunning of herring-sized schooling fishes with tail-slapping behaviors is proposed by Domenici (2001) as an adaptive strategy by killer whales to push the fluke portion of their bodies to higher speeds and accelerations more appropriately scaled to their prey than are whole-body attacks.

In both Antarctic and Pacific Northwest waters, prey specialization between killer whale populations occurs. In Washington and British Columbia coastal waters, resident killer whale populations feed almost exclusively on fishes, whereas nonresident or transient groups prey principally on marine mammals, especially harbor seals and sea lions (Bigg *et al.*, 1987; Ford *et al.*, 1998). Sightings of foraging resident pods correlate strongly with seasonal variations in their most common prey, salmon and steelhead. During months when salmon are unavailable, residents switch to feeding on herring or bottom fish (Felleman *et al.*, 1991) and may disperse northward several hundred kilometers. In contrast, the prey of transient groups is available year-round, and despite their label, these whales tend to remain in a small area for extended periods of time. Table 12.1 summarizes some important differences between these two populations of killer whales.

Different populations of killer whales use different foraging strategies, even when in close proximity to each other. These strategies reflect different prey types and possibly different pod traditions. Resident whales typically swim in a flank formation, possibly to maximize prey detection, although individual hunting is sometimes seen at the periphery of the main group. Resident pods appear to prefer foraging at slack tidal periods when salmon tend to aggregate. Echolocation and other vocalizations are common when resident pods feed on fish. Vocalizations are less frequent in transient groups preying on marine mammals. The latter are more likely to exhibit avoidance responses. Relative to residents, transients use short and irregular echolocation trains composed of clicks that appear structurally variable and low in intensity, more closely resembling random noise (Barrett-Lennard *et al.*, 1996). Yablokov (1966) has suggested that the contrasting black and white coloration patterns of killer whales facilitate visual coordination and communication within feeding groups. Transient groups commonly encircle their prey, taking turns hitting it with their flippers and flukes (Felleman *et al.*, 1991). This extended handling of prey may continue for 10–20 minutes before the prey is killed and consumed. When hunting alone, transient whales often employ repeated percussive tail-slaps or ramming actions to kill their prey before consuming it (Figure 12.21). Foraging by transient groups occurs more frequently at high tides when more of their pinniped prey are in the water and vulnerable to predation.

Table 12.1. A Comparison of Foraging-Related Differences Between Transient and Resident Killer Whales of the Pacific Northwest (Modified from Baird *et al.*, 1992)

Character	Residents	Transients
Group size	Large (3–80)	Small (1–15)
Dive pattern	Short and consistent	Long and variable
Temporal occurrence	With salmon runs	Unpredictable
Foraging areas	Deep water	Shallow water
Vocalize when hunting?	Frequently	Less frequently
Prey type	Salmon and other fish	Mammals
Relative prey size	Small	Large
Prey sharing	Usually no	Usually yes

Figure 12.21.   Transient killer whale attacking a Dall's porpoise in Chatham Strait, Alaska. (Courtesy of R. Baird.)

An interesting consequence of this prey specialization is the pattern of indirect trophic interactions that have evolved between these two populations of orcas. Resident whales compete directly for fish that are also the prey of pinniped species consumed by transient killer whales. Any increase in the carrying capacity of residents will have the effect of decreasing the carrying capacity of transients, because residents compete with the pinniped prey of transients for available fish resources. However, an increase in the carrying capacity of transients will have the opposite effect on residents by reducing competitive pressures from pinnipeds on fish resources (Baird *et al.*, 1992). These indirect trophic interactions and resulting trophic level efficiencies suggest that residents should be more numerous than transients, and estimates of the sizes of the two populations support this concept (Bigg *et al.*, 1987).

A similar dichotomy exists in Antarctic killer whales, with an offshore population consisting of groups of 10–15 animals each preying mostly (nearly 90%, based on content analyses of 785 stomachs) on other species of marine mammals. Inshore Antarctic populations occur in groups 10 times as large and feed almost exclusively on fishes (Berzin and Vladimirov, 1983).

Differences in the foraging behaviors of the two overlapping populations of killer whales in the Pacific Northwest have persisted through time, with each group expressing different vocalizations (see Chapter 11) and different social group sizes as well as different prey preferences and foraging behaviors. A similar discreteness likely exists between inshore and offshore Antarctic killer whale populations, where differences in the distribution, foraging behavior, group size, and even average body size prompted Soviet researchers to propose separate species status for the two populations (Berzin and Vladimirov, 1983). Baird *et al.* (1992) suggested that the two populations in the Pacific

Northwest are in the process of speciation. However, this picture may become more complicated as more is learned about a third group of killer whales occupying offshore waters from California to Alaska (Hoelzel *et al.*, 1998). Currently, little is known about the feeding habits or lifestyles of these offshore killer whales.

Killer whales are cooperative foragers and take advantage of their social group to create the ability, probably unique among marine mammals, to capture prey larger than themselves. The degree of foraging cooperation is variable, depending on the type of prey being attacked, but when large prey is involved hunting tactics can involve cooperative prey encirclement and capture, division of labor during an attack (Tarpy, 1979; Frost *et al.*, 1992), and sharing of prey after capture (Lopez and Lopez, 1985). In a study by Lopez and Lopez (1985), individual killer whales were seen repeatedly swimming up sloping beaches in Patagonia (Argentina) at high tide to capture hauled-out southern elephant seals and sea lions. This hunting strategy, termed intentional stranding, has also been observed among juvenile female killer whales in the capture of elephant seal pups at the Croizet Archipelago (Guinet and Bouvier, 1995). These studies suggest that adults teach this behavior to juveniles and that apprenticeship likely plays an important part in the transfer of hunting techniques from one generation to the next.

Although diets and prey capture strategies have been described for whales, less attention has been given to the energetic costs associated with foraging. Energy is expended through movement during prey searching and capture. Small cetaceans display a variety of foraging methods that vary in energetic costs and deplete their oxygen reserves in accordance with metabolic demands. For example, swimming at high speed, although an energetically costly strategy when searching for prey, might be advantageous in capturing prey that is patchy in distribution and ephemeral in time (Williams *et al.*, 1996).

12.4.6. Evolution of Prey Capture Mechanisms and Strategies Among Whales

Archaic whales (i.e., protocetids) had simple heterodont teeth. *Pakicetus* could probably shear and grind its prey as well as snap the jaw shut during capture (Gingerich and Russell, 1990). The feeding morphology of *Ambulocetus* has been compared to that of crocodilians. Both have long snouts, pointed teeth, and strong jaw adductors (especially the temporalis muscle), and they do not masticate their food. It has been suggested that *Ambulocetus* was an ambush predator that caught with its jaws large struggling fish and aquatic reptiles that were trapped in shallow water (Thewissen *et al.*, 1996). Basilosaurids showed a greater range of tooth and jaw form as well as body size (in comparison to protocetids) and may have exploited offshore or deeper water prey (Fordyce *et al.*, 1995).

The tooth and rostrum structure in the earliest mysticetes (e.g., aetiocetids) is consistent with tooth facilitated filter feeding. It has been suggested that filter feeding in mysticetes evolved in response to the availability of new food that became available when new oceanic circulation patterns associated with the creation of the Southern Ocean and the final breakup of Gondwana were created. Long distance migration perhaps evolved at the same time, allowing seasonal high latitude feeding to alternate with breeding in thermally less stressful temperate-tropical latitudes. Baleen-bearing mysticetes had evolved by the late Oligocene, 4–5 Ma after toothed mysticetes appeared in the fossil record. "Cetotheres" and balaenopterids, the latter group appearing in the late Miocene, possess cranial features (i.e., broad rostrum and elevated posterior ridge at the terminus of the coronoid process of the lower jaw; the latter is associated with the frontomandibular stay in extant balaenopterids), which suggest they represent successive stages in the development of engulfment feeding (Kimura, 2002; McGowen, personal communication). Early Miocene balaenids with their narrow arched rostra indicate that the skim feeding method of modern balaenids had already evolved by this time. The origin of the benthic suction feeding method of gray whales is not known because there is no record of this group before the Pleistocene.

The earliest known odontocetes probably used echolocation to hunt single prey items. Fordyce (1980) suggested that, similar to mysticete feeding modes, echolocation evolved in response to changing food resources, ocean circulation, and continental positions. Tooth number and form vary widely among odontocetes. A reduced role for teeth in processing food is shown by convergent tooth loss in diverse groups including some sperm, beaked, pygmy sperm, pilot whales, and narwhals.

A feeding adaptation that is unique in the record thus far for cetaceans is displayed in the extinct taxon *Odobenocetops* (Muizon, 1993a, 1993b; Muizon *et al.*, 2002). Two species of Odobenocetops lived during the early Pliocene along coastal Peru. These animals had large, down-turned, asymmetrical tusks and blunt snouts. Strong muscle scars on the anterior edge of the premaxillae suggest the presence of powerful upper lips. This feature together with the deeply vaulted palate and absence of teeth suggests a convergent feeding adaptation with the walrus for eating benthic mollusks using suction (Figure 12.22).

12.5. Feeding Specializations of Sirenians

12.5.1. Teeth

In the manatee there are usually five to seven functional teeth in each upper and lower jaw at one time. It has been estimated that from 20 to 30 teeth per jaw are possible within the life span of a given individual (Figure 12.23). The cheekteeth are **brachyodont,** (short-crowned), and are enameled but lack cementum. There are two vestigial incisors in each jaw at birth, but these are later resorbed (Harrison and King, 1980).

It was thought that tooth replacement in the manatee occurred throughout the life of the individual but there is evidence that this may not be the case (Husar, 1977). The teeth are replaced from the rear and, as the anterior teeth wear from excessive amounts of sand and grit in the manatee's normally vegetable diet, they fall out. Domning and Magnor (1977) confirmed observations that the increase in solid food intake after weaning, and hence more chewing, acts as a mechanical stimulus for the initiation and continuation of tooth-row movement. Although often and misleadingly compared to tooth replacement in elephants, this combination of horizontal movement with an apparent limitless supply of supernumerary (extra) molars is found in no other mammal, including the dugong. The elephant has a fixed number of teeth, which are replaced by a similar mechanism, but once an elephant has completely worn out its teeth it can no longer crush its food.

Dugongs, in contrast to manatees, have a fixed dental formula of I1/0, C0/0, P0/0, M2-3/2-3, with vertical replacement similar to most other mammals. Similar to manatees, two milk incisors are present in juveniles. The deciduous incisors are small and do not erupt but are instead resorbed. In males, the deciduous incisors are lost around the time of tusk eruption and their tooth sockets disappear as the tusks expand. In females, small

Figure 12.22.   *Odobenocetops peruvianus.* **(a)** Skull. **(b)** Life reconstruction. (From Muizon, 1993a, 1993b; life reconstruction courtesy of M. Parrish.)

partially resorbed incisors may persist until the animal is about 30 years old (Nishiwaki and Marsh, 1985). The permanent incisors (tusks) develop in both sexes but only erupt in males and some old, probably postreproductive females. In some old females the tusks erupt and wear, presumably because they have reached the bases of the premaxillae and cannot grow posteriorly any further. Male tusks erupt when the animal is 12–15 years old. Adult male tusks may be up to 15 cm long, but because they protrude from the mouth only a couple of centimeters, they do not seem to be particularly useful as

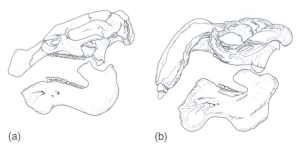

Figure 12.23.   Lateral views of skull of **(a)** manatee and **(b)** dugong illustrating differences in teeth and snout deflection. (From Werth, 2000a.)

weapons. Up to three pairs of vestigial incisors and one pair of canines may occur. However, apart from the erupted tusks, only the cheekteeth are functional.

In situ replacement of the milk dentition does not occur in dugongs. During the life of the animal there are a total of four to six cheekteeth on each side of the upper and lower jaws but these are never simultaneously all erupted and in use (Nishiwaki and Marsh, 1985). Three premolars are erupted at birth, but all the roots of the anterior teeth are resorbed, causing them to fall out. Their sockets then become occluded with bone without eruption of permanent premolars. The molars progressively erupt during growth. The entire process continues until molars 2 and 3 (and occasionally molar 1) are present in each quadrant of old animals. Molars 2 and 3 have persistent pulp cavities and continue to grow throughout life so that the total occlusal area of the cheekteeth is maintained and even increased after the anterior cheekteeth have been lost. The pattern of cheektooth replacement can be used to approximate the age of dugongs; however, more precise age estimates can be obtained by counting dentinal growth layers (Nishiwaki and Marsh, 1985).

12.5.2. Anatomy and Physiology of Digestion in Sirenians

Large parotid glands are the only salivary glands in the dugong. Submaxillary salivary glands are prominent in the West Indian manatee (Quiring and Harlan, 1953) but sublingual glands are small (Murie, 1872). The most characteristic feature of the sirenian tongue is the form of the taste buds; in the dugong they are contained in pits and in the manatee they occur in swellings (Yamasaki *et al.*, 1980). Tonsils are absent in sirenians (Hill, 1945).

The digestive tracts of both the dugong and the manatee are extremely long (Lomolino and Ewel, 1984). Detailed descriptions of the manatee stomach can be found in Langer (1988) and Reynolds and Rommel (1996). In both the manatee and dugong the stomach is a simple sac, differing from the typical mammalian stomach by the presence of a discrete cardiac gland that opens into the main sac via a single, small aperture (Marsh *et al.*, 1977; Reynolds and Rommel, 1996). The cardiac region of the Florida manatee stomach contains parietal, chief, and neck cells and secretes acid, mucus, and pepsin; its cellular composition corresponds to that of other mammals. The West African manatee lacks chief and parietal cells, except in the cardiac gland (Reynolds and Rommel, 1996, and references cited therein). The cardiac gland in the dugong contains all the chief cells and most of the parietal cells associated with the stomach (Marsh *et al.*,

1977). Reynolds and Rommel (1996) have shown that the Florida manatee has certain morphological adaptations for production of copious mucus, such as stratified lining of the pyloric stomach, and submucosal mucous glands in the greater curvature of the stomach. Other sirenians apparently lack these adaptations.

The duodenum of the manatee and dugong possesses large duodenal ampullae and paired duodenal diverticula (Figure 12.24). Marsh *et al.* (1977) suggested that the size of the ampullae and diverticula allow the passage of large volumes of digesta from the stomach. The cecum and large intestine, the contents of which may weigh twice that of the stomach contents, have been shown to be the principal areas of absorption of the fiber portion of the diet (Murray *et al.*, 1977). In the adult dugong, the large intestine is up to 30 m long, about twice as long as the small intestine. This compares with the manatee's large intestine that has been measured approximately 20 m long (references cited in Lanyon and Marsh, 1995). In comparison with other hindgut fermenting herbivores (e.g., the elephant has a 10 m hindgut), sirenians have longer hindguts (Lanyon and Marsh, 1995). Both the stomach shape and intestine size of the dugong and manatee provide evidence that they are nonruminant herbivores with a greatly enlarged hindgut that has a rich microflora, which enables them to digest cellulose and other fibrous carbohydrates (Nishiwaki and Marsh, 1985; Lanyon and Marsh, 1995). Both groups employ a digestive strategy atypical of hindgut fermenters; low-fiber material is retained for extended periods within the long hindgut and it is almost completely digested

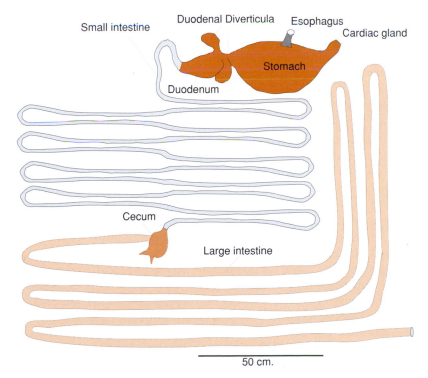

Figure 12.24. Diagram of a manatee digestive tract, drawn to scale. (Adapted from Reynolds and Rommel, 1996.)

(Lanyon and Marsh, 1995). Similarities in the gross anatomy of the gastrointestinal tracts exist between sirenians and their close relatives, elephants, especially regarding the specialized cardiac stomach (cardiac gland in sirenians) and the cecum (Reynolds and Rommel, 1996).

The liver of the dugong has three main lobes and is positioned against the almost horizontal diaphragm. A long common bile duct and a small gallbladder occur in the dugong (Harrison and King, 1980). By contrast, a relatively large gallbladder is present in the manatee. The dugong pancreas is small, compact, and lobulated (Husar, 1975).

12.5.3. Feeding Strategies

Even though sea grass meadows are very productive, sirenians and sea turtles are the only large herbivores to commonly graze on them. A wide variety of tropical and subtropical sea grasses are consumed (Hydrocharitaceae and Potamogetonaceae). Algae also are eaten but only in very small amounts if sea grasses are abundant (Nishiwaki and Marsh, 1985). Dugongs in subtropical Moreton Bay, Australia, appear to selectively consume sessile benthic invertebrates including ascidians (sea squirts) and polychaete worms. Nutritional stress, particularly the lack of dietary nitrogen, caused by strong seasonal patterns in the abundance of sea grasses, may explain the omnivory of the Moreton Bay animals, which live at the extreme southern edge of the dugongs' range (Preen, 1995). In contrast to the specialized sea grass grazing of dugongs, manatees are more generalized feeders, and they graze on a large variety of coastal and freshwater vegetation, including true grasses (Poaceae) in addition to sea grasses, floating freshwater plants (*Hydrilla* and water hyacinths), and even the leaves and shoots of emergent mangroves.

The degree of snout deflection in manatees and dugongs appears to be correlated with the degree of specialization for bottom feeding (see Figure 12.23). The strongly downturned snout of the dugong causes its mouth to open almost straight downward, and it is virtually an obligate bottom feeder subsisting on sea grasses less than 20 cm high. Manatees, in contrast, have only relatively slight deflection, and are generalists that can feed at any level in the water column from bottom to the surface, including the ability to take floating vegetation (Domning, 1977).

Sirenians appear to be the only marine mammals that use the snout in a prehensile manner. In a study of facial musculature of the Florida manatee (Marshall *et al.*, 1998a, 1998b, 2003) it was found that the short, muscular snout (especially the upper lip of the **rostral** or **oral disc**) is covered with modified vibrissae that achieve a prehensile function bringing vegetation to the mouth (Figure 12.25). Similarly, the rostral disc of the dugong is also used in feeding. During feeding, dugongs gouge tracks into the bottom sediments (Anderson and Birtels, 1978). It is apparent that the tracks represent the action of the snout while grubbing sea grasses from the bottom (Figure 12.26). Each track follows a serpentine course and appears to represent the continuous feeding effort of a single dive. The feeding trails suggest that the disc may be forced against the substrate while muscular contractions of its surface move the heavy bristles in a raking action, which extracts sea grasses from the bottom and gathers them toward the mouth. Although male dugongs have erupted tusks, they do not appear to be used in feeding. In addition to their peg-like molars located far back in the jaws, chewing also occurs by horny, grinding plates that cover the posterior surfaces of the tongue and mandible (Figure 12.27), adapted for crushing cellulose-rich vegetation.

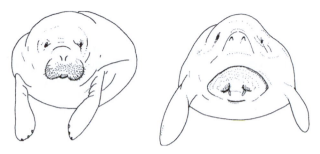

Figure 12.25.   Anterior view of oral discs of manatee (*left*) and dugong (*right*). (From Werth, 2000a.)

Figure 12.26.   Photo of dugong feeding tracks. (From Anderson and Birtels, 1978, courtesy of P. Anderson.)

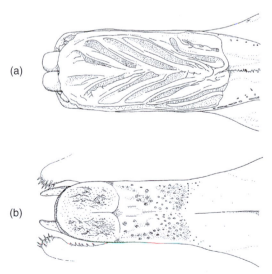

(a)

(b)

Figure 12.27.   Occlusal surfaces of oral grinding plates of **(a)** dugong and **(b)** Steller's sea cow. (From Werth, 2000a, 2000b.)

12.5.4. Evolution of Sirenian Feeding

Hypotheses of diet and feeding evolution among sirenians are discussed by Domning (2001 and earlier references cited therein) and MacFadden *et al.* (2004). Prorastomids and the earliest dugongids living 45–30 Ma were characterized by relatively flat rostra indicating little of the deflection of later sirenians and were likely selective browsers feeding on aquatic plants and to a lesser degree on sea grasses. Increased foraging on sea grasses is suggested for protosirenids with a rostral deflection of approximately 35 degrees. Oligocene faunas comprised as many as six apparently sympatric sirenian species. Miocene members of the *Metaxytherium-Halitherium* lineage (17–8 Ma) had strongly down-turned snouts (as much as 75 degrees in *Metaxytherium floridanum*) and small to medium sized tusks indicating that they consumed seagrass leaves and smaller rhizomes. The dugongine lineage in particular evolved large blade-like tusks presumably used to dig up larger rhizomes of seagrasses.

Among members of the hydrodamaline lineage, Steller's sea cow showed a number of unique cold water adaptations, including larger body size (body length double that of other sea cows), an extraordinarily thick epidermis, inability to dive, and loss of teeth and phalanges. Domning (1977) suggested that these adaptations reflected a change in diet away from bottom feeding to feeding on plants growing at the surface (and mainly on algae growing in exposed rocky shallows). He correlated the loss of phalanges as a specialization for surface and rough-water feeding and suggests that the stout claw-like forelimbs were adapted to grasp or fend against rocks and pull the animal forward while grazing in the shallows during bottom feeding. The relative ease of chewing algae compared to stringier sea grasses probably led to loss of the teeth. Although "overharvesting" by humans is usually blamed for the ultimate demise of Steller's sea cow, an intriguing explanation for their small, restricted populations was proposed by Anderson (1995) that involved a competitive interaction between Steller's sea cow and the kelp-sea urchin-sea otter relationship. According to this relationship, described in more detail later in this chapter, high-latitude shallow-water kelp communities are prolific only where sea otters are sufficiently abundant to limit sea urchin populations. Anderson (1995) proposed that if the sea cow food supply was otter dependent, then the niche into which Steller's sea cow evolved was "precariously balanced" on sea otter carnivory. The aboriginal hunting of sea otters, urchin grazing, and replacement of shallow-water kelps by deep-water species could have destroyed sea cow foraging areas, thereby limiting their populations prior to their discovery in the early 1700s.

In freshwater habitats of South America manatees adapted to an abrasive diet and increased tooth wear that resulted from their exploitation of a new high silica-containing resource, the true grasses, during the late Miocene (Domning, 1982). More recently manatees have adapted to a more diverse herbivorous diet.

12.6. Feeding Specializations of Other Marine Mammals

12.6.1. Sea Otters

Unlike their fish-eating relatives of freshwater lakes and streams, sea otters preferentially forage on benthic invertebrates, particularly sea urchins, gastropod and bivalve mollusks, and crustaceans. The dental formula of sea otters is I3/2, C1/1, P3/3, M1/2, differing from that of most other carnivores in having a reduced number of lower incisors.

In addition, the carnassials and molars are modified from the slicing role of terrestrial carnivores to blunt crushing teeth suitable for breaking up exoskeletons and shells of invertebrate prey (Figure 12.28). Captured prey is manipulated with the forepaws or is held temporarily in loose skin pouches in the armpits. For larger, heavier-shelled prey, otters will sometimes carry a rock to the surface, place it on their belly, and then use it to crack open the hard shells while they float on their backs (Figure 12.29). Otters have exceptionally high metabolic rates and correspondingly high food consumption rates: typical adult otters in Alaskan waters daily consume prey equal to 23–37% of their body weight daily (Costa, 1978; Kvitek *et al.*, 1993; Kvitek and Bretz, 2004).

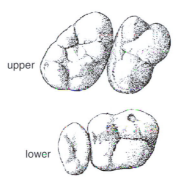

upper

lower

Figure 12.28. Occlusal surfaces of the upper (fourth premolar and first molar) and lower (first and second molars) teeth of a sea otter, with rounded cusps and a complete lack of shearing ability characteristic of most other carnivorans. (Adapted from Vaughn, 1986.)

Figure 12.29. Sea otter at sea surface with cracking rock. (Courtesy of P. Colla.)

Sea otters are completely restricted to shallow (generally <35 m depth) coastal waters of the North Pacific Ocean, where substantial populations of invertebrates inhabit rocky or sandy substrates. Sea otters are most commonly affiliated with near-shore canopy-forming kelps (principally the brown algae *Macrocystis* or *Nereocystis*), although large numbers of otters have been observed as much as 30 km offshore (Kenyon, 1969). In portions of the sea otter range, where their numbers have recovered or where they have been reintroduced (see Chapter 15), sea otters are considered a key-stone predator, playing a pivotal role in kelp bed communities by keeping populations of herbivorous invertebrates, mostly grazing sea urchins, low in number and allowing kelp plants to establish and thrive (Duggins, 1980). In the southern part of their historical California and northern Baja California range, healthy kelp forests persisted even though sea otters have been absent for more than a century. There, water temperature variations, storm surges, or other disturbances may play important roles in the dynamics of these otter-free near-shore kelp communities.

In a novel study of the influences of sea otters on marine plant–herbivore interactions, Steinberg *et al*. (1995) examined contrasting biotic assemblages between the northern and southern hemispheres. The northeast Pacific Ocean assemblage has been influenced by the effect of sea otter predation on herbivorous invertebrates, probably since the late Miocene. The temperate Australasia assemblage, in contrast, while containing similar guilds of plants and animals, lacks a predator comparable to the sea otter. Steinberg *et al*. (1995) suggested that in the northern hemisphere kelp forests evolved in a three-tiered food web, with sea otters and their recent ancestors at the top, herbivores in the middle, and macroalgae as the autotrophs. In such a food web, herbivores are limited by predation, thus protecting the autotrophs from intense disturbance via herbivory. In contrast, in the southern hemisphere where kelp is limited by herbivory, these autotrophs are left unprotected. These patterns led Steinberg *et al*. (1995) to predict that autotrophs defend against their herbivores in Australasia but not in the North Pacific. This prediction has been substantiated by the discovery of higher concentrations of secondary metabolites as defenses against herbivory in brown algae in the southern hemisphere compared to the northern hemisphere. In addition, there is evidence that herbivores in the southern regions have evolved differing abilities to cope with chemical defenses (Estes, 1996, and references cited therein). Results of this study suggest that sea otter predation of sea urchins, by freeing North Pacific autotrophs from the need to defend themselves, has been a key factor in the evolution of a marine flora that is vulnerable to herbivory. Kvitek and Bretz (2004) have shown that different secondary metabolites from phytoplankton, especially saxitoxin, provide butter clams *(Saxidomus giganteus)* some refuge from sea otter predation by accumulating saxitoxins in their tissues.

Ralls *et al*. (1995) monitored instrumented sea otters in California and found extensive variation in the foraging patterns of individual otters with respect to dive duration and success rate, surface interval length, size, and species of prey consumed, foraging bout length, and the degree of difference between diurnal and nocturnal foraging patterns. Juvenile males forage further offshore over deeper water than otters in other age and sex classes and they also tended to make longer dives. It is possible that they consume larger prey but this was not confirmed because they could not be observed from shore. In contrast, juvenile females foraged closer to shore and made shorter dives. Juvenile females also tended to have longer foraging bouts with short surface intervals, suggesting to Ralls *et al*. (1995) that they were perhaps capturing small prey of lower caloric value; they often feed on kelp crabs, turban snails, and mussels.

12.6.2. Polar Bears

Polar bears have 38–42 teeth and a dental formula of I 3/3, C1/l, P2-4/2-4, M2/3. The premolars and molars of polar bears are more jagged and sharper than those of other bears, reflecting their rapid evolutionary shift toward carnivory from the flatter grinding teeth of their more omnivorous relatives (Stirling, 2002). The principal prey species of polar bears throughout their range is the ringed seal. Polar bears hunt this species on both fjord and open sea ice; it is particularly important in the spring diet of polar bears. Ringed seals are captured at their breathing holes in the ice as well as in their lairs (small snow caves ringed seals use for resting during the winter and giving birth in spring). Female polar bears with new cubs emerge from the winter den at the time when ringed seals are giving birth to their young in lairs; successful hunting of ringed seals at this time is critical for the mother polar bears that have fasted for months in the den. In fact, most of the annual energy budget of polar bears comes from ringed seals, and most of this energy is consumed in spring and early summer. Polar bears employ still-hunting at seal holes, and they also stalk hauled out seals. Lairs are usually entered by crashing in the roof by jumping forcefully with the front paws (Figure 12.30), catching the occupant before it escapes through the hole in the ice. Polar bears also hunt other ice-living seals such as bearded, ribbon, spotted, harp, and perhaps young hooded seals.

The extent of bear predation on walruses varies in different areas of the Arctic, with higher rates of predation in the Canadian Arctic than in the Bering Sea. Bears frequently fail to make a kill by stalking hauled-out walrus herds, yet manage to frighten herds into the water. In the ensuing stampede, walrus pups are sometimes injured or isolated, making them easier prey for the bears (Riedman, 1990). Hunting of cetaceans by polar bears is almost certainly on an opportunistic basis; when belugas and narwhals are restricted to small areas of open water (polynyas) to breathe in ice-covered areas they can be taken following successive injuries inflicted when they surface. Waterfowl and seabirds are also taken by polar bears (Amstrup and DeMaster, 1988). During periods on land they will eat seaweed, eggs and chicks of ground nesting birds and even a single incident of a polar bear catching and eating a reindeer calf has been documented (Derocher et al., 2000). They will also eat carrion when it is available.

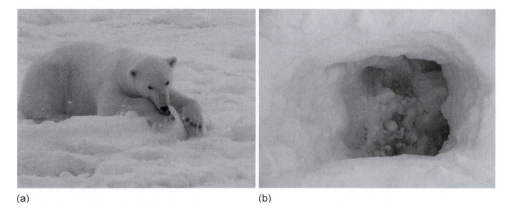

(a) (b)

Figure 12.30. Polar bear breaking into a ringed seal den **(a)**; ringed seal den after prey has been removed **(b)**. (Courtesy of Kit Kovacs.)

An interesting aspect of the foraging ecology of polar bears, indeed of bears in general, is their ability to fast for prolonged periods of time (Atkinson *et al.*, 1996). Pregnant polar bears are without food the entire period in the den, when giving birth and nursing young cubs. Other age groups undoubtedly also encounter circumstances that prevent them from hunting quite routinely for shorter periods. Polar bears in western Hudson Bay are forced to stay onshore in summer when the ice melts. During this ice-free period, they have little or no access to food. For up to 4 months they are stranded on shore until late autumn when the ice forms again and provides a hunting platform. Although most polar bears (except pregnant females) do not occupy dens regularly, there are data (i.e., urea-to-creatinine ratio as an indicator of protein loss through catabolism of body fat) that suggest that individuals are able to adopt an efficient state of protein conservation similar to that of denning bears during the winter. Such a metabolic fasting response might be employed during the mating season in spring when adult males forgo feeding to compete for access to females. Results of a study of fasting in polar bears suggests that among ursids in general, the ability to minimize protein loss during fasts may be largely dependent on the relative fatness of the animal at the start of a fast, which is in turn dependent on the extraordinary magnitude of fat stores that some individuals can accumulate in seasons when food is abundant (Atkinson *et al.*, 1996).

12.7. Summary and Conclusions

Marine mammal diets and foraging strategies are direct consequences of patterns of primary productivity, although most species prey on relatively large animals several trophic levels removed from the primary producers. The cheekteeth of pinnipeds and odontocetes are typically homodont with single pointed cusps adapted for feeding on fish and squid. Walruses employ a suction feeding strategy using the tongue as a piston to suck clams out of their shells. Tusks are used in social display rather than for feeding. Bearded seals also use suction feeding and hydraulic jetting when excavating benthic prey. Crabeater seals and a few other species filter feed on krill and have some specialization of their teeth for this feeding modality. The seasonal migration of the elephant seal, one of the best-studied feeding patterns among pinnipeds, involves an annual double migration with prolonged periods of offshore feeding. A generalist (pierce) feeding strategy appears to be the ancestral prey capture strategy in pinnipeds. A specialized suction feeding strategy evolved independently is several pinniped lineages,

Mysticetes employ rows of baleen plates to filter feed plankton and small fish from the water. They employ three different feeding strategies: skimming, engulfment, and suction feeding. Balaenids and neobalaenids pursue skimming by moving slowly through the water collecting prey on the baleen as water is expelled by the tongue. Balaenopterids employ engulfment feeding in which the mouth is enlarged by expansion of the throat grooves and invagination of the tongue to create a large ventral cavity. Eschrichtiids engage in benthic suction feeding. Large mysticetes (e.g., humpbacks, gray whales) typically feed at high latitudes during the summer and then migrate long distance to low latitudes in winter months.

Odontocete foraging strategies are varied. Some odontocetes, characterized by reduced numbers of teeth (e.g., beaked and sperm whales), employ a suction feeding strategy in which throat grooves and muscles of the tongue function in retraction and distension of the floor of the mouth. Sponge carrying has been proposed as a foraging specialization in some dolphins. Transient and resident killer whales not only employ different foraging strategies but also feed on different prey. The evolution of prey capture mechanisms and strategies in whales has been related to the availability of new food resources when oceanic circulation patterns were altered. This led to baleen filter feeding in mysticetes and to echolocation in odontocetes. Cetaceans are characterized by multi-chambered stomachs similar to ruminant artiodactyls, their close living relatives.

As the only herbivorous marine mammals, sirenians have modified short, broad-crowned teeth. Tooth replacement in manatees is unique among mammals. Teeth follow a "conveyor belt" replacement mechanism and are replaced throughout life. The degree of snout deflection in manatees and dugongs is correlated with the degree of specialization for bottom feeding on sea grasses or feeding in the water column or on floating vegetation. The dugong, with its strongly down-turned snout, is a bottom feeder, in contrast to manatees, which have relatively slight snout deflection and feed at all levels of the water column. Dugongs employ a rostral disc as a grazing mechanism; it gouges tracks into bottom sediments during feeding. The evolution of the conveyor belt replacement in manatees is related to an abrasive diet and increased tooth wear that resulted from exploitation of a new resource, sea grasses, approximately 10 Ma. The relatively long large intestine of sirenians is related to the high fiber content of their diets.

Sea otters possess blunt teeth for crushing the shells of their invertebrate prey. They are considered keystone predators and play a pivotal role in kelp bed communities by keeping populations of grazing sea urchins low. Study of the influences of sea otters on marine plant–herbivore interactions reveal that in the North Pacific sea otter predation may have been a key factor in the evolution of a marine flora that is vulnerable to herbivory. Polar bears are the most carnivorous ursids. Their feeding ecology differs from other marine mammals in that hunting occurs primarily in air; the annual cycle of polar bears is punctuated by periods of prolonged fasting, particularly in areas that have only seasonally available ice.

12.8. Further Reading

A general introduction to pinniped diet and foraging strategies is summarized by Riedman (1990) and also by Bowen and Siniff (1999; for cetaceans, see Evans (1987) and Gaskin (1982); for dugongs consult Anderson and Birtels (1978) and see Hartman (1979) for manatees. The anatomy and physiology of digestion for pinnipeds is summarized in King (1983); for cetaceans consult Slijper (1979) and for sirenians see Reynolds and Odell (1991). Adam and Berta (2002) review the evolution of feeding strategies in pinnipeds and see Fordyce *et al.* (1995) and Thewissen *et al.* (1996) for cetaceans. For sirenians see Domning (1982).

References

Adam, P. J., and A. Berta (2002). "Evolution of Prey Capture Strategies and Diet in the Pinnipedimorpha (Mammalia, Carnivora)." *Oryctos 4:* 83–107.

Amstrup, S. C., and D. P. De Master (1988). Polar bear, *Ursus maritimus. In* "Selected Marine Mammals of Alaska" (J. W. Lentfer, ed.), pp. 39–56. Mar. Mamm. Comm., Washington, DC.

Anderson, P. K. (1995). "Competition, Predation, and the Evolution and Extinction of Steller's Sea Cow, *Hydrodamalis gigas*." *Mar. Mamm. Sci. 11:* 391–394.

Anderson, P. K., and A. Birtels (1978). "Behavior and Ecology of the Dugong, *Dugong dugon* (Sirenia): Observations in Shoalwater and Cleveland Bays." *Queensl. Aust. Wildl. Res. 5:* 1–23.

Antonelis, G. A., M. S. Lowry, C. H. Fiscus, B. S. Stewart, and R. L. DeLong (1994). Diet of the northern elephant seal. *In* "Elephant Seals" (B. J. Le Boeuf, and R. M. Laws, eds.), pp. 211–223. University of California Press, Berkeley, CA.

Atkinson, S. N., R. A. Nelson, and M. A. Ramsay (1996). "Changes in the Body Composition of Fasting Polar Bears *(Ursus maritimus):* The Effect of Relative Fatness on Protein Conservation." *Physiol. Zool. 69:* 304–316.

Baird, R. W., P. A. Abrams, and L. M. Dill (1992). "Possible Indirect Interactions Between Transient and Resident Killer Whales: Implications for the Evolution of Foraging Specializations in the Genus *Orcinus*." *Oecologia 89:* 125–132.

Bakker, M. A. G., R. A. Kastelein, and J. L. Dubbledam (1996). "Histology of the Grooved Ventral Pouch of the Minke Whale, *Balaenoptera acutorostrata,* with Special Reference to the Occurrence of Lamellated Corpuscles." *Can. J. Zool. 75:* 563–567.

Barrett-Lennard, L. G., J. K. B. Ford, and K. A. Heise (1996). "The Mixed Blessing of Echolocation: Differences in Sonar Use by Fish-Eating and Mammal-Eating Killer Whales." *Anim. Behav. 51:* 553–565.

Barros, N. B., E. C. M. Parsons, and T. A. Jefferson (2000). "Prey of Offshore Bottlenose Dolphins from the South China Sea." *Aquat. Mamm. 26:* 2–8.

Benoit, D., and W. D. Bowen. (1990). "Seasonal and Geographic Variation in the Diet of Gray Seals *(Halichoerus grypus)* in Eastern Canada." *Can. J. Fish. Aquat. Sci. Bull. 222:* 215–226.

Berzin, A. A., and V. L. Vladimirov (1983). "A New Species of Killer Whale (Cetacea, Delphinidae) from the Antarctic Waters." *Zool. Zh. 62:* 287–295.

Best, R. C. (1981). "The Tusk of Narwhal *(Monodon monoceros L.):* Interpretation of Its Function (Mammalia: Cetacea)." *Can. J. Zool. 59:* 2386–2393.

Bigg, M. A., G. M. Ellis, J. K. B. Ford, and K. C. Balcomb (1987). *Killer Whales: A Study of Their Identification, Genealogy and Natural History in British Columbia and Washington State.* Phantom Press & Publ., Nanaimo, BC, Canada.

Bonner, W. N. (1982). *Seals and Man. A Study of Interactions.* University of Washington Press, Seattle, WA.

Bonner, W. N. (1989). *Whales of the World.* Facts on File, New York.

Bowen, W. D., and D. B. Siniff (1999). Distribution, population biology, and feeding ecology of marine mammals. *In* "Biology of Marine Mammals" (J. E. Reynolds, and S. A. Rommel, eds.), pp. 423–484. Smithsonian Institute Press, Washington, DC.

Bryden, M. M. (1971). "Anatomical and Allometric Adaptations in Elephant Seals, *Mirounga leonina*." *J. Anat. 108:* 208.

Bryden, M. M. (1972). Growth and development of marine mammals. *In* "Functional Anatomy of Marine Mammals," Vol. 1 (R. J. Harrison, ed.), pp. 1–79. Academic Press, New York.

Bryden, M. M., and A. W. Erickson (1976). "Body Size and Composition of Crabeater Seals *(Lobodon carcinophagus),* with Observations on Tissue and Organ Size in Ross Seals *(Ommatophoca rossi)*." *J. Zool. London 179:* 235–247.

Burns, J. J. (1981). The bearded seal *(Elignathus barbatus,* Erxleben, 1777). *In* "Handbook of Marine Mammals" (S. H. Ridgway and R. J. Harrison, eds.), pp. 145–170. Academic Press, London.

Corkeron, P. J., and R. C Connor (1999). "Why Do Baleen Whales Migrate?" *Mar. Mamm. Sci. 15:* 1228–1245.

Costa, D. P. (1978). "The Sea Otter. Its Interaction with Man." *Oceanus 21:* 24–30.

Costa, D. P. (1993). "The Relationship Between Reproductive and Foraging Energetics and the Evolution of the Pinnipedia." *Symp. Zool. Soc. London 66:* 293–314.

Cowan, D. F., and T. L. Smith (1995). "Morphology of the Complex Lymphoepithelial Organs of the Anal Canal ("Anal Tonsil") in the Bottlenose Dolphin, *Tursiosp truncatus*." *J. Morphol. 223:* 263–268.

Cowan, D. F., and T. L. Smith (1999). "Morphology of the Lymphoid Organs of the Bottlenose Dolphin, *Tursiops truncates*." *J. Anat. 194:* 505–517.

Darling, J. D. (1984). Gray whales off Vancouver Island, British Columbia. *In* "The Gray Whale, *Eschrichtius robustus* (Lilljeborg, 1861)" (M. L. Jones, S. L. Swartz, and S. Leatherwood, eds.), pp. 267–288. Academic Press, New York.

Darling, J. D., K. E. Koegh, and T. E. Steeves (1998). "Gray Whale *(Eschrichtius robustus)* Habitat Utilization and Prey Species off Vancouver Island, B.C." *Mar. Mamm. Sci. 14:* 692–720.

Davis, R. W., L. A. Fuiman, T. M. Williams, S. O. Collier, W. P. Hagey, S. B. Kanatous, S. Kohin, and M. Hornoing (1999). "Hunting Behavior of a Marine Mammal Beneath the Antarctic Fast Ice." *Science 283:* 993–996.

Davis, R. W., L. A. Fuiman, T. M. Williams, M. Horning, and W. Hagey (2003). "Classification of Weddell Seal Dives Based on 3 Dimensional Movements and Video-Recorded Observations." *Mar. Ecol. Prog. Ser. 264:* 109–122.

DeLong, R. L., and B. S. Stewart (1991). "Diving Patterns of Northern Elephant Seal Bulls." *Mar. Mamm. Sci. 7:* 369–384.

Derocher, A. E., Ø. Wiig, and G. Bangjord (2000) "Predation of Svalbard Reindeer by Polar Bears." *Polar Biol. 23:* 675–678.

Doidge, D. W., and J. P. Croxall (1985). Diet and energy budget of the Antarctic fur seal *Arctocephalus gazella* at South Georgia. *In* "Antarctic Nutrient Cycles and Food Webs" (W. R. Siegfried, P. R. Condy, and R. M. Laws, eds.), pp. 543–550. Springer-Verlag, Berlin.

Domenici, P. (2001). "The Scaling of Locomotor Performance in Predator-Prey Encounters: From Fish to Killer Whales." *Comp. Biochem. Physiol. A 131:* 169–182.

Domning, D. P. (1977). "An Ecological Model for Late Tertiary Sirenian Evolution in the North Pacific Ocean." *Syst. Zool. 25:* 352–362.

Domning, D. P. (1982). "Evolution of Manatees: A Speculative History." *J Paleontol. 56:* 599–619.

Domning, D. P. (2001). "Sirenians, Sea Grasses, and Cenozoic Ecological Change in the Caribbean." *Palaeogeogr. Palaeoclim. Palaeoecol. 166:* 27–50.

Domning, D. P., and D. Magnor (1977). "Taxa se substitiucao horizontal de dentes no peixe-boi." *Acta Amazon. 7:* 435–438.

Duggins, D. O. (1980). "Kelp Beds and Sea Otters: An Experimental Approach." *Ecology 61:* 447–453.

Dunham, J. S., and D. A. Duffus (2002). "Diet of Gray Whales *(Eschrichtius robustus)* in Clayoquot Sound, British Columbia, Canada." *Mar. Mamm. Sci. 18:* 419–437.

Eastman, J. T., and R. E. Coalson (1974). The digestive system of the Weddell seal, *Leptonychotes weddelli*-a review. *In* "Functional Anatomy of Marine Mammals" (R. J. Harrison, ed.), 2nd ed., pp. 253–320. Academic Press, London.

Estes, J. A. (1996). The influence of large, mobile predators in aquatic food webs: Examples from sea otters and kelp forests. *In* "Aquatic Predators and Their Prey" (S. P. R. Greenstreet, and M. L. Tasker, eds.), pp. 65–72. Blackwell Science, London.

Estes, J. A., M. T. Tinker, T. M. Williams, and D. F. Doak (1998). "Killer Whale Predation on Sea Otters Linking Oceanic and Nearshore Ecosystems." *Science 282:* 473–476.

Evans, G. H. P. (1987). *The Natural History of Whales and Dolphins*. Facts On File, New York.

Fay, F. H. (1982). Ecology and biology of the Pacific walrus, *Odobenus rosmarus divergens* Illiger. *North Am. Fauna Ser.*, U.S. Dep. Int., Fish Wildl. Serv., No. 74.

Fay, F. H. (1985). "*Odobenus rosmarus.*" *Mamm. Spec. 238:* 1–7.

Felleman, F. L., J. R. Heimlich-Boran, and R. W. Osborne (1991). The feeding ecology of killer whales *(Orcinus orca)* in the Pacific Northwest. *In* "Dolphin Societies" (K. Pryor, and K. S. Norris, eds.), pp. 113–147. University of California Press, Berkeley, CA.

Fiscus, C. H., and G. A. Baines (1966). "Food and Feeding Behavior of Steller and California Sea Lions." *J. Mammal. 47:* 195–200.

Ford, J. K. B., G. M. Ellis, L. G. Barrett-Lennard, A. B. Morton, R. S. Palm, and K. C. Balcomb III (1998). "Dietary Specialization in Two Sympatric Populations of Killer Whales *(Orcinus orca)* in Coastal British Columbia and Adjacent Waters." *Can. J. Zool. 76:* 1456–1471.

Ford, J., and D. Ford (1986). "Narwhal: Unicorn of the Arctic Seas." *Nat. Geogr. 169:* 354–363.

Fordyce, R. E. (1980). "Whale Evolution and Oligocene Southern Ocean Environments." *Palaeogeogr. Palaeoclim. Palaeoecol. 31:* 319–336.

Fordyce, R. E., L. G. Barnes, and N. Miyazaki (1995). "General Aspects of the Evolutionary History of Whales and Dolphins." *Island Arc 3:* 373–391.

Frost, K. J, and L. F. Lowry (1986). "Sizes of Walleye Pollock, *Theragra chalcogramma,* Consumed by Marine Mammals in the Bering Sea." *Fish. Bull. 84:* 192–197.

Frost, K. J., R. B. Russell, and L. F. Lowry (1992). "Killer Whales, *Orcinus orca,* in the Southeastern Bering Sea: Recent Sightings and Predation on Other Marine Mammals." *Mar. Mamm. Sci. 8:* 110–119.

Gaskin, D. E. (1982). *The Ecology of Whales and Dolphins*. Heinemann, London.

Gentry, R. L., G. L. Kooyman, and M. E. Goebel (1986). Feeding and diving behavior of northern fur seals. *In* "Fur Seals: Maternal Strategies on Land and at Sea" (R. L. Gentry, and G. L. Kooyman, eds.), pp. 61–78, Princeton University Press, Princeton, NJ.

Gilmore, R. M. (1960). "A Census of the California Gray Whale." *U.S. Fish Wildl. Serv., Spec. Sci. Rep. Fish. 342:* 1–30.

Gingerich, P. D., and D. E. Russell (1990). "Dentition of Early Eocene *Pakicetus* (Mammalia, Cetacea)." *Contrib. Mus. Paleontol. Univ. Mich. 28:* 1–20.

Guinet, C., and J. Bouvier (1995). "Development of Intentional Stranding Hunting Techniques in Killer Whale *(Orcinus orca)* Calves at Croizet Archipelago." *Can. J. Zool. 73:* 27–33.

Hain, J. H. W., S. L. Ellis, R. D. Kenney, P. J. Clapham, B. K. Gray, M. T. Weinrich, and I. G. Babb (1995). "Apparent Bottom Feeding by Humpback Whales on Stellwagen Bank." *Mar Mamm. Sci. 11:* 464–479.

Harrison, R. J., F. R. Johnson, and B. A. Young (1970). "The Oesophagus and Stomach of Dolphins *(Tursiops, Delphinus, Stenella)*." *J. Zool., London 160:* 377–390.

Harrison, R. J., and J. E. King (1980). *Marine Mammals,* 2nd ed. Hutchinson Co. Ltd., London.

Hartman, D. S. (1979). "Ecology and Behavior of the Manatee *(Trichechus manatus)* in Florida." *Am. Soc. Mamm. Spec. Publ. 5:* 1–153.

Hay, K. A., and A. W. Mansfield (1989). *Narwhal-Monodon monoceros* Linnaeus, 1758. *In* "Handbook of Marine Mammals," vol. 4 (S. H. Ridgway, and R. Harrison, eds.), pp. 145–176. Academic Press, London.

Helm, R. C. (1983). "Intestinal Length of Three California Pinniped Species." *J. Zool. London 199:* 297–304.

Herwing, R. P., and J. T. Staley (1986). "Anaerobic Bacteria from the Digestive Tract of North Atlantic Fin Whales *(Balaenoptera physalus)*." *FEMS Microbiol. Ecol. 38:* 361–371.

Herzing, D. L., and B. R. Mate (1984). Gray whale migrations along the Oregon coast. *In* "The Gray Whale, *Eschrichtius robustus* (Lilljeborg, 1861)" (M. L. Jones, S. L. Swartz, and S. Leatherwood, eds.), pp. 289–308. Academic Press, New York.

Heyning, J. E. (1984). "Functional Morphology Involved in Intraspecific Fighting of the Beaked Whale, *Mesoplodon carlhubbsi*." *Can. J. Zool. 62:* 1645–1654.

Heyning, J. E., and J. G. Mead (1996). "Suction Feeding in Beaked Whales: Morphological and Observational Evidence." *Contrib. Sci. Los Angeles Mus. Nat. Hist. 464:* 1–12.

Hill, W. C. O. (1945). "Notes on the Dissection of Two Dugongs." *J. Mammal. 26:* 153–175.

Hinga, K. R. (1978). "The Food Requirements of Whales in the Southern Hemisphere." *Deep-Sea Res. 25:* 569–577.

Hjelseth, A. M., M. Andersen, I. Gjertz, and C. Lydersen (1999). "Feeding Habits of Bearded Seals *(Erignathus barbatus)* from the Svalbard Area, Norway." *Polar. Biol. 21:* 186–193.

Hobson, K. A., and J. L. Sease (1998). "Stable Isotope Analyses of Tooth Annuli Reveal Temporal Dietary Records: An Example Using Steller Sea Lions." *Mar. Mamm. Sci. 14:* 116–129.

Hobson, K.A., and H. E. Welch (1992). "Determination of Trophic Relationships Within a high Arctic Food Web Using $\delta^{13}12C$ and $\delta^{15}del15N$ Analysis." *Mar. Ecol. Prog. Ser. 84:* 9–18.

Hoelzel, A. R., M. Dalheim, and S. J. Stern (1998). "Low Genetic Variation Among Killer Whales *(Orcinus orca)* in the Eastern North Pacific and Genetic Differentiation Between Foraging Specialists." *J. Hered. 89:* 121–128.

Hooper, S. N., M. Paradis, and R. G. Ackman (1973). "Distribution of Trans-6-Hexadecenoic Acid, 7-Methyl-7-Hexadecenoic Acid and Common Fatty Acids in Lipids of the Ocean Sunfish *Mola mola*." *Lipids 8:* 509–516.

Hoyt, E. (1984). *The Whale Called Killer.* E. P. Dutton, New York.

Husar, S. (1975). "A Review of the Literature of the Dugong *(Dugong dugon)*." *Res. Rep. U.S. Fish Wildl. Ser. 4:* 1–30.

Husar, S. (1977). "*Trichechus inunguis*." *Mamm. Spec.* 72: 1–4.

Imbrie, J., A. McIntyre, and T. C. Moore, Jr. (1983). The ocean around North America at the last glacial maximum. *In* "Late Quaternary Environments of the United States" (S. C. Porter, ed.), pp. 230–236. University of Minnesota Press, Minneapolis.

Iverson, S. J. (1993). "Milk Secretion in Marine Mammals in Relation to Foraging: Can Milk Fatty Acids Predict Diet?" *Symp. Zool. Soc. London 66:* 263–291.

Jaquet, N., and H. Whitehead (1999). "Movements, Distribution and Feeding Success of Sperm Whales in the Pacific Ocean, Over Scales of Days and Tens of Kilometers." *Aquat. Mamm. 25:* 1–13.

Kamiya, T. (1962). "On the Intramural Cystic Gland of the Cetacean." *Acta Anat. Nipponica 37:* 339–350.

Kastelein, R. A., J. L. Dubbledam, and M. A. G. de Bakker (1997a). "The Anatomy of the Walrus Head *(Odobenus rosmarus)*. Part 5: The Tongue and Its Function in Walrus Ecology." *Aquat. Mamm. 23:* 29–47.

Kastelein, R. A., N. M. Gerrits, and J. L. Dubbledam (1991). "The Anatomy of the Walrus Head *(Odobenus rosmarus)*. Part 2: Description of the Muscles and of Their Role in Feeding and Haul-Out Behavior." *Aquat. Mamm. 17:* 156–180.

Kastelein, R. A., C. Staal, A. Terlouw, and M. Muller (1997b). Pressure changes in the mouth of a feeding harbour porpoise *(Phocoena phocoena)*. In "The Biology of the Harbour Porpoise" (A. J. Read, P. R. Wiepkema, and P. E. Nachtigall, eds.), pp. 279–291. De Spil Publishers, Woerden, The Netherlands.

Kawamura, A. (1980). "A Review of Food of Balaenopterid Whales." *Sci. Rep. Whales Res. Inst. 32:* 155–197.

Kenyon, K. W. (1969). *The Sea Otter in the North Pacific Ocean,* North Am. Fauna, No. 68. U.S. Dept. Interior, Bureau Sport Fisheries and Wildlife, Washington, DC.

Kimura, T. (2002). "Feeding Strategy of an Early Miocene Cetothere from the Toyama and Akeyo Formations, Central Japan." *Paleontol. Res.* 6: 179–189.

King, J. E. (1983). *Seals of the World,* 2nd ed. Cornell University Press, Ithaca, NY.

Kirsch, P. E., S. J. Iverson, and W. D. Bowen (2000). "Effect of a Low-Fat Diet on Body Composition and Blubber Fatty Acids of Captive Juvenile Harp Seals *(Phoca groenlandica)*." *Physiol. Biochem. Zool. 73:* 45–59.

Krockenberger, M. B., and M. M. Bryden (1994). "Rate of Passage of Digesta Through the Alimentary Tract of Southern Elephant Seals *(Mirounga leonina)* (Carnivora: Phocidae)." *J. Zool. London 234:* 229–237.

Kvitek, R., and C. Bretz (2004) "Harmful Algal Bloom Toxins Protect Bivalve Populations from Sea Otter Predation." *Mar. Ecol. Prog. Ser. 271:* 233–243.

Kvitek, R. G., C. E. Bowlby, and M. Staedler (1993). "Diet and Foraging Behavior of Sea Otters in Southeast Alaska." *Mar. Mamm. Sci. 9:* 168–181.

Lambertson, R. H. (1983). "Internal Mechanism of Rorqual Feeding." *J. Mammal 64*(1): 76–88.

Lambertson, R. H., and R. J. Hintz (2001). "Roraqual Paradox Solved by Discovery of Novel Craniomandibular Articulation." *J. Morph. 248:* 253.

Lambertsen, R. H., and R. J. Hintz (2004). "Maxillomandibular Cam Articulation Discovered in North Atlantic Minke Whale." *J. Mammal. 85:* 446–452.

Lambertsen, R. H., K. J. Rasmussen, W. C. Lancaster, and R. J. Hinz (2005). "Functional Morphology of the Mouth of the Bowhead Whale and Its Implications for Conservation." *J. Mammal. 86:* 342–352.

Lambertsen, R. H., N. Ulrich, and J. Straley (1995). "Frontomandibular Stay of Balaenopteridae: A Mechanism for Momentum Recapture During Feeding." *J. Mammal. 76:* 877–899.

Langer, P. (1988). *The Mammalian Herbivore Stomach,* Fischer, New York.

Langer, P. (1996). "Comparative Anatomy of the Stomach of the Cetacea. Ontogenetic Changes Involving Gastric Proportions-Mesenteries-Arteries." *Z. Saugetierk. 61:* 140–154.

Lanyon, J. M., and H. Marsh (1995). "Digesta Passage Time in the Dugong." *Aust. J. Zool. 43:* 119–127.

Lavigne, D., W. Barchard, S. Innes, and N. A. Oritsland (1982). Pinniped bioenergetics. *In* "Mammals of the Sea," vol. 5., pp. 191–235. Food and Agriculture Organization, Rome.

Lavigne, D., S. Innes, R. E. Stewart, and G. A. J. Worthy (1985). An annual energy budget for northwest harp seals. *In* "Marine Mammals and Harp Seals" (J. R. Beddington, R. J. H. Beverton, and D. M. Lavigne, eds.), pp. 319–336. George Allen and Unwin, London.

Laws, R.M. 1984. Seals. *In* "Antarctic Ecology" (R. M. Laws, ed.), pp. 621–716. Academic Press, London.

Lawson, J. W., Gg. B. Stenson, and D. G. Mckinnon (1994). "Diet of Harp Seals *(Phoca groenlandica)* in Divisions 2J and 3KL During 1991-93." *NAFO Sci. Counc. Stud 21:* 143–154.

Leatherwood, S., and R. R. Reeves (1983). *The Sierra Club Handbook of Whales and Dolphins.* Sierra Club Books, San Francisco.

Le Boeuf, B. J., Y. Naito, A. C. Huntley, and T. Asage (1989). "Prolonged, Continuous, Deep Diving by Northern Elephant Seals." *Can. J. Zool. 67:* 2514–2519.

Levermann, N., A. Galatius, G. Ehlme, S. Rysgaard, and E. W. Born (2003). "Feeding Behaviour of Free-Ranging Walruses with Notes on Apparent Dextrality of Flipper Use." *BMC Ecol. 3:* 9–24.

Lewis, J. S., and W. W. Schroeder (2003). "Mud Plume Feeding, a Unique Foraging Behavior of the Bottlenose Dolphin in the Florida Keys." *Gulf of Mexico Sci. 21:* 92–97.

Lockyer, C. (1981a). "Growth and Energy Budgets of Large Baleen Whales from the Southern Hemisphere." *FAO Fish. Ser. (5) 3:* 379–487.

Lockyer, C. (1981b). "Estimates of Growth and Energy Budgets for the Sperm Whale, *Physeter catadon.*" *FAO Fish. Ser. (5) 3:* 489–504.

Lomolino, M. W., and K. C. Ewel (1984). "Digestive Efficiencies of the West Indian Manatee *(Trichechus manatus)*." *Florida Sci. 47:* 176–179.

Lopez, J. C., and D. Lopez (1985). "Killer Whales *(Orcinus orca)* of Patagonia, and Their Behavior of Intentional Stranding While Hunting Nearshore." *J. Mammal. 66:* 181–183.

Lowry, M. S., C. W. Oliver, and C. Macky. (1990). "Food Habits of California Sea Lions, *Zalophus californianus,* at San Clemente Island, California, 1981-86." *Fish. Bull. 88:* 509–521.

MacFadden, B. J., P. Higgins, M. T. Clementz, and D. S. Jones (2004). "Diets, Habitat Preferences, and Niche Differentiation of Cenozoic Sirenians from Florida: Evidence for Stable Isotopes." *Paleobiology 30:* 297–324.

Marsh, H., G. E. Heinsohn, and A. Spain (1977). The stomach and duodenal diverticula of the Dugong *(Dugong dugon)*. *In* "Functional Anatomy of Marine Mammals" (R. J. Harrison, ed.), pp. 271–295. Academic Press, London.

Marshall, C. D., L. A. Clark, and R. L. Reep (1998b). "The Muscular Hydrostat of the Florida Manatee *(Trichechus manatus latirostris):* A Functional Morphologic Model of Perioral Bristle Use." *Mar. Mamm. Sci. 14:* 290–303.

Marshall, C. D., G. D. Huth, V. M. Edmonds, D. L. Halin, and R. L. Reep (1998a). "Prehensile Use of Perioral Bristles During Feeding and Associated Behaviors of the Florida Manatee *(Trichechus manatus latirostris).*" *Mar. Mamm. Sci. 14*: 274–289.

Marshall, C. D., H. Maeda, M. Iwata, M., and S. A. Furuta, F. Rosas, and R. L. Reep. (2003). "Orofacial Morphology and Feeding Behaviour of the Dugong, Amazonian, West African and Antillean Manatees (Mammalia: Sirenia): Functional Morphology of the Muscular-Vibrissal Complex." *J. Zool. London 259:* 245–260.

Martensson, P.-E., E. S. Nordoy, E. B. Messelt, and A. S. Blix (1998). "Gut Length, Food Transit Time and Diving Habit in Phocid Seals." *Polar Biol. 20:* 213–217.

McConnell, B. J., and M. A. Fedak (1996). "Movements of Southern Elephant Seals." *Can. J. Zool. 74:* 1485–1496.

Merrick, R. L., and T. R. Loughlin (1997). "Foraging Behavior of Adult Female and Young-of-the-Year Steller Sea Lions in Alaskan Waters." *Can. J. Zool. 75:* 776–786.

Milette, L. L., and A. W. Trites (2003). "Maternal Attendance Patterns of Steller Sea Lions *(Eumetopias jubatus)* from Stable and Declining Populations in Alaska." *Can. J. Zool. 81:* 340–348.

Muizon, C. de (1993a). "Walrus-Like Feeding Adaptation in a New Cetacean from the Pliocene of Peru." *Nature 365:* 745–748.

Muizon, C de (1993b). "*Odobenocetops peruvianus:* Una remarcable convergencia de adaptación alimentaria entre morosa y delfin." *Bull. Inst. Fr. Etudes Andine 22:* 671–683.

Muizon, C. de, D. P. Domning, and D. R. Ketten (2002). "*Odobenocetops peruvianus*, the Walrus-Convergent Delphinoid (Mammalia: Cetacea) from Their Early Pliocene of Peru." *Smithson. Contr. Paleobiol. 93:* 223–262.

Murie, J. (1872). "On the Form and Structure of the Manatee." *Trans. Zool. Soc. London 8:* 127–202.

Murray, R., H. Marsh, G. E. Heinsohn, and A. V. Spain (1977). "The Role of the Midgut Cecum and Large Intestine in the Digestion of Sea Grasses by the Dugong (Mammalia: Sirenia)." *Comp. Biochem. Physiol. 56A:* 7–10.

Nemoto, T. (1959). "Food of Baleen Whales with Reference to Whale Movements." *Sci. Rep. Whales Res. Inst. 14:* 149–290.

Nerini, M. (1984). A review of gray whale feeding ecology. *In* "The Gray Whale, *Eschrichtius robustus* (Lilljeborg, 1861)" (M. L. Jones, S. L. Swartz, and S. Leatherwood, eds.), pp. 423–450. Academic Press, New York.

Nilssen, K.T., T. Haug, V. Potelov, V.A. Stasenkov, and Y.K. Timoshenko. (1995). "Food Habits of Harp Seals *(Phoca groenlandica)* During Lactation and Moult in March-May in the Southern Barents Sea and White Sea." *ICES J. Mar. Sci. 52:* 33–41.

Nishiwaki, M., and H. Marsh (1985). Dugong. *In* "Handbook of Marine Mammals" (S. H. Ridgway and R. Harrison, eds.), pp. 1–32. Acad. Press, New York.

Norris, K. S., and B. Mohl (1983). "Can Odontocetes Debilitate Prey with Sound?" *Am. Nat. 122:* 85–104.

Nøttestad, L., and T. Similä (2001). "Killer Whales Attacking Schooling Fish: Why Force Herring from Deep Water to the Surface?" *Mar. Mamm. Sci. 17:* 343–352.

Oliver, J. S., P. N. Slattery, E. F. O'Connor, and L. F. Lowry (1983). "Walrus, *Odobenus rosmarus,* Feeding in the Bering Sea: A Benthic Perspective." *Fish. Bull.* 81: 501–512.

Olsen, M. A., E. S. Nordoy, A. S. Blix, and S. D. Mathiesen (1994). "Functional Anatomy of the Gastrointestinal System of the Northeastern Atlantic Minke Whale *(Balaenoptera acutorostrata).*" *J. Zool. London 234:* 55–74.

Orton, L. S., and P. F. Brodie (1987). "Engulfing Mechanics of Fin Whales." *Can. J. Zool. 65:* 2898–2907.

Panigada, S., M. Zanardelli, S. Canese, and M. Jahoda. (1999). How Deep Can Baleen Whales Dive? *Mar. Ecol. Prog. Ser. 187:* 309–311.

Parrish, F. A., M. P. Craig, T. J. Ragen, G. J. Marshall, and B. M. Buhleier (2000). "Identifying Diurnal Foraging Habitat of Endangered Hawaiian Monk Seals Using a Seal-Mounted Video Camera." *Mar. Mammal Sci. 16:* 392–412.

Perryman, W. L., M. A. Donahue, J. L. Laake, and T. E. Martin (1999). "Diel Variation in Migration Rates of Eastern Pacific Gray Whales Measured with Thermal Imaging Sensors." *Mar Mamm. Sci. 15:* 426–445.

Pierce, G. J., P. R. Boyle, J. Watt, and M. Grisley (1993). "Recent Advances in Diet Analysis of Marine Mammals." *Symp. Zool. Soc. London 66:* 241–261.

Pike, G. C. (1962). "Migration and Feeding of the Gray Whale *(Eschrichitus gibbosus)*." *J. Fish. Res. Bd. Can. 19:* 815–838.

Pivorunas, A. (1979). "The Feeding Mechanisms of Baleen Whales." *Am. Sci.* 67: 432–440.

Poole, M. M. (1984). Migration corridors of gray whales along the central California coast, 1980-1982. *In* "The Gray Whale, *Eschrichtius robustus* (Lilljeborg, 1861)" (M. L. Jones, S. L. Swartz, and S. Leatherwood, eds.), pp. 389–408. Academic Press, New York.

Preen, A. (1995). "Diet of Dugongs: Are They Omnivores?" *J. Mammal. 76:* 163–171.

Pyle, P., M. J. Schramm, C. Keiper, and S. D. Anderson (1999). "Predation on a White Shark *(Carcharodon carcharias)* by a Killer Whale *(Orcinus orca)* and a Possible Case of Competitive Displacement." *Mar. Mamm. Sci. 15:* 563–568.

Quiring, D. P., and C. F. Harlan (1953). "On the Anatomy of the Manatee." *J. Mamm. 34:* 192–203.

Ralls, K., B. B. Hatfield, and D. B. Siniff (1995). "Foraging Patterns of California Sea Otters as Indicated by Telemetry. *Can. J. Zool. 73:* 523–531.

Reed, J. Z., D. J. Tollit, P. M. Thompson, and W. Amos (1997). "Molecular Scatology: The Use of Molecular Genetic Analysis to Assign Species, Sex and Individual Identity to Seal Faeces." *Mol. Ecol. 6:* 225–234.

Reilly, S. B. (1981). Population Assessment and Population Dynamics of the California Gray Whale, *Eschrichtius robustus*. Ph.D. thesis, University of Washington, Seattle.

Reynolds, J. E., III, and D. E. Odell (1991). *Manatees and Dugongs*. Facts on File, New York.

Reynolds, J. E., III, and S. A. Rommel (1996). "Structure and Function of the Gastrointestinal Tract of the Florida Manatee, *Trichechus manatus latirostris*." *Anat. Rec*. 245: 539–558.

Rice, D. W., and A. A. Wolman (1971). "The Life History and Ecology of the Gray Whale *(Eschrichtius robustus)*." *Am. Soc. Mammal. Spec. Publ. 3:* 1–42.

Rice, D. W., and A. A. Wolman (1990). "The Stomach of *Kogia breviceps*." *J. Mammal. 71:* 237–242.

Riedman, M. (1990). *The Pinnipeds: Seals, Sea Lions, and Walruses*. University of California Press, Berkeley.

Riedman, M., and J. Estes (1990). "The Sea Otter *(Enhydra lutris)*: Behavior, Ecology and Natural History." *U.S, Fish Wildl. Serv., Biol. Rep*. No. 90.

Robson, B. W., M. E. Goebel, J. D. Baker, R. R. Ream, T. R. Loughlin, R. C. Francis, G. A. Antonelis, and D. P. Costa (2004). "Separation of Foraging Habitat Among Breeding Sites of a Colonial Marine Predator, the Northern Fur Seal *(Callorhinus ursinus)*." *Can. J. Zool. 82:* 20–29.

Rodhouse, P. G., T. R. Arnbom, M. A. Fedak, J. Yeatman, and A. W. A. Murray (1992). "Cephalopod Prey of the Southern Elephant Seal, *Mirounga leonina* L." *Can. J. Zool. 70:* 1007–1015.

Rossbach, K. A., and D. L. Herzing, (1997). "Underwater Observations of Benthic-Feeding Bottlenose Dolphins *(Tursiops truncatus)* near Grand Bahama Island, Bahamas." *Mar. Mamm. Sci. 13:* 498–504.

Schumacher, U., P. Mein, J. Plotz, and U. Welsch (1995). "Histological, Histochemical, and Ultrastructural Investigations on the Gastrointestinal System of Antarctic Seals: Weddell Seal *(Leptonychotes weddellii)* and Crabeater Seal *(Lobodon carcinophagus)*." *J. Morphol. 225:* 229–249.

Sekiguchi, K., and P. B. Best (1992). "New Information on the Feeding Habits and Baleen Morphology of the Pygmy Right Whale *Caperea marginata*." *Mar. Mamm. Sci. 8:* 288–293.

Sharpe, F. A., and L. M. Dill (1997). "The Behavior of Pacific Herring Schools in Response to Artificial Humpback Whale Bubbles." *Can. J. Zool. 75:* 725–730.

Slijper, E. J. (1979). *Whales*. Hutchison, London.

Slip, D. J. (1995). "The Diet of Southern Elephant Seals *(Mirounga leonina)* from Heard Island." *Can. J. Zool. 73:* 1519–1528.

Smith, S. J., S. J. Iverson and W. D. Bowen (1997). "Fatty Acid Signatures and Classification Trees: New Tools for Investigating the Foraging Ecology of Seals." *Can. J. Fish. Aquat. Sci. 54:* 1377–1386.

Smolker, R., A. Richards, R. Conner, J. Mann, and P. Berggren (1997). "Sponge Carrying by Dolphins (Delphinidae, *Tursiops sp.*): A Foraging Specialization Involving Tool Use?" *Ethology 103:* 454–465.

Steinberg, P. D., J. A. Estes, and F. C. Winter (1995). "Evolutionary Consequences of Food Chain Length in Kelp Forest Communities." *Proc. Natl. Acad. Sci. U.S.A. 92:* 8145–8148.

Stewart, B. S. (1997). "Ontogeny of Differential Migration and Sexual Segregation in Northern Elephant Seals." *J. Mammal. 78:* 1101–1116.

Stewart, B. S., and R. L. DeLong (1993). "Seasonal Dispersion and Habitat Use of Foraging Northern Elephant Seals." *Symp. Zool. Soc. London 66:* 179–194.

Stewart, B. S., and R. L. DeLong (1994). Postbreeding foraging migrations of northern elephant seals. *In* "Elephant Seals" (B. J. Le Boeuf, and R. M. Laws, eds.), pp. 290–309. University of California Press, Berkeley.

Stewart, B. S., and R. L. DeLong (1995). "Double Migrations of the Northern Elephant Seal, *Mirounga angustirostris.*" *J. Mammal.* 76: 196–205.

Stirling, I. (1988). *Polar Bears.* University of Michigan Press, Ann Arbor.

Stirling, I. (2002). Polar Bears and Seals in the Eastern Beaufort Sea and Amundsen Gulf: A Synthesis of Population Trends and Ecological Relationships over Three Decades." *Arctic 55:* 59–76.

Sumich, J. L. (1986). "Latitudinal Distribution, Calf Growth and Metabolism, and Reproductive Energetics of Gray Whales, *Eschrichtius robustus.* Ph.D. thesis, Oregon State University, Corvallis.

Sumich, J. L. (2001). "Growth of Baleen of a Rehabilitating Gray Whale Calf." *Aquat. Mamm. 27:* 234–238.

Swingle, W. M., S. G. Barco, T. D. Pitchford, W. A. McLellan, and D. A. Pabst (1993). "Appearance of Juvenile Humpback Whales Feeding in the Nearshore Waters of Virginia." *Mar. Mamm. Sci. 9:* 309–315.

Takahashi, K., and F Yamasaki (1972). "Digestive Tract of Ganges Dolphin, *Platanista gangetica.* II. Small and Large Intestines." *Okajimas Folia Anat. Jpn. 48:* 427–452.

Tarpley, R. (1985). "Gross and Microscopic Anatomy of the Tongue and Gastrointestinal Tract of the Bowhead Whale *(Balaena mysticetus).* Ph.D dissertation, Texas A and M University, College Station, Texas.

Tarpy, C. (1979). "Killer Whale Attack!" *Nat. Geogr. 155:* 542–545.

Thewissen, J. G. M., S. I. Madar, and S. T. Hussain (1996). "*Ambulocetus natans,* an Eocene Cetacean (Mammalia) from Pakistan." *CFS Cour Forschungsinst. Senckenberg 191:* 1–86.

Tomilin, A. G. (1957). *Animals of the USSR and Neighboring Countries,* vol. 9. Cetaceans. Izdvo AN SSSR, Moscow.

Vaughn, T. A. (1986). *Mammalogy,* 3rd ed. W.B. Saunders, Philadelphia.

Wathne, J. A., T. Haug, and C. Lydersen (2000). "Prey Preferences and Niche Overlap of Ringed Seals *Phoca hispida* and Harp Seals *P. groenlandica* in the Barents Sea." *Mar. Ecol. Prog. Ser. 194:* 233–239.

Weinrich, M. T., M. R. Schilling, and C. R. Belt (1992). "Evidence for Acquisition of a Novel Feeding Behaviour: Lobtail Feeding in Humpback Whales, *Megaptera novaeangliae.*" *Anim. Behav. 44:* 1059–1072.

Werth, A. (2000a). "A Kinematic Study of Suction Feeding and Associated Behavior in the Long-Finned Pilot Whale, *Globicephala melas* (Traill)." *Mar. Mamm. Sci. 16:* 299–314.

Werth, A. (2000b). Feeding in marine mammals. *In* "Feeding" (K. Schwenk, ed.), pp. 487–526. Academic Press, San Diego, CA.

Williams, T. M., L. A. Fuiman, M. Horning, and R. W. Davis (2004). "The Cost of Foraging by a Marine Predator, the Weddell Seal *Leptonychotes weddellii*: Pricing by the Stroke." *J. Exp. Biol. 207:* 973–982.

Williams, T. M., S. F. Shippe, and M. J. Rothe (1996). Strategies for reducing foraging costs in dolphins. *In* "Aquatic Predators and Their Prey" (S. P. R. Greenstreet, and M. L. Tasker, eds.), pp. 4–9. Blackwell Science, Oxford.

Yablokov, A. V. (1966). *Variability of Mammals.* NSF/Smithsonian Institution, Springfield, VA. NTIS.

Yablokov, A. V., V. M. Bel'kovich, and V I. Botisov (1972). *Whales and Dolphins.* Nauka, Moscow.

Yamasaki, F., S. Komatsu, and T. Kamiya (1980). "A Comparative Study on the Tongues of Manatee and Dugong (Sirenia)." *Sci. Rep. Whales Res. Inst. 32:* 127–144.

13

Reproductive Structures, Strategies, and Patterns

13.1. Introduction

Successful reproduction over the life span of individuals, or the transmission of genetic information from parents to offspring, can be measured in terms of reproductive fitness. Individuals with high fitness produce more offspring than less fit individuals and, through natural selection, transfer more copies of their genes to subsequent generations. Like other mammals, the exaggerated difference in the costs of gamete production between male and female marine mammals (sperm is abundant and cheap; eggs are rare and expensive) has set the stage for the evolution of very different roles in the reproductive process for males and females of the same species. Female mammals, in addition to producing relatively expensive eggs, bear all the costs of pregnancy, lactation, and most if not all other costs associated with postnatal care. Consequently, males are free to invest their energy in activities associated with controlling access of other males to females, competing with other males in other ways (e.g., sperm competition), or mating with as many females as possible.

All groups of marine mammals share several features of reproduction that stem from their placental mammalian heritage. These basic conditions of reproduction are reflected in the strategies for care and nourishment of the young and in the resulting life histories of these animals (discussed in this chapter and the following chapters). Sex determination is based on male **heterogamy** (males have XY sex chromosomes) and female **homogamy** (females have XX sex chromosomes). Male-to-female ratios are very close to 1:1 (see Fredga, 1988, for mammalian exceptions). In contrast to the variety of sex determination mechanisms employed by some other vertebrates groups, male heterogamy of mammals has constrained the mechanics of mammalian sex determination into an evolutionary dead end. Sex ratios are almost always very close to 1:1, even though other sex ratios might be adaptive in some reproductive conditions, such as moderate to intense polygyny (see Bull and Charnov, 1985, for a review of this topic). Males that are completely excluded from reproduction by the actions of other polygynous males have zero reproductive fitness, and male heterogamy disallows the possibility of gender transformation (as is possible in some other vertebrate groups) to achieve higher fitness as females.

In all placental mammals, embryonic and fetal development occurs within the mother's uterus during extended **gestation** periods. Pregnancy is maintained by the production of chorionic gonadotropin, a placental hormone that blocks further ovulation until birth (parturition) occurs. At birth, most young marine mammals are precocious, capable of swimming or in the case of seals either moving on ice or land or in some pinniped species swimming very shortly after birth. For a time after birth, ranging from 4 days in hooded seals (Bowen *et al.*, 1985) to several years in some otariids, walruses, polar bears, odontocetes, and sirenians, mothers produce fat-rich milk for their developing, nutritionally dependent offspring. Neonates, being necessarily smaller than their mothers, have high surface-to-volume ratios with thinner fat or blubber layers and their heat-loss rate is therefore higher than that of adults (see Chapter 9). Therefore, species at the smaller end of the range of observed sizes of marine mammals tend to produce relatively larger offspring (relative to their mother's sizes; Figure 13.1).

The reproductive organs and associated structures of marine mammals are typically mammalian, with some specialization to accommodate body streamlining and, in pinnipeds, to protect testes and nipples during terrestrial locomotion.

Structural specializations for reproduction are described in the following section. We also examine the variety of mating systems seen among marine mammals. In this chapter the term "breeding" includes two different components of the reproductive process: **mating** (mate selection and copulation) and **parturition** (calving in whales and sirenians; pupping in pinnipeds and sea otters). Obviously, females are involved in both, whereas males are only involved in mating. Mating systems of mammals are often correlated with a species' broader social organization, and marine mammals are no exception in this regard. For many marine mammal species reproductive activities anchor one end of extended annual migrations. Consequently, for mysticete cetaceans and some pinnipeds, mating and parturition typically occur in the same locality at nearly the same time of the year. For other marine mammals, such as sea otters (Jameson and Johnson, 1993) and

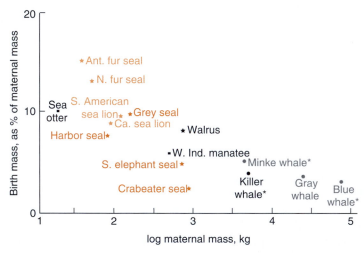

Figure 13.1. Birth mass, as a percentage of maternal mass for 15 species of marine mammals, spanning several orders of magnitude in body mass. Sources listed in Tables 13.1 and 13.5 and Kovacs and Lavigne, 1986, 1992a. *Weights estimated from length and girth measurements.

many odontocetes (Barlow, 1984), the timing of breeding is spread more broadly through the year with only a degree of seasonality. These species often remain in limited home ranges for both mating and parturition. Consequently, some discussion of social organization in a broader context is also addressed in this chapter.

13.2. Anatomy and Physiology of the Reproductive System

13.2.1. Reproductive Structures

Male—In phocid seals the testes are inguinal and lie lateral to the penis (Figure 13.2). In otariids, as in other carnivores, the testes lie outside the inguinal ring, a condition termed scrotal. Walrus testes are situated outside the muscular abdominal wall in the blubber lateral to the base of the penis but exhibit a tendency toward a scrotal arrangement during the mating season (Fay, 1982) or when hauled out in warm weather during the summer. Pinnipeds possess a **baculum,** or penis bone, which is the ossified anterior end of the corpus cavernosum (Figure 13.3). The baculum of the walrus and phocid seals is large relative to that of otariids (King, 1983), but otariids have a more complex shape relative to phocids or the walrus (Morejohn, 1975). The testes are ovoid, circular in cross section, and twice as long as they are broad. The weight of each testis increases with age early in life, with a rapid increase associated with the onset of sexual maturity.

Sperm are produced in the testes and then pass into the epididymis, which as in all mammals are long, coiled tubes that lie alongside each testis. The sperm mature as they pass along this tube and are stored near its end, the tail of the epididymis. The ductus deferens, which arises from the epididymis, passes, together with the testicular ligament and blood vessels, through an opening into the abdominal cavity. The sperm pass from the short ductus deferens to the urethra and then out of the penis during ejaculation. The prostate gland, which contributes fluid that nourishes the sperm, surrounds the neck of the bladder (Laws and Sinha, 1993).

The only conspicuous external difference between the sexes in cetaceans is the distance between the anal and genital slits that, in males, is approximately 10% of the body length, but in females the genital slit appears to be continuous with the anal slit and has a

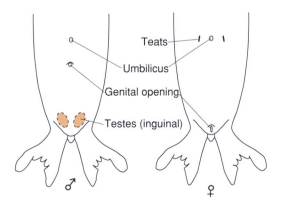

Figure 13.2.   External sexual differences of pinnipeds. (Redrawn from Laws and Sinha, 1993.)

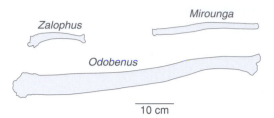

Figure 13.3.   Bacula of adult males of three pinniped genera.

mammary slit situated on either side (Figure 13.4). In all cetaceans, the testes are located
in the abdominal cavity, slightly behind and to the side of the kidneys. They are elongated, cylindrical, and oval or round in cross section. The testicle-weight-to-body-
weight ratio of cetaceans (Figure 13.5) is very high in some species, among the greatest
recorded for mammals. Odontocete whales in general have testes that are 7–24 times
larger than would be predicted based on body size. Right whales have testes than can
weigh up to 900 kg in combination, comprising almost 10% of their body mass
(Atkinson, 2002). The typical mammalian accessory glands are represented only by the
prostate, which surrounds the urogenital canal.

Neither cetaceans nor sirenians possess a baculum. The penis is totally concealed in
the body when in the retracted state (Figure 13.6). The fibroelastic penis of large baleen
whales reaches 2.5–3 m in length and up to 30 cm in diameter (Slijper, 1966). In all
cetaceans, the penis is erected by muscle fibers and not by vasodilation. Both the fibroelastic penis and associated penis retractor muscle appear to be synapomorphies shared
by artiodactyls and cetaceans (Pabst *et al.*, 1998). Based on anatomical constraints, copulation is assumed to take place belly-to-belly in all cetaceans, although it is rarely
observed.

The testes in manatees and dugongs are abdominal and the seminal vesicles are
large (Figure 13.7). The nonglandular prostate of manatees is composed of erectile
muscular tissue (Caldwell and Caldwell, 1985); the prostate is absent in the dugong.

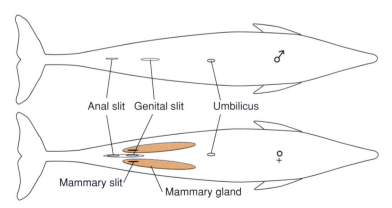

Figure 13.4.   External sexual differences of cetaceans.

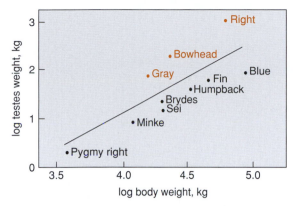

Figure 13.5. General relationship between body size and testis size of ten species of mysticetes. The three species with relatively larger testes (positioned above the diagonal line) are assumed to produce proportionally larger quantities of sperm with which to compete with the sperm of other males. (Redrawn from Brownell and Ralls, 1986.)

Bulbourethral-type glandular tissue is dispersed among the muscles close to the base of the penis in dugongs (Nishiwaki and Marsh, 1985).

Female—The ovaries in pinnipeds lie close to the posterolateral wall of the abdominal cavity near the kidneys and are enclosed in large peritoneal bursae (Figure 13.8). The size and shape of the ovaries depends on the age and reproductive status of the female. During pregnancy the follicle undergoes major physiological changes and develops into a **corpus luteum,** which produces hormones that maintain the placenta for the duration of gestation. After birth, the corpus luteum degenerates into a small, hard structure termed a **corpus albicans**. In marine mammals the duration of a recognizable corpus albicans varies between species and individuals; remnants of the corpus albicans may persist for several years in some species (Laws and Sinha, 1993). The uterus is bicornate (two horned) as in most mammals, but posteriorly the horns join to form a short uterus. The vagina is long, and it is separated from the vestibule by a fleshy hymeneal fold. Both the anus and vestibule open into a common furrow, which is surrounded by muscular tissue

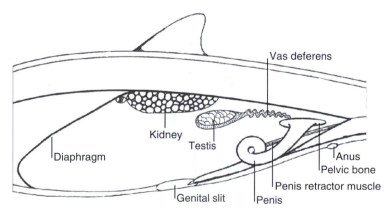

Figure 13.6. Cetacean male reproductive system. (Adapted from Slijper, 1966.)

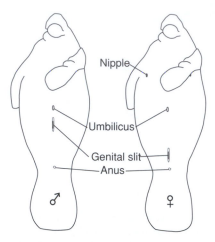

Figure 13.7. External sexual differences of sirenians. (Adapted from Reynolds and Odell, 1991.)

that acts as a sphincter. An os clitoridis is present in pinnipeds, but it is very small and its appearance even within a single species is irregular (King, 1983).

In cetaceans, the two ovaries lie in the abdominal cavity (Figure 13.9); in toothed whales the ovaries are held within well-developed deep ovarian pouches, but in baleen whales they are more exposed. Ovaries of baleen whales are oval, elongated, and convoluted and in toothed whales the ovaries are spherical and usually smooth. Ovaries of mature whales resemble a bunch of grapes in which the "grapes" are protruding follicles of different stages of development. For details of the morphology and histology of mysticete ovaries see Perrin and Donovan (1984). Follicular development is described in Laws (1962), Slijper (1966), and Harrison (1969) for fin whales and in Rice and Wolman (1971) for gray whales.

A peculiarity of most whales is that the corpora albicans remain for the entirety of an animal's life (Figure 13.10), providing a record of past ovulations. This makes it possible

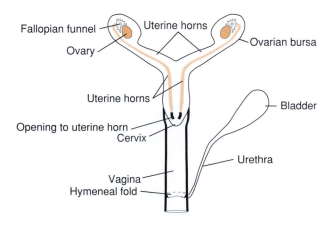

Figure 13.8. Pinniped female reproductive system. (Adapted from King, 1983.)

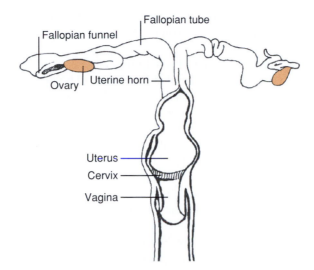

Figure 13.9. Cetacean female reproductive system.

to examine the reproductive history and estimate ages of individual whales; each corpus albicans represents one ovulation (although not necessarily a pregnancy; see also Chapter 14). In mysticetes unlike several odontocetes including the phocoenids in which the left ovary usually predominates and is more active (at least initially), both ovaries are fully functional and able to ovulate from the onset of sexual maturity. The uterus of

Figure 13.10. Left ovary of a Pacific whitesided dolphin showing corpora albicantia. (From Harrison and Bryden, 1988.)

cetaceans is bicornate as in pinnipeds. The vagina is covered with longitudinal and transverse folds termed "pseudocervices," which is a feature shared by artiodactyls and cetaceans (Pabst *et al.*, 1998).

Dugong ovaries are very large (up to 15 cm long), flattened, and ellipsoidal. Each lies in a peritoneal pouch hidden in the dorsal abdominal wall posterior to the kidney and lateral to the ureter (Nishiwaki and Marsh, 1985). The ovaries of manatees consist of large masses of bead-like spherules (Figure 13.11) and lack a heavy capsular coat (Caldwell and Caldwell, 1985). The uterus of sirenians is bicornate. The dugong vagina is a long straight tube with a very capacious lumen that possesses a raised shield-like region of keratinized material in its vault into which the cervix projects. The clitoris possesses a large conical glans (Nishiwaki and Marsh, 1985).

The ovary of the sea otter is a roughly lens-shaped, compressed oval with simple or complexly branched surface fissures (Estes, 1980). Changes in ovary form and in the gross anatomy and histology of the bicornate uterus through various stages of estrus development and through pregnancy are described by Sinha *et al.* (1966).

The size and shape of the mammary glands differ between phocids and otariids. Otariids and the walrus possess four mammary teats, located on the ventral abdominal wall anterior and posterior to the umbilicus. The bearded seal and the monk seals also have four teats, whereas all other phocids have only two abdominal teats adjacent to or posterior to the umbilicus (see Figure 13.2; King, 1983; Stewart and Stewart, 2002). The nipples of all pinnipeds are retractile.

The mammae of cetaceans consist of long, flat glands located along the belly on either side of the urogenital fold (see Figure 13.4). Their ducts open to a single nipple on each

Figure 13.11.   Multiple corpora *(arrows)* on the ovary of a mature Florida manatee. (From Marmontel, 1995.)

side (retained within mammary slits), and both are retracted in a skin pouch at rest (Yablokov *et al.*, 1972). The thickness (depth) of the glands depends on the sexual status of the female. The immature gland in blue and fin whales is thin (about 2.5 cm thick) and thickens with the onset of sexual maturity. The depth and color of the glands are important criteria in assessing the reproductive status of an individual female (Lockyer, 1984).

The mammary glands of sirenians consist of a single teat on each side located in the axilla (armpit) almost at the posterior border of the flipper (see Figure 13.7; Harrison and King, 1980).

Sea otters have two functional mammae, and polar bears have four. In both species, this may be a function of the evolution towards smaller litter sizes in the marine taxa (further discussed in Chapter 14; Derocher, 1990; Stewart and Stewart, 2002). Supernumerary teats (i.e., females with five or six teats) are not uncommon in polar bears.

13.2.2. Ovulation and Estrus

Reproduction in marine mammals is characterized by a cycle of events regulated by both the nervous and endocrine systems and especially by a reciprocal feedback regulation between reproductive hormones and the organs that secrete them. For many species this cycle is superimposed upon or entrained into annual patterns of migration between areas used for reproduction and those used for foraging (described later in this chapter and in Chapter 12).

The ovulation cycle of marine mammals includes a period of follicle growth and release of an ovum (usually only one) from an ovarian follicle and the subsequent development of a corpus luteum at the site of the ruptured follicle. Both the follicle and the corpus luteum contribute hormones necessary to prepare the uterus for implantation of the egg if fertilization does occur. The term **estrus** defines both the time period of maximum reproductive receptivity of female mammals as well as the associated receptive behavior. Estrus occurs simultaneously with and immediately following ovulation, but the timing varies among marine mammal groups. Phocids and otariids have a **postpartum estrus** in which estrus closely follows parturition, when there is a rapid regression of the corpus luteum from the previous pregnancy in one ovary, accompanied by rapid growth of follicles in the opposite ovary; one of those follicles ovulates to initiate the next cycle. Most otariids mate between 4–14 days after parturition before they go to sea, although a few species mate up to month after parturition, following the resumption of foraging trips (e.g., Steller sea lions) (Boness, 2002). The time delay between parturition and estrus and mating in phocids is more variable, but estrus is always near the end of lactation or soon after weaning has occurred (Riedman, 1990). Walruses are **polyestrous**; females enter one estrus period about 4 months postpartum when males are not producing sperm, then again about 6 months later when males are actively producing sperm and fertilization can occur. Phocids and otariids are thought to be monestrous, with only one estrus cycle each year (Riedman, 1990). However, this has been challenged.

Estrus and mating is more difficult to document in cetaceans than in pinnipeds, and so less is known of the timing of these reproductive events for most species. Gestation periods usually last between 11 and 18 months, and almost all cetaceans exhibit multiyear reproductive cycles (Table 13.1), with mating and parturition usually separated by at least one, and sometimes 2 or more years. High latitude odontocetes, such as beluga whales, and all migratory mysticetes have well-defined seasonal periods of estrus and

Table 13.1. Birth Weights, Total Gestation Periods, and Reproductive Intervals in Some Marine Mammals

Species	Approximate Birth Mass (kg)	Total Gestation (mo)	Reproductive Interval (yrs)	Source
Pinnipeds				
Phocids				
Harbor seal	11	9.5–12.5	1	Bowen (1991)[*]
Grey seal	16	11	1	Bowen (1991)[*]
N. elephant seal	40	11	1	Bowen (1991)[*]
Otariids				
N. fur seal	6	12	1	Bowen (1991)[*]
CA. sea lion	8	11	1	Bowen (1991)[*]
Antarctic fur seal	6	12	1	Bowen (1991)[*]
Walrus	50	14–16	2–4	Sease and Chapman (1988)
Cetaceans				
Odontocetes				
Dall's porpoise	na	7–11.4	na	Gaskin et al. (1984)[*]
Harbor porpoise	na	8–11	na	Gaskin et al. (1984)[*]
Bottlenose dolphin	na	12	na	Perrin and Reilly (1984)[*]
Spinner dolphin	na	9.5–10.7	na	
Killer whale	200	16–17	5	Ford et al. (1994)
Sperm whale	1050	14.5–16.5	3	Lockyer (1981)
Mysticetes				
Minke whale	230	10	1–2	Mitchell (1986)
Gray whale	920	11–13	2	Sumich (1986a,b)
Blue whale	2500	11–12	2–3	Rice (1986)
Sirenians				
Manatee	30	12–13	2–5	Reynolds and Odell (1991)
Dugong	na	13.9	3–7	Marsh (1995)
Other Marine Mammals				
Sea otter	2	6–7	1	Jameson and Johnson (1993)
Polar bear	0.7	8	2–4	Stirling (1988); Derocher and Stirling (1995)

Symbols: na=not available
[*]See for original sources.

mating (Evans, 1987), although mating-like behaviors can be observed in many of these species throughout the year.

In mysticetes, some uncertainty exists concerning the timing of ovulation (and subsequent conception) for many species. For example, in an historical examination of gray whale ovaries, Rice and Wolman (1971) found that all ($n = 28$) mature southbound female gray whales not carrying near-term fetuses had recently ovulated and were presumably in very early stages of pregnancy. These animals were collected off central California, so the timing of ovulation and conception was assumed by Rice and Wolman (1971) to have occurred during earlier, more northerly portions of the southern migration. Vigorous and extensive courting and mating behaviors of adults, presumably with the goal of fertilization, are observed easily during the latter portion of their southern migration and especially for the month or so spent in winter "breeding" lagoons at the southern terminus of this migration. If female gray whales have already ovulated and are pregnant before entering winter lagoons, our estimate of the timing of conception derived from direct histological examination is at odds with evidence based on field observations of copulatory behavior.

Manatees and dugongs are **polyovular,** producing numerous corpora lutea per pregnancy (see Figure 13.11). Such large numbers of corpora may be necessary to produce sufficient progesterone to maintain pregnancy (Marmontel, 1995).

In sea otters mating can take place with or without the formation of pair bonds. If the pair does remain together, the period is typically from 1–4 days. Sexual receptivity of females ends soon after pair-bond dissolution, suggesting that estrus averages 3–4 days (Riedman and Estes, 1990).

Field and captive observations indicate that polar bears are induced ovulators, with ovulation occurring only when the female is stimulated by the presence and activity of a male. Mating occurs in the spring but implantation of the blastocyst is delayed until autumn. At the time of implantation females stop feeding and enter dens in which they spend the winter without food and water (Ramsay and Stirling, 1988). Polar bears have reproductive rates that are low when compared to terrestrial mammals. The average reproductive interval for females is 2–3 years or sometimes even 4 years. Low reproductive output is due to late maturation, small litters, and long interbirth intervals (Derocher and Stirling, 1995).

13.2.3. Pregnancy and Birth

The duration of gestation in placental mammals ranges from 18 days in jumping mice to 22–24 months in elephants (Vaughan, 1999). In terrestrial species, gestation duration is roughly related to the size of the fetus; larger fetuses require longer gestation periods to develop. For some marine mammals that do not have postpartum estrus, both parturition and fertilization occur in the same geographic location and at the same time of year, but in successive years, and gestation periods for most species are between 11 and 17 months in duration regardless of body size (Table 13.1). Annual patterns are very practical for mammals that disperse broadly across wide ranges of ocean outside the breeding period. However, keeping fairly rigid annual cycles means that some animals must slow down their developmental cycle and others must escalate the timing of development. Pinnipeds, sea otters, and polar bears have seasonally delayed implantation extended to fit an annual cycle, whereas many large cetaceans must accelerate fetal development to complete gestation within 1 year.

All species of pinnipeds give birth out of water, and some also mate when on shore, so gestation periods can be determined by observing known or identified individuals at least intermittently from the date of copulation to subsequent birth. However, cetaceans and sirenians copulate and give birth in water, usually out of human sight. Observations of cetacean births in the wild are few (for descriptions of gray whale births, see Balcomb, 1974; Mills and Mills, 1979; for right whales, see Best, 1981). Furthermore, our attempts to visually determine when fertilization actually occurs are complicated by copulatory behavior by some cetaceans that serves nonreproductive social purposes rather than insemination (Johnson and Norris, 1994).

The postpartum estrus that characterizes most species of pinnipeds establishes a time frame of nearly 1 year between copulation and birth, even when an entire year is not required to complete fetal development. Adjustments to fit gestation periods of less than 1 year into an annual time frame are accomplished by a reproductive phenomenon known as **seasonal delayed implantation** (see Sandell, 1990, for a review of this topic). In seasonal delayed implantation, the zygote undergoes several cell divisions to form a hollow ball of a few hundred cells, the blastocyst, which then lies undifferentiated in the female's uterus for periods of a few weeks to several months, depending on the species (Table 13.2). Following this delay, the blastocyst becomes implanted into the inner wall of the uterus, a placental connection develops between the embryo and the uterine wall, and normal embryonic growth and development resume through the remainder of the gestation period (Figure 13.12). Delayed implantation substantially extends the developmental gestation period allowing mating to take place when adults are aggregated in rookeries rather than when dispersed at sea. By providing flexibility in the length of gestation, this reproductive strategy places parturition and mating into a relatively brief period of time and allows the young to be born when conditions are optimal for their survival. The critical period for mating is not necessarily immediately following birth but may in some species be associated with the timing of weaning, or the end of the postnursing fast, when food must be available to the inexperienced youngsters. The physiological and hormonal mechanisms controlling delayed implantation are not well understood. Also, because most marine mammals are at sea and inaccessible when blastocyst implantation occurs, the duration of the delay in some species has not been established with certainty (Riedman, 1990). However, delayed implantation is known or thought, to occur in all species of pinnipeds, sea otters, and polar bears as well as in several species of terrestrial mammals but not in sirenians or cetaceans. The reproductive strategy of delayed implantation requires precise timing of implantation in order to achieve the previously noted benefits. Temte (1994) reported that year-to-year precision in the mean date of birth is common for colonies of many pinniped species and he also suggested significant latitudinal variation in the timing of birth for several species.

Table 13.2. Durations of Developmental Gestations and Delayed Implantations of Some Marine Mammals, Which, Together, Equal the Total Gestation Durations Listed in Table 13.1

Species	Developmental Gestation (mo)	Delayed Implantation (mo)	Source
Pinnipeds			
Phocids			
Harbor seal	8–9.5	1.5–3	Riedman (1990)[*]
Gray seal	7.8	3.4	Riedman (1990)[*]
N. elephant seal	7	4	Riedman (1990)[*]
Otariids			
N. fur seal	8	4	Riedman (1990)[*]
CA. sea lion	8	3	Riedman (1990)[*]
Antarctic fur seal	8	4	Riedman (1990)[*]
Walrus	10–11	4–5	Riedman (1990)[*]
Other Marine Mammals			
Sea otter	4	2–3	Jameson and Johnson (1993)
Polar bear	4	4	Stirling (1988)

[*]Original sources listed in Riedman (1990), Jameson and Johnson (1993), and Stirling (1988).

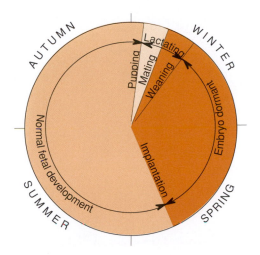

Figure 13.12. Generalized pinniped annual reproductive cycle, including durations of delayed implantation (*dark*) and developmental gestation (*light*). (Adapted from Sumich, 1996.)

The precision and latitudinal variation in the timing of birth in the harbor seal have been shown to be a response to photoperiod shifts (Temte, 1994).

Single-offspring births typify all species of marine mammals except polar bears. Confirmed multiple births are rare for pinnipeds or cetaceans in the wild, and only a few cases of captive-born "twins" and two verified cases of "triplets" (both by the same grey seal) have been reported (Spotte, 1982). The eventual success of seal pups born as a litter in the wild is not known but the large energy costs of lactation likely prohibits successful weaning of more than one pup, and hence serves as strong selection pressure against multiple birthing.

The gestation period of mysticetes ranges between 10 and 13 months (Table 13.1, Figure 13.13). Long-distance fasting migrations to low latitudes in winter months are typical of large mysticetes in both hemispheres, with both calving and presumed mating occurring in warm and often protected waters (Best, 1982). A general picture of the relationship between these migrations and reproductive timing and behavior is well-illustrated by gray whales. The migratory timing of gray whales has been clarified by Rice and Wolman (1971), Herzing and Mate (1984), Poole (1984), and Perryman *et al.* (1999). Collectively, these studies demonstrate that the annual migration of gray whales exists as two superimposed patterns related to the reproductive states of adult females (Figure 13.14). Each year one third to one half of the adult females are pregnant. These near-term pregnant females depart the Bering Sea at the start of the southbound migration 2 weeks before other gray whales (Rice and Wolman, 1971; dotted line, Figure 13.14). Nonpregnant adult females start moving southward later in company with adult males (although both groups partially overlap the earlier departure of pregnant females). These late-arriving whales are also the first to leave the lagoons (after approximately 30 days). While in the lagoons and adjacent coastal areas, most lactating cows with new calves maintain a spatial separation from other age and sex groups by occupying the inner, more protected reaches of the lagoons until the other animals depart (Norris *et al.*, 1983; Jones and Swartz, 1984). The early arriving females, accompanied by their calves,

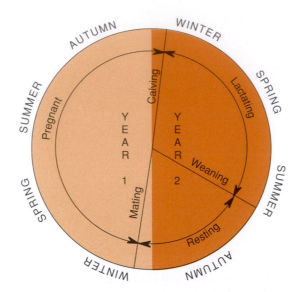

Figure 13.13.   Two-year reproductive cycle of fin and gray whales. (Adapted from Mackintosh, 1966.)

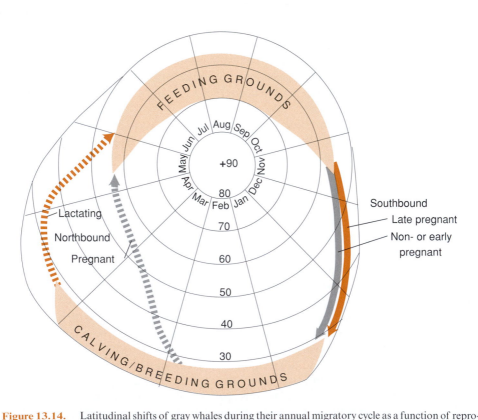

Figure 13.14.   Latitudinal shifts of gray whales during their annual migratory cycle as a function of reproductive status. (Adapted from Sumich, 1986b.)

are the last to leave Mexican calving lagoons the following spring (after about 80 days; Jones and Swartz, 1984).

Odontocetes differ from baleen whales in not having such extensive latitudinal migrations as a general trait. Gestation periods for odontocetes range from about 7 to 17 months (Table 13.1). There seems to be a relationship among the length of gestation, fetal growth rate, and calf size at birth among odontocetes, such that larger delphinids (e.g., killer whales and pilot whales) have longer gestation periods than do smaller taxa such as spinner or white-sided dolphins (Perrin and Reilly, 1984). Odontocetes typically produce single offspring that are physically well developed but socially undeveloped. Extended periods of postnatal social learning are crucial elements of the normal development of odontocetes (see Herzing, 1997).

Gestation in manatees lasts about 1 year. Calves are weaned about 1 year after birth but may remain with their mothers for another year before becoming independent (Marmontel, 1995). Thus, the calving interval of females is usually at least 3 years. Gestation in dugongs is estimated to be 13.9 months with long interbirth intervals ranging from about 3 to 7 years (Marsh, 1995).

After a gestation period of about 6 months, female sea otters give birth to a single pup in late winter or early spring (Jameson and Johnson, 1993). Birth may occur on land or in the water. It appears that the seasonal timing of pupping has evolved so as to occur with the period of maximum food supply, which permits maximal pup growth. Similarly gestation in polar bears is short in comparison to other marine mammals. Cubs are born in late December or early January, and the family emerges from the den in March, at the time of peak pupping for ringed seals. The cubs remain with their mother for about 2.5 years making interbirth intervals several years.

13.3. Mating Systems

Marine mammals exhibit considerable variation in the social arrangements surrounding breeding, but they all fall into two major groupings—polygyny and promiscuity. Some pinnipeds (e.g., elephant seals) are extremely **polygynous**, with a successful male mating with dozens of females during a single breeding season. Marine mammals, like all mammals, are predisposed to polygyny because of the disparity between the two sexes in terms of their basic reproductive physiology and anatomy (see Section 13.2). Males can produce vast numbers of energetically inexpensive gametes and are rarely involved in parental care activities. Females on the other hand have a fixed, low number of energetically expensive gametes that can be produced during their life times; they bear all the energetic costs of pregnancy and lactation and most or all other forms of postnatal parental care as well. Consequently, sexual selection pressure on males is fundamentally different from that on females (Emlen and Oring, 1977). Male mating strategies normally involve attempting to mate with as many partners as is possible each reproductive season, but the risk of being excluded from producing any offspring is high. In contrast, female mating strategies involve low risk of reproductive exclusion but result in the production of a single offspring per reproductive cycle, or a small litter (1–3 cubs) in the case of polar bears. In species in which large body size can confer a reproductive advantage to members of one sex, the conflict between the reproductive best interests of males and females of the same species can be associated with **sexual dimorphism** between the sexes in body size (Figure 13.15). Sexual selection can also drive the appearance of sexual ornamentation (usually in males), differences in color, shape, scent, and very different

Figure 13.15.   Male (right) and female (left) northern elephant seal, showing extreme sexual dimorphism.

behavior patterns of adult males and females, including different vocal repertoires (Ralls and Mesnick, 2002). Some cetaceans, some phocids and the sirenians have **promiscuous** mating systems, in which several sexually active adult males are associated with a group of estrous females for a variable period of time. Other cetaceans, most of the pinnipeds, sea otters, and polar bears exhibit various forms of polygyny. Some polygynous pinnipeds exhibit classical territorial or male dominance phenomena and it is the extremes among these species that make mating systems of pinnipeds most notable. However, there are also various alternate strategies employed by some species, or individuals, which do not conform to the classical territorial-male systems. Our ability to generalize about marine mammal mating and social organization, especially for cetaceans, is limited by a significant lack of knowledge, particularly with respect to species that are dispersed, pelagic dwellers.

13.3.1. Pinniped Mating Systems

Pinniped mating systems range from promiscuous systems, in which it is likely that members of both sexes mate with more than one partner during a breeding season, to the most extreme levels of polygyny displayed among mammals. In the latter grouping, a very small number of adult breeding males service all of the receptive female population in a given year. For example, individual northern fur seal males have been documented to have up to 100+ females in their harems. The basic dichotomy of a warm-blooded animal feeding at sea and thus having aquatic adapations (such as flippers) and yet having to come onto a solid substrate, where some of these same adaptations result in limited mobility, in order to give birth and care for young sets up an evolutionary predisposition for extreme polygyny (Figure 13.16; Bartholomew 1970; Stirling 1975, 1983; Le Boeuf, 1986, 1991; Boness, 1991). The degree to which species, or populations, of pinnipeds

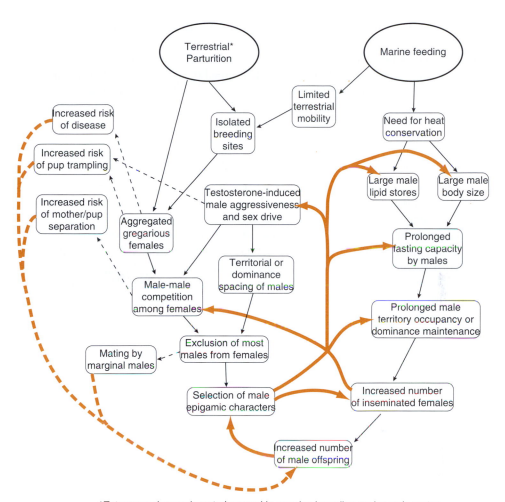

*Extreme polygyny is not observed in species breeding on ice or in water.

Figure 13.16. A simplified model of the evolution of polygyny in pinnipeds, with emphasis on male characteristics. *Black arrows* relate the major attributes of typical polygynous pinnipeds; the *colored arrows* indicate positive (*solid*) and negative (*dashed*) feedback loops. (Adapted from Bartholomew, 1970.)

exhibit polygyny is largely determined by how tightly females cluster during breeding and to what extent males can limit the access of other males to females. The physical structure of the birthing habitat (including accessibility of the water), aquatic versus terrestrial mating, and female intraspecific aggression (and hence spacing) all play a role in defining the number of females that a male can mate with during a breeding season.

Approximately one half of pinniped species give birth and mate on land. Rookeries are normally located on islands or on isolated mainland beaches and sandbars not easily accessible to large terrestrial predators. Consequently, available rookery space is limited and females aggregate densely. Because mother–pup separations may take place frequently on a crowded rookery, there are many opportunities for **alloparenting** or adoptive behavior. Because rookery sites are geographically fixed and the time of breeding remains quite constant from year to year, males can anticipate the return of females and can compete for breeding territories (in otariids) or establish dominance hierarchies (in elephant seals and land-breeding grey seals) prior to the return of the females. Females can depend on the predictable presence of adequate pupping sites and on the presence of high-quality breeding males. All pinniped species that give birth and mate on land are polygynous and strongly sexually dimorphic with males showing obvious secondary sexual characteristics (Stirling, 1975, 1983; Ralls and Mesnick, 2002). For example, adult male northern elephant seals in prime breeding condition are 5–6 times larger than adult females, and the largest southern elephant seal males may weigh 10 times as much as their female partners (Le Boeuf and Laws, 1994). In addition to their larger size, male elephant seals have additional secondary sexual characteristics including an elongated proboscis, enlarged canine teeth, and thick cornified skin on the neck. The inflatable proboscis is arguably the most conspicuous feature of adult males of this species (Plate 6). In females there is no nasal enlargement. The development of the proboscis in males begins at about 2 years of age and its full size is attained by about 8 years of age. The protuberance is an enlargement of the nasal cavity internally divided by the nasal septum. Inflation is attained through muscular action and blood pressure with minor assistance from the respiratory system. When inflated it forms a high, bilobed cushion on top of the snout with its anterior tip long enough to hang down in front of the open mouth. The snorts and other vocalizations are thus secondarily deflected downwards into the open mouth and pharynx, which act as a resonating chamber. Huge neck shields with thick fur ruffs are typical secondary sexual characters of mature otariid males, in addition to their large body size (see Miller, 1991).

In the territorial system employed by most otariids, competition revolves around the establishment and defense of a fixed space that meets the needs of females and their young. This type of mating system is referred to as **resource defense polygyny.** Several characteristics are common to those species that exhibit this form of mating system: (1) territorial males arrive before females; (2) females become receptive in close geographical proximity to where they give birth; (3) the rookery substrate is stable but discontinuous, facilitating its defense; and (4) there are relatively few aggressive interactions among males once established (i.e., those males that are not large enough to maintain or capture a territory are seen infrequently and are usually peripheral to the main breeding activity (Boness, 1991). The breeding behavior of northern fur seals is typical of this type of system and much of our understanding of the details regarding how this system functions come from a long-term study of northern fur seals on the Probilof Islands by Gentry (1998). Adult male northern fur seals usually maintain an established, specific territory for the duration of their breeding life (Figure 13.17). Territorial defense behavior is most frequently a two-male display using vocalizations and threat displays rather than physical combat, but males will fight violently at times. Males copulate more if they have larger body sizes, larger territories in good locations, the ability to fast for longer duration, longer tenure in a territory, and a higher frequency of territorial defense (Figure 13.16).

There are variations on the general territorial theme in some otariid species. For example, male Steller sea lions are present in the breeding areas before they are top candidates for breeding. Typically they start out in poor territories with associated poor reproductive success but gradually improve their competitive edge over a period of a few years (Gisiner, 1985). Once a male Steller sea lion succeeds in acquiring and holding a territory, he develops relationships with territorial neighbors that remain in place for some years (Higgins et al., 1988). It appears that the social connections and the memory and cognitive skills implied by having them may be as important as physical attributes in

Figure 13.17. Northern fur seal harem with territorial male. (Courtesy of J. Harvey.)

enhancing a male's reproductive success over several years. Another variant is seen in the Juan Fernandez fur seal, in which females do not remain sedentary throughout their stays on land during lactation. High temperatures contribute to both male and female movements to the water. Some males hold exclusively aquatic territories in areas when females predictably congregate. Other males hold either land-locked or shoreline territories where females give birth and nurse their young (Francis and Boness, 1991).

Polygynous, land birthing and mating phocids such as the elephant seals do not defend geographically fixed breeding territories as do otariids but, rather, they establish dominance hierarchies among groups of associating males within aggregations of females. This type of mating system often is referred to as **female (or harem) defense polygyny.** Similar to otariid males, elephant seals fast for the duration of the breeding season (a period of over 3 months for northern elephant seals), and the increased lipid stores associated with large male body mass confer clear advantages in successfully withstanding these extended periods without food. The most dominant (or alpha) male in a hierarchy defends nearby females against incursions by subordinate males. At Año Nuevo Island on the central California coast, elephant seal breeding behavior has been studied intensively for more than 3 decades (Le Boeuf and Laws, 1994). In areas crowded with females, an alpha male can control exclusive access to about 50 females. Because more females require even larger areas of the rookery, it becomes more difficult for an alpha male to exclude competing males from all of the females in larger groups. Even so, fewer than 10% of males manage to mate at all during their lifetimes, whereas very successful males mate with 100 or more females (Le Boeuf and Reiter, 1988). In male northern elephant seals, copulatory success is positively correlated with male size (Haley *et al.*, 1994), although this is not the case when controlling for age (Clinton and Le Boeuf, 1993). Thus, it is likely that other factors such as fighting experience influence actual mating success (Modig, 1996).

All of the territorial or hierarchical systems described previously are subject to cheater strategies by subdominant males. In some species nonterritorial males will intercept females on their way out to the ocean. Another strategy used by subdominant males is sneaking into large harems in which alpha males are more likely to be distracted. Females usually move away from these encroaching males or vocally protest mounting attempts by subdominant males, thus attracting the notice of higher ranking males (Boness *et al.*, 1982). Perhaps one of the most notable male-cheater strategies occurs in southern and Australian sea lions. Nonterritorial males in these two species make coordinated group raids into territories of breeding males. The defending male cannot deal with all of the raiders simultaneously, so some are able to forcibly mate with females or abduct them in the temporary confusion (Marlow, 1975; Campagna *et al.*, 1988; Cassini and Vila, 1990). A female's position within a group, especially a large group, determines her vulnerability to marginal males. Older or more dominant females near the center of a harem will be more likely to mate with the dominant male, leaving the females that are pushed to the edge of the harem more exposed to mating forays by subdominant males. Thus, satellite strategies serve to reinforce tight female aggregations (see Figure 13.16). Other sorts of "break-downs" in the system have also been suggested in the literature. Heath (2002) suggests that female California sea lions perform active mate choice, copulating in areas away from the pupping sites where they spend most of their time. Thus, the number of copulations a male accomplishes may be unrelated to the number of females in his territory in this species, although a high degree of polygyny is still attained with females exhibiting a high degree of unity in their selection of mates. Heath (2002) suggests that high temperatures at the breeding sites, and thus the need for routine access to water, has driven the evolution of this alternate system in this otariid species.

Three additional phocid species give birth on land but mate in the water, harbor seals, Hawaiian monk seals, and Mediterranean monk seals. The mating systems of these species are not uniformly well documented, but all three are thought to be polygynous at low levels. Male harbor seals appear to occupy selected display sites along routes frequently traveled by females for foraging, where they perform vocal displays to attract mates (Hanggi and Schusterman, 1994; Van Parijs *et al.*, 1997). Males initiate contact, but a female will rebuff some males while accepting another (Allen, 1985). Small groups of male harbor seals have been filmed off the California coast with a dominant singer being closely attended by other silent males, but the significance of these groups is not known. A low level of polygyny has been documented in harbor seals via microsatellite DNA markers in the small population breeding on Sable Island (Coltman *et al.*, 1998). Mobbing behavior by overly aggressive Hawaiian monk seal males attempting to mate with females is a phenomenon that has occurred in recent decades (Alcorn, 1984). Some young adult female and juvenile seals have been injured as a result of this behavior

(Johanos *et al.*, 1990) and wounded females are thought to be at increased risk of shark predation. A male-biased sex ratio is thought to be the source of this phenomenon, but how this situation initially arose is not known (Le Boeuf and Mesnick, 1990).

Approximately one third of pinnipeds give birth on ice and all of these species mate in the water. The ice breeders include: Weddell, leopard, Ross, and crabeater seals in the southern oceans and spotted, ringed, ribbon, bearded, hooded, Baikal, Caspian, harp, and hooded seals as well as walruses in the north. Some of these species are not well studied with regard to their mating behavior because of the logistical difficulties of following underwater behavior, particularly from an ice platform. However, among the species for which we have data, a variety of mating systems and patterns of sexual dimorphism are evident. Harp seals are close to being monomorphic and are thought to be promiscuous, breeding in very open floating pack-ice. Ribbon seals likely exhibit similar patterns given the similar body sizes of the sexes and the habitat similarities shared with harp seals. Hooded seals occupy a similar habitat to harp seals, but because of the exceptionally short lactation period in this species, and the fact that females aggregate very loosely and tend to stay some distance from the water when possible, males are able to achieve a degree of polygyny and are markedly sexual dimorphic in body size (Kovacs 1990, 1995; McRae and Kovacs, 1994). Hooded seal males attend a female (and her pup) until she is prepared to wean her offspring and mate. The pair then goes to the water together, but within a day or two the male is back in the breeding patch trying to establish association with another female. Males feed little if at all during the short, intense breeding season (Kovacs *et al.*, 1996).

Similar to elephant seals, male hooded seals also display cephalic ornamentation associated with attracting females and combating other males. An enlargement of the nasal cavity forms an inflatable hood on top of the head in adult male hooded seals (Plate 3). It reaches its maximum size in sexually mature bulls, which use it for courtship display. The hood starts developing at about 4 years of age and increases in size with age and increasing body size. By age 13, males possess large hoods. The hood can be inflated with air when the animal is disturbed or during the pupping and mating season. The hood is divided by the nasal septum and is lined by a continuation of the mucous membrane of the nose. The skin of the hood is very elastic, and the hairs that are present are shorter and grow less thickly than those on other parts of the body. No blubber is found in the hood; it is replaced by elastic tissue. There are muscle fibers around the nostrils where they form annular constrictor muscles that trap exhaled air within the hood. Hooded seals also possess the ability to blow (usually from the left nostril) a red balloon-shaped structure formed from the extensible membranous part of the internasal septum (See Plate 3). The seal produces this balloon by closing the right nostril, taking air into the hood, but then keeping the hood partially deflated by muscular action and pushing the air down the passage with the closed nostril. The pressure causes the elastic internasal septum to bulge and eventually to appear as a balloon through the open nostril of the other side. Similar to hooded seals, spotted seals and crabeater seals also appear in trios on the ice during the mating season and may have a **mate guarding** (serial monogamy) system similar to hooded seals, but no specific data are available on these species (but see Bengtsson and Siniff, 1981; Burns, 2002).

Ice-breeding Weddell seals, ringed seals, bearded seals, and walruses are also known or are suspected to achieve a degree of polygyny despite being affiliated with ice for birthing and water for mating. Male Weddell seals defend an underwater territory beneath a haul-out hole used by multiple females (Thomas and Keuche, 1982; Bartsh *et al.*, 1992; Harcourt *et al.*, 1998, 2000). Male ringed seals are thought to maintain a territory that overlaps with the breathing hole/lair complexes of several females (Smith and Hammill, 1981). They fight with one another, inflicting injuries that are concentrated around the hind-flippers and tail. Male bearded seals perform elaborate vocalizations at sites that they hold for a period of years, depending somewhat on ice conditions (Van Parijs *et al.*, 2001, 2004). The dispersed females are highly mobile and presumbly choose males based on qualities demonstrated by the position of their maritorries or some quality of their vocal displays.

Male walruses either display a **lek**-like system or perform female-defense polygyny from the water. Pacific walruses have been described as having a lek-like mating system. During the breeding season, female walruses and calves haul-out on ice or rest in the water while one or more adult males station themselves in the water alongside a herd and perform complex underwater displays, producing a series of amazingly bell-like or gong-like sounds while swimming in set patterns. The elaborate display behaviors of males allow females an opportunity to assess the quality of the males before selecting a mate. A different mating system has been described for the Atlantic walrus. Sjare and Stirling (1996) suggested that males performed female-defense polygyny rather than a lek-like system at the Dundas Island **polynya** (area of open water surrounded by stable fast ice) in the Canadian high Arctic. A large mature male had exclusive access to a group of several females for 1–5 days at a time. One or two males spent significantly more time with the herd and it was not evident that female preference was important in determining which male became the attending male (Figure 13.18). When attending the herd, a male continuously repeated a complex, stereotyped underwater song (see Figure 11.17). Other sexually mature males in the area behaved as silent herd members or vocal satellite males, or in some cases performed both of these roles (Sjare and Stirling, 1996). The herd attendance profile of males suggests that males maximize their reproductive success by roving between groups of females, a breeding strategy that also is characteristic of male sperm whales (Whitehead, 1993). In both subspecies of walruses males use the tusks extensively in their displays, and the tusks of males are significantly longer and more

Figure 13.18.   Herd of female and young walruses along the edge of an Arctic polynya. An attending male is stationed just under the water surface near the polynya edge and is singing. (Courtesy of I. Stirling/B. Sjare.)

robust in adult males compared to females (see Miller, 1975; Fay, 1982). Variability of the sea ice habitat is probably one of the factors contributing to the differences in the mating system and breeding behavior of Pacific and Atlantic walruses.

The degree of polygyny exhibited in ice breeders is reduced compared to land breeding species because of the difficulty of gaining, or restricting, access to females in the water or in physically unstable environments such pack ice (Stirling, 1975, 1983). The extensive distribution of sea ice promotes dispersal of breeding females while still providing protection from most predators and easy access to water. However, the temporary and unpredictable nature of the location, extent, and break-up time of pack ice in particular restricts the breeding season to a short time period. Several of the ice breeding seals exhibit reversed sexual dimorphism, with females being larger than males. It has been suggested for Weddell seals that the smaller size of the male in comparison to the female might make the male a more agile swimmer and therefore more attractive as a potential mate (Le Boeuf, 1991). This may also be the case for bearded seals, another species in which females are larger than males, which copulate aquatically. However, other aquatically mating species such as harbor seals and ringed seals exhibit dimorphism with males being larger, so there is not yet a completely satisfying explanation for the patterns of dimorphism exhibited among phocid seals. This is particularly true given findings such as those reported in Coltman *et al*. (1999) that suggest male mating success in harbor seals is not related to either body size or reproductive effort. Many of the ice breeding seals have developed complex underwater vocalizations to advertise attractiveness and mating receptivity (see Chapter 11), including harp seals, Weddell seals, and bearded seals (e.g., Van Parijs *et al*., 2001, 2003, 2004), which may play important roles in mate selection.

Grey seals are a particularly fascinating species in terms of their reproductive biology. They breed on both land and ice in different parts of their range and mate on land, on ice, or in water. In both birthing environments females cluster so males can control access by other males to some degree, but females do come and go from the water at many of the sites used by this species and hence have the potential to encounter and mate with males that do not appear in the established hierarchies in the breeding areas. At a few sites, including the top of Rona and Sable Island, female grey seals remain ashore throughout lactation, but at least at Rona, paternity is not tightly correlated with shore-based observations of copulatory success (Wilmer *et al*., 1999), and the social structure of breeding grey seals appears to have layers of complexity that are still not well understood (e.g., Pomeroy *et al*., 2000; Lidgard *et al*., 2001). A study by Amos *et al*. (1995) showed that females tended to produce several pups over a series of years that were sired by the same male, although often not the attendant male, suggesting that mating seals might be capable of recognizing each other in successive years, and that mating during the females' estrus can occur away from the "breeding" beaches. Male grey seals breeding in pack-ice areas do establish hierarchies, and the largest males tend to guard holes that females use to access the sea as well, but satellite strategies are clearly also used by other males that manage a degree of breeding success (Tinker *et al*., 1995; Lidgard *et al*., 2004).

Some recent applications of molecular genetics techniques that have explored paternity in polygynous seals suggest that territorial males may not father as high a proportion of pups as previously thought based on observations of mating on land (e.g., Wilmer *et al*., 1999; Gemmell *et al*., 2001; Lidgard *et al*., 2004), whereas other studies confirm precisely what is seen in harem systems (e.g., Fabiani *et al*., 2004). Thus our understanding and explanation of pinniped mating systems is likely to change somewhat in the future. Molecular techniques applied to phylogenetic questions are also changing some of our perceptions. Three decades ago the evolution of ice breeding among phocid seals was hypothesized to result from phocids having descended from an ice-breeding ancestor (Stirling, 1975, 1983). Reexamination of this and a related hypothesis (i.e., multiple origins for ice breeding; Bonner, 1984, 1990) in the context of a phylogenetic framework (Costa, 1993; Perry *et al.*, 1995) has suggested alternative hypotheses.

13.2.2. Cetacean Mating Systems

All cetaceans mate and give birth aquatically. Consequently, our understanding of these events in most cetaceans is very limited. Our knowledge of cetacean reproduction comes mainly from observations of small odontocetes in captivity, anatomical studies from exploited species (for whom large numbers of carcasses have been available), and from a handful of species in which direct underwater observations of courting, mating, and calving behavior are facilitated by clear water. With few exceptions, notably cooperative feeding by humpback whales, mysticetes do not group closely together, school, or form pods. This probably reflects their need to feed more or less individually because restricted areas cannot sustain the daily food requirements of a group. Most mysticetes show little evidence of persistent social behavior beyond mother–calf bonds that disappear at weaning (e.g., Valsecchi *et al.*, 2002). Unlike the mysticetes, odontocetes exist in social groups (e.g., schools, units, and pods, see Gaskin *et al.*, 1984). Of varying sizes whose structures persist longer than any individual member's life span, although it is not uncommon in some species for adult males to live much of their adult lives alone. Nonreproductive copulatory behavior complicates the study of mating systems in some species of odontocetes that use copulation as a social bonding behavior (e.g., Caldwell and Caldwell, 1972; Saayman *et al.*, 1979). Similar to the situation regarding pinnipeds, our knowledge of cetacean social behavior and mating systems is dominated by data on a few species from a handful of long-term studies (especially humpback whales, bottlenose dolphins, killer whales and sperm whales) and a few insightful molecular studies that have explored paternity and kinship (e.g., Duffield and Wells, 1991; Amos *et al.*, 1993a; Clapham and Palsbøll, 1997). Advances in satellite telemetry are currently also slowly but steadily expanding our knowledge of at-sea distribution patterns and social dynamics of cetaceans.

Many odontocete species live in structured social groups, or **schools**, characterized by long-term associations among individuals. The size of schools varies with species and location. Until quite recently it was assumed that adult males within the most stable social grouping sired the young within these odontocete groups, but this has repeatedly been found not to be the case. Enduring social relationships among female lineages appear to dominate the social structure of toothed whales, although sons also have prolonged associations with their mothers in species such as the killer whale. Odontocete mating systems appear to be a mix of promiscuous and polygynous systems, in which mobile males traveling among different groups of females, associating with individuals or groups for brief periods, appear to be a consistent pattern across a diverse array of species (Wells *et al.*, 1999; Mesnick and Ralls, 2002). Male–male competition seems to be a common element of mating among toothed whales, where many species are markedly sexually dimorphic or bear scar patterns from the teeth of their conspecifics. However, cooperative behavior among small coalitions of males also seems to be a strategy within some species.

Bottlenose dolphins inhabit near-shore waters of many temperate and warm coastlines, making some populations relatively easy to observe. One of the longest term studies of free-ranging bottlenose dolphins is that of the Sarasota Bay community on the west coast of Florida initiated, in 1970 (Irvine and Wells, 1972; Wells *et al.*, 1987; Scott *et al.*, 1989). This dolphin community contains about 100 individuals that live in small schools of socially interacting individuals. The groups are organized on the basis of age, sex, familial relationship, and reproductive condition. Although the composition of schools may fluctuate even over the course of a day, many associations are relatively long term. The mother–calf bond is close and may persist for many years. Membership of female bands is also relatively stable, and subadult and adult males may form lasting associations among themselves as well. Bottlenose dolphins probably mate promiscuously. In early adolescence some males form coalitions with one or two other adult males that last for years and may even be life long. These male coalitions exist as subgroups within larger schools and forage, swim, and court females together. However, single males as well as members of long-term male coalitions both sire offspring (Wells *et al.*, 1987; Duffield and Wells, 1991; Krutzen *et al.*, 2004). In the Sarasota Bay dolphin community, males typically start to sire offspring from the time they are in their late twenties and continue for about a decade. Over this time span, individual males produce offspring with several females, and the calves of any one female are fathered by different males. There is little aggression towards females by males in this population. This is not the case for bottlenose dolphins in Shark Bay, Australia, where coalitions of male bottlenose dolphins aggressively mate-guard and forcibly herd adult females, sometimes abducting them from other male coalitions (Connor *et al.*, 1992, 1996, 1999). Wilson (1995) suggested that male bottlenose dolphins do not form coalitions in the Moray Firth, but the reasons for such a difference in the mating system of this species at this location is unknown.

Observational studies of spinner dolphins in the clear waters of Hawaii by Norris *et al.* (1994) and in the Bahamas by Herzing (1997) indicate that spinners aggregate in schools of up to 100 individuals. Within each school there are subgroups of up to a dozen individuals whose members maintain close association with each other by remaining close and synchronizing their movements. Most commonly the animals occur in a staggered arrangement, or **echelon formation** (Figure 13.19), similar to that seen in flying formations of birds (Norris and Johnson, 1994). It has been suggested that this spatial arrangement might be part of a signaling system that allows organization and transmission of information throughout the school. Spinner dolphin societies are characterized by minimal sexual dimorphism and a high degree of cooperation among individuals. Mating is seasonal and apparently promiscuous, with more than one male achieving intromission with the same female in rapid succession; these males are often members of a coalition (Johnson and Norris, 1994).

Killer whales appear to live in several different types of social groupings. Transient groups are highly mobile and feed on other marine mammals, whereas resident populations occupy quite set home ranges and are largely fish eaters. A third variant is recognized, pelagic killer whales, but little is known about them. The social structure and mating behavior of resident killer whales is best documented. These killer whales exist in small stable social units knows as **pods.** Pods are characterized by specific dialects (see Chapter 11) and foraging strategies (see Chapter 12) as well as by their individual members. Killer whale pods are **matrilineal** social groups, each typically consisting of an older mature female, her male and female offspring, and the offspring of the second generation's mature females.

Figure 13.19.   Spinner dolphins in echelon formation.

Mature males remain with the pod into which they were born, and movement or exchange of individuals among pods has not been documented (Bigg *et al.*, 1990). Mating behavior has rarely been observed, yet genetic studies support the notion that male killer whales do not mate with their closely related pod members. Instead it is thought that interpod mating occurs when pods encounter each other in shared foraging or resting areas (Ford *et al.*, 1994). Genetic studies of long-finned pilot whales suggest a similar system regarding mating. It appears that males mate with multiple females and move between stable matrilineal pods but avoid mating with close relatives within their own natal social group (Amos *et al.*, 1991, 1993b).

Sperm whales occur in associations of varying sizes in all oceans (Whitehead, 2003, and references cited therein). Gregarious adult females and immature whales occur throughout tropical and subtropical latitudes year-round (Figure 13.20). These female/immature whale groups, or **units** (Whitehead *et al.*, 1991), are structurally equivalent to pods of killer whales and, like killer whales, appear to be matrilineal in structure (Whitehead *et al.*, 1991). Sperm whale social units average about a dozen whales and include closely related adult females and their offspring of both sexes (Whitehead and Waters, 1990). Unlike killer whales, however, male sperm whales leave these female groups at 4–5 years of age and gradually travel to higher and higher latitudes as they grow, so adult males are not permanent members of the female units. It is these large, solitary, physically mature males that participate in most matings, searching out estrus females as they move from one female unit to another (Whitehead, 1993). Adult male sperm whales are about twice the length of females and exhibit the greatest sexual size difference among cetaceans. The roving strategy of adult males may be assisted by long-distance vocal communication between traveling males and female social groups.

The tight social units seen among sperm whale females may be related to the need for a cooperative care system for calves that are too young to accompany their mothers during deep foraging dives (described in Chapter 12). Shared communal care of calves by

Figure 13.20. Group of sperm whales with an adult male at the top.

other adult females, also referred to as "babysitting" (Whitehead, 1996, and references cited therein), provides protection from predators as well as the possibility for communal nursing while the mother is foraging (Whitehead *et al.*, 1991). Whitehead (1996) reports that babysitting by sperm whales is a form of alloparental care. Such communal care behavior is likely to be selected for only if individuals within a sperm whale unit are closely related and remain together for extended time periods, perhaps even entire lifetimes. The benefits of this behavior may have been instrumental in the evolution of sociality in female sperm whales (Whitehead, 1996). Among female and immature sperm whales, two other levels of social organization, "groups" of associated units and even larger "aggregations" of associated, but unrelated, groups have been described (Whitehead and Kahn, 1992). Unlike units, temporal and geographic differences are observed in the sizes of groups and aggregations, probably caused by variations in the distribution and abundance of food.

Little is known about the mating behavior of narwhals and white whales, but both of these species live in societies that are age and sex structured. Groups are normally quite small, but huge aggregations are seen occasionally in both species. Adult male herds are autonomous for much of the year, traveling separately through areas also used by herds containing females and their calves and juvenile animals. The narwhal tusk is one of the most elaborate secondary sexual characteristics in mammals; it is a spiraled, erupted left canine that can be up to 3 m long. Only males normally possess the tusk, and they are frequently seen in a crossed tusk display on the summering grounds (Figure 13.21). It is thought that males establish a dominance hierarchy in this manner (Heide-Jørgensen, 2002).

Figure 13.21.   Tusks of two competing male narwhals. (Courtesy of M.-P. Heide-Jorgensen.)

Results of comparative studies of the social systems of odontocetes are beginning to emerge. For example, sperm whales have been described as having a social system similar to elephants and bottlenose dolphin societies have been compared with chimpanzees (Connor *et al.*, 1998 and references cited therein). Phenomena such as alloparenting have been documented for several species and may be a common feature of odontocete social groups. Future work based on long-term studies of identified individuals will allow us to extend these comparisons to a wider range of species.

Baleen whales are not as highly social as odontocetes. They occur in much smaller groups and in many species the only consistent, recognized social unit is the cow–calf pair. However, under some circumstances baleen whales do form loose aggregations that normally consist of 2–6 whales, although larger groups of more than 20 have been observed (Lockyer, 1981, 1984, and references cited therein). Most baleen whale species that have been studied appear to be promiscuous breeders and the relative size of the testes and penises of some species (Figure 13.23) suggest that **sperm competition** may be an important determinant of reproductive success in some species (right whales, bowhead, and gray whales; also see Brownell and Ralls, 1986). However, other species exhibit direct male–male competition, mate-guarding, and other characteristics of a polygynous system (Wells *et al.*, 1999; Mesnick and Ralls, 2002). The mating behavior of most baleen whales is intimately tied to a life cycle of feeding, migration, then breeding (described in greater detail in Chapter 12), although bowheads deviate from this pattern somewhat and out-of-season copulations are common among right whales. All mysticetes show reversed sexual size dimorphism in which females are the larger of the two sexes. This may allow the females to accommodate the energetic costs associated with rapid fetal growth and lactation (Clapham, 1996).

Humpback whales are the most studied baleen whales in terms of their social behavior. They occur in small, unstable groups on both summer feeding and winter breeding grounds (see Clapham, 1996, for a review). A principle feature of the mating behavior of this species is the performance by males of songs of extended duration (described in Chapter 11). Singing occurs mainly on the winter breeding grounds but songs are also occasionally heard on the summer feeding grounds. Many of the features of the song for a particular geographic area are shared by all of the whales in the region, but the song of the group changes through time. During the winter breeding season female humpbacks (even those with new calves) are often attended by an adult male, which has been termed a principal **escort** (Glockner-Ferrari and Ferrari, 1984; Medrano *et al.*, 1994). These males that attend a female are most likely waiting for mating opportunities or guarding the female after having copulated. Other males, termed secondary escorts, are sometimes present but do not normally play an active role with the pair. Principal escorts vigorously rebuff approaches by other males (Mobley and Herman, 1985; Clapham *et al.*, 1992). Defensive or competitive behaviors (Figure 13.22) include ramming, thrashing of the flukes, and using trumpet blows and bubble curtains or streams (see references cited in Weinrich, 1995; Pack *et al.*, 1998). Baker and Herman (1984) have found that the incidence of these agonistic behaviors increases with increasing density of whales in the breeding areas. This mating system, in which males display (sing) or directly compete for access to females, has been described by Clapham (1996) as being a floating-lek. It is not known whether a female mates with more than one partner in a season, but molecular analyses show that they are mated by different males between years (Clapham and Palsbøll, 1997).

Most gray whales mate in coastal or lagoon waters at the southern end of their annual migration. A persistent, yet unsubstantiated, myth describes the gray whale as typically mating in trios consisting of a receptive female and two males. Although courting trios are sometimes observed, the sexes of participants are usually unknown. Based on above-surface observations of courting activities, mating appears to be promiscuous, although mating success of individual males is impossible to establish from direct observations of their very active and vigorous courting encounters (Figure 13.23). Courting/mating groups including as many as 17 individuals have been reported (Jones and Swartz, 1984); however, the intensity of the physical interactions during these courting bouts precludes determining such basic features as gender and interwhale contact patterns or even accurate determinations of group sizes.

Right whales show sexual activity throughout the year, although calving occurs in a fixed annual cycle that suggests conception is restricted to the winter. Multiple males are simultaneously involved in stimulating and sequentially mating with a female. However, male–male aggression levels are low, suggesting that sperm competition is the means by which males compete with one another. Their enormous testicles also support this suggestion.

Little is known about the other mysticete whales with respect to mating. Even winter distributions are vague or unknown for many populations.

13.3.3. Sirenian Mating Systems

Manatees are solitary animals for most of the year. The few social interactions apart from mother–calf pairs observed appear to center on reproduction, or at least some form of sexual behavior (O'Shea, 1994). Mating and calving can occur throughout the year, but mating herds are most frequent in the warm season months in highly seasonal environments. Seasonality in the occurrence of newborn calves has been reported for West Indian manatees and for the more tropical dugongs (Reynolds and Odell, 1991). Male manatees have home ranges that overlap those of several females and they spend a considerable amount of time patrolling in search of estrus females. When a female becomes receptive she also becomes much more wide-ranging than is normal, straying well beyond her usual home range. Mating herds form around these females and scramble competition ensues that can involve frenzied pushing and shoving, with the female promiscuously mating with multiple partners. A manatee cow attracts as many as

(a)

(b)

Figure 13.22.   Views of humpback whale males competing **(a)** above surface and **(b)** subsurface. (Courtesy of Phil Colla.)

Figure 13.23. Surface view of a group of gray whales courting in a coastal lagoon in Baja California.

17 males that compete for copulation in these mating groups (Figure 13.24; see also Hartman, 1979; Bengtson, 1981; Rathbun and O'Shea, 1984). In the West Indian manatee a female can be accompanied by such a group for 3 to 4 weeks although she is receptive to mating for only a day or two. Rathbun *et al.* (1995) reported that participatory roles and ages of competing males were without distinct patterns, although it seems likely that large strong males would dominate in pushing competitions.

Figure 13.24. Courting/mating group of Florida manatees with focal female in the lead. (Courtesy of T. O'Shea, Sirenia Project, U. S. Geological Survey.)

Dugongs sometimes form large herds but are more often encountered in groups of less than a dozen; many individuals may be solitary (Andersen *et al.*, 2001). Less is known about dugong mating than for manatees because they often occupy quite turbid water where observations are difficult. Because dugongs are not conspicuously sexually dimorphic, the sexes of participants in social interactions are difficult to distinguish (Wells *et al.*, 1999). The sexual behavior of dugongs seems to differ from manatees primarily in their more intense physical competition for females during a much shorter period of time (Preen, 1989), at least in eastern Australia. Mating dugong herds only persist for a day or two. Dugongs, with scars presumably made by the tusks of other male dugongs during courtship battles, are frequently seen at the surface lunging at other dugongs in what are interpreted to be male–male fights. Only dugong males have erupted tusks, so it is assumed that they may be important in a social context. Breeding is thought to take place over a period of 4–5 months. Less male aggression and rather different mating behaviors have been described for dugongs by Anderson (1997). He suggests that dugongs occupying Shark Bay, Australia, exhibit a lek strategy in which males defend display territories and females appear in the cove only in the context of courtship and mating.

13.3.4. Mating Systems of Other Marine Mammals

Sea otters are polygynous breeders, and copulation occurs exclusively in the water. Copulation takes place repeatedly during a short period of consortship, after which the male and female separate. During mating a male grasps the female's nose in his mouth and rolls at the surface to achieve intromission. Distinctive scars are seen on many females as a result of this behavior (see Estes and Bodkin, 2002). When not breeding, sea otters segregate by gender, with adult males occupying annual home ranges at the extremes of the distribution range of females. During the spring breeding season, adult males leave these fringe areas and establish breeding territories (usually about 1 km long, running parallel to shore) within female areas (Jameson, 1989). As is typical of polygynous mammals, juvenile male sea otters in both California and Alaska tend to disperse farther than females. Juvenile males ultimately move to male areas, whereas juvenile females remain within the general areas occupied by their mothers (Ralls *et al.*, 1996). Territorial boundaries and preferred resting sites within territories are related to patterns of kelp canopy cover; some males exhibit strong fidelity for the same territory over several successive years. It is not known if nonterritorial males are occasionally successful at inseminating females.

Polar bears are also polygynous and sexually dimorphic with males being almost twice the size of females (Stirling, 2002). Female bears, like their dispersed prey, (e.g., ringed seals), are quite widely spread spatially and male bears can only attend one female at a time (Wiig *et al.*, 1992). Females typically do not mate when accompanied by cubs of the year. Therefore, only about one third of the adult females mate in any 1 year. Mating occurs in the spring. Mature male polar bears travel extensively during this time in search of receptive females. The breeding condition of females may be advertised by **pheromones** deposited on the ice via urine or from footpad secretions (Stirling, 1988). Whatever the mechanism, a male recognizes the scent of an estrus female and will follow her tracks (sometimes over distances up to 100 km) until she is located. Physical competition among males for available females is intense, based on the frequency of broken teeth in older males and scarring patterns accumulated with age in males (Stirling, 1988).

A successful male will remain with the female for several days, repeatedly testing her receptivity for mating. During peak receptivity the pair mate repeatedly over several days. Polar bears are induced ovulators so repeated copulation is required for ovulation and subsequent fertilization. Following mating the male then moves on in search of additional partners, while the pregnant female resumes hunting to increase fat stores through the summer before entering maternity dens dug in snow in November. Only pregnant polar bears enter deep snow dens; mothers with older offspring and all other bears remain active throughout the year and only build temporary snow beds for resting (Ramsay and Stirling, 1988). Maternity dens occur both on land and in multiyear pack ice. It is unclear whether pack ice denning is a result of increased human disturbance in traditional denning areas ashore.

13.4. Lactation Strategies

In placental mammals, provisioning offspring until they are nutritionally independent is the most energetically expensive segment of reproduction for most species. With very few exceptions, care and feeding of marine mammal offspring is the sole responsibility of the mother. It is in the best interest of each offspring to extract for its own use as much energy as possible from its mother, even at the expense of her future reproductive success. The genetic self-interests of the mother, on the other hand, may be better served if some of her resources are withheld from current offspring so they can be invested in later offspring to improve her own lifetime reproductive fitness (for a broader review and discussion of these and other genetic conflicts, see Hurst *et al.*, 1996).

Pinnipeds, with their predictable use of traditional breeding sites on land or ice, have been more amenable to studies of maternal investment strategies than have other groups of marine mammals. Three lactation strategies, each with several variations, are apparent in marine mammals: (1) fasting, (2) foraging cycle, and (3) aquatic nursing (Oftedal *et al.*, 1987; Boness and Bowen, 1996; Lydersen and Kovacs, 1999; Table 13.3). The fasting strategy, which is extremely uncommon in terrestrial mammals, characterizes some phocid seals and most large mysticetes. The foraging cycle strategy that is common to all otariid species, as well as members from all other marine mammal groups, is similar to terrestrial mammals that must leave their young for periods of time while they forage. However, the time frame is somewhat different than the norm for terrestrial species among some pinnipeds. The length of time that female pinnipeds spend away from their pups foraging varies between a few hours in some of the phocid species that feed while lactating up to almost 2 weeks in the most extreme otariid species. Among otariids a few

Table 13.3. Lactation Straegies in Marine Mammals with Examples from the Pinnipeds

Feature	Fasting (e.g., Northern Elephant seal)	Foraging Cycle (e.g., Antarctic fur seal)	Aquatic Nursing (walrus)
1. Duration of fasting	All of lactation	Variable (a few days)	Short (hours–days)
2. Duration of lactation	Short (~4 weeks)	Intermediate (~4 months)	Long (~2–3 yrs)
3. Fat content of milk	High (55%)	Intermediate + (40%)	Low (20%)
4. Pups/calves forage during later lactation	No	No	Yes

Original sources listed in Riedman (1990), Jameson and Johnson (1993), and Stirling (1988).

days is common and the average length of time away from shore is related to the distance females must travel to find adequate prey. The aquatic nursing strategy is seen in odontocete cetaceans, sirenians, and the walrus (Table 13.3), in which young remain with their mothers wherever they go and nurse at sea, as well as on ice and land in the case of walruses.

13.4.1. Fasting Strategy

The fasting strategy is exhibited in an extreme form by only a few phocid species among pinnipeds. In hooded seals (Kovacs and Lavigne, 1992b; Lydersen et al., 1997), northern and southern elephant seals (Arnbom, 1994; Costa et al., 1986), and some land-breeding populations of grey seals (Fedak and Anderson, 1982; Iverson et al., 1993; Reilly et al., 1996) mothers remain out of water for the entire duration of a relatively short lactation period, ranging from 4 days in hooded seals to 4–5 weeks in elephant seals (Table 13.4) and do not have access to food. During their lactation periods lipids are mobilized from maternal body tissues, converted to high-fat milk, and transferred to the large and rapidly growing offspring. Weaning in these phocids is abrupt, being marked by the departure of the mother from the breeding area.

Northern elephant seals have been studied most intensely regarding their reproductive behavior, among fasting phocids. Pregnant females of this species come ashore on rookery beaches from December to early March and give birth soon thereafter. The peak of the pupping season occurs in late January (Le Boeuf, 1974), and pups at birth weigh about 8% of their mother's mass. Females fast through their 4-week lactation period, remaining on the beach near their pups. Bouts of suckling, each lasting several minutes, occur three to four times each day, with the duration of suckling periods increasing throughout lactation. Elephant seal milk is quite low (about 10%) in fat in the early stages of lactation but increases sharply in fat through the first half of lactation reaching over 50% fat (Figure 13.25). The gradual substitution of fat for water is probably an adaptation that allows the mother to conserve water late in her fast (Riedman, 1990); the shift in milk composition, with increasing fat content over the nursing period likely also parallels the pup's ability to digest heavier food and to build its increasing blubber store prior to weaning. Pups weigh about 40 kg at birth and gain an average of 4 kg each day prior to weaning (Costa et al., 1986). Most of this growth occurs as an increase in pup adipose tissue, with little increase in lean body mass. Overall, nearly 60% of the total energy expenditure of a lactating elephant seal mother is in the form of milk for her pup. At weaning, most elephant seal pups have quadrupled their birth weights; however, a few may have increased their birth weights as much as seven-fold by sneaking milk from other lactating females after their own mothers have departed to feed at sea (Reiter et al., 1978). Weaned elephant seal pups do not immediately commence foraging for their own food. Rather, they remain on the pupping beaches for a postweaning fast lasting an average of 8–10 weeks. Factors affecting the duration of the postweaning fast likely include body weight at weaning (an indication of the magnitude of the pup's lipid reserves) and availability of appropriate prey in near-shore waters. A study of skeletal development in fasting northern elephant seal pups indicates that despite their fast they continue development during this period prior to going to sea (Patterson-Buckendahl et al., 1994). Only when pups have lost about 30% of their weaned body mass (and are at least 3 months old) do they go to sea to feed for the first time (Reiter et al., 1978).

Table 13.4. Some Features of Lactation in Marine Mammals

Species	Lactation Length (Weeks)	Milk % Fat	% Protein	Source
Pinnipeds				
Phocids				
Hooded seal	<1	61	na	Boness and Bowen, 1996[*]
Harbor seal	3.4	50	9	Boness and Bowen, 1996[*]
Harp seal	1.7	57	na	Boness and Bowen, 1996[*]
Gray seal	2.3	60	7	Boness and Bowen, 1996[*]
N. elephant seal	4	54	5–12	Boness and Bowen, 1996[*]
Weddell seal	8	48	na	Boness and Bowen, 1996[*]
Otariids				
Australian sea lion	60–72	26–37	10	Higgins and Gass (1993); Gales et al., (1996)
CA. sea lion	43	44	na	Boness and Bowen, 1996[*]
Steller sea lion	47	24	na	Boness and Bowen, 1996[*]
Antarctic fur seal	17	42	17	Boness and Bowen, 1996[*]
Galapagos fur seal	77	29	na	Boness and Bowen, 1996[*]
N. fur seal	18	42	14	Costa and Gentry (1986)
Walrus	100+	14–32	5–11	Riedman (1990)[*]
Cetaceans				
Odontocetes				
Harbor porpoise	32	46	11	Gaskin et al. (1984)[*]; Oftedal (1997)
Bottlenose dolphin	76	14	12–18	Pilson and Waller (1970); Oftedal (1997)[*]
Spinner dolphin	60–76	26	7	Pilson and Waller (1970)
Sperm whale	100	15–35	8–10	Lockyer (1981); Oftedal (1997)
Mysticetes				
Fin whale	24–28	17–51	4–13	Lockyer (1984); White (1953)
Minke whale	20–24	24	14	Best (1982)
Gray whale	28–32	53	6	Swartz (1986); Rice and Wolman (1971); Zenkovich (1938)
Blue whale	24–28	35–50	11–14	Lockyer (1984); Gregory et al., (1955)
Humpback	40–44	33–39	13	Lockyer (1984)[*]; Glockner-Ferrari and Ferrari (1984)
Sirenians				
Manatee	52	na	na	Reynolds and Odell (1991)
Dugong	78	na	na	Marsh (1995)
Other Marine Mammals				
Sea otter	20–30	21–26	9–12	Jameson and Johnson (1993); Riedman and Estes (1990)
Polar bear	130	17–36	9–13	Oftedal and Gittleman (1989); Derocher et al. (1993)

Symbols: [*]see for original sources listed.

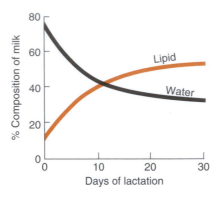

Figure 13.25.   Changes in the water and lipid components of northern elephant seal milk through lactation. (Adapted from Riedman, 1990.)

The other phocids that exhibit the fasting strategy follow the same basic patterns as northern elephant seals, but the time frame of both nursing and the postnursing pup fast are shorter. In the extreme, pack-ice breeding hooded seals have compressed lactation into a 3- to 4-day period during which pups grow from a birth mass of 24 kg to a weaning mass of approximately 50 kg (Bowen *et al.*, 1985; Oftedal *et al.*, 1993; Lydersen *et al.*, 1997). This amazing doubling of their body mass is achieved through drinking more than 10 liters of milk per day that has an average fat content of 60%. Extreme levels of immobility and on-demand nursing every few hours from a mother that is routinely located less than 1 m away leads to hooded seal neonates depositing about 70% of the energy they ingest as body tissue (see Lydersen and Kovacs, 1999). The fasting strategy is not a strict species-specific character for all of the phocids that exhibit the phenomenon. Many grey seals that give birth on land fast through lactation, but grey seals that utilize sea ice as their birth substrate (and some land breeders that have easy access to the water) forage during lactation to varying degrees among individuals (Lydersen *et al.*, 1995).

As has been already been mentioned previously, most baleen whales are characterized by long migrations between breeding grounds (where they give birth and lactate, but feed little if at all) and high-latitude summer foraging areas. For these migratory mysticetes, the occurrence of winter fasting during reproduction must be evaluated within the framework of the life history of the individual, which would be expected to vary by individual circumstances of age, sex, and reproductive condition. Acoustic data indicate that many North Atlantic fin whales migrate south into food-poor waters in winters, yet others remain at high latitudes through the winter. Pregnant female gray whales clearly commence fasting at the start of their southward winter migration (Rice and Wolman, 1971) and feed little or not at all in their low-latitude birthing locations. Calves are fed on fat-rich milk and grow quickly, and they are weaned during their first year of life, during or shortly after their first northward migration (Sumich, 1986a). A similar phenomenon is thought to occur for most of the migratory mysticetes.

Bowheads offer an interesting exception to this general strategy. They calve at high latitudes in late spring or summer, commence feeding again not long afterward, and have a long lactation period (Oftedal, 1997). This begs the question why? If bowhead can successfully remain in arctic waters during early lactation, why do other mysticetes go to such lengths to migrate to low latitudes for birthing? The answer might be the special feeding niche and extreme high-Arctic adaptations bowheads have evolved through time that permit them alone to do something exceptional. Or, the answer may reflect a more general and indirect ecological factor such as predation effects. Corkeron and Connor (1999) hypothesized that baleen whales migrate to southern waters to avoid killer whale predation. Bowhead whales manage to escape this predator by going into dense ice, where the killer whales with their high dorsal fins cannot readily pursue them.

13.4.2. Foraging Cycle Strategy

In the foraging cycle (or attendance) strategy, mothers normally fast only for a few days following birth, if they fast at all. Then the mothers begin to forage, usually leaving their pups on shore or on the ice, although some species such as harbor seals mothers are often accompanied by their young at sea. Phocid seals that forage during lactation do so during periods of only a few hours (Lydersen and Kovacs, 1999), so exhibit typical carnivore nursing behavior. The lactation cycles in otariids are more unusual; they involve days at a time spent in one or the other state (ashore nursing the pup or at sea foraging). The duration of foraging trips largely depends on the distance to foraging areas and on food abundance, but they are also loosely correlated with female body size (Costa, 1991, 1993; Higgins and Gass, 1999). Milk intake patterns in all otariid pups are intermittent, with a period of intense suckling being followed by fasting until their mothers return. Otariid milk is generally less energy-dense than that of phocids but varies in composition with trip duration and time ashore (e.g., Gales *et al.*, 1996; Trillmich, 1996). Lactation periods of otariids last from several months to a few years, depending on the species (Table 13.4). Subpolar species such as the northern and Antarctic fur seals have the shortest lactation periods and temperate and tropical species have much longer ones. As the lactation period progresses, most otariid pups begin to enter the water and supplement their milk diet with near-shore prey and the weaning process is a gradual one. However, this pattern does not follow for two species, the northern fur seal and Antarctic fur seal. In these species weaning is abrupt at about 4 months of age and shows very little variation. Many of the young appear to wean themselves by leaving the colony while the mothers are at sea. In general, flexibility in the timing of weaning is typical for temperate and tropical fur seals and sea lions. This flexibility may be a means by which mothers are able to buffer pups from unpredictable variations in food availability while pups are gradually increasing their own foraging skills and becoming independent (Trillmich, 1996).

Multiyear studies of Antarctic fur seals (Costa and Trillmich, 1988; Costa, 1991; Arnould *et al.*, 1996) illustrate a classical otariid foraging cycle approach to lactation and maternal care. Pregnant Antarctic fur seals give birth within a week of coming ashore at a breeding rookery (Costa *et al.*, 1985). At parturition, the percentage of body fat of Antarctic fur seal females is about one half that of elephant seal females (22% vs 40%; Boness and Bowen, 1996). Antarctic fur seals remain with their pups for about a week before returning to forage at sea. The foraging trips, during which the pup remains ashore, last 4–9 days each and the female will make 15–20 of these trips during lactation. After each trip, the pup is suckled several times a day for the 2 days that the mother remains at the rookery before her next foraging trip. Antarctic fur seal milk is about 40% fat (Costa *et al.*, 1985), a high fat content for an otariid. Antarctic fur seal pups weigh about 6 kg at birth and grow to 15–17 kg at weaning. This is one of the smallest birth masses among pinnipeds, yet the ratio of pup mass to maternal mass is almost twice as great in this species compared to elephant seals (15% vs 8%; see Figure 13.1). The pattern of pup growth through the 4-month lactation period is intermittent, reflecting the feast and fast pattern of milk availability. Unlike most other otariids, Antarctic fur seal pups do not begin to forage during the latter stages of lactation. Instead, the pup initiates weaning when it enters the sea to forage on its own, usually when its mother is out on a foraging trip herself. Because Antarctic fur seal pups (and other otariids) do not experience a strict postweaning fasting period they avoid the extensive postweaning mass losses characteristic of many phocid species.

Although it is often stated that phocid seals employ a fasting strategy, this is an overstatement of reality (see previous discussion). Most phocid mothers do some foraging during lactation, although their generally large body sizes allow them to cover the costs of lactation using on-board resources to a much greater degree than otariids. Species at the small end of the phocid size range all use a foraging cycle lactation strategy, but some large phocid species also engage in this strategy. For example, ringed seals and harbor seals both forage on a daily basis through lactation commencing a few days after the birth of their pups (Lydersen, 1995, Bowen *et al.*, 2001). Bearded seals, the largest of the northern phocids, also resume feeding early in lactation (Lydersen and Kovacs, 1999) and spend half of their time away from their pups during the nursing period (Krafft *et al.*, 2000). Likewise, Weddell seal mothers forage during lactation despite being large Antarctic phocids (Thomas and DeMaster, 1983). All of these species also show less abrupt weaning phenomena than the species that fast during lactation, and their young spend a considerable amount of time in the water prior to weaning, and commence foraging for themselves to some degree in most if not all species. Midsized phocids show considerable variability within the species that have been studied. For example, individual harp seal mothers vary dramatically in their food ingestion rates during lactation (Lydersen and Kovacs, 1993, 1996).

13.4.3. Aquatic Nursing Strategy

The third maternal care strategy, the aquatic nursing strategy, in which young are fed at sea and mothers forage during nursing is seen in all odontocete cetaceans, in the sirenians, and also in the walrus. A few days after birth, walrus calves go to sea with their mothers and join the herd (Sease and Chapman, 1988). Some time thereafter the calves accompany the adult animals at sea on foraging trips. Although little is known about at-sea mother–calf behavior, it is thought that young calves are left on the surface attended by some females or juvenile animals, until they are old enough to dive with their mothers, at approximately 5 months of age. The length of lactation in this species is long. Their dependency on milk decreases as their foraging skills are acquired and most calves are weaned by 2 years of age although they often remain in close association with their mothers for a much longer period of time (Fay, 1982). Walrus calves gain considerable foraging experience while still with the mother. Walruses are the only pinniped that can suckle while at sea (Miller and Boness, 1983); a skill facilitated by adaptations for suction feeding throughout life.

Odontocete whales have long lactation periods (Table 13.4). They differ from mysticetes in that most female odontocetes feed throughout lactation and weaning takes place when calves are from 1 to 3 years (Oftedal, 1997). The process of suckling in all cetaceans involves the tip of the tongue wrapping around the nipple from below and pushing the nipple against the hard palate. The base of the tongue then moves up and down rhythmically, producing suction which pulls the milk toward the throat (Logan and Robson, 1971). It has been suggested that the presence of marginal papillae on the tongues of neonate and juvenile beaked whales may assist the tongue in forming a tight seal while nursing and later in life may assist in holding prey against the palate while letting water escape between the papillae (Kastelein and Dubbledam, 1990; Heyning and Mead, 1996).

Manatee calves are weaned approximately 1 year after birth (Figure 13.26; Reynolds and Odell, 1991). Data on the length of lactation for dugongs are sparse, but the available data indicate that lactation is long, lasting at least 1.5 years, although dugong calves start eating sea grass soon after birth (Marsh, 1995).

Sea otter pups remain dependent on their mothers for about 6 months, and weaning is quite gradual (Jameson and Johnson, 1993). Pups are nursed at the surface while their mothers float on their backs. Pups remain on the surface, usually on floating kelp beds while their mother dive, until they learn to dive themselves.

Polar bears have exceptionally long lactation periods (~2.5 years) in comparison to other terrestrial carnivores (Oftedal and Gittleman, 1989). Polar bear cubs are born in a very altricial state and remain confined in the den consuming fat-rich milk for 2 to 4 months following birth, depending on the latitude of the den (Derocher et al., 1993). By the time they leave their dens, pups have increased their birth weight of less than 1 kg to 10–15 kg. Mean litter size at birth is not known, but it is probably close to two in most populations (Amstrup and DeMaster, 1988); females normally have one or two cubs although three and even four in a litter has been recorded. Polar bear cubs occasionally remain with their mothers for up to 3.5 years (Ramsay and Stirling, 1988; see also Chapter 14), although 2.5 years is more the norm.

Figure 13.26.   Manatee nursing. (Courtesy of U.S. Fish and Wildlife Service.)

13.5. Reproductive Patterns

13.5.1. Life-History Traits and Lactation Strategies

Phocid pups are larger at birth than are otariid pups because phocid mothers in most species are larger (62-kg ringed seals to 504-kg northern elephant seals) than otariid mothers (27-kg Galapagos fur seal to 273-kg Steller sea lion; Costa, 1991; Bowen, 1991). However, on a mass-specific basis the pinnipeds are quite similar (Kovacs and Lavigne, 1986a, 1992a). Likewise, the weaning mass of pinniped neonates based on maternal size is quite similar. However, the larger absolute body mass of phocid mothers permits the storage of sufficient lipid reserves to invest heavily in offspring during a short lactation period, with or without supplemental feeding. Otariid females, in contrast, lack the body mass necessary to bring ashore stored lipids reserves that are sufficient to support their reproductive investment during lactation. Relative to their body masses, otariid females make much larger total reproductive investments of energy, protein, and time in individual offspring than do phocid females (Table 13.5). The cyclical, extended, foraging pattern of lactating otariids allow for more flexibility in the duration of lactation than is possible for phocids. However, it can also potentially limit production potential (see Chapter 14). Such flexibility may serve as a buffer when food is scarce. The longer lactation period of otariids, the walrus and on a somewhat different time scale the foraging-cycle phocids, also permits offspring to develop swimming and foraging skills, while still depending on their mothers for energy.

The seasonally unstable nature of pack ice has selected for the extremely short lactation periods in ice-breeding phocid seals. Hooded seal pups suckle for only 4 days, and harp seals, ice-breeding grey seals, and bearded seals nurse for only 12, 17, and 18–24 days, respectively (Bowen *et al.*, 1994; Haller *et al.*, 1996; Lydersen and Kovacs, 1999) before they are weaned and forced to rely on the lipids accumulated while suckling.

Table 13.5. The Relationship between Maternal Mass, Lactation Duration, Pup Growth Rate, and Maternal Investment in Offspring in the Form of Protein and Energy in Some Pinnipeds

Species	Maternal (Mass, kg.)	Lactation, (Weeks)	Pup Mass Gain (Kg/day)	Maternal Investment Energy, MJ/kg	Protein, g/kg
Phocids					
Harbor seal	84	3.4	0.8	9	na
Harp seal	135	1.7	2.3	6	2
Hooded seal	179	<1	7.1	4	1
Gray seal	207	2.5	2.8	7	37
Weddell seal	447	8	2	7	32
N. elephant seal	504	4	4	5	36
Otariids					
N. fur seal	37	18	0.08	22	159
Ant. fur seal	39	17	0.11	31	192
CA. sea lion	88	43	0.13	36	202
Steller sea lion	273	47	0.38	26	204
Walrus					
Pacific	738	98	0.41	na	na
Atlantic	655	89	0.50	na	na

Original sources listed in Costa (1991), Boness and Bowen (1996), and Kovacs and Lavigne (1992).

Factors such as body size and correlated fasting ability (implying energetic limits) as well as ecological factors such as foraging options during lactation may also play a role in determining the actual duration of lactation (Trillmich, 1996). In any case, the brevity of lactation in phocids is compensated for by females providing large quantities of fat-rich, energy-dense milk that is transferred to pups at a high rate (see Oftedal *et al.*, 1996; Lydersen and Kovacs, 1999), which results in rapid growth by pups (e.g., Kovacs and Lavigne, 1986a, 1992a; Bowen, 1991). Other factors such as predation levels undoubtedly also play a role in determining the aquatic nature of some phocid pups and of course pup activity levels and the time spent in water have energetic consequences (Lydersen and Kovacs, 1999).

It is obvious that maternal care in marine mammals is very costly energetically. Mothers give birth to large offspring and subsequently transfer significant amounts of resources to their young. However, to assess the ultimate consequences of relative reproductive investment, changes in fitness must be measured. Fitness costs could include lowered future fertility or increased risk of mortality to a breeding female (Stearns, 1989). Data on fitness costs of reproduction in marine mammals exists only for only a few species, all of them pinnipeds. The northern elephant seal has been studied quite extensively in this regard but evidence for fitness costs of different levels of reproductive investment is mixed (e.g., Reiter *et al.*, 1981; Huber, 1987; Le Boeuf and Reiter, 1988). Mortality costs associated with females breeding at younger ages have been documented on Año Nuevo but not on the Farallon Islands. Data on fertility rates suggest costs are incurred at the Farallons but not on Año Nuevo. Information on the costs of reproduction in terms of reduced size at birth, growth, or survival to adulthood of pups of mothers that have pups in consecutive years versus those that skip at least a year also exist for two otariid species, the Galapagos fur seal and the Antarctic fur seal (Trillmich, 1986; Croxall *et al.*, 1988; Boyd *et al.*, 1995; Arnould *et al.*, 1996). For the Galapagos fur seal, the cost of extending the temporal investment in a pup is high; females are often still accompanied by a yearling in the next breeding season and they therefore either have a reduced birth rate or give birth and lose the newborn because of competition for milk by the older sibling (Trillmich, 1996). However, for the Galapagos fur seal, no increased mortality due to reproduction was measured in females, perhaps due to small sample sizes and high variability between years (Trillmich, 1996). Data for the Antarctic fur seal indicate that sustaining high fertility has substantial mortality costs. Boyd *et al.* (1995) suggested that lowered fertility of old Antarctic fur seals may be the result of one of two strategies whereby female seals either reproduce early and nearly continuously every year but die relatively young or reproduce every other year and survive longer. This would lead to an accumulation of low fertility females in old age classes and may explain the apparent decrease in pregnancy rate of older fur seal females without postulating a species-typical senescence.

The difference in maternal investment costs between otariids and phocids was briefly explored by Arnould (1997), who noted that the cost of reproduction in terms of mortality risk was much higher in the Antarctic fur seals compared to the northern elephant seal. Assuming higher juvenile survival in Antarctic fur seals (based on data for the northern fur seal), Arnould (1997) proposed that Antarctic fur seals may trade off the high cost of maternal investment for higher offspring survival. Estimates of juvenile survival in the Antarctic fur seal and other pinniped species are needed to test this hypothesis. Similarly, a recent review by Schulz and Bowen (2004) concluded that the precision and distribution of taxonomic data on basic maternal and offspring life-history traits of pinnipeds must be improved substantially before interspecific correlations can be used to effectively explore adaptive functions of the life-history traits in pinnipeds as a group. Such analyses are even further in the future for cetaceans.

13.5.2. Different Investment in Male and Female Offspring

Sexual selection theory predicts that parents (or, in the case of marine mammals, the mother) should either alter sex ratios such that more offspring of the gender with the least variability in reproductive success are favored or invest more in the gender whose reproductive success has the greatest variability (Trivers, 1972; Maynard-Smith, 1980). The first option is not available to any marine mammal because sex determination based on male heterogamy limits sex ratios to something very near 1:1. However, selective retention of male versus female fetuses may be related to their relative sizes and female body condition in highly dimorphic species (Arnbom *et al.*, 1994). For species with promiscuous mating habits and little gender size dimorphism, such as harp seals, little advantage can be won from investing more heavily in one gender than the other, and no differences in pup weight or growth rate are observed (Kovacs, 1987). In highly polygynous species, females rarely have difficulty securing a mate; if they survive to return to their breeding rookery, they can participate in reproduction. However, most males of the same species never will gain access to receptive females. Only a small percentage of males will succeed in breeding, and a few of these will be very successful. To improve their own fitness, mothers could capitalize on this high variability in potential reproductive success of their male offspring to enhance that male's chances of becoming a larger, more competitive, and ultimately more successful breeder when mature. Although some studies have suggested that some polygynous species of pinnipeds provide evidence of differential investment (Reiter *et al.*, 1978; Kovacs and Lavigne, 1986b; Ono and Boness, 1996), other studies have not found differences in investment in male and female offspring in polygynous, sexually dimorphic species (Kretzmann *et al.*, 1993, Smiseth and Lorentsen, 1995; Arnould *et al.*, 1996; Arbom *et al.*, 1997; Wilkinson and Van Aarde, 2001). More information on mass at weaning versus sex specific survival rates and perhaps also mass at birth and weaning versus final body size, particularly in males, are needed to explore these ideas further.

13.5.3. Evolution of Reproductive Patterns in Pinnipeds

Costa (1993) reflected on the evolution of reproduction and feeding energetics in phocids and otariids; he suggests that phocids are adapted to low metabolic expenditures and do not work as hard as otariids while foraging. He also suggested that the otariid breeding pattern (i.e., numerous short duration feeding trips that maximize energy and nutrient delivery to the pup) is ancestral, being most closely linked to their terrestrial common ancestors. According to his model, as ancestral otariids refined their foraging skills, they began to increase the duration of their foraging trips and the distance they traveled from shore. Such changes required a greater energy return per trip and an increased reliance on stored maternal reserves. This pattern might have been followed by the extinct desmatophocids, common ancestors of phocids, which evolved large body size. Large body size is a critical factor in reducing the need for regular feeding during lactation. Costa (1993) also suggested that basal phocids, the monachines, which occupied the Hawaiian Islands, were able to pass through the Central American seaway into the

Caribbean because they utilized highly dispersed prey with a high energy return per trip. Otariids either could not pass through the seaway or could not reproduce in the Caribbean due to its inherent low productivity. Once in the Atlantic, phocids dispersed easily in the absence of competition from otariids. On reaching high latitudes, their large bodies, and concomitant large fat stores, permitted a short lactation interval that in turn enabled phocids to breed on relatively stable ice. Once ice-breeding was initiated, the lactation interval was further shortened as phocids extended their range to pack ice. More rigorous testing of this model is required that incorporates recent phylogenetic advances.

13.6. Summary and Conclusions

The reproductive structures of marine mammals are similar to those of other mammals. A peculiarity of whales is that the corpora albicans remain visible for the rest of the animal's life, providing a record of past ovulations in the ovary. This makes it possible to examine the reproductive history of individual whales; each corpus albicans representing one ovulation (although not necessarily a pregnancy). The timing of ovulation and estrus varies among marine mammals. Phocids and otariids have a postpartum estrus which establishes an annual cycle with copulation in one breeding season and birth in the next. Adjustments to fit gestation periods of less than 1 year to the 1-year cycle are accomplished via delayed implantation. This combined reproductive strategy places parturition and mating in a relatively brief period of time when these animals are congregated at traditional breeding sites and allows the young to be born when conditions are optimal for their survival. Cetaceans, and some pinnipeds, show multiyear reproductive cycles with mating and parturition usually separated by at least 1, and sometimes 2 or more, years. Gestation in most marine mammals is approximately 1 year, although active gestation in polar bears is only a matter of a few months.

The mating systems of marine mammals include promiscuity and polygyny. Among pinnipeds, all otariids and many species of phocids are polygynous. Almost all pinniped species that breed on land are extremely polygynous and strongly sexually dimorphic. Because polygynous males must compete for reproductive control of females (or resources important to them), competition revolves around either the establishment and defense of breeding territories (resource defense polygyny) or the establishment of dominance hierarchies (female or harem defense polygyny). Pinnipeds that mate in water or on ice (walrus and ice seals) typically show a reduced level of polygyny that is explained, in part, by the difficulty of defending a resource or access to females in an unstable environment. In some ice-breeding species females might exhibit mate choice. Lek or lek-like mating strategies have been suggested for the Pacific walrus and at least two otariid species.

Mating in cetaceans and sirenians is typically promiscuous, although strategies such as mate guarding do exist. Odontocetes exist in social groups (e.g., schools, units, and pods), whereas mysticetes are more solitary. Humpback whales, spinner dolphins, bottlenose dolphin, and killer whales are among the best studied cetaceans with respect to social behavior. Killer whales live in pods that are matrilineal social groups, characterized by their specific dialects, foraging strategies, and individual members. Sperm whales occur in female and immature groups, or units, that are visited by breeding males. The tight bonding among females might be facilitated by the need to have calves cared for at the surface by some group members while mothers forage at great depths. The basic social unit in mysticetes is a cow and a calf. The larger testes of gray, bowhead, and right whales have been interpreted as evidence of sperm competition. Maternal care in mysticetes is shorter than that in odontocetes. The sexual behavior of dugongs and manatees involves scramble competition for access to an estrus female; dugongs in some areas exhibit quite violent competition between males for females, whereas in other areas they seem to use a lek strategy. Sea otters and polar bears are polygynous breeders.

Three maternal strategies characterize marine mammals: fasting (e.g., some phocid seals, mysticetes), foraging cycle (e.g., otariid seals, some phocids), and aquatic nursing (e.g., walruses, odontocetes, sirenians, and sea otters). The fasting stategy involves a short lactation period during which the pup is fed very fat-rich milk and grows quickly. The foraging strategy, generally speaking, involves a longer period of lactation during which pups receive milk that is lower in fat content, and neonatal growth is slower. The aquatic nursing strategy is variable in length from months to years depending on the species, and thus neonatal growth and milk fat-contents are also variable across the diverse groups utilizing this strategy. Polar bears exhibit a typical bear lactation strategy, but the maternal care period is long.

Evidence of fitness costs associated with different reproductive strategies in female seals is mixed. Similarly, there are mixed results concerning differential investment in male and female offspring. However, generally speaking the required data are lacking for rigorous testing of life-history phenomenon among marine mammals. A model for the evolution of reproductive patterns in pinnipeds proposes that the foraging-cycle pattern (i.e., numerous short duration feeding trips that maximize energy and nutrient delivery to the pup) is ancestral. According to this model, the fasting strategy resulted from an increased body size, seen in the extinct desmatophocids, which enabled a reliance on stored maternal reserves. Further evolution of the fasting strategy involved the initiation of ice breeding and, as a consequence, a shortened lactation interval.

13.7. Further Reading

For a summary of the pinniped reproductive system see Laws and Sinha (1993) and references cited therein. The reproductive system of cetaceans is described in Slijper (1966), Harrison (1969), Perrin and Reilly (1984), and Lockyer (1984); for sirenians see Hill (1945), Harrison (1969), and O'Shea *et al.* (1995). The evolution of maternal care patterns in pinnipeds is summarized by Kovacs and Lavigne (1986a, 1986b, 1992a) and Boness and Bowen (1996); see also Trillmich (1996) and Boyd (1998). Articles by Oftedal and coworkers summarize lactation in pinnipeds (e.g., Oftedal *et al.*, 1987) and cetaceans (Oftedal, 1997). Schulz and Bowen (2004) review the quality of available data and knowledge gaps regarding piniped lactation strategies. The behavioral ecology of neonatal pinnipeds is reviewed in Bowen (1991). Pinniped mating systems are summarized by Boness (1991) and Le Boeuf (1991). The classic paper on the evolution of pinniped mating systems is Bartholomew (1970). Lydersen and Kovacs (1999) provide a detailed energetic and behavioural summary of lactation strategies of five North Atlantic ice-breeding phocids. A general introduction to cetacean mating systems can be found in Evans (1987) and see papers in Mann *et al.* (2002). Manatee and dugong mating behaviors are discussed in Rathbun *et al.* (1995) and Anderson (1997).

References

Alcorn, D. J. (1984). *The Hawaiian Monk Seal on Laysan Island, 1982*. NOAA Tech. Mem. NMFS-SNFC NMFSS WFC-42, National Marine Fisheries Service, Honolulu, HI.

Allen, S. G. (1985). "Mating Behavior in the Harbor Seal." *Mar. Mamm. Sci. 1:* 84–87.

Amos, W., J. Barrett, and G. A. Dover (1991). "Breeding System and Social Structure in the Faroese Pilot Whales as Revealed by DNA Fingerprinting." *Rep. Int. Whal. Commn. Spec. Issue 13:* 255–270.

Amos, B., C. Schlotterer, and D. Tautz (1993b). "Social-Structure of Pilot Whales Revealed by Analytical DNA Profiling." *Science 260:* 670–672.

Amos, B., S. Twiss, P. P. Pomeroy, and S. Anderson (1995). "Evidence for Mate Fidelity in the Gray Seal." *Science 268:* 897–899.

Amos, W., S. Twiss, P. Pomeroy, and S. S. Anderson (1993a). "Male Mating Success and Paternity in the Gray Seal, *Halichoerus grypus*: A Study Using DNA Fingerprinting." *Proc. R. Soc. London 252:* 199–207.

Amstrup, S. C., and D. P. De Master (1988). Polar bear, *Ursus maritimus. In* "Selected Marine Mammals of Alaska" (J. W. Lentfer, ed.), pp. 39–56. Marine Mamm. Comm., Washington, DC.

Anderson, P. K. (1997). "Shark Bay Dugongs in Summer. 1. Lekking." *Behaviour 134:* 433–462.

Anderson, P. K., J. M. Packard, G. B. Rathbun, D. P. Domning, and R. Best (2001). Dugong and manatees. *In* "The New Encyclopedia of Mammals" (D. MacDonald, ed.), pp. 278–287. Oxford University Press, Oxford.

Arnbom, T. (1994). "Maternal Investment in Male and Female Offspring in the Southern Elephant Seal." Ph.D. Thesis, Stockholm University, Stockholm.

Arnbom, T., M. A. Fedak, and I. L. Boyd (1997). "Factors Affecting Maternal Expenditure in Southern Elephant Seals During Lactation." *Ecology 78:* 471–483.

Arnbom, T., M. A. Fedak, and P. Rothery (1994). "Offspring Sex-Ratio in Relation to Female Size in Southern Elephant Seals, *Mirounga leonina*." *Behav. Ecol. Sociobiol. 35:* 373–378.

Arnould, J. P. Y. (1997). "Lactation and the Cost of Pup-Rearing in Antarctic Fur Seals." *Mar Mamm. Sci. 13:* 516–526.

Arnould, J. P. Y., I. L. Boyd, and D. G. Socha (1996). "Milk Consumption and Growth Efficiency in Antarctic Fur Seal *(Arctocephalus gazella)* Pups." *Can. J. Zool. 74:* 254–266.

Atkinson, S. (2002). Male reproductive systems. *In* "Encyclopedia of Marine Mammals" (W. F. Perrin, B. Würsig, and J. G. M. Thewissen, eds.), pp. 700–704. Academic Press, San Diego.

Baker, C. S., and L. M. Herman (1984). "Aggressive Behavior Between Humpback Whales *(Megaptera novaeagliae)* on the Hawaiian Wintering Grounds." *Can. J. Zool. 62:* 1922–1937.

Balcomb, K. C., III (1974). "The Birth of a Gray Whale." *Pac. Disc. 27:* 28–31.

Barlow, J. (1984). "Reproductive Seasonality in Pelagic Dolphins (*Stenella* spp.): Implications for Measuring Rates." *Rep. Intern. Whal. Commn., Spec. Issue 6:* 191–198.

Bartholomew, G. A. (1970). "A Model for the Evolution of Pinniped Polygyny." *Evolution 24:* 546–559.

Bartsh, S. S., S. D. Johnston, and D. B. Siniff (1992). "Territorial Behavior and Breeding Frequency of Male Weddell Seals *(Leptonychotes weddellii)* in Relation to Age, Size and Concentration of Serum Testosterone and Cortisol." *Can. J. Zool. 70:* 680–692.

Bengtson, J. L. (1981). "Ecology of Manatees *(Trichechus manatus)* in the St. Johns River, Florida. Ph.D. thesis, University of Minnesota, Minneapolis.

Bengtson, J. L., and D. B. Siniff (1981). "Reproductive Aspects of Female Crabeater Seals, *Lobodon carcinophagus,* Along the Antarctic Peninsula." *Can J. Zool. 59:* 92–102.

Best, R. C. (1981). "The Status of Right Whales *(Eubalaena glacialis)* off South Africa, 1969-1979." *Rep. So. Afr. Sea Fish. Inst. 123:* 1–44.

Best, P. B. (1982). "Seasonal Abundance, Feeding, Reproduction, Age and Growth in Minke Whales off Durban (with Incidental Observations from the Antarctic)." *Rep. Int. Whal. Commn. 32:* 759–786.

Bigg, M. A., P. F. Olesiuk, G. M. Ellis, L. K. B. Ford, and K. C. Balcomb (1990). "Social Organization and Genealogy of Resident Killer Whales *(Orcinus orca)* in the Coastal Waters of British Columbia and Washington State." *Rep. Int. Whal. Commn., Spec. Issue 12:* 383–406.

Boness, D. J. (1991). Determinants of mating systems in the Otariidae (Pinnipedia). *In* "The Behaviour of Pinnipeds" (D. Renouf, ed.), pp. 1–44. Chapman & Hall, New York.

Boness, D. J. (2002). Estrus and estrous behaviour. *In* "Encyclopedia of Marine Mammals" (W. F. Perrin, B. Würsig, and J. G. M. Thewissen, eds.), pp. 395–398. Academic Press, San Diego.

Boness, D. J., S. S. Anderson, and C. R. Cox (1982). "Functions of Female Aggression During the Pupping and Mating Season of Gray Seals, *Halichoerus grypus* (Fabricius)." *Can. J. Zool. 60:* 2270–2278.

Boness, D. J., and W. D. Bowen (1996). "The Evolution of Maternal Care in Pinnipeds." *BioScience 46:* 645–654.

Bonner, W. N. (1984). Lactation strategies in pinnipeds: Problems for a marine mammalian group. *In* "Physiological Strategies in Lactation" (M. Peaker, R. G. Vernon, and C. H. Knight, eds.), pp. 253–272. Academic Press, London.

Bonner, W. N. (1990). *The Natural History of Seals.* Christopher Helm, London.

Bowen, W. D. (1991). Behavioral ecology of pinniped neonates. *In* "The Behaviour of Pinnipeds" (D. Renouf, ed.), pp. 66–127. Capman & Hall, New York.

Bowen, W. D., S. J. Iverson, D. J. Boness, and O. T. Oftedal (2001). "Foraging Effort, Food Intake and Lactation Performance Depend on Maternal Mass in a Small Phocid Seal." *Funct. Ecol. 15:* 325–334.

Bowen, W. D., O. T. Oftedal, and D. J. Boness (1985). "Birth to Weaning in Four Days: Remarkable Growth in the Hooded Seal, *Cystophora cristata.*" *Can. J. Zool. 63:* 2841–2846.

Bowen, W. D., O. T. Oftedal, D. J. Boness, and S. J. Iverson (1994). "The Effect of Maternal Age and Other Factors on Birth Mass in the Harbor Seal." *Can. J. Zool. 72:* 8–14.

Boyd, I. L. (1998). "Time and Energy Constraints in Pinniped Lactation." *Am. Nat. 152:* 717–728.

Boyd, I. L., K. Reid, and R. M. Bevan (1995). "Swimming Speed and Allocation of Time During the Dive Cycle in Antarctic Fur Seals." *Anim. Behav. 50:* 769–784.

Brownell, R. L. J., and K. Ralls (1986). "Potential for Sperm Competition in Baleen Whales." *Rep. Int. Whal. Commn., Spec. Issue 8:* 97–112.

Bull, J. J., and E. L. Charnov (1985). "On Irreversible Evolution." *Evolution 39:* 1149–1155.

Burns, J. J. (2002). Harbor and spotted seals. *In* "Encyclopedia of Marine Mammals" (W. F. Perrin, B. Würsig, and J. G. M. Thewissen, eds.), pp. 552–560. Academic Press, San Diego.

Caldwell, D. K., and M. C. Caldwell (1972). *The World of the Bottlenose Dolphin.* Lippincott, New York.

Caldwell, D. K., and M. C. Caldwell (1985). Manatees. *In* "Handbook of Marine Mammals," Vol. 3 (S. Ridgway, ed.), pp. 33–66. Academic Press, New York.

Campagna, C., B. J. Le Boeuf, and H. L. Cappozzo (1988). "Group Raids: A Mating Strategy of Male Southern Sea Lions." *Behavior 105:* 224–246.

Cassini, M. H., and B. L. Vila (1990). "Male Mating Behavior of the Southern Sea Lion." *Bull. Mar. Sci. 46:* 555–559.

Clapham, P. J. (1996). "The Social and Reproductive Biology of Humpback Whales: An Ecological Perspective." *Mamm. Rev. 26:* 27–49.

Clapham, P. J., and P. J. Palsbøll (1997). "Molecular Analysis of Paternity Shows Promiscous Mating in Female Humpback Whales (*Megaptera novaeangliae* Borowski). *Proc. R. Soc. Lond. B 264:* 95–98.

Clapham, P. J., P. J. Palsbøll, D. K. Mattila, and O. Vasquez (1992). "Composition and Dynamics of Humpback Whales Competitive Groups in the West-Indies." *Behaviour 122:* 182–194.

Clinton, W. L., and B. J. Le Boeuf. (1993). "Sexual Selection's Effects on Male Life History and the Pattern of Male Mortality." *Ecology 74:* 1884–1892.

Coltman, D. W., W. D. Bowen, and J. M. Wright (1998). "Male Mating Success in an Aquatically Mating Pinniped, the Harbour Seal *(Phoca vitulina),* Assessed by Microsatellite DNA Markers." *Mol. Ecol. 7:* 627–638.

Coltman, D. W., W. D. Bowen, and J. M. Wright (1999). "A Multivariate Analysis of Phenotype and Paternity in Male Harbor Seals, *Phoca vitulina,* at Sable Island, Nova Scotia." *Behav. Ecol. 10:* 169–177.

Connor, R. C., M. R. Heithaus, and L. M. Barre (1999). "Superalliance of Bottlenose Dolphins." *Nature 397:* 571–572.

Connor, R. C., J. Mann, P. L. Tyack, and H. Whitehead (1998). "Social Evolution in Toothed Whales." *Trends Ecol. Evol. 13:* 228–232.

Connor, R. C., A. F. Richards, R. A. Smolker, and J. Mann (1996). "Patterns of Female Attractiveness in Indian Ocean Bottlenose Dolphins." *Behaviour 133:* 37–69.

Connor, R. C., R. A. Smolker, and A. F. Richards. (1992). Dolphin alliances and coalitions. *In* "Coalitions and Alliances in Humans and Other Animals" (A. H. Harcourt, and F. B. M. De Waal, eds.), pp. 415–443. Oxford University Press, Oxford, UK.

Corkeron, P. J., and R. C. Connor. (1999). "Why Do Baleen Whales Migrate?" *Mar. Mamm. Sci. 15:* 1228–1245.

Costa, D. P. (1991). Reproductive and foraging energetics of pinnipeds; implications for life history patterns. *In* "The Behaviour of Pinnipeds" (D. Renouf, ed.), pp. 299–344. Chapman & Hall, London.

Costa, D. P. (1993). "The Relationship Between Reproductive and Foraging Energetics and the Evolution of the Pinnipedia." *Symp. Zool. Soc. London 66:* 293–314.

Costa, D. P., and R. L. Gentry (1986). Free-ranging energetics of northern fur seal. *In* "Fur Seals: Maternal Strategies on Land and at Sea" (R. L. Gentry, and G. L. Kooyman, eds.), pp. 79–101. Princeton University Press, Princeton, NJ.

Costa, D. P., B. J. Le Boeuf, A. C. Huntley, and C. L. Ortiz (1986). "The Energetics of Lactation in the Northern Elephant Seal, *Mirounga angustirostris*." *J. Zool. London 201:* 21–33.

Costa, D. P., P. H. Thorson, J. G. Herpolsheimer, and J. P. Croxall (1985). "Reproductive Bioenergetics of the Antarctic Fur Seal." *Antarct. J. U.S. 20:* 176–177.

Costa, D. P., and F. Trillmich (1988). "Mass Changes and Metabolism During the Perinatal Fast: Comparison Between Antarctic *(Arctocephalus gazella)* and Galapagos Fur Seals *(Arctocephalus galapagoensis)*." *Physiol. Zool. 61:* 160–169.

Croxall, J. P., T. S. McCann, P. A. Prince, and P. Rothery (1988). Reproductive performance of seabirds and seals at South Georgia and Signy Island, South Orkney Islands, 1976-1987: Implications for southern ocean monitoring studies. *In* "Antarctic Ocean and Resources Variability" (D. Sahrhage, ed.), pp. 261–285. Springer-Verlag, Heidelberg.

Derocher, A. E. (1990). "Supernumerary Mammae and Nipples in the Polar Bear *(Ursus maritimus)*." *J. Mammal. 71:* 236–237.

Derocher, A. E., D. Andriashek, and J. P. Y. Arnould (1993). "Aspects of Milk Composition in Polar Bears." *Can. J. Zool. 71:* 561–567.

Derocher, A. E., and I. Stirling (1995). "Temporal Variation in Reproduction and Body Mass of Polar Bears in Western Hudson Bay." *Can. J. Zool. 73:* 1657–1665.

Duffield, D. A., and R. S. Wells (1991). "The Combined Application of Chromosome, Protein and Molecular Data for the Investigation of Social Unit Structure and Dynamics in *Tursiops truncatus*." *Rep. Int. Whal. Commn. Spec. Issue 13:* 155–159.

Emlen, S. T., and L. W. Oring (1977). "Ecology, Sexual Selection, and the Evolution of Mating Systems." *Science 197:* 215–223.

Estes, J. A. (1980). "*Enhydra lutris*." *Mamm. Spec. 133:* 1–8.

Estes, J. A., and J. L. Bodkin (2002). Otters. *In* "Encyclopedia of Marine Mammals" (W. F. Perrin, B. Würsig, and J. G. M. Thewissen, eds.), pp. 842–858. Academic Press, San Diego.

Evans, G. H. P. (1987). *The Natural History of Whales and Dolphins*. Facts On File, New York.

Fabiani, A., F. Galimberti, S. Sanvito, and A. R. Hoelzel (2004). "Extreme Polygny Among Southern Elephant Seals on Sea Lion Island, Falkland Islands." *Behav. Ecol. 15:* 961–969.

Fay, F. H. (1982). "Ecology and Biology of the Pacific Walrus, *Odobenus rosmarus divergens* Illiger." North Am. Fauna Ser., U.S. Dep. Int., Fish Wildl. Serv., No. 74.

Fedak, M. A., and S. S. Anderson (1982). "The Energetics of Lactation: Accurate Measurements from a Large Wild Mammal, the Gray Seal *(Halichoerus grypus)*." *J. Zool. London 198:* 473–479.

Ford, J. K. B., G. M. Ellis, and K. C. Balcomb (1994). *Killer Whales: The Natural History and Genealogy of Orcinus orca in British Columbia and Washington State*. University of Washington Press, Seattle.

Francis, J. M., and D. J. Boness (1991). "The Effect of Thermoregulatory Behavior on the Mating System of the Juan Fernandez Fur Seal, *Arctocephalus philippii*." *Behaviour 119:* 104–126.

Fredga, K. (1988). "Aberrant Chromosomal Sex-Determining Mechanisms in Mammals, with Special Reference to Species with XY Females." *Phil. Trans. R. Soc. Lond. Ser B 322:* 83–95.

Gales, N. J., D. P. Costa, and M. Kretzmann (1996). "Proximate Composition of Australian Sea Lion Milk Throughout the Entire Supraannual Lactation Period." *Aust. J. Zool. 44:* 651–657.

Gaskin, D. E., G. J. D. Smith, A. P. Watson, W. Y. Yasui, and D. B. Yurik (1984). "Reproduction in the Porpoises (Phocoenidae): Implications for Management." *Rep. Int. Whal. Commn., Spec. Issue 6:* 135–147.

Gemmell, N. J., T. M. Burg, I. L. Boyd, and W. Amos (2001). "Low Reproductive Success in Territorial Male Antarctic Fur Seals *(Arctocephalus gazella)* Suggests the Existence of Alternative Mating Strategies." *Mol. Ecol. 10:* 451–460.

Gentry, R. L. (1998). *Behavior and Ecology of the Northern Fur Seal*. Princeton University Press, Princeton, NJ.

Gisiner, R. C. (1985). "Male Territoriality and Reproductive Behavior in the Steller Sea Lion, *Eumetopias jubatus*." Ph.D. thesis, University of California, Santa Cruz.

Glockner-Ferrari, D. A., and M. J. Ferrari (1984). "Reproduction in Humpback Whales, *Megaptera novaengliae,* in Hawaiian Waters." *Rep. Int. Whal. Commn. 6:* 237–242.

Gregory, M. E., S. K. Kon, S. J. Rowland, and S. Y. Thompson (1955). "The Composition of the Milk of the Blue Whales." *J. Dairy Res. 22:* 108–112.

Haley, M. P., C. J., Deutsch, and B. J. Le Boeuf (1994). "Size, Dominance and Copulatory Success in Male Northern Elephant Seals, *Mirounga angustirostris*." *Anim. Behav. 48:* 1249–1260.

Haller, M. A., K. M. Kovacs, and M. O. Hammill. (1996). "Maternal Investment by Fast-Ice Breeding Grey Seals *(Halichoerus grypus)*." *Can. J. Zool. 74:* 1531–1541.

Hanggi, E. B., and R. J. Schusterman (1994). "Underwater Acoustic Displays and Individual Variation in Male Harbour Seals, *Phoca vitulina*." *Anim. Behav. 48:* 1275–1283.

Harcourt, R. G., M. A. Hindell, and J. R. Waas (1998). "Under Ice Movements and Territory Use in Free-Ranging Weddell Seals During the Breeding Season. *N.Z. Nat. Sci. 23:* 72–73.

Harcourt, R. G., M. A. Hindell, and J. R. Waas (2000). "Three-Dimensional Dive Profiles of Free-Ranging Weddell Seals." *Polar Biol. 23:* 479–487.

Harrison, R. J. (1969). Endocrine organs: Hypophysis, thyroid, and adrenal. *In* "The Biology of Marine Animals" (H. T. Andersen, ed.), pp. 349–390. Academic Press, New York.

Harrison, R. J., and J. E. King (1980). *Marine Mammals,* 2nd ed. Hutchinson and Co., Ltd. London.

Hartman, D. S. (1979). Ecology and behavior of the manatee (*Trichechus manatus*) in Florida. Am. Soc. Mammal., Special Publ. No. 5, Shippenburg, PA.

Heath, C. B. (2002). California, Galapagos, and Japanese sea lions. *In* "Encyclopedia of Marine Mammals" (W. F. Perrin, B. Würsig, and J. G. M. Thewissen, eds.), pp. 180–186. Academic Press, San Diego.

Heide-Jørgensen, M. P. (2002). Narwhal. *In* "Encyclopedia of Marine Mammals" (W. F. Perrin, B. Würsig, and J. G. M. Thewissen, eds.), pp. 783–787. Academic Press, San Diego.

Herzing, D. L. (1997). "The Life History of Free-Ranging Atlantic Spotted Dolphins *(Stenella frontalis)*: Age Classes, Color Phases and Female Reproduction." *Mar. Mamm. Sci. 13:* 576–595.

Herzing, D. L., and B. R. Mate (1984). Gray whale migrations along the Oregon coast. *In* "The Gray Whale, *Eschrichtius robustus* (Lilljeborg, 1861)" (M. L. Jones, S. L. Swartz, and S. Leatherwood, eds.), pp. 289–308. Academic Press, New York.

Heyning, J. E., and J. G. Mead (1996). "Suction Feeding in Beaked Whales: Morphological and Observational Evidence." *Contrib. Sci. Los Angeles Mus. Nat. Hist. 464:* 1–12.

Higgins, L. V., D. P. Costa, A. C. Huntley, and B. J. Le Boeuf (1988). "Behavior and Physiological Measurements of the Maternal Investment in the Steller Sea Lion, *Eumetopias jubatus*." *Mar. Mamm. Sci. 4:* 44–58.

Higgins, L. V., and L. Gass (1993). "Birth to Weaning: Parturition, Duration of Lactation and Attendance Cycles of Australian Sea Lions *(Neophoca cinerea)*." *Can. J. Zool. 71:* 2047–2055.

Hill, W. C. O. (1945). "Notes on the Dissection of Two Dugongs." *J. Mammal. 26:* 153–175.

Huber, H. R. (1987). "Natality and Weaning Success in Relation to Age of First Reproduction in Northern Elephant Seals." *Can. J. Zool. 65:* 1311–1316.

Hurst, L. D., A. Allan, and B. O. Bengtson (1996). "Genetic Conflicts." *Q. Rev. Biol. 71:* 317–471.

Irvine, B., and R. S. Wells (1972). "Results of Attempts to Tag Atlantic Bottlenosed Dolphins *(Tursiops truncatus)*." *Cetology 13:* 1–5.

Iverson, S. J., W. D. Bowen, D. J. Boness, and O. T. Oftedal (1993). "The Effect of Maternal Size and Milk Energy Output on Pup Growth in Gray Seals *(Halichoerus grypus)*." *Physiol. Zool. 66:* 61–88.

Jameson, R. J. (1989). "Movements, Home Range, and Territories of Male Sea Otters off Central California." *Mar Mamm. Sci. 5:* 159–172.

Jameson, R. J., and A. M. Johnson (1993). "Reproductive Characteristics of Female Sea Otters." *Mar. Mamm. Sci. 9:* 156–167.

Johanos, T. C., B. L. Becker, M. A. Brown, B. K. Choy, L. M. Hiruki, R. E. Brainard, and R. L. Westlake (1990). The Hawaiian monk seal on Laysan Island. 1988. NOAA Tech. Mem. NMFSSWMFS-151, National Marine Fisheries Service, Honolulu, HI.

Johnson, C. M., and K. S. Norris (1994). Social behavior. *In* "The Hawaiian Spinner Dolphin" (K. S. Norris, B. Wursig, R. S. Wells, and M. Wursig, eds.), pp. 243–286. University of California Press, Berkeley.

Jones, M. L., and S. L. Swartz (1984). Demography and phenology of gray whales (*Eschrichtius robustus*) in Laguna San Ignacio, Baja California Sur, Mexico. *In* "The Gray Whale Whale, *Eschrichtius robustus* (Lilljeborg, 1861)" (M. L. Jones, S. L. Swartz, and S. Leatherwood, eds.), pp. 309–374. Academic Press, New York.

Kastelein, R. A., and J. L. Dubbledam (1990). "Marginal Papillae on the Tongue of the Harbour Porpoise *(Phocoena phocoena)*, Bottlenose Dolphin *(Tursiops truncatus)* and Commerson's Dolphin *(Cephalorhynchus commersonii)*." *Aquat. Mamm. 15:* 158–170.

King J. E. (1983). *Seals of the World.* Comstock, Ithaca, NY.

Kovacs, K. M. (1987). "Maternal Behavior and Early Behavioral Ontogeny of Harp Seals, *Phoca groenlandica*." *Anim. Behav. 35:* 844–855.

Kovacs, K. M. (1990). "Mating Strategies in Male Hooded Seals (*Cystophora cristata*)." *Can. J Zool. 68:* 2499–2502.

Kovacs, K. M. (1995). Harp and hooded seal reproductive behaviour and energetics—a case study in the determinants of mating systems in pinnipeds. *In* "Whales, Seals, Fish and Man" (A. S. Blix, L. Walløe, and Ø. Ulltang, eds.), pp. 329–335. Elsevier Sci. BV, Amsterdam.

Kovacs, K. M., and D. M. Lavigne (1986a). "Maternal Investment and Neonatal Growth in Phocid Seals." *J. Anim. Ecol. 55:* 1035–1051.

Kovacs, K. M., and D. M. Lavigne (1986b). "Growth of Grey Seal *(Halichoerus grypus)* Neonates: Differential Maternal Investment in the Sexes." *Can. J. Zool. 64:* 1937–1943.

Kovacs, K. M., and D. M. Lavigne (1992a). "Maternal Investment in Otariid Seals and Walruses." *Can. J. Zool. 70:* 1953–1964.

Kovacs, K. M., and D. M. Lavigne (1992b). "Mass Transfer Efficiency Between Hooded Seal *(Cystophora cristata)* Mothers and Their Pups in the Gulf of St. Lawrence." *Can. J. Zool. 70:* 1315–1320.

Kovacs, K. M., C. Lydersen, M. O. Hammill, and D. M. Lavigne (1996). "Reproductive Effort of Male Hooded Seals *(Cystophora cristata)*." *Can. J. Zool. 74:* 1521–1530.

Krafft, B. A., C. Lydersen, K. M. Kovacs, I. Gjertz, and T. Haug (2000). "Diving Behaviour of Lactating Bearded Seals *(Erignathus barbatus)* in the Svalbard Area." *Can. J. Zool. 78:* 1408–1418.

Kretzmann, M. B., D. P. Costa, and B. J. Le Boeuf (1993). "Maternal Energy Investment in Elephant Seal Pups: Evidence for Sexual Equality?" *Am. Nat. 141:* 466–480.

Krutzen, M., L. M. Barre, R. C. Connor, and J. Mann (2004). "'O Father: Where Art Thou?'—Paternity Assessment in an Open Fission-Fusion Society of Wild Bottlenose Dolphins *(Tursiops sp.)* in Shark Bay, Western Australia." *Mol. Ecol. 13:* 1975–1990.

Laws, R. M. (1962). Some effects of whaling on the southern Stocks of baleen whales. *In* "The Exploitation of Natural Animal Populations" (R. F. Le Cren, and M. W. Holdgate, eds.), pp. 137–158. Wiley, New York.

Laws, R. M., and A. A. Sinha (1993). Reproduction. *In* "Antarctic Seals" (R. M. Laws, ed.), pp. 228–267. Cambridge University Press, Cambridge, UK.

Le Boeuf, B. J. (1974). "Male-Male Competition and Reproductive Success in Elephant Seals." *Am. Zool. 14:* 163–176.

Le Boeuf, B. J. (1986). "Sexual Strategies of Seals and Walruses." *New Sci. 1491:* 36–39.

Le Boeuf, B. J. (1991). Pinniped mating systems on land, ice, and in water: Emphasis on the Phocidae. *In* "The Behaviour of Pinnipeds" (D. Renouf, ed.), pp. 45–65. Chapman & Hall, London.

Le Boeuf, B. J., and R. M. Laws (1994). Elephant seals: An introduction to the genus. *In* "Elephant Seals" (B. J. Le Boeuf, and R. M. Laws, eds.), pp. 1–26. University of California Press, Berkeley.

Le Boeuf, B. J., and S. Mesnick (1990). "Sexual Behavior of Male Northern Elephant Seals. I. Lethal Injuries to Adult Females." *Behaviour 16:* 143–162.

Le Boeuf, B. J., and J. Reiter (1988). Lifetime reproductive success in northern elephant seals. *In* "Reproductive Success" (T. H. Clutton-Brock, ed.), pp. 344–362. University of Chicago Press, Chicago.

Lidgard, D. C., D. J. Boness, and W. D. Bowen (2001). "A Novel Mobile Approach to Investigating Mating Tactics in Male Grey Seals *(Halichoerus grypus)*." *J. Zool. London 255:* 313–320.

Lidgard, D. C., D. J. Boness, W. D. Bowen, J. I. McMillan, and R. C. Fleisher (2004). "The Rate of Fertilization in Male Mating Tactics of the Polygynous Grey Seal." *Mol. Ecol. 13:* 3543–3548.

Lockyer, C. (1981). "Growth and Energy Budgets of Large Baleen Whales from the Southern Hemisphere." *FAO Fish. Ser. 3:* 489–504.

Lockyer, C. (1984). "Review of Baleen Whale (Mysticeti) Reproduction and Implications for Management." *Rep. Int. Whal. Commn. Spec. Issue 6:* 27–50.

Logan, F. D., and F. D. Robson (1971). "On the Birth of a Common Dolphin (*Delphinus delphinus* L.) in Captivity. *Zool. Garten (Leipzig) 40*(3): 115–124.

Lydersen, C. (1995). Energetics of pregnancy, lactation and neonatal development in ringed seals (*Phoca hispida*). *In* "Whales, Seals, Fish and Man" (A. S. Blix, L. Walløe, and Ø. Ulltang, eds.). pp. 319–327. Elsevier Sci. BV, Amsterdam.

Lydersen, C., M. O. Hammill, and K. M. Kovacs (1995). "Milk Intake, Growth and Energy Consumption in Pups of Ice-Breeding Grey Seals *(Halichoerus grypus)* from the Gulf of St. Lawrence, Canada." *J. Comp. Physiol. B 164:* 585–592.

Lydersen, C., and K. M. Kovacs. (1993). "Diving Behaviour of Lactating Harp Seals *(Phoca groenlandica)* from the Gulf of St. Lawrence, Canada." *Anim. Behav. 46:* 1213–1221.

Lydersen, C., and K. M. Kovacs. (1996). "Energetics of Lactation in Harp Seals *(Phoca groenlandica)* from the Gulf of St Lawrence, Canada." *J. Comp. Physiol. 166:* 295–304.

Lydersen, C., and K. M. Kovacs. (1999). "Behaviour and Energetics of Ice-Breeding, North Atlantic Phocid Seals During the Lactation Period." *Mar. Ecol. Prog. Ser. 187:* 265–281.

Lydersen, C., K. M. Kovacs, and M. O. Hammill (1997). "Energetics During Nursing and Early Postweaning Fasting in Hooded Seal *(Cystophora cristata)* Pups from the Gulf of St Lawrence, Canada." *J. Comp. Physiol. B 167:* 81–88.

Mackintosh, N. A. (1966). The distribution of southern blue and fin whales. *In* "Whales, Dolphins and Porpoises" (K. S. Norris, ed.), pp 125–144. University of California Press, Berkeley.

Mann, J., R. C. Connor, P. L. Tyack, and H. Whitehead (2002). *Cetacean Societies: Field Studies of Dolphins and Whales*. University of Chicago Press, Chicago.

Marlow, B. J. (1975). "The Comparative Behaviour of the Australasian Sea Lions, *Neophoca cinerea* and *Phocarctos hookeri* (Pinnipedia: Otariidae)." *Mammalia 39:* 159–230.

Marmontel, M. (1995). Age and reproduction in female Florida manatees. *In* "Population Biology of the Manatee" (T. J. O'Shea, B. B. Ackerman, and H. F. Percival, eds.), pp. 98–119. Tech. Rep. 1. Natl. Biol. Ser., Washington, DC.

Marsh, H. (1995). The life history, pattern of breeding, and population dynamics of the dugong. *In* "Population Biology of the Manatee" (T. J. O'Shea, B. B. Ackerman, and H. F. Percival, eds.), pp. 56–62. Tech. Rep. 1. Natl. Biol. Ser., Washington, DC.

Maynard-Smith, J. (1980). "A New Theory of Sexual Investment." *Behav. Ecol. Sociobiol. 7:* 247–251.

McRae, S. B., and K. M. Kovacs (1994). "Paternity Exclusion by DNA Fingerprinting, and Mate Guarding in the Hooded Seal, *Cystophora cristata*." *Mol. Ecol. 3:* 101–107.

Medrano, L., M. Salinas, I. Salas, P. Ladronde Guevara, and A. Aguayo (1994). "Sex Identification of Humpback Whales, *Megaptera novaeangliae*, on the Wintering Grounds of the Mexican Pacific Ocean." *Can. J. Zool. 72:* 1771–1774.

Mesnick, S. L. and K. Ralls (2002). Mating systems. *In* "Encyclopedia of Marine Mammals" (W. F. Perrin, B. Würsig, and J. G. M. Thewissen, eds.), pp. 726–733. Academic Press, San Diego.

Miller, E. H. (1975). "Walrus Ethology. 1. Social-Role of Tusks and Applications of Multidimensional Scaling." *Can. J. Zool. 53:* 590–613.

Miller, E. H. (1991). Communication in pinnipeds, with special reference to non-acoustic signalling. *In* "The Behaviour of Pinnipeds" (D. Renouf, ed.), pp. 128–235. Chapman & Hall, New York.

Miller, E. H., and D. J. Boness (1983). "Summer Behavior of Atlantic Walruses *Odobenus rosmarus rosmarus* (L.) at Coats Island, N.W.T. (Canada)." *Z. Säugetierkunde 48:* 298–313.

Mills, J. G., and J. E. Mills (1979). "Observations of a Gray Whale Birth." *Bull. S. Calif. Acad. Sci. 78:* 192–196.

Mitchell, E. (1986). Firmer whales. *In* "Marine Mammals of Eastern North Pacific and Arctic Waters" (D. Haley, ed.), pp. 49–55. Pacific Search Press, Seattle, WA.

Mobley, J. R. J., and L. M. Herman (1985). "Transience of Social Affiliations Among Humpback Whales *(Megaptera novaeangliae)* on the Hawaiian Wintering Grounds." *Can. J. Zool. 63:* 762–772.

Modig, A. O. (1996). "Effect of Body Size on Male Reproductive Behavior in the Southern Elephant Seal." *Anim. Behav. 51:* 1295–1306.

Morejohn, G. V. (1975). "A Phylogeny of Otariid Seals Based on Morphology of the Baculum." *Rapp. P-v. Réun. Cons. Int. Explor. Mer. 169:* 49–56.

Nishiwaki, M., and H. Marsh (1985). Dugong, *Dugong dugon* (Mueller, 1776). *In* "Handbook of Marine Mammals. Vol. 3. The Sirenians and Baleen Whales" (S. H. Ridgway, and R. Harrison, eds.), pp. 1–31. Academic Press, London.

Norris, K. S., and C. M. Johnson (1994). Locomotion. *In* "The Hawaiian Spinner Dolphin" (K. S. Norris, B. Würsig, R. S. Wells, and M. Würsig, eds.), pp. 201–205. University of California Press, Berkeley.

Norris, K. S., B. Villa-Ramirez, G. Nichols, B. Würsig, and K. Miller (1983). Lagoon entrance and other aggregations of gray whales (*Eschrichtius robustus*). *In* "Communication and Behavior of Whales" (R. Payne, ed.), pp. 259–293. Westview Press, Boulder, CO.

Norris, K. S., B. Würsig, R. S. Wells, and M. Würsig (eds.) (1994). *The Hawaiian Spinner Dolphin*. University of California Press, Berkeley.

Oftedal, O. T. (1997). "Lactation in Whales and Dolphins: Evidence of Divergence Between Baleen- and Toothed-Species." *J. Mammary Gland Biol. and Neoplasia 2:* 205–230.

Oftedal, O. T., D. J. Boness, and R. A. Tedman (1987). The behavior, physiology and anatomy of lactation in the Pinnipedia. *In* "Current Mammalogy," Vol. 1 (H. Genoways, ed.), pp. 175–245. Plenum Press, New York.

Oftedal, O. T, W. D. Bowen, and D. J. Boness (1993). "Energy Transfer by Lactating Hooded Seals and Nutrient Deposition in Their Pups During the Four Days from Birth to Weaning." *Physiol. Zool. 66:* 412–436.

Oftedal, O. T., W. D. Bowen, and D. J. Boness (1996). "Lactation Performance and Nutrient Deposition in Pups of the Harp Seal, *Phoca groenlandica,* on Ice Floes off Southeast Labrador." *Physiol. Zool. 69:* 635–657.

Oftedal, O. T, and J. L. Gittleman (1989). Patterns of energy output during reproduction in carnivores. *In* "Carnivore Behavior, Ecology, and Evolution" (J. L. Gittleman, ed.), pp. 355–379. Cornell University Press, Ithaca, NY.

Ono, K. A., and D. J. Boness (1996). "Sexual Dimorphism in Sea Lion Pups: Differential Maternal Investment, or Sex-Specific Differences in Energy Allocation?" *Behav. Ecol. Sociobiol. 38:* 31–41.

O'Shea, T. J. (1994). "Manatees." *Sci. Am. 273:* 66–72.

O'Shea, T. J., B. B. Ackerman, and H. F. Percival (eds.) (1995). *Population Biology of the Florida Manatee.* Natl. Biol. Serv. Tech. Rep. No. 1.

Pabst, D. A., S. A. Rommel, and W. A. McLellan (1998). Evolution of the thermoregulatory function in the cetacean reproductive systems. *In* "The Emergence of Whales: Patterns in the Origin of Cetacea" (J. G. M. Thewissen, ed.), pp. 379–397. Plenum Press, New York.

Pack, A. A., D. R. Salden, M. J. Ferrari, D. A. Glockner-Ferrari, L. M. Herman, H. A. Stubbs, and J. M. Straley (1998). "Male Humpback Whale Dies in Competitive Group." *Mar. Mamm. Sci. 14:* 861–873.

Patterson-Buckendahl, P, S. H. Adams, R. Morales, W. S. S. Jee, C. E. Cann, and C. L. Ortiz (1994). "Skeletal Development in Newborn and Weanling Northern Elephant Seals." *Am. J. Physiol. 267:* R726–R734.

Perrin, W. F., and G. P. Donovan (eds.) (1984). "Report of the Workshop on Reproduction of Whales, Dolphins and Porpoises." *Rep. Int. Whal. Commn. Spec. Issue 6:* 1–24.

Perrin, W. F., and S. B. Reilly (1984). "Reproductive Parameters of Dolphins and Small Whales of the Family Delphinidae." *Rep. Int. Whal. Commn. Spec. Issue 6:* 97–125.

Perry, E. A., S. M. Carr, S. E. Bartlett, and W. S. Davidson (1995). "A Phylogenetic Perspective on the Evolution of Reproduction Behavior in Pagophilic Seals of the Northwest Atlantic as Indicated by Mitochondrial DNA Sequences." *J. Mammal. 76:* 22–31.

Perryman, W. L., M. A. Donahue, J. L. Laake, and T. E. Martin (1999). "Diel Variation in Migration Rates of Eastern Pacific Gray Whales Measured with Thermal Imaging Sensors." *Mar. Mamm. Sci. 15:* 426–445.

Pilson, M. E., and D. W. Waller (1970). "Composition of Milk from Spotted and Spinner Porpoises." *J. Mammal. 51:* 74–79.

Pomeroy, P. P., S. D. Twiss, and P. Redman (2000). "Philopatry, Site Fidelity and Local Kin Associations Within Grey Seal Breeding Colonies." *Ethology 106:* 899–919.

Poole, M. M. (1984). Migration corridors of gray whales along the central California coast, 1980-1982. *In* "The Gray Whale, *Eschrichtius robustus* (Lilljeborg, 1861)" (M. L. Jones, S. L. Swartz, and S. Leatherwood, eds.), pp. 389–408. Academic Press, New York.

Preen, A. (1989). "Observations of Mating Behavior in Dugongs." *Mar. Mamm. Sci. 5:* 382–386.

Ralls, K., T. C. Eagle, and D. B. Siniff (1996). "Movement and Spatial Use Patterns of California Sea Otters." *Can. J. Zool. 74:* 1841–1849.

Ralls, K. and S. L. Mesnick (2002). Sexual dimorphism *In* "Encyclopedia of Marine Mammals" (W. F. Perrin, B. Würsig and J. G. M. Thewissen, eds.), pp. 1071–1078. Academic Press, San Diego.

Ramsay, M. A., and I. Stirling (1988). "Reproductive Biology of Female Polar Bears *(Ursus maritimus)*." *J. Zool. London 214:* 601–634.

Rathbun, G. B., and T. J. O'Shea (1984). The manatee's simple social life. *In* "The Encyclopedia of Mammals" (D. Macdonald, ed.), pp. 300–301. Facts on File, New York.

Rathbun, G. B., J. P. Reid, R. K. Bonde, T. J. O'Shea, and J. A. Powell (1995). Reproduction in free-ranging Florida manatees. *In* "Population Biology of the Florida Manatee" (T. J. O'Shea, B. B. Ackerman, and H. F. Percival, eds.), pp. 135–156. U. S. Dept. of the Interior, Washington, DC.

Reilly, J. J., M. A. Fedak, D. H. Thomas, W. A. A. Coward, and S. S. Anderson (1996). "Water Balance and the Energetics of Lactation in Gray Seals *(Halichoerus grypus)* as Studied by Isotopically Labelled Water Methods." *J Zool. London 238:* 157–165.

Reiter, J., K. J. Panken, and B. J. Le Boeuf (1981). "Female Competition and Reproductive Success in Northern Elephant Seals. *Anim. Behav. 29:* 670–687.

Reiter, J., N. L. Stinson, and B. J. Le Boeuf (1978). "Northern Elephant Seal Development: The Transition from Weaning to Nutritional Independence." *Behav. Ecol. Sociobiol. 3:* 337–367.

Reynolds, J. E., III, and D. E. Odell (1991). *Manatees and Dugongs.* Facts on File, New York.

Rice, D. W. (1986). Gray whale. *In* "Marine Mammals of Eastern North Pacific and Arctic Waters" (D. Haley, ed.), pp. 2–71. Pacific Search Press, Seattle, WA.

Rice, D. W., and A. A. Wolman (1971). "The Life History and Ecology of the Gray Whale *(Eschrichtius robustus)*." *Am. Soc. Mammal. Spec. Publ. 3:* 1–142.

Riedman, M. (1990). *The Pinnipeds: Seals, Sea Lions, and Walruses*. University of California Press, Berkeley.

Riedman, M., and J. Estes (1990). "The Sea Otter *(Enhydra lutris):* Behavior, Ecology and Natural History. *U.S. Fish Wild. Biol. Rep. 90*(14): 1–126.

Saayman, G. S., C. K. Tayler, and D. Bower (1979). The socioecoogy of humpback dolphins (*Sousa sp.*). *In* "Behavior of Marine Animals, Vol. 3 Cetaeans" (H. E. Winn and B. L. Olla, eds.), pp. 165–226. Plenum Press, New York.

Sandell, M. (1990). "The Evolution of Seasonal Delayed Implantation." *Q. Rev. Biol. 65:* 23–42.

Schulz, T. M., and W. D. Bowen (2004). "Pinniped Lactation Strategies: Evaluation of Data on Maternal and Offspring Life History Traits." *Mar. Mamm. Sci. 20:* 86–114.

Scott, M. D., R. S. Wells, and A. B. Irvine (1989). A long-term study of bottlenose dolphins on the west coast of Florida. *In* "The Bottlenose Dolphin" (S. Leatherwood, and R. R. Reeves, eds.), pp. 235–244. Academic Press, Orlando, FL.

Sease, J. L., and D. G. Chapman (1988). Pacific walrus—*Odobenus rosmarus divergens. In* "Selected marine Mammals of Alaska," pp. 17–38. Mar. Mamm. Comm., Washington, DC.

Sinha, A. A., C. H. Conaway, and K. W. Kenyon (1966). "Reproduction in the Female Sea Otter." *J. Wildl. Manage. 30:* 121–130.

Sjare, B., and I. Stirling (1996). "The Breeding Behaviour of Atlantic Walruses, *Odobenus rosmarus rosmarus*, in the Canadian High Arctic." *Can. J. Zool. 75:* 897–911.

Slijper, E. J. (1966). Functional morphology of the reproductive system in Cetacea. *In* "Whales, Dolphins and Porpoises (K. Norris, ed.), pp. 277–319. University of California Press, Berkeley.

Smiseth, P. T., and S.-H. Lorentsen (1995). "Evidence of Equal Maternal Investment in the Sexes in the Polygynous and Sexually Dimorphic Grey Seal *(Halichoerus grypus)."* *Behav. Ecol. Sociobiol. 36:* 145–150.

Smith, T. G., and M. O. Hammill (1981). "Ecology of the Ringed Seal, *Phoca hispida* in Its Fast Ice Breeding Habitat." *Can. J. Zool. 59:* 966–981.

Spotte, S. (1982). "The Incidence of Twins in Pinnipeds." *Can. J. Zool. 60:* 2226–2233.

Stearns, S. C. (1989). "Trade-Offs in Life History Evolution." *Funct. Ecol. 3:* 259–268.

Stewart, R. E. A., B. E. Stewart (2002). Female reproductive systems. *In* "Encyclopedia of Marine Mammals" (W. F. Perrin, B. Wursig, and J. G. M. Thewissen, eds.), pp.422–428. Academic Press, San Diego, CA.

Stirling, I. (1975). "Factors Affecting the Evolution of Social Behavior in the Pinnipedia." *Rapp. P-v. Réun. Cons. Int. Explor Mer 169:* 205–212.

Stirling, I. (1983). The social evolution of mating systems in pinnipeds. *In* "Advances in the Study of Mammalian Behavior" (J. F. Eisenberg, and D. G. Kleiman, eds.), pp. 489–527. Spec. Publ. No. 7, Am. Soc. Mammal, Shippenburg, PA.

Stirling, I. (1988). *Polar Bears*. University of Michigan Press, Ann Arbor, MI.

Stirling, I. (2002). Polar bears. *In* "Encyclopedia of Marine Mammals" (W. F. Perrin, B. Würsig and J. G. M. Thewissen, eds.), pp. 945–948. Academic Press, San Diego, CA.

Sumich, J. L. (1986a). "Growth in Young Gray Whales *(Eschrichtius robustus)."* *Mar. Mamm. Sci. 2:* 145–152.

Sumich, J. L. (1986b). "Latitudinal Distribution, Calf Growth and Metabolism, and Reprodcutive Eneretics of Gray Whales, *Eschrichtius robustus."* PhD thesis, Oregon State University, Corvallis.

Sumich, J. L. (1996). *An Introduction to the Biology of Marine Life*. Wm. C. Brown, Dubuque, IA.

Sumich, J. L., and I. Show (1999). "Aerial Survey and Photogrammetric Comparisons of Southbound Migrating Gray Whales in the Southern California Bight, 1998-1990." *Rep. Int. whal. Commn. Spec. Issue*

Swartz, S. L. (1986). "Gray Whale Migratory, Social and Breeding Behavior." *Rep. Int. Whal. Comm. Spec. Issue 8:* 207–229.

Temte, J. L. (1994). "Photoperiod Control of Birth Timing in the Harbour Seal." *J. Zool. London 233:* 369–384.

Thomas, J. A., and D. P. Demaster (1983). "Diel Haul-Out Patterns of Weddell Seal *(Leptonychotes weddelli)* Females and Their Pups." *Can. J. Zool. 61:* 2084–2086.

Thomas, J. A., and V. Keuche (1982). "Quantitative Analysis of Weddell Seal *(Leptonychotes weddellii)* Underwater Vocalizations in McMurdo Sound, Antarctica." *J. Acoust. Soc. Am. 72:* 1730–1738.

Tinker, T. M., K. M. Kovacs, and M. O. Hammill (1995). "The Reproductive Behaviour and Energetics of Male Grey Seals *(Halichoerus grypus)* Breeding on a Landfast Ice Substrate." *Behav. Ecol. Sociobiol. 36:* 159–170.

Trillmich, F. (1986). Attendance behavior of Galapagos fur seals. *In* "Fur Seals: Maternal Strategies on Land and at Sea" (R. L. Gentry, and G. L. Kooyman, eds.), pp. 168–185. Princeton University Press, Princeton, NJ.

Trillmich, F. (1996). Parental investment in pinnipeds. *In* "Parental Care: Evolution, Mechanisms, and Adaptive Significance" (J. S. Rosenblatt, and C. T. Snowdon, eds.), pp. 533–577. Academic Press, San Diego, CA.

Trivers, R. L. (1972). Parental investment and sexual selection. *In* "Sexual Selection and the Descent of Man, 1871-1971" (B. Campbell, ed.), pp. 136–179. Aldine, Chicago, IL.

Valsecchi, E., P. Hale, P. Corkeron, and W. Amos (2002). "Social Structure in Migrating Humpback Whales *(Megaptera novaeangliae)*." *Mol. Ecol. 11:* 507–518.

Van Parijs, S. M., K. M. Kovacs, and C. Lydersen (2001). "Temporal and Spatial Distribution of Male Bearded Seal Vocalizations—Implications for Mating System." *Behaviour 138:* 905–922.

Van Parijs, S. M., C. Lydersen, and K. M. Kovacs (2003). "Vocalisations and Movements Suggest Alternate Mating Tactics in Male Bearded Seals." *Anim. Behav.* 65: 273–283.

Van Parijs, S. M., C. Lydersen, and K. M. Kovacs (2004). The effects of ice cover on the behavioural patterns of aquatic mating male bearded seals. *Anim. Behav. 68:* 89-96.

Van Parijs, S. M., P. M. Thompson, D. J. Tollit, and A. Mackay (1997). "Distribution and Activity of Male Harbour Seals During the Mating Season." *Anim. Behav. 54:* 35–43.

Vaughan, T. A. (1999). *Mammalogy,* 4th ed. Brooks/Cole Publishers, Pacific Grove, CA.

Weinrich, M. (1995). "Humpback Whale Competitive Groups Observed on a High-Latitude Feeding Ground." *Mar. Mamm. Sci. 11:* 251–254.

Wells, R. S., D. J. Boness, and G. B. Rathbun (1999). Behavior. *In* "Biology of Marine Mammals" (J. E. Reynolds III, and S. A. Rommel, eds.), pp. 324–422. Smithsonian Institute Press, Washington, DC.

Wells, R. S., M. D. Scott, and A. B. Irvine (1987). The social structure of free-ranging bottlenose dolphins on the west coast of Florida. *In* "Current Mammalogy," Vol 1 (H. H. Genoways, ed.), pp. 235–244. Plenum Press, New York.

White, J. C. D. (1953). "Composition of Whale's Milk." *Nature 171:* 612.

Whitehead, H. (1993). "The Behaviour of Mature Male Sperm Whales on the Galapagos Islands Breeding Grounds." *Can J. Zool. 71:* 689–699.

Whitehead, H. (1996). "Babysitting, Dive Synchrony, and Indications of Alloparental Care." *Behav. Evol. Sociobiol. 38:* 237–244.

Whitehead, H., and B. Kahn (1992). "Temporal and Geographic Variation in the Social Structure of Female Sperm Whales." *Can. J. Zool. 70:* 2145–2149.

Whitehead, H., and H. Waters (1990). "Social Organization and Population Structure of Sperm Whales off the Galapagos Islands, Ecuador (1985 and 1987)." *Rep. Int. Whal. Comm. Spec. Issue 12:* 1–440.

Whitehead, H., S. Waters, and T. Lyrholm (1991). "Social Organization of Female Sperm Whales and Their Offspring: Constant Companions and Casual Acquaintances." *Behav. Ecol. Sociobiol. 29:* 385–389.

Wiig, Ø., I. Gjertz, R. Hansson, and J. Thomassen (1992). "Breeding Behaviour of Polar Bears in Hornsund, Svalbard." *Polar Rec. 28:* 157–159.

Wilkinson, I. D., and R. J. Van Aarde (2001). "Investment in Sons and Daughters by Southern Elephant Seals, *Mirounga leonina*, at Marion Island." *Mar. Mamm. Sci. 17:* 873–887.

Wilmer, J. W., P. J. Allen, P. P. Pomeroy, S. D. Twiss, and W. Amos (1999). "Where Have All the Fathers Gone? An Extensive Microsatellite Analysis of Paternity in the Grey Seal *(Halichoerus grypus)*." *Mol. Ecol. 8:* 1417–1429.

Wilson, D. R. B. (1995). "The Ecology of Bottlenose Dolphins in the Moray Firth, Scotland: A Population at the Northern Extreme of the Species' Range." Ph.D. thesis, University of Aberdeen, Aberdeen, Scotland.

Yablokov, A. V., V. M. Bel'kovich, and V. I. Botisov (1972). *Whales and Dolphins*. Nauka, Moscow.

Zenkovich, B. A. (1938). "Milk of Large-Sized Cetaceans." *Dokl. Akad. Nauk. SSSR 16:* 203–205.

14

Population Structure and Dynamics

14.1. Introduction

Attempts to understand the ecology of marine mammals and the ecological roles they play in marine ecosystems necessitates knowledge of their abundance and the trends in their numbers. Yet, despite the high levels of interest in these animals, few good estimates exist for population sizes of marine mammals. Even within individual species abundance is usually only known for some populations within the global range of the species. The reasons why abundance of marine mammal populations is difficult to assess are numerous and most of them are tightly linked to the distribution patterns and natural behavior of marine mammals. Many marine mammal populations are broadly dispersed much of the year and virtually all species spend considerable amounts of their time underwater and are therefore unavailable to normal census methods. Hence, estimation methods for almost all marine mammal species must include estimates for unseen portions of a population under study. In general the abundance of exploited populations of marine mammals is better known than unexploited populations, because economic and management interests result in greater efforts at population assessment. Additionally, pinniped population sizes are generally better known that those of cetaceans or sirenians, because at least the reproductive segment of most pinniped populations congregates annually at traditional breeding sites on land or on ice to give birth and mate, permitting enumeration of pups, breeding adults, or both age groups.

Although our knowledge of population sizes is limited for marine mammals in general, some characteristics of populations of these animals can provide valuable information regarding potential responses of the populations and likely trends in their numbers. The life-history characteristics of individuals in a population and basic parameters that characterize populations are fundamental to understanding dynamics of populations through time and they also provide essential information directly for marine mammal management (also see Chapter 15).

A mammal **population** is defined as group of interbreeding individuals of the same species; a population usually occupies a defined geographical area although populations of a species can overlap geographically for parts of the year and yet not interbreed. In rare cases a population can consist of all the living members of a closely confined species, but usually a species is comprised of a number of semiautonomous populations spread across the species range. The reproductive behavior of members of a population creates a gene pool common to them but isolated to varying degrees from the gene pools of other populations. Genetic change in a population can result from emigration, immigration, genetic drift, and selection. Management of many species of marine mammals in the past has been practiced on a geographic basis and hence management agencies have often focused on **stocks,** which are segments of populations (usually defined geographically, not biologically) that are subject to commercial exploitation.

14.2. Abundance and Its Determination in Marine Mammals

Not surprisingly given the taxonomic and ecological diversity within marine mammals, the range of body sizes displayed, in addition to their respective histories with respect to commercial exploitation, abundance varies enormously among marine mammals. Abundances range from severely endangered species such as the baiji and the Mediterranean monk seal, with only a few or a few hundred individuals world-wide, respectively, to species that number in the millions, such as crabeater seals, harp seals, and some of the oceanic dolphins (Table 14.1). The current status and population structure of some species, particularly large baleen whales, but also sea otters, elephant seals, and other species is influenced heavily by past commercial exploitation. Some populations were extirpated and other species were reduced via hunting to small fragments of the original stocks (see Brownell *et al.*, 1989).

In simplistic terms, determining the abundance of marine mammal populations is approached in two basic ways: (1) total population counts (census) and (2) counting a sample of the population that is then extrapolated to represent the whole population (see

Table 14.1. Estimates of Worldwide Abundances of Selected Marine Mammals

Species	Estimated Abundance	Source
Pinnipeds		
Crabeater seal	10,000,000–15,000,000	Bengtson (2002)
Harp seal	7,000,000	Lavigne (2002)
Saima seal	200	Sipila and Hyvärinen (1998)
Antarctic fur seal	3,000,000	Gentry (2002)
Hooker's sea lion	11,100–14,000	Gales and Fletcher (1999)
Guadalupe fur seal	>7,000	Arnould (2002)
Walrus	230,000	Kovacs (2005)
Mediterranean monk seal	350–450	Gilmartin and Forcada (2002)
Cetaceans		
Pantropical spotted dolphin	Low millions	LeDuc (2002)
Baiji	A few dozen	Kaiya (2002)
Minke whales (combined)	935,000+	Gambell (1999)
Gray whale	26,000+	Jones and Swartz (2002)
North Atlantic right whale	<300	Reynolds *et al.* (2002)
Sirenians		
Florida manatee	3,300	Reynolds and Powell (2002)
Dugong	>85,000	Marsh (2002)
Sea otter	100,000	Bodkin *et al.* (1995)
Polar bear	21,000–28,000	Derocher *et al.* (1998)

Garner *et al.*, 1999). Total population counts are rare for marine mammals because of the difficulties of enumerating all members of a population. Bigg *et al.* (1990) counted the resident killer whale population off the coast of British Columbia and Washington by identifying individuals using natural markings. Similar methods have been used to census North Atlantic right whales (Kenny, 2002) and a few other species that retain individual markings throughout their lives (see Section 14.3). Indices of abundance, such as assessments of catch per unit effort for dispersed populations subject to harvesting provides information that can be used to assess relative trends in abundance. However, the most common assessment methods for marine mammals involve counting a sample of the population and then applying some assumptions with a model to extend the count and estimate the whole population. Methods of counting in these assessments include transect surveys (distance sampling), mark-recapture, migration counts, and colony counts (Buckland and York, 2002).

Transect surveys for marine mammals are done from ships or aircraft and involve line, strip, or cue counts. Lines or strips can be randomly selected or placed on systematic grids (often based on relative probabilities of locating animals in different habitat types within the area of the survey) throughout the study area. The number of animals counted and their spatial arrangement on and around the line are used to model the population size (Buckland *et al.*, 1993a). Zig-zag designs are often employed in marine mammal counts to minimize time spent off-line in the survey process. This census method is commonly used to determine abundance of cetaceans (e.g., Laake *et al.*, 1997; Hammond *et al.*, 2002). Strip transects are similar to line surveys, but all animals within a strip of a specified width are counted. For this method to provide reasonable results all animals within the designated areas must be detectable by the survey. This method has been used to estimate pup production in harp (Stenson *et al.*, 1993) and hooded seals (Bowen *et al.*, 1987) as well as assess ringed seal abundance during their annual molting period (Reeves, 1998). Cue surveys do not census individual animals but rather count some cue from them, such as blows of some cetacean species; cue density per unit time is subsequently converted into animal density.

Mark-recapture methods are useful primarily for assessing populations that aggregate at some specific location(s) each year. The method basically involves marking a large number of individuals in a population (branding or tagging see Section 14.3) and then at some future point in time sampling the population in some manner (recapture, resighting, or harvest) and using the proportion marked versus unmarked to assess population size. Many assumptions are made about the probability of recapture of individual animals, and many refinements to mark-recapture methods have been developed over time (see Seber, 1982). Pinniped populations have been assessed using these methods with considerable success (Chapman and Johnson, 1968; Siniff *et al.*, 1977; Bowen and Sergeant, 1983; York and Kozloff, 1987), and attempts have been made in the past with cetacean populations, but these have met with limited success. The problems of using photographic images of natural marks to assess unequal capture probabilities in capture-recapature studies of populations have been addressed for sperm whales by Whitehead (2001), for northern bottlenose whale by Gowans and Whitehead (2001), and by Stevick *et al.* (2001) for humpback whales. Migration counts can yield reliable indexes and assessments for some species, which travel along narrow migration corridors (often coastal). For example, gray whale sightings and frequencies of bowhead vocalizations along migration routes have been used to assess abundance (Buckland *et al.*, 1993b).

Visual or photographic counts of the number of animals at colony sites can provide indexes of abundance, despite the fact that the whole population is not present at the site, and some proportion of the animals are usually in the water (e.g., harbor seal assessment; Thompson and Harwood, 1990). Such counts can be useful if they are supplemented with appropriate behavioral data that allow counts to be corrected to adjust for missing animals (spread in pupping through the season or sex and age patterns in the timing of molt, or diel patterns of foraging, etc.).

Catch-per-unit effort methods, adapted from fisheries biology, have also been used to estimate marine mammal abundance from harvests (see Bowen and Siniff, 1999).

14.3. Techniques for Monitoring Populations

When individual members of populations can be recognized and identified over extended time periods, several life-history characteristics can be determined and some parameters of the populations to which these animals belong can be deduced. Currently, the most widely applied techniques for repeated identification of individual noncaptive marine mammals are flipper tagging, photo-identification, radio-tagging, and genetic identification. Individual pinnipeds or sea otters tagged as pups with easily visible plastic tags can be identified repeatedly when hauled out (Figure 14.1; Table 14.2), or even

Figure 14.1.   Weaned northern elephant seal pup with flipper tags and a hot brand on the left flank used for long-term identification and to estimate rates of flipper tag loss. (Courtesy of B. Stewart, Hubbs-Sea World Research Institute.)

Table 14.2. A Survey of a Variety of 'Tags' used on Noncaptive Marine Mammals

Type of Tag	Method of Attachment	Species	Dive features			Image		Geoposition	Audio	I.D.	Other	Example of Application
			Duration	Depth	Profile	Video	Still					
CTD	penetrant	White whale	x	x	x			x		x	salinity conductivity	Lydersen et al., 2002
SRDL	adhesive	Ringed seals	x	x	x			x		x	water temp.	Lydersen et al., 2004
VHF	penetrant	Gray whale	x					x				Harvey and Mate, 1984
SL	penetrant	Bowhead whale						x				Mate et al., 2000
VHF w/TDR	suction cup	Beaked whale	x	x	x			x				Baird et al., 2004
Acoustic transponder	implanted	Sperm whale	x	x	x			x		x		Watkins et al., 1993
VHF w/TDR	implanted	Sperm whale	x	x	x			x		x		Watkins et al., 2002
TDR w/audio	suction cup	Sperm whale	x	x	x				x			Madsen et al., 2002
D-tag	suction cup	Sperm whale	x	x	x			3-D acceler.	x		direction	Miller et al., 2004
Flipper tag	penetrant	Weddell seal										Testa and Rothery, 1992
SL w/TDR	adhesive	Weddell seal	x	x				x		x		Burns and Castellini, 1998
TDR w/speed	adhesive	Weddell seal	x	x	x	x		3-D location	x			Davis et al., 2003
Head tag	adhesive	Harbor seal								x		Hall et al., 2000
SL	adhesive	Harbor seal						x				Lowry et al., 2001
TDR w/digital camera	adhesive	Fur seal	x	x	x		x					Hooker et al., 2002
VHF w/flipper tag	implanted	Sea otter						x				Siniff and Ralls, 1991
Transponder chip	implanted	Sea otter								x		Thomas et al., 1987
Transponder chip	implanted	Manatee								x		Wright et al., 1998

VHF = very high frequency radio; TDR = time/depth recorder; SL = Satellite link while resting at the sea surface, and movement patterns, home ranges or territories, growth rates, longevity, and reproductive success can sometimes be determined from observations associated with repeated sightings of the same individual.

14.3.1. Photographic Identification

Life history and population information can be established for some species of pinnipeds, cetaceans, and manatees using photographic techniques to identify individuals based on patterns of scars, natural pigmentation, callosities, or barnacle patches on the skin. In its simplest form, the technique involves photographing a visually unique part of the subject animal, such as its head or flukes, then collecting and cataloging these images so that they may be compared to photographs of the same animal taken at another time and/or place. The complex and separate genealogies of Northwest Pacific killer whale populations (see Figure 11.26) and North Pacific and North Atlantic humpbacks and North Atlantic right whales are based on an extensive collection and cataloging system of photographic images of individual whale pigmentation and scar patterns (Figure 14.2).

Some of the photographic image collections used for individual identification purposes include up to 10,000 images representing only a few thousand whales. As the numbers of images of these whales continue to grow, new and faster methods for digitizing images and computer-based retrieval and matching methods have been developed (Mizroch et al., 1990; Hillman et al., 2003). High-resolution digital video or still digital images are beginning to replace film-based photography in the field (see Mizroch, 2003 and Markowitz et al., 2003 for a discussion of the relative merits of each). Large-format aerial photographs have also been used successfully to evaluate body size, reproductive status, and body fatness of migrating gray whales (Perryman and Lynn, 2002).

14.3.2. Radio and Satellite Telemetry

The location, behavior, and even identity of individual marine mammals are sometimes difficult to monitor because they travel long distances at sea, remain submerged for prolonged periods of time, and are simply difficult to identify individually in crowded groupings. These limitations on individual behavioral studies are being overcome with a variety of electronic monitoring or recording devices, including behavioral and physiological recorders and a variety of very high frequency (VHF) and ultra high frequency (UHF) radio transmitters for data telemetry and position determination. All of these technologies have benefited from advances in Earth-orbit satellite communications, electronic miniaturization, information storage, and battery power capacities. These technologies have resulted in improved quality of data, reduced instrument package size, better instrument protection from sea water and hydrostatic pressure, and improved methods of attaching instruments to animals (Stewart et al., 1989; Stevick et al., 2002).

Attaching recorders and transmitters to free-ranging pinnipeds usually requires their capture and physical restraint or chemical immobilization (DeLong and Stewart, 1991). Substantial progress has been made in the development of immobilizing drugs for terrestrial mammals (including polar bears). However, the use of compounds such as ketamine hydrochloride on pinnipeds is relatively new and is still considered experimental (Erickson and Bester, 1993), and such compounds cannot be used at all on cetaceans or sirenians. Consequently, physical restraint is often preferred for smaller

(a)

(b)

Figure 14.2.   Individual identification photographs of two killer whales from A pod, northeastern Pacific **(a)**, and a humpback whale near Isla Socorro, eastern North Pacific **(b)**. (Courtesy of J. Jacobsen.)

animals (Figure 14.3), which can be handled safely without chemical immobilization (Gentry and Casanas, 1997).

Once an animal is immobilized or restrained, several methods of attaching instrument packages have been employed successfully, ranging from fast-setting adhesives on pinniped guard hairs (Figure 14.4; DeLong and Stewart, 1991) to body harnesses on gray whale calves (Norris and Gentry, 1974), tail tethers on manatees (Reid *et al.*, 1995), tusk mounts on walruses (Gjertz *et al.*, 2001), surgical implantation in sea otters (Ralls *et al.*, 1989), and mechanical attachments to dorsal fins of odontocetes (Mate *et al.*, 1995). For radio transmitters, the choice of attachment sites is limited to the dorsal surface, as signal transmission will not occur when the transmitter antenna is submerged. Nontransmitting recorders, such as the time-depth data archiving recorders (e.g., Le Boeuf *et al.*, 1988; DeLong and Stewart, 1991), have more flexibility with respect to attachment site but their use is constrained by the requirement to relocate and recapture

Figure 14.3. Temporary net restraint of harbor seals for measuring and tag attachment. (Courtesy of J. Harvey.)

the subject animal at the end of the study period to retrieve the recorder package. Data recorders have been successfully deployed on pinniped species that predictably return to the same breathing holes in the ice on completion of diving (i.e., Weddell seals; Kooyman *et al.*, 1980) or to the same rookery beaches after a foraging trip at sea (fur seals and both species of elephant seals; DeLong *et al.*, 1992; Walker and Boveng, 1995). TDRs and other data recorders (see Chapter 10) have also been deployed on more mobile species, but equipment losses tend to be higher as is relocation effort (e.g., Lydersen and Kovacs, 1993; Krafft *et al.*, 2000). Signals from low power and relatively inexpensive VHF transmitters can be monitored with portable directional receiving antennas from shore, ships, or aircraft within line-of-sight range of the transmitter (usually less than 50 km). Even without specialized sensors, these radio tags can provide useful information regarding the location, foraging and haul-out patterns (of pinnipeds), frequency and duration of dives, and swimming velocities.

To avoid the complications inherent in capturing or restraining large cetaceans, Watkins and Schevill (1977) pioneered the use of small projectile-fired radio transmitters to track sperm whales. Since those early studies, modifications of delivery and attachment systems for unrestrained animals have led to radio transmitters being attached using skin-penetrating devices (Ray *et al.*, 1978; Mate and Harvey, 1984; Mate *et al.*, 1995) and nonpenetrating suction cup attachments (Figure 14.5; Stone *et al.*, 1994). Devices have also been applied using cross-bows or long attachment poles (Figure 14.6).

In general terms, each of these systems is intended to record and either transmit or store information about some aspects of an animal's behavior, physiology, or

Figure 14.4.   Harbor seal with microprocessor-based TDR glued to newly molted dorsal pelage. Rear flipper tags are also visible. (Courtesy of B. Stewart, Hubbs-Sea World Research Institute.)

Figure 14.5.   Suction cup attachment of a transmitter to a humpback whale. (Courtesy of J. Goodyear.)

Figure 14.6. VHF transmitter being applied to a gray whale by pole. (Courtesy of S. Ludwig.)

geographical position (and, if monitored over time, its pattern of dispersal or migration). Some systems, such as the acoustic time-depth transmitters employed by Lydersen (1991) on ringed seals to obtain information on diving and haul-out activities, have a limited range, but are an effective tool to monitor subjects in difficult under-ice conditions. At the other extreme, satellite-linked transmitters and Global Positioning System (GPS) navigational receivers give relatively precise, global-scale monitoring and position determination. Powerful UHF transmitters attached to a wide variety of marine mammals (platform transmitting terminals [PTTs] or satellite relay data loggers [SRDLs]), monitored by satellites such as the polar-orbiting Argos system, are revolutionizing our ability to document at-sea movements and behaviour of marine mammals (e.g., Mate *et al.*, 1995; Westgate and Read, 1998; Lagerquist *et al.*, 2000; Bennett *et al.*, 2001; Suydam *et al.*, 2001; Deutsch *et al.*, 2003; Mauritzen *et al.*, 2003; Burns *et al.*, 2004; Laidre *et al.*, 2004). Many of these studies are providing new insight regarding population boundaries as well as exchange and overlap. Even environmental conditions are now being measured by on-board sensors (e.g., Lydersen *et al.*, 2002, 2004; Costa *et al.*, 2003; Hooker and Boyd, 2003). Vast amounts of data can be collected, even on animals in remote or inhospitable (to humans) environments.

14.3.3. Molecular Genetic Techniques

Naturally occurring variations in an organism's DNA, resulting from its tendency to mutate in a regular, or at least predictable, pattern, are being used to establish individual identification, gender, parentage (especially paternity), and also population sizes and boundaries (Dover, 1991). The techniques involved include older, low resolution methods of protein analysis through allozyme electrophoresis, cytogenetics (chromosome studies), and DNA–DNA hybridization, as well as a variety of newer, high resolution techniques for analysis of variation in DNA including restriction fragment length polymorphisms (RFLPs), multilocus DNA fingerprinting, direct DNA sequencing, and microsatellite analyses (Table 14.3). These methods are discussed in detail elsewhere (e.g., Hillis *et al.*, 1996; Dizon *et al.*, 1997). In the following discussion we focus on the

Table 14.3. Application of Various Molecular Genetic Techniques to Problems in Population Biology

Problem	Population Structure	Geographic Variation	Paternity Testing	Mating Systems	Individual Relatedness
Allozymes	+	+	m	+	m
Chromosomes	m	m	−	m	−
DNA sequence analysis (mtDNA)	+	+	+	+	+
Restriction analysis (RFLP)	+	m	+	m	m
DNA profiling (mini- and microsatellites, fingerprinting)	+, −	+, m	+, +	+, −	+, m

inappropriate use of technique; +, appropriate use of technique; m, marginally appropriate or appropriate under limited circumstances. Based on Hillis *et al.*, 1996.

application of these different techniques to the analysis of the individual identification and population structure in marine mammals and discuss the diversity of issues that may be addressed using molecular techniques. For additional information see reviews of this topic by Hoelzel (1993, 1994) and Hoelzel *et al.* (2002a).

14.3.3.1. Allozymes

Allozyme electrophoresis, a procedure for separating proteins of different molecular sizes and electrical charges that therefore have different migration rates in electric fields, is the simplest, most versatile, and least expensive of the techniques for detecting levels of genetic variation within and between populations. The resolution of this technique is low, because only protein-coding regions of DNA can be evaluated and only a small proportion of the changes in those regions will cause a detectable change in the mobility of the protein. Allozymes are visualized by chemical staining the electrophoretic gel after migration. Differences in mobility and the presence of allozyme variants are indicated by different band positions on the gel for **homozygotes** and multiple bands for **heterozygotes.** These different band positions represent the allozymes and thus gene frequencies at the loci being evaluated.

The most extensive analysis of genetic variation based on allozyme variation within and between baleen whale populations to date is that conducted by Wada and Numachi (1991). They surveyed a total of 17,925 whales (including fin, sei, minke, and Bryde's whales) from 15 locations. With the exception of Bryde's whale, which has a tropical/subtropical distribution, these species are distributed in all major oceans and in both hemispheres.

14.3.3.2. Chromosomes

Similarities and differences in the size and appearance of chromosomes (chromosome heteromorphisms) have often been used in the study of gene maps, population and species differences, and reproductive isolation. In cetaceans, chromosomes have been used to examine patterns of chromosome variability within and among species. In population studies, chromosome heteromorphism analysis makes it possible to investigate relationships among animals of known associations and to establish patterns of reproductive exchange among marine mammal groups. For example, using this method the parentages of known social groups of bottlenose dolphin have been determined (Duffield and Chamberlin-Lea, 1990; Duffield and Wells, 1991).

14.3.3.3. DNA Sequence Analysis

DNA changes over time in a finite number of ways, including substitution of one of the four nucleotides bases for another, deletion of one or more bases, insertion of one or more bases, and the duplication of a segment of DNA. These changes can be quantified through direct analysis of DNA sequences. DNA can be extracted from a variety of tissues including muscle, skin, blood, bone, hair, and internal organs. Once extracted, a specific sequence of DNA can then be amplified (copied). This is done by the polymerase chain reaction method (PCR) by which a highly specific DNA region is simultaneously copied and isolated many times. The capacity of PCR to amplify and isolate DNA means that only very small amounts of tissue are required and can be obtained from small biopsy darts or even from collected patches of shed skin. The rate and pattern of nucleotide substitutions differ among genomic regions, rendering each region more or less appropriate for population genetic analysis. The genomic target region to survey is chosen on this basis. The characteristics or changes in specific genomic regions can then be appropriately matched to a particular technique.

14.3.3.3.1. *Mitochondrial DNA (mtDNA) Analysis*

Several aspects of population structure and dynamics can be addressed with studies of mtDNA sequence data. Such studies usually begin with an assessment of genetic diversity from which geographic range descriptions can then be mapped. Migration rates between population ranges or their subdivisions can be estimated. Estimates of effective population size also can be made from measures of genetic diversity. But perhaps one of the most powerful uses of mtDNA data has been in the determination of species identity and/or stock structure in many endangered or managed species.

MtDNA evolves 5–10 times faster than nuclear genomic DNA and, because of this rapid rate, polymorphisms are more likely to be detectable than in proteins or nuclear DNA. MtDNA is maternally inherited, thus it represents only matriarchial phylogeny, allowing a direct assessment of the dispersal patterns of females. The almost exclusive (although some exceptions are known; see Avise, 1996 for a summary) maternal pattern of inheritance of mtDNA makes it more sensitive to fluctuations in population size. Essentially, if more females are present in a population, more variation in mtDNA should exist. For example, Hoelzel and Dover (1991a) estimated the long-term effective population size of minke whales to be 400,000 based on mtDNA variation in three populations.

Analysis of mtDNA variation also can provide information on genetic variation among populations for use in determining stock identifications for management purposes. Results of studies of mtDNA variation in three major humpback whale lineages (i.e., North Atlantic, North Pacific, and Southern Ocean populations) revealed high levels of genetic variation among populations (e.g., Baker *et al.*, 1990, 1993, 1994; Baker and Medrano-Gonzalez, 2002). It has been suggested that this distribution of humpback clades (and maternally directed fidelity) is consistent with colonization of new feeding grounds following the retreat of the last ice age (Baker and Medrano-Gonzalez, 2002). Studies of control region (part of the mtDNA genome) diversity in other cetacean species, including killer whales (Hoelzel and Dover, 1991b; Hoelzel *et al.*, 2002b) and harbor porpoises (Rosel *et al.*, 1999), indicate low genetic diversity (in all killer whales and the Northeast Atlantic population of harbor porpoises) that might be caused by historical bottlenecks (see later).

In pinniped species, patterns of mtDNA variation in the harbor seal revealed that populations in the Pacific and Atlantic Oceans are highly differentiated (Stanley *et al.*, 1996). These data are consistent with an ancient isolation of populations in both oceans due to the development of polar sea ice about 2–3 Ma. In the Atlantic and Pacific, populations appear to have been colonized from west to east, with the European population showing the most recent common ancestry. The results of Stanley *et al.* (1996) were used to define a hierarchy of population units for ranking conservation priorities. In another study, analysis of mtDNA variation in southern hemisphere fur seals revealed distinctly different patterns of molecular evolution and population substructure among four congeneric species (Lento *et al.*, 1994, 1997). Other population subdivisions for southern fur seals were suggested by Wynen *et al.* (2001). These molecular data have been used to argue for reevaluation of "species" definitions that, in turn, has important implications for management. Similar studies of mtDNA variation in populations of California sea lions (Maldonado *et al.*, 1995), Steller sea lions (Bickham *et al.*, 1996; Truijillo *et al.*, 2004), Atlantic and Pacific walruses (Cronin *et al.*, 1994; Anderson *et al.*, 1998), harbor porpoises (Rosel *et al.*, 1999), belugas (Brown-Gladden *et al.*, 1997; O'Corry-Crowe *et al.*, 1997), sea otters (Cronin *et al.*, 1996), and polar bears (Cronin *et al.*, 1991) have concluded that mtDNA is particularly useful in stock identification and conservation management. For sirenians, which are less mobile than cetaceans and pinnipeds, mtDNA studies have identified strong geographic structure among the West Indian manatee populations (Garcia-Rodriquez *et al.*, 1998). Other species that exhibit morphologic variation have been shown to possess considerable genetic differentiation using both mtDNA and nuclear DNA markers (e.g., near-shore and offshore forms of some species of the bottlenose dolphin; Hoelzel *et al.*, 1998).

Analysis of genetic variation also can provide important information on the demographic history of some species, including the extent of inbreeding within a population. Several marine mammal populations that were overhunted experienced severe population reductions within short time periods known as **population bottlenecks**. Among pinnipeds the northern elephant seal provides provides a classic example of a population bottleneck. Hoelzel *et al.* (1993) and Halley and Hoelzel (1996) used mtDNA variation and census data to calculate the size and duration of the population bottleneck experienced by the northern elephant seal. Confidence limits of 95% indicated a bottleneck event of fewer than 30 seals for less than 20 years or a single year bottleneck of fewer than 20 seals. Historical evidence supports the latter hypothesis of a 1-year bottleneck. Results of a study that considered genetic diversity in pre- and postexploitation northern elephant seal populations suggested that preexploitation populations had more genetic diversity than modern populations (Weber *et al.*, 2000). Similar results were found for pre–fur trade populations and modern populations of sea otters (Larson *et al.*, 2002). Low levels of genetic diversity among Hawaiian monk seals, also hunted to near extinction, have been attributed to a population bottleneck. In contrast to northern elephant seals the inability of Hawaiian monk seals to recover from exploitation has been explained as the result of low reproductive success among females (Kretzmann *et al.*, 1997). In the case of the vaquita, low genetic variability coupled with its restricted range suggest a persistent effect of small effective population size or a founder event (Rosel and

Rojas-Bracho, 1999). Other species reported to have gone through bottlenecks have maintained genetic diversity including humpback whales, Guadalupe fur seals, Antarctic fur seals, and San Juan fur seals (see Hoelzel *et al.,* 2002 for original references).

The mtDNA sequencing and other molecular evidence has also been used to infer hybridization between various marine mammal species both in captivity and in the wild. Hybrids among wild pinnipeds have been reported for Antarctic and subantarctic fur seals (Goldsworthy *et al.,* 1999) and harp and hooded seals (Kovacs *et al.,* 1997) and among cetaceans including fin and blue whales (Arnason *et al.,* 1991; Spilliaert *et al.,* 1991; Berube and Aguilar, 1998), and Dall's and harbor porpoises (Baird *et al.,* 1998; Willis *et al.,* 2004). In some cases evidence for hybridization is supported by morphology (i.e., fin and blue whales, harp and hooded seal) and skin pigment patterns (Dall's and harbor porpoises). In other cases morphologic evidence exists for hybridization in the absence of molecular evidence (e.g., dusky dolphin and southern right whale dolphin and several otariid species—e.g., Brunner, 1998, 2002; Yazdi, 2002).

14.3.3.3.2. Restriction Site Analysis

Restriction enzymes (originally isolated from bacteria) cut double-stranded DNA at particular nucleotide sequences into fragments, usually four, five, or six nucleotides long. A restriction enzyme will cleave the DNA wherever the particular recognition sequence (or restriction site) of the enzyme occurs. The result is a series of "restriction fragments" of the DNA. Variation in the lengths of these fragments between individuals indicates variation in the presence or absence of the restriction sites, and tests to examine length differences are termed RFLPs. RFLP analyses begin by cutting DNA from each individual with one or more restriction enzymes, then separating the resulting fragments by gel electrophoresis, and scoring the visualized bands of size-sorted fragments. The frequency distribution of RFLP bands among individuals in a population is then used to describe the population structure. This method can be used with great efficiency to screen populations or species for specific changes in sequence.

Dizon *et al.* (1991) compared mtDNA RFLP data among different populations of spinner dolphins (i.e., eastern, whitebelly, and pantropical). Results of this study showed high levels of variation within and between populations of eastern and whitebelly forms, indicating significant genetic interchange between them despite large differences in morphology. In another study, Dowling and Brown (1993) investigated mtDNA RFLP variation among bottlenose dolphins from the eastern Pacific and western North Atlantic. Isolation of Atlantic versus Pacific populations and possibly the existence of two or more populations in the western North Atlantic were the principal conclusions of this study. Boskovic *et al.* (1996) studied grey seal populations in the North Atlantic and Baltic. There were no shared haplotypes between eastern and western Atlantic populations and no evidence of separation between Gulf of St. Lawrence and Sable Island animals within Canada. Population structure and history of Antarctic and subantarctic fur seal colonies in relation to the effects of commercial sealing has also been explored using RFLPs in combination with other techniques (Wynen *et al.,* 2000).

14.3.3.3.3. DNA Fingerprinting and Microsatellite Analysis

DNA profiling, a commonly used technique, is based on a very high rate of change seen in repetitive DNA regions such as **microsatellites** and **minisatellites.** Researchers employ DNA profiling to investigate variation in minisatellite loci using single locus or multilocus probes, which measure variability at one or more locations in the nuclear genome.

The resulting barcode-like bands that constitute a DNA profile or fingerprint can be used to determine three major things: (1) identity, (2) parentage, and (3) relatedness. DNA profiles can be used to determine identity in the same way that traditional fingerprints are used to identify individuals uniquely. DNA fingerprints can be used to determine parentage because approximately 50% of the bands in the offspring come from the mother and the remaining bands come from the father. Thus it is possible to exclude individuals as parents if bands in offspring are not present in the profiles of either parent (Figure 14.7). Alternatively, a probabilistic statement may be made regarding the likelihood of parentage. This is done by determining a band-sharing coefficient for all possible pairs of sires and offspring and dams and offspring. Pairs with the highest band-sharing coefficients are the most likely to be primary relatives. It is this capability that gives DNA fingerprinting its greatest potential. DNA fingerprinting also can be used to determine nonparental relatedness. In many species, the band-sharing coefficient between two siblings, like that between a parent and offspring, is greater than 50%. The proportion of band-sharing between two unrelated individuals can be as low as 20–30%. Thus the degree of relatedness between two individuals can be estimated by their degree of band-sharing based on a survey of band-sharing in individuals of known relatedness.

The use of DNA profiling to study population structure, mating systems, and reproductive behavior has largely involved cetaceans. The use of DNA profiling to test paternity

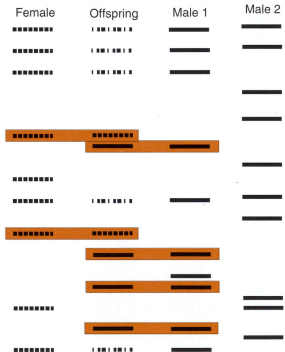

Figure 14.7.   Representation of paternity testing by DNA fingerprinting. Offspring bands were derived from the mother (*even dashes*), the father (*solid lines*), or both (*uneven dashes*). Male 2 may be eliminated as the father because he and the offspring share no bands. (Redrawn from Hoelzel, 1993.)

can be especially time consuming when many pairwise comparisons have to be made. Such is the case in the study of paternity among polygynous species. To test paternity, it is usually necessary to collect samples from both mother and offspring, as well as the set of potential sires. However, even the identification of potential sires requires the long-term recognition of individuals and observations of associations between the female and potential mates at the time of conception. One example where the field observation data are this detailed is for a bottlenose dolphin study on the Gulf coast of Florida (Duffield and Wells, 1991). Results from this study indicate that males moving through the home ranges of matrifocal (female related) groups achieve matings through temporary associations. Exclusive male access to females or groups within the community was not observed.

In a related study of paternity in pods of long-finned pilot whales captured in the Faroese drive fishery, Amos *et al.* (1991) found that for only 4 out of 34 pods did males sire members of their pod. It was assumed in most cases that the entire pod had been captured and sampled. Paternal exclusions were achieved using the multilocus DNA fingerprint probe. A pair of bands apparently representing a single locus within the multilocus pattern was used to assess the probability that mating was random. Amos *et al.* (1991) concluded that males often successfully mate with two or more females, and generally not with females within their own pod. Similar findings have been made recently for other cetacean species (see Chapter 13).

There are several reasons why DNA profiling has not been as widely used to study pinnipeds as it has to study cetaceans. The reproductive behavior of many pinnipeds can be observed directly, whereas in cetaceans direct observation is difficult. Pinnipeds do not form small social groups or exhibit obvious cooperative behavior as is often found in cetacean social groups. DNA profiling has, however, revealed important information about pinniped social structure, as reviewed by Boness *et al.* (1993), and is increasingly being used to assess earlier assumptions made about social and population structure within pinnipeds. For example, at least two studies suggest that female harbor seals may be choosing mates and that males positioned in the water just off the beach do not simply have exclusive mating rights to females that move through their territories (Harris *et al.*, 1991; Perry, 1993). Using DNA fingerprinting, McRae and Kovacs (1994) excluded the possibility that attending males within hooded seal trios (consisting of an adult female, her pup, and an attending adult male) were the fathers of the pups, establishing either that they did not remain paired from one breeding season to the next or that, like grey seals, attending males do not necessarily fertilize all of the females they attend.

Several other pinniped researchers have used analysis of hypervariable minisatellite loci to investigate general questions of inherent levels of genetic variability. Minisatellites are among the most variable DNA sequences yet discovered and are best suited for indicating individual relatedness. Additionally, in populations with reduced genetic variability, minisatellites often are the only markers able to detect useful polymorphisms (Burke *et al.*, 1996). DNA profiles were examined in the northern elephant seal and compared to the harbor seal (Lehman *et al.*, 1993). Results indicated that this species lacks genetic variability, with roughly 90% of alleles being shared among all individuals tested. In contrast, harbor seals from the eastern Pacific possess much greater levels of variation at these loci. In other studies, Amos (1993) and Amos *et al.* (1995) used analysis of multi- and single-locus profiles to assign paternity in colonies of grey seals on North Rona Island, Scotland. Long-term mate fidelity in grey seals was also studied (Amos *et al.*, 1995), with results indicating a surprisingly high degree of mate fidelity for this polygynous species (discussed in Chapter 13).

In addition to minisatellites, another class of repetitive DNA, microsatellite loci, has been used as an alternative to traditional methods of individual recognition because these loci are permanent and exist in all individuals. The frequency distribution of microsatellite alleles at a single or a few loci can be used to characterize population substructure. Statistical analysis of allelic combinations at several loci can be used to make individual identifications. In a study of humpback whales, analysis of microsatellite loci allowed the unequivocal identification of individuals (Palsbøll *et al.*, 1997). In addition to revealing local and migratory movements of humpbacks, genetic tagging allowed the first estimates of animal abundance based on genotypic data (Palsbøll *et al.*, 1997). Microsatellite markers were used to determine paternity in humpback whales, confirming observations of a promiscuous mating system (Clapham and Palsbøll, 1997). In a study of long-finned pilot whale pod structure, Amos (1993) used microsatellite loci to confirm results of an earlier study (Amos *et al.*, 1991) reporting that pilot whales exist in matrifocal kin groups. Microsatellite data also were used to study the genetic relationship between Canadian polar bear populations (Paetkau *et al.*, 1995, 1999). Considerable genetic variation was detected among most populations. Minimal genetic structure in several polar populations indicated that, despite the long-distance seasonal movements undertaken by polar bears, gene flow among some local populations is restricted. In another study, the genetic variation in sperm whales measured by analysis of microsatellite markers offered insight into the patterns of kinship, suggesting that females form permanent social units based on one or several matrilines (Richard *et al.*, 1996). A study based on microsatellites showed no genetic structure among sperm whales and indicated that males but not females frequently breed in different ocean basins (Lyrholm *et al.*, 1999).

The use of multiple markers and integrative approaches that combine molecular genetic data with field studies and various other disciplines (i.e., morphology) has increased our understanding of the population genetics of various species. Identification of population structure is a requisite to the development and implementation of future conservation and management plans for any marine mammal species, as well as for understanding basic biology of the species such as patterns of movement of males and females within ocean basins.

14.4. Population Structure and Dynamics

Life-history characteristics of individual members of a population influence (or control) population size and the patterns of change through time. Population growth rates (either positive or negative) result from the interplay between factors that promote birth and survival and those that promote mortality, in addition to emigration and immigration. **Natality** is a population parameter that describes the rate at which offspring are produced in a population; it is determined by the collective **birth rates** of individual females in a population, which in turn depend on the age at which sexual maturity is reached, how many young are born in a reproductive episode, how often females reproduce, and when in their lives they terminate reproduction. **Mortality** describes the rate of death in a population. The balance between natality and mortality establishes patterns of **survivorship** and the **age distribution** of a population (Figure 14.8).

All marine mammals have relatively large body sizes and long life spans, which make them slow to mature and reproduce. Relatively few offspring are produced, and these are

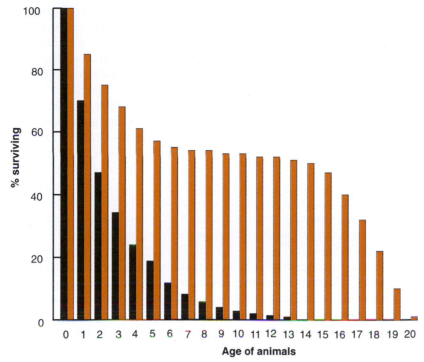

Figure 14.8. Hypothetical age distributions of an r-selected (*black bars*) and k-selected (*colored bars*) population. These populations also differ markedly in patterns of mortality, mean age, maximum age, and survivorship.

intensely nurtured. Hence, juvenile mortality is relatively low and animals that reach sexual maturity tend to live into old age before dying. In population biology terms, marine mammals are classic examples of **k-selected species** (see Figure 14.8); species that have evolved to maintain relatively stable population sizes through time, at or near the **carrying capacity** of their environment (MacArthur and Wilson, 1967).

Although marine mammals are k-selected species, and hence buffered to some extent from interannual variability in their numbers, their abundances do change through time. Population dynamics is the study of how and why populations vary in abundance through time. This involves intrinsic factors (i.e., birth rate, growth rate, and longevity) as well as extrinsic factors (i.e., disease and natural toxins in the marine environment, interspecific competition, and predation).

When interpreting population parameters and life-history characteristics of marine mammals, it must be remembered that many populations experienced commercial exploitation that devastated their numbers; some populations have recovered to their preexploitation levels recently, but many are still in a state of recovery, and yet others show little sign of recovery.

14.4.1. Birth and Pregnancy Rates

The number of offspring a female marine mammal produces is determined by her frequency of pregnancy (which is never more than once per year; Table 13.1), by the duration of her reproductive lifetime, and in polar bears, by the number of cubs produced during each pregnancy (all other marine mammals normally produce a single offspring per successful pregnancy). In some species, birth rates of individual females are known to change over their lifetimes. Sexual maturity in most marine mammals precedes physical maturity by several years, allowing for substantial body mass increase after a female is capable of her first reproduction. Huber *et al.* (1991) found that younger (and presumably smaller) sexually mature female elephant seals from the Farallon Islands were more likely than older females to "skip" a year by forgoing a pregnancy, and were more likely to "skip" in later years if they had matured early. However, at nearby Año Nuevo Island, no significant age-specific variation in natality was reported by Le Boeuf and Reiter (1988). In species such as elephant seals that fast for part or all of their lactation, skipping a year may provide a small female the time necessary to acquire additional body mass between successive pregnancies (Sydeman and Nur, 1994). Similar "skips" are suggested for newly mature female gray whales whose body masses are likely near the minimum size capable of supporting the energetic cost of pregnancy and lactation while fasting (Sumich, 1986a).

A gradual decline in pregnancy rates in older animals has been documented in short-finned pilot whales (Marsh and Kasuya, 1991). In this species, killer whales (Ford *et al.*, 1994), and likely in other odontocetes, pregnancy rates decline with increasing age of females. Female short-finned pilot whales are known to live to maximum ages of at least 63 years, yet no individual older than 36 years has been found to be pregnant; Marsh and Kasuya (1991) proposed that all females in their study population over 40 years of age (about 25% of all females) had ceased to ovulate. Even so, many were found with sperm in their reproductive tracts, indicating a continuation of mating well into their postreproductive years. The functional role of these postreproductive females is unclear. Norris and Pryor (1991) suggest two possible roles that can be viewed as mechanisms to increase these animals' reproductive fitness through the investment of energy, time, or information to their direct descendants or other close kin: (1) in stable, long-duration matrilineal or **matrifocal** (aggregations of matrilines) social groups of odontocetes (i.e., pilot whales, sperm whales, and killer whales), the old members of the group are always females, and these old females may serve as valuable repositories and transmitters of cultural information, and (2) nurturing responsibilities such as babysitting and fostering (including occasional nursing of other females' young) have been reported for sperm whales (Whitehead, 1996, 2004), and, as these species are all deep diving whales, foraging mothers could benefit from the presence of other older and closely related females to tend their calves. Reproductive senescence of this type has not been observed in pinnipeds. Pistorius and Bester (2002) explored the impact of age on reproductive rates in southern elephant seals and found no evidence of senescence (also see Sergeant, 1966). Once female seals are through the first few years of reproduction, they seem to have consistently high pregnancy rates to the end of their lives, with most mature females pregnant each reproductive period (e.g., Bjørge 1992; Pitcher *et al.*, 1998; Lydersen and Kovacs, 2005). Polar bears exhibit smaller litter sizes and other reductions in reproductive performance with increasing age beyond their prime reproductive years (Derocher and Stirling, 1994).

14.4.2. Growth Rates

The growth rates of suckling marine mammal neonates vary tremendously among species (also see Chapter 13). In terms of absolute growth, average rates of mass gain range over several orders of magnitude, from less than 0.1 kg/day in fur seals (Table 14.4) to more than 5 kg/day during the 4-day lactation of hooded seals (Bowen *et al.*, 1985) to more than 100 kg/day in blue whale calves (Rice, 1986). Presumably, absolute rates of mass gain increase even more between weaning and sexual maturity; however, far fewer data on postweaning growth rates are available. Relative rates of growth also can be compared using the time required to double in body mass (Table 14.4). A clear distinction between phocids and otariids is once again apparent, with the doubling time of harbor seal masses being intermediate between typical phocids and typical otariids.

Although body masses vary substantially on a seasonal basis in those species that experience prolonged postweaning or seasonal fasts, body length tends to increase regularly until physical maturity is reached. The pattern of length increase with age usually is sigmoidal in shape and is asymptotic (McLaren, 1993). Several commonly used exponential equations are employed to describe and model the pattern of body length increase with increasing age (Brody, 1968; Richards, 1959).

Two features of individual growth patterns of marine mammals stand out. First, growth to physical maturity continues for several years after reaching sexual maturity (see discussion later), so old sexually mature individuals are often much larger than younger, but still sexually mature, individuals of the same gender. Second, in polygynous and sexually dimorphic species of pinnipeds and odontocetes, patterns of male growth exhibit a delay in the age of sexual maturity to accommodate a period of accelerated growth into body sizes much larger than those of females (Figure 14.9). Before polygynous males can compete successfully for breeding territories or establish a high dominance rank, they must achieve a body size substantially larger than that of females. The longer wait to sexual maturity comes at a substantial cost reflected in overall mortality. The life history of male northern elephant seals is geared toward high mating success late

Table 14.4. Birth Mass Doubling Times and Rates of Body Mass Increase for Various Species of Marine Mammals

Species	Approximate Time to Double Birth Mass (days)	Rate of Birth Mass Increase (kg/day)	Source
Pinnipeds			
Phocids			
Harbor seal	18	0.6	Costa (1991); Bowen (1991)[*]
Grey seal	9	2.7	Costa (1991); Bowen (1991)[*]
N. elephant seal	10	3.2	Costa (1991); Bowen (1991)[*]
Otariids			
N. fur seal	85	0.07	Costa (1991); Bowen (1991)[*]
CA sea lion	79	0.13	Costa (1991); Bowen (1991)[*]
Antarctic fur seal	62	0.08	Costa (1991); Bowen (1991)[*]
Cetaceans			
Mysticetes			
Gray whale	60	16	Sumich (1986)
Blue whale	25	108	Gambell (1979); Rice (1986)
Other Marine Mammals			
Polar bear	10	0.1	Stirling (1988)

na=not available

[*]See for original sources.

in life, although the chance of living to an age of high mating success is low because male mortality is relatively high. On average, the highest mating success in males occurs at 12–13 years of age, but male survivorship to this age is approximately 1% of males born (Clinton and Le Boeuf, 1993). In phocids in which the sexes are approximately the same size and in mysticetes, growth of both sexes is approximately the same until the age of sexual maturity, when females usually outpace males to achieve a slightly larger body size at sexual maturity (Rice and Wolman, 1971; Lockyer, 1984; Kovacs and Lavigne, 1986). Studies of growth rates in mysticetes in particular have been hampered by a scarcity of old and physically mature individuals in heavily exploited species and by a general lack of any mature animals whose ages are known with certainty.

14.4.3. Age of Sexual Maturity

Marine mammals are considered sexually mature when they are capable of producing gametes (sperm and eggs). As previously noted and indicated in Figure 14.9 and in Table 14.5, there exists a male-to-female dichotomy in the age of sexual maturity for sexually dimorphic species. Sexual maturity of females occurs around 4 years of age in northern elephant and grey seals and 6 years of age in males. Walrus males mature about 4 years

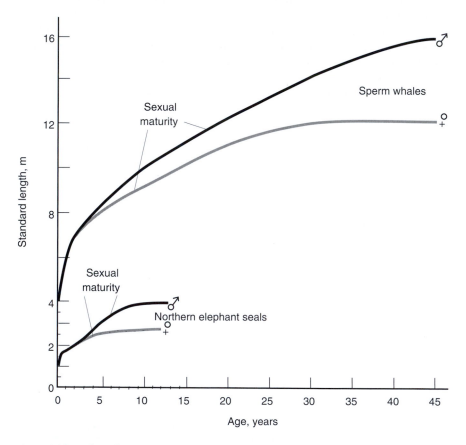

Figure 14.9.   Growth curves for male and female sperm whales adapted from Lockyer, 1981 and northern elephant seals adapted from Clinton, 1994.

later than females (Table 14.5). The average age of sexual maturity for baleen whales ranges from 10 years in the blue whale to 7–14 years in minke whales (Lockyer, 1984; Table 14.5), with no apparent differences between genders. Among odontocetes, estimates of sexual maturity range from 4–5 years in common dolphins to 14 years in killer whales (Table 14.5) with few gender-based age differences. The more dimorphic sperm whale requires about 3 years longer for males to reach sexual maturity compared to females.

The average age of sexual maturity in some species appears sensitive to density-related competition for space or food. In crowded and commercially unexploited southern elephant seal breeding colonies, females mature at an average age of 6 years, whereas on recently colonized and uncrowded rookeries, females reach puberty at 2 years of age (Carrick *et al.*, 1962). Similar reductions in age of maturity have been demonstrated in southern hemisphere fin and sei whales when their populations were reduced by commercial hunting, and in minke whales as the numbers of their larger mysticete food competitors were dramatically reduced during the first half of the 20th century (Figure 14.10; Lockyer, 1984; Kato and Sakuramoto, 1991). Lockyer (1984) estimated that by 1970, the overall mysticete whale density was reduced over the entire Antarctic summer

Table 14.5. Approximate Ages of Sexual Maturity for Various Species of Marine Mammals

Species	Mean Age at Sexual Maturity		Source
	Females	Males	
Pinnipeds			
Phocids			
Harbor seal	2–7	3–7	Riedman (1990)[*]
Grey seal	3–5	6	Riedman (1990)[*]
N. Elephant seal	4	6	Riedman (1990)[*]
Otariids			
N. fur seal	3–7	5	Riedman (1990)[*]
CA sea lion	4–5	4–5	Riedman (1990)[*]
Antarctic fur seal	3–4	3–4	Riedman (1990)[*]
Walrus	5–6	9–10	Riedman (1990)[*]
Cetaceans			
Odontocetes			Perrin and Reilly (1984)
Bottlenose dolphin	~12	~11	Perrin and Reilly (1984)
Striped dolphin	9	9	Perrin and Reilly (1984)
Common dolphin	2–6	3–7	Perrin and Reilly (1984)
L.fin pilot whale	6–7	12	Perrin and Reilly (1984)
Killer whale	14	12–14	Ford *et al.* (1994)
Sperm whale	9	12	Lockyer (1981)
Mysticetes			
Gray whale	9	9	Rice and Wolman (1971)
Blue whale	10	10	Rice (1986)
Minke whale	7–14	7–14	Lockyer (1984)
Sirenians			
Manatee	6–10, 12.6	6–10	Reynolds and Odell (1991); Marmontel (1995)
Dugong	9.5	9–10	Marsh (1995)
Other Marine Mammals			
Sea otter	4	6–7	Jameson (1989); Jameson and Johnson (1993)
Polar bear	4	6	Stirling (1988)

*See for original sources.

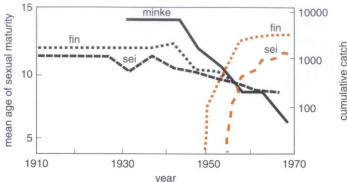

Figure 14.10. Changes in sexual maturity for three species of Antarctic balaenopterids (*black lines*), and the cumulative harvests in the same whaling areas (*colored lines*). (From Lockyer, 1984; and Kato and Sakuramoto, 1991.)

feeding area to about 15% of its pre-1920 level. Whether the effects of these and other whaling-induced changes in resource competition may have influenced interpregnancy intervals of these species is discussed by Clapham and Brownell (1996).

Both sexes of manatees typically become sexually mature between 6 and 10 years of age (Reynolds and Odell, 1991), although an average age of 12.6 years for female Florida manatees was reported by Marmontel (1995). In dugongs, the age of sexual maturity reported in a female in one study was 9.5 years and ranged between 9 and 10 years for males (Marsh, 1995; Table 14.5). Sexual maturity in female Californian and Alaskan sea otters occurred at an average age of 4 years (Riedman and Estes, 1990; Jameson and Johnson, 1993). Both sexes of polar bears become sexually mature at an average age of 4–6 years and are reproductively senescent by age 20 (Ramsay and Stirling, 1988).

14.4.4. Age Determination and Longevity

The determination of an individual's age is fundamental to evaluating its reproductive contribution to the population. Techniques for age determination of noncaptive marine mammals vary widely among species and among investigators. Several methods of tagging or marking animals, such as the metal Discovery projectile tags (Figure 14.11) used widely in the first half of the 20th century, have provided minimum estimates of the ages of some whales. However, this technique works only if the tagged animal is killed at some later time to recover the embedded tag. Consequently, Discovery tags have been used successfully only with species subject to extensive harvesting and a reasonable likelihood of tag recovery; they are not currently applied to any species.

Another, more reliable method involves determining an individual's age using teeth and bones (Scheffer, 1950; Laws, 1962). As an animal grows, incremental layers accumulate in several hard tissues, especially teeth and bones. These incremental growth layers become a record of the life history of the individual animal, analogous to rings in tree trunks; counting these layers is now widely used to establish ages of individual marine mammals. This technique assumes that teeth or other hard parts can be obtained (teeth are preferred over bone), that incremental growth layers can be discerned, and that the time interval represented by a single growth increment can be independently established. Teeth typically are obtained from dead specimens or from captive or temporarily

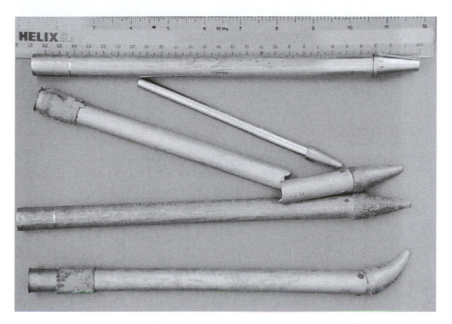

restrained individuals of all groups of marine mammals except mysticetes. Procedures for enhancing the visibility of the growth increments usually include thinly slicing and polishing teeth, then etching or staining the polished surface to better resolve the growth layers (reviewed in Perrin and Myrick, 1980). Each countable unit of repeating incremental growth layers contains at least one change in tissue density, hardness, or opacity, and is referred to as a **growth layer group (GLG);** Figure 14.12). In most species so far examined, each GLG is thought to represent an annual increment (e.g., Perrin and Myrick, 1980).

In mysticetes, GLGs of bony tissues such as tympanic bullae and skull bones have been studied with limited success, with the maximum number of GLGs providing only a minimum estimate of age (Klevezal' *et al.*, 1986). More commonly, alternating light and dark zones in the large waxy ear plugs of mysticetes have been used (Figure 14.13). Several studies have established that one GLG, consisting of one light and one adjacent dark band, represents an annual increment (Laws and Purves, 1956; Roe, 1967; Rice and Wolman, 1971; Sumich, 1986b), although only slightly more than one half of the ear plugs collected in each of two studies of gray whales by Rice and Wolman (1971) and Blokhin and Tiupeleyev (1987) were considered readable. Additionally, in older animals, there often is some loss of earplug laminae deposited in an animal's early years, and this technique is usually not considered a reliable estimator of age much beyond the onset of sexual maturity.

For cetacean species that have experienced extended periods of commercial hunting and collection of anatomical specimens (e.g., sperm whales and large mysticetes), reproductive organs can provide some evidence of age as well as of reproductive history. As mentioned in the previous chapter, the reproductive history of the individual female can

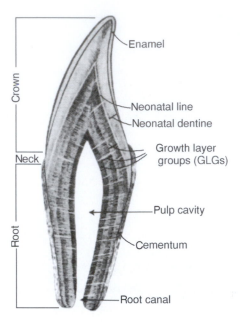

Crown

Enamel

Neck

Neonatal line
Neonatal dentine

Growth layer
groups (GLGs)

Root

Pulp cavity

Cementum

Root canal

Figure 14.12.   Polished cross-section of a dolphin tooth showing major structural features and well-defined growth layer groups. (From Perrin and Myrick, 1980.)

be read from an examination of the ovaries and a count of the total number of corpora albicantia. If the time interval between successive pregnancies can be independently established, and the age at sexual maturity is known (see previous section), an estimate of age can be derived. For example, a large and physically mature gray whale female reported by Rice and Wolman (1971) to have 34 corpora albicantia is estimated to have been 75–80 years old when she died [(34 corpora) × (2 year inter-pregnancy interval) + (8–9 years to sexual maturity)], and she was pregnant when she was killed.

For most species of cetaceans, estimates of average or maximum life spans are simply unavailable, either because the species has never experienced intensive commercial

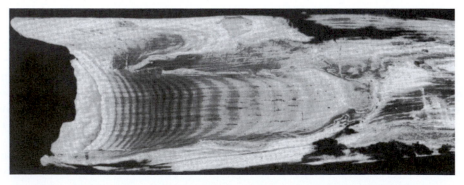

Figure 14.13.   Polished cross-section of a fin whale ear plug showing annual growth layers. (From Lockyer, 1984.)

exploitation and the tissues needed to establish age have not been available or because populations of long-lived species with extensive histories of commercial exploitation (including most species of mysticetes) currently express age distributions that have been biased by the harvesting of larger and older animals, leaving relatively youthful populations with few individuals expressing maximum ages (Sumich and Harvey, 1986).

Seals and polar bears are routinely aged using growth layers in the teeth, with good accuracy for most species. A few generalizations concerning pinniped longevity can be made. Females, especially those of polygynous species, tend to live longer than males, most of whom do not survive even to the delayed age of sexual maturity (Bigg, 1981). Southern elephant seal females have been known to pup successfully at 24–25 years, and individual wild ringed and grey seals have been known to exceed 40 years of age (Bonner, 1971; Lydersen and Gjertz, 1987).

14.4.5. Natural Mortality

Common causes of natural mortality (human-induced mortality is discussed in Chapter 15) in marine mammal populations include disease, interspecific competition, and predation. In general terms, age-specific mortality rates are a function of a population's age distribution and of its age-specific mortality patterns. Most mammalian populations exhibit U-shaped mortality-to-age curves, with high mortality rates in young age classes followed by several years of low mortality in middle-aged individuals and increasing mortality rates in old-age classes (Caughley, 1966). As noted previously, however, several species of large cetaceans that have experienced episodes of population size reduction due to commercial hunting within the last century have yet to recover to their preexploitation population sizes, and these relatively youthful populations exhibit little of the natural mortality characteristic of older animals. Consequently, mortality rates of early year classes are better documented and understood than are rates of older age classes for these animals. First-year mortality rates exceeding 50% are not uncommon and have been documented for several pinnipeds (Barlow and Boveng, 1991; Harcourt, 1992) and cetaceans (Sumich and Harvey, 1986; Aguilar, 1991). Preweaning mortality in pinnipeds is variable and often high because of the physical trauma associated with living in crowded rookery conditions. During early colonization of Año Nuevo Island, northern elephant seal pup mortality averaged 34% while on the rookery, and it increased to 63% by 1 year of age. Most (about 60%) of the mortality occurring on these rookeries results from injuries inflicted by adults (Le Boeuf *et al.*, 1994). Postrookery survivorship estimates showed no relation to gender or body mass of pups at weaning. However, in southern sea lions, pups living in colonies had substantially lower preweaning mortality rates than do pups reared by sequestered or isolated mothers (Campagna *et al.*, 1992).

14.4.5.1. Disease

Disease is a frequent cause of death in marine mammal populations; infectious agents include bacteria, viruses, fungi, and protozoa, and animal parasites (Cowan, 2002; Harwood, 2002). Marine mammals can contract cancer, tuberculosis, herpes, arthritis, and other diseases that might not immediately be thought of as wildlife diseases. Bacterial infection is thought to be the main cause of disease and disease related deaths in marine mammals, especially in captivity (Howard *et al.*, 1983). Sometimes bacterial agents appear in the form of outbreaks, such as the one that occurred in 1998 in an endangered New Zealand sea lion population (Gales *et al.*, 1999 cited in Cowan, 2002), resulting in mortalities of over 50% of pups during their first 2 months of life. Other viruses known or suspected to infect marine mammals include those causing papillomas, genital warts, and plaques on penile or vaginal epithelia, conditions that may interfere with successful reproduction. Viruses can also cause significant mortality levels, sometimes at epidemic proportions in marine mammal populations (Harwood, 2002). Herpes infections cause skin lesions and pneumonia and herpes was linked to a major outbreak of cancer in California sea lions (Gulland *et al.*, 1996, Lipscomb *et al.*, 2000). Protozoan and acanthocephalan induced diseases and cardiac disease were identified by Kreuder *et al.* (2003) as additional common causes of mortality in southern sea otters.

Morbilliviruses (Figure 14.14), which cause distemper in dogs and measles in humans, are known to be serious pathogens of cetaceans and pinnipeds (Hall, 1995; Kennedy, 1998). In early 1988, harbor seals of the North Sea and adjacent Baltic Sea experienced a massive die-off of 17,000–20,000 animals. The deaths accounted for 60–70% of harbor seals in some local populations. The virus that caused this mortality event, phocine distemper virus (PDV; Osterhaus and Vedder, 1988; Harwood and Grenfell, 1990; Grenfell *et al.*, 1992), suppresses immune system responses, leaving animals susceptible to secondary bacterial infections. A second outbreak of this virus, somewhat less dramatic than the first, struck European harbor seals again in 2002 (Harding *et al.*, 2002). Two other morbilliviruses that affect marine mammals have been recognized, porpoise morbillivirus (PMV) and dolphin morbillivirus (DMV). The latter was responsible for large-scale mortalities in striped dolphins in the Mediterranean in 1990 (Duignan *et al.*, 1992). Antibodies to morbilliviruses have been detected in Florida manatees, polar bears, and a wide variety of small cetaceans, indicating that they have had exposure to the virus (e.g., Duignan *et al.*, 1995a, 1995b; Cowan, 2002). Some of these species may be dead-end hosts, allowing sufficient viral replication to elicit an antibody response but not enough to transmit the disease.

Massive seal die-offs are not new events in marine mammal populations (see Geraci *et al.*, 1999 and Domingo *et al.*, 2002 for reviews) . Historical records indicate that harbor seal die-offs with symptoms like those seen in 1988 have occurred around Great

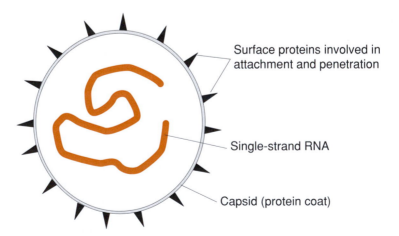

Figure 14.14.   Simplified general structure of morbillivirus. (Redrawn from Hall, 1995.)

Britain at least four times in the past 200 years. Also, in 1918, harbor seal populations around Iceland succumbed in large numbers to "pneumonia". Another virus-induced die-off of harbor seals occurred in New England in 1979–1980. In 1955, more than 60% of the crabeater seals around the Antarctic died of an unidentified viral infection. Although most of these seal deaths were caused by a virus (possibly the same virus), we do not yet understand what initiates the rapid spread and increased virulence leading to a large-scale die-off of infected individuals. Several hypotheses have been proposed, including crowding and increased resource competition resulting from rapidly increasing population sizes, increasing levels of environmental contaminants in food, toxic dinophyte or diatom blooms (Boesch *et al.*, 1997; Bargu *et al.*, 2002), and increasing water temperatures (Lavigne and Schmitz, 1990). For each of these hypotheses, some evidence exists to support its claim as the factor that caused the seal populations to suddenly become susceptible to the virus in 1988.

Phycotoxins such as **domoic acid** produced by marine algae have resulted in mass mortalities of marine mammals in several documented cases. In May and June of 1998 over 400 California sea lions died of domoic acid poisoning. A similar event occurred in 2002 (February–August) and in 2003 (April–June) along the central and southern California coast with reported strandings of hundreds of marine animals including common dolphins, and California sea lions (Marine Mammal Commission, 2003). Similar toxins have been implicated in the deaths of Hawaiian monk seals, humpback whales and manatees, as well as a recent mass die-off of Mediterranean monk seals (see Harwood, 2002).

Several populations of the southern elephant seal have been found to have low genetic variability at immune response loci (i.e., major histocompatibility complex [MHC]), as measured by RFLPs (see Section 14.3.3.3), in comparison to terrestrial mammals (Slade, 1992). Similar observations have been reported for fin whales. Together these results have been interpreted as indicating that at least some marine mammals are exposed to relatively few pathogens, and, consequently, there is less selection pressure to maintain population-based diversity at immune response loci. Reduced exposure to pathogens, which are thought to be diffusely concentrated in the marine environment, would also mean that marine mammal populations might be susceptible to occasional pathogen-induced mass mortalities (e.g., morbilliviruses discussed previously). Additionally, low levels of genetic variation have been implicated as increasing the vulnerability of a species to infectious diseases (Kretzmann *et al.*, 1997, and references cited therein), and many marine mammal populations have experienced bottlenecks and subsequent reductions in genetic diversity (see Section 14.3).

14.4.5.2. Interspecific Competition

Competition among marine mammal species and perhaps also between marine mammals and other marine predators likely has impacts on marine mammal populations, but definitive evidence of competition is generally lacking. It has been suggested that the virtual extirpation of walruses in Svalbard permitted an expansion of the bearded seal population in this region (Weslawski *et al.*, 2000) and that the rapid expansion of grey seals on Sable Island, Nova Scotia, has caused severe declines in the harbor seal population on the island (Bowen *et al.*, 2003). Similar claims have been made regarding the decline in blue and other large baleen whales allowing the minke whale and crabeater seal populations to increase in the Antarctic or a related, somewhat reversed, claim that the current minke whale abundance in the Antarctic is restricting the recovery of other baleen whales (Wade, 2002). However, the effects of competition are difficult to demonstrate with certainty in complex marine systems.

14.4.5.3. Predators

The strong tendency of pinnipeds to haul-out on ice or islands limits the impact of terrestrial predators on these populations. However, opportunistic terrestrial predators of pinnipeds (especially pups) include wolves, dogs, foxes, jackals, hyenas, and pumas (summarized by Riedman, 1990); some avian predation on pinniped pups has also been documented (e.g., Lydersen and Smith, 1989). Predation by polar bears on arctic ice seals was discussed in Chapter 12. Other marine predators that can have serious effects on some pinniped populations include adult male sea lions and leopard seals, killer whales (also described in Chapter 12), and several species of large sharks. White sharks have been observed killing and feeding on small odontocetes and scavenging on carcasses of large cetaceans (Long and Jones, 1996, and references cited therein). Greenland sharks (Lucas and Stobbo, 2000; Bowen *et al.*, 2003) and at least four other shark species (tiger, white-tip, bull, and gray) have been reported to prey on pinnipeds (Riedman, 1990). Long *et al.* (1996) summarized pinniped records of great white shark bites and provided inferences of annual, seasonal, and geographic trends and variations in shark-pinniped predatory interactions over a 23-year period. White sharks also kill California sea otters (Ames *et al.*, 1996), although it is not known whether they actually eat them. Hiruki *et al.* (1992, 1993) and Westlake and Gilmartin (1990) have described the impact of wounds inflicted by tiger and white-tip sharks on the reproductive success of female Hawaiian monk seals. Klimley *et al.* (1996) documented the predatory behavior of white sharks on young northern elephant seals (described earlier by Tricas and Mc Cosker, 1984, and Ainley *et al.*, 1985); a stealthy approach along the sea bottom in shallow areas where seals enter the water is followed by attack. White sharks are characterized by strong countershading patterns, and their approach must be difficult for seals to detect when they are swimming at the surface. The bleeding seal is carried underwater where it is held tightly in the shark's jaws until it dies of blood loss. Stomach content analyses indicate that white sharks prefer pinnipeds or whales to other prey such as birds or sea otters. This selective preference for marine mammals with extensive lipid stores may be necessary to satisfy the demands of maintaining elevated muscle temperatures and high growth rates in the cool temperate waters where their attacks on pinnipeds are concentrated (Ainley *et al.*, 1985).

14.5. Summary and Conclusions

The broad distribution and mostly underwater behavior of marine mammals makes estimating the size of populations challenging. There are two basic approaches for determining marine mammal abundances: total population (census) counts or counts of a sample of individuals and extrapolation to the whole population. Techniques for the repeated identification of marine mammals include flipper tagging, photo-identification, radio and satellite telemetry, and a variety of molecular genetic methods (i.e., analysis of chromosomes, allozymes, and DNA sequences). Molecular genetic techniques are now widely used to establish individual identification and gender, parentage (especially paternity), and population sizes and boundaries. This genetic information has proven particularly useful in stock identification and management.

Marine mammals are k-selected species with life-history patterns based on low birth and death rates that have evolved to maintain relatively stable population sizes at or near carrying capacity. Growth rates vary considerably between species over several orders of magnitude. Among common causes of natural mortality in marine mammal populations are predators (especially sharks and killer whales), parasites, diseases (e.g., morbilliviruses, toxic dinoflagellate blooms), and trauma.

14.6. Further Reading

For a review of pinniped life-history characteristics see Costa (1991), and for cetaceans see Perrin and Reilly (1984) and Lockyer (1984). The population dynamics of sirenians are summarized in O'Shea and Hartley (1995). A collection of edited papers on the application of molecular genetics to problems of marine mammal population biology are in Boyd (1993) and Dizon *et al.* (1997). Bowen and Siniff (1999) review distribution and population biology of marine mammals and McLaren and Smith (1985) review the population ecology of seals. Garner *et al.* (1999) review marine mammal survey and population assessment methods, and Dizon *et al.* (1997) provide a review of molecular genetics of marine mammals.

References

Aguilar, A. (1991). "Calving and Early Mortality in the Western Mediterranean Striped Dolphin, *Stenella coeruleoalba*." *Can. J. Zool. 69:* 1408–1412.

Ainley, D. G., R. P. Henderson, H. R. Huber, R. J. Boekekheide, S. G. Allen, and T. L. McElroy (1985). "Dynamics of White Shark/Pinniped Interactions in the Gulf of the Farallones." *South. Calif. Acad. Sci. Mem. 9:* 109–122.

Ames, J. A., J. J. Geibel, F. E. Wendell, and C. A. Pattison (1996). White shark-inflicted wounds of sea otters in California, 1968–1992. *In* "Great White Sharks: The Biology of *Carcharodon carcharias*" (A. P. Klimley, and D. G. Ainley, eds.), pp. 309–319. Academic Press, San Diego, CA.

Amos, B. (1993). "Use of Molecular Probes to Analyze Pilot Whale Pod Structure: Two Novel Approaches." *Symp. Zool. Soc. London 66:* 33–48.

Amos, B., J. Barrett, and G. A. Dover (1991). "Breeding Behaviour of Pilot Whales Revealed by DNA Fingerprinting." *Heredity 67:* 49–55.

Amos, B., S. Twiss, P. P. Pomeroy, and S. Anderson (1995). "Evidence for Mate Fidelity in the Gray Seal." *Science 268:* 1897–1899.

Andersen, L. W., E. W. Born, I. Gjertz, Ø. Wiig, L. E. Holm, and C. Bendixen (1998). "Population Structure and Gene Flow of the Atlantic Walrus *(Odobenus rosmarus rosmarus)* in the Eastern Atlantic Arctic Based on Mitochondrial DNA and Microsatellite Variation." *Mol. Ecol. 7:* 1323–1336.

Arnason, U., R. Spilliaert, A. Palsdottir, and A. Arnason (1991). "Molecular Identification of Hybrids Between the Two Largest Whale Species, the Blue Whale *(Balaenoptera musculus)* and the Fin Whale *(B. physalus)*." *Hereditas 115:* 183–189.

Arnould, J. P. Y. (2002). Southern fur seals (*Arctocephalus* spp.). *In* "Encyclopedia of Marine Mammals" (W. F. Perrin, B. Würsig and J. G. M.Thewissen, eds.), pp. 1146–1151. Academic Press, San Diego, CA.

Baird, R. W., D. J. McSweeney, A. D. Ligon, and D. L. Webster (2004). "Tagging Feasibility and Diving of Cuvier's Beaked Whales *(Ziphius cavirostris)* and Blainville's Beaked Whales *(Mesoplodon densirostris)* in Hawai'i." Report prepared under Order No. AB133F-03-SE-0986 to the Hawaii Wildlife Fund, Volcano, HI.

Baird, R. W., P. M. Willis, T. J. Guenther, P. J. Wilson, and B. N. White (1998). "An Inter-Generic Hybrid in the Family Phocoenidae." *Can. J. Zool. 76:* 198–204.

Baker, C. S., and L. Medrano-Gonzalez (2002). Worldwide distribution and diversity of humpback whale mitochondrail lineages *In* "Molecular and Cell Biology of Marine Mammals" (C. J. Pfeiffer, ed.), pp. 84–99. Krieger Publ., Melbourne, FL.

Baker, C. S., S. R. Palumbi, R. H. Lambertsen, M. T. Weinrich, J. Calambokidis, and S. O'Brien (1990). "Influence of Seasonal Migration on Geographic Distribution of Mitochondrial DNA Haplotypes in Humpback Whales." *Nature 344:* 238–240.

Baker, C. S., A. Perry, J. L. Bannister, M. T. Weinrich, R. B. Abernathy, J. Calambokidis, J. Lien, R. H. Lambertsen, J. Urban Ramirez, O. Vasquez, P. J. Clapham, A. Ailing, S. J. O'Brien, and R. S. Palumbi (1993). "Abundant Mitochondrial DNA Variation and World-Wide Population Structure in Humpback Whales." *Proc. Natl. Acad. Sci. U.S.A. 90:* 8239–8243.

Baker, C. S., R. W. Slade, J. L. Banister, R. B. Abernathy, M. T. Weinrich, J. Lien, J. Urban Ramirez., P. Corkeron, J. Calambokidis, O. Vasquex, and S. R. Palumbi (1994). "Hierarchical Structure of Mitochondrial DNA Gene Flow Among Humpback Whales *Megaptera novaeangliae,* Worldwide." *Mol. Ecol. 3:* 313–327.

Bargu, S., C. L. Powell, S. L. Coale, M. Busman, G. J. Doucette, and M. W. Silver. (2002). "Krill: A Potential Vector for Domoic Acid in Marine Food Webs." *Mar. Ecol. Prog. Ser. 237:* 209–216.

Barlow, J., and P. Boveng (1991). "Modeling Age-Specific Mortality for Marine Mammal Populations." *Mar. Mamm. Sci. 7:* 50–65.

Bengtson, J. L. (2002). Crabeater seal *(Lobodon carcinophaga). In* "Encyclopedia of Marine Mammals" (W. F. Perrin, B. Würsig, and J. G. M. Thewissen, eds.), pp. 302–304. Academic Press, San Diego, CA.

Bennett, K. A., B. J. McConnell, and M. A. Fedak (2001). "Diurnal and Seasonal Variations in the Duration and Depth of the Longest Dives in Southern Elephant Seals *(Mirounga leonina)*: Possible Physiological and Behavioural Constraints." *J. Exp. Biol. 204:* 649–662.

Berube, M., and A. Aguilar (1998). "A New Hybrid Between a Blue Whale, *Balaenoptera musculus* and a Fin Whale, *B. physalus*: Frequency and Implications of Hybridization." *Mar. Mamm Sci. 14:* 82–98.

Bickham, J. W., J. C. Patton, and T. R. Loughlin (1996). "High Variability for Control-Region Sequences in a Marine Mammal: Implications for Conservation and Biogeography of Steller Sea Lions." *J. Mammal. 77:* 95–108.

Bigg, M. A. (1981). Harbour seal—*Phoca vitulina* and *P. largha. In* "Handbook of Marine Mammals," vol. 2 (S. H. Ridgway, and R. J. Harrison, eds.), pp. 1–27. Academic Press, London.

Bigg, M. A., P. F. Olesiuk, G. M. Ellis, J. K. B Ford, and K. C. Balcomb III (1990). "Social Organization and Genealogy of Resident Killer Whales *(Orcinus orca)* in the Coastal Waters of British Columbia and Washington State." *Rep. Int. Whal. Commn. Spec. Issue 12:* 383–405.

Bjørge, A. (1992). "The Reproductive Biology of the Harbor Seal, *Phoca vitulina L.*, in Norwegian Waters." *Sarsia 77:* 47–51.

Blokhin, S. A., and P. A. Tiupeleyev (1987). "Morphological Study of the Earplugs of Gray Whales and the Possibility of Their Use in Age Determination (SC/38/PS 18)." *Rep. Int. Whal. Commn. 37:* 341–345.

Bodkin, J. L., R. J. Jameson, and J. A. Estes (1995). Sea otters in the North Pacific Ocean. *In* "Our Living Resources: A Report to the Nation on the Distribution, Abundance and Health of U.S. Plants, Animals and Ecosystems" (E. T. LaRoe III, G. S. Farris, C. E. Puckett, P. D.Doran, and M. J. Mac, eds.), pp 353–356. Department of the Interior, Natl. Biol. Serv., Washington, DC.

Boesch, D. F., D. M. Anderson, R. A. Homer, S. E. Shumway, P. A. Tester, and T. E. Whitledge (1997). Harmful algal blooms in coastal waters: Options for prevention, control, and mitigation. NOAA Coastal Ocean Program Decision Analysis Series No. 10.

Boness, D. J., W. D. Bowen, and J. M. Francis (1993). "Implications of DNA Fingerprinting for Mating Systems and Reproductive Strategies of Pinnipeds." *Symp. Zool. Soc. London 66:* 61–93.

Bonner, W. N. (1971). "An Aged Seal *(Halichoerus grypus)." J. Zool., London 164:* 261–262.

Boskovic, R., K. M. Kovacs, M. O. Hammill, and B. N. White (1996). "Geographic Distribution of Mitochondrial DNA Haplogypes in Grey Seals *(Halichoerus grypus)." Can. J. Zool. 74:* 1787–1796.

Bowen, W. D. (1991). Behavioural ecology of pinniped neonates. *In* "The Behaviour of Pinnipeds" (D. Renouf, ed.), pp. 66–127. Chapman & Hall, London.

Bowen, W. D., S. L. Ellis, S. J. Iverson, and D. J. Boness (2003). "Maternal and Newborn Life-History Traits During Periods of Contrasting Population Trends: Implications for Explaining the Decline of Harbour Seals *(Phoca vitulina),* on Sable Island." *J. Zool. London 261:* 155–163.

Bowen, W. D., R. A. Myers, and K. Hay (1987). "Abundance Estimation of a Dispersed, Dynamic Population: Hooded Seals *(Cystophora cristata)* in the Northwest Atlantic." *Can. J. Fish. Aquat. Sci. 44:* 282–295.

Bowen, W. D., O. T. Oftedal, and D. J. Boness (1985). "Birth to Weaning in Four Days: Remarkable Growth in the Hooded Seal, *Cystophora cristata." Can. J. Zool. 63:* 2841–2846.

Bowen, W. D., and D. E. Sergeant. (1983). "Mark-Recapture Estimates of Harp Seal Pup *(Phoca groenlandica)* Production in the Northwest Atlantic." *Can. J. Fish. Aquat. Sci. 40:* 728–742.

Bowen, W. D., and D. B. Siniff (1999). Distribution, population biology, and feeding ecology of marine mammals. *In* "Biology of Marine Mammals" (J. E. Reynolds III, and S. A. Rommel, eds.), pp. 423–484. Smithsonian Institute Press, Washington, DC.

Boyd, I. L. (ed.) (1993). Marine Mammals: Advances in Behavioural and Population Biology. *Symp. Zool. Soc. London* 66, Clarendon Press, Oxford, UK.

Brody, S. (1968). *Bioenergetics and Growth*. Hafner, New York.

Brownell, R. L., K. Ralls, and W. F. Perrin (1989). "The Plight of the Forgotten Whales." *Oceanus 32:* 5–11.

Brown-Gladdenm J. G., M. M. Ferguson, and J. W. Clayton (1997). "Matriarchial Genetic Population Structure of North American Beluga Whales *Delphinapterus leucas* (Cetacea: Monodontidae)." *Mol. Ecol. 6:* 1033–1046.

Brunner, S. (1998). "Cranial Morphometrics of the Southern Fur Seals *Arctocephalus forsteri* and *A. pusillus* (Carnivora: Otariidae)." *Aust. J. Zool. 46:* 67–108.

Brunner, S. (2002). "A Probable Hybrid Sea Lion—*Zalophus californianus* X *Otaria byronia*." *J. Mammal. 83:* 135–144.

Buckland, S. T., D. R. Anderson, K. P. Burnham, and J. L. Laake. (1993). *Distance Sampling: Estimating Abundance of Biological Populations*. Chapman & Hall, London.

Buckland, S. T., J. M. Breiwick, K. L. Cattanach, and J. L. Laake (1993). "Estimated Population-Size of the California Grey Whale." *Mar. Mamm. Sci. 9:* 235–249.

Buckland, S. T., and A. E. York (2002). Abundance Estimation *In* "Encyclopedia of Marine Mammals" (W. F. Perrin, B. Würsig, and J. G. M.Thewissen, eds.), pp. 1–6. Academic Press, San Diego, CA.

Burke, T., O. Hanotte, and I. Van Piljen (1996). Minisatellite analysis in conservation genetics. *In* "Molecular Genetic Approaches in Conservation" (T. B. Smith, and R. K. Wayne, eds.), pp. 251–277. Oxford University Press, Oxford.

Burns, J. M., and M. A. Castellini (1998). "Dive Data from Satellite Tags and Time-Depth Recorders: A Comparison in Weddell Seal Pups." *Mar. Mamm. Sci. 14:* 750–764.

Burns, J. M., D. P. Costa, M. A. Fedak, M. A. Hindell, C. J. A. Bradshaw, N. J. Gales, B. McDonald, S. J. Trumble, and D. E. Crocker (2004). "Winter Habitat Use and Foraging Behavior of Crabeater Seals Along the Western Antarctic Peninsula." *Deep-Sea Res. Part II Tropical Stud. Oceanogr. 51:* 2279–2303.

Campagna, C., B. J. Le Boeuf, M. Lewis, and C. Bisoli (1992). "The Fisherian Seal: Equal Investment in Male and Female Pups in Southern Elephant Seals." *J. Zool. 226:* 551–561.

Caughley, G. (1966). "Mortality Patterns in Mammals." *Ecology 47:* 906–918.

Chapmann, D. G., and A. M. Johnson (1968). "Estimation of Fur Seal Populations by Randomized Sampling." *Trans. Am. Fish. Soc. 97:* 264–270.

Clapham, P. J., and R. L. J. Brownell (1996). "The Potential for Interspecfic Competition in Baleen Whales." *Rep. Int. Whal. Comm. 46:* 361–367.

Clapham, P. J., and P. J. Palsbøll (1997). "Molecular Analysis of Paternity Shows Promiscuous Mating in Female Humpback Whales (*Megaptera novaeangliae,* Borowski)." *Proc. R. Soc. London Ser. B 264:* 95–98.

Clinton, W. L. (1994). Sexual selection and growth in male northern elephant seals. *In* "Elephant Seals" (B. J. Le Boeuf and R. M. Laws, eds.), pp. 154–168. University of California Press, Berkeley, CA.

Clinton, W. L., and B. J. Le Boeuf (1993). "Sexual Selection's Effects on Male Life History and the Pattern of Male Mortality." *Ecology 74:* 1884–1892.

Costa, D. P. (1991). Reproductive and foraging energetics of pinnipeds; implications for life history patterns. *In* "The Behavior of Pinnipeds" (D. Renouf, ed.), pp. 299–344. Chapman & Hall, London.

Costa, D. P., J. Klinck, E. Hofman, J. M. Burns, M. A. Fedak and D. E. Crocker (2003). "Marine Mammals as Ocean Sensors." *Integr. Comp. Biol. 43:* 920–920.

Cowan, D. F. (2002). Pathology *In* "Encyclopedia of Marine Mammals (W. F. Perrin, B. Würsig, and J. G. M.Thewissen, eds.), pp. 883–890. Academic Press, San Diego, CA.

Cronin, M. A., S. C. Amstrup, G. Garner, and E. R. Vyse (1991). "Intra- and Interspecific Mitochondrial DNA Variation in North American Bears *(Ursus)*." *Can. J. Zool. 69:* 2985–2992.

Cronin, M. A., J. Bodkin, B. Ballachey, J. Estes, and J. C. Patton (1996). "Mitochondrial Variation Among Subspecies and Populations of Sea Otters *(Enhydra lutris)*." *J. Mammal. 77:* 546–557.

Cronin, M. A., S. Hills, E. W. Born, and J. C. Patton (1994). "Mitochondrial DNA Variation in Atlantic and Pacific Walruses." *Can. J. Zool. 72:* 1035–1043.

Davis, R. W., L. A. Fuiman, T. M. Williams, M. Horning, and W. Hagey (2003). "Classification of Weddell Seal Dives Based on 3-Dimensional Movements and Video-Recorded Observations." *Mar. Ecol. Prog. Ser. 264:* 109–122.

DeLong, R. L., and B. S. Stewart (1991). "Diving Patterns of Northern Elephant Seal Bulls." *Mar Mamm. Sci.* 7: 369–384.

DeLong, R. L., B. S. Stewart, and R. D. Hill (1992). "Documenting Migrations of Northern Elephant Seals Using Day Length." *Mar Mamm. Sci. 8:* 155–159.

Derocher, A. E., G. W. Garner, N. J. Lunnn, and Ø. Wiig (eds.) (1998). "Polar Bears." Proceedings of the Twelfth Working Meeting of the IUCN/SSC Polar Bear Specialist Group, 3-7 February 1997, Oslo, Norway. IUCN Species Survival Commn. Paper No. 19.

Derocher, A. E., and I. Stirling (1994). "Age-Specific Reproductive Performance of Female Polar Bears *(Ursus maritimus)*." *J. Zool. London 234:* 527–536.

Deutsch, C. J., J. P. Reid, R. K. Bonde, D. E. Easton, H. I. Kochman, and T. J. O'Shea (2003). "Seasonal Movements, Migratory Behavior, and Site Fidelity of West Indian Manatees Along the Atlantic Coast of the United States." *Wild. Monogr. 151:* 1–77.

Dietz, R., and M. P. Heide-Jorgensen (1995). "Movements and Swimming Speed of Narwhals, *Monodon monoceros,* Equipped with Satellite Transmitters in Melville Bay, Northwest Greenland." *Can. J. Zool. 73:* 2106–2119.

Dizon, A. E., S. J. Chivers, and W. F Perrin (eds.) (1997). Molecular Genetics of Marine Mammals, *Spec. Publ. No. 3, Society for Marine Mammalogy.* Allen Press, Kansas.

Dizon, A. E., S. O. Southern, and W. F. Perrin (eds.) (1991). "Molecular Analysis of mtDNA Types in Exploited Populations of Spinner Dolphins *(Stenella longirostris)*." *Rep. Int. Whal. Commn. Spec. Issue 13:* 183–202.

Domingo, M., S. Kennedy, and M.-F. Van Bressem (2001). Marine mammal mass mortalities. *In* "Marine Mammals Biology and Conservation"(P. G. H. Evans, and J. A. Raga, eds.), pp. 425–456. Kluwer Publishing, New York.

Dover, G. A. (1991). "Understanding Whales with Molecular Probes." *Rep. Int. Whal. Commn. Spec. Issue 13:* 29–38.

Dowling, T. E., and W. M. Brown (1993). "Population Structure of the Bottlenose Dolphin as Determined by Restriction Endonuclease Analysis of Mitochondrial DNA." *Mar Mamm. Sci. 9:* 138–155.

Duffield, D. A., and J. Chamberlin-Lea (1990). Use of chromosome heteromorphisms and hemoglobins in studies of bottlenose dolphin populations and paternities. *In* "Bottlenose Dolphin" (S. Leatherwood, and R. R. Reeves, eds.), pp. 609–622. Academic Press, San Diego, CA.

Duffield, D. A., and R. Wells (1991). "The Combined Application of Chromosome, Protein, and Molecular Data for the Investigation of Social Units Structure and Dynamics in *Tursiops truncatus*." *Rep. Int. Whal. Commn. Spec. Issue 13:* 155–170.

Duignan, P. J., J. R. Geraci, J. A. Raga, and N. Calzada (1992). "Pathology of Morbillivirus Infection in Striped Dolphins *(Stenella coeruleoalba)* from Valencia and Murcia, Spain." *Can. J. Vet. Res. 56:* 242–248.

Duignan, P. J., C. House, J. R. Geraci, N. Duffy, B. K. Rima, M. T. Walsh, G. Early, D. J. St. Aubin, S. Sadove, H. Koopman, and H. Rhinehart (1995b). "Morbillivirus Infection in Cetaceans of the Western Atlantic." *Vet. Microbiol. 44:* 241–249.

Duignan, P. J., C. House, M. T. Walsh, T. Campbell, G. D. Bossart, N. Duffy, P. J. Fernandes, B. K. Rima, S. Wright, and J. R. Geraci (1995a). "Morbillivirus Infection in Manatees." *Mar. Mamm. Sci. 11:* 441–451.

Erickson, A. W., and M. N. Bester (1993). Immobilization and capture. *In* "Antarctic Seals" (R. M. Laws, ed.), pp. 46–88. Cambridge University Press, New York.

Ford, J. K. B., G. M. Ellis, and K. C. Balcomb (1994). *Killer Whales: The Natural History and Genealogy of Orcinus orca in British Columbia and Washington State.* University of Washington Press, Seattle.

Gales, N. J., and D. J. Fletcher (1999). "Abundance, Distribution and Status of the New Zealand Sea Lion, *Phocarctos hookeri*." *Wildl. Res. 26:* 35–52.

Gambell, R. (1979). "The Blue Whale." *Biologist 26:* 209–215.

Gambell, R. (1999). The International Whaling Commission and the contemporary whaling debate *In* "Conservation and Management of Marine Mammals" (J. R. Twiss, Jr., and R. R. Reeves, eds), pp 179–198. Smithsonian Inst. Press, Washington, DC.

Garcia-Rodriguez, A. I., B. W. Bowen, D. Domning, A. A. Mignucci-Giannoni, M. Marmontel, R. A. Montoya-Ospina, B. Morales-Vela, M. Rudin, R. K. Bonde, and P. M. McGuire (1998). "Phylogeography of the West Indian Manatee *(Trichechus manatus):* How Many Populations and How Many Taxa?" *Mol. Ecol. 7:* 1137–1149.

Garner, G.W., S. C. Amstrup, J. L. Laake, B. F. J. Manly, L. L. McDonald, and D. G. Robertson (eds.) (1999). *Marine Mammal Survey and Assessment Methods.* A. A. Balkema, Rotterdam, Netherlands.

Gentry, R. L. (2002). Eared seals. *In* "Encyclopedia of Marine Mammals" (W. F. Perrin, B. Würsig and J. G. M.Thewissen, eds.), pp. 348–351. Academic Press, San Diego, CA.

Gentry, R. L., and V. R. Casanas (1997). "A New Method for Immobilizing Otariid Neonates." *Mar. Mamm. Sci. 13:* 155–156.

Geraci, J. R., J. Harwood, and V. J. Lounsbury (1999). Marine Mammal die-offs: causes, investigations and issues. *In* "Conservation and Management of Marine Mammals" (J. R. Twiss, Jr., and R. R. Reeves, eds.), pp. 367–395. Smithsonian Institute Press, Washington, DC.

Gilmartin, W. G., and J. Forcada (2002). Monk seals (*Monachus monachus, M. tropicalis,* and *M. schauins-landi*) *In* "Encyclopedia of Marine Mammals" (W. F. Perrin, B. Würsig, and J. G. M.Thewissen, eds.), pp. 756–758. Academic Press, San Diego, CA.

Gjertz, I., D. Griffiths, B. A. Krafft, C. Lydersen, and Ø. Wiig (2001). "Diving and Haul-Out Patterns of Walruses *Odobenus rosmarus* on Svalbard." *Polar Biol. 24:* 314–319.

Goldsworthy, S. D., D. J. Boness, and R. C. Fleisher (1999). "Mate Choice Among Sympatric Fur Seals: Female Preference for Conphenotypic Males." *Behav. Ecol. Sociobiol. 45:* 253–267.

Gowans, S., and H. Whitehead (2001). "Photographic Identification of Northern Bottlenose Whales (*Hyperoodon ampullatus*): Sources of Heterogeneity from Natural Marks." *Mar. Mamm. Sci. 17:* 76–93.

Grenfell, B. T., M. E. Lonergan, and J. Harwood (1992). "Quantitative Investigations of the Epidemiology of Phocine Distemper Virus (PDV) in European Common Seal Populations." *Sci. Tot. Environ. 115:* 15–29.

Gulland, F. M. D., J. G. Trupkiewicz, T. R. Spraker, and L. J. Lowenstine (1996). "Metastatic Carcinoma of Probable Transitional Cell Origin in 66 Free-Living California Sea Lions (*Zalophus californianus*) 1979 to 1994." *J. Wildl. Dis. 32:* 250–258.

Hall, A. J. (1995). "Morbilliviruses in Marine Mammals." *Trends Microbiol.* 7: 4–9.

Hall, A. J., S. Moss, and B. McConnell. 2000. "A New Tag for Identifying Seals." *Mar. Mamm. Sci. 16:* 254–257.

Halley, J., and A. R. Hoelzel (1996). Simulation models of bottleneck events in natural populations. *In* "Molecular Genetic Approaches in Conservation" (T. B. Wayne, and R. K. Wayne, eds.), pp. 347–364. Oxford University Press, Oxford, UK.

Hammond, P. S., P. Berggren, H. Benke, D. L. Borchers, A. Collet, M. P. Heide-Jorgensen, S. Heimlich, A. R. Hiby, M. F. Leopold, and N. Øien (2002). "Abundance of Harbour Porpoise and Other Cetaceans in the North Sea and Adjacent Waters." *J. Appl. Ecol. 39:* 361–376.

Harcourt, R. (1992). "Factors Affecting Early Mortality in the South American Fur Seal (*Arctocephalus australis*) in Peru: Density-Related Effects and Predation." *J. Zool. London 226:* 259–270.

Harding, K. C., T. Härkönen, and H. Caswell (2002). "The 2002 European Seal Plague: Epidemiology and Population Consequences." *Ecol. Letters 5:* 727–732.

Harris, A. S., J. S. F. Young, and J. M. Wright (1991). "DNA Fingerprinting of Harbour Seals (*Phoca vitulina concolor*): Male Mating Behavior May Not Be a Reliable Indicator of Reproductive Success." *Can. J. Zool. 69:* 1862–1866.

Harvey, J. T., and B. R. Mate. 1984. Dive characteristics and movements of radio-tagged gray whales in San Ignacio, Baja California Sur, Mexico. *In* "The Gray Whale, *Eschrichtius robustus* (Lilljeborg, 1861)" (eds.), pp. 561–575, Academic Press, San Diego, CA.

Harwood, J. (2002). Mass die-offs. *In* "Encyclopedia of Marine Mammals" (W. F. Perrin, B. Würsig, and J. G. M.Thewissen, eds.), pp. 724–726. Academic Press, San Diego, CA.

Harwood, J., and B. Grenfell (1990). "Long Term Risks of Recurrent Seal Plagues." *Mar Pollut. Bull. 21:* 284–287.

Hillis, D. M., C. Moritz, and B. K. Mable (1996). *Molecular Systematics,* 2nd ed. Sinauer Associates, Sunderland, MA.

Hillman, G. R., B. Wursig, G. A. Gailey, N. Kehtarnavaz, A. Drobyshevsky, B. N. Araabi, H. D. Tagare, and D. W. Weller (2003). "Computer-Assisted Photo-Identification of Individual Marine Vertebrates: A Multi-Species System." *Aquat. Mamm. 29:* 117–123.

Hiruki, L. M., W. G. Gilmartin, B. L. Becker, and I. Stirling (1993). "Wounding in Hawaiian Monk Seals (*Monachus schauinslandi*)." *Can. J. Zool. 71:* 458–468.

Hiruki, L. M., I. Stirling, W. G. Gilmartin, T. C. Johanos, and B. L. Becker (1992). "Significance of Wounding to Female Reproductive Success in Hawaiian Monk Seals (*Monachus schauinslandi*) at Laysan Island." *Can. J. Zool. 71:* 469–474.

Hoelzel, A. R. (1993). "Genetic Ecology of Marine Mammals." *Symp. Zool. Soc. London 66:* 15–29.

Hoelzel, A. R. (1994). "Genetics and Ecology of Whales and Dolphins." *Annu. Rev. Ecol. Syst. 25:* 377–399.

Hoelzel, A. R., and G. A. Dover (1991a). "Mitochondrial d-Loop DNA Variation Within and Between Populations of the Minke Whale (*Balaenoptera acutorostrata*)." *Rep. Int. Whal. Comm., Spec. Issue 13:* 171–183.

Hoelzel, A. R., and G. A. Dover (1991b). "Genetic Differentiation Between Sympatric Killer Whale Populations." *Heredity 66:* 191–195.

Hoelzel, A. R., S. D. Goldsworthy, and R. C. Fleisher (2002). Population genetic structure. *In* "Marine Mammal Biology" (A. R. Hoelzel, ed.), pp. 325–352. Blackwell Publishing, Oxford, UK.

Hoelzel, A. R., J. Halley, C. Campagna, T. Arnbom, B. Le Boeuf, S. J. O'Brien, K. Ralls, and G. A. Dover (1993). "Elephant Seal Genetic Variation and the Use of Simulation Models to Investigate Historical Population Bottle-Necks." *J. Hered. 84:* 443–449.

Hoelzel, A. R. A. Natoli, M. E. Dalheim, C. Olavarria, R. W. Baird, and N. A. Black (2002b). "Low Worldwide Genetic Diversity in the Killer Whale *(Orcinus orca):* Implications for Demographic History." *Proc. R. Soc. Lond. B 269:* 1467–1473.

Hoelzel A. R., C. W. Potter, and P. Best (1998). "Genetic Differentiation Between Parapatric `Nearshore' and `Offshore' Populations of the Bottlenose Dolphin." *Proc. R. Soc. London B 265:* 1–7.

Hooker, S. K., and I. L. Boyd (2003). "Salinity Sensors on Seals: Use of Marine Predators to Carry CTD Data Loggers." *Deep-Sea Res. Part I – Oceanogr. Res. Papers 50:* 927–939.

Hooker, S. K., I. L. Boyd, M. Jessopp, O. Cox, J. Blackwell, P. L. Boveng, and J. L. Bengtson (2002). "Monitoring the Prey-Field of Marine Predators: Combining Digital Imaging with Data-Logging Tags." *Mar. Mamm. Sci. 18:* 680–697.

Howard, E. B., Britt, J. O., G. K. Matsumoto, R. Itahara, and C. Nagano. (1983). Bacterial diseases. *In* "Pathobiology of Marine Mammals" (E. B. Howard, ed.), pp. 69–118. CRC Press, Boca Raton, FL.

Huber, H. R., A. C. Rovetta, L. A. Fry, and S. Johnston (1991). "Age-Specific Natality of Northern Elephant Seals at the South Farallon Islands, California." *J. Mammal. 72:* 525–534.

Jameson, R. J. (1989). "Movements, Home Range, and Territories of Male Sea Otters off Central California." *Mar Mamm. Sci. 5:* 159–172.

Jameson, R. J., and A. M. Johnson (1993). "Reproductive Characteristics of Female Sea Otters." *Mar Mamm. Sci. 9:* 156–167.

Jones, M.L., and S. L. Swartz (2002). Gray whale *(Eschrichtius robustus). In* "Encyclopedia of Marine Mammals" (W. F. Perrin, B. Würsig, and J. G. M.Thewissen, eds.), pp. 524–536. Academic Press, San Diego, CA.

Kaiya, Z. (2002). Baiji. *In* "Encyclopedia of Marine Mammals" (W. F. Perrin, B. Würsig, and J. G. M.Thewissen, eds.), pp. 58–61. Academic Press, San Diego, CA.

Kato, H., and K. Sakuramoto (1991). "Age at Sexual Maturity of Southern Minke Whales: A Review and Some Additional Analyses." *Rep. Int. Whal. Commn. 41:* 331–337.

Kennedy, S. (1998). "Morbillivirus Infections in Aquatic Mammals." *J. Comp. Path. 119:* 201–225.

Kenny, R. D. (2002). North Atlantic, North Pacific, and Southern Right Whales (*Eubalaena glacialis, E. japonica,* and *E. australis*). *In* "Encyclopedia of Marine Mammals" (W. F. Perrin, B. Würsig, and J. G. M.Thewissen, eds.), pp. 806–813. Academic Press, San Diego, CA.

Klevezal', G. A., L. I. Sukhovskaya, and S. A. Blokhin (1986). "Age Determination of Baleen Whales from Bone Layers." [OT: Opredelenie vozrasta usatykh kitov po godovym sloyam v kosti]. *Zool. Zh. 65:* 1722–1730.

Klimley, A. P., P. Pyle, and S. D. Anderson (1996). The behavior of white sharks and their pinniped prey during predatory attacks. *In* "Great White Sharks: The Biology of *Carcharodon carcharias*" (A. P. Klimley, and D. G. Ainley, eds.), pp. 175–192. Academic Press, San Diego, CA.

Kooyman, G. L., E. A. Wahrenbrock, M. A. Castellini, R. W. Davis, and E. E. Sinnett (1980). "Aerobic and Anaerobic Metabolism During Voluntary Diving in Weddell Seals: Evidence of Preferred Pathways from Blood Chemistry and Behavior." *J. Comp. Physiol. 138:* 335–346.

Kovacs, K. M., and D. M. Lavigne (1986). "Maternal Investment and Neonatal Growth in Phocid Seals." *J. Anim. Ecol. 55*(3): 1035-1051.

Kovacs, K. M. (ed.) (2005). "Birds and Mammals of Svalbard." *Polarhandbok* no. 13, Norweigian Polar Institute, Tromso, Norway.

Kovacs, K. M., C. Lydersen, M. O. Hammill, B. N. White, P. J. Wilson, and S. Malik (1997). "A Harp Seal X Hooded Seal Hybrid." *Mar. Mamm. Sci. 13:* 460–468.

Krafft, B. A., C. Lydersen, K. M. Kovacs, I. Gjertz, and T. Haug (2000). "Diving Behaviour of Lactating Bearded Seals *(Erignathus barbatus)* in the Svalbard Area." *Can. J. Zool. 78:* 1408–1418.

Kretzmann, M., W. G. Gilmartin, A. Meyer, G. P. Zegers, S. R. Fain, B. F. Taylor, and D. P. Costa (1997). "Low Genetic Variability in the Hawaiian Monk Seal." *Conserv. Biol. 11:* 482–490.

Kreuder, C., M. A. Miller, D. A. Jessup, L. J. Lowenstine, M. D. Harris, J. A. Ames, T. E. Carpenter, P. A. Conrad, and J. A. K. Mazet (2003). "Patterns of Mortality in Southern Sea Otters *(Enhydra lutris nereis)* from 1998–2001." *J. Wildl. Dis. 39:* 495–509.

Laake, J. L., J. Calambokidis, S. D. Osmek, and D. J. Rugh (1997). "Probability of Detecting Harbor Porpoise from Aerial Surveys: Estimating g(0)." *J. Wildl. Manage. 61:* 63–75.

Lagerquist, B. A., K. M. Stafford, and B. R. Mate (2000). "Dive Characteristics of Satellite-Monitored Blue Whales *(Balaenoptera musculus)* off the Central California Coast." *Mar. Mamm. Sci. 16:* 375–391.

Laidre, K. L., M.-P. Heide-Jørgensen, M. L. Logdson, R. C. Hobbs, P. Heagerty, R. Dietz, O. A. Jørgensen, and M. A. Treble (2004). "Seasonal Narwhal Habitat Associations in the High Arctic." *Mar. Biol. 145:* 821–831.

Larson, S., R. Jameson, M. Etinier, M. Fleming, and P. Bentzen (2002). "Loss of Genetic Diversity in Sea Otters *(Enhydra lutris)* Associated with the Fur Trade of the 18th and 19th Centuries." *Mol. Ecol. 11:* 1899–1903.

Lavigne, D. M. (2002). Harp seals (*Pagophilus groenlandicus*) *In* "Encyclopedia of Marine Mammals" (W. F. Perrin, B. Würsig, and J. G. M.Thewissen, eds.), pp. 560–562. Academic Press, San Diego, CA.

Lavigne, D. M., and O. J. Schmitz (1990). "Global Warming and Increasing Population Densities: A Prescription for Seal Plagues." *Mar. Pollut. Bull. 21:* 280–284.

Laws, R. M. (1962). "Age Determination of Pinnipeds with Special Reference to Growth Layers in Teeth." *Zeits. Säug. 27:* 129–146.

Laws, R. M., and P. E. Purves (1956). "The Ear Plug of the Mysticeti as an Indication of Age with Special Reference to the North Atlantic Fin Whale *(Balaenoptera physalus* Linn.)." *Nor. Hvalfangst-Tid. 45:* 413–425.

Le Boeuf, B. J., D. P. Costa, A. C. Huntley, and S. D. Feldkamp (1988). "Continuous, Deep Diving in Female Northern Elephant Seals, *Mirounga angustirostris.*" *Can. J. Zool. 66:* 416–458.

Le Boeuf, B. J., P. Morris, and J. Reiter (1994). Juvenile survivorship of northern elephant seals. *In* "Elephant Seals" (B. Le Boeuf, and R. M. Laws, eds.), pp. 121–136. University of California Press, Berkeley, CA.

Le Boeuf, B. J., and J. Reiter (1988). Lifetime reproductive success in northern elephant seals. *In* "Reproductive Success" (T. H. Clutton-Brock, ed.), pp. 344–362. University of Chicago Press, Chicago.

LeDuc, R. (2002). Delphinids, Overview. *In* "Encyclopedia of Marine Mammals" (W. F. Perrin, B. Würsig, and J. G. M.Thewissen, eds.), pp. 310–316. Academic Press, San Diego, CA.

Lehman, N., R. K. Wayne, and B. S. Stewart (1993). "Comparative Levels of Genetic Variability in Harbor Seals and Northern Elephant Seals as Determined by Genetic Fingerprinting." *Symp. Zool. Soc. London 66:* 49–60.

Lento, G., M. Hadden, G. K. Chambers, and C. S. Baker (1997). "Genetic Variation of Southern Hemisphere Fur Seals: *Arctocephalus* spp.: Investigation of Population Structure and Species Identity." *J. Hered. 88:* 202–208.

Lento, G. M., R. H. Mattlin, G. K. Chambers, and C. S. Baker (1994). "Geographic Distribution of Mitochondria Cytochrome b DNA Haplotypes in New Zealand Fur Seals *(Arctocephalus forsteri)*." *Can. J. Zool. 72:* 293–299.

Lipscomb, T. P., D. P. Scott, R. L. Garber, A. E. Krafft, M. M. Tsai, J. H. Lichy, J. F. Taubenberger, F. Y. Schulman, and F. M. D. Gulland (2000). "Common Metastatic Carcinoma of California Sea Lions *(Zalophus californianus):* Evidence of Genital Origin and Association with Novel Gammaherpesvirus." *Vet. Path. 37:* 609–617.

Lockyer, C. (1984). "Review of Baleen Whale (Mysticeti) Reproduction and Implications for Management." *Rep. Int. Whal. Comm., Spec. Issue 6:* 27–50.

Long, D. J., K. D. Hanni, P. Pyle, J. Roletto, R. E. Jones, E. Pyle, and R. Bandar (1996). White shark predation on four pinniped species in central California waters: Geographic and temporal patterns inferred from wounded carcasses. *In* "Great White Sharks: The Biology of *Carcharodon carcharias"* (A. P. Klimley, and D. G. Ainley, eds.), pp. 263–291. Academic Press, San Diego, CA.

Long, D. J., and R. E. Jones (1996). White shark predation and scavenging on cetaceans in the eastern North Pacific Ocean. *In* "Great White Sharks: The Biology of *Carcharodon carcharias"* (A. P. Klimley, and D. G. Ainley, eds.), pp. 293–308. Academic Press, San Diego, CA.

Lowry, L. F., K. J. Frost, J. M. VerHoef, and R. A. DeLong (2001). "Movements of Satellite-Tagged Subadult and Adult Harbor Seals in Prince William Sound, Alaska." *Mar. Mamm. Sci. 17:* 835–861.

Lucas, Z., and W. T. Stobo (2000). "Shark-Inflicted Mortality on a Population of Harbour Seals *(Phoca vitulina)* at Sable Island, Nova Scotia." *J. Zool. London 252:* 405–414.

Lydersen, C. (1991). "Monitoring Ringed Seal *(Phoca hispida)* Activity by Means of Acoustic Telemetry." *Can J. Zool. 69:* 1178–1182.

Lydersen, C., and I. Gjertz (1987). "Population Parameters of Ringed Seals (*Phoca hispida* Schreber, 1775) in the Svalbard Area." *Can. J. Zool. 65:* 1021–1027.

Lydersen, C., and K. M. Kovacs (1993). "Diving Behaviour of Lactating Harp Seal, *Phoca groenlandica,* Females from the Gulf of St Lawrence, Canada." *Anim. Behav. 46:* 1213–1221.

Lydersen, C., and K. M. Kovacs (2005). "Growth and Population Parameters of the World's Northernmost Harbour Seals *Phoca vitulina* Residing in Svalbard, Norway." *Polar Biol. 28:* 156–163.

Lydersen, C., O. A. Nost, K. M. Kovacs, and M. A. Fedak (2004). "Temperature Data from Norwegian and Russian Waters of the Northern Barents Sea Collected by Free-Living Ringed Seals." *J. Mar. Syst. 46:* 99–108.

Lydersen, C., O. A. Nost, P. Lovell, B. J. McConnell, T. Gammelsrod, C. Hunter, M. A. Fedak, and K. M. Kovacs (2002). "Salinity and Temperature Structure of a Freezing Arctic Fjord—Monitored by White Whales *(Delphinapterus leucas)*." *Geophys. Res. Letters 29:* Art. No 2119.

Lydersen, C., and T. G. Smith (1989). "Avian Predation on Ringed Seal *Phoca hispida* Pups." *Polar Biol. 9:* 489–490.

Lyrholm, T., O. B. Leimar, and U. Gyllensten (1999). "Sex-Biased Dispersal in Sperm Whales: Contrasting Mitochondrial and Nuclear Genetic Structure of Global Populations." *Proc. R. Soc. Lond. B 266:* 347–354.

MacArthur, R. H., and E. O. Wilson (1967). *The Theory of Island Biogeography*. Princeton University Press, Princeton, NJ.

Madsen, P. T., R. Payne, N. U. Kristiansen, M. Wahlberg, I. Kerr, and B. Mohl (2002). "Sperm Whale Sound Production Studied with Ultrasound Time/Depth-Recording Tags." *J. Exp. Biol. 205:* 1899–1906.

Maldonado, J. E., F. Orta Davila, B. S. Stewart, E. Geffen, and R. K. Wayne (1995). "Intraspecific Genetic Differentiation in California Sea Lions *(Zalophus californianus)* from Southern California and the Gulf of California." *Mar Mamm. Sci. 11:* 46–58.

Marine Mammal Commission (2003). Ann. Rep. Mar. Mamm. Commn., Mar. Mamm. Commn., Washington, DC.

Markowitz, T. M., A. D. Harlin, and B. Würsig (2003). "Digital Photography Improves Efficiency of Individual Dolphin Identification: A Reply to Mizroch." *Mar. Mamm. Sci. 19:* 608–612.

Marmontel, M. (1995). Age and reproduction in female Florida manatees. *In* "Population Biology of the Florida Manatee" (T. J. O'Shea, B. B. Ackerman, and H. F. Percival, eds.), pp. 98–119. Inf. Tech. Rep. No.1, Natl. Biol. Ser., Washington, DC.

Marsh, H. (1995). The life history, pattern of breeding, and population dynamics of the dugong. *In* "Population Biology of the Florida Manatee" (T. J. O'Shea, B. B. Ackerman, and H. F. Percival, eds.), pp. 56–62. Inf. Tech. Rep. No.1, Natl. Biol. Ser., Washington, DC.

Marsh, H. (2002). Dugong *(Dugong dugon)*. *In* "Encyclopedia of Marine Mammals" (W. F. Perrin, B. Würsig, and J. G. M.Thewissen, eds.), pp. 344–347. Academic Press, San Diego, CA.

Marsh, H., and T. Kasuya (1991). An overview of the changes in the role of a female pilot whale with age. *In* "Dolphin Societies: Discoveries and Puzzles" (K. Pryor, and K. S. Norris eds.), pp. 281–285. University of California Press, Berkeley, CA.

Mate, B. R., and J. L. Harvey (1984). Ocean movements of radio-tagged whales. *In* "The Gray Whale, *Eschrichtius robustus* (Lilljeborg, 1861)" (M. L. Jones, S. Leatherwood, and S. L. Swartz, eds.), pp. 577–589. Academic Press, New York.

Mate, B. R., G. K., Krutzikowsky, and M.Windsor, (2000). Satellite-Monitored Movements of Radio-Tagged Bowhead Whales in the Beaufort and Chukchi Seas During the Late-Summer Feeding Season and Fall Migration." *Can. J. Zool. 78:* 1168–1181.

Mate, B. R., K. A. Rossbach, S. L. Nieukirk, R. S. Wells, A. B. Irvine, M. D. Scott, and A. J. Read (1995). "Satellite-Monitored Movements and Dive Behavior of a Bottlenose Dolphin *(Tursiops truncatus)* in Tampa Bay, Florida." *Mar Mamm. Sci. 11:* 452–463.

Mauritzen, M., A. E. Derocher, O. Pavlova, and Ø Wiig (2003). "Female Polar Bears, *Ursus maritimus,* on the Barents Sea Drift Ice: Walking the Treadmill." *Anim. Behav. 66:* 107–113.

McLaren, I. A. (1993)."Growth in Pinnipeds." *Biol. Rev. Cambridge Philos. Soc. 68:* 1–79.

McLaren, I. A., and T. G. Smith (1985). "Population Ecology of Seals—Retrospective and Prospective Views." *Mar. Mamm. Sci. 1:* 5483.

McRae, S. B., and K. M. Kovacs (1994). "Paternity Exclusion by DNA Fingerprinting, and Mate Guarding in the Hooded Seal, *Cystophora cristata*." *Mol. Ecol. 3:* 101–107.

Miller, P. J. O., M. P. Johnson, P. L. Tyack, and E. A. Terray (2004). "Swimming Gaits, Passive Drag and Buoyancy of Diving Sperm Whales *Physeter macrocephalus*." *J. Exp. Biol. 207:* 1953–1967.

Mizroch, S. A. (2003). "Digital Photography Improves Efficiency of Individual Dolphin Identification: A Reply to Markowitz et al." *Mar. Mamm. Sci. 19:* 612–614.

Mizroch, S. A. J. A. Beard, and M. Lynde (1990). "Computer Assisted Photo-Identification of Humpback Whales." *Rep. Int. Whal. Comm., Spec. Issue 12:* 63–70.

Norris, K. S., and R. L. Gentry (1974). "Capture and Harnessing of Young California Gray Whales, *Eschrichtius robustus*." *Mar. Fish. Rev. 36:* 58–64.

Norris, K. S., and K. Pryor (1991). Some thoughts on grandmothers. *In* "Dolphin Societies: Discoveries and Puzzles" (K. Pryor, and K. S. Norris, eds.), pp. 287–289. University of California Press, Berkeley, CA.

O'Corry-Crowe, G. M., R. S. Suydam, A. Rosenberg, K. J. Frost, and A. E. Dizon (1997). "Phylogeography, Population Structure and Dispersal Patterns of the Beluga Whale *Delphinapterus leucas* in the Western Nearctic Revealed by Mitochondria DNA." *Mol. Ecol. 6:* 955–970.

O'Shea, T. J., and W. C. Hartley (1995). Reproduction and early-age survival of manatees at Blue Spring, upper St. Johns River, Florida. *In* "Population Biology of the Florida Manatee" (T. J. O'Shea, B. B. Ackerman, and H. F. Percival, eds.), pp. 157–170. Inf. Tech. Rep. No.1, Natl. Biol. Ser., Washington, DC.

Osterhaus, A. D. M. E., and E. J. Vedder (1988). "Identification of Virus Causing Recent Seal Deaths." *Nature 335:* 20.

Paetkau, D., S. C. Amstrup, E. W. Born, W. Calvert, A. E. Derocher, G. W. Garner, F. Messier, I. Stirling, M.K. Taylor, Ø. Wigg, and C. Strobeck (1999). "Genetic Structure of the World's Polar Bear Populations." *Mol. Ecol. 8:* 1571–1584.

Paetkau, D., W. Calvert, I. Stirling, and C. Strobeck (1995). "Microsatellite Analysis of Population Structure in Canadian Polar Bears." *Mol. Ecol. 4:* 347–354.

Palsbøll, P. J., J. Allen, M. Bérubé, P. J. Clapham, T. P. Feddersen, P. S. Hammond, R. R. Hudson, H. Jorgensen, S. Katona, A. Holm Larsen, F. Larsen, J. Lein, D. K. Mattlla, J. Sigurjónsson, R. Sears, T. Smith, R. Sponer, P. Stevick, and N. Olen (1997). "Genetic Tagging of Humpback Whales." *Nature 388:* 767–769.

Perrin, W. F., and A. C. Myrick (1980). "Growth of Odontocetes and Sirenians. Problems in Age Determination." *Rep. Int. Whal. Commn. Spec. Issue 3:* 1–229.

Perrin, W. F., and S. B. Reilly (1984). "Reproductive Parameters of Dolphins and Small Whales of the Family Delphinidae." *Rep. Int. Whal. Comm. Spec. Issue 6:* 97–134.

Perry, E. A. (1993). "Mating System of Harbour Seals. Ph.D. thesis, Memorial University, Newfoundland.

Perryman, W. L., and M. S. Lynn (2002). "Evaluation of Nutritive Condition and Reproductive Status of Migrating Gray Whales *(Eschrichtius robustus)* Based on Analysis of Photogrammetric Data." *J. Cetacean Res. Manage. 4:* 155–164.

Pistorius, P. A., and M. N. Bester (2002). "A Longitudinal Study of Senescence in a Pinniped." *Can. J. Zool. 80:* 395–401.

Pitcher, K. W., D. G. Calkins, and G. W. Pendleton (1998). "Reproductive Performance of Female Steller Sea Lions: An Energetic-Based Reproductive Strategy?" *Can. J. Zool. 76:* 2075–2083.

Ralls, K., D. B. Siniff, T. D. Williams, and V B. Kuechle (1989). "An Intraperitoneal Radio Transmitter for Sea Otters." *Mar Mamm. Sci. 5:* 376–381.

Ramsay, M.A. and I. Stirling (1988). "Reproductive Biology of Female Polar Bears *(Ursus maritimus)*." *J. Zool. 214:* 601–634.

Ray, G. C., E. D. Mitchell, D. Wartzok, V. M. Kozicki, and R. Maiefski (1978). "Radio Tracking of a Fin Whale *(Balaenoptera physalus)*." *Science 202:* 521–524.

Reeves RR. (1998). "Distribution, Abundance and Biology of Ringed Seals *(Phoca hispida):* An Overview." *NAMMCO Sci. Publ. 1:* 9–45.

Reid, J. P., R. K. Bonde, and T. J. O'Shea (1995). Reproduction and mortality of radio-tagged and recognizable manatees on the Atlantic coast of Florida *In* "Population Biology of the Florida Manatee" (T. J. O'Shea, B. B. Ackerman, and H. F. Percival, eds.), pp. 171–190. Inf. Tech. Rep. No.1, Natl. Biol. Ser., Washington, DC.

Reynolds, J. E., III, D. P. DeMaster, and G.K. Silber (2002). Endangered species and populations *In* "Encyclopedia of Marine Mammals" (W. F. Perrin, B. Würsig, and J. G. M. Thewissen, eds.), pp. 373–382. Academic Press, San Diego, CA.

Reynolds, J. E., III, and D. K. Odell (1991). *Manatees and Dugongs.* Facts on File, New York.

Reynolds, J. E., III, and J.A. Powell (2002). Manatees *(Trichechus manatus, T. senegalensis,* and *T. inunguis). In* "Encyclopedia of Marine Mammals" (W. F. Perrin, B. Würsig, and J. G. M.Thewissen, eds.), pp. 709–720. Academic Press, San Diego, CA.

Rice, D. W. (1986). Gray whale. *In* "Marine Mammals of Eastern North Pacific and Arctic Waters" (D. Haley, ed.), pp. 62–71. Pacific Search Press, Seattle, Washington.

Rice, D. W., and A. Wolman (1971). "The Life History and Ecology of the Gray Whale *(Eschrichtius robustus)*." *Am. Soc. Mammal. Spec. Pub. 3:* 1–142.

Richard, K. R., M. C. Dillon, H. Whitehead, and J. M. Wright (1996). "Patterns of Kinship in Groups of Free-Living Sperm Whales *(Physeter macrocephalus)* Revealed by Multiple Molecular Genetic Analyses." *Proc. Natl. Acad. Sci. U. S. A. 93:* 8792–8795.

Richards, F. J. (1959). "A Flexible Growth Function for Empirical Use." *J. Exp. Botany 10:* 290–300.

Riedman, M. (1990). *The Pinnipeds: Seals, Sea Lions, and Walruses.* University of California Press, Berkeley, CA.

Riedman, M. L., and J. A. Estes (1990). The sea otter *(Enhydra lutris):* Behavior, ecology and natural history. Biol. Rep. No. 90(14). US. Fish and Wildl. Ser., Washington, DC.

Roe, H. S. J. (1967). "Rate of Lamina Formation in the Ear Plug of the Fin Whale." *Nor. Hvalfangst-Tid. 56:* 41–45.

Rosel, P. E., A. E. Dizon, and M. G. Haygood (1995). "Variability in the Mitochondrial Control Region in Populations of the Harbor Porpoises, *Phocoena phocoena,* on Interoceanic and Regional Scales." *Can. J. Fish. Aquat. Sci. 52:* 1210–1219.

Rosel, P.E., and L. Rojas-Bracho (1999). "Mitochondrial DNA Variation in the Critically Endangered Vaquita *Phocoena sinus* Norris and MacFarland, 1958." *Mar. Mamm. Sci. 15:* 990–1003.

Scheffer, V. B. (1950). "Growth Layers on the Teeth of Pinnipedia as an Indication of Age." *Science 112:* 309–311.

Seber, G. A. F. (1982). *The Estimation of Animal Abundance and Related Parameters,* 2nd ed. Macmillan, New York.

Sergeant, D. E. (1966). "Reproductive Rates of Harp Seals, *Pagophilus groenlandicus* (Erxleben)." *J. Fish. Res. Bd. Canada 23:* 757–766.

Siniff,D. B., D. P. DeMaster, R. J. Hofman, and L. L. Eberhardt. (1977). "An Analysis of the Dynamics of a Weddell Seal Population." *Ecol. Monogr. 47:* 319–335.

Siniff, D. B. and K. Ralls (1991). "Reproduction, Survival and Tag Loss in California Sea Otters." *Mar. Mamm. Sci. 7:* 211–229.

Sipila, T., and H. Hyvärinen (1998). "Status and Biology of Saimaa *(Phoca hispida saimensis)* and Ladoga *(Phoca hispida lodogensis)* Ringed Seals." *NAMMCO Sci. Publ. 1:* 46–62.

Slade, T. W. (1992). "Limited MHC Polymorphism in the Southern Elephant Seal: Implications for MHC Evolution and Marine Mammal Population Biology." *Proc. R. Soc. Lond. B 249:* 163–171.

Spilliaert, G. Vikingsson, U. Arnason, A. Palsdottir, J. Sigurjonsson, and A. Arnason (1991). Species Hybridization Between a Female Blue Whale *(Balaenoptera musculus)* and a Male Fin Whale *(B. physalus):* Molecular and Morphological Documentation." *J. Hered. 82:* 269–274.

Stanley, H., F. S. Casey, J. M. Carnahan, S. Goodman, J. Harwood, and R. K. Wayne (1996). "Worldwide Patterns of Mitochondrial DNA Differentiation in the Harbor Seal *(Phoca vitulina)." Mol. Biol. Evol. 13:* 368–382.

Stenson, G. B., R. A. Myers, M. O. Hammill, I. Ni, W. G. Warren, and M.C. S. Kingsley (1993). "Pup Production of Harp Seals, *Phoca groenlandica*, in the Northwest Atlantic." *Can. J. Fish. Aquat. Sci. 50:* 2429–2439.

Stevick, P. T., B. J McConnell, and P. S. Hammond (2002). Patterns of movement. *In* "Marine Mammal Biology: An Evolutionary Approach" (A. R. Hoelzel, ed.), pp. 185–216. Blackwell Publishing, Oxford, UK.

Stevick, P. T., P. J. Palsbøll, T. D. Smith, M. V. Bravington, and P. S. Hammond. 2001. "Errors in Identification Using Natural Markings: Rates, Sources and Effects on Capture-Recapture Estimates of Abundance." *Can. J. Fish Aquat. Sci. 58:* 1861–1870.

Stewart, B. S., S. Leatherwood, P. K. Yochem, and M.-P. Heide-Joergensen (1989). "Harbor Seal Tracking and Telemetry by Satellite." *Mar. Mamm. Sci. 5:* 361–375.

Stirling, I. (1988). *Polar Bears.* University of Michigan Press, Ann Arbor.

Stone, G., J. Goodyear, A. Hutt, and A. Yoshinaga (1994). "A New Non-Invasive Tagging Method for Studying Wild Dolphins." *Mar. Technol. Soc. J. 28:* 11–16.

Sumich, J. L. (1986a). "Latitudinal Distribution, Calf Growth and Metabolism, and Reproductive Energetics of Gray Whales, *Eschrichtius robustus*." Ph.D. dissertation. Oregon State University, Corvallis.

Sumich, J. L. (1986b). "Growth in Young Gray Whales *(Eschrichtius robustus)." Mar. Mamm. Sci. 2:* 145–152.

Sumich, J. L., and J. T. Harvey (1986). "Juvenile Mortality in Gray Whales *(Eschrichtius robustus)." J. Mammal. 67:* 179–182.

Suydam, R. S., L. F. Lowry, K. J. Frost, G. M. O'Corry-Crowe, and D. Pikok (2001). "Satellite Tracking of Eastern Chukchi Sea Beluga Whales into the Arctic Ocean." *Arctic 54:* 237–243.

Sydeman, W., and J. N. Nur (1994). Life history strategies of female northern elephant seals. *In* "Elephant Seals" (B. J. Le Boeuf, and R. M. Laws, eds.), pp. 137–153. University of California Press, Berkeley.

Testa, J. W., and P. Rothery (1992). "Effectiveness of Various Cattle Ear Tags for Weddell Seals." *Mar. Mamm. Sci. 8:* 344–353.

Thomas, J. A., L. H. Cornell, B. E. Joseph, T. D. Williams, and S. Dreischman (1987). "An Implanted Transponder Chip Used as a Tag for Sea Otters *(Enhydra lutris)*." *Mar. Mamm. Sci. 3:* 271–274.

Thompson, P. M., and J. Harwood (1990). "Methods for Estimating the Population Size of Common Seals *Phoca vitulina.*" *J. Appl. Ecol. 27:* 924–938.

Tricas, T. C., and J. E. Mc Cosker (1984). "Predatory Behavior of the White Shark *(Carcharodon carcharias)*." *South. Calif. Acad. Sci. Mem. 9:* 81–91.

Trujillo, R. G., T. R. Loughlin, N. J. Gemmell, J. C. Patton, and J. W. Bickham (2004). "Variation in Microsatellites and mtDNA Across the Range of the Steller Sea Lion, *Eumetopias jubatus.*" *J. Mamm. 85:* 338–346.

Wada, S., and K. I. Numachi (1991). "Allozyme Analysis of Genetic Differentiation Among the Populations and Species of the *Balaenoptera.*" *Rep. Intern. Whal. Commn., Spec. Issue 13:* 125–154.

Wade, P. R. 2002. Population dynamics. *In* "Encyclopedia of Marine Mammals" (W. F. Perrin, B. Würsig, and J. G. M. Thewissen, eds.), pp. 974–982. Academic Press, San Diego, CA.

Walker, B. G., and P. L. Boveng (1995). "Effects of Time-Depth Recorders on Maternal Foraging and Attendance Behavior of Antarctic Fur Seals *(Arctocephalus gazella)*." *Can. J. Zool. 73:* 1538–1544.

Watkins, W. A., M. A. Daher, N. A. DiMarzio, A. Samuels, D. Wartzok, K. M. Fristrup, P. W. Howey, and R. R. Maiefski. (2002). "Sperm Whale Dives Tracked by Radio Tag Telemetry." *Mar. Mamm. Sci. 18:* 55–68.

Watkins, W. A., M. A. Daher, K. M. Fristrup, T. J. Howald, and G. N. Disciara (1993). "Sperm Whales Tagged with Transponders and Tracked Underwater by Sonar." *Mar. Mamm. Sci. 9:* 55–67.

Watkins, W. A., and W. E. Schevill (1977). *The Development and Testing of a Radio Whale Tag,* Ref. No. 77-58. WHOI, Woods Hole, MA.

Weber, D. S., B. S. Stewart, J. C. Garza, H. T. Prins (2000). "An Empirical Genetic Assessment of the Severity of the Northern Elephant Seal Population Bottleneck." *Curr. Biol. 10:* 1287–1290.

Weslawski, J. M., L. Hacquebord, L. Stempniewicz, and M. Malinga (2000). "Greenland Whales and Walruses in the Svalbard Food Web Before and After Exploitation." *Oceanologia 42:* 37–56.

Westgate, A. J., and A. J. Read (1998). "Applications of New Technology to the Conservation of Porpoises." *Mar. Techn. Soc. J. 32:* 70–81.

Westlake, R. L., and W. G. Gilmartin (1990). "Hawaiian Monk Seal Pupping Locations in the Northwestern Hawaiian Islands." *Pac. Sci. 44:* 366–383.

Whitehead, H. (1996). "Babysitting, Dive Synchrony, and Indications of Alloparental Care." *Behav. Evol. Sociobiol. 38:* 237–244.

Whitehead, H. (2001). "Direct Estimation of Within-Group Heterogeneity in Photo-Identification of Sperm Whales." *Mar. Mamm. Sci. 17:* 718–728.

Whitehead, H. (2004). *Sperm Whales.* University of Chicago Press, Chicago.

Willis, P. M., B. J. Crespi, L. M. Dill, R. W. Baird, and M. B. Hanson (2004). "Natural Hybridization Between Dall's porpoises *(Phocoenoides dalli)* and the Harbour Porpoises *(Phocoena phocoena)*." *Can. J. Zool. 82:* 828–834.

Wright, I. E., S. D. Wright, and J. M. Sweat (1998). "Use of Passive Integrated Transponder (PIT) Tags to Identify Manatees *(Trichechus manatus latirostris)*." *Mar. Mamm. Sci. 14:* 641–645.

Wynen, L. P., S. D. Goldsworthy, S. M. Insley, M. Adams, J. W. Bickham, J. Francis, J. P. Gallo, A. R. Hoelzel, P. Majluf, R. W. G. White, and R. Slade (2001). "Phylogenetic Relationships Within the Eared Seals (Otariidae:Carnivora), with Implications for the Historical Biogeography of the Family." *Mol. Phylogenet. Evol. 21:* 270–284.

Wynen, L. P., S. D. Goldsworthy, C. Guinet, M. N. Bester, I. L. Boyd, G. J. G. Hofmeyr, R. G. White, and R. Slade (2000). "Postsealing Genetic Variation and Population Structure of Two Species of Fur Seal *(Arctocephalus gazella* and *A. tropicalis)*." *Mol. Ecol. 9:* 299–314.

Yazdi , P. (2002). "A Possible Hybrid Between the Dusky Dolphin *(Lagenorhynchus obscurus)* and the Southern Right Whale Dolphin *(Lissodelphis peronii)*." *Aquat. Mamm. 28:* 211–217.

York, A. E., and P. Kozloff. (1987). "On the Estimation of Numbers of Northern Fur Seal, *Callorhinus ursinus,* Pups on St. Paul Island." *Fish. Bull. 85:* 367–375.

15

Exploitation and Conservation

15.1. Introduction

In this final chapter we consider both the commercial exploitation and conservation of marine mammals. The overexploitation of whales, seals, sea cows, and sea otters has resulted in seriously reduced population sizes, the extinction of several species, and the endangered status of several others. Recognition of marine mammals as crucial natural resources and valued ecosystem components that require protection has resulted in the establishment of several international legal frameworks for their conservation. This framework monitors many human-induced marine mammal mortalities, such as those resulting from net entanglements, incidental fisheries takes, and environmental contaminants. Consideration is also given to the effect of ecotourism activities on marine mammals. Finally, we evaluate our progress at conserving and protecting marine mammals and set goals for the future.

15.2. Commercial Exploitation of Marine Mammals

Marine mammals have been exploited for centuries for subsistence purposes by coastal aboriginal human populations on all continents except Antarctica. Pinnipeds at rookery or haul-out sites have been especially vulnerable to hunters foraging for food, clothing, and shelter materials. Midden remains of seals, sea lions, and near-shore cetaceans date back as far as 8500 years (Krupnik, 1984; O'Leary, 1984; Beland, 1996). Historical evidence indicates that cetaceans were abundant in the North Sea and in the English Channel during the Middle Ages and earlier. Right whales and perhaps gray whales were likely hunted regularly in this area from at least the 9th century onward; harbor porpoises and other small cetaceans were hunted in this area before the 16th century (Smet, 1981). The impact of aboriginal hunting pressure on marine mammal populations was apparently localized and small in scale in comparison to later industrialized commercial hunting.

456

15.2.1. Historical Practices

The history of commercial exploitation of marine mammal populations, including whaling, sealing, or ottering, represent tragic examples of how not to manage renewable living resources. Beginning with the Basques in the 12th century, who took North Atlantic right whales, and late 16th century Japanese, who took coastal gray whales (Omura, 1984), the story of "commercial" exploitation of marine mammals has been one of repeated overharvesting and serious reductions in population sizes. Over-exploitation is in part responsible for the extinction of the Steller's sea cow, the Caribbean monk seal, and the Atlantic gray whale, and for the endangered status of several other species including the northern right whale, the Atlantic bowhead population, and nearly all populations of manatees (Hofman, 1995), dugongs, and river dolphins.

Sea otters, all species of fur seals, harp and hooded seals, both species of elephant seals, and walruses were heavily harvested for skins and oil beginning in the early 19th century (Scammon, 1874; Busch, 1985; Figure 15.1), but these hunters left few records of the numbers of animals taken. By the late 19th century, California sea otters and northern elephant seals were thought to be extinct. Other species, including southern hemisphere fur seals, were depleted (Shaughnessy, 1982; Riedman, 1990).

In the early 19th century, "Yankee" whalers from New England extended their reach into the Pacific and Indian Oceans in a widening search for sperm, right, bowhead, and gray whales that swam slowly and could be killed with hand harpoons and lances (Figure 15.2). By the end of the 19th century, right whales in both hemispheres, as well as bowhead and gray whales, had been severely depleted. The North Atlantic right whale remains the most endangered of the great whales, numbering fewer than 300 animals, despite almost complete protection for over 100 years (Donovan, 1995). In 1851 in *Moby Dick,* Herman Melville questioned the future that all whales faced ". . . the thousand harpoons and lances darting along all the continental coasts; the moot point is, whether Leviathan can long endure so wide a chase, and so remorseless a havoc" (Melville, 1851).

15.2.2. Current Practices

The era of modern commercial whaling was initiated in the 1860s with the invention of the cannon-fired, explosive-head harpoon. In combination with the development of faster steam-powered catcher boats, whalers could for the first time take large num-

Figure 15.1.   Eighteenth century Aleutian Island hunters chasing sea otters. (From Scammon, 1874.)

Figure 15.2.   Nineteenth century whaling scene with bowhead whales being taken with shoulder-fired bomb lance gun. (From Scammon, 1874.)

bers of the faster swimming rorqual whales (initially blue and fin whales). These whales had previously been ignored by whalers with hand harpoons, because they were too fast to be overtaken in sail- or oar-powered boats. At first, whaling for rorquals was conducted from land stations and the number that could be exploited was limited. The first Antarctic whaling station was established in 1904 on South Georgia; 195 whales were taken that year. By 1913, there were 6 land stations and 21 floating factories and the total catch was 10,760 whales (Donovan, 1995).

The invention of the stern slipway in 1925 allowed pelagic factory ships to haul harpooned whales aboard for processing at sea (Figure 15.3) rather than depending on shore stations and the kill of large rorquals rose dramatically. From a harvest of 176 blue whales in 1910, the annual take climbed to over 37,000 whales (mostly blues) by 1931, when 41 factory ships were working in Antarctic waters (Figure 15.4). After this peak year blue whales became increasingly scarce, and catches declined steadily until the catch was commercially insignificant by the mid-1950s. In 1966, only 70 blue whales were killed in the entire world's oceans. Only then, when substantial numbers could no longer turn a profit, was the hunting of blue whales banned in the southern hemisphere.

Pelagic whaling operations effectively ceased during World War II. The war caused a world shortage in the supply of whale oil, which encouraged several nations to renew their pelagic whaling activities at the end of World War II. It was in this immediate post-war setting that discussions were held in London in 1945 and in Washington in 1946 on the international regulation of whaling (the results of these discussions are outlined in the next section). Meanwhile, without large populations of blue whales to exploit,

Figure 15.3. Stern view of an Antarctic whaling factory ship *(Southern Venturer)* showing slipway for pulling harpooned whales to the processing deck. (Courtesy of Southhampton Oceanography Centre.)

pelagic whalers switched to the smaller, more numerous fin whales, catches of which sky-rocketed to over 25,000 whales each year for most of the 1950s (see Figure 15.4; Laws, 1962). By 1960, the fin whale catch began to plummet, and whaling pressure was diverted to the even smaller sei whale. The total southern hemisphere sei whale population

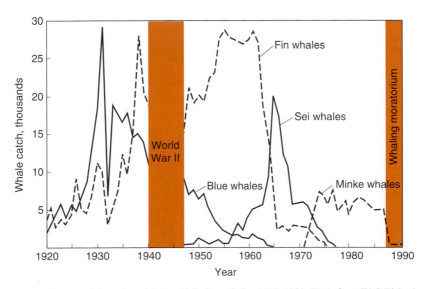

Figure 15.4. Commercial catches of Antarctic baleen whales, 1920–1990. (Data from FAO Fisheries Catch and Landing Statistics.)

probably never exceeded 60,000; one third of that population was killed in 1965 alone. By the late 1960s, sei whales had followed their larger relatives to commercial extinction, and the whaling effort shifted to the even smaller minke whale. Minke whales remained the principal target species of pelagic whalers until pelagic harvesting of baleen whales ceased during a moratorium initiated in 1986 (discussed later).

Some harvests of cetaceans continue at present, including the Faroe Island drive fisheries of pilot whales and Japanese harvests of small coastal odontocetes, and sperm whales and three species of mysticetes that are taken under International Whaling Commission (IWC)-sanctioned "scientific research whaling" under permits issued by the Japanese government (Clapham *et al.*, 2003). Norway has a commercial harvest of minke whales, in which some hundreds of animals are taken annually. Pinniped populations are not immune to commercial pressures either. The Namibian government continues to harvest Cape fur seals, primarily as a fisheries protection measure (Wickens, 1995). Commercial harvests take place in Canada for harp and hooded seals, and grey seal hunting is encouraged by a variety of local and governmental subsidy programs. For harp seals, the Canadian quota was increased in 2003 to 350,000 animals (2003–2005 total quota of 975,000, with annual quotas up to 350,000 in 2 of the 3 years). This quota is above the calculated sustainability limits for this population. This management strategy was initiated to reduce the harp seal population because it is assumed that harp seals are a significant factor impeding the recovery of Atlantic cod stocks in the western North Atlantic (Marine Mammal Commission, 2003; DFO, 2003). Norway and Russia also have commercial hunts for harp and hooded seals, but at levels well below currently calculated sustainable limits.

15.3. Legal Framework for Marine Mammal Conservation and Protection

15.3.1. International

The United States has entered into agreements regarding the conservation or protection of marine mammals with more than 30 countries. Additionally, several other international agreements and conventions have been established that involve conservation of marine mammal populations or their habitats. Some of the more important ones are listed with their date of enactment in Table 15.1.

Before 1900, several agreements were made to govern the taking of seals and fur seals, mostly regarding land-based operations in the Antarctic. In 1911, the International Convention on Pribilof fur seals banned the pelagic harvesting of fur seals, although harvesting of young bachelor males continued on land for much of the 20th century (Gentry and Kooyman, 1986). After the fur seal Convention lapsed in 1984, commercial harvesting of northern fur seals effectively ceased (Bonner, 1994).

The post-World War II history of modern whaling is essentially that of the history of the IWC. In 1946, the International Convention for the Regulation of Whaling was signed, and in 1948 it was ratified, establishing the IWC. This convention set a precedent for international regulation of natural resources by giving conservation equal billing with the economics of whaling. Its stated aim was "to provide for the proper conservation of whale stocks and thus make possible the orderly development of the whaling industry." The IWC was originally established to regulate Antarctic whaling, to make

Table 15.1. Some International Agreements and Treaties Affecting Marine Mammals, with Their Date of Inception or Ratification (Wallace, 1994 and Hedley, 2000[*])

1911	International Convention on Pribilof fur seals
1935	League of Nations Convention for the Regulation of Whaling
1940	Convention on Nature Protection and Wild Life Preservation in the Western Hemisphere
1946	International Convention for the Regulation of Whaling
1949	Convention for the Establishment of an Inter-American Tropical Tuna Commission
1957	Interim Convention on Conservation of North Pacific Fur Seals
1958	Convention on the High Seas
	Convention on the Territorial Sea and the Contiguous Zone
	Convention on Fishing and Conservation of the Living Resources of the High Seas
	Convention on the Continental Shelf
1959	The Antarctic Treaty
1964	Convention for the International Council for the Exploration of the Sea
1969	International Convention on Civil Liability for Oil Pollution Damage
1970	United Kingdom, Conservation of Seals Act
1972	Convention for the Conservation of Antarctic Seals
1973	Convention on International Trade in Endangered Species of Wild Fauna and Flora (CITES)
	International Convention for the Prevention of Pollution from Ships
	Agreement on the Conservation of Polar Bears
1974	Bern Convention—Convention on the Conservation of European Wildlife and Natural Habitats
1979	Convention on the Conservation of Migratory Species of Wild Animals
1980	Convention on the Conservation of Antarctic Marine Living Resources
1982	United Nations Convention on the Law of the Sea
1983	Eastern Pacific Ocean Tuna Fishing Agreement
1989	Convention on the Prohibition of Fishing with Long Driftnets in the South Pacific
1989–1991	United Nations General Assembly Resolutions on Large-Scale Pelagic Driftnet Fishing and Its Impacts on the Living Marine Resources of the World's Oceans and Seas (Non-binding)
1990	Agreement on the Conservation of Seals in the Wadden Sea
	International Convention on Oil Pollution Preparedness, Response and Co-operation
1992	Convention on Biological Diversity
	Agreement on the Conservation of Small Cetaceans of the Baltic and North Seas
	Agreement on Cooperation in Research, Conservation and Management of Marine Mammals in the North Atlantic
	Agreement to Reduce Dolphin Mortality in the Eastern Tropical Pacific Tuna Fishery
1992	OSPAR—Oslo and Paris Convention for the Protection of the Marine Environment of the North-East Atlantic

[*]See Hedley (2000) for a brief overview of the prodigious amount of information now available on the Internet relating to the international law of the sea, international fisheries, and marine mammals.

recommendations on quotas and minimum sizes, and to ban the taking of females with calves. It has since been extended to cover all commercial pelagic whaling activities of member nations and it serves as a venue for discussions of coastal whaling issues and small cetacean conservation practices as well. The IWC recognized the need for scientific advice and established an advisory Scientific Committee the members of which are nominated by member governments. The IWC was intended to "encourage, recommend, and organize" studies and to "collect, analyze, study, appraise, and disseminate" information (Article IV); however, it lacked both inspection and enforcement powers (Wallace, 1994). Any member government could object to any decision with which it did not agree and excuse itself from the limitations of that decision. This, along with the right of member nations to unilaterally issue permits to catch whales for scientific purposes, has severely limited the authority of the IWC to enforce its own recommendations and regulations.

Early IWC management procedures were based on the use of the **Blue Whale Unit** (BWU) as a means of setting catch quotas. In terms of oil yield, it was considered that one blue whale was equal to 2 fin, 2.5 humpback, or 6 sei whales (based on their relative oil yields; Figure 15.5; Andresen, 1993), with no distinction given to the actual species being harvested. In 1962–1963, a call from the Scientific Committee for quotas by species was rejected but the total quota was reduced from 15,000 to 10,000 BWUs and humpback whales were protected throughout the southern hemisphere. In 1964–1965, blue and humpback whales were given complete protection and Norway, the Netherlands, and the United Kingdom ceased their Antarctic whaling operations, leaving only Japan and the Soviet Union actively engaged in the pelagic harvesting of baleen and sperm whales (Tonnessen and Johnsen, 1982).

Reductions in target population sizes, new national and international regulations, and changing market demands, as well as changing attitudes in many nations concerning the ethics of killing large whales for profit, all contributed to the demise of large-scale pelagic whaling. A new IWC management procedure was instituted in 1974, based on the recognition that management procedures should apply to geographically localized stocks rather than to species. This new procedure also recognized a protected-stock category that permitted no harvesting. In 1979, a proposal to end pelagic whaling for all species except minke whales was adopted and a sanctuary was declared for the Indian Ocean outside the Antarctic. By 1982, the IWC had agreed to a deferred pause (a "moratorium") in all commercial whaling beginning in 1986, with a responsibility for conducting a comprehensive assessment of large whale stocks by 1990. Japan and Norway both continued to kill whales under scientific research permits, although the IWC passed several resolutions calling on Norway and Japan to cease their whaling operations. Following stock assessments of minke whales and other large baleen whales in the North East Atlantic, Norway objected formally to the moratorium on all whaling and resumed commercial exploitation of minke whales in the North Atlantic in 1993, although the moratorium had not and still has not been lifted (Hofman, 1995).

In 1994, the IWC accepted a revised procedure for estimating the number of whales that could be taken without causing the affected population to be reduced below its max-

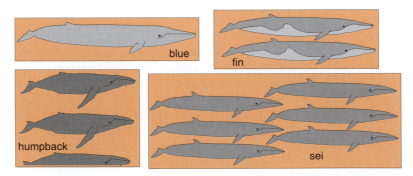

Figure 15.5.   BWU equivalents for four species of baleen whales: I BWU = I blue whale = 2 fin whales = 2.6 humpback whales = 6 sei whales.

imum net productivity. This **revised management procedure (RMP)** has been evaluated using computer simulations, but if commercial whaling is resumed it is clear that independent monitoring of the affected population will be required to verify that RMP is working (Hofman, 1995).

The 1973 Convention on International Trade in Endangered Species of Wild Fauna and Flora (CITES) created a mechanism to regulate the international trade of endangered plant and animal species, including numerous species of marine mammals. Species listed in CITES Appendix I include all species deemed to be threatened with extinction that are, or may be, affected by international trade policies or activities. Species listed in CITES Appendix II include species that, although not currently threatened with extinction, may become so unless international trade is restricted or regulated by participating governments (Wallace, 1994). Marine mammal species currently listed on either CITES Appendix I or II are listed in Table 15.2. Despite cooperation that has resulted from resolutions adopted by CITES and the IWC, incidents of illegal trade in whale products (i.e., meat and blubber) have continued to take place. However, there are now techniques to recognize the source species and even the stock-identity in many cases of such tissues. Among the recently developed techniques that allow fast and reliable identification of commercial whale products for sale in retail markets is liquid chromatographic analysis of water-soluble proteins in skeletal muscles (Ukishima *et al.*, 1995, and references cited therein). Molecular genetic analyses of whale products found in retail market places in Japan and Korea has revealed a surprising diversity of whale and dolphin species (Baker and Palumbi, 1994; Baker *et al.*, 1996; Baker *et al.*, 2002). These same molecular techniques have been used to make species identifications of beaked whales incidentally taken in the California drift gillnet fishery (Henshaw *et al.*, 1997).

Table 15.2. Current Status of Marine Mammal Species Listed as Endangered or Threatened by the U.S. Endangered Species Act (ESA), the U.S. Marine Mammal Protection Act (MMPA), or the Convention on International Trade in Endangered Species of Wild Fauna and Flora (CITES) (Braham, 1992; Marine Mammal Commission, 1997)

	Pop. Size	% of Initial/ Trend	Legal Status
Pinnipeds			
Hawaiian monk seal	1200–1800	decr.	Endangered (ESA); App. I (CITES)
Mediterranean monk seal	(450–525)	na	Endangered (ESA); App. I (CITES)
Southern elephant seal	na	na	App. II (CITES)
Siamaa seal	200–250	na	Endangered (ESA); App. I (CITES)
Steller sea lion	87,000	25%/decr.	Endangered (ESA); App. I (CITES)
N. fur seal	1,020,000	decr.	Depleted-MMPA; App. II (CITES)
Guadalupe fur seal	na	na	Endangered (ESA); App. I (CITES)
All other fur seals	na	na	App. II (CITES)
Cetaceans			
Right whales	>2000	2%	Endangered (ESA); App. I (CITES)
No. Pacific	200		
Bowhead whale	<600	<15%	Endangered (ESA); App. I (CITES)
Pygmy right whale	na	na	

Blue whale	<9000	<5%	Endangered (ESA); App. I (CITES)
N. Atlantic Ocean	100–555	<50%	
N. Pacific Ocean	1600	33%	
N. Indian Ocean	na	na	
S. Indian Ocean	5000	50%	
Antarctic	450–1900	<1%	
Humpback whale	>10,000	8%	Endangered (ESA); App. I (CITES)
Sei whale	25,000	<24%	Endangered (ESA); App. I (CITES)
Fin whale	119,000	<26%	Endangered (ESA); App. I (CITES)
Minke whale	725,000	na	App. I (CITES); W. Greenland-App.II
Bryde's whale	90,000	na	App. I (CITES)
Gray whale	22,000	100%/incr.	removed ESA; App. I (CITES)
East N. Pacific	22,000	100%/incr.	removed ESA; App. I (CITES)
West N. Pacific	<100	na	Endangered (ESA): App. I (CITES)
Sperm whale	1,810,000	65%	Endangered (ESA): App. I (CITES)
N. Atlantic Ocean	190,000	58%	
N. Pacific Ocean	930,000	74%	
S. Oceans	780,000	66%	
Chinese R. dolphin	100–300	na	
Indus and Ganges Susu	na	na	Endangered (ESA): App. I (CITES)
Beaked whales	na	na	Endangered (ESA): App. I (CITES)
Bottlenose whale	na	na	App. I (CITES)
Tucuxi	na	na	App. I (CITES)
Humpbacked dolphins	na	na	App. I (CITES)
Finless porpoise	na	na	App. I (CITES)
Vaquita	<500	na	Endangered (ESA) : App. I (CITES)
Sirenians			
Dugong	na	na	App. I (CITES)
Amazonian manatee	na	na	App. I (CITES)
West Indian manatee	2300 (U.S. pop.)	na	App. I (CITES)
West African manatee	na	na	App. I (CITES)
Other marine mammals			
Polar bear	na	na	App. I (CITES)
California sea otter	2200	na	App. I (CITES)

na = not available, decr. = decreasing, incr. = increasing. Population figures in parentheses are rough estimates.

The Convention for the Conservation of Antarctic Marine Living Resources (CCAMLR), ratified in 1980, was the first international agreement to approach the management of marine resources from an ecosystem perspective. Commercial ventures in the Antarctic, including whale, fish, or krill harvesting, must be consistent with the goals of the Convention, namely the "maintenance of the ecological relationships between harvested, dependent, and related populations of Antarctic marine living resources and the restoration of depleted populations" (Beddington and May, 1982). This was the first in a slowly growing trend of national and international efforts to recognize the integrity of large marine ecosystems (LMEs). To date, about 50 LMEs have been identified, mostly in coastal waters, as ecological systems that respond to external stresses such as overexploitation, environmental fluctuations (such as El Niño-Southern Oscillation), or large-scale pollution problems (Alexander, 1993).

In 1982, the United Nations Conference on the Law of the Sea adopted a draft "Law of the Sea (LOS) Treaty" that was to go into effect 12 months after ratification (Craven and Schneiden, 1989). Over 2 decades later, ratification of the treaty has still not taken place; it requires support from 60 nations to enter into force. However, many nations have unilaterally adopted major elements of the LOS Treaty. The LOS treaty includes many of the key features found in the United States Fisheries Conservation and Management Act. In particular, it internationalizes the concept of 200-mile-wide (324 km) exclusive economic zones (EEZs), granting coastal nations sovereign rights with respect to natural resources (including fishing), scientific research, and environmental preservation. Additionally, the LOS Treaty obliges nations to prevent or control marine pollution, to promote the development and transfer of marine technology to developing nations, and to settle peacefully disputes arising from the exploitation of marine resources.

The imposition of 200-mile-wide EEZs by essentially all coastal nations of the world since 1982 has dramatically changed the assumption of responsibility for management and conservation of marine mammals because most species of marine mammals spend all, or at least a critical portion of their life cycles (either for feeding or for reproduction) within 200 miles of some nation's shoreline. The LOS Treaty is notably silent regarding the Antarctic upwelling area, for national claims to territory on the Antarctic continent are not recognized. Consequently, the CCAMLR provides the fundamental legal framework for management issues regarding Antarctic marine mammal species.

15.3.2. United States

The **Marine Mammal Protection Act** (MMPA) of 1972 established a moratorium on the taking of marine mammals in U.S. waters and on importing marine mammals and marine mammal products into the United States. It applies to the activities of U.S. citizens and U.S. flagged vessels in U.S. territorial seas or its EEZ. The moratorium does not apply to Indians, Aleuts, or Inuits in coastal Alaska who hunt marine mammals for subsistence (see later) or for making and selling handicrafts. Under a permit system, the act also allows the taking and importing of marine mammals for scientific research, for education and public display, and for incidental catches occurring in the course of commercial fishing operations, such as the purse-seine tuna fishery (Hofman, 1989). In addition to its regulatory aspects, the MMPA established the U.S. Marine Mammal Commission (consisting of three scientists appointed by the President), a Committee of Scientific Advisors on Marine Mammals (appointed by the Commission), and a Marine Mammal Health and Stranding Response Program, which includes the marine mammal stranding networks discussed later.

The MMPA and the U.S. Endangered Species Act (ESA) of 1973 were the first U.S. legislative acts to recognize the values of nonconsumptive uses of protected species. Protected species were listed as either *endangered* (in danger of becoming extinct over a significant portion of its range) or *threatened* (likely to become endangered). A threatened population is automatically considered *depleted* under the MMPA. A depleted population is defined as one that contains fewer individuals than its **optimum sustainable population (OSP)** level, where OSP has subsequently been interpreted as a population level between 60 and 100% of carrying capacity (Gerrodette and DeMaster, 1990). Criteria were established by the ESA for protection of threatened and endangered species, which included the effects of overutilization, habitat modification, or

destruction, disease or predation, inadequate regulations, or other natural or anthropogenic factors. All large commercially hunted whale species were initially assigned
endangered status (Braham, 1992). Only one, the Northeast Pacific gray whale population, has been removed from the list because of subsequent population recovery (*Federal
Register*, 1994).

The development of criteria to be used for delisting threatened species is a topic of
considerable current interest (e.g., Ralls *et al.*, 1996). The ESA specifies that the recovery
plan developed for each species shall include "objective, measurable criteria which, when
met, would result in a determination . . . that the species be removed from the list."
Although there have been several proposals to improve the ESA by providing quantitative guidance in the form of specific probabilities of extinction within some time frame,
experiences of the Southern Sea Otter Recovery Team indicated that these guidelines
should not be overly rigid and should allow flexibility for specific situations. This recovery team concluded that the most important consideration in a recovery plan is to
appoint a team of individuals that is both technically well qualified and unconstrained
by pressures from management agencies (Ralls *et al.*, 1996).

15.3.3. Native Subsistence Harvesting

The IWC establishes catch limits for native subsistence whaling in the United States and
other nations. In 2002, the IWC agreed to 2003–2007 catch limits for several whale stocks
subject to native subsistence whaling including 19 fin whales and 187 Minke whales
(taken by Greenland natives) and 4 humpback whales (taken by natives of St. Vincent
and the Grenadines), and 280 individuals from the Bering-Chukchi-Beaufort Sea stock
of bowhead whales (taken by Alaskan Inuit and native peoples of Chukotka).
Additionally, annual catches of up to 140 gray whales for subsistence purposes by the
Makah tribe of Washington and Russian natives were authorized (Marine Mammal
Commission, 2003). One gray whale was taken by the Makah in 1999, but none have
been taken since. The Makah tribe has been legally enjoined by U.S. authorities from
continued whaling until a new Endangered Species Environmental Impact Assessment
and a MMPA waiver, currently under litigation, are issued (see NOAA Fisheries Office
of Protected Resources website for current information regarding this issue).

15.4. Incidental Taking of Marine Mammals

15.4.1. Entanglements

The incidental, unintentional capture of cetaceans and pinnipeds in pelagic gill nets
(both set and drift types) and pelagic long-lines is a global problem, occurring wherever
marine mammals and nets occur together. Entanglement in trap or pot lines is also common (Perrin *et al.*, 1994). Little behavioral information exists regarding when and how
entanglements in nets occur, but it probably includes components of nondetection,
curiosity, and possibly social organization as well. The Scientific Committee of the IWC
estimates that numerous pinnipeds and as many as 300,000 cetaceans are killed annually
by entanglement in fishing gear (Burns and Wandesforde Smith, 2002).

Available evidence indicates that several stocks of small cetaceans likely cannot sustain the mortality rates occurring due to trap and passive net entanglement. These

include (with estimates of the number killed annually) the vaquita in the Gulf of California (30–40), the baiji in the Yangtze River, the Indo-Pacific humpbacked dolphin (7–8), bottlenose dolphins (30–40) of the South African Natal coast, Mediterranean Sea striped dolphins (5000–10,000), harbor porpoises of the western North Atlantic (5000–8600), eastern South Pacific dusky dolphins (1800–1900), North Pacific northern right whale dolphins (19,000), Mediterranean Sea sperm whales (20–30), and several thousand small Mediterranean Sea odontocetes (Perrin *et al.*, 1994; www.nmfs.noaa.gov).

In addition to entanglement in actively tended nets, an unknown (and probably unknowable) number of marine mammals are also killed by discarded or lost fishing gear (Figure 15.6). The shift since 1940 from the use of natural to synthetic fibers for nets, lines, and other fishing gear has led to a large increase in the quantity of lost and discarded fishing gear because these materials are so resistant to degradation. Gill net fisheries for squid in the central North Pacific Ocean, with over 1 million kilometers of nets set annually, constitute a large potential source of derelict gear. These nets, if lost, often continue to drift for years, taking a continuing toll on populations of marine mammals and birds.

Even after nets and lines drift ashore, these derelict materials can continue to kill. One survey of Antarctic fur seals on Bird Island, South Georgia, indicated that even on that remote island, nearly 1% of the seals were entangled in synthetic debris (Croxall *et al.*, 1990). Most of the entangled seals were wearing "neck collars" made of plastic strapping (59%) or fishing line and net material (29%; Figure 15.7). A large proportion of these seals exhibited signs of physical injury, and an unknown number of animals presumably

Figure 15.6.   Net entangled gray whale beached on the Oregon coast.

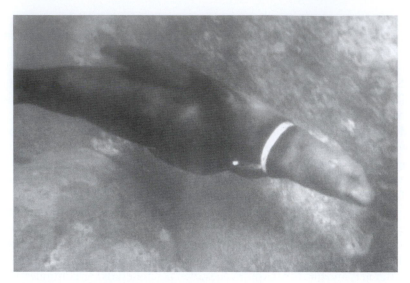

Figure 15.7.   Photo of sea lion with neck wrapped in line.

died after such entanglement. Between 1985 and 1995, the Marine Entanglement Research Program in the United States supported work to assess marine debris, including ways to monitor debris levels, inform the public about problems and solutions, reduce derelict fishing gear, and encourage international efforts to address marine debris pollution (Marine Mammal Commission, 1997; Walsh, 1998).

15.4.2. Tuna-Dolphin Interactions

The behavioral association between yellowfin tuna and several species of dolphins (particularly *Stenella attenuata, S. longirostris,* and *Delphinus delphis*) in the eastern tropical Pacific (ETP) is not well understood. However, it has long been used by the tuna purse seine fishery to locate and catch yellowfin tuna (Lennert and Hall, 1996); small quantities of skipjack and bigeye tuna are also taken in this way. When tuna seiners sight dolphin schools and yellowfin tuna, nets are set around all of the animals and the net circle is then tightened and closed, trapping both tuna and dolphins (Figure 15.8). Dolphins killed in these sets are easily counted; however, an additional unobserved mortality of dependent calves separated from their mothers is estimated to be about 10% of observed deaths (Archer *et al.*, 2004).

Although it was commercial whaling that initially brought about the enactment of the MMPA in the United States, it was the incidental takes of dolphins in the ETP tuna purse seine fishery that became symbolic of the problems confronting marine mammals. By 1970, 200,000–300,000 dolphins were being killed each year in tuna purse seine operations. Throughout the 1960s and into the early 1970s, the U.S. fleet dominated this fishery and was responsible for more than 80% of the dolphin mortality (Young *et al.*, 1993). Reduction of incidental mortality of dolphins in this fishery has been a primary focus of the MMPA since it was enacted in 1972. Initially, the MMPA gave commercial fisheries a 2-year exemption from the general permit system. During this period, U.S. and foreign vessels in the tuna purse seine fishery killed 424,000 dolphins in 1972 and 265,000 in 1973

Figure 15.8.   Modern tuna purse seine, with dolphins escaping (bottom of image) during backdown
                   procedure. (Courtesy of W. Perryman.)

(Marine Mammal Commission, 1991). In 1974, the National Marine Fisheries Service
(NMFS), which is responsible for administering the MMPA for these species, allowed
the U.S. tuna purse seine fleet to take an unlimited number of dolphins until December
31, 1975. This permit was challenged in court and was invalidated but the annual kill
exceeded 108,000 (Marine Mammal Commission, 1991).

   In 1977, NMFS issued a 3-year general permit with annual dolphin quotas set to
decline from 52,000 in 1978 to 31,000 in 1980. Accompanying this quota were requirements for onboard observers and several fishing gear modifications and procedural
restrictions such as the use of fine-mesh net panels and new back-down procedures (see

Figure 15.8) to release dolphins captured within nets (National Research Council, 1992). During these years, dolphin mortality dropped to less than 50% of the quota allotment (Marine Mammal Commission, 1991).

In 1980, NMFS issued another general permit through until 1984 including a total annual dolphin quota of 20,500. The permit also imposed additional observer requirements and other regulations governing fishing gear and techniques. By 1984 the U.S. tuna purse seine fleet had declined from 94 vessels in 1980 to approximately 42 vessels (IATTC, 1981). This decline was due, in part, to United States vessels moving to and fishing under the flags of other nations to evade MMPA regulations and to avoid the high operating and labor costs in the United States. The international tuna fleet, which is now larger than the U.S. fleet, began to contribute heavily to dolphin mortality in the ETP. In fact, the combined U.S. and foreign estimated kill levels for most of the 1980s exceeded the combined kill estimates of the late 1970s (Marine Mammal Commission, 1992). In 1986, for example, 133,174 dolphins were killed in the tuna fishery (Young *et al.*, 1993).

To address foreign takes of dolphins, Congress amended the MMPA in 1984 to require that each nation exporting tuna to the United States had to adopt a regulatory program governing the incidental taking of marine mammals in the course of tuna fishing that is comparable to that of the U.S. fleet. These nations also had to ensure that the average rate of incidental takes by the vessels of that nation are comparable to the average rate of the incidental take of marine mammals by United States vessels in the course of tuna purse seining. These amendments also required intermediary nations exporting tuna to the United States to provide proof that they have acted to prohibit the importation of tuna from those nations prohibited from exporting tuna directly to the United States and to establish an onboard observer program that would give reliable estimates of the average rate of incidental take (Young *et al.*, 1993; Hall, 1998).

Another conservation effort has focused on identifying new methods of catching yellowfin tuna without the incidental capture of dolphins. The U.S. Secretary of Commerce was instructed to have the National Academy of Sciences (NAS) perform an independent review of alternatives. The NAS recommendations included (Young *et al.*, 1993):

- Research into incentives to improve captain performance
- Research on tracking flotsam-associated yellowfin, fish aggregating devices, and monitoring of radiotagged dolphins
- Research on the nocturnal behavior of tuna and dolphins
- Use of satellites to locate aggregations of tuna
- Assessment of the impact on tuna populations if purse seining for tuna associated with dolphin is discontinued

The U.S. Dolphin Protection Consumer Information Act (DPCIA) was passed in 1990, legislating the "Dolphin Safe" certification program for tuna not caught by intentionally deploying purse seine nets around dolphins. At the same time, the three major U.S. tuna canners announced that they would no longer purchase tuna caught by encircling dolphins and that they would develop a tracking and labeling system for their tuna products. This system would enable the federal government to track tuna labeled as "dolphin safe" back to the harvesting vessel to verify whether the product was properly labeled.

The result of the passage of the DPCIA was a decrease in the number of U.S. vessels setting nets on dolphins and a decline in the domestic incidental U.S. kill from 5083 in 1990 to 812 in 1991, to fewer than 500 in 1992 and to 0 in 1996 (Marine Mammal Commission, 1997). The International Dolphin Conservation Program Act of 1992 (finalized in 1997) authorized the Secretary of State to enter into international agreements to establish a global moratorium on the intentional encirclement of dolphins with purse seine nets during tuna fishing operations.

The United Nations adopted a resolution in December 1991 calling for a global moratorium on all large-scale driftnet fishing on the high seas. The moratorium became effective at the end of 1992 (Hofman, 1995). Collectively, this agreement and others have reduced ETP dolphin mortality in domestic and foreign tuna purse seining operations from over 130,000 animals in 1986 to about 1500 by 2003 (Marine Mammal Commission, 2003) (Figure 15.9), and reduced relative annual mortality of all dolphin stocks involved in this fishery to less than 0.2% (Lennert and Hall, 1996). Currently, both the eastern Pacific spinner dolphin and northeastern Pacific spotted dolphin population sizes are considered depleted but stable at about 560,000 and 730,000 animals, respectively.

15.4.3. Pinniped-Fishery Interactions

The tuna-dolphin problem led to consideration of the impacts of other fisheries on marine mammals within the United States (see Hall and Donovan, 2001). In 1994 amendments to the MMPA included a new section that specifically addressed interactions between pinnipeds and fishery resources. An important feature was its provision allowing states the authority to lethally remove individual pinnipeds identified as affecting certain salmonid stocks provided that certain conditions were met (Marine Mammal Commission, 1997). Perhaps the best example of such an interaction is that between California sea lions and spawning steelhead salmon in Seattle, Washington. The number of steelhead running through the Ballard Locks declined from nearly 3000 in the early 1980s to just 42 in the 2001 run. During that time there was an increase in the number of California sea lions congregating near the locks and preying on the steelhead. Because of the ineffectiveness of previous measures taken, in 1995 the Washington Department of Fish and Wildlife was granted authority until 1997 (later extended through 2006) to lethally take individually identifiable California sea lions preying on steelhead migrating through the Ballard locks. To date, no sea lions have been killed: instead, several were

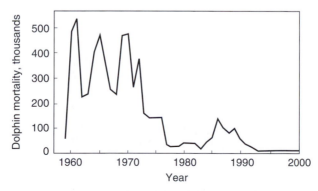

Figure 15.9.	Annual dolphin mortality in the tropical Pacific Ocean tuna purse seine fishery, excluding numbers of calves separated from their mothers and presumed lost. (Redrawn from National Research Council, 1992.)

marked, transported, and released to a different location. In addition an acoustic array installed to deter sea lions who might approach the locks has resulted in no sea lions being observed foraging on the steelhead at the locks since then (Marine Mammal Commission, 2002, 2003).

15.4.4. Acoustic Deterrence

The number of entanglements and deaths of marine mammals can be reduced by the use of active acoustic alarms, or **pingers,** to warn animals about the presence of nets. Field experiments off the New England coast tested the effectiveness of underwater acoustic pingers, which emitted a broad-band signal with a frequency of 10 kHz, well within the hearing range of harbor porpoises and harbor seals. Results indicated that acoustic alarms reduced the incidental catch of harbor porpoises by an order of magnitude (e.g., Kraus *et al.*, 1997; Bordino *et al.*, 2002; Barlow and Cameron, 2003). These results have not yet been fully replicated elsewhere, which suggests that pinger effectiveness studies warrant further research (Dawson *et al.*, 1998). Slightly different devices, referred to as acoustic harassment devices, have been used to deter pinnipeds from approaching nets and aquaculture sites. However, it has been discovered that the level of noise pollution associated with these devices may negatively impact marine mammal populations (Johnston and Woodley, 1998), and some species habituate quickly to acoustic scaring devices and may actually use them to locate potential food sources.

15.5. Environmental Contaminants

The fact that environmental contaminants pose risks to marine mammal populations is widely acknowledged. The high trophic position occupied by many marine mammal species and the tendency of all marine mammals to maintain a significant blubber layer for their thermoregulatory and energetic needs leaves them particularly vulnerable to toxic compounds that are persistent in the environment and tend to bioaccumulate through food webs (biomagnification)—particularly lipophilic substances. Persistent organic pollutants (POPs), hydrocarbons, metals, and other compounds have been found in high concentrations in the tissues of various marine mammal species and have been associated with organ and skeletal anomalies, impaired reproduction and immune function, various disease outbreaks, and a single incidence of acute poisoning of a small harbor seal colony (Kannan *et al.*, 1997; Cowan, 2002). Marine mammal populations in all corners of the globe are affected by contaminants including Arctic and Antarctic regions; far from the sources of contamination because these compounds are transported over long distances by the atmospheric and ocean currents (AMAP, 1998). Numerous studies have suggested that exposure to pollutants has an impact on marine mammal populations, mainly on reproduction and mortality, but few have demonstrated a direct relationship. This is not surprising given the complex nature of marine ecosystems and the logistical and ethical difficulties in performing effects studies on marine mammals.

Two "case studies" and numerous laboratory investigations carried out on laboratory animals strongly suggest the possibility of negative effects of exposure to contaminants in marine mammals. A host of studies conducted over a 20+ year period on beluga whales in the St. Lawrence River Estuary has documented heavy contamination of the whales with mercury, lead, polychlorinated biphenyls (PCBs), dichlorodiphenyl-trichloroethane (DDT), Mirex, and other pesticides and unusually high incidences of lesions, tumors, ulcers, and conditions that are often associated with impaired immunity in other mammals (e.g., Martineau *et al.*, 1994, 2002; De Guise *et al.*, 1995; Beland, 1996; McKinney *et al.*, 2004). Although there are inherent problems with comparing contaminant levels and incidents of disease between stranded animals and samples obtained in other ways (see Hobbs *et al.*, 2003; Kingsley, 2002), the suggestion that St. Lawrence beluga have been compromised by industrial pollutants is compelling. Major efforts to limit environmental release of toxins along the River in recent decades, as well as contaminant clean-up efforts, might be paying off for this population. There are some indications that the St. Lawrence beluga population might be increasing (Kingsley, 1999; but also see Martineau, 2002).

Heavy pollution loads over a period of decades in the Baltic Sea have been suggested to be the cause of pathological impairments, including reproductive disturbances, which have resulted in depressed reproductive capacity, and immune disruption in ringed seals and grey seals living in the region (e.g., Helle *et al.*, 1983; Mattson *et al.*, 1998, Nyman *et al.*, 2002, 2003). Declining levels of DDT and PCB in marine mammal tissues in the Baltic (and in other areas) are now being detected, following marked reductions in their global use in recent decades (AMAP, 1998; Nyman *et al.*, 2002), and there is evidence that both grey seal and ringed seal stocks have shown signs of recovery (Harding and Härkönen, 1999).

Catastrophic oil spills such as the *Torrey Canyon* (1967), *Santa Barbara* (1969), *Amoco Cadiz* (1978), *Exxon Valdez* (1989), *Brear* (1993), and *Prestige* (2002) engender a concern for the marine environment in a manner that no invisible contaminant can. Spilled oil floats on seawater, providing a constant reminder of its presence until it is washed ashore, sinks, or evaporates. Large volumes of oil suffocate benthic organisms by clogging their gills and filtering structures or fouling their digestive tracts. Marine birds and mammals suffer heavily as their feathers or fur become oil-soaked and matted, and they lose insulation and buoyancy. Longer-term detrimental effects result from ingestion of oil while grooming or feeding, and from inhalation of hydrocarbon vapors.

On March 22, 1989, the supertanker *Exxon Valdez* ran aground on Bligh Reef in Alaska's Prince William Sound (Figure 15.10). The grounding punched holes in 8 of the 11 cargo tanks and 3 of the 7 segregated ballast tanks. The result was the largest oil spill to date in United States waters (242,000 barrels or nearly 40,000,000 liters). The spill occurred in an area noted for its rich assemblages of seabirds, marine mammals, fish, and other wildlife. It was one of the most pristine stretches of coastal waters in the United States, with specially designated natural preserves such as the Kenai Fjords and the shoreline of Katmai National Park.

The area of the spill, because of its gravel and cobble beaches, was particularly sensitive to oil. In places, the thick, tarry crude oil penetrated over a meter below the beach surface. High winds, waves, and currents in the days following the accident quickly spread oil over 26,000 km^2. The toll on wildlife was devastating. Over 33,000 dead birds were recovered, and many sea otters, seals, sea lions, and other marine mammals, as well as many fish, were killed. Traditional fishing and cultural activities of the native communities in Prince William Sound were halted for the year, creating enormous social and economic costs. The biological impact of the *Exxon Valdez* oil spill on marine mammal populations has been summarized by Loughlin (1994).

Figure 15.10.   The supertanker *Exxon Valdez* spilling crude oil after colliding with Bligh Reef in Prince William Sound, Alaska, 1989. (Courtesy of J. Harvey.)

In November 2002, the even larger supertanker *Prestige* split apart about 200 km off the northwest coast of Spain and lost much of its cargo of oil. The seas at the site of the *Prestige* spill were very rough, so little could be done but watch the oil wash ashore, eventually, to form a 1-m deep foamy mixture of oil and sea water referred to as "chocolate mousse." The volume of oil released from the *Prestige* was greater than that from the *Exxon Valdez* and, because of higher water temperatures along the Spanish coast, was also more toxic. The slick threatened one of Europe's most picturesque and wildlife-rich coasts, with more than a thousand Spanish and French beaches contaminated with spilled *Prestige* oil. Local fisheries have been destroyed, and current studies indicate that the environmental damage and the clean-up and remediation costs will be substantially greater than those incurred in the *Exxon Valdez* spill. In addition to the direct adverse effects of oil pollution on marine mammals, a lot of concern has been expressed about other possible effects of offshore oil industry operations. Sounds produced by ships, seismic exploration, offshore drilling, and other activities may interfere with communication, breeding, and feeding behaviors causing marine mammals to abandon or avoid these areas (Hofman, 1995). For example, in a comparison of the behavior of bowhead whales in regions of different amounts of human activity, Richardson *et al.* (1995a) noted that the west Arctic bowheads, with greater exposure to human activities, made longer dives, spent a lower percentage of time at the surface, and less frequently raised their flukes, presumably to make them less conspicuous when at the surface.

The impacts of large spills are clearly visible. However, there is also serious cause for concern about the effects that chronic low-level exposure to pollutants can have on marine mammals (e.g., Jenssen, 1996).

15.6. Single Beachings vs Mass Strandings

Beached and stranded marine mammals have long been a focus of curiosity and scientific interest. Many of the original descriptions of species included in this book, as well as the establishment of an important and extensive series of base-line measurements of anatomical features, mortality patterns, disease, toxic contamination loads, and other population indicators, have been based on examinations of beached and stranded marine mammals. The terms "beached" and "stranded" often are used interchangeably in regard to marine mammals. However, for the purpose of this discussion, we will follow the definitions of Hofman (1991), using the term "beached" to refer to any dead marine mammal that washes up on shore, and "stranded" to refer more restrictively only to live cetaceans or sirenians that swim onto or are unintentionally trapped onshore by waves or receding tides. For shorelines with high human usage, a live pinniped, sea otter, or polar bear found ashore may be considered stranded if it is seen as a potential threat to human health or safety. Otherwise, in the United States, it is considered by the NMFS to be "in its element," even if it is ill or injured, and therefore is not stranded (Dierauf, 1990). Accordingly, because death usually occurs at sea, beached animals almost always occur individually, whereas strandings can occur singly (Figure 15.11) or as a group or **mass stranding** (Figure 15.12), defined by Klinowska (1985) as three or more individuals of the same species stranded in the same general area at approximately the same time. Mass stranding events usually involve social odontocetes including pilot whales *(Globicephala)* and Atlantic white-sided dolphins

Figure 15.11.   Beached minke whale on the Oregon coast.

Figure 15.12.   Mass stranding of sperm whales on the Oregon coast, 1979.

(Lagenorhynchus). Stranding frequency varies geographically and population sizes and nearness to shore are among the factors that are thought to influencing stranding patterns.

15.6.1. Hypotheses Regarding Mass Strandings

For a review of the occurrences and causes of marine mammal mass strandings see Geraci *et al.* (1999) and Walsh *et al.* (2001). The occurences of several recent mass strandings of beaked whales have been correlated with military uses of LFA sonar systems (described in Chapter 11). Additional explanations for odontocete mass strandings include the presence of debilitating but nonlethal parasitic infestations of the respiratory tracts, brain, or middle ear; group-wide bacterial or viral infections (e.g., morbillivirus discussed in Chapter 14); panic flight responses to predators (including humans); cohesive social bonds causing the entire herd to follow one strander; and near-shore disorientation of echolocation or geomagnetic signals used for navigation. The latter explanation is based on an assumption that odontocetes are capable of detecting and responding to the Earth's magnetic field direction or intensity or both. Biological magnetic detectors (small crystals of a magnetic form of iron oxide, magnetite) have been found in diatoms, intertidal mollusks (Frankel, 1986), and sharks and rays (Kalmijn, 1977) and have been suggested for insects and birds (Gould, 1980) and fin whales (Walker *et al.*, 1992). Magnetic material has been found in the brains, bone, blubber, and muscle of the bottlenose dolphin, Cuvier's beaked whale, Dall's porpoise, and the humpback whale (Bauer *et al.*,

1985). Magnetite crystals are thought to continually orient themselves in line with the earth's magnetic field, just like a compass. By sensing changes in the orientation of these crystals, the host animal is thought to be able to determine the direction in which it is traveling, a useful ability during extended open-ocean migrations. Normally, the natural magnetic force fields run north to south at an even intensity. In some places, however, the field is distorted by certain types of geological formations, such as those rich in iron. Such distortions are called geomagnetic anomalies. Because some live strandings of whales occur in areas where geomagnetic anomalies are present, it has been suggested (Klinowska 1985, 1986, 1988) that these strandings may be the result of navigational mistakes on the part of whales attempting to use their magnetic sense.

Indirect and experimental evidence for this hypothetical explanation of mass strandings comes from analyses of cetacean stranding patterns on the United Kingdom (Klinowska, 1985, 1986, 1988) and eastern U.S. coastlines (Kirschvink *et al.*, 1986; Kirschvink, 1990; Walker *et al.*, 1992). This hypothesis, however, has been questioned by a number of workers. Hui (1994) evaluated sightings of free-ranging common dolphins and demonstrated only an association with bottom topography patterns. He found no support for the hypothesis that free-ranging dolphins use magnetic intensity gradients as navigational pathways. An analysis of a comparably sized sample of New Zealand single and mass strandings by Brabyn and Frew (1994) found no apparent relationship between mass stranding locations and either geomagnetic contours or magnetic minima, and there was no evidence that whales appeared to be avoiding magnetic gradients.

Using confusion of echolocation signals or disturbances of geomagnetic patterns to explain mass strandings relies on viewing strandings as accidental consequences of navigational mistakes when animals are near shore. It also requires that potential stranders rely predominantly or exclusively on that single sensory mode for navigational cues, while ignoring contradictory information from other sensory modes such as vision. More information regarding the health status of individual animals involved in strandings that fit the geomagnetic hypothesis may help to clarify this issue.

15.6.2. Stranding Networks and Strandings as a Source for Study Specimens

The MMPA provided the legal basis to establish six regional stranding networks on the U.S. coastline and an amendment to the MMPA in 2000 established the Marine Mammal Rescue and Assistance Act administered by the Fish and Wildlife Service (FWS) and NMFS to provide financial assistance (2001–2003) to stranding networks (Marine Mammal Commission, 2001). These stranding networks in the United States and those in other nations are summarized by Gulland *et al.* (2001). The development and activities of stranding networks are reviewed by Wilkinson and Worthy (1999) and Gulland *et al.*, 2001). The MMPA provided each network with appropriate authority and resources to respond to beachings and strandings of marine mammals for the purposes of necropsy and tissue salvage of beached animals and of rescue and/or rehabilitation of live stranded animals (Dierauf, 1990; Hofman, 1991). These networks have become valuable (and sometimes the only) sources of tissues from rare or legally protected marine mammal species.

15.7. Ecotourism

15.7.1. Whale Watching

Whale watching is an ancient pastime but a relatively youthful industry. Only in the past 4 decades has there been a defined and growing market for large numbers of people to gain very close contact with whales for educational or ecotourist pursuits. These activities range from casual observations of migrating, feeding, or courting whales to intensive and prolonged observations or interactions from boats, aircraft, or swimming in the water. In 1991, an estimated 4 million people worldwide went whale watching. By 1994 this had increased to 5.4 million, with total revenues estimated to be more than $500 million (Hoyt, 1995), and over 50 countries and territories offered whale watching tours (IFAW, 1995). Some of the better known whale-watching sites are listed in Table 15.3. In the Canary Islands alone, whale-watching activities target 16 species of odontocetes, are conducted from large ferry-like vessels often carrying over 1000 passengers/day, and generate 12 million ε annually.

The International Whaling Commission (IWC) confronts new responsibilities in overseeing the conservation of whale stocks with respect to nonconsumptive use of cetacean populations. According to Barstow (1986), these nonconsumptive uses include benign research using remote sensing or noninvasive techniques, habitat protection in the form of sanctuaries or refuges, cultural valuation involving aesthetic appreciation of cetaceans through the arts and emphasizing whales as a unique educational resource, and finally, recreational whale watching.

A basic question facing the IWC and local management agencies of individual stocks exposed to whale watching activities concerns the impact of these activities on the normal behavior of the target animals. Do whale watching activities constitute harassment? In the terminology of the MMPA, harassment was defined in 1994 as any activity that has the potential to injure or substantially disrupt the normal behavior of a marine mammal. There is no doubt that some whale or seal watching, as well as some benign research activities (Figure 15.13), result in obvious low-level harassment. Some studies have examined the effects of the presence of boat traffic and noise on baleen whales (reviewed by Richardson *et al.*, 1995b and see also Patenaude *et al.*, 2002), and a few have studied influences on smaller odontocetes (reviewed by Buckstaff, 2004). Much less information is available on the effects of tourism on pinnipeds (reviewed by Richardson *et al.*, 1995b; Constantine, 1998). The dominant behavioral reactions of

Table 15.3. Target Species and Locations of Commercial Whale-Watching Activities

Target species	Location(s)
Bottlenose dolphins	United States, Australia, and Japan
Spinner and spotted dolphins	Caribbean Islands, Brazil, and Hawaii
Killer whales	Norway, British Columbia, Puget Sound, Antarctica, Patagonia
Sperm whales	Canary Is., Azores Is., S. Island of New Zealand, Norway
Humpback whales	Hawaii, Alaska, Australia, Japan, and New England
Gray whales	British Columbia to Baja California
Right whales	Patagonia

Figure 15.13. A close encounter with a gray whale (sampling expired lung gases) in its winter breeding lagoon.

cetaceans reported in these studies were short-term effects such as an increase in swimming speed, spatial avoidance, increase in breathing synchrony, and changes in diving behavior (Janik and Thompson, 1996; Nowacek *et al.*, 2001; Hastie *et al.*, 2003; Buckstaff, 2004; Constantine *et al.*, 2004). Such short-term disruptions could cause longer-term changes in the behavior of a population. However, evidence of harassment is not always clear-cut. Salden (1988) noted that Hawaiian humpback cows and calves may be deserting traditional winter resting areas near the Maui shore in favor of waters 3–4 km offshore, presumably in response to increased whale watching activities. However, no evidence has been found of a decline in the relative encounter rate of individual humpbacks, humpback pods, and calves. In the southern California Bight, where whale watching of migrating gray whales is an important winter component of the sportfishing industry, Sumich and Show (1999) demonstrated a definite shift of migratory routes from coastal to offshore locations, even though this population continued to increase in size.

For humpback whales in Hawaii and gray whales in California, their normal (or at least historical) behavior has changed, but it is not at all certain that the changes are consequences of whale watching activities. Behavioral studies need to be conducted (e.g., Buckstaff, 2004) alongside more detailed research on individual survival and reproductive rates and movement patterns to assess whether boat traffic has a significant impact at the population level. Ultimately, a critical question we need to address is what trade-offs are we willing to accept for the presumed educational, recreational, and commercial benefits of whale watching?

15.7.2. Dolphin Swim and Feeding Programs

Other social interactions between humans and marine mammals include captive and wild dolphin swimming and feeding programs. The bottlenose dolphin is the species typically involved in human/dolphin interactions. Several facilities in the United States and elsewhere conduct swim with dolphin programs in captivity. Quantitative behavioral studies of bottlenose dolphins in these swim with dolphin programs revealed that human swimmers and dolphins both have risky encounters in some cases (Samuels and Spradlin, 1995; Samuels *et al.*, 2000). Risky social interactions (e.g., aggressive, submissive, or sexual behavior) occurred at higher rates when encounters between dolphins and swimmers were not directly controlled by staff. In controlled situations, trainers diminished the potential for dolphin distress and swimmer injury. The long-term effects of swim programs on dolphin behavior are unknown and require tracking the behavior of individual dolphins over several years and analyzing comparative quantitative behavioral data (i.e., Samuels and Spradlin, 1995; Samuels *et al.*, 2000) with that from non-swim dolphins in zoo and aquarium environments and in the wild.

Although swimming with or feeding wild dolphins is illegal in the United States by an amendment to the MMPA (Marine Mammal Commission, 1994), there are a number of opportunities in the United States to participate in these activities (e.g., spinner dolphins in Kealakekua Bay in Hawaii; Würsig *et al.*, 1995). Study of the responses of wild bottlenose dolphins to the presence of swimmers in the Bay of Islands, New Zealand revealed that dolphins changed their behavior one third to one half of the time that they were approached by the operator's boat. Dolphin response varied with swimmer placement. The "line abreast" strategy for placing swimmers in the water resulted in the lowest rate of avoidance by the dolphins but there are some data to suggest that dolphins may modify their responses with time (Constantine and Baker, 1997; Constantine, 2001). It is possible to swim with whales other than dolphins in a few countries (e.g., humpback whales in Tonga, sperm whales in the Galapagos and Azores Islands, and pilot whales and beaked whales off the Canary Islands; Constantine, 1999). Among the countries allowing the feeding of wild dolphins is Australia. Monkey Mia in Western Australia is one of the best known areas with a history of human and dolphin interactions (Conner and Smolker, 1985). Significant differences in the behavior of wild and provisioned dolphins have been reported, including an increased mortality rate for calves born to provisioned mothers (references cited in Constantine, 1999). At another feeding program at Tangalooma, Moreton Bay Island, Queensland, forceful contact (e.g., pushy behavior) between humans and wild dolphins during feeding sessions was reported (Orams *et al.*, 1996). The number of dolphins attending a particular feeding significantly increased the pushiness as did the presence of adult males at a feeding.

15.8. Progress and the Future

As we begin the 21st century, we can look back on some successes in our attempts to manage and conserve populations of marine mammals. Northern elephant seals and Pacific humpbacks and East Pacific gray whales have all made dramatic recoveries in the past century. Populations of several large baleen whales in the southern hemisphere are also increasing (e.g., southern hemisphere blue whales; Branch *et al.*, 2004). The global population of polar bears is currently being managed in a sustainable fashion, despite hunting taking place in many parts of their range (Lunn *et al.*, 2002). Baltic Sea harbor seal numbers have increased after protection was introduced for them in the 1960s and 1970s (Heide-Jørgensen and Härkönen, 1988) and similar recoveries have taken place in some walrus populations (Born *et al.*, 1995).

Some sea otter populations have been rescued from the brink of extinction; the recovery plan for the California sea otter is an interesting case study. The small size and distribution of the California sea otters and the growing risk of oil spills as a result of increasing tanker traffic in their geographic range resulted in recognition of the fact that the best way to minimize the threat from oil spills would be to encourage expansion of the population into a greater portion of its preexploitation range. However, it was recognized that range expansion of the sea otter could impact commercial and recreational abalone and other fisheries that had developed in the absence of sea otters. Despite this latter concern, the U.S. Fish and Wildlife Service undertook a translocation program to establish a sea otter colony at San Nicholas Island in the California Channel Islands. Although only about 24% of the original number of otters removed from the mainland population and released on the island remain at the island colony at present (Marine Mammal Commission, 2003), valuable information was obtained for evaluating whether translocation programs are viable alternatives for conservation.

Few large-scale commercial harvests of marine mammals take place today and marine mammals currently enjoy substantial empathy from the public. However, in spite of the widespread awareness of and concern for the welfare of marine mammals, some marine mammal populations (and even species) are still at risk, some as a result of direct interactions with humans or as a result of the impacts humans have on the world's oceans.

Since 1990, a 90% crash in Alaska's Aleutian Island populations of the sea otter has been documented, and declines have taken place in some Alaskan harbor seal colonies as well as precipitous declines in Steller sea lion numbers (Figure 15.14). The sea otter decline has been attributed to increased predation by killer whales (Estes *et al.*, 1998), which are presumed to be eating more sea otters in response to decreases that have taken place in populations of their preferred prey, sea lions and seals (Figure 15.15). Although the causes of these collapses of various marine mammal populations in this region are poorly understood, two competing hypotheses provide starting points for understanding the linkages between these widely disparate populations. The top-down forcing hypothesis, elucidated by Springer *et al.* (2003), proposes that the decimation of populations of large whales by the commercial whaling industry forced their major predators, killer whales, to shift their predatory efforts to smaller pinnipeds and sea otters. Arguments for this hypothesis are based on the timing of the collapses, known diets, and observed foraging behaviors of killer whales and their prey and on energetic modelling considerations. The alternate bottom-up forcing hypothesis proposes instead that Steller sea lion populations and other pinnipeds have collapsed because their prey populations (principally herring, pollock, and capelin) were overfished or were negatively affected by climate regime shifts (NRC, 2003). According to this hypothesis, although killer whale predation may be contributing to the decline of pinniped and sea otter populations (Figure 15.15), predation from this source cannot adequately explain the observed pinniped and sea otter population collapses (Wade *et al.*, 2003).

Marine mammal species listed by the MMPA as depleted, or by the ESA as endangered, or in CITES Appendix I or II (see Table 15.2) include species from the large baleen whales, whose populations have been reduced by direct harvesting, to small odontocetes

Figure 15.14.   Steller sea lion rookery beach photographed on the same day of the year in **(a)** 1969 and **(b)** 1987. (Courtesy of National Marine Mammal Laboratory, NMFS, NOAA.)

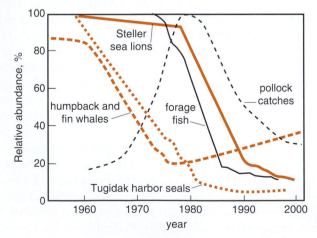

Figure 15.15.   Trends of population changes for several species of marine mammals (*color*) and forage fish (*solid black*) and commercial catch of pollock (*dashed black*) in the Gulf of Alaska, 1960–2000. A similar decline for Steller sea lions is documented for the Bering Sea. (Adapted from National Research Council, 2003.)

such as the vaquita, whose small populations continue to decline due to incidental mortality in fishing nets (Hohn *et al.*, 1996), to manatees in Florida, which experience high mortality because of recreational boat activity. With the human population of the world expected to increase by about 1.3% each year, the future of the seas as a source of protein and as a place for recreation and aesthetic pursuits will depend on a better under-standing of the consequences of our intrusions on the workings of marine ecosystems. The history of our involvement with marine mammal populations has repeatedly demon-strated a lack of practical awareness of the fundamental effects our interventions cause in these natural systems. As we begin the 21st century, we can hopefully develop a sense of stewardship toward the world's oceans and their inhabitants that is reflected in our per-sonal as well as our political decisions. Only by better understanding the fate we are shap-ing for the world's ocean can we expect to have healthy populations of marine mammals to be witnessed with fascination and respect by the children of those reading this book.

15.9. Summary and Conclusions

Overexploitation has resulted in the extinction of three marine mammal species: Steller's sea cow, the Atlantic gray whale, and the Caribbean monk seal. Despite establishment of legal frameworks to address exploitation, many other species have been and continue to be affected by human activities such as bycatch in commercial fisheries, net entangle-ments, and environmental contaminants. Information about mortality patterns, disease, and levels of environmental contaminants have been obtained from beached and stranded marine mammals as well as studies designed to address these issues in wild pop-ulations or via laboratory tissue studies. It is clear that the presence of boat traffic, as well as swimming and feeding activities with whales and dolphins, alters, at least in the short term, their normal behavior patterns, although the long-term effects of these ecotourism activities are unknown. The recovery of the northern elephant seal, the Pacific hump-back, and the California sea otter indicates that some of our conservation efforts have been successful. The future of marine mammal conservation requires better information on the population status and ecological relationships of marine mammals with their ecosystems as well as greater understanding of the effects of human-induced activities.

15.10. Further Reading

Marine mammal conservation is the subject of multiauthored volumes by Simmonds and Hutchinson (1996), Reeves and Twiss (1999), Clapham *et al.* (1999), and see also Annual Reports of the Marine Mammal Commission especially for U.S. legislation affecting marine mammal conservation. The *Journal of Cetacean Research and Management* pub-lishes papers about the conservation and management of whales, dolphins, and por-poises, especially topics taken up by the IWC Scientific Committee. The effects of the *Exxon Valdez* oil spill on marine mammals is reviewed in Loughlin (1994). See Richardson *et al.* (1995a) for a summary of the effects of noise (i.e., boat traffic, offshore drilling, seismic profiling) on marine mammals and Reijnders *et al.* (1999) and Vos *et al.* (2003) for a summary of the effects of persistent contaminants on marine mammals.

Among marine mammal conservation groups/agencies with Internet addresses are the following:

http://www.nmfs.gov—headquarters of National Marine Fisheries Service (NMFS), the office primarily responsible for marine mammal conservation with links to other NMFS offices.

http://elfi.com/csihome.html—Cetacean Society International, volunteer organization involved in conservation of wild whales and dolphins world-wide.

http://www.iucn.org—World Conservation Union (IUCN), environmental biologists active in global cetacean conservation.

References

Alexander, L. M. (1993). "Large Marine Ecosystems." *Mar. Policy 17:* 186–198.

AMAP (Arctic Monitoring and Assessment Programme) (1998). AMAP Assessment Report: Arctic Pollution Issues. AMAP, Oslo, Norway.

Andresen, S. (1993). "The Effectiveness of the International Whaling Commission." *Arctic 46:* 108–115.

Archer, F., T. Gerrodette, S. Chivers, and A. Jackson (2004). "Annual Estimates of the Unobserved Incidental Kill of Pantropical Spotted Dolphin *(Stenella attenuata attenuata)* Calves in the Tuna Purse-Seine Fishery of the Eastern Tropical Pacific." *Fish. Bull. 102:* 233–244.

Baker, S. C., F. Cipriano, and S. R. Palumbi (1996). "Molecular Genetic Identification of Whale and Dolphin Products from Commercial Markets in Korea and Japan." *Mol. Ecol. 5:* 671–685.

Baker, C. S., M. L. Dalebout, and G. M. Lento (2002). "Gray Whale Products Sold in Commercial Markets Along the Pacific Coast of Japan." *Mar. Mamm. Sci. 18:* 295–300.

Baker, S. C., and S. R. Palumbi (1994). "Which Whales Are Hunted? A Molecular Genetic Approach to Monitoring Whaling." *Science 265:* 1538–1539.

Barlow, J. and G. A. Cameron (2003). "Field Experiments Show that Acoustic Pingers Reduce Marine Mammal By-Catch in the California Drift Net Fishery." *Mar. Mamm. Sci. 19:* 265–283.

Barstow, R. (1986). "Non-Consumptive Utilization of Whales." *Ambio 15:* 155–163.

Bauer, G. B., M. Fuller, A. Perry, J. R. Dunn, and J. Zoeger (1985). Magnetoreception and biomineralization of magnetite in cetaceans. *In* "Magnetic Biomineralization and Magnetoreception in Organisms" (J. L. Kirschvink, D. S. Jones, and B. J. McFadden, eds.), pp. 489–508. Plenum Press, New York.

Beddington, J. R., and R. M. May (1982). "The Harvesting of Interacting Species in a Natural Ecosystem." *Sci. Am.* November: 62–69.

Beland, P. (1996). "The Beluga Whales of the St. Lawrence River." *Sci. Am.* May: 58–65.

Bonner, W. N. (1994). *Seals and Sea Lions of the World*. Facts on File, New York.

Bordino, P. S., S. Kraus, D. Albareda, A. Fazio, A. Palmerino, M. Mendez, and S. Botta (2002). "Reducing Incidental Mortality of Franciscana Dolphin *Pontoporia blainvillei* with Acoustic Warning Devices Attached to Fishing Nets." *Mar. Mamm. Sci. 18:* 833–842.

Born, E. W., I. Gjertz, and R. R. Reeves (1995). *Population Assessment of Atlantic Walrus*. Meddelelser, No. 138., Norsk Polarinstitutt, Oslo, Norway.

Brabyn, M., and R. V. C. Frew (1994). "New Zealand Herd Stranding Site Do Not Relate to Geomagnetic Topography." *Mar Mamm. Sci. 10:* 195–207.

Braham, H. (1992). "Endangered Whales: Status Report." *Natl. Mar Fish. Serv*.

Branch, T. A., K. Matsuoka, and T. Miyashita (2004). "Evidence for Increases in Antarctic Blue Whales Based on Bayesian Modelling." *Mar. Mamm. Sci. 20:* 726–754.

Buckstaff, K.C. (2004). "Effects of Watercraft Noise on the Acoustic Behavior of Bottlenose Dolphins, *Tursiops truncatus,* in Sarasota Bay, Florida." *Mar. Mamm. Sci. 20:* 709–725.

Burns, W. C. G., and G. Wandesforde-Smith. (2002). "The International Whaling Commission and the Future of Cetaceans in a Changing World." *Rev. Eur. Comm. Int. Environ. Law 11:* 199–210.

Busch, B. C. (1985). *The War Against the Seals. A History of the North American Seal Fishery*. McGill Queen's University Press, Kingston, Ont.

Calkins, D. G., D. C. McAllister, K. W. Pitcher, and G. W. Pendelton (1999). "Steller Sea Lion Status and Trend in Southeast Alaska: 1979–1997." *Mar. Mamm. Sci. 15:* 462–477.

Clapham, P. J., P. Berggren, S. Childerhouse, N. A. Friday, T. Kasuya, L. Kell, K. H. Kock, S. Manzanilla-Naim, G. N. Di Sciara, W. F. Perrin, A. J. Read, R. R. Reeves, E. Rogan, L. Rojas-Bracho, T. D. Smith,

M. Stachowitsch, B. L. Taylor, D. Thiele, P. R. Wade, and R. L. Brownell (2003). "Whaling as a Science." *Bioscience 53:* 210–212.

Clapham, P. J., S. B. Young, and R. L. Brownell, Jr. (1999). "Baleen Whales: Conservation Issues and the Status of the Most Endangered Populations." *Mamm. Rev. 29:* 35–60.

Conner, R. C., and R. Smolker (1985). "Habituated Dolphins *(Tursiops* sp.) in Western Australia." *J. Mammal. 66:* 398–400.

Constantine, R. (1999). *Effects of Tourism on Marine Mammals.* New Zealand Dept. Conserv. Rep., Sci. Conserv. No. 104, Wellington, NZ.

Constantine, R. (2001). "Increased Avoidance of Swimmers by Wild Bottlenose Dolphins *(Tursiops truncatus)* Due to Long-Term Exposure to Swim with Dolphin Tourism." *Mar. Mamm. Sci. 17:* 689–702.

Constantine, R., and C. S. Baker (1997). "Monitoring the Commercial Swim-with-Dolphin Operations in the Bay of Islands." *N. Z. Dep. Conserv. Sci. Conserv. Ser. 56:* 1–59.

Constantine, R., D. H. Brunton, and T. Dennis (2004). "Dolphin-Watching Tour Boats Change Bottlenose Dolphin *(Tursiops truncates)* Behaviour." *Biol. Conserv. 117:* 299–307.

Cowan, D. F. (2002). Pathology. *In* "Encyclopedia of Marine Mammals" (W. F. Perrin, B. Würsig, and J. G. M. Thewissen, eds.), pp. 883–890. Academic Press, San Diego, CA.

Craven, J. P., and J. Schneiden (eds.) (1989). The international implication of extended maritime jurisdiction in the Pacific. *Proceedings of Law of the Sea, 21st Annual Conference,* p. 650.

Croxall, J. P, S. Rodwell, and I. L. Boyd (1990). "Entanglement in Man-Made Debris of Antarctic Fur Seals at Bird Island, South Georgia." *Mar. Mamm. Sci. 6:* 221–233.

Dawson, S. M., A. Read, and E. Slooten (1998). "Pingers, Porpoises and Power: Uncertainties with Using Pingers to Reduce Bycatch of Small Cetaceans." *Biol. Conserv. 84:* 141–146.

De Guise, S., D. Martineau, P. Beland, and M. Fournier (1995). "Possible Mechanisms of Action of Environmental Contaminants on St. Lawrence Beluga Whales *(Delphinapterus leucas)."* *Environ. Health Perspect. 103:* 73–77.

DFO (Department of Fisheries and Oceans Canada) (2003). *Facts about Seals 1003.* Fisheries and Oceans Canada, Ottawa.

Dierauf, L. A. (1990). Marine mammal stranding networks. *In* "CRC Handbook of Marine Mammal Medicine: Health, Disease, and Rehabilitation," pp. 667–672. CRC Press, Boca Raton, FL.

Donovan, G. P. (1995). *The International Whaling Commission and the Revised Management Procedure. Additional Essays on Whales and Man.* High North Alliance, N-8390 Reine i Lofoten, Norway.

Estes, J. A., M. T. Tinker. T. M. Williams, and D. F. Doak (1998). "Killer Whale Predation on Sea Otters Linking Oceanic and Nearshore Ecosystems." *Science 282:* 473–476.

Federal Register (1993). "Protected Species Special Exception Permits." *NOAA. 58:* 53320–53364.

Frankel, R. B. (1986). "Magnetic Skeletons in Davy Jones' Locker." *Nature 320:* 575.

Fromberg, A., M. Cleemann, and L. Carlsen (1999). "Review on Persistent Organic Pollutants in the Environment of Greenland and Faroe Islands." *Chemosphere 38:* 3075–3093.

Gentry, R. L., and G. L. Kooyman (1986). Introduction. *In* "Fur Seals: Maternal Strategies on Land and at Sea" (R. L. Gentry, and G. L. Kooyman, eds.), pp. 3–27. Princeton University Press, Princeton, NJ.

Gerrodette, T., and D. P. DeMaster (1990). "Quantitative Determination of Optimum Sustainable Population Level." *Mar Mamm. Sci. 6:* 1–16.

Gould, J. L. (1980). "The Case for Magnetic Sensitivity in Birds and Bees (Such As It Is)." *Am. Sci. 68:* 256–267.

Gulland, F. M. D., L. A. Dierauf, and T. K. Rowles (2001). Marine mammal stranding networks. *In* "CRC Handbook of Marine Mammal Medicine," 2nd ed. (L. Dierauf, and F. M. D. Gulland eds.), pp. 45–67. CRC Press, Boca Raton, FL.

Hall, M. A. (1998). "An Ecological View of the Tuna-Dolphin Problem: Impacts and Trade-Offs." *Rev. Fish Biol. Fish. 8:* 1–34.

Harding, K. C., and T. J. Härkönen (1999). "Development in the Baltic Grey Seal *(Halichoerus grypus)* and Ringed Seal *(Phoca hispida)* Populations During the 20th Century." *Ambio 28:* 619–627.

Hastie, G. D., B. Wilson, L. H. Tufft, and P. M. Thompson (2003). "Bottlenose Dolphins Increase Breathing Synchrony in Response to Boat Traffic." *Mar. Mamm. Sci. 19:* 74–84.

Hedley, C. (2000). "The Law of the Sea and the Internet: A Resource Guide with Special Reference to the Conservation and Management of Marine Living Resources." *Int. J. Mar. Coastal Law 15:* 567–579.

Heide-Jørgensen, M. P., and T. J. Härkönen (1988). "Rebuilding Seal Stocks in the Kattegat-Skagerrak." *Mar Mamm. Sci. 4:* 231–246.

Helle, E., H. Hyvarinen, H. Pyssalo, and K. Wickstrom (1983). "Levels of Organochlorine Compounds in an Inland Seal Population in Eastern Finland." *Mar. Pollut. Bull. 14:* 256–260.

Henshaw, M. D., R. G. Le Due, S. J. Chivers, and A. E. Dizon (1997). "Identifying Beaked Whales (Family Ziphiidae) Using mtDNA Sequences." *Mar. Mamm. Sci. 13:* 487–495.

Hobbs, K. E., d. C. g. Muir, R. Michaud, P. Beland, R. J. Letcher, and R. J. Norstrom (2003). "PCBs and Organochlorine Pesticides in Blubber Biopsies from Free-Ranging St. Lawrence River Estuary Beluga Whales *(Delphinapterus leucas)*, 1994–1998." *Environ. Pollut. 122:* 291–302.

Hofman, R. J. (1989). "The Marine Mammal Protection Act: A First of Its Kind Anywhere." *Oceanus 32:* 21–28.

Hofman, R. J. (1991). History, goals, and achievements of the regional marine mammal stranding networks in the United States. *In* "Marine Mammal Strandings in the United States. Proceedings of the Second Marine Mammal Stranding Workshop, Miami, Florida, December 3–5, 1987" (John E. Reynolds III, and Daniel K. Odell, eds.), pp. 7–16. NTIS number: PB91-173765.

Hofman, R. J. (1995). "The Changing Focus of Marine Mammal Conservation." *Trends Ecol. Evol. 10:* 462–465.

Hohn, A. A., A. J. Read, S. Fernandez, O. Vidal, and L. T. Findley (1996). "Life History of the Vaquita, *Phocoena sinus* (Phocoenidae, Cetacea)." *J. Zool. London 239:* 235–251.

Holt, S. (2003). "Is the IWC Finished as an Instrument for the Conservation of Whales and the Regulation of Whaling?" *Mar. Poll. Bull. 46:* 924–926.

Hoyt, E. (1995). *The Worldwide Value and Extent of Whale Watching: 1995*. Report to the Whale and Dolphin Conservation Society, England.

Hui, C. A. (1994). "Lack of Association Between Magnetic Patterns and the Distribution of Free-Ranging Dolphins." *J. Mammal. 75:* 399–405.

IATTC (Inter-American Tropical Tuna Commission) (1981). *Annual Report*. La Jolla, CA.

IFAW (1995). *International Fund for Animal Welfare (IFAW) and Tethys European Conservation—Report of the Workshop on the Scientific Aspects of Managing Whale Watching*. IFAW, Tethys and European Conservation, Montecastello de Vibio, Italy.

Janik, V. M., and P. M. Thompson (1996). "Changes in Surfacing Patterns of Bottlenose Dolphins in Response to Boat Traffic." *Mar Mamm. Sci. 12:* 597–601.

Jenssen, B. M. (1996). "An Overview of Exposure to, and Effects of, Petroleum Oil and Organochlorine Pollution in Grey Seals *(Halichoerus grypus)*." *Sci. Tot. Environ. 186:* 109–118.

Johnston, D. W., and T. H. Woodley (1998). "A Survey of Acoustic Harassment Device (AHD) Use in the Bay of Fundy, NB, Canada." *Aquat. Mamm. 24:* 51–61.

Kalmijn, A. J. (1977). "The Electric and Magnetic Sense of Sharks, Skates, and Rays." *BioScience 20:* 45–52.

Kannan, K., K. Senthilkumar, B. G. Loganathan, S. Takahashi, D. K. Odell, and S. Tanabe (1997). "Elevated Accumulation of Tributyltin and Its Breakdown Products in Bottlenose Dolphins *(Tursiops truncatus)* Found Stranded Along the U.S. Atlantic and Gulf Coasts." *Environ. Sci. Technol. 31:* 296–301.

Kingsley, M. C. S. (1999). Population indices and estimates for the belugas of the St. Lawrence estuary. Can. Tech. Rep. Fish. Aqu. Sci. No. 2266.

Kingsley, M. C. S. (2002). "Comment on Martineau et al. 1999. Cancer in Beluga Whales from the St. Lawrence Estuary, Quebec, Canada: A Potential Biomarker of Environmental Contamination." *Mar. Mamm. Sci. 18:* 572–574.

Kirschvink, J. L. (1990). "Geomagnetic Sensitivity in Cetaceans: An Update with Live Stranding Records in the United States." *Life Sci. 196:* 639–650.

Kirschvink, J. L., A. E. Dizon, and J. A. Westphal (1986). "Evidence from Stranding for Geomagnetic Sensitivity in Cetaceans." *J. Exp. Biol. 120:* 1–24.

Klinowska, M. (1985). "Cetacean Live Stranding Dates Relate to Geomagnetic Disturbances." *Aquat. Mamm. 11:* 109–119.

Klinowska, M. (1986). The cetacean magnetic sense-evidence from strandings. *In* "Research on Dolphins," (M. M. Bryden, and R. Harrison, eds.), pp. 401–432. Clarendon Press, Oxford, UK.

Klinowska, M. (1988). "Cetacean 'Navigation' and the Geomagnetic Field." *J. Navig. 41:* 52–71.

Kraus, S. D., A. J. Read, A. Solow, K. Baldwin, T. Spradlin, E. Anderson, and J. Williamson (1997). "Acoustic Alarms Reduce Porpoise Mortality." *Nature 388:* 525.

Krupnik, I. I. (1984). Gray whales and the aborigine of the Pacific Northwest: The history of aboriginal whaling. *In* "The Gray Whale, *Eschrichtius robustus* (Lilljeborg, 1861)" (M. L. Jones, S. L. Swartz, and S. Leatherwood, eds.), pp. 103–120. Academic Press, New York.

Laws, R. M. (1962). Some effects of whaling on the southern stocks of baleen whales. *In* "The Exploitation of Natural Animal Populations" (R. F. LeCren, and M. W. Holdgate, eds.), pp. 137–158. Wiley, New York.

Lennert, C., and M. A. Hall (1996). "Estimates of Incidental Mortality of Dolphins in the Eastern Pacific Ocean Tuna Fishery in 1994." *Rep. Int. Whal. Commn. 46:* 555–557.

Loughlin, T. R. (ed.) (1994). *Impacts of the Exxon Valdez Oil Spill on Marine Mammals.* Academic Press, San Diego, CA.

Loughlin, T. R., A. S. Perlov, and V. A. Vladimirov (1992). "Range-Wide Survey and Estimation of Total Number of Steller Sea Lions in 1989." *Mar. Mamm. Sci. 8:* 220–239.

Lunn, N. J., S. Schliebe, and E. W. Born (2002). Polar bears. *In* "Proceeding of the 13th Working Meeting of the IUCN/SSC Polar Bear Specialist Group, 23-28 June 2001, Nuuk, Greenland." Occas. Pap. IUCN Species Surv. Commn. No. 26.

Marine Mammal Commission (1991). *Annual Report to Congress.* Marine Mammal Commission, Washington, DC.

Marine Mammal Commission (1992). *Annual Report to Congress.* Marine Mammal Commission, Washington, DC.

Marine Mammal Commission (1994). *Annual Report to Congress.* Marine Mammal Commission, Washington, DC.

Marine Mammal Commission (1997). *Annual Report to Congress.* Marine Mammal Commission, Washington, DC.

Marine Mammal Commission (2002). *Annual Report to Congress.* Marine Mammal Commission, Washington, DC.

Marine Mammal Commission (2003). *Annual Report to Congress.* Marine Mammal Commission, Washington, DC.

Marine Mammal Commission (2004). *Annual Report to Congress.* Marine Mammal Commission, Washington, DC.

Martineau, D., S. De Guise, M. Fournier, L. Shugart, C. Girard, A. Lagace, and P. Beland (1994). "Pathology and Toxicology of Beluga Whales from the St Lawrence Estuary, Quebec, Canada—Past, Present and Future." *Sci. Tot. Environ. 154:* 201–215.

Martineau, D., K. Lemberger, A. Dallaire, P. Michel, P. Beland, P. Labelle, and T. P. Lipscomb (2002). "St. Lawrence Beluga Whales, the River Sweepers?" *Environ. Health Perspect. 110:* A562–A564.

Mattson, M., H. Raunio, O. Pelkonen, and E. Helle (1998). "Elevated Levels of Cytochrome P4501A (CYP1A) in Ringed Seals from the Baltic Sea." *Aquat. Toxicol. 43:* 41–50.

McKinney, M. A., A. Arukwe, S. De Guise, D. Martineau, P. Beland, A. Dallaire, S. Lair, M. Lebeauf, and R. J. Letcher (2004). "Characterization and Profiling of Hepatic Cytochromes P450 and phase II Xewmobiotic-Metabolizing Enzymes in Beluga Whales *(Delphinapterus leucas)* from the St. Lawrence River Estuary and the Canadian Arctic." *Aquat. Toxicol. 69:* 35–49.

Melville, H. (1851). *Moby Dick.* Dutton, New York.

National Research Council (USA) (1992). *Dolphins and the Tuna Industry.* National Academy Press, Washington, DC.

National Research Council (USA) (2003). *Decline of the Steller Sea Lion in Alaskan Waters.* National Academy Press, Washington, DC.

Nowacek, S. M., R. S. Wells, and A. Solow (2001). "Short-Term Effects of Boat Traffic on Bottlenose Dolphins, *Tursiops truncatus,* in Sarasota Bay, Florida." *Mar. Mamm. Sci. 17:* 673–688.

Nyman, M., J. Koistinen, M. L. Fant, T. Vartiainen, and E. Helle (2002). "Current Levels of DDT, PCB and Trace Elements in the Baltic Ringed Seals *(Phoca hispida baltica)* and Grey Seals *(Halichoerus grypus)."* *Environ. Pollut. 119:* 399–412.

Nyman, M., M. Bergknut, M. L. Fant, H. Raunio, M. Jestoi, C. Bengs, A. Murk, J. Koistinen, C. Backman, O. Pelkonen, M. Tysklind, T. Hirvi, and E. Helle (2003). "Contaminant Exposure and Effects in Baltic Ringed and Grey Seals as Assessed by Biomarkers." *Mar. Environ. Res. 55:* 73–99.

Odell, D. K., E. D. Asper, J. Baucom, and L. H. Cornell (1980). "A Recurrent Mass Stranding of the False Killer Whale, *Pseudorca crassidens,* in Florida." *Fish. Bull. 78:* 171–177.

O'Leary, B. L. (1984). Aboriginal whaling from the Aleutian Islands to Washington State. *In* "The Gray Whale, *Eschrichtius robustus* (Lilljeborg, 1861)" (M. L. Jones, S. L. Swartz, and S. Leatherwood, eds.), pp. 79–102. Academic Press, New York.

Omura, H. (1984). History of gray whales in Japan. *In* "The Gray Whale, *Eschrichtius robustus* (Lilljeborg, 1861)" (M. L. Jones, S. L. Swartz, and S. Leatherwood, eds.), pp. 27–77. Academic Press, New York.

Orams, M. B., G. J. E. Hill, and A. J. Baglioni, Jr. (1996). ""Pushy" Behavior in a Wild Dolphin Feeding Program at Tangalooma, Australia." *Mar. Mamm. Sci. 12:* 107–117.

O'Shea, T. J., R. R. Reeves and A. K. Long (eds.) (1999). *Marine Mammals and Persistent Ocean Contaminants.* Marine Mammal Commission, Bethesda, MD.

Patenaude, N. J., W. J. Richardson, M. A. Smultea, W. R. Kosk, G. W. Miller, B. Würsig, and C. R. Green, Jr. (2002). "Aircraft Sound and Disturbance to Bowhead and Beluga Whales During Spring Migrations in the Alaskan Beaufort Sea." *Mar. Mamm. Sci. 18:* 309–335.

Perrin, W. F., G. P. Donovan, and J. Barlow (eds.) (1994). "Gillnets and Cetaceans." *Rep. Int. Whal. Commn. Spec. Issue 15:* .

Ralls, K., D. P. DeMaster, and J. A. Estes (1996). "Developing a Criterion for Delisting the Southern Sea Otter Under the U.S. Endangered Species Act." *Conserv. Biol. 10:* 1528–1537.

Reeves, R. R. and J. R. Twiss (eds.) (1999). *Conservation and Management of Marine Mammals.* Smithsonian Institution Press, Washington, DC.

Reijnders, P. J. H., and A. Aguilar (2002). Pollution and marine mammals. *In* "Encyclopedia of Marine Mammals" (W. F. Perrin, B. Würsig, and J. G. M. Thewissen, eds.), pp. 948–957. Academic Press, San Diego.

Reijnders, P. J. H., A. Aguilar, and G. P. Donovan, eds. (1999). "Chemical Pollutants and Cetaceans." *J. Cetacean Res. Manage. Spec. Issue 1.*

Richardson, W. J., K. Finley, G. W. Miller, R. A. Davis, and W. R. Koski (1995a). "Feeding, Social and Migration Behavior of Bowhead Whales, *Balaena mysticetus,* in Baffin Bay vs. the Beaufort Sea-Regions with Different Amounts of Human Activity." *Mar Mamm. Sci. 11:* 1–45.

Richardson, W. J., C. R. Greene, C. I. Malme, D. H. Thompson, S. E. Moore, and B. Würsig (1995b). *Marine Mammals and Noise.* Academic Press, San Diego, CA.

Riedman, M. (1990). *The Pinnipeds: Seals, Sea Lions, and Walruses.* University of California Press, Berkeley.

Salden, D. R. (1988). "Humpback Whale Encounter Rates Offshore of Maui, Hawaii." *J. Wildl. Manage. 52:* 301–304.

Samuels, A., and T. R. Spradlin (1995). "Quantitative Behavioral Study of Bottlenose Dolphins in Swim-with-Dolphin Programs in the United States." *Mar. Mamm. Sci. 11:* 520–544.

Samuels, A., L. Bejder, and S. Heinrich (2000). "A Review of the Literature Pertaining to Swimming with Wild Dolphins." *Rept. Mar. Mamm. Comm.* Silverspring, MD, T74463123, 57p.

Scammon, C. M. (1874). *The Marine Mammals of the Northwestern Coast of North America.* Dover, New York (republished in 1968).

Shaughnessy, P. D. (1982). "The Status of the Amsterdam Island Fur Seal." *Mamm. Seas 4:* 411–421.

Simmonds, M. P., and J. D. Hutchinson (1996). *The Conservation of Whales and Dolphins.* John Wiley and Sons, New York.

Smet. W. M. A. D. (1981). "Evidence of Whaling in the North Sea and English Channel During the Middle Ages." *Mamm. Seas 3:* 301–309.

Springer, A. M., J. A. Estes, G. B. Van Vilet, T. M. Williams, D. F. Doak, E. M. Danner, K. A. Forney, and K. A. Pfister (2003). "Sequential Megafaunal Collapse in the North Pacific Ocean: An Ongoing Legacy of Industrial Whaling?" *Proc. Natl. Acad. Sci. 100:* 12223–12228.

Steller Sea Lion Recovery Team (1994). Recovery Plan for the Steller Sea Lion *(Eumetopias jubatus).* Report to the Office of Protected Resources. NOAA Natl. Mar. Fish. Ser., Place.

Sumich, J., and L. I. Show (1999). "Aerial Survey and Photogrammetric Comparisons of Southbound Migrating Gray Whales in the Southern California Bight, 1988–1990." *Rep. Int. Whal. Commn. Spec. Issue.*

Tonnessen, J. N., and A. O. Johnsen (1982). *The History of Modern Whaling.* [Translated from Norwegian by R. I. Christophersen.] C. Hurst & Company, London, Australian National University Press, Canberra.

Ukishima, Y., Y. Sakane, A. Fukuda, S. Wada, and S. Okada (1995). "Identification of Whale Species by Liquid Chromatographic Analysis of Sarcoplasmic Proteins." *Mar. Mamm. Sci. 11:* 344–361.

Vos, J. G., G. D. Bossart, M. Fournier, and T. J O'Shea (2003). *Toxicology of Marine Mammals.* Taylor and Francis, New York.

Wade, P. et al. (2003). Commercial whaling and "whale killers": A reanalysis of evidence for sequential megafaunal collapse in the North Pacific. XV Biennial Conference on the Biology of Marine Mammals. Greensboro, North Carolina.

Walker, M. M., J. L. Kirschvink, G. Ahmed, and A. E. Dizon (1992). "Evidence that Fin Whales Respond to the Geomagnetic Field During Migration." *J. Exp. Biol. 171:* 67–78.

Wallace, R. L. (1994). *The Marine Mammal Commission Compendium of Selected Treaties, International Agreements, and Other Relevant Documents on Marine Resources, Wildlife, and the Environment,* vols. 1–3. Marine Mammal Commission, Washington, DC.

Walsh, V. M. (1998). "Eliminating Driftnets from the North Pacific Ocean: U.S.–Japanese Cooperation in the International North Pacific Fisheries Commission, 1953–1993." *Ocean Dev. Int. Law* 29: 295–322.

Watson, A. (1994). "Polydactyly in a Bottlenose Dolphin, *Tursiops truncatus.*" *Mar Mamm. Sci. 10:* 93–100.

Wickens, P. (1995). "Namibian Sealing Debacle: An Environmental Disaster and Managerial Blunder." *Afr. Wildl. 49:* 6–7.

Wilkinson, D., and G. A. J. Worthy (1999). Marine Mammal Stranding Networks. *In* Conservation and Management of Marine Mammals (J. R. Twiss Jr. and R. R. Reeves, eds.) pp. 396–411, Smithsonian Inst. Press, Washington, D.C.

Würsig, B., K. Barr, R. Constantine, K. Dudzinski, A. Forest, and S. Yin (1995). Swim-with-dolphin programs: Monitoring results and needs. *Abstr 11th Bienn. Conf. Biol. Mar Mamm.*, Orlando, FL.

Young, N. M., S. Iudecello, K. Evans, and D. Baur (1993). *The Incidental Capture of Marine Mammals in U.S. Fisheries.* Center for Marine Conservation.

APPENDIX

Classification of Marine Mammals

The classification schemes used here follow those proposed by Berta and Wyss (1994) for pinnipeds, Fordyce and Barnes (1994) for cetaceans, and Domning (1994) for sirenians and desmostylians. For the sea otter and polar bear refer to the text. An additional important reference is a review of the taxonomy and distribution of marine mammals provided by Rice (1998). Certain higher taxonomic categories and ranks are not used although hierarchical levels are conveyed by indentation. Phylogenetic systematic studies of marine mammals published in recent years have altered our traditional concepts of included groups and taxa. There continues to be disagreement regarding specific phylogenetic relationships, the validity of particular taxa, and the name and rank of higher taxonomic categories. The higher level classification of various marine mammal taxa proposed by McKenna and Bell (1997) as part of a larger classification system for mammals incorporates both fossil and recent taxa. Detailed review of this scheme, particularly as it relates to marine mammals, is beyond the scope of this book and will require further examination. The most significant changes to traditional higher taxonomic arrangements advocated by McKenna and Bell (1997) are provided under remarks. Emphasis is placed on providing definitions, diagnoses, content (included genera and species), distributional and fossil record data for monophyletic groups at the "family" level. The definition of a taxon is based on ancestry and taxonomic membership and diagnosis is a list of shared derived characters for a particular taxon. The geographic distributions of extant taxa are based on Jefferson *et al.* (1993), Reeves *et al.* (1992), and Riedman (1990) for pinnipeds and sirenians and Leatherwood and Reeves (1983) for cetaceans. Quotation marks are used for taxa whose monophyly is questioned, and a dagger designates extinct taxa. The time scale used (Figure 1.1) follows Harland *et al.* (1990) and Berggren *et al.* (1995) and the geographic coverage scheme (Figure 1) and age correlations for fossil taxa follows Deméré (1994b).

491

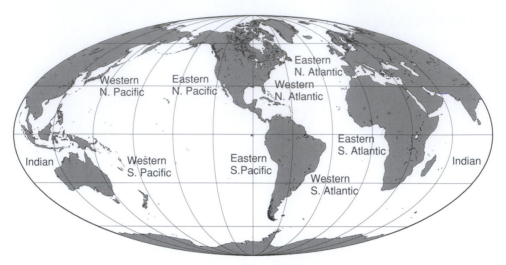

Figure 1.   World map with ocean basins discussed in text. (From Smith *et al.*, 1994.)

CARNIVORA

PINNIPEDIA (Illiger, 1811)

Diagnosis—Exclusive of fossil taxa *Enaliarctos†* and *Pteronarctos†:* large infraorbital foramen; anterior palatine foramina anterior of maxilla-palatine suture; round window large with round window fossula; postglenoid foramen vestigial or absent; enlarged jugular foramen; basal whorl of cochlea enlarged; processus gracilis and anterior lamina of malleus reduced; M1-2 reduced in size relative to premolars; M1-2 cingulum reduced or absent; M1 metaconid reduced or absent; greater and lesser humeral tuberosities enlarged; short robust humerus; shallow olecranon fossa; digit I on manus emphasized; digit I and V on the pes emphasized; short ilium; strong medial inclination of femoral condyles; large cochlear aqueduct; middle ear cavity and external auditory meatus with distensible cavernous tissue; deciduous dentition reduced; number of lower incisors reduced; fore flipper claws short; manus, digit V intermediate phalanx strongly reduced; embrasure pit p4-M1 shallow or absent; mastoid process close to paroccipital process, the two connected by a high continuous ridge; maxilla makes significant contribution to the orbital wall, lacrimal fuses early and does not contact jugal; pit for tensor tympani absent; 13 lingual cingulum absent; M1-2 trigonid suppressed; supinator ridge on humerus absent; olecranon process laterally flattened, expanded distal half; metacarpal heads smooth with flat phalanges, hinge-like articulations; pes long and flattened, metatarsal shafts with flattened heads, hinge-like articulations; pubic symphysis unfused; five lumbar vertebrae; fovea for teres femoris ligament strongly reduced or absent; greater femoral trochanter large and flattened.

Definition—The monophyletic group including the common ancestor of the Otariidae and all of its descendants including the Odobenidae, Phocidae, and *Pinnarctidion†, Desmatophoca†,* and *Allodesmus†.*

Distribution—Pinnipeds occur in all of the world's oceans (Figures 2–6).

Fossil history—Pinnipeds are recognized in the fossil record beginning in the early early Miocene (23 Ma)-Pleistocene, North Pacific (North America, Japan); late Miocene-Pleistocene, eastern South Pacific (South America); late middle Miocene-Pleistocene, North Atlantic (western Europe, North America); latest Miocene-early Pliocene, western South Pacific (Australasia); early Pliocene, eastern South Atlantic (South Africa) (Deméré *et al.*, 2003).

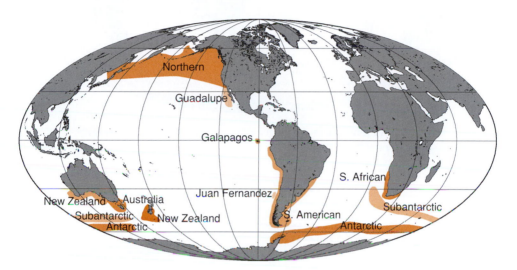

Figure 2.   Distribution of fur seals. (Based on Riedman, 1990.)

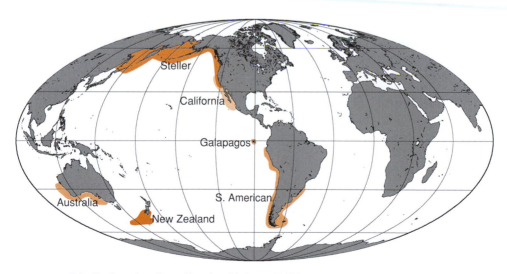

Figure 3.   Distribution of sea lions. (Based on Riedman, 1990.)

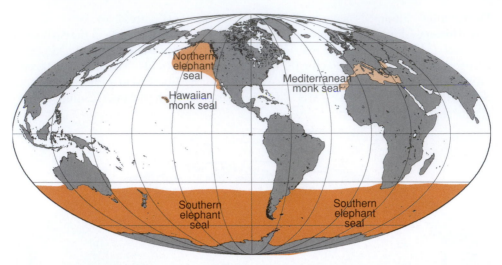

Figure 4.   Distribution of "monachines" (monk and elephant seals). (Based on Riedman, 1990.)

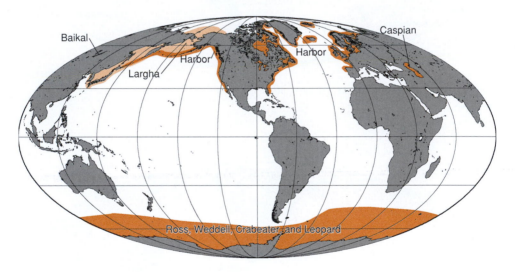

Figure 5.   Distribution of some phocine seals. (Based on Riedman, 1990.)

The earlier diverging pinnipedimorphs *Enaliarctos†* and *Pteronarctos†* are known from the late Oligocene through the early Miocene (27–16 Ma), North Pacific (North America, Japan; Deméré, 1994b). Basal pinnipedimorphs are reviewed by Berta (1991, 1994).

Remarks—McKenna and Bell (1997) divide pinnipeds into two families, the Otariidae and the Phocidae; the latter includes walruses as a subfamily. These families together

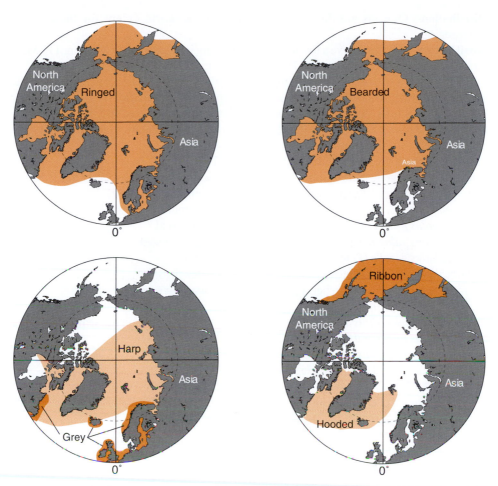

Figure 6.   Distribution of Arctic phocine seals (ringed, harp and ribbon seals). (Based on Riedman, 1990.)

with a number of extinct taxa (i.e., *Pachycynodon†, Kolponomos†, Allocyon†, Enaliarctos†,* and *Pteronarctos†*) are grouped in the superfamily Phocoidea.

Content—Composed of three families containing all living pinnipeds. A number of fossil genera are included, among which are several not belonging to modern families (i.e., *Pinnarctidion†, Allodesmus†* and *Desmatophoca†*).

OTARIIDAE (Gill, 1866)

Diagnosis—Supraorbital processes large and shelf-like, particularly among adult males; secondary spine on the scapula; uniformly spaced pelage units; trachea with anterior bifurcation of bronchi (Berta and Wyss, 1994).

Definition—The monophyletic group including the common ancestor of *Pithanotaria†* and all of its descendants including *"Thalassoleon"†, Hydrarctos†, Arctocephalus, Callorhinus, Otaria, Zalophus, Eumetopias, Neophoca,* and *Phocarctos.*

Distribution—Otariids are distributed throughout the world except for the extreme polar regions.

Fossil history—Otariids have a fossil record beginning in the early late Miocene (11 Ma)-Pleistocene, eastern North Pacific; Pleistocene, eastern South Pacific (Peru).

Extant genera with a fossil record include *Neophoca,* early Pleistocene, South Pacific (New Zealand); *Arctocephalus,* late Pleistocene, South Pacific (Africa) and late Pleistocene, South America; *Callorhinus,* late Pliocene, eastern North Pacific (California, Japan); *Eumetopias,* Pleistocene, eastern North Pacific (Japan) and eastern South Pacific (South America).

Extinct genera are reviewed by Repenning and Tedford (1977).

Content—Seven genera containing 14–16 recent species and 2–5 subspecies. See Brunner (2004) for a different interpretation of otariid taxonomy.

Arctocephalus australis (Zimmerman, 1783)—South American fur seal
Arctocephalus forsteri (Lesson, 1828)—New Zealand fur seal
Arctocephalus galapagoensis (Heller, 1904)—Galapagos fur seal
Arctocephalus gazella (Peters, 1875)—Antarctic fur seal
Arctocephalus philippii (Peters, 1866)—Juan Fernandez fur seal
Arctocephalus pusillus (Schreber, 1775)—South African fur seal
    Arctocephalus pusillus pusillus
    Arctocephalus pusillus doriferus (Wood Jones, 1925)—Australian fur seal
Arctocephalus townsendi (Merriam, 1897)—Guadalupe fur seal
Arctocephalus tropicalis (Gray, 1872)—subantarctic fur seal
Callorhinus ursinus (Linnaeus, 1758)—northern fur seal
Eumetopias jubatus (Schreber, 1776)—Steller sea lion
Neophoca cinerea (Péron, 1816)—Australian sea lion
Otaria byronia (Blainville, 1820)—southern sea lion
Phocarctos hookeri (Gray, 1844)—Hooker's sea lion or New Zealand sea lion
Zalophus californianus (Lesson, 1828)—California sea lion
    Zalophus californianus californianus
    Zalophus californianus wollebacki (Sivertsen, 1953)—Galapagos sea lion
    Zalophus californianus japonicus (Peters, 1866)—Japanese sea lion

ODOBENIDAE (Allen, 1880)

Diagnosis—Pterygoid strut broad and thick; P4 protocone shelf strong and posterolingually placed with convex posterior margin; m1 talonid absent; calcaneum with prominent medial tuberosity (Deméré and Berta, 2001).

Definition—The monophyletic group containing the common ancestor of *Neotherium†* and all of its descendants including *Aivukus†, Alachtherium†, Dusignathus†, Gomphotaria†, Imagotaria†, Odobenus, Pliopedia†, Pontolis†, Proneotherium†, Prorosmarus†, Protodobenus†, Prototaria†* and *Valenictus†* (Deméré, 1994a).

Distribution—Modern walruses inhabit the shallow, circumpolar Arctic coasts.

Fossil history—Walruses are known as far back as the early middle Miocene (16 Ma)-Pleistocene, eastern North Pacific (North America); early Pliocene-Pleistocene, North Atlantic (western Europe); late Pleistocene, eastern North Pacific (Japan) (Deméré, 1994b).

Extinct genera are reviewed by Deméré (1994a) and Kohno *et al.* (1995).

Content—One genus and species and 2–3 subspecies are recognized.

Odobenus rosmarus (Linnaeus, 1758)—walrus

    Odobenus rosmarus divergens (Illiger, 1815)—Pacific walrus

    Odobenus rosmarus rosmarus (Linneaus, 1758)—Atlantic walrus

    Odobenus rosmarus laptevi (Chapskii, 1940)

PHOCIDAE (Gray, 1821)

Diagnosis—Lack of the ability to draw the hind limbs forward under the body due to a massively developed astragalar process and greatly reduced calcaneal tuber; lack of an alisphenoid canal; lack of an auditory meatus and prefacial commissure; pachyostotic mastoid region; pit for insertion of tympanohyal anterior to stylomastoid foramen; greatly inflated heavily ossified caudal entotympanic; transversely directed basilar cochlear whorl; opening of cochlear fenestra outside the tympanic cavity to form a cochlear foramen; strongly everted ilia; insertion for the ilial psoas major muscle.

Definition—The monophyletic group containing the common ancestor of *Leptophoca†*, and all of its descendants including *Acrophoca†*, *Homiphoca†*, *Piscophoca†*, "*Monachus*", *Mirounga*, *Hydrurga*, *Leptonychotes*, *Lobodon*, *Ommatophoca*, *Erignathus*, *Cystophora*, *Halichoerus*, *Histriophoca*, *Pagophilus*, and *Phoca* (includes *Pusa*).

Distribution—Phocids are distributed throughout the world (see Figures 4–6).

Fossil history—A record of phocids from the late Oligocene (29–23 Ma) has recently been reported (Koretsky and Sanders, 2002). However, the stratigraphic provenance of the specimen is uncertain. Phocids are well documented from the middle Miocene-Pleistocene, western North Atlantic (western Europe), eastern North Atlantic (Maryland and Virginia); early Pliocene-Pleistocene, South Pacific (Africa); middle Miocene-Pleistocene, South America.

Extinct genera are reviewed by Muizon (1981, 1982, 1984), Hendey and Repenning (1972), and Cozzuol (1996).

Content—Thirteen genera, 19 recent species and 9 subspecies. Taxonomy follows Rice (1998) but see Stanley *et al.* (1996) for *Phoca vitulina*.

Cystophora cristata (Nilsson, 1841)—Hooded seal

Erignathus barbatus (Erxleben, 1777)—Bearded seal

    Erignathus barbatus barbatus

    Erignathus barbatus nauticus (Pallas, 1811)

Halichoerus grypus (Fabricus, 1791)—grey or Atlantic seal

    Halichoerus halichoerus grypus

    Halichoerus halichoerus macrorhynchus (Hornschuch and Schilling, 1851)

Histriophoca fasciata (Zimmermann, 1783)—ribbon seal

Hydruga leptonyx (Blainville, 1820)—leopard seal

Leptonychotes weddelli (Lesson, 1826)—Weddell seal

Lobodon carcinophagus (Hombron and Jacquinot, 1842)—crabeater seal

Mirounga angustirostris (Gill, 1866)—northern elephant seal

Mirounga leonina (Linnaeus, 1758)—southern elephant seal

Monachus monachus (Hermann, 1779)—Mediterranean monk seal

Monachus schauinslandi (Matschie, 1905)—Hawaiian monk seal

Monachus tropicalis (Gray, 1950)—West Indian monk seal
Ommatophoca rossi (Gray, 1844)—Ross seal
Pagophilus groenlandicus (Erxleben, 1777)—harp seal
Phoca caspica (Gmelin, 1788)—Caspian seal
Phoca hispida botnica (Gmelin, 1788)
Phoca hispida hispida (Schreber, 1775)—ringed seal
Phoca hispida ladogensis (Nordquist, 1899)
Phoca hispida ochotensis (Pallas, 1811)
Phoca hispida saimensis (Nordquist, 1899)
Phoca largha (Pallas, 1811)—spotted seal
Phoca sibirica (Gmelin, 1788)—Baikal seal
Phoca vitulina (Linnaeus, 1758)—harbor seal
   Phoca vitulina concolor (De Kay, 1842)
   Phoca vitulina mellonae—freshwater harbor seal
   Phoca vitulina richardsi (Gray, 1864)—Pacific harbor seal
   Phoca vitulina stejnegeri (Allen, 1902)
   Phoca vitulina vitulina

CETACEA (Brisson, 1762)

Diagnosis—Mastoid process of petrosal not exposed posteriorly; pachyosteosclerotic bulla; bulla articulates with the squamosal via a circular entoglenoid process; fourth upper premolar protocone absent; fourth upper premolar paracone height twice that of the first upper molar (Geisler, 2001).

Definition—The monophyletic group including the common ancestor of *Pakicetus†* and all of its descendants including *Ambulocetus†,* other archaeocetes, Odontoceti, and Mysticeti (Thewissen, 1994).

Distribution—Cetaceans are distributed throughout the world's oceans.

Fossil history—Cetaceans are recognized in the fossil record beginning in the early Eocene (about 53–54 Ma) in the eastern Tethys (India and Pakistan); middle Eocene, central Tethys (Egypt); late Eocene, western Tethys (southeastern United States, Australasia); and extending through the Pleistocene.

Remarks—McKenna and Bell (1997) recognize the Cetacea as a suborder in the Order Cete (including Mesonychidae and two other families).

Content—Composed of two suborders containing all living cetaceans and fossil taxa including *Pakicetus†, Ambulocetus†,* and other archaeocetes (see text).

ODONTOCETI (Flower, 1867)

Diagnosis—Nasals elevated above the rostrum; frontals higher than nasals; premaxillary foramen present; maxillae overlay supraorbital process (Geisler and Sanders, 2003).

Definition—The monophyletic group including the common ancestor of the extinct Agorphidae† and all of its descendants including the archaic families Albeirodontidae†, Dalpiazinindae†, Eurhinodelphidae†, Kentriodontidae†, Squalodelphidae†, Squalodontidae†, and Odobenocetopsidae† and modern families

Delphinidae, Iniidae, Kogiidae, Lipotidae, Monodontidae, Phocoenidae, Physeteridae, Platanistidae, Pontoporiidae, and Ziphiidae.

Fossil history—The earliest odontocetes are known from the late Oligocene (29.3–23 Ma)-Pleistocene of western North Atlantic (southeastern United States), Paratethys (Europe, Asia), eastern North Pacific (Japan), and western South Pacific (New Zealand); early early Miocene-Pleistocene, western South Atlantic (South America), and eastern North Pacific (Oregon and Washington) (Deméré, 1994b).

DELPHINIDAE (Gray, 1821) (incl. Holodontidae Brandt, 1873; Hemisyntrachelidae Slijper, 1936)

Definition—The monophyletic group including the common ancestor of *Albireo†* and all of its descendants including *Cephalorhynchus, Delphinus, Feresa, Globicephala, Grampus, Lagenorhynchus, Lissodelphis, Orcaella, Orcinus, Pepenocephala, Pseudorca, Sotalia, Sousa, Stenella,* and *Tursiops*.

Distribution—Widely distributed in all oceans of the world.

Fossil history—The oldest delphinids known are from the latest Miocene (11–7 Ma) Pleistocene, western North Atlantic (Virginia and North Carolina); late Miocene-Pleistocene, western North Pacific (New Zealand, Japan); early Pliocene, eastern South Pacific (Peru); late Pleistocene, eastern North Atlantic (western Europe) (Deméré, 1994b).

Globicephala and *Pseudorca* are known from the late Pleistocene (790–10,000 yrs.), eastern North Atlantic (Florida; Morgan, 1994); *Delphinus* and *Tursiops* are known from late Pleistocene, eastern North Atlantic (western Europe; Van der Feen, 1968); *Delphinus* is known from the late Pliocene, New Zealand (Fordyce, 1991; McKee, 1994).

Extinct genera are reviewed by Fordyce and Muizon (2001), and Cozzuol (1996; South American record).

Content—17 genera and 36 recent species are recognized. Heyning and Perrin (1994) review the taxonomy of *Delphinus;* Jefferson and Van Waerebeek (2002) update the taxonomic status of common dolphins).

Cephalorhynchus commersonii (Lacépéde, 1804)—Commerson's dolphin
    Cephalorhynchus commersonii commersonii
    Cephalorhynchus commersonii subsp.
Cephalorhynchus eutropia (Gray, 1846)—black dolphin
Cephalorhynchus heavisidii (Gray, 1828)—Heaviside's dolphin
Cephalorhynchus hectori (van Beneden, 1881)—Hector's dolphin
    Cephalorhynchus hectori hectori (Baker *et al.*, 2002)—South Island Hector's dolphin
    Cephalorhynychus hectori maui
Delphinus capensis (Gray, 1828)—long-beaked common dolphin
    Delphinus capensis capensis
    Delphinus capensis tropicalis (van Bree, 1971)—Arabian common dolphin
Delphinus delphis (Linnaeus, 1758)—common dolphin
Feresa attenuata (Gray, 1874)—pygmy killer whale
Globicephala macrorhynchus (Gray, 1846)—short-finned pilot whale
Globicephala melas (Traill, 1809)—long-finned pilot whale
    Globicephala melas edwardii
    Globicephala melas melas
    Globicephala melas subsp.

Grampus griseus (G. Cuvier, 1812)—Risso's dolphin
Lagenodelphis hosei (Fraser, 1956)—Fraser's dolphin
Lagenorhynchus acutus (Gray, 1828)—Atlantic white-sided dolphin
Lagenorhynchus albirostris (Gray, 1846)—white-beaked dolphin
Lagenorhynchus australis (Peale, 1848)—Peale's dolphin
Lagenorhynchus cruciger (Quoy and Gaimard, 1824)—hourglass dolphin
Lagenorhynchus obliquidens (Gill, 1865)—Pacific white-sided dolphin
Lagenorhynchus obscurus (Gray, 1828)—dusky dolphin
   Lagenorhynchus obscurus fitzroyi
   Lagenorhynchus obscurus obscurus
   Lagenorhynchus obscurus subsp.
Lissodelphis borealis (Peale, 1848)—northern right whale dolphin
Lissodelphis peronii (Lacépéde, 1804)—southern right whale dolphin
Orcaella brevirostris (Gray, 1866)—Irrawaddy dolphin
Oracella heinsohni (Beasley *et al.*, 2005)
Orcinus orca (Linnaeus, 1758)—killer whale
Peponocephala electra (Gray, 1846)—melon-headed whale
Pseudorca crassidens (Owen, 1846)—false killer whale
Sotalia fluviatilis (Gervais, 1853)—tucuxi
Sousa chinensis (Kukenthal, 1892)—Indo-Pacific hump-backed dolphin
Sousa plumbea (Cuvier, 1829)—Indian humpback dolphin
Sousa teuszii (Kukenthal, 1892)—Atlantic hump-backed dolphin
Stenella attenuata (Gray, 1846)—spotted dolphin
   Stenella attenuata subsp. A of Perrin (1975)
   Stenella attenuata subsp. B of Perrin (1975)
   Stenella attenuata graffmani (Lonnberg, 1934)
Stenella clymene (Gray, 1850)—short-snouted spinner dolphin
Stenella coeruleoalba (Meyen, 1833)—striped dolphin
Stenella frontalis (Curier, 1829)—Atlantic spotted dolphin
Stenella longirostris (Gray, 1828)-long-snouted spinner dolphin
   Stenella longirostris orientalis (Perrin, 1990)
   Stenella longirostris centroamericana (Perrin, 1990)
Steno bredanensis (Lesson, 1828)—rough-toothed dolphin
Tursiops aduncus (Ehrenberg, 1833)—Indian Ocean bottlenose dolphin
Tursiops truncatus (Montagu, 1821)—bottlenose dolphin

PHYSETERIDAE (Gray, 1821)

Definition—The monophyletic group including the common ancestor of *Apenophyseter†, "Aulophyseter"†, Diaphorocetus†, Helvicetus†, Idiorophus†, Orycterocetus†, Physeterula†, Placoziphius†, Preaulophyseter†, Prophyseter†, Scaldicetus†, Thalassocetus†,* and all of its descendants including *Physeter.*
Distribution—Widely distributed in all oceans of the world, avoiding only the polar pack ice in both hemispheres.
Fossil history—The record of physeterids goes back at least to the early early Miocene (23.3–21.5 Ma), Mediterranean (Italy), Parathys (Europe), and western South Atlantic (Argentina); late early Miocene, western South Pacific (Australasia), eastern

North Pacific (central California), and western North Pacific (Japan); early middle Miocene, western North Atlantic (Maryland and Virginia); late middle Miocene-late Miocene, eastern North Atlantic (western Europe); late middle Miocene-late Pliocene, eastern North Atlantic (Florida); late late Miocene, eastern South Pacific (Peru); late Pliocene, eastern North Pacific (California), and perhaps earlier if *Ferecetotherium†* is included from the late Oligocene (29–23 Ma; Deméré, 1994b).

Extinct genera are reviewed by Fordyce and Muizon (2001) see Cozzuol (1996) for South American record).

Remarks—McKenna and Bell (1997) include the pygmy sperm whale as a subfamily.

Content—One genus and species are recognized.

Physeter macrocephalus (Linnaeus, 1758) (syn. *Physeter catadon*)—sperm whale

KOGIIDAE (Gill, 1871; Miller, 1923)

Definition—The monophyletic group including the common ancestor of *Kogiopsis†*, *Scaphokogia†* and all of its descendants including *Kogia*.

Distribution—Distributed in all oceans of the world (temperate, subtropical and tropical waters), with the range of *Kogia breviceps* recorded as in warmer seas (Handley, 1966).

Fossil history—The fossil record of kogiids goes back to the early late Miocene (8.5–5.2 Ma), eastern South Pacific (Peru); early Pliocene, western North Atlantic (Florida); late Pliocene, Mediterranean (Italy).

Extinct genera are reviewed by Fordyce and Muizon (2001).

Remarks—see Physeteridae.

Content—One genus and two recent species are recognized.

Kogia breviceps (de Blainville, 1838)—pygmy sperm whale

Kogia simus (Owen, 1866)—dwarf sperm whale

ZIPHIIDAE (Gray, 1865) (including Choncziphiidae Cope, 1895; Hyperoodontidae Gray, 1866)

Definition—The monophyletic group including the common ancestor of *Choneziphius†*, *Cetorhynchus†*, *Ninoziphius†*, *Paleoziphius†*, *Squaloziphius†*, *Ziphiorostrum†*, and all of its descendants including *Berardius*, *Hyperoodon*, *Mesoplodon*, *Tasmacetus*, and *Ziphius*.

Distribution—Temperate and tropical waters of all oceans, some species located far offshore in deep waters (exceptions *Berardius bairdii* from North Pacific and *Mesoplodon bidens* from cold temperature to subarctic North Atlantic).

Fossil history—The record of ziphiids extends back to the early early Miocene (23.3–21.5 Ma), eastern North Pacific (Oregon and Washington); later middle Miocene (14.6–11 Ma), eastern North Atlantic (western Europe); early Pliocene, North America; middle-late Miocene, western North Pacific (Japan); latest Miocene-early Pliocene, Australasia, western North Atlantic (Virginia and North Carolina, Florida), and eastern South Pacific (Peru); late Pliocene, Mediterranean (Italy), and Australasia (Deméré, 1994b).

Mesoplodon is known from the late middle Miocene (14.6–11 Ma), eastern North Atlantic (western Europe) and late Pleistocene-Holocene (western South Pacific, Argentina).

Extinct genera are reviewed by Fordyce and Muizon (2001) and Cozzuol (1996; South American record).

Remarks—As noted by McKenna and Bell (1997) although the name Hyperoodontidae has priority over Ziphiidae, the latter has been in universal use for more than a century.

Content—Five genera and 19 recent species.

Berardius arnuxii (Duvernoy, 1851)—Arnoux's beaked whale

Berardius bairdii (Stejneger, 1883)—Baird's beaked whale

Hyperoodon ampullatus (Forster, 1770)—northern bottlenose whale

Hyperoodon planifrons (Flower, 1862)—southern bottlenose whale

Indopacetus pacificus (Longman, 1926)—Indo-Pacific beaked whale, Longman's beaked whale

Mesoplodon bidens (Sowerby, 1804)—Sowerby's beaked whale

Mesoplodon bowdoini (Andrews, 1908)—Andrews' beaked whale

Mesoplodon carlhubbsi (Moore, 1963)—Hubb's beaked whale

Mesoplodon denisrostris (de Blainville, 1817)—Blainville's beaked whale

Mesoplodon europaeus (Gervais, 1855)—Gervais' beaked whale

Mesoplodon ginkgodensis (Nishiwaki and Kamiya, 1958)—ginkgo-toothed beaked whale

Mesoplodon grayi (von Haast, 1896)—Gray's beaked whale

Mesoplodon hectori (Gray, 1871)—Hector's beaked whale

Mesoplodon layardii (Gray, 1865)—strap-toothed beaked whale

Mesoplodon mirus (True, 1913)—True's beaked whale

Mesoplodon perrini (Dalebout *et al.*, 2002)—Perrin's beaked whale

Mesoplodon peruvianus (Reyes *et al.*, 1991)—Peruvian beaked whale

Mesoplodon stejnegeri (True, 1885)—Stejneger's beaked whale

Mesoplodon traversii (Gray, 1874)—Spade toothed whale

Tasmacetus shepardi (Oliver, 1937)—Tasman beaked whale

Ziphius cavirostris (Cuvier, 1823)—Cuvier's beaked whale

PLATANISTIDAE (Gray, 1863)

Definition—The monophyletic group including the common ancestor of *Pomatodelphis†, Zarachis†,* and all of its descendants including *Platanista*.

Distribution—Indus and Ganges River drainages of Pakistan and India.

Fossil history—Fossil plantanistids are known from the late early Miocene (21.5–16.3 Ma), western North Atlantic (Florida); early middle Miocene, western North Atlantic (Maryland and Virginia, Florida).

The stratigraphic occurrence of platanistids is further discussed in Gottfried *et al.* (1994), Morgan (1994), and Hamilton *et al.* (2001).

Content—A single genus and Recent species and two subspecies are recognized following Perrin and Brownell (2001).

Platanista gangetica

   Platanista gangetica gangetica (Roxburgh, 1801)—Ganges susu

   Platanista gangetica minor (Owen, 1853)—Indus susu

INIIDAE (Flower, 1867) (includes Pontplanodidae Ameghino, 1894; Saurocetidae Ameghino, 1891; Saurodelphidae, Abel, 1905).

Definition—The monophyletic group including the common ancestor of *Goniodelphis†, Ischyrorhynchus†, Saurocetes†* and all of its descendants including *Inia*.

Distribution—Amazon and Orinoco River drainages in central and northern South America.

Fossil history—Iniids are known from early late Miocene (10.4–6.7 Ma), western South Atlantic (Argentina, Uruguay); early Pliocene, western North Atlantic (Florida).

Fossil iniids are reviewed in Fordyce and Muizon (2001) and see Cozzuol (1996) for South American record.

Content—A single genus and two Recent species and three subspecies are recognized (Best and da Silva, 1989).

Inia geoffrenesis (de Blainville, 1817)—boto; Amazon river dolphin

   Inia geoffrensis humboldtiana (Pilleri and Gihr, 1977)

   Inia geoffrensis geoffrensis (Van Bree and Robineau, 1973)

   Inia geoffrensis bolivienis (D'Orbigny, 1834)

LIPOTIDAE (Zhou Qian and Li, 1979)

Definition—The monophyletic group including the common ancestor of *Prolipotes†* and all of its descendants including *Lipotes*.

Distribution—Confined to Yangtze (Changjiang) River drainage on the Chinese mainland.

Fossil history—A single extinct genus, *Prolipotes,* is reported from the Neogene of China (Zhou *et al.,* 1984) has not been confirmed as a lipotiid.

Content—A single genus and species is recognized.

Lipotes vexillifer (Miller, 1918)—baiji, Yangtse river dolphin

PONTOPORIIDAE (Gill, 1871; Kasuya, 1973) (=Stenodelphinae True, 1908)

Definition—The monophyletic group including the common ancestor of *Brachydelphis†, Parapontoporia†, Piscorhynchus†, Pliopontes†, Pontistes†* and all of its descendants including *Pontoporia*.

Distribution—Restricted to coastal central Atlantic waters of South America.

Fossil history—Fossil record of pontoporiids extends back to the late Miocene (10.4–6.7 Ma), western South Atlantic (Argentina); late late Miocene, eastern North Pacific (California, Baja California), and eastern South Pacific (Peru); early Pliocene, western North Atlantic (Virginia and North Carolina), eastern North Pacific (California), and eastern South Pacific (Peru).

Pontoporia is reported from the early late Miocene (10.4–6.4 Ma), western South Atlantic (Argentina).

Extinct genera are reviewed by Fordyce and Muizon (2001) and see Cozzuol (1996) for South American record.

Content—A single genus and species is recognized.

Pontoporia blainvillei (Gervais and d'Orbigny, 1844)—franciscana, La Plata dolphin

MONODONTIDAE (Gray, 1821; Miller and Kellogg, 1955)

Definition—The monophyletic group including the common ancestor of *Denebola†* and all of its descendants including *Monodon* and *Delphinapterus*.

Distribution—The beluga and narwhal are confined to Arctic and subarctic waters.

Fossil history—Monodontids are known in the fossil record as far back as early late Miocene (10.4–6.7 Ma); late late Miocene, eastern North Pacific (California); late late Miocene, eastern North Pacific (Baja California); latest Miocene-early Pliocene, western North Atlantic (Virginia and North Carolina) and eastern North Pacific (California); late Pliocene, eastern North Pacific (California); late Pleistocene, eastern North Atlantic (western Europe) and Arctic Ocean (northern Alaska).

Extinct genera are reviewed by Fordyce and Muizon (2001).

The extant genus *Delphinapterus* is known from the western North Atlantic (Virginia and North Carolina; Whitmore, 1994). Extant species include *Monodon monoceras* and *Delphinapterus leucas* late Pleistocene, western North Atlantic (Canada; Harrington, 1977).

Content—Two genera and two species are recognized.

Delphinapterus leucas (Pallas, 1776)—white whale or beluga
Monodon monoceros (Linnaeus, 1758)—narwhal

PHOCOENIDAE (Gray, 1825; Bravard, 1885)

Definition—The monophyletic group including the common ancestor of *Australithax†, Lomacetus†, Piscolithax†, Piscorhynchus†, Salumiphocoena†,* and all of its descendants including *Neophocoena, Phocoena,* and *Phocoenoides*.

Distribution—Phocoenids are distributed throughout the world.

Fossil history—Phocoenids are known in the fossil record from early late Miocene-late late Miocene (10.4–6.7 Ma), eastern North Pacific (California); late late Miocene, eastern North Pacific (Baja California), and eastern South Pacific (Peru); early Pliocene, western North Atlantic (Virginia and North Carolina); early-late Pliocene, eastern North Pacific (California); late Pleistocene, eastern North Atlantic (western Europe), and western North Atlantic (Canada).

Extinct genera are reviewed by Fordyce and Muizon (2001) and see Cozzuol (1996) for South American record.

The extant species *Phocoena phocoena* is known from the late Pleistocene, eastern North Atlantic (western Europe; Van der Feen, 1968) and western North Atlantic (Canada; Harrington, 1977).

Content—Three genera, six species and three subspecies are recognized (Rosel *et al.*, 1995).

Neophocaena phocaenoides (Cuvier, 1829)—finless porpoise

Phocoena (Australophocoena) dioptrica (Lahille, 1912; Barnes, 1985)—spectacled
  porpoise
Phocoena phocoena (Linnaeus, 1758)—harbour porpoise
  Phocaena phocoena phocoena
  Phocoena phocoena vomerina (Gill, 1865)
  Phocoena phocoena relicta
  Phocoena phocoena subsp.
Phocoena sinus (Norris and McFarland, 1958)—vaquita
Phocoena spinipinnis (Burmeister, 1865)—Burmeister's porpoise
Phocoenoides dalli (True, 1885)—Dall's porpoise

MYSTICETI (Flower, 1864)

Diagnosis—Lateral margin of maxillae thin; descending process of maxilla present as infraorbital plate; posterior portion of vomer exposed on basicranium and covering basisphe-
noid/basioccipital suture; basioccipital crest wide and bulbous; mandibular symphysis
unfused with only a ligamental or connective tissue attachment, marked by anteroposterior
groove; mandibular symphysis short with large boss dorsal to groove; dorsal aspect of
mandible curved laterally and baleen plates numerous and thin. (Deméré *et al.*, in press).
Definition—The monophyletic group including the common ancestor of the extinct
Aetiocetidae† and all of its descendants including the archaic families Llanocetidae†,
Mammalodontidae†, "Cethotheriidae"† and modern families Balaenidae,
Balaenopteridae, Eschrichtiidae, and Neobalaenidae.
Fossil history—The earliest mysticetes are known from the late Eocene-early Oligocene
(34–33 Ma), Antarctica; late Oligocene-Pleistocene, eastern North Pacific (California),
western North Pacific (Japan, Australasia); early early Miocene-Pleistocene, western
South Atlantic; early middle Miocene, western North Atlantic; late middle Miocene-
Pleistocene, eastern North Atlantic (western Europe) (Deméré *et al.*, in press).

BALAENIDAE (Gray, 1825)

Diagnosis—Lack of dorsal fin; right and left baleen racks separated by anterior cleft;
long palatine with overlap of pterygoids nearly reaching pterygoid fossa; supraorbital
process of frontal abruptly depressed below vertex (Deméré *et al.*, in press).
Definition—The monophyletic group including the common ancestor of *Balaenula†*
and all of its descendants including *Balaena* and *Eubalaena*.
Distribution—Bowheads are circumpolar in the Arctic and right whales are known in
temperate waters of both hemispheres.
Fossil history—Balaenids are known in the fossil record from early late Miocene (23 Ma),
western North Atlantic (Florida); late late Miocene, eastern North Pacific (California);
late Miocene-early Pliocene, western South Pacific (Australasia); early Pliocene, western
North Atlantic (Virginia and North Carolina), eastern North Pacific (California), and
western North Pacific (Japan); late Pliocene, eastern North Pacific (California), eastern
North Pacific (Florida), and Mediterranean (Italy) (Deméré, 1994b).

The genus *Balaena* is reported from the early Pliocene (5.2–3.4 Ma), western North Atlantic (Virginia and North Carolina); late Pliocene, Mediterranean (Italy); early Pleistocene, eastern North Pacific (Oregon).

Extant species *Balaena mysticetus* is reported from the late Pleistocene, North Atlantic (Canada); *Eubalaena glacialis,* late Pleistocene, western North Atlantic (Florida).

Remarks—The taxonomic history of fossil balaenids has been reviewed by McLeod *et al.* (1993).

Content—Two genera and three recent species are recognized.

Balaena mysticetus (Linnaeus, 1758)—bowhead whale

Eubalaena (=Balaena) glacialis (Miiller, 1776)—northern right whale

E. (= Balaena) australis (Desmoulins, 1822)—southern right whale

ESCHRICHTIIDAE (Ellerman and Morrison-Scott, 1951) (=Rhachianectidae Weber, 1904).

Diagnosis—Mandibular symphysis short with large boss dorsal to groove; mandibular condyle directed posterodorsally; dorsal humps present and baleen plates few and thick (Deméré *et al.*, in press).

Definition—The monophyletic group including San Diego Natural History Museum 95017 and "*Balaenoptera" gastaldi, Eschrichtius,* and all of its descendants (Deméré *et al.*, in press).

Distribution—Three or more stocks of gray whales are recognized: North Atlantic stock (or stocks), apparently exterminated in recent historical times (perhaps as late as the seventeenth or eighteenth century); a Korean or western Pacific stock hunted at least until 1966, and now close to extinction; and a California or eastern Pacific stock (Leatherwood and Reeves, 1983).

Fossil history—The genus *Eschrichtius* is known from the late Pleistocene, western North Atlantic (Florida), eastern North Pacific (California; Deméré *et al.*, in press).

Remarks—see Balaenopteridae.

Content—One genus and recent species.

Eschrichtius robustus (Lilljeborg, 1861)—gray whale

BALAENOPTERIDAE (Gray, 1864)

Diagnosis—Numerous throat grooves that extend past the throat region (Barnes and McLeod, 1984).

Definition—The monophyletic group including the common ancestor of *Idiocetus†* and all of its descendants including *Balaenoptera* and *Megaptera* (Demére *et al.*, in press).

Distribution—Humpback, minke, fin, and blue whales are widely distributed in all oceans. Sei whales and Bryde's whales are usually found in tropical and warm temperate waters and not in polar regions as are other rorquals.

Fossil history—Balaenopterids are known from the early late Miocene (12 Ma), eastern North Pacific (California); late late Miocene, western North Atlantic (Maryland and Virginia), eastern North Pacific (southern California, Baja California), western North Pacific (Japan), and eastern South Pacific (Peru); latest Miocene-early Pliocene, Australasia; early Pliocene, eastern North Atlantic (western Europe), western North Atlantic (Virginia and North Carolina), western North Atlantic (Florida), eastern

North Pacific (California), western North Pacific (Japan), and eastern South Pacific (Peru); late Pliocene, Mediterranean (Italy), western North Atlantic (Florida), and eastern North Pacific (California, western South Pacific, Australasia); early Pleistocene, western North Atlantic (western Europe); late Pleistocene, western North Atlantic (Canada) and eastern North Pacific (California).

The genus *Megaptera is* known from the early late Miocene, eastern North Pacific (central California); late Miocene-early Pliocene (Australasia); early Pliocene, western North Atlantic (Virginia and North Carolina); late Pliocene, western North Atlantic (Florida); late Pleistocene, western North Atlantic (Florida) and western North Atlantic (Canada). Sei whale *(Balaenoptera borealis)* is reported from the late Pleistocene, western North Pacific (Japan). The humpback whale *(Megaptera novaeangliae)* is known from the late Pleistocene, western North Atlantic (Florida) and western North Atlantic (Canada) (Deméré, 1994b).

Extinct genera are reviewed by (Deméré *et al.*, in press).

Remarks—McKenna and Bell (1997) include the gray whale as a subfamily.

Content—Two genera containing six or seven species are recognized.

Balaenoptera acutorostrata (Lacépéde, 1804)—minke whale

    Balaenoptera acutorostrata acutorostrata

    Balaenoptera acutorostrata scammoni (Deméré, 1986)

    Balaenoptera acutorostrata subsp.

Balaenoptera bonaerensis (Burmeister, 1867)—Antarctic minke whale

Balaenoptera borealis (Lesson, 1828)—sei whale

    Balaenoptera borealis borealis

    Balaenoptera borealis schlegelli (Flower, 1865)

Balaenoptera brydei (Olsen, 1913)—Eden's whale

Balaenoptera edeni (Anderson, 1878)—Bryde's whale

Balaenoptera musculus (Linnaeus 1758)—blue whale

    Balaenoptera musculus musculus

    Balaenoptera musculus indica

    Balaenoptera musculus brevicauda (Ichihara, 1966)—pygmy blue whale

    Balaenoptera musculus intermedia

Balaenoptera omurai (Wada *et al.*, 2003)—Omura's whale

Balaenoptera physalus (Linnaeus, 1758)—fin whale

Megaptera novaeangliae (Borowski, 1781)—humpback whale

NEOBALAENIDAE (Gray, 1874; Miller, 1923)

Diagnosis—Palate with maxilla window not reaching to posterior margin (Deméré *et al.*, in press).

Definition—The monophyletic group including the common ancestor of *Caperea* and all of its descendants.

Distribution—Known only in temperate waters of southern hemisphere (including Tasmania, coasts of South Australia, New Zealand, and South Africa).

Fossil history—No fossil species are known.

Content—One genus and species is recognized.

Caperea marginata (Gray, 1846)—pygmy right whale

SIRENIA (Illiger, 1811)

Diagnosis—External nares retracted and enlarged, reaching to or beyond the level of the anterior margin of the orbit; premaxilla contacts frontal; sagittal crest absent; pl-5, MI-3, or secondarily reduced from this condition by loss of anterior premolars; mastoid inflated and exposed through orbital fenestra; ectotympanic inflated and drop-like; pachyostosis and os-teosclerosis present in skeleton (Domning, 1994).

Definition—The monophyletic group including the last common ancestor of *"Prorastomus"*† and all of its descendants including *"Protosiren"*†, the Trichechidae, and the "Dugongidae" (Donming, 1994).

Distribution—Sirenians occur in the Indo-Pacific (dugongs) and the southeastern United States, Caribbean, and the Amazon drainage of South America (manatees).

Fossil history—Sirenians are recognized in the fossil record beginning in the early Eocene (56.5–50 Ma), western Tethys (Jamaica); middle Eocene, eastern Tethys (Pakistan and India), central Tethys (Egypt), and western Tethys (southeastern United States), late Eocene, central Tethys (Egypt, Europe); and extending through to the Recent.

Remarks—McKenna and Bell (1997) include sirenians in the suborder Tethytheria along with desmostylians and elephants. They propose the Order Uranotheria to include the suborders Hyracoidea, Embrithopoda (extinct rhinoceros-like mammals) and Tethytheria.

Content—Composed of two families containing all living sirenians.

DUGONGIDAE (Gray, 1821)

Diagnosis—Ventral extremity of jugal lies approximately under posterior edge of orbit, but forward of jugal's postorbital process (if present); ventral border of horizontal ramus of mandible not tangent to angle; development of triangular flukes (Domning, 1994).

Definition—Paraphyletic family includes *Bharatisiren†, Caribosiren†, Crenatosiren†, Corytosiren†, Dioplotherium†, Dusisiren†, Eosiren†, Eotheroides†, "Halitherium"†, Hy-drodamalis†, Metaxytherium†, Prototherium†, Rytiodus†,* and *Xenosiren†.*

Distribution—The dugong is widely distributed in shallow coastal bays of the Indo-Pacific region.

Fossil history—Dugongids are reported from middle Eocene (50–38.6 Ma), central Tethys; late Eocene, central Tethys (Egypt); late Oligocene, Paratethys (Caucasia) and central Tethys (Europe); early early Miocene, western South Atlantic (Argentina); late early Miocene, Paratethys (Austria and Switzerland), western North Atlantic (Florida), and western South Atlantic (Brazil); early middle Miocene, western North Atlantic (Maryland and Virginia); late middle Miocene, Mediterranean (Italy), eastern North Atlantic (western Europe), western North Atlantic (Florida), and eastern North Pacific (California, Baja California); early late Miocene, western North Atlantic (Florida), eastern North Pacific (California), and western North Pacific (Japan); late late Miocene, easter North Pacific (southern California), and western North Pacific (Japan); early Pliocene, western North Atlantic (Florida), eastern North Pacific (California), and western North Pacific (Japan); late Pliocene, Mediterranean (Italy), western North Atlantic (Florida), and eastern North Pacific (California); early Pleistocene, western South Pacific

(Australasia); late Pleistocene, eastern North Pacific (Alaska to central California) and western North Pacific (Japan; Deméré, 1994b).

The extant species *Dugong dugon,* may also occur in the Pleistocene (Domning, 1996). **Content**—One extant genus and Recent species is recognized. Taxonomic history of dugongids is reviewed by Domning (1996).

Dugong dugon (Miiller, 1776)—dugong

TRICHECHIDAE (Gill, 1872) (1821)

Diagnosis—External auditory meatus very broad and shallow, wider anteroposteriorly than high, reduction of neural spines; possibly tendency to enlargement and (at least in *Trichechus*) anteroposterior elongation of thoracic centra (Domning, 1994).

Definition—Most recent common ancestor of *Miosiren †* and all of its descendants including *Anomotherium†, Potamosiren†, Ribodon†,* and *Trichechus* (Domning, 1994).

Distribution—Members of this family inhabit the Gulf coast of Florida *(T. manatus manatus),* the Caribbean and South American coast *(T. manatus latirostris),* the Amazon Basin *(T. inunguis)* and West Africa *(T. senegalensis).*

Fossil history—Trichechids are reported from the late Miocene, western South Pacific (Argentina), and eastern South Pacific (Brazil and Peru); early Pliocene, western North Atlantic (Virginia and North Carolina); late Pleistocene, western North Atlantic (Virginia, Florida (Deméré, 1994b); (Cozzuol, 1996; South American record).

The Antillean manatee *Trichechus manatus* is reported from the late Pleistocene, Western North Atlantic (Florida); Deméré, 1994b).

Content—One extant genus and three species are recognized. Domning and Hayek (1986) describe morphologic distinction of subspecies.

Trichechus inunguis (Natterer in von Pelzeln, 1883)—Amazonian manatee

    T. manatus manatus (Linnaeus, 1758)—Antillean manatee

    T. manatus latirostris (Harlan, 1824; Hatt, 1934)—Florida manatee

Trichechus senegalensis (Link, 1795)—West African manatee

DESMOSTYLIA (Reinhart, 1953)

Diagnosis—Lower incisors transversely aligned; enlarged passage present through squamosal from external auditory meatus to roof of skull; roots of P1 fuse; paroccipital process enlarged (Ray *et al.,* 1994).

Definition—The monophyletic group containing the common ancestor of *Behemotops†* and all of its descendants including: *Paleoparadoxia†, Cornwallius†,* and *Desmostylus†* (Ray *et al.,* 1994).

Distribution—Late Oligocene-Miocene of North America and Japan.

Fossil history—Desmostylians are known from late Oligocene, North America; late Oligocene or early Miocene, North America; late early and early middle Miocene, Japan; late Oligocene, North America; middle Miocene, North America (Domning, 1996).

Content—Contains no living genera.

References

Baker, A. N., A. N. H. Smith, and F. Pichler (2002). "Geographical Variation in Hector's Dolphin: Recognition of a New Subspecies of *Cephalorhynchus hectori*." *J. R. Soc. N. Z. 32*: 713–727.

Barnes, L. G., and S. A. McLeod (1984). The fossil record and phyletic relationships of gray whales. *In* "The Gray Whale *Eschrichtius robustus*" (M. L. Jones, S. L. Schwartz, and S. Leatherwood, eds.), pp. 3–32. Academic Press, New York.

Berggren, W. A., D. V. Kent, C. C. Swisher, Jr., and M. P. Aubry (1995). A Revised Cenozoic Geochronology and Chronostratigraphy. *In* "Geochronology, Time Scales and Global Stratigraphic Correlations" (W. A. Berggren *et al.*, eds.), pp. 129–212. SEPM Special Publication, No. 54.

Berta, A. (1991). "New *Enaliarctos†* (Pinnipedimorpha) from the Oligocene and Miocene of Oregon and the Role of "Enaliarctids" in Pinniped Pylogeny." *Smithson. Contrib. Paleobiol. 69:* 1–33.

Berta, A. (1994). "New Specimens of the Pinnipediform *Pteronarctos* from the Miocene of Oregon." *Smithson. Contrib. Paleobiol. 78:* 1–30.

Berta, A., and A. R. Wyss (1994). "Pinniped Phylogeny." *Proc. San Diego Soc. Nat. Hist. 29:* 33–56.

Best, R. C. and V. M. F. da Silva (1989). Biology, status and conservation of *Inia geoffrensis* in the Amazon and Orinoco River basins. *In* "Biology and Conservation of River Dolphins" (W. F. Perrin, R. L. Brownell Jr., K. Zhou, and J. Liu, eds.), pp. 23–24. IUCN Species Survival Commission Occasional Paper no. 3. IUCN, Gland Switzerland.

Brunner, S. (2004). "Fur Seals and Sea Lions (Otariidae): Identification of Species and Taxonomic Review." *Systematics Biodiversity 1:* 339–439.

Cozzuol, M. A. (1996). "The Record of the Aquatic Mammals in Southern South America." *Milnchner Geowiss. Abh. 30:* 321–342.

Dalebout, M., J. G. Mead, C. S. Baker, A. N. Baker, and A. Van Helden (2002). "A New Species of Beaked Whale *Mesoplodon perrini* sp.n. (Cetacea: Ziphiidae) Discovered Through Phylogenetic Analyses of Mitochondrial DNA Sequences." *Mar. Mamm. Sci. 18:* 577–608.

Deméré, T. A. (1994a). "The Family Odobenidae: A Phylogenetic Analysis of Fossil and Living Taxa." *Proc. San Diego Soc. Natl. Hist. 29:* 99–123.

Deméré, T. A. (1994b). "Phylogenetic Systematics of the Family Odobenidae (Mammalia; Carnivora) with Description of Two New Species from the Pliocene of California and a Review of Marine Mammal Paleofaunas of the World." Unpublished Ph.D. thesis, University of California, Los Angeles.

Deméré, T. A. and A. Berta (2001). "A Re-Evaluation of *Proneotherium repenningi* from the Middle Miocene Astoria Formation of Oregon and Its Position as a Basal Odobenid (Pinnipedia: Mammalia)." *J. Vert. Paleontol. 21:* 279–310.

Deméré T.A., A. Berta, and P. J. Adam (2003). "Pinnipedimorph Evolutionary Biogeography." *Bull. Amer. Mus. Nat. Hist. 279:* 32–76.

Deméré, T. A., A. Berta, and R. McGowen (in press). "The Taxonomic and Evolutionary History of Fossil and Modern Balaenopteroid Mysticetes." *J. Mamm. Evol.*

Domning, D. P. (1994). "A Phylogenetic Analysis of the Sirenia." *Proc. San Diego Soc. Natl. Hist. 29:* 177–190.

Domning, D. P. (1996). "Bibliography and Index of the Sirenia and Desmostylia." *Smithson. Contrib. Paleobiol. 80:* 1–611.

Domning, D. P., and L. C. Hayek (1986). "Interspecific and Intraspecific Morphological Variation in Manatees (Sirenia: *Trichechus*)." *Mar Mamm. Sci. 2:* 87–144.

Fordyce, E. (1991). The fossil vertebrate record of New Zealand. *In* "Vertebrate Paleontology of Australasia" (P. Vickers-Rich, J. M. Monaghan, R. F Baird, and T. Rich, eds.), pp. 1191–1314. Pioneer Design Studio, Melbourne, Australia.

Fordyce, E., and L. G. Barnes (1994). "The Evolutionary History of Whales and Dolphins." *Annu. Rev. Earth Planet Sci. 22:* 419-455.

Fordyce, E., and C. de Muizon (2001). Evolutionary history of cetaceans: A review. *In* "Secondary Adaptation of Tetrapods to Life in Water" (J.-M. Mazin, and V. de Buffrenil, eds.), pp. 169–233. Verlag, F. Pfeil, Munchen, Germany.

Geisler, J. H. (2001). "New Morphological Evidence for the Phylogeny of Artiodactyla, Cetacea and Mesonychidae." *Am. Mus. Novit. 3344:* 1–53.

Geisler, J. H., and A. E. Sanders (2003). "Morphological Evidence for the Phylogeny of the Cetacea." *J. Mamm. Evol. 10*(1/2): 23–29.

Gottfried, M. D., D. J. Bohaska, and F. C. Whitmore, Jr. (1994). "Miocene Cetaceans of the Chesapeake Group." *Proc. San Diego Soc. Natl. Hist. 29:* 229–238.

Hamilton, H. S. Caballero, A. Collins, and R. L. Brownell, Jr. (2001). "Evolution of River Dolphins." *Proc. R. Soc. London Ser B Biol Sci. 268:* 549–556.

Handley, C. O. Jr. (1966). A synopsis of the genus *kogia* (pygmy sperm whales). *In* "Whales, Dolphins and Porpoises" (K. S. Norris, ed.), pp. 62–69. University of California Press, Berkeley.

Harland, W. B., R. L. Armstrong, A. V. Cox, L. E. Craig, A. G. Smith, and D. G. Smith (1990). *A Geologic Time Scale-1989.* Cambridge University Press, New York.

Harrington, C. R. (1977). "Marine Mammals from the Champlain Sea and the Great Lakes." *Ann. N. Y. Acad. Sci. 288:* 508–537.

Hendey, Q. B., and C. A. Repenning (1972). "A Pliocene Phocid from South Africa." *Ann. So. Afr. Mus. 59*(4): 71–98.

Heyning, J. E., and W. F. Perrin (1994). "Evidence for Two Species of Common Dolphin (Genus *Delphinus*) from the Eastern North Pacific." *Contrib. Sci. Natl. Hist. Mus. L. A. Cty. 442.*

Jefferson, T. A., S. Leatherwood, and M. A. Webber. (1993) *Marine Mammals of the World.* FAO Species identification Guide. United Nations Environmental Programme, Food and Agriculture Organization of the United Nations, Rome.

Jefferson, T. A., and Van Waerebeek (2002). "The Taxonomic Status of the Nominal Dolphin Species *Delphinus tropicalis* Van Bree, 1971." *Mar. Mamm. Sci. 18:* 787–818.

Kohno, N., L. G. Barnes, and K. Hirota (1995). "Miocene Fossil Pinnipeds of the Genus *Prototaria* and *Neotherium* (Carnivora: Otariidae): Imagotariinae in the North Pacific Ocean: Evolution, Relationship and Distribution." *Island Arc 3:* 285–308.

Koretsky, I. and A. E. Sanders (2002). "Paleontology from the Late Oliogocene Ashley and Chandler Bridge Formations of South Carolina 1: Paleogene Pinniped Remains: The Oldest Known Seal (Carnivora: Phocidae)." *Smithson. Contrib. Palaeobiol. 93:* 179–183.

Leatherwood, S., and R. R. Reeves (1983). *The Sierra Club Handbook of Whales and Dolphins.* Sierra Club Books, San Francisco.

McKee, J. W. A. (1994). "Geology and Vertebrate Paleontology of the Tangahoe Formation, South Taranaki Coast, New Zealand." *Geol. Soc. New Zealand Misc. Publ.* 80B: 63–91.

McKenna, M. C., and S. K. Bell (1997). *Classification of Mammals above the Species Level.* Columbia University Press, New York.

McLeod, S. A., F. C. Whitmore, and L. G. Barnes (1993). Evolutionary relationships and classification. *In* "The Bowhead Whale" (J. J. Burns, J. J. Montague, and C. J. Cowles, eds.), pp. 45–70. Special Publication 2, Society of Marine Mammalogy, Allen Press, Lawrence, KS.

Morgan, G. S. (1994). "Miocene and Pliocene Marine Mammal Faunas from the Bone Valley Formation of Central Florida." *Proc. San Diego Soc. Natl. Hist. 29:* 239–268.

Muizon, C. de (1981). Les vertébrés fossiles de la formation Pisco (Pérou). *Recherche sur les grandes civilisations.* Mémoire no. 6, Instituts Francais d'etudes andines, editions A.D.P.F, Paris.

Muizon, C. de (1982). "Phocid Phylogeny and Dispersal." *Ann. So. Afr. Mus. 89:* 175–213.

Muizon, C. de (1984). Les vertébrés fossiles de la Formation Pisco (Pérou). *Deuxieme partie, les odontocees (Cetacea, Mammalia) du Pliocene inferieur de Sud-Sacaco. Recherche sur les grandes civilisations.* Institut Francais d'Etudes Andines, Mémoire 50: 1–188.

Perrin, W. F., and R. L. Brownell, Jr. (2001). "Appendix 1 (of Annex U). Update on the List of Recognized Species of Cetaceans." *J. Cetacean Res. Manage. 3*(Suppl.): 364–365.

Ray, C. E., D. P. Domning, and M. C. McKenna (1994). "A New Specimen of *Behemotops proteus* (Order Desmostylia) from the Marine Oligocene of Washington." *Proc. San Diego Soc. Natl. Hist. 29:* 205–222.

Reeves, R. R., B. S. Stewart, and S. Leatherwood (1992). *The Sierra Club Handbook of Seals and Sirenians.* Sierra Club Books, San Francisco, CA.

Repenning, C. A., and R. H. Tedford (1977). "Otarioid Seals of the Neogene." *Geol. Surv. Prof. Pap. (U.S.) 992:* 93.

Rice, D. W. (1998). *Marine Mammals of the World.* Special Publications 4, Society of Marine Mammalogy. Allen Press, Lawrence, KS.

Riedman, M. (1990). *The Pinnipeds.* University of California Press, Berkeley.

Rosel, P. E., A. E. Dizon, and M. G. Haygood (1995). "Variability of the Mitochondrial Control Region in Populations of the Harbour Porpoises, *Phocoena phocoena,* on Interoceanic and Regional Scales." *Can. J. Fish Aquat. Sci. 52:* 1210–1219.

Smith, A. G., D. G. Smith, and B. Funnell (1994). *Atlas of Mesozoic and Cenozoic Coastlines.* Cambridge University Press, Cambridge.

Stanley, H., S. Casey, J. M. Carnahan, S. Goodman, J. Harwood, and R. K. Wayne (1996). "Worldwide Patterns of Mitochondrial DNA Differentiation in the Harbor Seal *(Phoca vitulina)."* *Mol. Biol. Evol. 13:* 368-382.

Thewissen, J. G. M. (1994). "Phylogenetic Aspects of Cetacean Origins: A Morphological Perspective." *J. Mamm. Evol. 2:* 157–183.

Van der Feen, P. J. (1968). "A Fossil Skull Fragment of a Walrus from the Mouth of the River Scheldt (Netherlands)." *Bhdragen Tot de Dierkunde 38:* 23–30.

Wada, S. M., M. Oishi, and T. K. Yamada (2003). "A Newly Discovered Species of Living Baleen Whale." *Nature 426:* 278–281.

Whitmore, F. (1994). "Neogene Climatic Change and the Emergence of the Modern Whale Fauna of the North Atlantic Ocean." *Proc. San Diego Soc. Nat. Hist. 29:* 223–227.

GLOSSARY

Acoustic Thermometry of Ocean Climate Program (ATOC) A project that involved the installation of low-frequency sound generators off the Hawaiian and California coasts to determine the effect of these sound transmissions on marine mammals

Adaptation A particular biological structure, physiological process, or behavior that results from natural selection

Aerobic dive limit (ADL) The length of time that an animal can remain underwater without accumulating lactate in the blood

Age distribution The relative number of individuals of each age in a population

Allometry The variation in the relative rates of growth of various parts of the body in relation to the entire animal

Alloparenting A male or female that takes care of an offspring other than its own

Allozyme One of two or more alternate forms of a gene of an enzyme

Anaerobic Lacking oxygen

Ancestral A preexisting character or condition; also referred to as primitive

Antitropical Species pairs in which one member occupies the northern hemisphere and the other occupies the southern hemisphere

Apnea Breath holding

Apomorphic A condition of a character that is derived relative to a preexisting character of that condition; also referred to as derived

Archaeocetes An archaic group of whales

Area cladogram A hypothesized reconstruction of the geographic distribution of related organisms

Baculmn Bone in the penis of some marine mammals (sea otter, polar bear, and pinnipeds)

Baleen Plates of cornified epithelial tissue suspended from the upper jaw of mysticete whales, also known as whalebone

Basal metabolic rate Minimum caloric requirement needed to sustain life in a resting individual

Bayesian methods In phylogenetics, modeling tree topology, branch lengths, and substitution parameters as probability distributions

Bends A serious condition that results when a diver breathing pressurized air ascends too rapidly after spending a period of time at depth; occurs when nitrogen under high pressure escapes into the blood, joints, and nerve tissue causing pain, paralysis, and death unless treated by gradual decompression

Birth rates The ratio between births and individuals in a specified population

Blastocyst An embryonic stage in mammals consisting of a hollow ball of cells

Blow Exhaled air mixed with water and oil emitted by whales from the blowhole

Blubber A thick layer of fat that occurs in the deep layer of the skin in whales and other marine mammals

Blue whale unit An early management procedure for estimating catch quotas using one blue whale in terms of oil yield as a standard equal to 2 fin, 2.5 humpback, or 6 sei whales

Bootstrap analysis A technique for estimating which portions of a tree are well-supported and which portions are weak based on the percentage of replicates of a character matrix supporting that group

Brachyodont Low crowned teeth

Bradycardia A slow heart rate

Bremer support A technique for estimating which portions of a tree are more well-supported than others based on the number of extra steps you need to construct a tree where that clade is no longer present

Brown fat Specialized internal fat with a capacity for a high rate of oxygen consumption and heat production; allows newborn seals to generate heat efficiently; also known as brown adipose tissue (BAT)

Bubble-cloud feeding A strategy employed by humpback and other whales that involves emitting a train of air bubbles underwater that form a circular net; the whale surfaces in the center with mouth open to scoop up entrapped prey

Callosities Hard, thick, raised patches of skin found on the head of right whales; distributed near the mouth; the arrangement of these patches often is used to identify specific individuals

Carnivora An order of mammals that includes pinnipeds

Carrying capacity The maximum population that can be supported by existing resources

Cavum ventrale A hollow sac-like structure formed by the tongue that invaginates forming a ventral pouch in balanopterid whales; expansion of this pouch enlarges the capacity of the mouth and is used to engulf prey

Cenozoic era The last 65 million years of geologic time; marked by the diversification of mammals

Center of origin/dispersalist explanation Biogeographic origin of a group of organisms from a small area or center from which they dispersed

Cetacea An order of mammals that contains the whales, dolphins, and porpoises and their relatives

Cetology The study of cetaceans or whales

Characterization A list of distinguishing characters that includes shared derived and shared primitive characters and their taxonomic distributions

Characters Heritable morphologic, molecular, physiologic, or behavioral attributes of organisms

Character states Alternate forms of a character

Chevron bones Ventral processes on the posterior vertebrae that mark the beginning of the tail in cetaceans

Circumpolar Having a distribution around the poles

Clades Monophyletic groups of organisms

Cladistics A method of reconstructing the evolutionary history of organisms based on the shared possession of derived characters that provide evidence of a shared common ancestry

Cladogram A branching diagram that depicts relationships among organisms based on cladistic methods

Classification The arrangement of taxa into a hierarchy

Click A broad frequency sound of short duration

Codas Rhythmic patterns of clicks used by sperm whales for communication

Commensalism A relationship between species in which one species benefits and the other is unharmed

Connecting stomach The third of four compartments of the cetacean stomach between the main stomach and the pyloric stomach

Consensus tree Summary of multiple equally parsimonious trees

Convergence A similarity between organisms that is not the result of inheritance

Corpus albicans A permanent visible scar on the ovary that results from individual ovulations; a peculiarity of whales is that these scars are visible as rounded protuberances that remain for the rest of the animal's life

Corpus luteum The mass of tissue that is left after ovulation when a mature follicle ruptures from the ovary

Corridor A route of dispersal followed by organisms

Cosmopolitan Having a broad, wide-ranging distribution

Cost of transport The power required to move a given body mass at some velocity

Countercurrent heat exchange systems The opposite flow of adjacent fluids that maximize heat transfer rates

Crossover speed A velocity at which it is energetically advantageous to switch from submergence and move above the water surface

Cutaneous ridges Ridges located on the surface of the skin of many cetaceans that play a role in streamlining or tactile sensing

Deep scattering layer Concentrations of zooplankton that are found at depth during the day and that migrate upward to shallower water at night

Defined Recognizing a taxon based on its ancestry and taxonomic membership

Dental formula A convenient way of designating the number and arrangement of mammalian teeth (for example, I3/3, C1/1, P4/4, M3/3). The letters indicate incisors, canines, premolars, and molars. The numbers above the line indicate the number of teeth on one side of the upper jaw; those below the line indicate the number of teeth on one side of the lower jaw

Derived A character or condition that is changed from a preexisting character or condition

Desmostylians The only known extinct group of marine mammals; their close relatives are sirenians

Diagnosed The listing of shared derived characters and their taxonomic distribution for a group of organisms

Dialects Repetitions or calls that are characteristic of a particular geographic region

Diatoms A phylum of algae with glass-like walls that are important constituents of the base of the marine food chain

Digitigrade Walking on the digits (fingers and toes)

Diphyletic Having two separate ancestries

Disjunct The distribution of a species pair in which they are separated by a geographic barrier

Domoic acid A toxin produced by a number of marine diatoms

Echelon formation A staggered arrangement of animals (e.g., spinner dolphin societies) that may serve a signaling function

Echolocation The production of high frequency sound and its reception by reflected echoes; used by toothed whales to navigate and locate prey

El Niño-Southern Oscillation Oceanographic events that occur in the eastern Pacific Ocean every few years in which prevailing current patterns change, water temperature increases, and upwelling declines; results in reduced food resources for marine mammals

Embolism The obstruction of a blood vessel by a foreign object (e.g., air bubble)

Encephalization quotient A numeric comparison that considers brain size relative to body size

Endemic Confined to a particular geographic region

Energetics Quantitative assessment of resource acquisition and allocation by animals

Eocene An epoch of geologic time beginning approximately 55 Ma that marks the earliest record of cetaceans and sirenians

Escort One of several male humpbacks that attend a female. The principal escort usually is the one involved in mating; often secondary escorts are present that do not play an active role in mating

Estrus The period of sexual receptivity in females

Exaptation A different function for a structure than its original function

Extant Refers to living as opposed to extinct organisms

Facial and cranial asymmetry Bones and soft tissue in the odontocete skull that are of differing sizes; those on the right side are larger than those on the left side

Fast ice Sea ice attached to land

Female or harem defense polygyny Mating strategy in which males establish dominance hierarchies that confer priority access to females

Fineness Ratio The relationship between maximum body length and maximum body diameter

Flukes Horizontally flattened tail of cetaceans and sirenians

Forestomach The first compartment of a multichambered stomach; characteristic of whales and some ungulate mammals

Frequency (power) spectrum A graphic representation of sound pressure levels with frequency

Frictional drag Resistant component to force that results from interaction of the animal's surface with surrounding water

Frontomandibular stay Portion of the temporalis (jaw closing) muscle that optimizes the angle at which the mouth can be opened during engulfment feeding by balaenopterid whales

Fundic chamber The second of four compartments of the cetacean stomach; also referred to as the main stomach

Genetic drift Changes in the gene pool of a population due to chance

Geographic location dive time and depth recorders (GLTDRs) Instruments that record geographic location, duration, and depth of dive

Gestation The period between fertilization and birth

Global Positioning System (GPS) An instrument that is used to locate precise geographic position using satellite data while the user is in the field

Glove finger A hardened plug of tissue in the external ear canal of baleen whales

Growth layer group (GLG) A layer of tissue in bone or teeth that represents a regular interval of time (usually 1 year)

Guard hairs Outer coat of coarse protective hair found on most mammals

Hematocrit The percentage of red blood cells to total blood volume

Hemoglobin An iron-containing protein in red blood that transports oxygen

Heterodont A dentition in which teeth are differentiated into several types, such as incisors, canines, premolars, and molars, with different functions

Heterogamy Having different gametes (e.g., males have XY sex chromosomes)

Heterozygosity Containing two different alleles for a given trait

Homodont A dentition with all teeth very similar in form and function

Homogamy Having the same gametes (e.g., females have XX sex chromosomes)

Homology A similarity that is the result of inheritance

Homoplasy A similarity that is not the result of inheritance

Homozygotes Containing two identical alleles for a given trait

Hybridization An offspring of two different species

Hypercapnia The presence of excessive amount of carbon dioxide in the blood

Hyperosmotic Having a higher osmotic pressure than the surrounding environment

Hyperphalangy An increase in number of digits, seen in whales

Hypoosmotic Having a lower osmotic pressure than the surrounding environment

Hypoxia An oxygen deficiency that reaches the body's tissues

Induced drag Retardation of movement due to the presence of flukes or flippers

Induced ovulators In female polar bears the release of a mature follicle from the ovary (i.e., ovulation) occurs only when she is stimulated by the presence and activity of a male

Infrasonic Low frequency sounds below 18 Hz

Ingroup The group of organisms whose evolutionary relationships are the subject of investigation (see outgroup)

Introgression The transplantation of genes between species resulting from fertile hybrids mating successfully with one of the parent species

Ischemia A local decrease in blood flow resulting from its obstruction

Isotope dilution techniques Field method used to estimate the fat and fat-free components and rate of water intake in an animal. It involves administering a concentration of an isotope to an animal and subsequently measuring the concentration of this isotope in body tissues (e.g., blood)

Junk The region of the sperm whale head, below the spermaceti organ, filled with oil and connective tissue; called junk by whalers because of its lesser quantity of oil

Kolponomos An extinct bear-like carnivore that may have filled the sea otter niche during the middle Miocene

Krill Shrimp-like crustaceans of the family Euphausidae that form the primary food resource for baleen whales and certain pinnipeds (e.g., crabeater seal)

K-selected species Characterized by long life spans, low mortality rates, large body size, and production of few offspring that are intensively nurtured (see r-selected species)

Lactate Metabolic end product produced during anaerobic metabolism

Lek A mating strategy or courtship display in which males display to females

Lineages Ancestor-descendant populations of organisms through time

Lobtail feeding Strategy observed in humpbacks in which feeding is followed by a tail slap that may serve to aggregate prey

Low-frequency military sonars Devices that transmit low-frequency sound

Main stomach The second compartment in the multichambered stomach of cetaceans

Mapping Placing the distributions of nonheritable characters (e.g., geographic distribution, ecological association) on a particular tree topology

Marine Mammal Protection Act Federal regulation enacted by legislature in 1972 to protect marine mammals in U.S. waters

Marine mammals Those species of mammals that spend all or most of their existence in the water

Mariposia The practice of drinking seawater

Mass stranding Three or more individuals of the same species that intentionally swim or are unintentionally trapped ashore by waves or receding tides

Mate guarding Mating strategy that involves males displacing one another while attending nursing females

Mating Reproductive activity in animals; also referred to as breeding

Matrifocal Aggregations of matrilines

Matrilineal Descendants of a single female

Maximum likelihood In phylogenetics, a method for choosing a preferred tree among many possible trees based on probability-based distribution of data

Melon Fat-filled structure on the forehead of odontocetes that functions to focus sound

Mesonychids Group of extinct ungulates; they are hypothesized to be the closest relatives of whales

Microsatellites Very short repeated DNA sequences

Minisatellites Short repeated DNA sequences

Miocene An epoch of geologic time beginning approximately 23 Ma and ending 5 Ma

Monestrous Having only one estrus cycle each year

Monkey lips/dorsal bursa (MLDB complex) Soft tissue structures located in the nasal passages of odontocetes hypothesized as the mechanisms of sound production

Monogamous A mating strategy in which adult males mate with only one female

Monophyletic group A group of organisms that includes a common ancestor and all of its descendants

Morbillivirus Serious pathogen that affects marine mammals by suppressing immune system responses

Mortality The proportion of deaths in a population

Museau de singe Valve-like slit that opens into the right nasal passage of the sperm whale, homologous to the nasal plugs of other odontocetes

Mutualists Two species which both benefit from the relationship between them

Myoglobin Hemoglobin in the muscles

Nasal plugs Masses of connective tissue that when retracted open the nasal passages of odontocetes

Natality Birth rate

Nitrogen narcosis A state of euphoria and exhilaration that occurs when air is breathed under increased pressure that occurs with depth

Nodes Branching points on a cladogram

Nomenclature The formal system of naming taxa

Nonshivering thermogenesis Process by which an animal keeps warm by metabolizing brown fat to produce heat

Oligocene An epoch of geologic time beginning approximately 35 Ma and ending 23 Ma

Optimization *A posteriori* arguments as to how particular characters should be polarized given a particular tree topology

Optimum sustainable population (OSP) Population level between 60% of its carrying capacity and carrying capacity

Osmotic The flow of solvent through a semi-permeable membrane

Osteosclerotic Containing compact bone

Otarioid Nonmonophyletic group of pinnipeds that includes walruses, otariids, and their extinct relatives

Outgroup Taxon or group of organisms that is closely related but outside the group whose relationships are the subject of investigation

Outgroup comparison Procedure for determining the polarity (ancestral or derived) condition of a character; assumes that the character found in the outgroup is the ancestral condition for the group in question (ingroup)

Pachyostosis Thick dense bone

Pack ice Sea ice that is unattached to land

Paedomorphism A change in developmental timing such that adults retain juvenile characteristics

Pagophilic Seals that inhabit ice

Paleolithic That period of the Stone Age characterized by stone implements

Pan bone Thin area of bone in the lower jaw of odontocetes filled with a fat channel that is hypothesized as the sound receiver

Parallelism Characters that have evolved independently in two groups

Paraphyletic group A group of organisms that does not include the common ancestor and all of its descendants

Paraxonic An arrangement of digits in which the plane of symmetry passes between the third and fourth digits

Parsimony The principle that prefers the hypothesis with the fewest number of steps

Parturition Giving birth; calving in whales and pupping in pinnipeds and sea otters

Peripheral vasoconstriction Redistribution of blood such that some tissues receive more blood than other tissues

Peritympanic sinuses Air-filled sinuses surrounding the tympanic region of cetaceans

Pharyngeal pouches Air-filled sacs located in the throat of walruses that when struck with the front flippers produce distinctive sounds

Phenetic A method of classification based on overall similarity of the taxa

Pheromones Chemical signals that function in communication between organisms

Photic zone The uppermost zone in the water where sunlight permits photosynthesis

Phylogenetic systematics Method of determining evolutionary relationships among taxa based on shared possession of derived characters; also referred to as cladistics

Phylogenetic tree Branching diagram that represents the relationships of a group of organisms; sometimes distinguished from cladogram as an ecological reconstruction rather than a genealogical reconstruction

Phylogeny Evolutionary history of a group of organisms

Phytoplankton Microscopic algae floating near the surface that forms the base of the food chain

Pingers Acoustic sound devices that are used to prevent marine mammals from net entanglement

Pinnae External ear; reduced or absent in many marine mammals

Pinnipeds Monophyletic group that includes seals, sea lions, and walruses

Pleistocene An epoch of geologic time beginning approximately 1.6 Ma

Plesiomorphic Ancestral or primitive character

Pliocene An epoch of geologic time beginning approximately 5 Ma and ending 1.6 Ma

Pods Social units of certain cetaceans, particularly killer whales

Polychlorinated biphenols Industrial chemicals (i.e., organochlorides) that are toxic to many animals including marine mammals and are identified as pollutants in many coastal environments

Polydactyly Possession of several or many additional digits

Polyestrous Having more than one estrus cycle each year

Polygynous A mating strategy in which one male mates with more than one female during a breeding season

Polynya A large area of water in pack ice that remains open throughout the year

Polyovular Producing numerous corpora lutea each pregnancy

Polytomy A pattern of unresolved relationships among taxa

Population A group of organisms of the same species occupying a specific geographic region

Porpoising Leaping locomotion; an animal partially avoids the drag associated with surfacing by leaping above the water

Postpartum estrus Female sexual receptivity that occurs immediately following birth and is a characteristic of phocids and otariids

Precocious Born in an advanced state of development (i.e., able to walk and swim); in contrast to altricial, which refers to birth in an undeveloped state

Pressure drag Resistant force created by the need to displace an amount of water equal to an animal's largest cross-sectional diameter

Primary production Organisms such as plants able to produce their own food and that support other levels

Primitive The ancestral form of a character (see derived)

Promiscuous Mating system of many cetaceans and sirenians in which adult males randomly associate with females for variable periods of time

Pyloric stomach The final compartment in the multichambered stomach of cetaceans

Reniculate Refers to a kidney that is made up of small lobes or reniculi that is characteristic of pinnipeds, cetaceans, polar bears, and sea otters

Resident One of two populations of killer whales occupying coastal waters of the

Pacific Northwest. Resident populations are distinguished from transient populations by a number of foraging-related differences

Resource defense polygyny Type of mating system in which males exclude other males from access to females by defending breeding territories

Rete ridges Flap-like projections between the epidermis and dermis of the skin of cetaceans

Retia mirabilia Groups of blood vessels that form blocks of tissue on the dorsal wall of the thoracic cavity, extremities, or body periphery; they function as reservoirs for blood and as countercurrent heat exchange systems

Reversal Loss of a derived feature and reestablishment of an ancestral feature

Revised management procedure (RMP) Estimations of the number of whales that can be taken with low risk of causing the affected population to be reduced below its maximum net productivity

Reynolds number A widely used indicator of the forces acting on submerged bodies, approximated by the equation (body length) ×(swimming velocity)/(water viscosity)/(water density)

Rorquals Whales of the family Balaenopteridae; includes humpbacks, sei, minke, and fin whales

Rostral or oral disc Upper lip and mouth of manatees and dugongs; covered with vibrissae and used in feeding

Rostrum Anterior portion or beak region of the skull that is elongated in most cetaceans

R-selected species Characterized by rapid population increase (often in unpredictable, fluctuating environments) and production of many offspring with little parental investment

Schools Structured social groups observed in odontocetes characterized by long-term association among individuals

Seasonal delayed implantation Suspension of development of the embryo in some marine mammals (pinnipeds and sea otters) for several months

Sexual dimorphism External differences between males and females of a particular species

Signature whistles Narrow band frequency modulated sound produced by dolphins and hypothesized to function for individual recognition as well as communication among group members

Sirenia The only herbivorous order of marine mammals; includes manatees and dugongs

Sister group The closest relative to the ingroup

Sonogram A graphic representation of sound frequency with time

Speciation Splitting of a lineage resulting in the formation of two species from a single common ancestor

Speciation events Those factors such as emergence of geographic barriers that cause lineage splitting

Spectogram A graphic representation of sound waves per unit time

Sperm competition Mating strategy seen in some mysticetes in which copulating males attempt to displace or dilute the sperm of other males in an attempt to increase the probability of being the male for fertilize the female

Spermaceti Waxy fluid that fills the spermaceti organ; this is the sperm whale oil that was sought after by whalers

Spermaceti organ Elongate connective tissue sac in the forehead of sperm whales that contains spermaceti

Stocks Subdivision of a population of animals that is subject to commercial exploitation; term often used by management agencies

Subdermal connective tissue sheath Collagenous tissue layer that envelops the lateral surfaces of cetaceans and serves as an important anchor for muscles of the tail

Surfactants Substances lining lung surfaces that reduce surface tension

Survivorship The number of individuals of a group that are alive at each age

Synapomorphy Shared derived character, used as evidence of a shared common ancestry between taxa

Systematics Study of biological diversity that has as its emphasis the reconstruction of evolutionary histories among organisms

Tapetum lucidum A specialized layer behind the retina that reflects light and enables an animal to see better in darkness

Taphonomic Processes that affect an organism after death

Taxon A particular taxonomic group at a given rank

Taxonomy A branch of systematics that deals with the description, identification, and classification of species

Telemetry Techniques for measuring the distance of an object (e.g., animal) from an observer

Telescoping Change in the relationship of bones of the cetacean skull in response to the migration of the nasal opening to the top of the skull

Tethytheria The monophyletic group that includes proboscideans (elephants), sirenians, and extinct desmostylians

Thalassocnus Extinct aquatic sloth that lived during the early Pliocene

Thermoneutral Falling within a temperature range at which an animal can maintain its body temperature

Throat grooves Furrows in the throat region of balaenopterids that can open up like pleats during feeding to expand the mouth cavity

Time depth recorders Instruments that record the duration and depth of dives

Transient One of two populations of Pacific Northwest killer whales that usually do not reside in an area for an extended period of time (i.e., nonresidents) and that can be distinguished from resident populations by a number of foraging-related differences

Trophic levels The division of species in an ecological community based on their main food source

Tusks Enlarged incisor or canine teeth

Tympano-periotic complex Bones of the skull that make up the auditory bulla

Ultrasonic High-frequency sounds

Underfur hairs Short fine hairs that serve primarily for insulation

Units Social groups of females and immature sperm whales

Upwelling Areas off continental margins where circulation patterns bring nutrient rich water to the surface

Vicariance explanation An hypothesis to explain patterns of distribution among organisms that argues that organisms occur in an area because they evolved there

Wave drag Friction created by wave action

Wave riding Behavior seen in dolphins in which they ride the waves of a boat's wake

Whalebone Another name for baleen; term used by whalers

Whale lice Amphipod crustaceans (cyamids) found on the skin of large cetaceans

INDEX

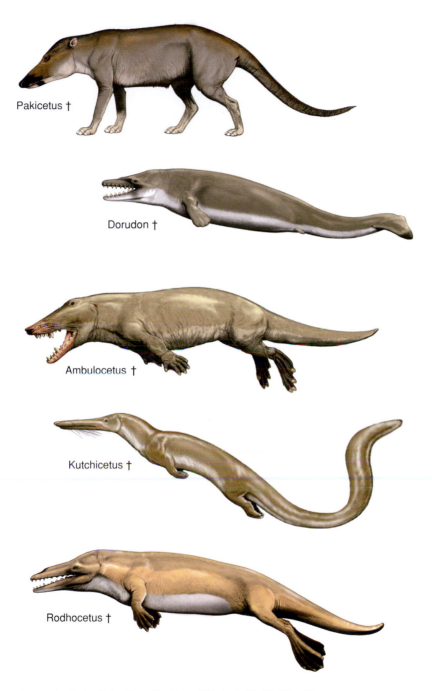

Pakicetus †

Dorudon †

Ambulocetus †

Kutchicetus †

Rodhocetus †

Reconstructions of archaic whales. † = extinct taxa. (Illustrated by Carl Buell.)

Plate 1.

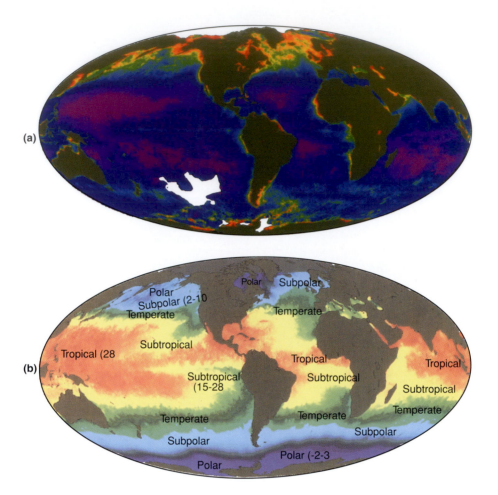

(a) Multi-annual mean distribution of sea-surface temperatures and marine climatic zones based on averaged July observations (NOAA SEAWifs image.) **(b)** Three-year mean of the distribution of marine primary production; low in the central gyres (less than 50 gC/m^2/year; magenta to deep blue), moderate (50–100 gC/m^2/year; light blue to green areas, and high in coastal zones and upwelling areas (greater than 100 gC/m^2/year; yellow, orange, and red). (Courtesy NASA/GSFC Space Data Computing Division.)

Plate 2.

Inflated hood of male hooded seal (*Cystophora cristata*). (Photo by Kit M. Kovacs and Christian Lydersen, NPI.)

Inflated nasal septum from left nostril of male hooded seal. (Photo by Kit M. Kovacs and Christian Lydersen, NPI.)

Adult bearded seal (*Erignathus barbatus*). (Photo by Bjorn Krafft.)

Plate 3.

Grey seal (*Haliochoerus grypus*). (Photo by Kit M. Kovacs and Christian Lydersen, NPI.)

Ringed seal (*Phoca hispida*) mother and pup. (Photo by Kit M. Kovacs and Christian Lydersen, NPI.)

Adult male walrus (*Odobenus rosmarus*). (Photo by Kit M. Kovacs and Christian Lydersen, NPI.)

Plate 4.

Leopard seal (*Hydruga leptonyx*). (Photo by Simon Allen.)

Leopard seal (*Hydruga leptonyx*). (Photo by Rob Williams.)

Weddell seal (*Leptonychotes weddelli*) with digital camera logger. (Photo by Katsufumi Sato.)

Plate 5.

Male Northern elephant seal (*Mirounga angustirostris*). (Photo by Tanguy de Tillesse.)

N. elephant seal with sleeping California sea lion (*Zalophus californianus*). (Photo by Lizabeth Bowen.)

Plate 6.

New Zealand fur seal (*Arctocephalus forsteri*) eating octopus. (Photo by Abigail Caudron.)

Antarctic fur seal (*Arctocephalus gazella*) mom and pup. (Photo by Abigail Caudron.)

Female Australian fur seal (*Arctocephalus pusillus*). (Photo by Simon Allen.)

Plate 7.

Male Subantarctic fur seal (*Arctocephalus tropicalis*). (Photo by Simon Allen.)

Male Northern fur seal (*Callorhinus ursinus*). (Photo by James Sumich.)

Sterller's sea lion (*Eumetopias jubatus*). (Photo by Nicholas Brown.)

Plate 8.

Bottlenose dolphins (*Tursiops truncatus*). (Photo by Kevin Robinson.)

Bottlenose dolphins (*Tursiops truncatus*). (Photo by Ben Chicoski.)

Commerson's dolphin (*Cephalorhynchus commersonii*). (Photo by Rob Williams.)

Plate 9.

Finless porpoise (*Neophocaena phocaenoides*). (Photo by Grant Abel.)

Amazon River dolphin or boto (*Inia goeffrensis*). (Photo by Elizabeth L. Zuniga.)

Beluga (*Delphinapterus leucas*). (Photo by John Morrissey.)

Plate 10.

Common dolphin (*Delphinus delphis*) and Atlantic spotted dolphin (*Stenella frontalis*). (Photos by Megan C. Mattson, NOAA Fisheries.)

False killer whales (*Pseudorca crassidens*). (Photo by Jeremy Kiszka.)

Plate 11.

Indian humpback dolphin (*Sousa plumbea*). (Photo by Jeremy Kiszka.)

Pilot whale (*Globicephala sp.*). (Photo by Megan C. Mattson, NOAA Fisheries.)

Killer whale (*Orcinus orca*) female and calf. (Photo by Rob Williams.)

Plate 12.

Blue whales (*Balaenoptera musculus*). (Photo by Ingrid Over-gard.)

Minke whale (*Balaenoptera acutorostrata*). (Photo by Danielle Dion, Quoddy Link Marine, Inc.)

Fin whale (*Balaenoptera physalus*). (Photo by Danielle Dion, Quoddy Link Marine, Inc.)

Plate 13.

Humpback whale (*Megaptera novaeangliae*). (Photo by Magaly Chambellant.)

Southern right whale (*Eubalaena australis*). (Photo by José Palazzo, IWC/Brasil.)

Bowhead whale (*Balaena mysticetus*). (Photo courtesy of NOAA.)

Plate 14.

Gray whale (*Eschrichtius robustus*) cow and calf. (Photo by James Sumich.)

Gray whale calf with baleen and rostral hairs visible. (Photo by Julie-Ann Kondor.)

Courting gray whale male. (Photo by Julie-Ann Kondor.)

Plate 15.

Sea otter (*Enhydra lutris*). (Photo by André Moura.)

Florida manatee (*Trichechus manatus*). (Photo courtesy Sirenia Project, USGS.)

Dugong (*Dugong dugon*). (Photo by Ben Chicoski.)

Plate 16.